Geophysical Monograph Series

Geophysical Monograph Series

- **212 The Early Earth: Accretion and Differentiation James Badro** *and Michael Walter (Eds.)*
- **213 Global Vegetation Dynamics: Concepts and Applications in the MC1 Model** *Dominique Bachelet and David Turner (Eds.)*
- **214 Extreme Events: Observations, Modeling and Economics** *Mario Chavez, Michael Ghil, and Jaime Urrutia-Fucugauchi (Eds.)*
- **215 Auroral Dynamics and Space Weather** *Yongliang Zhang and Larry Paxton (Eds.)*
- **216 Low-Frequency Waves in Space Plasmas** *Andreas Keiling, Dong- Hun Lee, and Valery Nakariakov (Eds.)*
- **217 Deep Earth: Physics and Chemistry of the Lower Mantle and** *Core Hidenori Terasaki and Rebecca A. Fischer (Eds.)*
- **218 Integrated Imaging of the Earth: Theory and Applications** *Max Moorkamp, Peter G. Lelievre, Niklas Linde, and Amir Khan (Eds.)*
- **219 Plate Boundaries and Natural Hazards** *Joao Duarte and Wouter Schellart (Eds.)*
- **220 Ionospheric Space Weather: Longitude and Hemispheric Dependences and Lower Atmosphere Forcing Timothy Fuller- Rowell,** *Endawoke Yizengaw, Patricia H. Doherty, and Sunanda Basu (Eds.)*
- **221 Terrestrial Water Cycle and Climate Change Natural and Human-Induced Impacts** *Qiuhong Tang and Taikan Oki (Eds.)*
- **222 Magnetosphere-Ionosphere Coupling in the Solar System** *Charles R. Chappell, Robert W. Schunk, Peter M. Banks, James L. Burch, and Richard M. Thorne (Eds.)*
- **223 Natural Hazard Uncertainty Assessment: Modeling and Decision Support** *Karin Riley, Peter Webley, and Matthew Thompson (Eds.)*
- **224 Hydrodynamics of Time-Periodic Groundwater Flow: Diffusion Waves in Porous Media** *Joe S. Depner and Todd C. Rasmussen (Auth.)*
- **225 Active Global Seismology** *Ibrahim Cemen and Yucel Yilmaz (Eds.)*
- **226 Climate Extremes** *Simon Wang (Ed.)*
- **227 Fault Zone Dynamic Processes** *Marion Thomas (Ed.)*
- **228 Flood Damage Survey and Assessment: New Insights from Research and Practice** *Daniela Molinari, Scira Menoni, and Francesco Ballio (Eds.)*
- **229 Water-Energy-Food Nexus – Principles and Practices** *P. Abdul Salam, Sangam Shrestha, Vishnu Prasad Pandey, and Anil K Anal (Eds.)*
- **230 Dawn–Dusk Asymmetries in Planetary Plasma Environments** *Stein Haaland, Andrei Rounov, and Colin Forsyth (Eds.)*
- **231 Bioenergy and Land Use Change** *Zhangcai Qin, Umakant Mishra, and Astley Hastings (Eds.)*
- **232 Microstructural Geochronology: Planetary Records Down to Atom Scale** *Desmond Moser, Fernando Corfu, James Darling, Steven Reddy, and Kimberly Tait (Eds.)*
- **233 Global Flood Hazard: Applications in Modeling, Mapping and Forecasting** *Guy Schumann, Paul D. Bates, Giuseppe T. Aronica, and Heiko Apel (Eds.)*
- **234 Pre-Earthquake Processes: A Multidisciplinary Approach to Earthquake Prediction Studies** *Dimitar Ouzounov, Sergey Pulinets, Katsumi Hattori, and Patrick Taylor (Eds.)*
- **235 Electric Currents in Geospace and Beyond** *Andreas Keiling, Octav Marghitu, and Michael Wheatland (Eds.)*
- **236 Quantifying Uncertainty in Subsurface Systems** *Celine Scheidt, Lewis Li, and Jef Caers (Eds.)*
- **237 Petroleum Engineering** *Moshood Sanni (Ed.)*
- **238 Geological Carbon Storage: Subsurface Seals and Caprock Integrity** *Stephanie Vialle, Jonathan Ajo-Franklin, and J. William Carey (Eds.)*
- **239 Lithospheric Discontinuities** *Huaiyu Yuan and Barbara Romanowicz (Eds.)*
- **240 Chemostratigraphy Across Major Chronological Eras** *Alcides N.Sial, Claudio Gaucher, Muthuvairavasamy Ramkumar, and Valderez Pinto Ferreira (Eds.)*
- **241 Mathematical Geoenergy:Discovery, Depletion, and Renewal** *Paul Pukite, Dennis Coyne, and Daniel Challou (Eds.)*
- **242 Ore Deposits: Origin, Exploration, and Exploitation** *Sophie Decree and Laurence Robb (Eds.)*
- **243 Kuroshio Current: Physical, Biogeochemical and Ecosystem** *Dynamics Takeyoshi Nagai, Hiroaki Saito, Koji Suzuki, and Motomitsu Takahashi (Eds.)*
- **244 Geomagnetically Induced Currents from the Sun to the Power Grid** *Jennifer L. Gannon, Andrei Swidinsky, and Zhonghua Xu (Eds.)*
- **245** *Shale: Subsurface Science and Engineering Thomas Dewers, Jason Heath, and Marcelo Sánchez (Eds.)*
- **246 Submarine Landslides: Subaqueous Mass Transport Deposits From Outcrops to Seismic Profiles** *Kei Ogata, Andrea Festa, and Gian Andrea Pini (Eds.)*
- **247 Iceland: Tectonics, Volcanics, and Glacial Features** *Tamie J. Jovanelly*
- **248 Dayside Magnetosphere Interactions** *Qiugang Zong, Philippe Escoubet, David Sibeck, Guan Le, and Hui Zhang (Eds.)*
- **249 Carbon in Earth's Interior** *Craig E. Manning, Jung-Fu Lin, and Wendy L. Mao (Eds.)*
- **250 Nitrogen Overload: Environmental Degradation, Ramifications, and Economic Costs** *Brian G. Katz*
- **251 Biogeochemical Cycles: Ecological Drivers and Environmental Impact** *Katerina Dontsova, Zsuzsanna Balogh-Brunstad, and Gaël Le Roux (Eds.)*
- **252 Seismoelectric Exploration: Theory, Experiments, and Applications** *Niels Grobbe, André Revil, Zhenya Zhu, and Evert Slob (Eds.)*
- **253 El Niño Southern Oscillation in a Changing Climate** *Michael J. McPhaden, Agus Santoso, and Wenju Cai (Eds.)*
- **254 Dynamic Magma Evolution** *Francesco Vetere (Ed.)*
- **255 Large Igneous Provinces: A Driver of Global Environmental and Biotic Changes** *Richard. E. Ernst, Alexander J. Dickson, and Andrey Bekker (Eds.)*
- **256 Coastal Ecosystems in Transition: A Comparative Analysis of the Northern Adriatic and Chesapeake Bay** *Thomas C. Malone, Alenka Malej, and Jadran Faganeli (Eds.)*
- **257 Hydrogeology, Chemical Weathering, and Soil Formation** *Allen Hunt, Markus Egli, and Boris Faybishenko (Eds.)*
- **258 Solar Physics and Solar Wind** *Nour E. Raouafi and Angelos Vourlidas (Eds.)*
- **259 Magnetospheres in the Solar System** *Romain Maggiolo, Nicolas André, Hiroshi Hasegawa, and Daniel T. Welling (Eds.)*
- **260 Ionosphere Dynamics and Applications** *Chaosong Huang and Gang Lu (Eds.)*
- **261 Upper Atmosphere Dynamics and Energetics** *Wenbin Wang and Yongliang Zhang (Eds.)*
- **262 Space Weather Effects and Applications** *Anthea J. Coster, Philip J Erickson, and Louis J Lanzerotti (Eds.)*

Space Physics and Aeronomy Collection Volume 1
Geophysical Monograph 258

Solar Physics and Solar Wind

Nour E. Raouafi
Angelos Vourlidas
Editors

Yongliang Zhang
Larry J. Paxton
Collection Editors in Chief

This Work is a co-publication of the American Geophysical Union and John Wiley and Sons, Inc.

AGU
ADVANCING EARTH AND SPACE SCIENCE

WILEY

This edition first published 2021

Published under the aegis of the AGU Publications Committee

Brooks Hanson, Executive Vice President, Science
Carol Frost, Chair, Publications Committee
For details about the American Geophysical Union visit us at www.agu.org

Registered Office
John Wiley & Sons, Inc., 111 River Street, Hoboken, NJ 07030, USA

Editorial Office
111 River Street, Hoboken, NJ 07030, USA

For details of our global editorial offices, customer services, and more information about Wiley products visit us at www.wiley.com.

Wiley also publishes its books in a variety of electronic formats and by print-on-demand. Some content that appears in standard print versions of this book may not be available in other formats.

Library of Congress Cataloging-in-Publication Data

Names: Raouafi, Nour E., editor. | Vourlidas, Angelos, editor.
Title: Solar physics and solar wind / Nour E. Raouafi, Angelos Vourlidas, editors.
Description: Hoboken, NJ : Wiley, 2021. | Series: Geophysical monograph series | Includes bibliographical references and index.
Identifiers: LCCN 2020047581 | ISBN 9781119507536 (hardback) | ISBN 9781119815488 (adobe pdf) | ISBN 9781119815471 (epub)
Subjects: LCSH: Sun. | Solar wind.
Classification: LCC QB521 .A88 2021 | DDC 523.7–dc23
LC record available at https://lccn.loc.gov/2020047581

Cover Design: Wiley
Cover Image: © NOAA National Environmental Satellite, Data, and Information Service (NESDIS)

Set in 10/12pt Times New Roman by SPi Global, Pondicherry, India

CONTENTS

LIST OF CONTRIBUTORS

Olga Alexandrova
Laboratoire d'Etudes Spatiales et d'Instrumentation en Astrophysique
Observatoire de Paris
Université PSL
CNRS
Sorbonne Université
Université de Paris
Meudon, France

Alessandro Bemporad
INAF Osservatorio Astrofisico di Torino
Turin, Italy

Luca Bertello
National Solar Observatory
Boulder, Colorado, USA

Christina M. S. Cohen
Space Radiation Laboratory
California Institute of Technology
Pasadena, California, USA

Serena Criscuoli
National Solar Observatory
Boulder, Colorado, USA

Mausumi Dikpati
High Altitude Observatory
Boulder, Colorado, USA

Cooper Downs
Predictive Science Inc.
San Diego, California, USA

Aleida K. Higginson
NASA Goddard Space Flight Center
Greenbelt, Maryland, USA

Rachel Howe
School of Physics and Astronomy
University of Birmingham
Birmingham, United Kingdom

James A. Klimchuk
NASA Goddard Space Flight Center
Greenbelt, Maryland, USA

Michael Lavarra
Institut de Recherche en Astrophysique et Planétologie
Toulouse, France

Benoit Lavraud
Laboratoire d'Astrophysique de Bordeaux
Université de Bordeaux
CNRS, B18N
Pessac, France

Gang Li
Department of Space Science
University of Alabama in Huntsville
Huntsville, Alabama, USA

Mark Linton
Space Science Division
Naval Research Laboratory
Washington, D.C., USA

Noé Lugaz
University of New Hampshire
Durham, New Hampshire, USA

Glenn M. Mason
Johns Hopkins University Applied Physics Laboratory
Laurel, Maryland, USA

Lorenzo Matteini
Department of Physics
Imperial College London
London, UK

Ineke De Moortel
School of Mathematics and Statistics
University of St Andrews
St Andrews, United Kingdom;
and
Rosseland Centre for Solar Physics
University of Oslo
Norway

Susanna Parenti
Université Paris-Saclay
CNRS
Institut d'Astrophysique Spatiale
Orsay, France

Gordon Petrie
National Solar Observatory
Boulder, Colorado, USA

Viviane Pierrard
Belgian Institute for Space Aeronomy
Brussels, Belgium

Rui Pinto
Laboratoire Dynamique des Etoiles, des (Exo)planètes et de leur Environnement (LDE3)
Astrophysics Division (DAp/AIM)
Saclay Nuclear Research Centre (CEA Saclay)
Gif-sur-Yvette, France;
and
Institut de Recherche en Astrophysique et Planétologie
Toulouse, France

Jiong Qiu
Montana State University
Bozeman, Montana, USA

Nour E. Raouafi
Johns Hopkins University Applied Physics Laboratory
Laurel, Maryland, USA

Fabio Reale
Dipartimento di Fisica & Chimica
Universitá di Palermo
Palermo, Italy

Alexis P. Rouillard
Institut de Recherche en Astrophysique et Planétologie
Toulouse, France

Eduardo Sanchez-Diaz
Institut de Recherche en Astrophysique et Planétologie
Toulouse, France

Albert Y. Shih
NASA Goddard Space Flight Center
Greenbelt, Maryland, USA

Alphonse C. Sterling
NASA Marshall Space Flight Center
Huntsville, Alabama, USA

Barbara J. Thompson
NASA Goddard Space Flight Center
Greenbelt, Maryland, USA

Nicholeen Viall
NASA Goddard Space Flight Center
Greenbelt, Maryland, USA

Christian Vocks
Leibniz Institute for Astrophysics Potsdam
Potsdam, Germany

Angelos Vourlidas
Johns Hopkins University Applied Physics Laboratory
Laurel, Maryland, USA

Linghua Wang
School of Earth and Space Sciences
Peking University
Beijing, China

David F. Webb
Institute for Scientific Research
Boston College
Chestnut Hill, Massachusetts, USA

Yihong Wu
Leibniz Institute for Astrophysics Potsdam
Potsdam, Germany

PREFACE

The upcoming decade will mark a turning point in solar and heliospheric research. The spacecraft fleet comprising the Heliophysics Systems Observatory will be augmented by a set of unique space missions and ground-based telescopes that will greatly expand our knowledge of the heliosphere. The Parker Solar Probe (PSP) was launched in August 2018, sixty years after it was conceived as humanity's first mission to enter and study a stellar atmosphere from "within." The PSP will approach closer to the Sun than any spacecraft before to explore the solar atmosphere as close as 8.86 solar radii above the surface. The PSP has completed 6 of its 24 scheduled orbits as of this writing. The data recorded during the first six orbits show an unprecedented view of the nascent solar wind. We expect more discoveries as the spacecraft flies ever closer to the Sun. The Solar Orbiter (SolO), launched in February 2020, will approach within 60 solar radii but from an orbit inclined by up to 34° out of the ecliptic. SolO will give us the first unprojected images of the solar poles and measure the magnetic fields in these regions. The largest ground-based telescope ever built, the 4-m Daniel K. Inouye Solar Telescope (DKIST), will be operational next year. The DKIST will reveal solar structures as small as 20–30 km in diameter. In the near future, the 4-m European Solar Telescope (EST) will also become operational.

The observations and measurements from the PSP and SolO and those we will get from the DKIST, EST, and other future missions and telescopes will not only help answer long-standing scientific questions but also lead to significant discoveries and open new avenues for exploration. These measurement capabilities will write a new chapter of space and solar physics. So, it is a good time now to take a look at the status of the solar and heliospheric research. This book presents seven chapters that cover most aspects of solar and heliospheric physics.

Important technological inventions, along with essential advances in mathematics and physical theories made over the last few centuries, led to fundamental solar and astronomical discoveries. These discoveries created new puzzles that form the main pillars and axes of solar and heliospheric research. The invention of the telescope and the discovery of sunspots by Galileo in the 16th century represent a turning point in solar physics research. Soon after, the 11-year solar cycle was revealed with its long- and short-term variability. The observation of the strongest flare in recorded history by Carrington in 1858 was another important milestone and the building block for solar activity and space weather. The exact nature of these phenomena remained hidden until the discovery of solar magnetism by George Hale (1908). He stated in his *The Astrophysical Journal* paper: "The present paper describes an attempt to enter one of the new fields of research opened by this recent work with the spectroheliograph," which is a testament to the fact that scientific advances go hand in hand with technology. In the next half- century, three major phenomena were discovered: the coronal heating problem, the solar wind and its acceleration, and coronal mass ejections. All of these phenomena are driven by the magnetic field, which is generated deep in the convection zone

The Sun remains the most observed and studied star in the Universe. Yet many of the processes that govern its behavior are not fully understood. The solar dynamo and convection, coronal heating, the acceleration of the solar wind, flares, coronal mass ejections, and solar energetic particles are all outstanding examples of challenging solar and heliospheric problems. Solar and heliospheric research has undergone a renaissance in the last decades. Since the advent of the Space Age, significant strides have been made in our understanding of the Sun along with breakthrough discoveries. Important advances in theory and computing power are keeping pace, allowing access to fundamental physical processes that are otherwise impossible to explore because of the complexity of the underlying theories. These advances have helped us better understand the challenges facing us when we try to comprehend how the Sun and its corona work and also what it takes to obtain the measurements needed to make progress. Figure I.1 is an illustration of the integrated ground–space–theory system, which is a prerequisite to overcoming the hurdles we were facing for decades.

Understanding the Sun is essential not only because we live in its extended atmosphere (i.e., the corona and the solar wind) but also because it is the only star we can study in detail. The knowledge we gain from observing the Sun and its environment provides insights into other worlds that may harbor life like our Earth.

This book provides an overview of solar physics and the advances made over the past few decades as well as the challenging problems that remain. The seven chapters cover the solar interior, the atmosphere, magnetism and radiation, plasma heating and acceleration, and solar activity. It is a comprehensive view of how our star works and what is needed to understand it better. We hope that this reference work will help researchers in other fields, young scientists, teachers, students, and the public to familiarize themselves with the status of the field as of 2019.

Nour E. Raouafi and Angelos Vourlidas
Johns Hopkins University Applied Physics Laboratory
USA

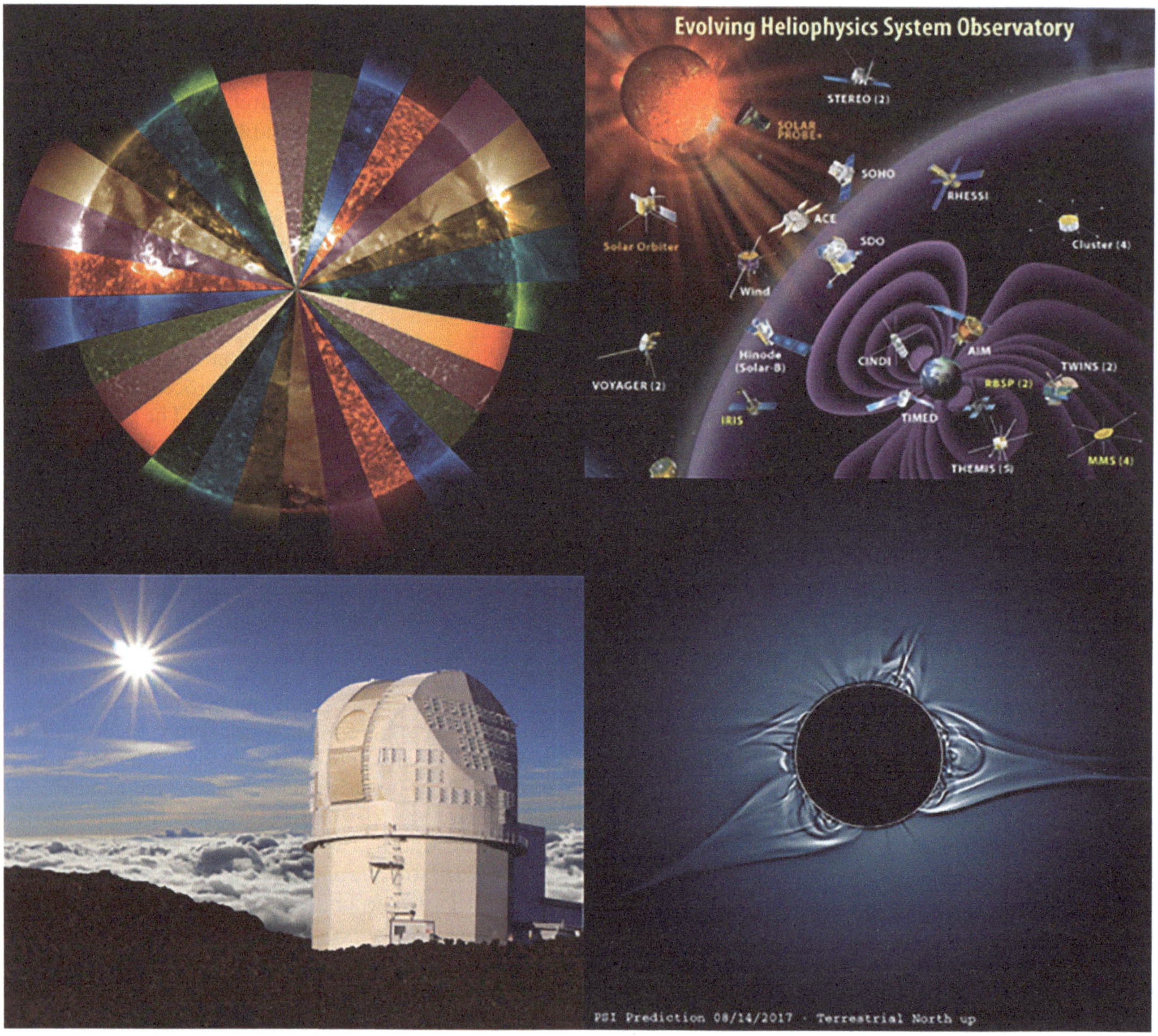

Figure I.1 Illustration of the complex solar environment and the recent observational and modeling advances that will lead to breakthrough insights into challenging phenomena that have been puzzling scientists for decades. Top left: Slice view showing the Sun at different wavelengths. Top right: The fleet of heliospheric space missions. Bottom left: Image of the 4-m DKIST solar telescope. Bottom right: Simulation of the structure of the solar corona for the 2017 eclipse. (*Source:* Reproduced with permission from Predictive Science Inc.)

1

The Solar Wind

Alexis P. Rouillard[1], Nicholeen Viall[2], Viviane Pierrard[3], Christian Vocks[4], Lorenzo Matteini[5], Olga Alexandrova[6], Aleida K. Higginson[2], Benoit Lavraud[7], Michael Lavarra[1], Yihong Wu[4], Rui Pinto[1,8], Alessandro Bemporad[9], and Eduardo Sanchez-Diaz[1]

ABSTRACT

The solar corona produces a continuous outflow of plasma influencing the entire solar system and beyond to over a hundred of astronomical units. Due to the Sun's proximity the source regions of that wind can be observed remotely with great precision and the escaping plasma can be measured directly in situ. Space missions launched over the past six decades have exploited this fantastic plasma laboratory to reveal the very rich physical processes at play as the solar wind forms and expands into space. This chapter provides a brief synthesis of these observational discoveries. It covers a broad range of topics, from the large-scale imaging and detailed spectroscopic observations of the wind source regions to the dynamical evolution of interplanetary plasma down to kinetic scales. Despite the great advances accomplished in solar wind research over the past decades, many mysteries remain concerning the origins of the solar winds. These mysteries are currently being addressed with a new fleet of extraordinary missions en route to our star.

1.1. INTRODUCTION

One of the current mysteries in heliophysics is the heating of the solar atmosphere to temperatures that are orders of magnitude hotter than the solar surface. As a result of this heating, the Sun cannot contain its atmosphere, and a continual outflow of plasma streams out from the solar corona to interplanetary space and beyond. For the debate surrounding the exact physical mechanisms of the heating of the corona, we direct the reader to Chapter 2. We here discuss the physical mechanisms behind the formation and propagation of the solar wind that are not yet well understood.

The whole volume of space influenced by the solar wind is called the heliosphere, and its size is in part modulated by the solar wind ram pressure. The solar wind extends from the corona to well beyond a hundred astronomical units (AU) to a termination shock. Beyond that shock, the solar wind slows down abruptly in response to the pressure of the interstellar medium, and the plasma becomes compressed and more turbulent until it reaches a zone where it can no longer push back the

[1] *Institut de Recherche en Astrophysique et Planétologie, Toulouse, France*

[2] *NASA Goddard Space Flight Center, Greenbelt, Maryland, USA*

[3] *Belgian Institute for Space Aeronomy, Brussels, Belgium*

[4] *Leibniz Institute for Astrophysics Potsdam, Potsdam, Germany*

[5] *Department of Physics, Imperial College London, London, UK*

[6] *Laboratoire d'Etudes Spatiales et d'Instrumentation en Astrophysique, Observatoire de Paris, Université PSL, CNRS Sorbonne Université, Université de Paris, Meudon, France*

[7] *Laboratoire d'Astrophysique de Bordeaux, Université de Bordeaux, CNRS, B18N, Pessac, France*

[8] *Laboratoire Dynamique des Etoiles, des (Exo)planètes et de leur Environnement (LDE3), Astrophysics Division (DAp/AIM), Saclay Nuclear Research Centre (CEA Saclay), Gif-sur-Yvette, France*

[9] *INAF Osservatorio Astrofisico di Torino, Turin, Italy*

Space Physics and Aeronomy Collection Volume 1: Solar Physics and Solar Wind, Geophysical Monograph 258, First Edition.
Edited by Nour E. Raouafi and Angelos Vourlidas.

DOI: 10.1002/9781119815600.ch1

interstellar plasma. In situ measurements of these outer regions are sparse, and our exploration of this boundary layer has only just begun with the *Voyager* spacecraft. In particular, the global shape of the heliosphere is still not known because the *Voyager* has only measured a very small region of the heliospheric boundary.

Birkeland (1908) argued very early that a corpuscular emission from sunspots consisting of relativistic electrons must impact Earth's magnetic field and be deflected to the polar regions to create the aurora. For several decades, it was realized that particles could be emitted from the Sun during flares, but it was generally thought that the space around Earth was mostly empty or perhaps traversed by occasional streams of particles from the Sun (Chapman & Ferraro, 1931). The prevailing view was that the solar corona consists of a hot gas (possibly extending to 1 AU), in thermal and hydrostatic equilibrium, pulled back by the solar gravitational field (Chapman & Zirin, 1957). Detailed observational studies of comets by Biermann (1951) showed that a subset of their tails cannot be accelerated by radiation pressure alone but may also respond to material flowing out from the Sun's atmosphere. He suggested that the passage of solar particles at the comet formed an ion tail and that these particles must have a very high speed relative to the comet in order to align the tail in the Sun's direction. Parker (1958) built on these observations and realized that the high temperature of the corona can provide enough energy to force coronal plasma to accelerate from subsonic to supersonic speeds. He demonstrated that the hydrostatic approach predicted too high kinetic pressure at infinity and that a continuous radial expansion of solar gas must act to reduce the coronal pressure. This was the first theory describing the continual expansion of what we now call the *solar wind*.

In this model, a dominant force affecting coronal particles and pushing them outward is induced by the thermal pressure gradient in the corona. Parker's original model assumed an isothermal corona, but subsequent models allowing for a varying temperature with radial distance confirmed that a supersonic wind can also form under such conditions. The presence of an outflowing supersonic wind ranging from 300 to 800 km/s was confirmed by plasma measurements from the *Luna* 1, 2, 3; *Venera* 1; and *Mariner 2* (e.g., Neugebauer & Snyder, 1962) and the numerous subsequent solar wind dedicated missions (e.g., Hundhausen & Gosling, 1976; Marsch et al., 1982). The Parker model provided an acceleration from subsonic speed in the coronal region to supersonic speeds of around 300–400 km/s typical of the slow solar wind. His model was unable, however, to explain fast solar wind speeds of 700–800 km/s without assuming unrealistic coronal hole temperatures in excess of 2×10^6K. As we shall show in this chapter, coronal and solar wind measurements suggest that additional physical mechanisms must be accounted for to produce a fast solar wind.

The solar wind has been measured in situ for several decades near 1 AU. Typical solar wind speeds near 1 AU range from 300 to 800 km/s, proton temperatures take values between 10^5 K in the slow wind to about 2×10^5 in the fast wind. There is now no doubt that the fast solar wind measured in situ originates near the center of coronal holes at the Sun (see Chapters 2 and 3). Coronal holes are cooler regions of the solar atmosphere that exhibit drops in Extreme UltraViolet (EUV) emissions. The cooler temperatures result from the significant escape of heat out in the solar wind. The fast solar wind streams out along open magnetic fields connecting regions deep inside coronal holes to the interplanetary medium. The origin of the slow wind is more complex and likely consists of multiple source components: these include transient releases from helmet streamers, plasma accelerating on rapidly expanding magnetic fields rooted at the boundary of coronal holes (through processes likely similar to that of the fast wind), and/or continual plasma exchanges between loops and open magnetic fields. We will describe these different components in the following sections.

1.2. OBSERVATIONS OF THE NASCENT SOLAR WIND

1.2.1. Remote-Sensing Observations of Coronal Heating and the Solar Wind

We begin our story on the solar wind with its formation in the solar corona; we first present remote-sensing observations that have provided important information on the conditions in which the winds are produced.

It is hard to observe the coronal source regions of the fast and slow solar winds in white-light images obtained routinely by coronagraphs. This is because the fast solar wind originates in very tenuous coronal holes, whereas the slow solar wind emerges from the vicinity of dense streamer loops that completely dominate coronal brightness. In contrast, spectroscopic observations provide more detailed information about the temperatures, flow velocities, and wave properties during the formation of the winds near their sources.

The *Solar and Heliospheric Observatory* (*SoHO*; Domingo et al., 1995) revolutionized how we observe the corona and the nascent solar wind. In particular, the Ultraviolet Coronagraph Spectrometer (UVCS/SoHO; Kohl et al., 1995) revealed that heavy ions such as O^{5+} and Mg^{9+} are heated hundreds of times more strongly than protons and electrons, and have very anisotropic kinetic temperatures (Antonucci et al., 2000; Kohl et al., 1997, 1998), meaning that temperatures measured in the direction perpendicular to the magnetic field are often much larger than

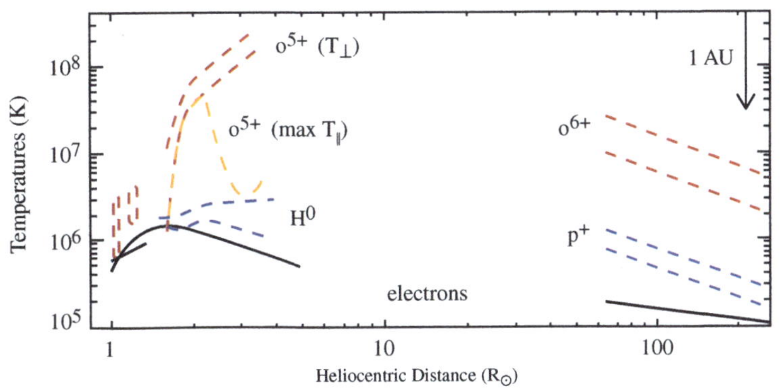

Figure 1.1 Radial evolution of solar wind temperatures from the corona to 1 AU. Indirect estimates of coronal temperatures are derived for three species (electrons, oxygen, and hydrogen) from an "empirical coronal model" that exploits SUMER (1–1.2$R_\odot$) and UVCS (1.5–4$R_\odot$) line widths compared with direct in situ measurements at distances greater than 60$R_\odot$) in the high-speed wind (Cranmer et al., 1999). The in situ data were assembled from *Helios*, *IMP*, *Ulysses*, and *Voyager* particle data, and double sets of curves denote rough lower and upper bounds on representative fast-wind values. The upper and lower limits correspond to extreme values of an assumed non-thermal component to the line broadening that is attributed to unresolved MHD wave motions along the line of sight. Temperatures of electrons (solid black), hydrogen (dotted), and oxygen (dashed) are shown. Oxygen ions correspond to O5+ in the corona but O6+ in the far solar wind, and coronal temperatures of neutral hydrogen are here compared with proton temperatures in the solar wind. Figure taken with permission from (Cranmer et al., 1999).

those parallel to the field (see Chapter 2). A comparison between different ion and electron temperatures derived from coronal observations and solar wind measurements are shown in Figure 1.1. The measured temperatures provide support to coronal heating mechanisms involving collisionless wave–particle resonances with frequencies of 10 Hz to 10 kHz. These waves could be damped to heat preferentially heavy ions (Cranmer et al., 1999; Tu & Marsch, 1997). It is not clear yet how such hypothetical waves can be generated from lower-frequency Alfvén waves typically emitted at minute periods. Some proposed mechanisms involve the turbulent cascade of magneto-plasma fluctuations from low to high frequencies (Hollweg, 2002). Then kinetic processes must occur to damp the small-scale fluctuations that have spectrally cascaded to oblique wavevectors. Proposed mechanisms include ion-cyclotron and Landau damping (Leamon et al., 1998). Nonlinear processes such as beam instabilities or mode conversion and damping have also been proposed.

Emission lines, and more specifically the dimming of certain lines measured via coronal spectroscopy, have been used to infer the outflow speed and density of the forming fast solar wind. This effect is most important for spectral lines in the UV range that have a significant component due to ion excitation via resonant scattering of chromospheric emission (followed by spontaneous emission). For some lines, such as O VI doublet 103.2 nm and 103.8 nm lines, and of the HI Lyman-α 121.6 nm and Lyman-β 102.5 nm lines, the collisional and radiative components can be separated (Marocchi et al., 2001). This separation can be used to investigate the dimming of the radiative component, which is due to the radial expansion of the emitting coronal atoms. This Doppler shifts the (narrow) exciting chromospheric profile with respect to the (broad) atomic absorption profile. As a result, the UV intensity of the resonantly scattered component of the line emission decreases with increasing outflow velocities (Hyder & Lites, 1970; Noci et al., 1987; Withbroe et al., 1982). These observations applied to the O VI doublet lines have shown that the fast solar wind becomes supersonic much closer to the Sun than the slow solar wind. The fast O^{5+} ions reach speeds in excess of 600 km/s within 4 solar radii from the solar surface (Antonucci et al., 2000). Identical techniques were also used to study flows in the vicinity of streamers, adjacent but not above the helmet, and found that the slow solar wind accelerates more slowly, with its outflow speed remaining below 200 km/s at least until 4 solar radii (Abbo et al., 2010; Strachan et al., 2000). Figure 1.2 presents a summary of these results.

A similar analysis of the Doppler dimming technique has been applied to the Lyman-α (121.6 nm) emission of neutral hydrogen. The latter moves outward in response to rapid charge-exchange coupling with the heated protons that eventually form the bulk of the solar wind. Neutral hydrogen therefore acts as a proxy for protons at heights up to 2.5 solar radii in coronal holes, and higher heights in more dense structures. In some coronal structures, He^+ can act as a proxy for alpha particles, which are also an important component of the solar corona. The

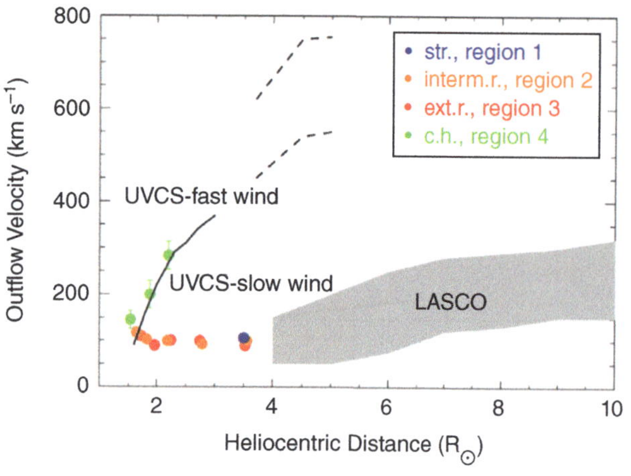

Figure 1.2 Outflow velocity (km/s) of the solar wind for the considered four regions as a function of the heliocentric distance (in solar radii). The gray band from 4 to 10 $R_\odot$ shows the range of outflow velocities for the slow wind obtained with LASCO (Sheeley et al., 1997). The solid curve up to 3 $R_\odot$ represents the values of the fast wind obtained from the UVCS data (Antonucci et al., 2000), and the dashed curves show the results by Telloni et al. (2007) of the fast wind velocity. The error bars (small in many cases) have been estimated based on the propagation of the statistical uncertainties of the observed OVI 1032 and 1037 line intensities. (*Source*: Reproduced from Abbo et al., 2010. © 2010, Elsevier.)

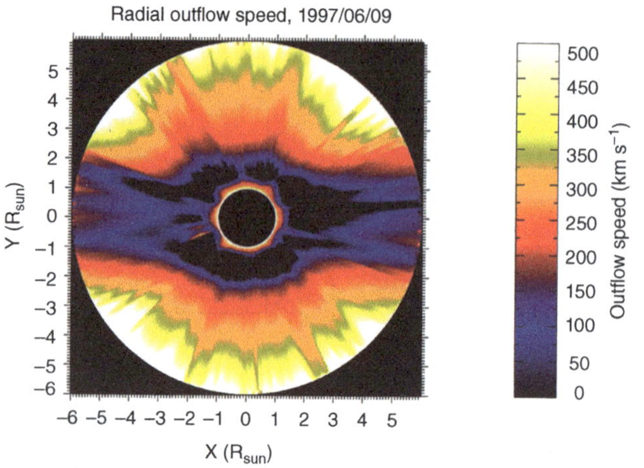

Figure 1.3 2D map of radial outflow velocity in the plane of the sky derived from the ratio between white-light and UV coronal emissions. The outer white region corresponds to altitudes where the Doppler dimming technique with the Ly spectral line cannot be applied anymore. (*Source*: Taken from Bemporad, 2017. © 2017, IOP Publishing.)

UVCS/SoHO instrument has provided H I Lyman-α spectral line data over 18 years. This makes it possible to study proton outflow speeds throughout the solar cycle, focusing on the coronal region sampled by the spectrometer field of view, such as coronal streamers (Susino et al., 2008; Zangrilli & Poletto, 2016) and coronal holes (Antonucci et al., 2000; Strachan et al., 1993; Teriaca et al., 2003).

In a recent analysis (Bemporad, 2017), UVCS daily Lyman-α synoptic data were combined to provide the first 2D images of coronal Lyman-α emission, representative of future data that will be acquired by the Metis coronagraph onboard Solar Orbiter (Antonucci et al., 2017). These have been directly combined with classical 2D coronagraphic images acquired in white light with LASCO to derive 2D maps of HI outflow speeds, with a technique originally described by Withbroe et al. (1982) that neglects line-of-sight integration effects. As pointed out by Bemporad (2017), because both the radiative component of Lyman-α emission and the white-light polarized emission depend on the electron density distribution integrated along the line of sight, this latter quantity can be simplified by directly taking the ratio between the two UV and white-light intensities.

Figure 1.3 shows an example of a 2D outflow velocity map, ranging between 1.5 and 4.0 $R_\odot$, that was obtained this way (Bemporad, 2017). The map shows an increasing speed with altitude from about 150–200 km/s in the equatorial regions to 400 km/s in the polar regions. As we shall see, these values are in agreement with the expected latitudinal distribution of slow and fast solar wind components during solar minimum, corresponding to equatorial regions, and mid-latitude and polar regions, respectively. Alternatively, as shown more recently by Dolei et al. (2018), line-of-sight integration effects can be fully taken into account (under some assumptions) by deriving electron densities from white-light coronagraphic images. From this, one can derive HI outflow speed maps by again exploiting the Doppler dimming technique discussed above.

The solar wind acceleration has also been measured via oscillations of the coronal plasma, identified as Alfvén waves in coronagraphic observations. Alfvén waves transverse to the plane-of-sky have been observed to be omnipresent in coronal holes within the first 0.3 $R_\odot$ of the solar atmosphere by the Coronal Multichannel Polarimeter (Tomczyk et al., 2016). Detailed analysis of off-limb Doppler shifts (Fe XIII) at high time cadence has furthermore revealed the presence of upward- and downward-propagating waves and, by comparing their respective phase speeds, allowed for the determination of the bulk flow speed profile of nascent fast wind flows (Morton et al., 2015, 2016). These observations have furthermore shown that the spectra of the oscillations show a predominant 1/f slope, reminiscent of solar wind measurements made beyond 0.3 AU by past space probes (e.g., Bavassano, Dobrowolny, Fanfoni et al., 1982; M. L. Goldstein et al., 1995). This supports the idea that the 1/f component of the oscillation spectra observed in the solar wind are already set in the low solar atmosphere and are advected outward (see, for example, Verdini et al., 2012). The statistical properties of fluctuations at magnetohydrodynamic (MHD) and kinetic scales are discussed further in Section 1.4.

1.2.2. Transient Coronal Outflows in the Nascent Solar Wind

The formation of the background solar wind, introduced in the previous section, is continually perturbed by the ejection of jets and small transients that form in the corona. Direct observations of these transient outflows in EUV and white-light images by the *STEREO* and *SoHO* spacecraft have provided new insights on the origin of mesoscale structures measured in situ in the solar wind.

Variable solar wind outflows in the form of plasmoids are continually released from helmet streamers in white-light (i.e., electron density) observations (e.g., Harrison et al., 2009; Rouillard, Davies, et al., 2010; Rouillard, Lavraud, et al., 2010; Rouillard et al., 2009; Sheeley et al., 1997; Sheeley et al., 2007; Wang et al., 1998; Wang et al., 2000). These plasmoids have been tracked from the corona, through the inner heliosphere, and in some cases out to 1 AU using heliospheric imagers (Rouillard, Davies, et al., 2010; Rouillard et al., 2011; Sheeley & Rouillard, 2010), showing in a direct way that some of the helmet streamer structures produced in the corona result in density structures measured in the inner heliosphere. Plasmoids (or "blobs") have been tracked from the tip of streamers where they typically form to several tens of solar radii, and analysis of their kinematic properties has confirmed that they are advected in the slow wind (Sheeley et al., 1997). In fact, this type of analysis has provided one of the rare kinematic measurements of the forming slow wind. It has revealed that a subset of the slow solar wind is released right above helmet streamers and accelerates over 20–30 solar radii to reach its terminal speed of about 300 km/s (this acceleration is shown as the gray area in Figure 1.2).

Helmet streamers form in the corona where magnetic fields of opposite polarity meet, and therefore a complex reconfiguration of the solar magnetic field is likely to occur due to magnetic reconnection. Magnetic reconnection can occur high up in the solar atmosphere (4–6 solar radii), and the collapse of newly formed magnetic loops can force the downward motion of coronal plasma. When densities are high enough, these plasma "inflows" are detected in coronagraphs in the vicinity of streamers and the coronal neutral line where the heliospheric current sheet (described next) forms (Wang et al., 2000). Multispacecraft studies using *SoHO* and *STEREO* images have recently shown that these inflows are associated with the release of density blobs in the slow solar wind (Sanchez-Diaz et al., 2017).

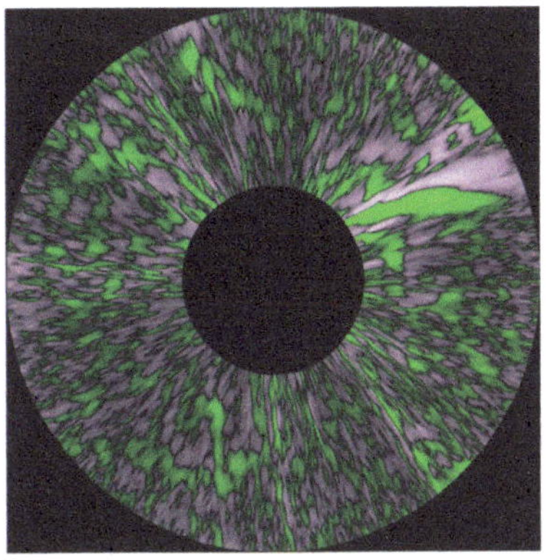

Figure 1.4 Propagating brightness fluctuations (green/black) derived from COR-2 observations during a dedicated deep-field campaign. The fluctuations are present at all azimuths and times, with a wide range of brightnesses and lateral sizes. The fluctuations appear with smaller scales than the Sheeley blobs (discussed in the text); such a blob is observed off the north-western limb in this image as the larger bright green feature. (*Source*: Taken from DeForest et al., 2018.)

An important property of density blobs and plasmoids released in the solar wind is their multi-scale and cyclic nature. Fourier analysis of brightness variations released from a highly tilted current sheet near solar maximum has shown that the large plasmoids are released from the Sun with characteristic time scales of about 19–20 hr (Sanchez-Diaz et al., 2017). Similar spectral analysis has revealed characteristic 90 min timescales embedded within the larger plasmoids (Kepko et al., 2016; Viall et al., 2010; Viall & Vourlidas, 2015). DeForest et al. (2018) used deep-field, high-cadence coronagraph observations to show that there is still more substructure (shown in Figure 1.4)—both time dynamic on scales smaller than 90 min, and time stationary, filamentary streamer structures down to the resolution limit. The time-dynamic characteristic scale sizes likely have different formation causes, such as inherent time scales due to the characteristics of coronal heating (Endeve et al., 2004), or waves (Pylaev et al., 2017).

Observations of the corona therefore show that highly dynamic streamers host transient processes associated with the likely continual emergence, redistribution, and removal of solar magnetic flux directly influencing the properties of the magnetic fields and particles of the solar wind (Owens et al., 2013; Sanchez-Diaz et al., 2016). Magnetic reconnection occurring at the boundary of coronal holes was suggested to occur in the 1980s to explain the rigid rotation of coronal holes (Wang et al., 1988). Theoretical considerations (Fisk, 1996) and sophisticated numerical MHD simulations have highlighted the inevitable, complex evolution that takes place all along the corona's open-closed field boundaries (Antiochos et al., 2007; Linker et al., 2011; Lionello et al., 2005; Titov et al., 2009; Titov et al., 2011). The activity hosted by helmet streamers are the clearest example, with many different scales of plasma and magnetic field signatures generated by magnetic reconnection (Higginson et al., 2017; Higginson & Lynch, 2018; Lionello et al., 2005; Török et al., 2009).

Heliospheric imaging has also shown that not all structures at MHD scales measured in the solar wind form in the corona. DeForest, Matthaeus, Viall, and Cranmer (2016) showed turbulent density fluctuations setting in around 30 solar radii away from the Sun. These fluctuations coexist with outflowing helmet streamer plasmoids and are advected with the solar wind.

1.3. MEASUREMENTS OF THE SOLAR WIND IN THE INNER HELIOSPHERE

1.3.1. Bulk Properties and Large-Scale Structures

The properties of the solar wind escaping the corona change throughout the solar cycle. This results from the evolving coronal magnetic topology that responds to the emergence and evolution of photospheric magnetic fields. This evolution alters both the magnetization and the bulk properties of the wind at different heliocentric latitudes. Spacecraft have measured in situ the properties of solar wind magnetic fields and particles at different locations in the heliosphere and over several decades. Figure 1.5 displays three dial plots showing the distribution of solar wind speeds with latitude measured by the *Ulysses* spacecraft during its three polar orbits. The sunspot number plotted below the dial plots is low in the left-hand and right-hand plots, marking the occurrence of solar minima (McComas et al., 2008). At these times, the Sun's magnetic field is quasi-dipolar, and the ambient solar wind is very clearly structured in at least two types of plasma flows. A fast wind is measured at high latitudes above coronal holes, and a more complex slow wind is measured at the low latitudes of solar streamers. As the solar cycle advances toward the activity maximum (middle panel), the polar coronal holes can disappear temporarily and the magnetic field evolves toward a non-dipolar structure. In response to this process, the bulk speed loses the ordered latitudinal structure shown in Figure 1.5 (left). At solar maximum, the large-scale spatial separation between the slow and fast solar wind is less clear as shown in the middle dial plot. Global numerical models of the solar wind show the clear link between these

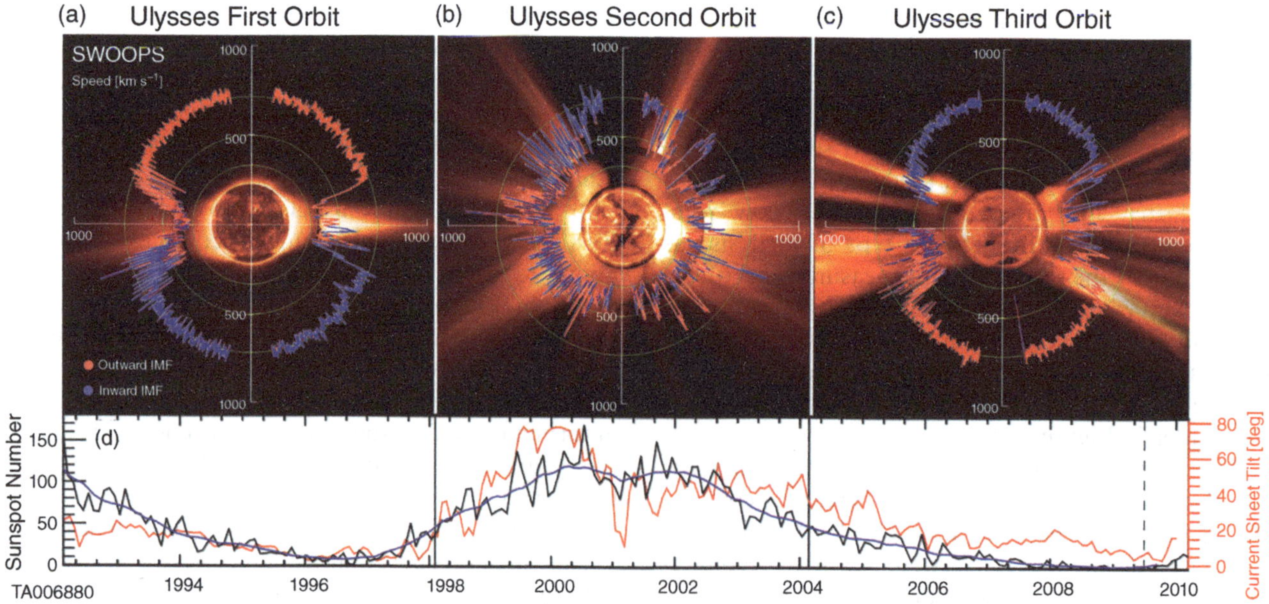

Figure 1.5 (a–c) Polar plots of the solar wind speed, colored by IMF polarity for Ulysses' three polar orbits to indicate measured magnetic polarity. (d) Contemporaneous values for the smoothed sunspot number (black) and heliospheric current sheet tilt (red), lined up to match Figures 1.1a–c. In Figures 1.1a–c, the solar wind speed is plotted over characteristic solar images for solar minimum for cycle 22 (17 August 1996), solar maximum for cycle 23 (7 December 2000), and solar minimum for cycle 23 (28 March 2006). From the center out, we blend images from the *Solar and Heliospheric Observatory* (*SoHO*) Extreme ultraviolet Imaging Telescope (Fe XII at 1950 nm), the Mauna Loa K coronagraph (700–950 nm), and the SoHO C2 white-light coronagraph. (*Source*: Image reproduced with permission from McComas et al., 2008, © 2013 John Wiley & Sons.)

changes in the coronal magnetic topology and wind streams (Linker et al., 2011; Oran et al., 2013; Pinto et al., 2011; Pinto et al., 2016; van der Holst et al., 2014).

The first models capable of describing the general properties of the solar wind assumed a thermally driven flow. These include Parker's original (1958) theory that assumed a constant coronal temperature as well as subsequent fluid models that allowed for thermal stratification and inhomogeneities. The latest fluid models account for more detailed energy injection and transport mechanisms (Linker et al., 2011; Lionello et al., 2014; Oran et al., 2013; Pinto & Rouillard, 2017; van der Holst et al., 2014). Fluid models are not able, for known coronal temperatures, to explain the high speeds of the fast solar wind without including additional physical processes than just the effect of the thermal pressure gradient. These processes could involve Alfvén waves with their induced turbulent wave pressure and Reynolds stresses that would contribute to further accelerate the solar wind (Chandran, 2018; Cranmer et al., 1999; Lionello et al., 2014; Oran et al., 2013). Kinetic solar wind models suggest that heated particles such as suprathermal electrons, ubiquitous in the solar wind, could also contribute to the acceleration of the wind by imposing an electric field on the ions, extracting them out of the corona to high speeds (Pierrard & Pieters, 2014). It is, however, very challenging to deal with these different types of processes altogether in a unified view, as they work on very different scales.

At solar minimum, the quasi-dipolar distribution of coronal magnetic field lines extends into two magnetic hemispheres of opposite polarities in the interplanetary medium. These hemispheres are separated by a neutral line called the heliospheric current sheet (HCS). A schematic of the global structure of the HCS is shown in Figure 1.6. The HCS is typically identified in situ as an abrupt change in the magnetic field direction and a 180° switch in the pitch angle of suprathermal electrons (above tens of eV). It is also commonly called a sector boundary (Liu et al., 2014). The HCS is usually surrounded by a high-density region called the Heliospheric Plasma Sheet (HPS) that may correspond to the heliospheric extension of coronal streamer rays (Winterhalter et al., 1994). The sector boundary and the magnetic field reversal are, at times, not collocated in situ (Owens et al., 2013). This is likely indicative of the dynamic effects, already mentioned, occurring in the solar corona near the source of sector boundaries leading to a local folding of the magnetic field that is not necessarily associated with the large-scale HCS (Owens et al., 2013). The HCS exerts only small latitudinal displacements during solar minimum, but can extend over a broad latitudinal range at solar maximum when solar active regions are more numerous at low latitudes. For weak to moderate solar activity, the topological shape of the HCS resembles that of a ballerina skirt, as illustrated in Figure 1.6.

The expansion of the solar wind plasma combined with the effect of solar rotation shapes the interplanetary magnetic field into Archimedean spirals rooted at the Sun with a radial position $r(\theta)=k\theta/(2\pi)$, where r is the radial coordinate, θ the azimuthal angle, and k is

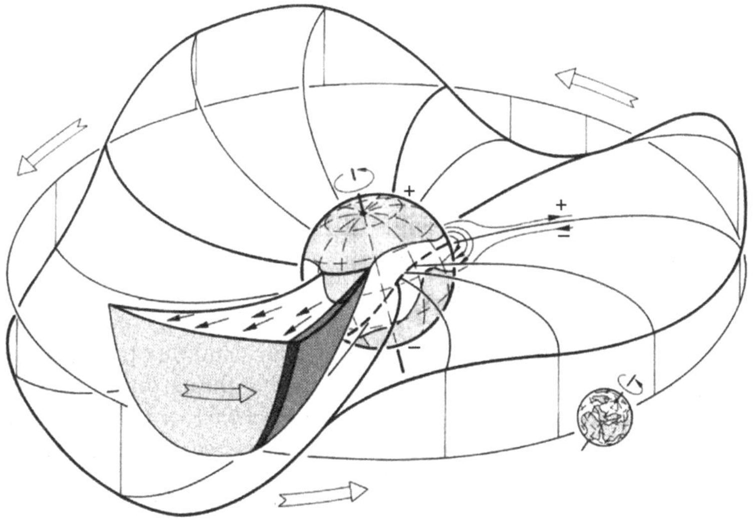

Figure 1.6 Configuration in the inner heliosphere of a "ballerina skirt" heliospheric current sheet extending above the streamer belt near solar minimum (for A > 0 solar magnetic field polarity, that is, outward field at the north pole), which lies ahead of a high-speed stream (drawn truncated at high latitudes) from an equatorward extension of a northern polar coronal hole. The dark shaded region is the interaction region. (*Source*: Image reproduced with permission from Schwenn, 1990, © 1990, Springer.)

the product of synoptic period of the Sun and the solar wind speed (Parker, 1958). These spirals are known as Parker spirals and, according to Maxwell's equations, plasma elements situated on a same Parker spiral share a common source region at the Sun. The measured dependence between the magnetic field angle with respect to the radial direction and solar wind speed agrees very well with the simple Parker spiral model. This is true in the Helios spacecraft measurements of the inner heliosphere (Bruno & Bavassano, 1997), of near 1 AU measurements taken over several decades (Borovsky, 2010), and further out along the *Pioneer* and *Voyager* orbits in the more distant outer heliosphere (Burlaga & Ness, 1993). Measurements of magnetic fields at high latitudes have been limited to those made by the *Ulysses* spacecraft. These measurements also confirm that the angle of the interplanetary magnetic field to the radial direction also agrees well with the Parker spiral model (Forsyth et al., 2002). Deviation from the average spiral orientations can be due to dynamic processes that occur near the source region of the solar wind or the onset and development of turbulent flows (Fisk, 1996). Both of these are discussed at length in the later sections of this chapter; they are likely important to place constraints on theories of coronal heating and solar wind formation.

1.3.2. Composition of the Solar Winds

Figure 1.7 presents solar wind parameters during the passage of a high-speed stream bound by two intervals of the slow wind. In situ measurements at 1 AU and beyond are not ideal to study the origin of the solar winds because the latter evolve greatly between the Sun and 1 AU. For example, the fast and slow winds interact during transit to 1 AU, and the distribution functions are strongly affected by a number of processes during transport from the Sun as discussed later in this chapter. In essence, the solar wind undergoes a number of processes that irreversibly erase information on their source regions during transit. Wind properties that remain unchanged during transit to 1AU are the abundance and ionization state of minor ions. They provide crucial information on the origins of these winds.

As an element rises in the solar atmosphere from the chromosphere to the hot corona, its ionization level increases due to radiative and collisional processes. This carries on until it reaches a height where radiative ionization is too weak and collisions are rare. Beyond that height, the charge state is said to be "frozen" and remains unchanged until it is measured in situ in the interplanetary medium. A measure of the charge state of heavy ions therefore provides unique information on the temperature at the source regions of the solar winds (Geiss et al., 1995). The charge state of an ion species will

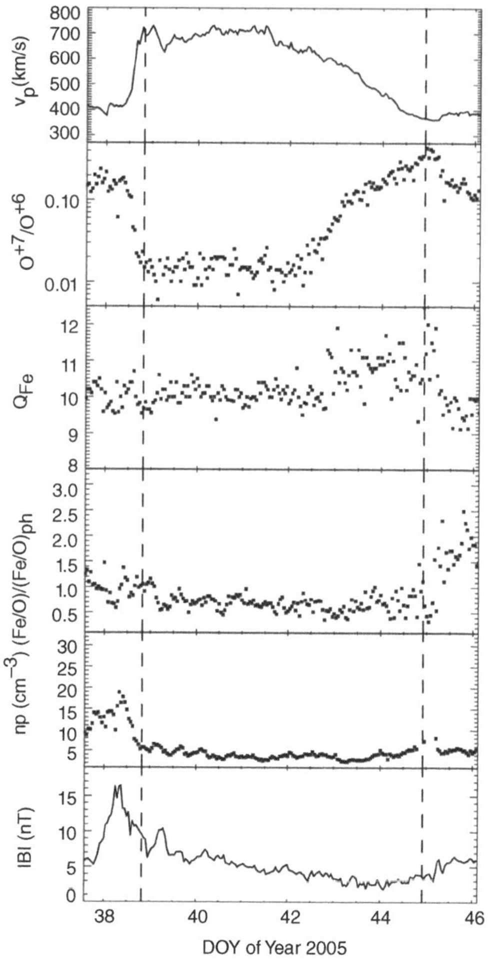

Figure 1.7 Typical time profiles of solar wind parameters for a selected solar wind interval. The SW parameters are (from top down) proton speed (vp), O+7/O+6 density ratio, average charge of Fe (QFe), Fe/O relative to its photospheric value (0.0646; Asplund et al., 2009, related to the first ionization potential, FIP, bias), proton density (np), and the magnitude of the associated interplanetary magnetic field (—B—). The selected SW interval is between the two vertical dashed lines. (*Source*: Image taken with permission from Ko et al., 2014. © 2014, IOP Publishing.)

increase with temperature at the collisional coronal base of a solar wind flux tube. Solar wind measurements have shown that the carbon and oxygen ions have lower charge states (07+/06+ or C5+/C6+) in the fast than in the slow wind (Geiss et al., 1995; Kasper et al., 2012). This is clearly seen in Figure 1.7 for the 07+/06+ ratio.

Electron temperatures prevailing in the source region of the fast solar wind are therefore lower than temperatures near the source region of the slow wind (temperature smaller than 1.2 MK). Near 1 AU, where ion temperatures of the fast and slow solar winds are higher than temperatures estimated from charge state ratios.

This suggests that sustained heating processes occur in the fast and slow solar wind in the corona well above the heights where ionization states are frozen (Lopez et al., 1986).

Sharp O7+/O6+ boundaries occur at both leading and trailing edges of high-speed streams, suggesting rapid transitions between different coronal sources (Borovsky & Denton, 2016; Burton et al., 1999). High-cadence measurements of the ionization state of heavy ions in the slow wind have recently revealed a high degree of variability, changing by an order of magnitude inside density structures (Kepko et al., 2016). This will be discussed later in the chapter.

The higher charge state of heavy ions measured in the slow wind necessarily results from higher temperatures and densities at the coronal base of open field lines channeling the slow wind. Numerical models of the solar coronal plasma and magnetic field have been used to study the origin(s) of the slow wind. They reveal that magnetic fields are generally stronger at the base of flux tubes channeling the slow wind (Wang et al., 2009). There is a clear statistical correlation between coronal regions threaded by strong magnetic fields in the form of loops or open magnetic fields and high-plasma temperatures (Schrijver et al., 2004). Therefore, one possible interpretation for the high charge-state ratios measured in the slow wind resides in strong local heating near its source region due to the strong magnetic fields (Wang et al., 2009). Such a strong heating at the coronal base would lead to a strong heat flux conducted down to the chromosphere and enhanced densities in the slow wind. In contrast, less evaporation is likely to occur near the cooler source region of the fast wind, leading to a more tenuous fast wind. This interpretation can explain the fairly constant mass flux measured in situ in the fast and slow solar winds.

An alternative theory for the origin of the high charge states measured in the slow wind suggests that plasma initially confined to coronal loops is released along open magnetic field lines (Fisk et al., 1998; Schwadron et al., 1999). Loop plasma would naturally be pushed to higher ionization states because loops host typically hotter and denser plasma than open fields of coronal holes (Schwadron et al., 1999). For this mechanism to work, loop plasma must find a way to be transferred to the open field, which would require magnetic reconnection to occur continually.

The fast and slow solar winds also differ in their abundance of elements with low First Ionisation Potential (FIP), that is, the minimal energy required to ionize an atom. This first ionization occurs in the chromosphere for most elements. Overall, the corona and the solar winds exhibit higher abundances of elements with low FIP than known photospheric abundances. The slow wind exhibits even greater abundances of low-FIP elements such as Si, Fe, and Mg than the fast solar wind (von Steiger et al., 2000; Zurbuchen et al., 1999) as seen in the fourth panel of Figure 1.7 and in Figure 1.8. Such high abundances of low-FIP elements are not measured remotely in coronal holes but are more typical of closed

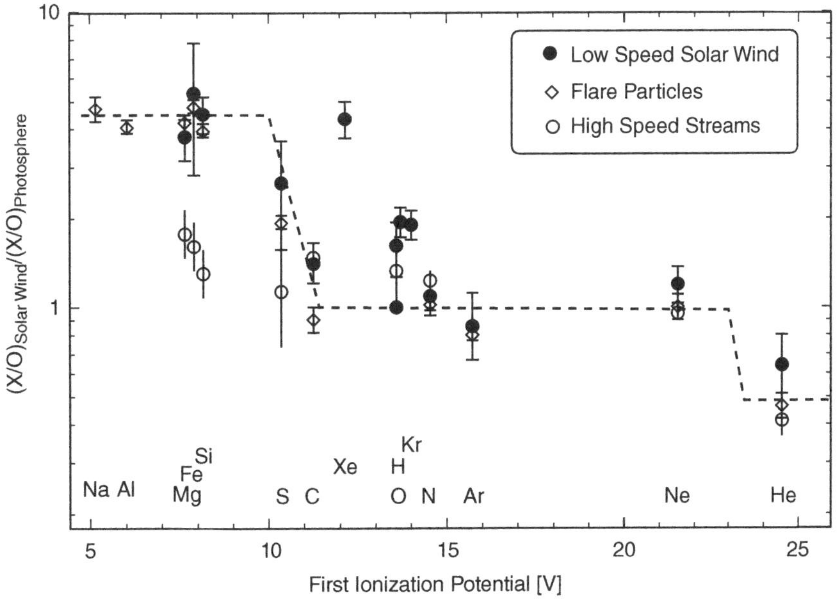

Figure 1.8 Element abundances as a function of the first ionization potential (FIP) in the average slow solar wind, fast solar wind. Abundances are given relative to oxygen and are normalized to photospheric abundances. (*Source*: This figure was taken from Geiss, 1998 with permission from SSR. © 1998, Springer Nature.)

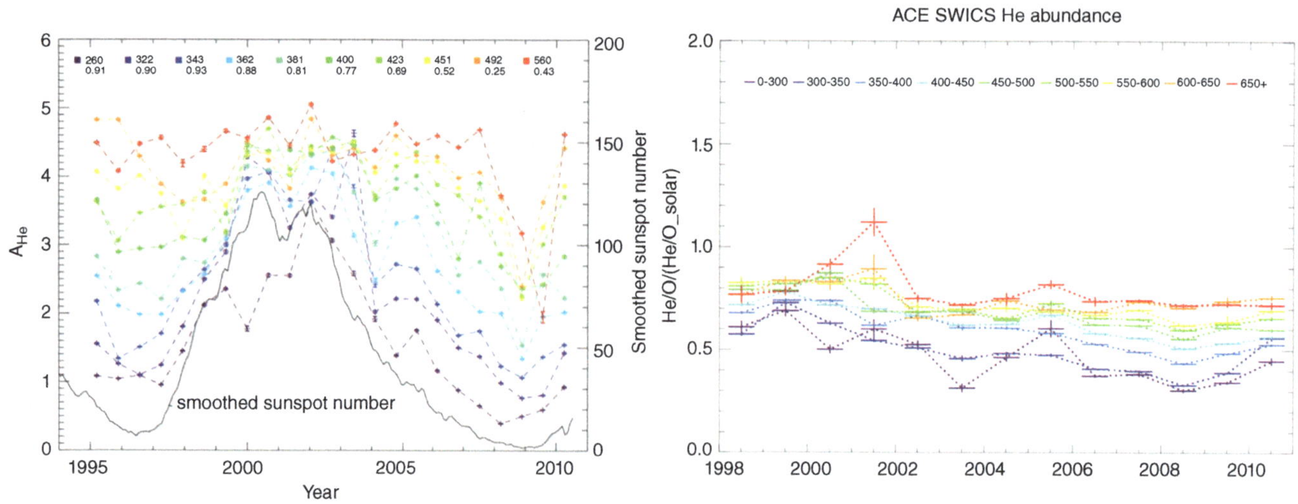

Figure 1.9 Left: The abundance of helium (computed as 100 times the density ratio of hydrogen) during 17 years of solar wind measurements by the Advanced Composition Explorer (ACE). The color coding indicates the solar wind speed. The monthly smoothed sunspot number is plotted as a black curve. Right: The abundance of helium relative to oxygen, measured by ACE/SWICS. The same trend of depletion with wind speed as for He/H is seen, but the solar cycle dependence is less pronounced. (*Source*: Images reproduced with permission from Kasper et al., 2012 and Rakowski & Laming, 2012. © 2012, IOP Publishing.)

coronal loops (Widing & Feldman, 2001). This supports the idea that the slow wind is supplied by plasma initially trapped on coronal loops and that are subsequently released in the wind (Fisk et al., 1998). In contrast, the fast wind carries fewer elements with low FIP (von Steiger et al., 2000). The composition of the fast solar wind agrees roughly with that measured in coronal holes.

The fast and slow winds can also differ in their abundance of elements with higher FIP such as helium. As shown in the left-hand panel of Figure 1.9 taken from Kasper et al. (2012), the helium abundance of the fast wind remains steady (4–6%) throughout the solar cycle. In contrast, the helium abundance in the slow wind varies from around 1% at solar minima to around 4–6% at solar maxima (Aellig et al., 2001; Kasper et al., 2012). Therefore, the slow wind carries less helium than the fast wind at solar minima, but both present similar abundances at solar maxima.

There is currently no accepted mechanism for the regulation of heavy ion abundances in the solar wind. All proposed mechanisms must necessarily occur where elements are first ionized in the chromosphere and fractionate elements in a mass-independent manner. Several mechanisms have been invoked in this region to modulate the transfer of heavy ions from the chromosphere to the corona. They include, for example, the effect of Coulomb collisions in high-temperature gradients (Bø et al., 2013) and the effect of MHD waves via a ponderomotive force (Laming, 2009, 2015). The latter mechanism, in particular, is able to explain a broad range of composition measurements in the fast and slow solar winds for reasonable conditions in the solar atmosphere.

1.3.3. Solar Wind Interaction Regions

The rotation of the Sun and its associated solar wind sources such as coronal holes can result in the radial alignment of the fast and slow solar wind. A kinematic gathering of the interplanetary plasma when the fast wind catches up the slow stream, inducing an increase in plasma density near the stream interface. When the solar wind behind is much faster than the solar wind ahead, it can compress the slow wind before it reaches 1AU (as seen in Figure 1.7). When a compression region is measured over at least two consecutive solar rotations it is called a Corotating Interaction Region (CIR), otherwise one refers to simply a Stream Interaction Region (SIR; L. Jian et al., 2006). The interaction between fast and slow solar wind that creates a CIR is illustrated in Figure 1.10. Detailed descriptions of CIRs can be found in Gosling and Pizzo, 1999, and in Pizzo, 1982. As seen in the left-hand schematic of Figure 1.10, the interaction region forms, like the background solar wind, an Archimedian spiral rooted at the Sun. In three dimensions, the stream interface can be titled as it responds to the tilt of the solar magnetic dipole. Figure 1.10 illustrates how plasma flows are typically deflected inside CIRs in latitude as well as in longitude (Pizzo, 1982).

CIRs and SIRs can be more complex structures than spirals; they are composed of a multitude of smaller

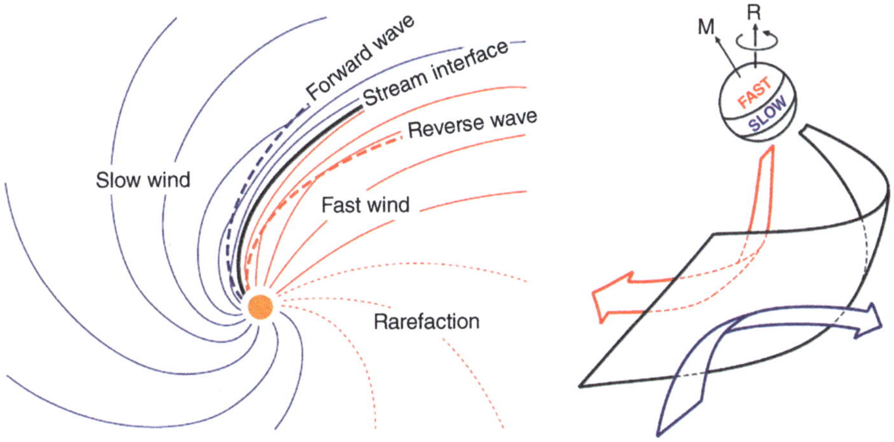

Figure 1.10 Two schematics illustrating CIRs where the fast solar wind catches up and compresses the slow solar wind. The left-hand schematic is a view from solar north, and the right-hand schematic illustrates the 3D interaction between flows when the stream interface is inclined relative to the north-south direction. (*Source*: Figure taken from Owens & Forsyth, 2013. © 2013, Springer Nature.)

interaction regions that merge during the wind propagation to 1 AU. This is a consequence of the variability of the slow wind released by helmet streamers (as discussed in Section 1.2.2). Strong dynamic effects associated with the rise in total pressure inside the SIRs and CIRs are most apparent beyond 1 AU, where forward/reverse shock pairs (Lee, 2000) can bound the interaction region.

The interaction of the fast and slow solar winds, particularly in SIRs and CIRs, reduces the range of velocities of the solar wind. The slow solar wind has typical slow speeds between 300 km/s and 450 km/s. Between 0.3 and 0.4 AU, where interaction regions have not yet formed to accelerate the slowest plasma, the slow solar wind exhibits speeds less than 300 km/s nearly 10% of the time. This very slow solar wind can have speeds as low as 200 km/s inside 0.7 AU, a speed never measured at 1 AU. Wind speeds less than 300 km/s are very seldom measured in situ at 1 AU but have been extensively observed in white-light images (Rouillard, Davies, et al., 2010). This very slow wind typically has lower temperatures and higher densities than the regular slow solar wind. The properties and the source of this VSSW as well as its solar cycle variability were analyzed by Sanchez-Diaz et al. (2016).

1.3.4. Mesoscale Structures

There is an abundance of structures in the solar wind that are above the kinetic scales, but well below the global scales of the heliosphere. These so-called mesoscale structures abound in the solar wind that fills the inner heliosphere and their in situ measurements provide insights on the formation mechanisms of the solar wind.

As we saw in section 1.2.2, evidence that at least some in situ mesoscale density structures originate within the solar corona, as the solar wind forms, is found in the remote imaging of the corona. The images recorded by *SoHO* and *STEREO* have provided a tracking of density fluctuations continuously at mesoscales (several 100 Mm) from the Sun to the interplanetary medium, where it is measured in situ. As we describe below, composition, electron strahl, velocity, magnetic field, plasma temperature, and plasma density measured in situ have also been used to begin to piece together a picture where time dynamics, such as reconnection, and spatial structure at the Sun combine to create mesoscale structure in the solar wind. In essence, the solar wind measured in situ is far from homogeneous and is most likely formed that way.

The solar wind measured near the HCS is one region where myriad mesoscale structures are found. Crooker et al. (1996, 1993) showed that the highly structured solar wind measurements associated with the HCS at 1 AU were not simply due to multiple crossings of a single wavy current sheet, but rather the sampling of a HCS comprising intertwined flux ropes. They suggested that some were formed in the solar corona as the result of transient activity. In a statistical survey of mesoscale flux ropes found in the solar wind, Cartwright and Moldwin (2010a) found them to by highly concentrated near the heliospheric current sheet. As previously discussed, *STEREO* has tracked structures of sufficient density variations from the Sun to the spacecraft making in situ measurements (Rouillard, Davies, et al., 2010; Rouillard, Lavraud, et al., 2010; Rouillard et al., 2011). This unequivocally confirms the association of many of the mesoscale structures as magnetic flux ropes measured at 1 AU at the HCS and "blobs" released from helmet streamers. We now discuss mesoscale structures measured in situ that have not yet been associated with specific coronal features because they could not be tracked *from the Sun* continuously in *remote imaging*. For these, only likely associations have been made with coronal structures observed separately near the Sun by using numerical simulations of the solar corona and wind.

As already shown in Figure 1.5, the slow solar wind and the HCS are generally associated with the helmet streamer structure in the solar corona (McComas et al., 1998). Gosling et al. (1981) showed helium abundance variations associated with the crossing of the HCS, confirming that variations associated with the HCS are of solar origin. Kilpua et al. (2009) identified in *STEREO* in situ data 17 different transient structures at the HCS, which they linked to time dynamics in helmet streamers, 7 of which had counter-streaming electrons, indicating that the structures were still connected at both ends back to the Sun. Kepko et al. (2016) identified a cyclic train of mesoscale structures around the HCS. They exhibited cyclic compositional changes, confirming a solar source. One of the structures was a flux rope with counter-streaming electrons, followed by a strahl dropout; the compositional changes indicate that magnetic reconnection in the corona created these structures.

Mesoscale structures are not restricted to the HCS. The slow solar wind includes the HCS, but can be observed as far as 30° away from the HCS (Burlaga et al., 1982), and is more generally associated with the boundary between field lines that are open to the heliosphere (coronal holes) and those that are close in the corona (streamers). Numerical modeling shows that open-closed boundaries can be a complex web of separatrices (the so-called S-web; Antiochos et al., 2011) where reconnection is likely to occur. During solar maximum, these separatrices can form a complex web, mapping to many locations away from the HCS in the heliosphere (Crooker et al., 2012; Crooker et al., 2014). Mesoscale structures in the slow solar wind outside of the HCS are observed in density as seen in Figure 1.11 taken from Viall et al. (2008), but thye are also observed in magnetic field (Borovsky, 2008), and composition (Viall et al., 2009).

It is thought that interchange reconnection could be a source of mesoscale structures perhaps forming at these modeled separatrices (Higginson et al., 2017). One signature expected when interchange reconnection occurs is that the electron strahl—which always flows away from the Sun—is observed to be in the opposite sense expected from the magnetic field direction (Crooker et al., 1996; Crooker et al., 2004; S. Kahler & Lin, 1994; S. W. Kahler et al., 1996), indicating that the magnetic field is locally folded back on itself. Owens et al. (2013) shows these inverted strahl signatures in the slow, dense solar wind at 1 AU associated both with helmet streamers, and with pseudostreamers, also associated with separatrices. Stansby and Horbury (2018) and Di Matteo et al. (2019) argue that signatures of interchange reconnection away from the HCS can be identified in *Helios* data inside of 1 AU. They identified mesoscale structures using density and showed concurrent temperature signatures, which are retained close to the Sun, strongly suggesting a solar source.

Mesoscale structures also occur in the fast wind. One prominent example is that of microstreams (Neugebauer, 2012; Neugebauer et al., 1995; Neugebauer et al., 1997). Microstreams are observed in the fast, polar solar wind with velocity fluctuations of ±35 km/s, last 6 hr or longer, have higher kinetic temperatures, higher proton flux, and slightly FIP enhanced compared to the rest of the fast solar wind. They are associated with large angle magnetic discontinuities and compositional changes that are consistent with a solar origin. Neugebauer (2012) showed that X-ray jets and the reconnection that causes

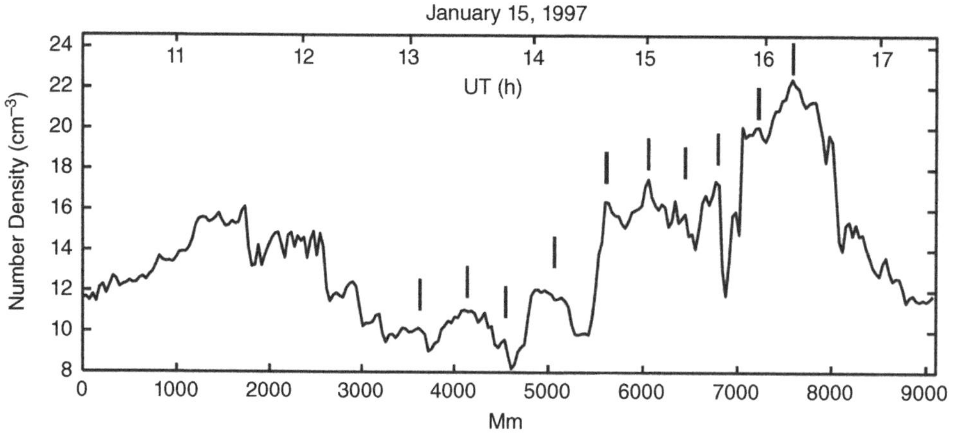

Figure 1.11 Solar wind number density data for 15 January 1997. Bottom *x*-axis is in radial-length scale steps, top *x*-axis shows the corresponding UT. Tick marks indicate a clear 400 Mm periodicity. (*Source*: Taken from Viall et al., 2008. © 2008, John Wiley and Sons.)

them are the most likely sources of microstreams, though they could also be related to polar plumes (Neugebauer et al., 1997; Poletto, 2015). The distinction is difficult, because jets and plumes are themselves related (Raouafi et al., 2016). Simulations support the plume–microstream connection (Velli et al., 2011), and the jet–microstream/ Alfvén wave connection (Karpen et al., 2017).

Horbury et al. (2018) found even smaller structures in *Helios* data at 0.3 AU, lasting tens of seconds to minutes, and reaching up to 1000 km/s. They are Alfvénic in nature, exhibiting large magnetic field deflections. These structures may form during jets from the chromosphere and/or low corona. Borovsky (2016) showed hours-long structures in the fast solar wind with large variations in number density, temperature, magnetic field strength, composition, electron strahl, and proton specific entropy, and also argue these mesoscale structures map to features in the solar corona. In contrast to the dynamic sources described above, Borovsky (2016) argues that these mesoscale structures are the result of relatively time stationary coronal flux tubes.

Pressure balances structures where the magnetic pressure balances the thermal pressure (Burlaga & Ogilvie, 1970) are also prevalent in the fast solar wind (Bavassano et al., 2004; Reisenfeld et al., 1999; Thieme et al., 1990). Unlike microstreams, McComas et al. (1996) showed that PBSs were not distinguishable from the rest of the fast solar wind, and may not be relics of transient coronal structure.

Mesoscale structures in the solar wind are an important part of the solar terrestrial connection, because they can drive magnetospheric dynamics. Often, mesoscale structures are cyclic, identified as discrete frequencies in plasma density (Di Matteo & Villante, 2017; Sanchez-Diaz et al., 2017; Viall et al., 2008) and dynamic pressure (Kepko & Spence, 2003; Kepko et al., 2002) Sometimes the structures exhibit periodicities in all plasma components (Stephenson & Walker, 2002). They are observed to directly drive global oscillations of the magnetosphere at the exact same frequencies (Kepko & Spence, 2003; Kepko et al., 2002; Viall et al., 2009; Villante et al., 2013), even by ground-based magnetometer on Earth (Villante et al., 2016) in radar oscillations in the high latitude ionosphere (Fenrich & Waters, 2008), polar UV imaging data (Liou et al., 2008), and even the equatorial ionosphere (Dyrud et al., 2008). MHD simulations have confirmed that cyclic solar wind dynamic pressure structures directly drive magnetospheric oscillations, and locations of field line resonance will even amplify the waves (Claudepierre et al., 2010; Hartinger et al., 2014).

The variations in the magnetic field of mesoscale structures in the ambient solar wind are important for an understanding of both their creation and their effect on the heliosphere and, in particular, on energetic particles. One fundamental scale size in the magnetic field on mesoscales is the correlation scale length perpendicular to the mean field. Crooker et al. (1982) measured this quantity at 1 AU and found it to have a characteristic size of 130 Mm during low magnetic field variance, and a characteristic length scale of 320 Mm during high variance. Collier, Slavin, and Lepping (2000) extended this analysis of magnetic field characteristic size scales at 1 AU and found that there is an additional scale size—the radius of curvature of the magnetic field—equal to 640 Mm.

The magnetic field changes that produce mesoscale length scales are either in the form of flux ropes (e.g., Feng et al., 2007; Moldwin et al., 1995; Moldwin et al., 2000) or flux tubes with magnetic field discontinuities at their boundaries (e.g., Bruno et al., 2001; Thieme et al., 1989). Magnetic flux ropes have a characteristic core magnetic field and magnetic rotation, and are often consistent with a force-free equilibrium (Burlaga, 1988; H. Goldstein, 1983). Flux ropes in large-scale ICMEs are a common, well-known example. There is a population of flux ropes lasting tens of minutes to hours, which are not associated with ICMEs, and instead occur with the ambient solar wind. Unlike ICMEs, the mesoscale flux ropes are not observed to be expanding due to overpressure, they do not exhibit ICME temperature depletion, they occur more often during solar minimum, and they occur in conjunction with the heliospheric current sheet. These observations make it clear that they are a slow wind feature, and not a continuation of a distribution of ICMEs to smaller scales (Cartwright & Moldwin, 2008, 2010b; Crooker et al., 1996; Moldwin et al., 1995, 2000).

Though flux ropes occur often in the slow solar wind, statistical measurements indicate that they do not make up a large portion, with occurrence rates of only six per year. There is considerable uncertainty about this rate, however, as different identification criteria can lead to vastly different event lists, and compressive Alfvén waves can be mistaken for flux ropes (Cartwright & Moldwin, 2008, 2010b; Feng et al., 2007; Higginson & Lynch, 2018).

Flux tube structures take the form of a discontinuous, planar change in the magnetic field and plasma parameters at the boundaries of the mesoscale structures. These are typically described as tangential and rotational discontinuities (Hudson, 1970), and are abundant throughout the solar wind, occurring even in the slow wind away from the HCS and in the fast wind (Neugebauer et al., 1984).

Mesoscale structures ahead of energetic particle events can influence energetic particle transport. *STEREO* observations of energetic particles have shown that variations appear to be linked to flux tubes with diameters of 6000 Mm (von Rosenvinge et al., 2009). Furthermore, even shorter fluctuations in the particle properties suggest that these flux tubes have sizes that extend down to 500 Mm. These intensity changes are similar to particle "dropouts" (Chollet et al., 2007; Mazur et al., 2000). Dropouts have been attributed to changes in the magnetic connectivity between the particle detector and the particle accelerator at the Sun due to flux tubes (at scales of 10–100 Mm) that meander or mix through the interplanetary medium (Chollet & Giacalone, 2011; Giacalone et al., 2000). Particle dropouts and heat flux loss often indicate changes within the underlying magnetic topology occurring with HCS crossings or local kinks in magnetic flux tubes (Borovsky, 2008; Crooker et al., 1982). We note that the majority of sharp changes in energetic particle populations during ambient solar wind conditions occur with changes in plasma and magnetic fields at tangential discontinuities (Neugebauer & Giacalone, 2015).

Whether the smaller flux tubes and flux ropes are created at the Sun and advected with the solar wind, or whether they are created in transit is still an open question. For example, Cartwright and Moldwin (2008) argued that flux ropes are generated by reconnection across the HCS in transit, and indeed plenty of reconnection is known to occur locally in the solar wind (Gosling, 2012), though not in direct association with a mesoscale flux rope. Magnetic reconnection in the solar wind is discussed in Section 1.3.5. On the other hand, Crooker et al. (1996) argued the flux ropes were formed at the Sun as the solar wind is created. Measurements of the flux ropes as a function of radial distance from the Sun shows that there are fewer at greater distances, and that they are larger with distance (Cartwright & Moldwin, 2010b), indicating merging or expansion rather than local creation.

One idea is that the solar wind is filled with coherent "fossil structure" flux tubes from the solar corona that advect with the solar wind (Bruno et al., 2001; Marsch & Tu, 1993; Thieme et al., 1989). Borovsky (2008) and Collier et al. (2000) both used geometric arguments (e.g., solar rotation rate, radial expansion of the solar wind) to relate the mesoscale flux tube structure observed at 1 AU to granules. A comparison of the occurrence distribution of flux tube sizes between in situ and observed solar features is shown in Figure 1.12. Neugebauer and Giacalone (2015) argued that tangential discontinuities were preexisting flux tube boundaries formed at the Sun based on the corresponding plasma parameters, and were not consistent with in-transit turbulence. However, it is unclear if a direct connection with granules in the photosphere is

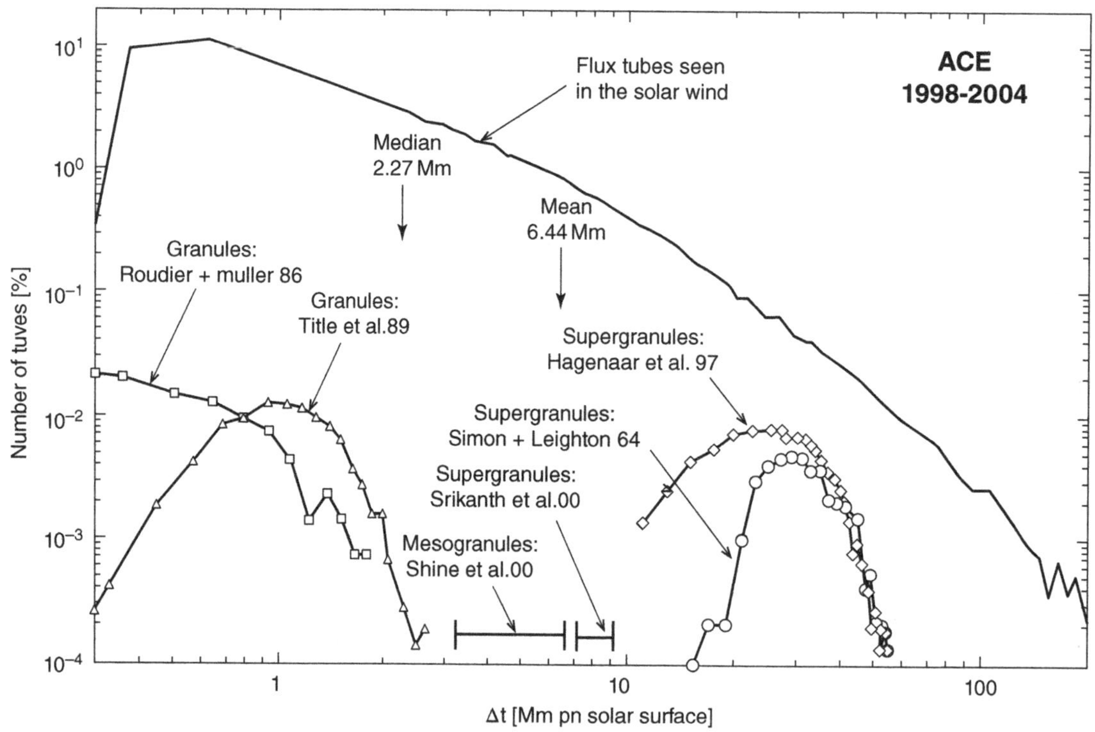

Figure 1.12 Occurrence distribution of flux tube sizes mapped to the solar surface is plotted as the black curve. Also plotted are distributions of solar granules and supergranules, and supergranule sizes obtained with high-resolution measurements are indicated with horizontal bars. (*Source*: Taken with permission from Borovsky, 2008. © 2008, John Wiley and Sons.)

possible, and simulations show that the boundaries of flux tubes created from granulation would not survive to 1 AU (Cranmer et al., 2013).

Some of the mesoscale structures, such as the characteristic magnetic field correlation lengths, may be related to turbulence. In general, it is known that the spectra of fluctuations in the solar wind have an inertial range at smaller scales, and at larger scales, follow a 1/f form that is a condition of low-frequency Alfvénic turbulence. The origin of the 1/f turbulent fluctuations is still debated but may itself originate from the corona (e.g., Matthaeus & Goldstein, 1986; Nicol et al., 2009) and footpoint stirring in the solar photosphere, with the inertial range being the transit turbulent decay. When compositional changes are associated with discontinuities and flux ropes, that is uncontroversial evidence that those structures were formed at the Sun (Borovsky, 2012; Borovsky & Denton, 2016; Kepko et al., 2016; Neugebauer, 2012; Viall et al., 2009). Structures without compositional change are ambiguous and could be formed either at the Sun or in transit (Owens et al., 2011).

We should point out the impact of mesoscale structures on global heliospheric structures. The variable coronal outflows imaged by coronagraphs and discussed in Section 1.2.2 indeed have repercussions on the structure of the solar wind. The standard picture of a CIR (Burlaga & Barouch, 1976; Lee, 2000; Pizzo, 1982) considers them as recurrent, uniform, and stable structures. In situ measurements made during the maximum phase of cycle 23 have revealed that interaction regions do not necessarily recur, a result of coronal hole reconfigurations and of the single-point nature of in situ measurements (L. Jian et al., 2006). Heliospheric imaging further reveals that individual CIRs or SIRs are by no means smooth compression regions distributed along a spiral but exhibit strong variability along the interaction region in response to several dynamic processes occurring at the Sun (Rouillard, Davies, et al., 2010). First, coronal holes can rapidly form and disappear during a solar rotation period, which leads to the appearance and disappearance of interaction regions in the interplanetary medium. Second, the density variations induced by the release of the small and large-scale streamer events discussed in the previous paragraph modifies the global structure of CIRs. In this regard, heliospheric imagery has provided new insights into the global structure of CIRs. It has been shown that in the inner heliosphere, CIRs are made up of compressed density structures (Rouillard, Lavraud, et al., 2010). The presence of strong pressure variations along the CIR surface suggests that shock formation will also be nonuniform with heliospheric location and time. Strong pressure enhancements would develop due to the compression by high-speed streams of small-scale transients. The shape of the CIR shock may therefore become a highly irregular surface beyond 1 AU. This could have implications for the formation of MIRs by the interaction of CIRs toward the outer regions of the heliosphere.

1.3.5. Magnetic Reconnection in the Solar Wind

Over the last couple of decades, magnetic reconnection has become a widely studied plasma process, ubiquitous in most if not all astrophysical contexts. Although it has long been known to occur at the Sun and in Earth's magnetosphere, only a few studies proposed that it should be occurring frequently in the solar wind and at the interfaces of CMEs during propagation (McComas et al., 1988; Moldwin et al., 1995, 2000; Riley et al., 2004; Wei et al., 2003). Despite these suggestions, it was only later that the first in situ observational evidence for magnetic reconnection in the solar wind was provided by Gosling et al. (2005). The evidence consisted in the observation of a velocity change (or jet) bounded by two discontinuities (cf. Figure 1.13) where the flow and field properties where consistent with the Walén relation (Hudson, 1970; Phan et al., 2006) but with opposite signs. The relation is of opposite sign at each discontinuity because the Alfvénic perturbation is propagating parallel to the magnetic field at one discontinuity (A1 in Figure 1.13) but anti-parallel at the other (A2).

Since then, a number of observational studies have shown that magnetic reconnection is indeed frequent in the solar wind. It has in particular been shown that (a) the magnetic reconnection exhausts may contain two oppositely streaming proton beams resulting from the mixing of the two populations from the inflow media (Gosling et al., 2005); (b) it often occurs at the HCS, where

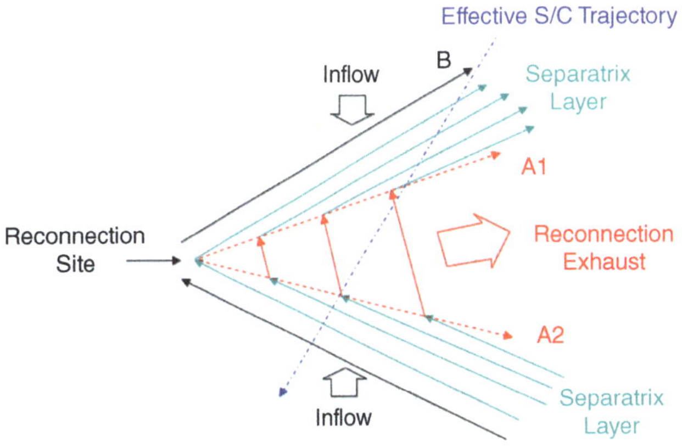

Figure 1.13 Highly idealized planar projection of a slightly asymmetric reconnection exhaust, not to scale, convecting with the nearly radial (from the Sun) solar wind flow. The dash-dot arrow shows an effective spacecraft trajectory through the exhaust. For simplicity, the oppositely directed exhaust is not shown. (*Source*: From Gosling & Szabo, 2008. © 2008 John Wiley and Sons.)

spacecraft typically observe counter-streaming electron populations resulting from the mixing of either strahl or halo electrons (Gosling, 2005; Lavraud et al., 2009), signaling field lines either closed or disconnected from the Sun; (c) it occurs also well inside of 1 AU; (d) reconnection X lines in the solar wind may extend over very large distances and be active for long periods of time (from multiple spacecraft observations), suggesting that the reconnection process is not inherently intermittent (Gosling et al., 2006; Lavraud et al., 2009; Phan et al., 2006); (e) reconnection occurs for a wide range of guide fields and plasma beta properties, but is more frequent for low plasma beta (Eriksson et al., 2009; Gosling & Phan, 2013); and (f) magnetic reconnection may be driven within or at the interfaces of CMEs (Gosling et al., 2005; Ruffenach et al., 2015; Wang et al., 2012), as well as at small flux ropes in the solar wind (Feng et al., 2011; Tian et al., 2010). Many other studies, in particular based on theory and simulations, have been conducted to understand the structure of the exhaust, its boundaries, and its evolution, but these are not summarized here. It is important to note, however, that the occurrence of reconnection at CME boundaries was early perceived as broadly important (McComas et al., 1988). Indeed, its occurrence implies that the magnetic structure of the CME will change over time during propagation, so that the outer magnetic shells of the structure are peeled off, or in other words, the CME is eroded (Dasso et al., 2007; Möstl et al., 2008; Ruffenach et al., 2012; Tian et al., 2010). Direct evidence for this process was studied in detail through both case and statistical studies in Ruffenach et al. (2012, 2015), demonstrating the frequent occurrence of CME erosion during propagation. The process was also studied through global simulations by Manchester et al. (2014) and Manchester, van der Holst, and Lavraud (2014). The importance of CME erosion by magnetic reconnection also comes from its potential effect on CME geoeffectiveness (Lavraud et al., 2014).

Other broader impacts of the frequent occurrence of magnetic reconnection in the solar wind come from the fact that it may affect solar wind heating, in particular near the Sun, as well as the generation of turbulence and associated intermittency, topics that are reviewed in the next sections.

1.4. WAVES AND TURBULENCE

1.4.1. Spectra of Solar Wind Fluctuations at All Scales

In situ measurements allow us to make detailed analysis of electromagnetic fluctuations in the solar wind plasma over a wide range of scales. Figure 1.14 (left) shows an example of a composite magnetic spectrum taken at 1 AU using the *ACE* and *Cluster* spacecraft (Kiyani et al., 2015). Fluctuations are organized in different frequency ranges, each with a different power law. The main central part (10^{-3}Hz $< f <$ 10^{-1}Hz) is characterized by a Kolmogorov $-5/3$ spectral slope; this corresponds to the inertial range of an MHD turbulent cascade that transports energy from large to small scales, as in hydrodynamics turbulence (Bruno & Carbone, 2013).

It should be recalled here that due to the high speed of the solar wind flow compared to the typical velocities associated to the motion of the plasma fluctuations (Taylor hypothesis), time frequencies measured in the spacecraft frame can be interpreted as spatial k-vectors in the plasma frame. It is then possible to study the 3D distribution of k-vectors in the inertial range using measurements at different angles with respect to the main field (Horbury et al., 2008; Saur & Bieber, 1999; Wicks et al., 2010); this suggests a quasi-2D distribution of the power, constituted by very elongated turbulent eddies along the magnetic field and a turbulent cascade that occurs preferentially for k-vectors perpendicular with respect to the magnetic field.

At larger scales, the spectrum of magnetic fluctuations is often characterized by a shallower slope, close to -1 (Bavassano, Dobrowolny, Mariani et al., 1982; Denskat & Neubauer, 1982); this range, called 1/f, is sometimes considered as the energy reservoir for the turbulent cascade, in analogy with hydrodynamics, although this analogy is not necessarily straightforward (Tu & Marsch, 1995) and more generally, the origin of the 1/f range is still under debate in the community (Chandran, 2018; Matteini et al., 2018; Matthaeus & Goldstein, 1986; Velli et al., 1989; Verdini et al., 2012).

At higher frequencies, around scales corresponding to the typical ion characteristic lengths (ion gyroradius ρ_i and inertial length d_i), the spectrum becomes steeper. This is expected when MHD breaks down, and kinetic physics starts to play a role. Moreover, fluctuations become more compressible at these scales, as a result of the transition from a regime where the electric field is controlled by ideal-MHD to a Hall-MHD regime (e.g., Alexandrova et al., 2008; Kiyani et al., 2013; Lacombe et al., 2017). As a consequence, the spectrum of the electric field, which follows that of the magnetic field at large MHD fluid scales, starts to depart at ion scales and display a shallower spectral slope at sub-ion scale, such that the ratio of electric-to-magnetic fluctuations increases linearly with f (Matteini et al., 2017).

At even higher frequencies, approaching electron plasma scales (f: 50–100Hz, corresponding to a few km at 1 AU), the magnetic spectrum of the background turbulence steepens further; see Figure 1.14 (right). We will discuss the background turbulence at electron scales in more detail below. Note that at these frequencies whistler wave activity is also sometimes observed, characterized by

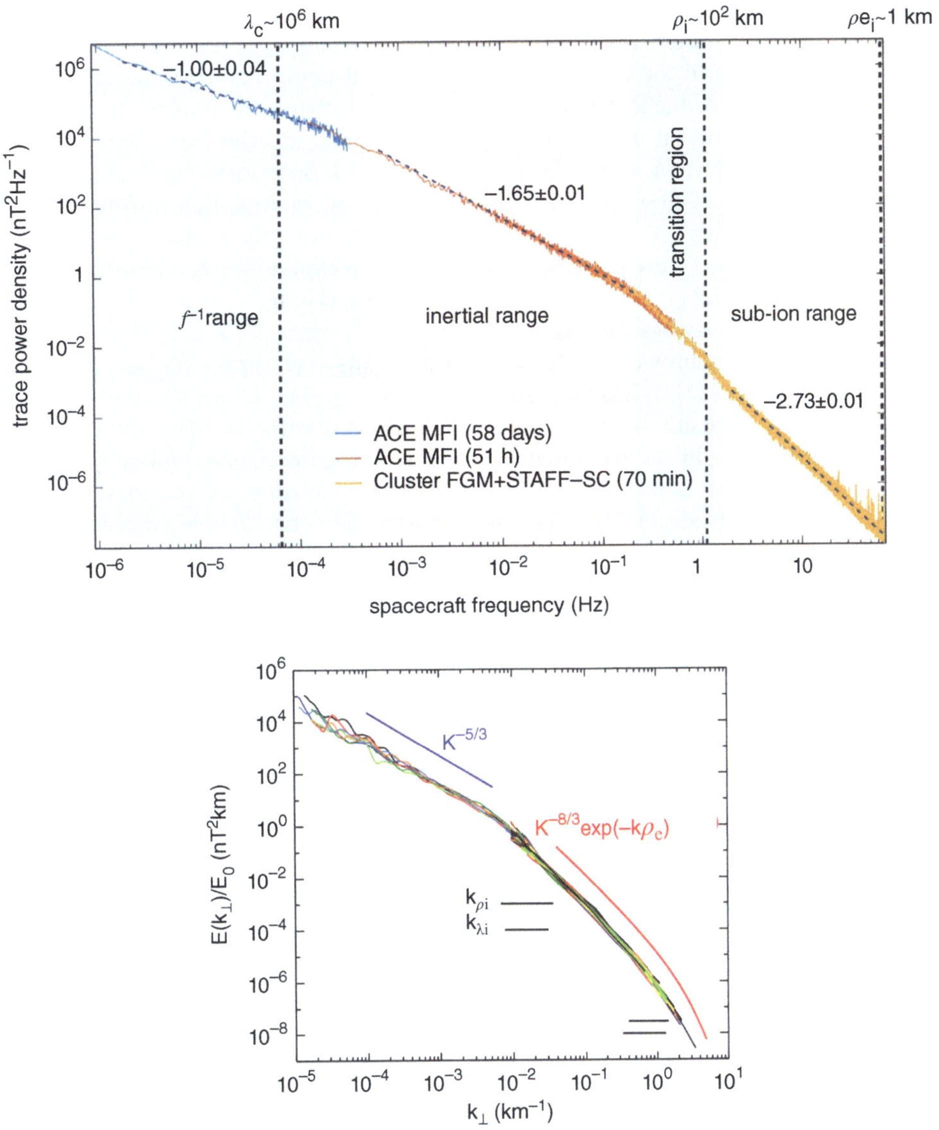

Figure 1.14 Top: Typical spectrum of magnetic field fluctuations in the solar wind from Kiyani et al. (2015). Measurements of ACE and Cluster at 1 AU from MHD to electron scales are shown. Doppler-shifted characteristic scales, such as the correlation length λ_c and the ion and electron Larmor radii $\rho_{i,e}$ are indicated by vertical dashed lines. Bottom: Superposed (more then 100) turbulent spectra of magnetic fluctuations under different plasma conditions from inertial to sub-electron scales as measured by Cluster (Alexandrova et al., 2012). (*Source*: Kiyani et al., 2015.)

right-handed circular polarization very distinct from that of the background turbulence (Kajdič et al., 2016a; Lacombe et al., 2014; O. W. Roberts et al., 2017; Stansby et al., 2016)

1.4.2. Alfvén Waves in the Fast and Slow Winds

At large scales, one of the most distinctive feature is the high degree of correlation between magnetic and velocity fluctuations, suggesting an Alfvénic origin and corresponding to a flux of waves moving away from the Sun (Belcher & Davis, 1971). Even though the spectrum of velocity fluctuations within the inertial range follows a shallower power law, $f^{-3/2}$, than the one governing magnetic fluctuations (Podesta et al., 2006; C. Salem et al., 2009). Alfvénic fluctuations are ubiquitous in the fast solar wind, where the correlation between velocity and magnetic field is typically very high, especially closer to the Sun. This is not the case in the slow wind, which is more irregular and compressible, and in general it does not display a high level of Alfvénic correlation. However, there are also periods of slow wind that appear highly

Alfvénic, suggesting then a common origin as the main fast wind coming from coronal holes (D'Amicis & Bruno, 2015). Consistent with that, all the properties of this Alfvénic slow wind match those of the fast wind, thus reinforcing the idea of a common origin of the plasma (D'Amicis et al., 2018). Interestingly, in situ measurements suggest that a higher fraction of slow solar wind is also Alfvénic closer to the Sun (Stansby et al., 2019); this is an aspect that will benefit from the future explorations of *Solar Orbiter* and *Parker Solar Probe*.

Alfvénic streams (fast and slow) always display a $1/f$ range at large scales (Bavassano, Dobrowolny, Mariani et al., 1982; Denskat & Neubauer, 1982). It has been suggested that these fluctuations originate in the corona (Matthaeus & Goldstein, 1986), could be generated by Alfvén wave reflection in the acceleration region due to strong density gradients (Velli et al., 1989; Verdini et al., 2012), by parametric instability (Chandran, 2018), or be a consequence of a saturation of the wave amplitude (Matteini et al., 2018).

Alfvénic periods in the solar wind are also characterized by a remarkably low plasma and magnetic field compressibility (Matteini et al., 2015). This means that fluctuations in the 1/f range, although large and comparable to the background magnetic field intensity, act mostly like directional changes rather than compressing the field. How this state is achieved is today not fully understood, but it is very well maintained by the plasma during expansion, as confirmed by observations by *Ulysses* beyond 1 AU. Another consequence of this state is that modulation of the magnetic field also implies (anti-correlated) local variations of the flow speed. This results in a spiky velocity profile in the fast solar wind (Matteini et al., 2014); the amplitude of the velocity enhancements tracks the Alfvén speed and is then largest close to the Sun. As mentioned earlier, Horbury et al. (2018) have suggested that the Alfvénic spikes observed in fast streams could be signatures of Alfvénic pulses injected by coronal jets surviving in interplanetary space (Karpen et al., 2017; M. A. Roberts et al., 2018). It has been argued that the effect of jets and velocity shears that may develop in the corona could significantly increase the strength of the radial component of the interplanetary magnetic field with radial distance away from the Sun (Lockwood et al., 2009a, 2009b). These kinematic effects could be the source of discrepancy between the open magnetic flux derived from numerical models of the solar corona and the magnetic field measured in situ (Lockwood et al., 2009a).

Below the $1/f$ range, and at scales smaller than a few hours, a turbulent inertial range is observed in all main fields (magnetic, electric, velocity and density); for example, see the review paper by Alexandrova et al. (2013). In the fast/Alfvénic wind, this corresponds to an almost incompressible MHD cascade, where fluctuations in density are small compared to magnetic and velocity fluctuations. The slow wind typically has a higher level of compressibility, suggesting a mixture of Alfvénic fluctuations and compressible structures, whose weight in the power spectrum also varies as a function of distance. Estimations of the turbulent cascade rate (Coburn et al., 2014; Sorriso-Valvo et al., 2007) confirm the existence of an energy flux through MHD scales, whose amplitude roughly matches the external energy input needed to explain ion temperature profiles in interplanetary space (Hellinger et al., 2011)

1.4.3. Solar Wind Fluctuations at Kinetic Scales

We now discuss the turbulent fluctuations at the end of the Kolmogorov inertial range and at smaller scales, that is, the *kinetic range* of the solar wind turbulence, sometimes called the *dissipation range* by analogy with the usual hydrodynamic (HD) turbulence.

Approaching ion scales (100 km at 1 AU), the MHD approximation is no more valid, and ions and electrons cannot be considered as one fluid. As protons are much heavier than electrons, in the vicinity of the ion inertial length λ_i, there is a separation into two fluids (Hall-MHD description). Arriving at the ion Larmor radius scale, ρ_i, ions should be considered as particles and not a fluid. The corresponding time scale is the ion cyclotron frequency f_{ci}. In the vicinity of f_{ci}, Alfvén waves become dispersive. In plasma with anisotropic ions, the Alfvén ion cyclotron (AIC) instability generates quasi-parallel AIC waves, visible in turbulent spectra as bumps or breaks at the vicinity of f_{ci} (e.g., L. K. Jian et al., 2014; Lion et al., 2016). The intermittent structures present in the inertial range, have their smallest spatial scales with the strongest gradients. For example, the thickness of the current sheets is observed to be several λ_i, or the cross section of magnetic vortices is about a few ρ_i (Perrone et al., 2016, 2017). These small-scale discontinuities have a distinct Fourier spectrum, f^{-2}, for current sheets and shocks, and f^{-4} for magnetic vortices (Alexandrova, 2008). Thus, the presence of different phenomena, like ion temperature instabilities (Bale et al., 2009; Hellinger et al., 2006) and the strongest gradients of the intermittent structures, is at the origin of the spectral variability observed in the frequency range covering ion scales~[0.1,3] Hz (C. W. Smith et al., 2006).

Between ion and electron scales, ~[3, 30] Hz at 1 AU, one observes another general power-law behavior of the magnetic spectrum (Alexandrova et al., 2009, 2012; Sahraoui et al., 2013). This suggests a small-scale turbulent cascade of the electron fluid (e.g., Alexandrova et al., 2008; Schekochihin et al., 2009). This cascade is more compressible than the MHD inertial range (Alexandrova et al., 2008; Hamilton et al., 2008; Kiyani et al., 2013; Leamon et al., 1998; C. S. Salem et al., 2012; Turner et al., 2011). The observed compressibility can be explained by the β-dependent compressibility of oblique kinetic Alfvén waves (KAWs; Cerri et al., 2017; Lacombe et al., 2017; C. S. Salem et al., 2012).

In the vicinity of the electron scales (1 km at 1 AU), ions are fully kinetic and electrons start to become kinetic, so no magneto-fluid motions are expected anymore. At these scales, the magnetic spectrum has an exponential shape similar to the dissipation range spectrum of the usual HD turbulence; see Figure 1.1 (right). This spectrum can be explained by the dissipation of KAWs via Landau damping (Howes et al., 2006, 2011; Schreiner & Saur, 2017).

In fact, the statistical properties described above, such as spectral shape and compressibility, are just averaged characteristics of the turbulent flow. The nature of turbulence at kinetic scales remains an open question. Even if KAWs can explain the observed compressibility and the exponential spectral shape, it is not clear if turbulence at these small scales is a mixture of such linear waves. To address the question of the nature of turbulent fluctuations, one should consider time-domain measurements and not only spectral properties. Few observations at sub-ion scales show the presence of intermittent coherent structures in the form of current sheets (Greco et al., 2016; Perri et al., 2012) and magnetic vortices (Jovanic et al., 2020 submitted). These observations suggest a non-homogeneous dissipation of the electromagnetic turbulent cascade in space that contradicts the homogeneous dissipation of KAWs via Landau damping.

It is possible that both phenomena, that is, a mixture of quasi-linear waves and coherent structures, coexist within the kinetic range of scales. This can be at the origin of a balance between linear and nonlinear effects (the so-called *critical balance*). To verify this point, one needs to find a way to estimate characteristic times at different scales of the kinetic range.

1.5. KINETIC PROCESSES AND HEATING IN THE SOLAR WIND

1.5.1. Evolution of Anisotropies of the Solar Wind Distribution Function

Particle distributions in interplanetary space have been studied over many decades with *Helios*, *Ulysses*, *Wind*, *Cluster*, and various other spacecraft, mainly near 1 AU, but with also some observations from 0.3 to 4 AU up to now. The observed particle distributions are not Maxwellian, but show distinguishable structures. The physical processes that regulate the shape of particle distribution functions in the super-Alfvénic solar wind include plasma heating/instabilities, particle collisions and adiabatic particle motion. The main constituents of the solar wind, protons, electrons and alpha particles, can develop highly anisotropic velocity distribution functions. In this chapter, we illustrate these effects by considering electrons only and refer the reader to other reviews of the field of 'ion kinetics' that is not discussed here (e.g. Matteini et al. 2012). Electron distributions usually consist of a thermal core, an extended suprathermal halo, and an anti-sunward beam or "strahl" (Lin, 1974; Pilipp et al., 1987). Core and halo distributions often show temperature anisotropies, that is, different temperatures $T_{\parallel}$ parallel to the background magnetic field, and $T_{\perp}$ perpendicular to it. The temperatures of electron distributions can be determined by fits with bi-Maxwellian or bi-kappa distributions to core and halo components, the latter after subtraction of the asymmetric strahl component (Ogilvie et al., 2000). The different components of the electron distributions are illustrated in Figure 1.16.

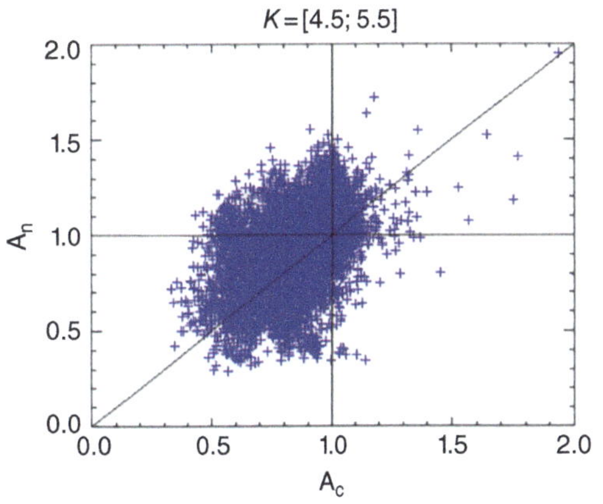

Figure 1.15 Electron temperature anisotropy of the halo ($A_h = T_{h\perp}/T_{h\parallel}$) as a function of the temperature anisotropy of the electron core ($A_c = T_{c\perp}/T_{c\parallel}$) observed in the solar wind with *Helios*, *Cluster*, and *Ulysses* from 0.7 to 1.5 AU for kappa values ranging between 4.5 and 5.5. (*Source*: From Pierrard et al., 2016. © 2016, Springer Nature.)

A naive expectation for an electron distribution in the expanding solar wind would be focusing into the magnetic field direction, caused by the mirror force in the expanding magnetic field geometry of the Parker spiral. This would lead to a pronounced temperature anisotropy $T_{\parallel} > T_{\perp}$. However, this is not observed, Figure 1.15 displays temperature anisotropies measured in the inner heliosphere and shows that even $T_{\perp} > T_{\parallel}$ is frequently measured in situ in the inner heliosphere. This observation alone already indicates that processes must be active in the solar wind that are capable of increasing $T_{\perp}$ relative to $T_{\parallel}$.

Early observations (Pilipp et al., 1990) have revealed that $T_{\parallel} > T_{\perp}$ is dominant in the fast solar wind, whereas $T_{\perp} > T_{\parallel}$ is more commonly found in slow solar wind streams. An extensive review of solar wind electron temperature anisotropies for solar distances between 0.3 AU and 4 AU has been provided by Pierrard et al. (2016). Core and halo electron temperature anisotropies, A_c and A_h, with $A = T_{\perp}/T_{\parallel}$, are generally correlated, but their

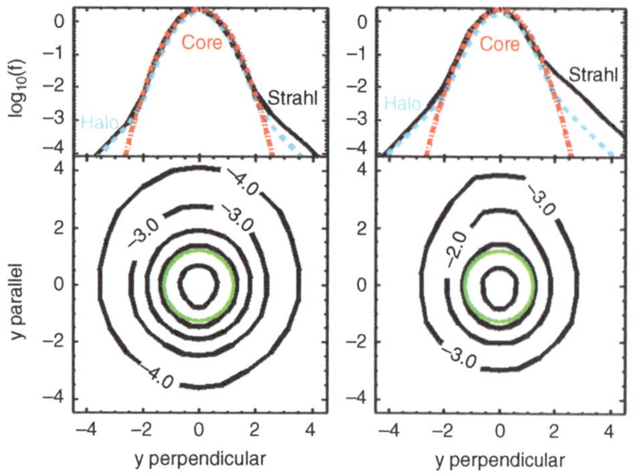

Figure 1.16 Typical electron velocity distribution function (VDF) observed by *WIND* at 1 AU in the low-speed (left) and in the high-speed (right) solar wind. Bottom panels: The plane of velocities (normalized to the thermal velocity) parallel and perpendicular to the interplanetary magnetic field. Top panels: Parallel (solid black line) and perpendicular (dashed blue line) cross section of the observed VDF. The dashed-dotted red line represents the Maxwellian distribution that well fits the core of observed VDF. (*Source*: From Pierrard, 2011. © 2011, Springer Nature.)

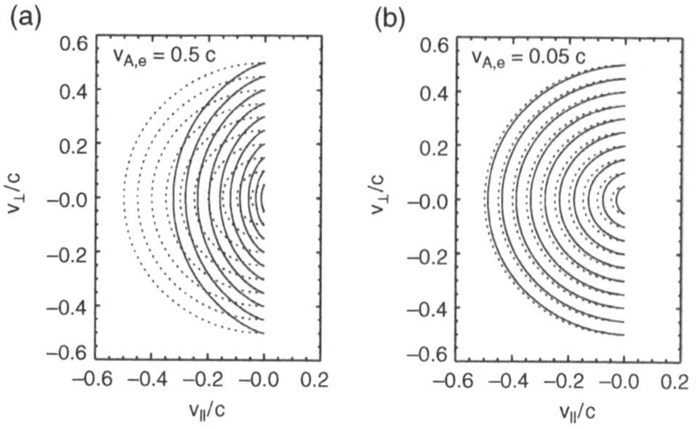

Figure 1.17 Kinetic shells formed in the electron VDF by resonant electron–whistler interaction (solid lines), and isolines of an undisturbed Maxwellian VDF (dotted lines). Shown are two plots for different values of the electron Alfvén speed, $v_{A,e} = 0.5c$ (left, (a)), and $v_{A,e} = 0.05c$ (right, (b)). (*Source*: From Vocks & Mann, 2003. © 2003, IOP Publishing.)

values are widely scattered between 0.5 and 2 (see Figure 1.15). Generally, $A_c < 1$ and $A_h < 1$, but instances with $A_c > 1$ and $A_h < 1$, and vice versa, are observed. The latter case, $A_c < 1$ and $A_h > 1$, then is much more common.

The maximum range of values for temperature anisotropies is constrained by the onset of plasma instabilities. $A > > 1$ values are unstable toward the generation of whistler waves, and $A < < 1$ values are subject to the firehose instability (Gary 1993). $A > > 1$

This picture is similar for solar wind proton distributions. They frequently show temperature anisotropies $T_\perp > T_\parallel$ plus an anti-sunward beam (see Marsch, 2012, for a review based on *Helios* data covering solar distances between 0.3 AU and 1 AU). The temperature anisotropy, A, shows a strong correlation with the observed ion cyclotron wave power, which indicates that the distributions are shaped by wave–particle interactions. As for electrons, the temperature anisotropies are constrained by thresholds for firehose and mirror instabilities, but are hardly observed near these limits. Observations suggest that temperature anisotropy instabilities may play a role in controlling the level of temperature anisotropy.

1.5.2. Wave–Particle Interactions, Kinetic Instabilities, and Collisions

The low density of the solar wind plasma could lead to the expectation that it could be described as collisionless; that is, Coulomb collisions are negligible. Nevertheless, Coulomb collisions do play a role in constraining temperature anisotropies (Pierrard et al., 2016; Pierrard et al., 2011; C. Salem et al., 2003; Štverák et al., 2008)

Non-Maxwellian particle distributions can become unstable toward numerous plasma instabilities (Štverák et al., 2008). Quasi-linear theory (Kennel & Engelmann, 1966) is a useful tool to study the effect of whistler waves on the evolution of solar wind electron distributions. It considers nonlinear effects between plasma waves and the fluctuations they cause in the plasma, and therefore diffusion terms are included in the Boltzmann–Vlasov equation. Solar wind models including the effects of whistler wave turbulence are capable of reproducing the observed characteristics of solar wind electrons (Pierrard et al., 2011; Vocks, 2012).

While whistler wave turbulence leads to waves propagating in all directions relative to the background magnetic field (Chen et al., 2010; Gary & Smith, 2009), even solar wind models that consider only the interaction between electrons and whistlers propagating parallel to the magnetic field result in electron distributions with a core, halo, and strahl (Vocks et al., 2005).

The quasi-linear interaction between parallel propagating whistlers and electrons leads to pitch-angle scattering in the reference frame of the waves, which leads to the formation of "kinetic shells" (Isenberg et al., 2001); see Figure 1.17. In interplanetary space, whistler wave phase speeds are not larger than electron thermal speeds, so there is little difference between the wave frame and the plasma rest frame. So whistler turbulence tends to lead to more pitch-angle diffusion. The observation of finite strahl widths already implies that some pitch-angle scattering mechanism must be active in the solar wind, because otherwise the conservation of magnetic momentum in the expanding magnetic field geometry of the solar corona and wind would focus the electrons into an extremely narrow beam.

If the whistler wave phase speed is much higher than electron thermal speeds, as is the case in the solar corona, then the kinetic shell formation not only leads to pitch-angle diffusion, but also deforms the distribution toward higher temperature anisotropies, $T_\perp > T_\parallel$. The diffusion process brings electrons from a phase-space location with low $v_\parallel$, $v_\perp = 0$ to a position with high $v_\perp$, $v_\parallel = 0$. In the plasma frame, this corresponds to a gain of kinetic energy. Therefore, this mechanism is capable of producing a suprathermal population out of an initially Maxwellian electron distribution (Vocks et al., 2008), thus forming a halo component with a power-law distribution.

1.5.3. Suprathermal Particles

Kinetic effects induce strong distortions of the VDFs in the thermal regime, but also by forming suprathermal populations that include electron strahl, non-thermal ion beams, and heavy ion differential streaming. In this chapter, we focus on the origin and important effects of suprathermal electrons and direct the reader to the review by Marsch et al. (2009) for a discussion of suprathermal ions.

As already mentioned, solar wind electron distributions show distinct particle populations; a thermal core; an extended suprathermal halo; the asymmetric, anti-sunward strahl; and an isotropic super-halo at higher energies (Lin, 1998). The thermal core can be described by a bi-Maxwellian distribution, but this does not work so well for the halo (Pilipp et al., 1987) component. Core and halo can be much better fitted by kappa distributions (Maksimovic et al., 1997), that show power-law suprathermal tails $\propto v^{-2(\kappa+1)}$. The κ parameter can reach values as low as 2. Note that $\kappa \rightarrow \infty$ corresponds to a Maxwellian, and for $\kappa \rightarrow 1.5$, the thermal energy, that is, the temperature, diverges.

An even better fit to observed electron data can be achieved if the core is additionally fitted by a (bi-) Maxwellian (Pierrard et al., 2016). The density of the halo is less than that of the core, but due to the higher halo temperature, their thermal energies are comparable (Maksimovic et al., 2000). A coronal origin of these suprathermal electrons is possible (Pierrard et al., 1999; Vocks et al., 2005), but whistler wave turbulence in the solar wind also plays a role in shaping these distributions (Yoon et al., 2015).

Pierrard et al. (2016) have studied the evolution of the suprathermal halo through the heliosphere between 0.3 AU and 4 AU. Temperature anisotropies were considered by applying bi-kappa fits to the electron halo population. As already found in earlier works (Maksimovic et al., 1997), the kappa value, and thus the strength of the suprathermal population, is not constant but varies in time with solar wind conditions.

Throughout the heliosphere, the kappa of the halo population generally decreases with increasing solar distance. Values of $\kappa = 7$ are frequently found closer than 1 AU, but rarely beyond 1 AU. A similar trend is found for kappa values around 5, and $\kappa = 3$ is hardly found below 1.5 AU, and more frequently beyond 1.5 AU. The strahl population can follow a different trend.

This reduction of kappa corresponds to an increase in kinetic temperature. Therefore, some mechanism for energizing these electrons has to be present in the solar wind.

The presence of non-thermal populations in many space plasmas suggests a universal mechanism for their creation. This concerns mainly electrons, but also the ions in the solar wind. Observations of electron suprathermal tails in the solar wind suggest their existence in the solar corona, because the electron mean free path in the solar wind is around 1 AU. The ion-charge measurements also confirm that suprathermal electrons can already be present in the corona. In any case, measurements made by Helios a radial distances as low as 0.3 AU confirm the presence of suprathermal electrons, even if they are characterized by a larger kappa index for the halo population at low radial distances (Pierrard et al., 2016). *Solar Orbiter* will have the advantage of measuring the distributions at low distances inside and outside the ecliptic plane. Up to now, the distributions at low distances have come from *Helios*, and those at large distances from *Ulysses*. Even if useful to study the average evolution with the radial distances, the observations were made at very different time periods (1970s for *Helios*, 1990s for *Ulysses*) and by different instruments measuring the distributions on different energy ranges.

Different scenarios have been suggested to generate the suprathermal particles (see Pierrard & Lazar, 2010, for a review). Turbulence associated with the long range of the Coulomb interactions has been proposed. Models including the generation of suprathermal electrons by resonant interaction with whistler waves in the solar corona and wind have been developed (Pierrard et al., 2011; Vocks & Mann, 2003). Recent scenarios use the kinetic theory to show that Kappa-like distributions could correspond to a particular thermodynamic equilibrium state. Kappa-like distributions can result as a consequence of the entropy generalization in nonextensive statistics (Leubner & Vörös, 2005), physically related to the long-range nature of the Coulomb potential, turbulence, and intermittency.

Suprathermal particles have important consequences for plasma fluctuations, resonant and nonresonant wave–particle acceleration, the heat flux, and plasma heating. These effects are well described by the kinetic approaches, where no closure requires the distributions to be nearly Maxwellian. The heating of the corona can be naturally explained by the velocity filtration, without depositing wave or magnetic field energy. Indeed, if suprathermal particles are present, their ratio over thermal ones increases as a function of altitude in an attraction field and automatically lead to an increase in temperature. This process works for electrons, but also protons and other ions if they are characterized by an enhanced population of suprathermal particles already in the low corona. The mass-to-charge ratio then determines the temperature of each ion species in good agreement with observations (Pierrard & Lamy, 2003). At larger radial distances (typically above 1.5 solar radii), the wind starts to blow, and the temperature reaches a maximum before starting to decrease. The presence of suprathermal electrons also increases the solar wind velocity as shown by Pierrard and Pieters (2014), in comparison to what is obtained with a simple thermal population characterized by a Maxwellian distribution.

1.6. THE EVOLUTION OF THE SOLAR WIND FROM THE INNER TO OUTER HELIOSPHERE

There are far fewer measurements of the solar wind in the outer heliosphere than in the inner heliosphere; however, measurements of the solar wind taken by the *Pioneer 10/11* and *Voyager 1/2* spacecraft showed that the heliosphere changes in two important ways. First, beyond 1 AU, solar wind interaction regions begin to dominate the heliosphere. Around 2–3 AU (Gosling et al., 1976; Hundhausen & Gosling, 1976; E. J. Smith & Wolfe, 1976), the dynamic compression of the slow wind by the fast wind creates a high-pressure region that expands in the reference frame of the solar wind, producing a forward fast shock and a reverse shock. The expansion of the compression region into the fast stream continually removes plasma from the high-speed stream to create larger, simpler structures. As the wind continues outward, the slower streams become more prominent, and the interaction regions broaden ahead of the fast wind.

At approximately 7 to 10 AU, by which time the fast and slow wind streams have given way to large interaction regions (Burlaga et al., 1995), successive interaction regions begin to interact themselves, forming "merged interaction regions." On average, there exists a single merged interaction region per solar rotation near 10 AU (Hundhausen & Gosling, 1976). Around 20 AU, these merged interaction regions and their associated shocks are replaced by corotating pressure enhancements that slowly form large pressure rings concentric to the Sun (Whang & Burlaga, 1990). These have been called "global merged interaction regions" and, together with merged interaction regions, play a significant role in the modulation of galactic cosmic rays (Burlaga et al., 1993; Wang et al., 2006).

The second important difference in the solar wind of the outer heliosphere is the existence of a new population of ions called "pick-up ions." Beyond several tens of AUs, these pick-up ions actually modify the properties of the solar wind and must be accounted for. A pick-up ion is created when a neutral atom, streaming into the heliosphere from the local interstellar medium, becomes ionized (Vasyliunas & Siscoe, 1976; Isenberg, 1987). The ionization processes include charge exchange with solar wind ions, photoionization from solar photons, and impact ionization from solar wind electrons. Out of these, charge exchange is the dominant interaction process, whereby a singly charged ion passes close to the neutral atom and captures one of its electrons.

Once ionized, the new pick-up ion starts to gyrate around the magnetic field. The ion starts with a very small velocity as compared with the solar wind flow (about 23 km/s). Therefore, the ion experiences a motional electric field caused by the moving solar wind plasma. The pick-up ion gains energy from this electric field until its guiding center reaches a velocity that is equivalent to the solar wind flow velocity, at which point it becomes part of the bulk flow and is convected outward with the expanding solar wind. After the pick-up, the ions experience gyrotropization and isotropization by either ambient or self-generated, low-frequency electromagnetic fluctuations in the solar wind plasma. In the spacecraft frame, this process appears as if those new ions are "picked up" by the solar wind flow.

We have not thoroughly grasped the transport of pick-up ions in the solar wind due to the lack of measurements of keV-ions in the outer heliosphere. The production of the suprathermal tails on the velocity distribution functions, in particular, remains a puzzle (Chalov et al., 2003; Fahr & Fichtner, 2011). It is believed that during their propagation through the outer heliosphere, pick-up ions undergo pitch-angle scattering and stochastic acceleration by interactions with different kinds of solar wind turbulences (Chalov et al., 1997). Pick-up ions may also play a significant role in mediating the solar wind interaction with the interstellar medium, possibly changing the structure of the heliosheath through charge exchange (Zank, 1999), but more measurements are needed.

1.7. OUTSTANDING QUESTIONS AND FUTURE PROSPECTS

We have decades of in situ measurements of the solar wind plasma, composition, and magnetic field within 1 AU, and we have decades of remote imaging and spectroscopy of the solar corona and young solar wind. NASA's *Parker Solar Probe* and ESA's *Solar Orbiter* missions provide an amazing opportunity to finally unite these disparate data sets and make a direct connection between the remotely observed corona and the "ground truth" in situ solar wind observations.

These missions will enable highly detailed studies of the mysterious solar wind heating. Not only will these provide new constraints on kinetic models of solar wind ions and elections and their interaction with plasma waves, but also direction comparison of the data with kinetic model results will be possible. Additionally, because together *Solar Orbiter* and the *Parker Solar Probe* will cover such a wide range of solar distances, it will be possible to study (a) the radial evolution of plasma turbulence in the expanding solar wind, (b) temperature anisotropy evolution for different particle species, and (c) to distinguish whether power-law suprathermal tails are formed due to local turbulent processes in the solar wind or have their origin in the solar corona. Similarly, we will be able to investigate the nature of the formation of in situ mesoscale structures, weighing the importance of turbulent processing or magnetic reconnection that occurs during transit against time-dynamic magnetic reconnection in the corona or stationary coronal streamer substructure.

Despite the many unanswered questions surrounding the formation and propagation of the solar wind, it is obvious that heliophysics is poised on the threshold of an exciting era of discovery.

ACKNOWLEDGMENTS

OA is supported by the French Centre National d'Etude Spatiales (CNES). VP and CV thank ISSI for the project on kappa distributions. APR, YU, BL, RP, and ML acknowledge support from the French space agency (Centre National des Etudes Spatiales; CNES; https://cnes.fr/fr) that funds activity in plasma physics data center (Centre de Données de la Physique des Plasmas; CDPP; http://cdpp.eu/) and the Multi Experiment Data & Operation Center (MEDOC; https://idoc.ias.u-psud.fr/MEDOC), and the space weather team in Toulouse (Solar-Terrestrial Observations and Modelling Service; STORMS; http://storms-service.irap.omp.eu/). This includes funding for the data mining tools AMDA (http://amda.cdpp.eu/), CLWEB (clweb.cesr.fr/), and the propagation tool (http://propagationtool.cdpp.eu). APR, YW, BL, RP, and ML also acknowledge financial support from ERC for the project SLOW_SOURCE - DLV-819189 as well as ANR project COROSHOCK No $ANR-18-ERC1-0006-01$.

REFERENCES

Abbo, L., Antonucci, E., Mikić, Z. et al. (2010, December). Characterization of the slow wind in the outer corona. *Advances in Space Research 46*: 1400–1408. https://doi.org/10.1016/j.asr.2010.08.008.

Aellig, M.R., Lazarus, A.J., and Steinberg, J.T. (2001). The solar wind helium abundance: Variation with wind speed and the solar cycle. *Geophysical Research Letters 28*: 2767–2770. https://doi.org/10.1029/2000GL012771.

Alexandrova, O. (2008, February). Solar wind vs magnetosheath turbulence and Alfvén vortices. *Nonlinear Processes in Geophysics 15*: 95–108.

Alexandrova, O., Carbone, V., Veltri, P., and Sorriso-Valvo, L. (2008, February). Small-scale energy cascade of the solar wind turbulence. *The Astrophysical Journal 674*: 1153–1157. https://doi.org/10.1086/524056.

Alexandrova, O., Chen, C.H.K., Sorriso-Valvo, L. et al. (2013, October). Solar wind turbulence and the role of ion instabilities. *Space Science Reviews 178*: 101–139. https://doi.org/10.1007/s11214-013-0004-8.

Alexandrova, O., Lacombe, C., Mangeney, A. et al. (2012, December). Solar wind turbulent spectrum at plasma kinetic scales. *The Astrophysical Journal 760*: 121. https://doi.org/10.1088/0004-637X/760/2/121.

Alexandrova, O., Saur, J., Lacombe, C. et al. (2009, October). Universality of solar-wind turbulent spectrum from MHD to electron scales. *Physical Review Letters 103* (16): 165003. https://doi.org/10.1103/PhysRevLett.103.165003.

Antiochos, S.K., DeVore, C.R., Karpen, J.T., and Mikić, Z. (2007, December). Structure and dynamics of the sun's open magnetic field. *The Astrophysical Journal 671*: 936–946. https://doi.org/10.1086/522489.

Antiochos, S.K., Mikić, Z., Titov, V.S. et al. (2011, April). A model for the sources of the slow solar wind. *The Astrophysical Journal 731*: 112. https://doi.org/10.1088/0004-637X/731/2/112.

Antonucci, E., Andretta, V., Cesare, S., Ciaravella, A., Doschek, G., Fineschi, S., et al. (2017, November). METIS, the Multi Element Telescope for Imaging and Spectroscopy: an instrument proposed for the solar orbiter mission. In *Society of photo-optical instrumentation engineers (spie) conference series* (Vol. *10566*, p. 105660L). doi: https://doi.org/10.1117/12.2308225.

Antonucci, E., Dodero, M.A., and Giordano, S. (2000, November). Fast solar wind velocity in a polar coronal hole during solar minimum. *Solar Physics 197*: 115–134. https://doi.org/10.1023/A:1026568912809.

Bavassano, B., Dobrowolny, M., Fanfoni, G. et al. (1982, June). Statistical properties of MHD fluctuations associated with high-speed streams from Helios-2 observations. *Solar Physics 78*: 373–384. https://doi.org/10.1007/BF00151617.

Bavassano, B., Dobrowolny, M., Mariani, F., and Ness, N.F. (1982, May). Radial evolution of power spectra of interplanetary Alfvenic turbulence. *Journal of Geophysical Research 87*: 3617–3622.

Bavassano, B., Pietropaolo, E., & Bruno, R. (2004, February). Compressive fluctuations in high-latitude solar wind. *Annales Geophysicae, 22*, 689–696. doi: 10 .5194/angeo-22-689-2004

Belcher, J.W. and Davis, L. Jr. (1971). Large-amplitude waves in the interplanetary medium, 2. *Journal of Geophysical Research Atmospheres 76*: 3534–3563.

Bemporad, A. (2017, September). Exploring the inner acceleration region of solar wind: A study based on coronagraphic UV and visible light data. *The Astrophysical Journal 846*: 86. https://doi.org/10.3847/1538-4357/aa7de4.

Biermann, L. (1951). Kometenschweife und solare Korpuskularstrahlung. *Zeitschrift für Astrophysik 29*: 274.

Birkeland, K. (1908). The Norwegian aurora polaris expedition, 1902–1903, Vol. I, Section 1, H Aschehoug, Oslo.

Bø, I.M.T., Esser, R., and Lie-Svendsen, Ø. (2013, May). Effect of Coulomb collisions on the gravitational settling of low and high first ionization potential elements. *The Astrophysical Journal 769*: 60. https://doi.org/10.1088/0004-637X/769/1/60.

Borovsky, J.E. (2008, August). Flux tube texture of the solar wind: Strands of the magnetic carpet at 1 AU? *Journal of Geophysical Research: Space Physics 113*: A08110. https://doi.org/10.1029/2007JA012684.

Borovsky, J.E. (2010, September). On the variations of the solar wind magnetic field about the Parker spiral direction. *Journal of Geophysical Research: Space Physics 115*: A09101. https://doi.org/10.1029/2009JA015040.

Borovsky, J. E. (2012, June). Looking for evidence of mixing in the solar wind from 0.31 to 0.98 AU. *Journal of Geophysical Research: Space Physics*, *117*, A06107. doi: 10.1029/2012JA017525.

Borovsky, J.E. (2016, June). The plasma structure of coronal hole solar wind: Origins and evolution. *Journal of Geophysical Research: Space Physics 121*: 5055–5087. https://doi.org/10.1002/2016JA022686.

Borovsky, J.E. and Denton, M.H. (2016, July). The trailing edges of high-speed streams at 1 AU. *Journal of Geophysical Research: Space Physics 121*: 6107–6140. https://doi.org/10.1002/2016JA022863.

Bruno, R. and Bavassano, B. (1997, September). On the winding of the IMF spiral for slow and fast wind within the inner heliosphere. *Geophysical Research Letters 24*: 2267. https://doi.org/10.1029/ 97GL02183.

Bruno, R., & Carbone, V. (2013, May). The solar wind as a turbulence laboratory. *Living Reviews in Solar Physics*, *10*, 2. doi: 10.12942/lrsp-2013-2.

Bruno, R., Carbone, V., Veltri, P. et al. (2001, October). Identifying intermittency events in the solar wind. *Planetary and Space Science 49*: 1201–1210. https://doi.org/10.1016/S0032-0633(01)00061-7.

Burlaga, L.F. (1988, July). Magnetic clouds and force-free fields with constant alpha. *Journal of Geophysical Research: Space Physics 93*: 7217–7224. https://doi.org/10.1029/JA093iA07p07217.

Burlaga, L.F. and Barouch, E. (1976, January). Interplanetary stream magnetism – Kinematic effects. *The Astrophysical Journal 203*: 257–267. https://doi.org/10.1086/154074.

Burlaga, L.F., Lepping, R.P., Behannon, K.W. et al. (1982, June). Large-scale variations of the interplanetary magnetic field – Voyager 1 and 2 observations between 1 and 5 AU. *Journal of Geophysical Research: Space Physics 87*: 4345–4353. https://doi.org/10.1029/JA087iA06p04345.

Burlaga, L.F., McDonald, F.B., and Ness, N.F. (1993, Jan). Cosmic ray modulation and the distant heliospheric magnetic field: Voyager 1 and 2 observations from 1986 to 1989. *Journal of Geophysical Research: Space Physics 98* (A1): 1–12. https://doi.org/10.1029/92JA01979.

Burlaga, L.F. and Ness, N.F. (1993, October). Large-scale distant heliospheric magnetic field: Voyager 1 and 2 observations from 1986 through 1989. *Journal of Geophysical Research: Space Physics 98*: 17451–17460. https://doi.org/10.1029/93JA01475.

Burlaga, L.F., Ness, N.F., and McDonald, F.B. (1995, August). Magnetic fields and cosmic rays in the distant heliosphere at solar maximum: Voyager 2 observations near 32 AU during 1990. *Journal of Geophysical Research: Space Physics 100*: 14763–14772. https://doi.org/10.1029/95JA01557.

Burlaga, L.F. and Ogilvie, K.W. (1970, November). Magnetic and thermal pressures in the solar wind. *Solar Physics 15*: 61–71. https://doi.org/10.1007/BF00149472.

Burton, M.E., Neugebauer, M., Crooker, N.U. et al. (1999, May). Identification of trailing edge solar wind stream interfaces: A comparison of Ulysses plasma and composition measurements. *Journal of Geophysical Research: Space Physics 104*: 9925–9932. https://doi.org/10.1029/JA104iA05p09925.

Cartwright, M.L. and Moldwin, M.B. (2008, September). Comparison of small-scale flux rope magnetic properties to large-scale magnetic clouds: Evidence for reconnection across the HCS? *Journal of Geophysical Research: Space Physics 113*: A09105. https://doi.org/10.1029/2008JA013389.

Cartwright, M.L. and Moldwin, M.B. (2010a, August). Heliospheric evolution of solar wind small-scale magnetic flux ropes. *Journal of Geophysical Research: Space Physics 115*: A08102. https://doi.org/10.1029/2009JA014271.

Cartwright, M.L. and Moldwin, M.B. (2010b, August). Heliospheric evolution of solar wind small-scale magnetic flux ropes. *Journal of Geophysical Research: Space Physics 115*: A08102. https://doi.org/10.1029/2009JA014271.

Cerri, S.S., Servidio, S., and Califano, F. (2017, September). Kinetic cascade in solar-wind turbulence: 3D3V Hybrid-kinetic simulations with electron inertia. *The Astrophysical Journal 846*: L18. https://doi.org/10.3847/2041-8213/aa87b0.

Chalov, S.V., Fahr, H.J., and Izmodenov, V. (1997, April). Spectra of energized pick-up ions upstream of the two-dimensional heliospheric termination shock. II. Acceleration by Alfvenic and by large-scale solar wind turbulences. *Astronomy and Astrophysics 320*: 659–671.

Chalov, S.V., Fahr, H.J., and Izmodenov, V.V. (2003, June). Evolution of pickup proton spectra in the inner heliosheath and their diagnostics by energetic neutral atom fluxes. *Journal of Geophysical Research: Space Physics 108*: 1266. https://doi.org/10.1029/2002JA009492.

Chandran, B.D.G. (2018, February). Parametric instability, inverse cascade and the range of solar-wind turbulence. *Journal of Plasma Physics 84* (1): 905840106. https://doi.org/10.1017/S0022377818000016.

Chapman, S. and Ferraro, V.C.A. (1931). A new theory of magnetic storms. *Terrestrial Magnetism and Atmospheric Electricity (Journal of Geophysical Research) 36*: 77. https://doi.org/10.1029/TE036i002p00077.

Chapman, S. and Zirin, H. (1957). Notes on the solar corona and the terrestrial ionosphere. *Smithsonian Contributions to Astrophysics 2*: 1.

Chen, C.H.K., Horbury, T.S., Schekochihin, A.A. et al. (2010, June). Anisotropy of solar wind turbulence between ion and electron scales. *Physical Review Letters 104* (25): 255002. https://doi.org/10.1103/PhysRevLett.104.255002.

Chollet, E. E., & Giacalone, J. (2011, February). Evidence of confinement of solar-energetic particles to interplanetary magnetic field lines. *The Astrophysical Journal*, *728*, 64. doi: 10 .1088/0004-637X/728/1/64.

Chollet, E.E., Giacalone, J., Mazur, J.E., and Al Dayeh, M. (2007, November). A new phenomenon in impulsive-flare-associated energetic particles. *The Astrophysical Journal 669*: 615–620. https://doi.org/10.1086/521670.

Claudepierre, S. G., Hudson, M. K., Lotko, W., Lyon, J. G., & Denton, R. E. (2010, November). Solar wind driving of magnetospheric ULF waves: Field line resonances driven by dynamic pressure fluctuations. *Journal of Geophysical Research: Space Physics, 115*, A11202. doi: https://doi.org/10.1029/2010JA015399.

Coburn, J.T., Smith, C.W., Vasquez, B.J. et al. (2014, May). Variable cascade dynamics and intermittency in the solar wind at 1 AU. *The Astrophysical Journal 786*: 52. https://doi.org/10.1088/0004-637X/786/1/52.

Collier, M.R., Slavin, J.A., and Lepping, R.P. (2000, June). IMF length scales and predictability: The two length scale medium. *International Journal of Geomagnetism and Aeronomy 2*: 3–16. https://doi.org/10.1029/2012JA017525.

Cranmer, S.R., Field, G.B., and Kohl, J.L. (1999, June). Spectroscopic constraints on models of ion cyclotron resonance heating in the polar solar corona and high-speed solar wind. *The Astrophysical Journal 518*: 937–947. https://doi.org/10.1086/307330.

Cranmer, S.R., van Ballegooijen, A.A., and Woolsey, L.N. (2013, April). Connecting the Sun's high-resolution magnetic carpet to the turbulent heliosphere. *The Astrophysical Journal 767*: 125. https://doi.org/10.1088/0004-637X/767/2/125.

Crooker, N.U., Antiochos, S.K., Zhao, X., and Neugebauer, M. (2012, April). Global network of slow solar wind. *Journal of Geophysical Research: Space Physics 117*: A04104. https://doi.org/10.1029/2011JA017236.

Crooker, N.U., Burton, M.E., Siscoe, G.L. et al. (1996, November). Solar wind streamer belt structure. *Journal of Geophysical Research 101*: 24331–24342. https://doi.org/10.1029/96JA02412.

Crooker, N.U., Kahler, S.W., Larson, D.E., and Lin, R.P. (2004, March). Large- scale magnetic field inversions at sector boundaries. *Journal of Geophysical Research: Space Physics 109*: A03108. https://doi.org/10.1029/2003JA010278.

Crooker, N.U., McPherron, R.L., and Owens, M.J. (2014, June). Comparison of interplanetary signatures of streamers and pseudostreamers. *Journal of Geophysical Research: Space Physics 119*: 4157–4163. https://doi.org/10.1002/2014JA020079.

Crooker, N.U., Siscoe, G.L., Russell, C.T., and Smith, E.J. (1982, April). Factors controlling degree of correlation between ISEE 1 and ISEE 3 interplanetary magnetic field measurements. *Journal of Geophysical Research: Space Physics 87*: 2224–2230. https://doi.org/10.1029/JA087iA04p02224.

Crooker, N.U., Siscoe, G.L., Shodhan, S. et al. (1993, June). Multiple heliospheric current sheets and coronal streamer belt dynamics. *Journal of Geophysical Research: Space Physics 98*: 9371–9381. https://doi.org/10.1029/93JA00636.

D'Amicis, R. and Bruno, R. (2015, May). On the origin of highly Alfvénic slow solar wind. *The Astrophysical Journal 805*: 84. https://doi.org/10.1088/0004-637X/805/1/84.

D'Amicis, R., Matteini, L., & Bruno, R. (2018, December). On slow solar wind with high Alfv\'enicity: From composition and microphysics to spectral properties. *arXiv e-prints*.

Dasso, S., Nakwacki, M.S., Démoulin, P., and Mandrini, C.H. (2007, August). Progressive transformation of a flux rope to an ICME. Comparative analysis using the direct and fitted expansion methods. *Solar Physics 244*: 115–137. https://doi.org/10.1007/s11207-007-9034-2.

DeForest, C.E., Howard, R.A., Velli, M. et al. (2018, July). The highly structured outer solar corona. *The Astrophysical Journal 862*: 18. https://doi.org/10.3847/1538-4357/aac8e3.

DeForest, C.E., Matthaeus, W.H., Viall, N.M., and Cranmer, S.R. (2016, September). Fading coronal structure and the onset of turbulence in the young solar wind. *The Astrophysical Journal 828*: 66. https://doi.org/10.3847/0004-637X/828/2/66.

Denskat, K.U. and Neubauer, F.M. (1982, April). Statistical properties of low- frequency magnetic field fluctuations in the solar wind from 0.29 to 1.0 AU during solar minimum conditions – HELIOS 1 and HELIOS 2. *Journal of Geophysical Research: Space Physics 87*: 2215–2223. https://doi.org/10.1029/JA087iA04p02215.

Di Matteo, S., Viall, N.M., Kepko, L. et al. (2019). Helios observations of quasiperiodic density structures in the slow solar wind at 0.3, 0.4, and 0.6 AU. *Journal of Geophysical Research: Space Physics 124* (2): 837–860. https://doi.org/10.1029/2018JA026182.

Di Matteo, S. and Villante, U. (2017, May). The identification of solar wind waves at discrete frequencies and the role of the spectral analysis techniques. *Journal of Geophysical Research: Space Physics 122*: 4905–4920. https://doi.org/10.1002/2017JA023936.

Dolei, S., Susino, R., Sasso, C. et al. (2018, May). Mapping the solar wind HI outflow velocity in the inner heliosphere by coronagraphic ultraviolet and visible-light observations. *Astronomy & Astrophysics (A&A) 612*: A84. https://doi.org/10.1051/0004-6361/201732118.

Domingo, V., Fleck, B., and Poland, A.I. (1995, December). The SOHO mission: An overview. *Solar Physics 162*: 1–37. https://doi.org/10.1007/BF00733425.

Dyrud, L.P., Behnke, R., Kepko, E.L. et al. (2008, July). Ionospheric ULF oscillations driven from above Arecibo. *Geophysical Research Letters 35*: L14101. https://doi.org/10.1029/2008GL034073.

Endeve, E., Holzer, T.E., and Leer, E. (2004, March). Helmet streamers gone unstable: Two-fluid magnetohydrodynamic models of the solar corona. *The Astrophysical Journal 603*: 307–321. https://doi.org/10.1086/381239.

Eriksson, S., Gosling, J.T., Phan, T.D. et al. (2009). Asymmetric shear flow effects on magnetic field configuration within oppositely directed solar wind reconnection exhausts. *Journal of Geophysical Research: Space Physics 114* (A7) https://doi.org/10.1029/2008JA013990.

Fahr, H.J. and Fichtner, H. (2011, September). Pick-up ion transport under conservation of particle invariants: How important are velocity diffusion and cooling processes? *Astronomy and Astrophysics 533*: A92. https://doi.org/10.1051/0004-6361/201116880.

Feng, H.Q., Wu, D.J., and Chao, J.K. (2007, February). Size and energy distributions of interplanetary magnetic flux ropes. *Journal of Geophysical Research: Space Physics 112*: A02102. https://doi.org/10.1029/2006JA011962.

Feng, H.Q., Wu, D.J., Wang, J.M., and Chao, J.W. (2011, March). Magnetic reconnection exhausts at the boundaries of small interplanetary magnetic flux ropes. *Astronomy & Astrophysics 527*: A67. https://doi.org/10.1051/0004-6361/201014473.

Fenrich, F.R. and Waters, C.L. (2008, October). Phase coherence analysis of a field line resonance and solar wind oscillation. *Geophysical Research Letters 35*: L20102. https://doi.org/10.1029/2008GL035430.

Fisk, L.A. (1996, July). Motion of the footpoints of heliospheric magnetic field lines at the Sun: Implications for recurrent energetic particle events at high heliographic latitudes. *Journal of Geophysical Research: Space Physics 101*: 15547–15554. https://doi.org/10.1029/96JA01005.

Fisk, L.A., Schwadron, N.A., and Zurbuchen, T.H. (1998, July). On the slow solar wind. *Space Science Reviews 86*: 51–60. https://doi.org/10.1023/A:1005015527146.

Forsyth, R.J., Balogh, A., and Smith, E.J. (2002, November). The underlying direction of the heliospheric magnetic field through the Ulysses first orbit. *Journal of Geophysical Research: Space Physics 107*: 1405. https://doi.org/10.1029/2001JA005056.

Gary, S.P. and Smith, C.W. (2009, December). Short-wavelength turbulence in the solar wind: Linear theory of whistler and kinetic Alfvén fluctuations. *Journal of Geophysical Research: Space Physics 114*: A12105. https://doi.org/10.1029/2009JA014525.

Geiss, J. (1998). Constraints on the FIP mechanisms from solar wind abundance data. *Space Science Reviews 85*: 241–252.

Geiss, J., Gloeckler, G., and von Steiger, R. (1995, April). Origin of the solar wind from composition data. *Space Science Reviews 72*: 49–60. https://doi.org/10.1007/BF00768753.

Giacalone, J., Jokipii, J.R., and Mazur, J.E. (2000, March). Small-scale gradients and large-scale diffusion of charged particles in the heliospheric magnetic field. *The Astrophysical Journal Letters 532*: L75 L78. https://doi.org/10.1086/312564.

Goldstein, H. (1983, November). On the field configuration in magnetic clouds. In *Nasa conference publication* (Vol. 228).

Goldstein, M. L., Roberts, D. A., & Matthaeus, W. H. (1995). Magnetohydrodynamic turbulence in the solar wind. *Annual Review of Astronomy and Astrophysics, 33*, 283–326. doi: 10.1146/annurev.aa .33.090195.001435.

Gosling, J.T. (2005). Magnetic disconnection from the Sun: Observations of a reconnection exhaust in the solar wind at the heliospheric current sheet. *Geophysical Research Letters 32* (5) https://doi.org/10.1029/2005GL022406.

Gosling, J.T. (2012, November). Magnetic reconnection in the solar wind. *Space Science Reviews 172*: 187–200. https://doi.org/10.1007/s11214-011-9747-2.

Gosling, J.T., Borrini, G., Asbridge, J.R. et al. (1981, July). Coronal streamers in the solar wind at 1 AU. *Journal of Geophysical Research 86*: 5438–5448. https://doi.org/10.1029/JA086iA07p05438.

Gosling, J.T., Eriksson, S., and Schwenn, R. (2006, October). Petschek-type magnetic reconnection exhausts in the solar wind well inside 1 AU: Helios. *Journal of Geophysical Research 111* (A10) https://doi.org/10.1029/2006JA011863.

Gosling, J.T., Hundhausen, A.J., and Bame, S.J. (1976, May). Solar wind stream evolution at large heliocentric distances – Experimental demonstration and the test of a model. *Journal of Geophysical Research 81*: 2111–2122. https://doi.org/10.1029/JA081i013p02111.

Gosling, J.T. and Phan, T.D. (2013, January). Magnetic reconnection in the solar wind at current sheets associated with extremely small field shear angles. *The Astrophysical Journal 763* (2): L39. https://doi.org/10.1088/2041-8205/763/2/L39.

Gosling, J.T. and Pizzo, V.J. (1999, July). Formation and evolution of corotating interaction regions and their three dimensional structure. *Space Science Review 89*: 21–52. https://doi.org/10.1023/A:1005291711900.

Gosling, J.T. and Szabo, A. (2008). Bifurcated current sheets produced by magnetic reconnection in the solar wind. *Journal of Geophysical Research: Space Physics 113* (A10).

Gosling, J.T., Skoug, R.M., McComas, D.J., and Smith, C.W. (2005, January). Direct evidence for magnetic reconnection in the solar wind near 1 AU. *Journal of Geophysical Research: Space Physics 110*: A01107. https://doi.org/10.1029/2004JA010809.

Greco, A., Perri, S., Servidio, S. et al. (2016, June). The complex structure of magnetic field discontinuities in the turbulent solar wind. *The Astrophysical Journal Letters 823*: L39. https://doi.org/10.3847/2041-8205/823/2/L39.

Hamilton, K., Smith, C.W., Vasquez, B.J., and Leamon, R.J. (2008, January). Anisotropies and helicities in the solar wind inertial and dissipation ranges at 1 AU. *Journal of Geophysical Research: Space Physics 113*: A01106. https://doi.org/10.1029/2007JA012559.

Harrison, R.A., Davis, C.J., and Davies, J.A. (2009, October). Pre-CME onset fuses – Do the STEREO heliospheric imagers hold the clues to the CME onset process? *Solar Physics 259*: 277–296. https://doi.org/10.1007/s11207-009-9417-7.

Hartinger, M.D., Welling, D., Viall, N.M. et al. (2014, October). The effect of magnetopause motion on fast mode resonance. *Journal of Geophysical Research: Space Physics 119*: 8212–8227. https://doi.org/10.1002/2014JA020401.

Hellinger, P., Matteini, L., Štverák, S., Trávníček, P. M., & Marsch, E. (2011, September). Heating and cooling of protons in the fast solar wind between 0.3 and 1 AU: Helios revisited. *Journal of Geophysical Research: Space Physics, 116*, 9105. doi: 10.1029/2011JA016674.

Higginson, A.K., Antiochos, S.K., DeVore, C.R. et al. (2017, May). Formation of heliospheric arcs of slow solar wind. *The Astrophysical Journal 840*: L10. https://doi.org/10.3847/2041-8213/aa6d72.

Higginson, A.K. and Lynch, B.J. (2018, May). Structured slow solar wind variability: Streamer-blob flux ropes and torsional Alfvén waves. *The Astrophysical Journal 859*: 6. https://doi.org/10.3847/1538-4357/aabc08.

Hollweg, J.V. (2002). Heating and acceleration of the solar wind in coronal holes: cyclotron resonances. *Advances in Space Research 30*: 469–469. https://doi.org/10.1016/ S0273-1177(02)00324-1.

Horbury, T. S., Forman, M., & Oughton, S. (2008, October). Anisotropic scaling of magnetohydrodynamic turbulence. *Physical Review Letters, 101* (17), 175005-+. doi: https://doi.org/10.1103/ PhysRevLett.101.175005.

Horbury, T. S., Matteini, L., & Stansby, D. (2018, August). Short, large-amplitude speed enhancements in the near-Sunfast solar wind. *Monthly Notices of the Royal Astronomical Society, 478*, 1980–1986. doi: 10 .1093/mnras/sty953.

Howes, G.G., Cowley, S.C., Dorland, W. et al. (2006, November). Astrophysical gyrokinetics: Basic equations and linear theory.

The Astrophysical Journal 651: 590–614. https://doi.org/10.1086/506172.

Howes, G.G., Tenbarge, J.M., Dorland, W. et al. (2011, July). Gyrokinetic simulations of solar wind turbulence from ion to electron scales. *Physical Review Letters 107* (3): 035004. https://doi.org/10.1103/PhysRevLett.107.035004.

Hudson, P.D. (1970, November). Discontinuities in an anisotropic plasma and their identification in the solar wind. *Planetary and Space Science 18*: 1611–1622. https://doi.org/10.1016/0032-0633(70) 90036-X.

Hundhausen, A.J. and Gosling, J.T. (1976, March). Solar wind structure at large heliocentric distances – an interpretation of Pioneer 10 observations. *Journal of Geophysical Research 81*: 1436–1440. https://doi.org/10.1029/JA081i007p01436.

Hyder, C.L. and Lites, B.W. (1970, September). Hα Doppler brightening and Lyman-a Doppler dimming in moving Hα Prominences. *Solar Physics 14*: 147–156. https://doi.org/10.1007/BF00240170.

Isenberg, P.A. (1987, February). Evolution of interstellar pickup ions in the solar wind. *Journal of Geophysical Research 92*: 1067–1073. https://doi.org/10.1029/JA092iA02p01067.

Isenberg, P.A., Lee, M.A., and Hollweg, J.V. (2001, April). The kinetic shell model of coronal heating and acceleration by ion cyclotron waves: 1. *Outward propagating waves. Journal of Geophysical Research: Space Physics 106*: 5649–5660. https://doi.org/10.1029/2000JA000099.

Jian, L., Russell, C.T., Luhmann, J.G., and Skoug, R.M. (2006, December). Properties of stream interactions at one AU during 1995 2004. *Solar Physics 239*: 337–392. https://doi.org/10.1007/s11207-006-0132-3.

Jian, L.K., Wei, H.Y., Russell, C.T. et al. (2014, May). Electromagnetic waves near the proton cyclotron frequency: STEREO observations. *The Astrophysical Journal 786*: 123. https://doi.org/10.1088/0004-637X/786/2/123.

Jovanović , D., Alexandrova , O., Maksimović , M., and Belic, M. (2020, June). Fluid Theory of Coherent Magnetic Vortices in High-β Space Plasmas. *The Astrophysical Journal 896* (1):8. https://doi.org/10.3847/1538-4357/ab8a45.

Kahler, S. and Lin, R.P. (1994, July). The determination of interplanetary magnetic field polarities around sector boundaries using E greater than 2 keV electrons. *Geophysical Research Letters 21*: 1575–1578. https://doi.org/10.1029/94GL01362.

Kahler, S.W., Crooker, N.U., and Gosling, J.T. (1996, November). The topology of intrasector reversals of the interplanetary magnetic field. *Journal of Geophysical Research 101*: 24373–24382. https://doi.org/10.1029/96JA02232.

Kajdič, P., Alexandrova, O., Maksimovic, M. et al. (2016a, December). Suprathermal electron Strahl widths in the presence of narrow-band whistler waves in the solar wind. *The Astrophysical Journal 833*: 172. https://doi.org/10.3847/1538-4357/833/2/172.

Kajdič, P., Alexandrova, O., Maksimovic, M. et al. (2016b, December). Suprathermal electron strahl widths in the presence of narrow-band whistler waves in the solar wind. *The Astrophysical Journal 833*: 172. https://doi.org/10.3847/1538-4357/833/2/172.

Karpen, J.T., DeVore, C.R., Antiochos, S.K., and Pariat, E. (2017, January). Reconnection-driven coronal-hole jets with gravity and solar wind. *The Astrophysical Journal 834*: 62. https://doi.org/10.3847/1538-4357/834/1/62.

Kasper, J.C., Stevens, M.L., Korreck, K.E. et al. (2012, February). Evolution of the relationships between helium abundance, minor ion charge state, and solar wind speed over the solar cycle. *The Astrophysical Journal 745*: 162. https://doi.org/10.1088/0004-637X/745/2/162.

Kennel, C.F. and Engelmann, F. (1966, December). Velocity space diffusion from weak plasma turbulence in a magnetic field. *Physics of Fluids 9*: 2377–2388. https://doi.org/10.1063/1.1761629.

Kepko, L. and Spence, H.E. (2003, June). Observations of discrete, global magnetospheric oscillations directly driven by solar wind density variations. *Journal of Geophysical Research: Space Physics 108*: 1257. https://doi.org/10.1029/2002JA009676.

Kepko, L., Spence, H.E., and Singer, H.J. (2002, April). ULF waves in the solar wind as direct drivers of magnetospheric pulsations. *Geophysical Research Letters 29*: 1197. https://doi.org/10.1029/ 2001GL014405.

Kepko, L., Viall, N.M., Antiochos, S.K. et al. (2016, May). Implications of L1 observations for slow solar wind formation by solar reconnection. *Geophysical Research Letters 43*: 4089–4097. https://doi.org/10.1002/2016GL068607.

Kilpua, E.K.J., Luhmann, J.G., Gosling, J. et al. (2009, May). Small solar wind transients and their connection to the large-scale coronal structure. *Solar Physics 256*: 327–344. https://doi.org/10.1007/s11207-009-9366-1.

Kiyani, K.H., Chapman, S.C., Sahraoui, F. et al. (2013, January). Enhanced magnetic compressibility and isotropic scale invariance at sub-ion Larmor scales in solar wind turbulence. *The Astrophysical Journal 763*: 10. https://doi.org/10.1088/0004-637X/763/1/10.

Kiyani, K. H., Osman, K. T., & Chapman, S. C. (2015). Dissipation and heating in solar wind turbulence: from the macro to the micro and back again. *Philosophical Transactions of the Royal Society A: Mathematical, Physical and Engineering Sciences*, *373*(2041), 20140155. doi: https://doi.org/10.1098/rsta.2014.01550.

Ko, Y.-K., Muglach, K., Wang, Y.-M., Young, P. R., & Lepri, S. T. (2014). Temporal evolution of solar wind ion composition and their source coronal holes during the declining phase of cycle 23. I. Low-latitude extension of polar coronal holes. *The Astrophysical Journal*, *787*(2), 121.

Kohl, J.L., Esser, R., Gardner, L.D. et al. (1995, December). The ultraviolet coronagraph spectrometer for the solar and heliospheric observatory. *Solar Physics 162*: 313–356. https://doi.org/10.1007/BF00733433.

Kohl, J.L., Noci, G., Antonucci, E. et al. (1998, July). UVCS/SOHO empirical determinations of anisotropic velocity distributions in the solar corona. *The Astrophysical Journal 501*: L127–L131. https://doi.org/10.1086/311434.

Kohl, J.L., Noci, G., Antonucci, E. et al. (1997, October). First results from the SOHO ultraviolet coronagraph spectrometer. *Solar Physics 175*: 613–644. https://doi.org/10.1023/A:1004903206467.

Lacombe, C., Alexandrova, O., and Matteini, L. (2017, October). Anisotropies of the magnetic field fluctuations at kinetic scales in the solar wind: cluster observations. *The*

Astrophysical Journal 848: 45. https://doi.org/10.3847/1538-4357/aa8c06.

Lacombe, C., Alexandrova, O., Matteini, L. et al. (2014, November). Whistler mode waves and the electron heat flux in the solar wind: Cluster observations. *The Astrophysical Journal 796*: 5. https://doi.org/10.1088/0004-637X/796/1/5.

Laming, J.M. (2009, April). Non-Wkb models of the first ionization potential effect: Implications for solar coronal heating and the coronal helium and neon abundances. *The Astrophysical Journal 695*: 954–969. https://doi.org/10.1088/0004-637X/695/2/954.

Laming, J.M. (2015, September). The FIP and inverse FIP effects in solar and stellar coronae. *Living Reviews in Solar Physics 12*: 2. https://doi.org/10.1007/lrsp-2015 -2.

Lavraud, B., Gosling, J.T., Rouillard, A.P. et al. (2009, May). Observation of a complex solar wind reconnection exhaust from spacecraft separated by over 1800 R E. *Solar Physics 256*: 379–392. https://doi.org/10.1007/s11207-009-9341-x.

Lavraud, B., Ruffenach, A., Rouillard, A.P. et al. (2014, January). Geo-effectiveness and radial dependence of magnetic cloud erosion by magnetic reconnection. *Journal of Geophysical Research: Space Physics 119*: 26–35. https://doi.org/10.1002/2013JA019154.

Lazar, M., Yoon, P.H., López, R.A., and Moya, P.S. (2018, January). Electromagnetic electron cyclotron instability in the solar wind. *Journal of Geophysical Research: Space Physics 123*: 6–19. https://doi.org/10.1002/2017JA024759.

Leamon, R.J., Smith, C.W., Ness, N.F. et al. (1998, March). Observational constraints on the dynamics of the interplanetary magnetic field dissipation range. *Journal of Geophysical Research Atmospheres 103*: 4775. https://doi.org/10.1029/97JA03394.

Lee, M.A. (2000, May). An analytical theory of the morphology, flows, and shock compressions at corotating interaction regions in the solar wind. *Journal of Geophysical Research 105*: 10491–10500. https://doi.org/10.1029/1999JA000327.

Leubner, M.P. and Vörös, Z. (2005, January). A nonextensive entropy approach to solar wind intermittency. *The Astrophysical Journal 618*: 547–555. https://doi.org/10.1086/425893.

Lin, R.P. (1974, June). Non-relativistic solar electrons. *Space Science Reviews 16*: 189–256. https://doi.org/10.1007/BF00240886.

Lin, R.P. (1998, July). WIND observations of suprathermal electrons in the interplanetary medium. *Space Science Reviews 86*: 61–78. https://doi.org/10.1023/A:1005048428480.

Linker, J.A., Lionello, R., Mikić, Z. et al. (2011, April). The evolution of open magnetic flux driven by photospheric dynamics. *The Astrophysical Journal 731*: 110. https://doi.org/10.1088/0004-637X/731/2/110.

Lion, S., Alexandrova, O., and Zaslavsky, A. (2016, June). Coherent events and spectral shape at ion kinetic scales in the fast solar wind turbulence. *The Astrophysical Journal 824*: 47. https://doi.org/10.3847/0004-637X/824/1/47.

Lionello, R., Riley, P., Linker, J.A., and Mikić, Z. (2005, May). The effects of differential rotation on the magnetic structure of the solar corona: Magnetohydrodynamic simulations. *The Astrophysical Journal 625*: 463–473. https://doi.org/10.1086/429268.

Lionello, R., Velli, M., Downs, C. et al. (2014, April). Validating a Time-dependent Turbulence-driven Model of the Solar Wind. *The Astrophysical Journal 784*: 120. https://doi.org/10.1088/0004-637X/784/2/120.

Liou, K., Takahashi, K., Newell, P.T., and Yumoto, K. (2008, August). Polar Ultraviolet Imager observations of solar wind-driven ULF auroral pulsations. *Geophysical Research Letters 35*: L16101. https://doi.org/10.1029/2008GL034953.

Liu, Y.C.-M., Huang, J., Wang, C. et al. (2014, November). A statistical analysis of heliospheric plasma sheets, heliospheric current sheets, and sector boundaries observed in situ by STEREO. *Journal of Geophysical Research: Space Physics 119*: 8721–8732. https://doi.org/10.1002/2014JA019956.

Lockwood, M., Owens, M., & Rouillard, A. P. (2009a, November). Excess open solar magnetic flux from satellite data: 1. Analysis of the third perihelion Ulysses pass. *Journal of Geophysical Research: Space Physics, 114* (A11), A11103. doi: 10.1029/2009JA014449.

Lockwood, M., Owens, M., & Rouillard, A. P. (2009b, November). Excess open solar magnetic flux from satellite data: 2. A survey of kinematic effects. *Journal of Geophysical Research: Space Physics, 114* (A11), A11104. doi: 10.1029/2009JA014450.

Lopez, R.E., Freeman, J.W., and Roelof, E.C. (1986, July). The relationship between proton temperature and momentum flux density in the solar wind. *Geophysical Research Letters 13*: 640–643. https://doi.org/10.1029/GL013i007p00640.

Maksimovic, M., Gary, S.P., and Skoug, R.M. (2000, August). Solar wind electron suprathermal strength and temperature gradients: Ulysses observations. *Journal of Geophysical Research: Space Physics 105*: 18337–18350. https://doi.org/10.1029/2000JA900039.

Maksimovic, M., Pierrard, V., and Riley, P. (1997). Ulysses electron distributions fitted with Kappa functions. *Geophysical Research Letters 24*: 1151–1154. https://doi.org/10.1029/97GL00992.

Manchester, W.B., Kozyra, J.U., Lepri, S.T., and Lavraud, B. (2014, July). Simulation of magnetic cloud erosion during propagation. *Journal of Geophysical Research: Space Physics 119* (7): 5449–5464. https://doi.org/10.1002/2014JA019882.

Manchester, W.B., van der Holst, B., and Lavraud, B. (2014, March). Flux rope evolution in interplanetary coronal mass ejections: The 13 May 2005 event. *Plasma Physics and Controlled Fusion 56* (6): 064006. https://doi.org/10.1088/0741-3335/56/6/064006.

Marocchi, D., Antonucci, E., and Giordano, S. (2001, February). Oxygen abundance in coronal streamers during solar minimum. *Annales Geophysicae 19*: 135–145. https://doi.org/10.5194/angeo-19-135-2001.

Marsch, E. (2012, November). Helios: Evolution of distribution functions 0.3-1 AU. *Space Science Reviews 172*: 23–39. https://doi.org/10.1007/s11214-010-9734-z.

Marsch, E., Schwenn, R., Rosenbauer, H. et al. (1982, January). Solar wind protons – Three-dimensional velocity distributions and derived plasma parameters measured between 0.3 and 1 AU. *Journal of Geophysical Research: Space Physics 87*: 52–72. https://doi.org/10.1029/JA087iA01p00052.

Marsch, E. and Tu, C.-Y. (1993, December). Modeling results on spatial transport and spectral transfer of solar wind

Alfvenic turbulence. *Journal of Geophysical Research: Space Physics 98*: 21. https://doi.org/10.1029/93JA02365.

Matteini, L., Alexandrova, O., Chen, C.H.K., and Lacombe, C. (2017, April). Electric and magnetic spectra from MHD to electron scales in the magnetosheath. *Monthly Notices of the Royal Astronomical Society 466*: 945–951. https://doi.org/10.1093/mnras/stw3163.

Matteini, L., Horbury, T.S., Neugebauer, M., and Goldstein, B.E. (2014). Dependence of solar wind speed on the local magnetic field orientation: Role of Alfvenic fluctuations. *Geophysical Research Letters 41*: 259–265. https://doi.org/10.1002/2013GL058482.

Matteini, L., Horbury, T.S., Pantellini, F. et al. (2015, March). Ion kinetic energy conservation and magnetic field strength constancy in multi-fluid solar wind Alfvénic turbulence. *The Astrophysical Journal 802*: 11. https://doi.org/10.1088/0004-637X/802/1/11.

Matteini, L., Stansby, D., Horbury, T.S., and Chen, C.H.K. (2018). On the 1/f spectrum in the solar wind and its connection with magnetic compressibility. *The Astrophysical Journal Letters 869*: L32.

Matteini, L., Hellinger, P., Landi, S. et al. (2012, November). Ion Kinetics in the solar wind: Coupling global expansion to local microphysics, *Space Science Review 172* (1-4): 373-396. https://doi.org/10.1007/s11214-011-9774-z.

Matthaeus, W. H., & Goldstein, M. L. (1986, July). Low-frequency 1/f noise in the interplanetary magnetic field. *Physical Review Letters*, *57*, 495–498. doi: 10.1103/PhysRevLett.57.495.

Mazur, J.E., Mason, G.M., Dwyer, J.R. et al. (2000, March). Interplanetary magnetic field line mixing deduced from impulsive solar flare particles. *The Astrophysical Journal Letters 532*: L79–L82. https://doi.org/10.1086/312561.

McComas, D.J., Bame, S.J., Barraclough, B.L. et al. (1998). Ulysses' return to the slow solar wind. *Geophysical Research Letters 25*: 1–4. https://doi.org/10.1029/97GL03444.

McComas, D.J., Ebert, R.W., Elliott, H.A. et al. (2008, September). Weaker solar wind from the polar coronal holes and the whole Sun. *Geophysical Research Letters 35*: L18103. https://doi.org/10.1029/2008GL034896.

McComas, D.J., Gosling, J.T., Winterhalter, D., and Smith, E.J. (1988, April). Interplanetary magnetic field draping about fast coronal mass ejecta in the outer heliosphere. *Journal of Geophysical Research: Space Physics 93* (A4): 2519–2526. https://doi.org/10.1029/JA093iA04p02519.

McComas, D.J., Hoogeveen, G.W., Gosling, J.T. et al. (1996, December). ULYSSES observations of pressure-balance structures in the polar solar wind. *Astronomy and Astrophysics 316*: 368–373.

Moldwin, M.B., Ford, S., Lepping, R. et al. (2000, January). *Small-scale magnetic flux ropes in the solar wind. 27*: 57–60. https://doi.org/10.1029/1999GL010724.

Moldwin, M.B., Phillips, J.L., Gosling, J.T. et al. (1995, October). Ulysses observation of a noncoronal mass ejection flux rope: Evidence of interplanetary magnetic reconnection. *Journal of Geophysical Research: Space Physics 100*: 19903–19910. https://doi.org/10.1029/95JA01123.

Morton, R. J., Tomczyk, S., & Pinto, R. (2015, July). Investigating Alfvenic wave propagation in coronal open-field regions. *Nature Communications, 6*. Retrieved 2015-07-27, from http://www.nature.com/ncomms/2015/150727/ncomms8813/full/ncomms8813.html doi: https://doi.org/10.1038/ncomms8813.

Morton, R. J., Tomczyk, S., & Pinto, R. F. (2016, September). A global view of velocity fluctuations in the corona below 1.3 R with CoMP. *The Astrophysical Journal, 828*, 89. Retrieved 2016-09-12, from http://adsabs.harvard.edu/abs/2016ApJ...828...89M doi: https://doi.org/10.3847/0004-637X/828/2/89.

Möstl, C., Miklenic, C., Farrugia, C. J., Temmer, M., Veronig, A., Galvin, A. B., et al. (2008, October). Two-spacecraft reconstruction of a magnetic cloud and comparison to its solar source. *Annales Geophysicae*, *26*(10), 3139–3152. doi: https://doi.org/10.5194/angeo-26-3139-2008.

Neugebauer, M. (2012, May). Evidence for polar X-ray jets as sources of microstream peaks in the solar wind. *The Astrophysical Journal 750*: 50. https://doi.org/10.1088/0004-637X/750/1/50.

Neugebauer, M., Clay, D.R., Goldstein, B.E. et al. (1984, July). A reexamination of rotational and tangential discontinuities in the solar wind. *Journal of Geophysical Research: Space Physics 89*: 5395–5408. https://doi.org/10.1029/JA089iA07p05395.

Neugebauer, M. and Giacalone, J. (2015, October). Energetic particles, tangential discontinuities, and solar flux tubes. *Journal of Geophysical Research: Space Physics 120*: 8281–8287. https://doi.org/10.1002/2015JA021632.

Neugebauer, M., Goldstein, B.E., McComas, D.J. et al. (1995, December). Ulysses observations of microstreams in the solar wind from coronal holes. *Journal of Geophysical Research: Space Physics 100*: 23389–23396. https://doi.org/10.1029/95JA02723.

Neugebauer, M., Ruzmaikin, A., & McComas, D. J. (1997, January). Wavelet analysis of the structure of microstreams in the polar solar wind. In S. R. Habbal (Ed.), *Robotic exploration close to the sun: Scientific basis* (Vol. *385*, p. 41–46). doi: https://doi.org/10.1063/1.51765.

Neugebauer, M. and Snyder, C.W. (1962, December). Solar plasma experiment. *Science 138*: 1095–1097. https://doi.org/10.1126/science.138.3545.1095-a.

Nicol, R.M., Chapman, S.C., and Dendy, R.O. (2009, October). Quantifying the anisotropy and solar cycle dependence of "1/f" solar wind fluctuations observed by advanced composition explorer. *The Astrophysical Journal 703*: 2138–2151. https://doi.org/10.1088/0004-637X/703/2/2138.

Noci, G., Kohl, J.L., and Withbroe, G.L. (1987, April). Solar wind diagnostics from Doppler-enhanced scattering. *The Astrophysical Journal 315*: 706–715. https://doi.org/10.1086/165172.

Ogilvie, K.W., Fitzenreiter, R., and Desch, M. (2000, December). Electrons in the low-density solar wind. *Journal of Geophysical Research: Space Physics 105*: 27277–27288. https://doi.org/10.1029/2000JA000131.

Oran, R., van der Holst, B., Landi, E. et al. (2013, December). A global wave-driven magnetohydrodynamic solar model with a unified treatment of open and closed magnetic field topologies. *The Astrophysical Journal 778*: 176. https://doi.org/10.1088/0004-637X/778/2/176.

Owens, M.J. and Forsyth, R.J. (2013). The heliospheric magnetic field. *Living Reviews in Solar Physics 10*.

Owens, M.J., Crooker, N.U., and Lockwood, M. (2013, May). Solar origin of heliospheric magnetic field inversions: Evidence for coronal loop opening within pseudostreamers. *Journal of Geophysical Research: Space Physics 118*: 1868–1879. https://doi.org/10.1002/jgra.50259.

Owens, M. J., Wicks, R. T., & Horbury, T. S. (2011, April). Magnetic discontinuities in the near-earth solar wind: Evidence of in-transit turbulence or remnants of coronal structure?, 269, 411-420. doi: https://doi.org/10.1007/s11207-010-9695-0.

Parker, E.N. (1958, November). Dynamics of the interplanetary gas and magnetic fields. *The Astrophysical Journal 128*: 664. https://doi.org/10.1086/146579.

Perri, S., Goldstein, M.L., Dorelli, J.C., and Sahraoui, F. (2012, November). Detection of small-scale structures in the dissipation regime of solar-wind turbulence. *Physical Review Letters 109* (19): 191101. https://doi.org/10.1103/PhysRevLett.109.191101.

Perrone, D., Alexandrova, O., Mangeney, A. et al. (2016, August). Compressive coherent structures at ion scales in the slow solar wind. *The Astrophysical Journal 826*: 196. https://doi.org/10.3847/0004-637X/826/2/196.

Perrone, D., Alexandrova, O., Roberts, O.W. et al. (2017, November). Coherent structures at ion scales in fast solar wind: Cluster observations. *The Astrophysical Journal 849*: 49. https://doi.org/10.3847/1538-4357/aa9022.

Phan, T.D., Gosling, J.T., Davis, M.S. et al. (2006, January). A magnetic reconnection X-line extending more than 390 Earth radii in the solar wind. *Nature 439* (7073): 175–178. https://doi.org/10.1038/nature04393.

Pierrard, V. (2011, February). *Solar wind electron transport: Interplanetary electric field and heat conduction. 172*: 315–324. https://doi.org/10.1007/s11214-011-9743-6.

Pierrard, V., & Lamy, H. (2003, September). The effects of the velocity filtration mechanism on the minor ions of the corona. *Solar Physics, 216*, 47–58. doi: 10.1023/ A:1026157306754.

Pierrard, V. and Lazar, M. (2010, November). Kappa distributions: Theory and applications in space plasmas. *Solar Physics 267*: 153–174. https://doi.org/10.1007/s11207-010-9640-2.

Pierrard, V., Lazar, M., Poedts, S. et al. (2016, August). The electron temperature and anisotropy in the solar wind. comparison of the core and halo populations. *Solar Physics 291*: 2165–2179. https://doi.org/10.1007/s11207-016-0961-7.

Pierrard, V., Lazar, M., and Schlickeiser, R. (2011, April). Evolution of the electron distribution function in the whistler wave turbulence of the solar wind. *Solar Physics 269*: 421–438. https://doi.org/10.1007/s11207-010-9700-7.

Pierrard, V., Maksimovic, M., and Lemaire, J. (1999, August). Electron velocity distribution functions from the solar wind to the corona. *Journal of Geophysical Research: Space Physics 104*: 17021–17032. https://doi.org/10.1029/1999JA900169.

Pierrard, V. and Pieters, M. (2014, December). Coronal heating and solar wind acceleration for electrons, protons, and minor ions obtained from kinetic models based on kappa distributions. *Journal of Geophysical Research: Space Physics 119*: 9441–9455. https://doi.org/10.1002/2014JA020678.

Pilipp, W.G., Miggenrieder, H., Montgomery, M.D. et al. (1987, February). Characteristics of electron velocity distribution functions in the solar wind derived from the HELIOS plasma experiment. *Journal of Geophysical Research: Space Physics 92*: 1075–1092. https://doi.org/10.1029/JA092iA02p01075.

Pilipp, W.G., Muehlhaeuser, K.-H., Miggenrieder, H. et al. (1990, May). Large-scale variations of thermal electron parameters in the solar wind between 0.3 and 1 AU. *Journal of Geophysical Research: Space Physics 95*: 6305–6329. https://doi.org/10.1029/JA095iA05p06305.

Pinto, R.F., Brun, A.S., Jouve, L., and Grappin, R. (2011, August). Coupling the solar dynamo and the corona: Wind properties, mass, and momentum losses during an activity cycle. *The Astrophysical Journal 737*: 72. https://doi.org/10.1088/0004-637X/737/2/72.

Pinto, R. F., Brun, A. S., & Rouillard, A. P. (2016, August). Flux-tube geometry and solar wind speed during an activity cycle. *Astronomy & Astrophysics, 592*, A65. Retrieved 2019-03-04, from https://www.aanda.org/articles/aa/abs/2016/08/aa28599-16/aa28599-16.html doi: https://doi.org/10.1051/0004-6361/ 201628599.

Pinto, R.F. and Rouillard, A.P. (2017, April). A multiple flux-tube solar wind model. *The Astrophysical Journal 838*: 89. https://doi.org/10.3847/1538-4357/aa6398.

Pizzo, V.J. (1982, June). A three-dimensional model of corotating streams in the solar wind. *III - Magnetohydrodynamic streams. Journal of Geophysical Research: Space Physics 87*: 4374–4394. https://doi.org/10.1029/ JA087iA06p04374.

Podesta, J. J., Roberts, D. A., & Goldstein, M. L. (2006, October). Power spectrum of small-scale turbulent velocity fluctuations in the solar wind. *Journal of Geophysical Research: Space Physics, 111* (A10), 10109-+. doi: 10.1029/2006JA011834.

Poletto, G. (2015, December). Solar coronal plumes. *Living Reviews in Solar Physics 12*: 7. https://doi.org/10.1007/lrsp-2015-7.

Pylaev, O.S., Zaqarashvili, T.V., Brazhenko, A.I. et al. (2017, May). Oscillation of solar radio emission at coronal acoustic cut-off frequency. *Astronomy and Astrophysics 601*: A42. https://doi.org/10.1051/0004-6361/201629218.

Rakowski, C.E. and Laming, J.M. (2012). On the origin of the slow speed solar wind: helium abundance variations. *The Astrophysical Journal 754* (1): 65.

Raouafi, N.E., Patsourakos, S., Pariat, E. et al. (2016, November). Solar coronal jets: Observations, theory, and modeling. *Space Science Reviews 201*: 1–53. https://doi.org/10.1007/s11214-016-0260-5.

Reisenfeld, D.B., McComas, D.J., and Steinberg, J.T. (1999). Evidence of a solar origin for pressure balance structures in the high-latitude solar wind. *Geophysical Research Letters 26*: 1805–1808. https://doi.org/10.1029/1999GL900368.

Riley, P., Linker, J., Lionello, R. et al. (2004, October). Fitting flux ropes to a global MHD solution: A comparison of techniques. *Journal of Atmospheric and Solar-Terrestrial Physics 66* (15–16): 1321–1331. https://doi.org/10.1016/j.jastp.2004.03.019.

Roberts, M.A., Uritsky, V.M., DeVore, C.R., and Karpen, J.T. (2018, October). Simulated encounters of the parker solar probe with a coronal-hole jet. *The Astrophysical Journal 866*: 14. https://doi.org/10.3847/1538-4357/aadb41.

Roberts, O.W., Alexandrova, O., Kajdič, P. et al. (2017, December). Variability of the magnetic field power spectrum in the solar wind at electron scales. *The Astrophysical Journal 850*: 120. https://doi.org/10.3847/1538-4357/aa93e5.

Rouillard, A. P., Davies, J. A., Lavraud, B., Forsyth, R. J., Savani, N. P., Bewsher, D., et al. (2010, April). Intermittent release of transients in the slow solar wind: 1. Remote sensing observations. *Journal of Geophysical Research: Space Physics, 115*, A04103. doi: 10.1029/2009JA014471.

Rouillard, A. P., Lavraud, B., Davies, J. A., Savani, N. P., Burlaga, L. F., Forsyth, R. J., et al. (2010, April). Intermittent release of transients in the slow solar wind: 2. In situ evidence. *Journal of Geophysical Research: Space Physics, 115*, A04104. doi: 10.1029/2009JA014472.

Rouillard, A. P., Savani, N. P., Davies, J. A., Lavraud, B., Forsyth, R. J., Morley, S. K., et al. (2009, May 01). A multispacecraft analysis of a small-scale transient entrained by solar wind streams. *Solar Physics, 256*(1), 307–326. Retrieved from https://doi.org/10.1007/s11207-009-9329-6 doi: 10.1007/s11207-009-9329-6.

Rouillard, A.P., Sheeley, N.R. Jr., Cooper, T.J. et al. (2011, June). The solar origin of small interplanetary transients. *The Astrophysical Journal 734*: 7. https://doi.org/10.1088/0004-637X/734/1/7.

Ruffenach, A., Lavraud, B., Farrugia, C.J. et al. (2015, January). Statistical study of magnetic cloud erosion by magnetic reconnection. *Journal of Geophysical Research: Space Physics 120*: 43–60. https://doi.org/10.1002/2014JA020628.

Ruffenach, A., Lavraud, B., Owens, M. J., Sauvaud, J.-A., Savani, N. P., Rouillard, A. P., et al. (2012, September). Multispacecraft observation of magnetic cloud erosion by magnetic reconnection during propagation: Magnetic Cloud Erosion. *Journal of Geophysical Research: Space Physics, 117*(A09101), 1-16. doi: 10.1029/2012JA017624.

Sahraoui, F., Huang, S.Y., Belmont, G. et al. (2013, November). Scaling of the electron dissipation range of solar wind turbulence. *The Astrophysical Journal 777*: 15. https://doi.org/10.1088/0004-637X/777/1/15.

Salem, C., Hubert, D., Lacombe, C. et al. (2003, March). Electron properties and Coulomb collisions in the solar wind at 1 AU: Wind observations. *The Astrophysical Journal 585*: 1147–1157. https://doi.org/10.1086/346185.

Salem, C., Mangeney, A., Bale, S.D., and Veltri, P. (2009, September). Solar wind magnetohydrodynamics turbulence: Anomalous scaling and role of intermittency. *The Astrophysical Journal 702*: 537–553. https://doi.org/10.1088/0004-637X/702/1/537.

Salem, C.S., Howes, G.G., Sundkvist, D. et al. (2012, January). Identification of kinetic Alfven wave turbulence in the solar wind. *The Astrophysical Journal Letters 745*: L9. https://doi.org/10.1088/2041-8205/745/1/ L9.

Sanchez-Diaz, E., Rouillard, A.P., Davies, J.A. et al. (2017, January). Observational evidence for the associated formation of blobs and raining inflows in the solar corona. *The Astrophysical Journal Letters 835*: L7. https://doi.org/10.3847/2041-8213/835/1/L7.

Sanchez-Diaz, E., Rouillard, A.P., Lavraud, B. et al. (2016, April). The very slow solar wind: Properties, origin and variability. *Journal of Geophysical Research: Space Physics 121*: 2830–2841. https://doi.org/10.1002/2016JA022433.

Saur, J. and Bieber, J.W. (1999, May). Geometry of low-frequency solar wind magnetic turbulence: Evidence for radially aligned Alfénic fluctuations. *Journal of Geophysical Research104*:9975–9988.https://doi.org/10.1029/1998JA900077.

Schekochihin, A.A., Cowley, S.C., Dorland, W. et al. (2009, May). Astrophysical Gyrokinetics: Kinetic and fluid turbulent cascades in magnetized weakly collisional plasmas. *The Astrophysical Journal 182*: 310–377. https://doi.org/10.1088/0067-0049/182/1/310.

Schreiner, A. and Saur, J. (2017, February). A model for dissipation of solar wind magnetic turbulence by kinetic Alfven waves at electron scales: Comparison with observations. *The Astrophysical Journal 835*: 133. https://doi.org/10.3847/1538-4357/835/2/133.

Schrijver, C.J., Sandman, A.W., Aschwanden, M.J., and De Rosa, M.L. (2004, November). The coronal heating mechanism as identified by full-sun visualizations. *The Astrophysical Journal 615*: 512–525. https://doi.org/10.1086/424028.

Schwadron, N.A., Fisk, L.A., and Zurbuchen, T.H. (1999, August). Elemental fractionation in the slow solar wind. *The Astrophysical Journal 521*: 859–867. https://doi.org/10.1086/307575.

Schwenn, R. (1990). Large-scale structure of the interplanetary medium. *Physics of the Inner Heliosphere I*: 99–181.

Sheeley, N.R. and Rouillard, A.P. (2010, May). Tracking streamer blobs into the heliosphere. *The Astrophysical Journal 715*: 300–309. https://doi.org/10.1088/0004 -637X/715/1/300.

Sheeley, N.R., Wang, Y.-M., Hawley, S.H. et al. (1997, July). Measurements of flow speeds in the corona between 2 and 30 R. *The Astrophysical Journal 484*: 472–478. https://doi.org/10.1086/304338.

Sheeley, N.R., Warren, H.P., and Wang, Y.-M. (2007, December). A streamer ejection with reconnection close to the Sun. *The Astrophysical Journal 671*: 926–935. https://doi.org/10.1086/522940.

Smith, C.W., Hamilton, K., Vasquez, B.J., and Leamon, R.J. (2006, July). Dependence of the dissipation range spectrum of interplanetary magnetic fluctuations on the rate of energy cascade. *The Astrophysical Journal Letters 645*: L85–L88. https://doi.org/10.1086/506151.

Smith, E.J. and Wolfe, J.H. (1976, March). Observations of interaction regions and corotating shocks between one and five AU – Pioneers 10 and 11. *Geophysical Research Letters 3*: 137–140. https://doi.org/10.1029/GL003i003p00137.

Sorriso-Valvo, L., Marino, R., Carbone, V. et al. (2007, September). Observation of inertial energy cascade in interplanetary space plasma. *Physical Review Letters 99* (11): 115001. https://doi.org/10.1103/PhysRevLett.99.115001.

Stansby, D. and Horbury, T.S. (2018, June). Number density structures in the inner heliosphere. *Astronomy & Astrophysics (A&A) 613*: A62. https://doi.org/10.1051/0004-6361/201732567.

Stansby, D., Horbury, T.S., Chen, C.H.K., and Matteini, L. (2016, September). Experimental determination of whistler wave dispersion relation in the solar wind. *The Astrophysical Journal Letters 829*: L16. https://doi.org/10.3847/2041-8205/829/1/L16.

Stansby, D., Horbury, T.S., and Matteini, L. (2019, January). Diagnosing solar wind origins using in situ measurements in the inner heliosphere. *Monthly Notices of the Royal Astronomical Society 482*: 1706–1714. https://doi.org/10.1093/mnras/sty2814.

Stephenson, J.A.E. and Walker, A.D.M. (2002, May). HF radar observations of Pc5 ULF pulsations driven by the solar wind. *Geophysical Research Letters 29*: 1297. https://doi.org/10.1029/2001GL014291.

Strachan, L., Kohl, J.L., Weiser, H. et al. (1993, July). A Doppler dimming determination of coronal outflow velocity. *The Astrophysical Journal 412*: 410–420. https://doi.org/10.1086/172930.

Strachan, L., Panasyuk, A.V., Dobrzycka, D. et al. (2000, February). Latitudinal dependence of outflow velocities from O VI Doppler dimming observations during the Whole Sun Month. *Journal of Geophysical Research: Space Physics 105*: 2345–2356. https://doi.org/10.1029/1999JA900459.

Štverák, S., Travnicek, P., Maksimovic, M. et al. (2008, March). Electron temperature anisotropy constraints in the solar wind. *Journal of Geophysical Research: Space Physics 113*: A03103. https://doi.org/10.1029/2007JA012733.

Susino, R., Ventura, R., Spadaro, D. et al. (2008, September). Physical parameters along the boundaries of a mid-latitude streamer and in its adjacent regions. *Astronomy & Astrophysics (A&A) 488*: 303–310. https://doi.org/10.1051/0004-6361:200809713.

Telloni, D., Antonucci, E., & Dodero, M. A. (2007, September). Outflow velocity of the $O+^5$ ions in polar coronal holes out to 5 R. *Astronomy & Astrophysics (A&A)*, *472*, 299–307. doi: 10.1051/ 0004-6361:20077083.

Teriaca, L., Poletto, G., Romoli, M., and Biesecker, D.A. (2003, May). The nascent solar wind: Origin and acceleration. *The Astrophysical Journal 588*: 566–577. https://doi.org/10.1086/368409.

Thieme, K.M., Marsch, E., and Schwenn, R. (1990, November). Spatial structures in high-speed streams as signatures of fine structures in coronal holes. *Annales Geophysicae 8*: 713–723.

Thieme, K.M., Schwenn, R., and Marsch, E. (1989). Are structures in high-speed streams signatures of coronal fine structures? *Advances in Space Research 9*: 127–130. https://doi.org/10.1016/0273-1177(89)90105-1.

Tian, H., Yao, S., Zong, Q. et al. (2010, September). Signatures of magnetic reconnection at boundaries of interplanetary small-scale magnetic flux ropes. *The Astrophysical Journal 720*: 454–464. https://doi.org/10.1088/0004-637X/720/1/454.

Titov, V.S., Forbes, T.G., Priest, E.R. et al. (2009, March). Slip-squashing factors as a measure of three-dimensional magnetic reconnection. *The Astrophysical Journal 693*: 1029–1044. https://doi.org/10.1088/0004-637X/693/1/1029.

Titov, V.S., Mikić, Z., Linker, J.A. et al. (2011, April). Magnetic topology of coronal hole linkages. *The Astrophysical Journal 731*: 111. https://doi.org/10.1088/0004-637X/731/2/111.

Tomczyk, S., Landi, E., Burkepile, J.T. et al. (2016, August). Scientific objectives and capabilities of the Coronal Solar Magnetism Observatory. *Journal of Geophysical Research: Space Physics 121*: 7470–7487. https://doi.org/10.1002/2016JA022871.

Török, T., Aulanier, G., Schmieder, B. et al. (2009, October). Fan-spine topology formation through two-step reconnection driven by twisted flux emergence. *The Astrophysical Journal 704*: 485–495. https://doi.org/10.1088/0004-637X/704/1/485.

Tu, C.-Y., & Marsch, E. (1995, July). Comment on "Evolution of energy-containing turbulent eddies in the solar wind" by W. H. Matthaeus, S. Oughton, D. H. Pontius, Jr., and Y. Zhou. *Journal of Geophysical Research*, *100*, 12323–12328. doi: https://doi.org/10.1029/95JA01103.

Tu, C.-Y. and Marsch, E. (1997, April). Two-fluid model for heating of the solar corona and acceleration of the solar wind by high-frequency Alfven waves. *Solar Physics 171*: 363–391. https://doi.org/10.1023/A:1004968327196.

Turner, A.J., Gogoberidze, G., Chapman, S.C. et al. (2011, August). Nonaxisymmetric anisotropy of solar wind turbulence. *Physical Review Letters 107* (9): 095002. https://doi.org/10.1103/PhysRevLett.107.095002.

van der Holst, B., Sokolov, I.V., Meng, X. et al. (2014, February). Alfvén Wave Solar Model (AWSoM): coronal heating. *The Astrophysical Journal 782*: 81. https://doi.org/10.1088/0004-637X/782/2/81.

Vasyliunas, V.M. and Siscoe, G.L. (1976, March). On the flux and the energy spectrum of interstellar ions in the solar system. *Journal of Geophysical Research 81*: 1247–1252. https://doi.org/10.1029/JA081i007p01247.

Velli, M., Grappin, R., and Mangeney, A. (1989, October). Turbulent cascade of incompressible unidirectional Alfven waves in the interplanetary medium. *Physical Review Letters 63*: 1807–1810. https://doi.org/10.1103/PhysRevLett.63.1807.

Velli, M., Lionello, R., Linker, J.A., and Mikić, Z. (2011, July). Coronal plumes in the fast solar wind. *The Astrophysical Journal 736*: 32. https://doi.org/10.1088/0004-637X/736/1/32.

Verdini, A., Grappin, R., Pinto, R., and Velli, M. (2012, May). On the origin of the 1/f spectrum in the solar wind magnetic field. *The Astrophysical Journal Letters 750*: L33. https://doi.org/10.1088/2041 -8205/750/2/L33.

Viall, N.M., Kepko, L., and Spence, H.E. (2008, July). Inherent length-scales of periodic solar wind number density structures. *Journal of Geophysical Research: Space Physics 113*: A07101. https://doi.org/10.1029/2007JA012881.

Viall, N.M., Kepko, L., and Spence, H.E. (2009, January). Relative occurrence rates and connection of discrete frequency oscillations in the solar wind density and dayside magnetosphere. *Journal of Geophysical Research: Space Physics 114*: A01201. https://doi.org/10.1029/2008JA013334.

Viall, N.M., Spence, H.E., and Kasper, J. (2009, December). Are periodic solar wind number density structures formed in the solar corona? *Geophysical Research Letters 36*: L23102. https://doi.org/10.1029/2009GL041191.

Viall, N.M., Spence, H.E., Vourlidas, A., and Howard, R. (2010, November). Examining periodic solar-wind density structures observed in the SECCHI heliospheric imagers. *Solar Physics 267*: 175–202. https://doi.org/10.1007/s11207-010-9633-1.

Viall, N. M., & Vourlidas, A. (2015, July). Periodic density structures and the origin of the slow solar wind. *The Astrophysical Journal*, *807*, 176. doi: 10 .1088/0004-637X/807/2/176.

Villante, U., Del Corpo, A., and Francia, P. (2013, January). Geomagnetic and solar wind fluctuations at discrete frequencies: A case study. *Journal of Geophysical Research: Space Physics 118*: 218–231. https://doi.org/10.1029/2012JA017971.

Villante, U., Di Matteo, S., and Piersanti, M. (2016, January). On the transmission of waves at discrete frequencies from the solar wind to the magnetosphere and ground: A case study. *Journal of Geophysical Research: Space Physics 121*: 380–396. https://doi.org/10.1002/2015JA021628.

Vocks, C. (2012, November). Kinetic models for whistler wave scattering of electrons in the solar corona and wind. *Space Science Reviews 172*: 303–314. https://doi.org/10.1007/s11214-011-9749-0.

Vocks, C. and Mann, G. (2003, August). Generation of suprathermal electrons by resonant wave-particle interaction in the solar corona and wind. *The Astrophysical Journal 593*: 1134–1145. https://doi.org/10.1086/376682.

Vocks, C., Mann, G., and Rausche, G. (2008, March). Formation of suprathermal electron distributions in the quiet solar corona. *Astronomy & Astrophysics (A&A) 480*: 527–536. https://doi.org/10.1051/0004-6361:20078826.

Vocks, C., Salem, C., Lin, R.P., and Mann, G. (2005, July). Electron Halo and Strahl formation in the solar wind by resonant interaction with whistler waves. *The Astrophysical Journal 627*: 540–549. https://doi.org/10.1086/430119.

von Rosenvinge, T.T., Richardson, I.G., Reames, D.V. et al. (2009, May). The solar energetic particle event of 14 December 2006. *Solar Physics 256*: 443–462. https://doi.org/10.1007/s11207-009-9353-6.

von Steiger, R., Schwadron, N.A., Fisk, L.A. et al. (2000, December). Composition of quasi-stationary solar wind flows from Ulysses/Solar Wind Ion Composition Spectrometer. *Journal of Geophysical Research: Space Physics 105*: 27217–27238. https://doi.org/10.1029/1999JA000358.

Wang, Y., Wei, F.S., Feng, X.S. et al. (2012, March). Variations of solar electron and proton flux in magnetic cloud boundary layers and comparisons with those across the shocks and in the reconnection exhausts. *The Astrophysical Journal 749* (1): 82. https://doi.org/10.1088/0004-637X/749/1/82.

Wang, Y.-M., Ko, Y.-K., and Grappin, R. (2009, January). Slow solar wind from open regions with strong low-coronal heating. *The Astrophysical Journal 691*: 760–769. https://doi.org/10.1088/0004-637X/691/1/760.

Wang, Y.M., Sheeley, J., and N. R., & Rouillard, A. P. (2006, Jun). Role of the Sun's nonaxisymmetric open flux in cosmic-ray modulation. *The Astrophysical Journal 644* (1): 638–645. https://doi.org/10.1086/503523.

Wang, Y.-M., Sheeley, N.R., Socker, D.G. et al. (1998, December). Observations of correlated white-light and extreme-ultraviolet jets from polar coronal holes. *The Astrophysical Journal 508*: 899–907. https://doi.org/10.1086/306450.

Wang, Y.-M., Sheeley, N.R., Socker, D.G. et al. (2000, November). The dynamical nature of coronal streamers. *Journal of Geophysical Research 105*: 25133–25142. https://doi.org/10.1029/2000JA000149.

Wang, Y.-M., Sheeley, N.R. Jr., Nash, A.G., and Shampine, L.R. (1988, April). The quasi-rigid rotation of coronal magnetic fields. *The Astrophysical Journal, Part 1* (327): 427–450. https://doi.org/10.1086/166205.

Wei, F., Liu, R., Fan, Q., and Feng, X. (2003, June). Identification of the magnetic cloud boundary layers. *Journal of Geophysical Research: Space Physics 108* (A6) https://doi.org/10.1029/2002JA009511.

Whang, Y.C. and Burlaga, L.F. (1990, December). Simulation of period doubling of recurrent solar wind structures. *Journal of Geophysical Research: Space Physics 95*: 20663–20671. https://doi.org/10.1029/JA095iA12p20663.

Wicks, R.T., Horbury, T.S., Chen, C.H.K., and Schekochihin, A.A. (2010). Power and spectral index anisotropy of the entire inertial range of turbulence in the fast solar wind. *Monthly Notices of the Royal Astronomical Society: Letters 407*: L31–L35. https://doi.org/10.1111/j.1745-3933.2010.00898.x.

Widing, K.G. and Feldman, U. (2001, July). On the rate of abundance modifications versus time in active region plasmas. *The Astrophysical Journal 555*: 426–434. https://doi.org/10.1086/ 321482.

Winterhalter, D., Smith, E.J., Burton, M.E. et al. (1994, April). The heliospheric plasma sheet. *Journal of Geophysical Research: Space Physics 99*: 6667–6680. https://doi.org/10.1029/93JA03481.

Withbroe, G.L., Kohl, J.L., Weiser, H., and Munro, R.H. (1982, March). Probing the solar wind acceleration region using spectroscopic techniques. *Space Science Reviews 33*: 17–52. https://doi.org/10.1007/BF00213247.

Yoon, P.H., Kim, S., and Choe, G.S. (2015, October). Steady-state model of solar wind electrons revisited. *The Astrophysical Journal 812*: 169. https://doi.org/10.1088/0004-637X/812/2/169.

Zangrilli, L. and Poletto, G. (2016, October). Evolution of active region outflows throughout an active region lifetime. *Astronomy & Astrophysics (A&A) 594*: A40. https://doi.org/10.1051/0004-6361/ 201628421.

Zank, G.P. (1999, July). Interaction of the solar wind with the local interstellar medium: A theoretical perspective. *Space Science Reviews 89*: 413–688. https://doi.org/10.1023/A:1005155601277.

Zurbuchen, T.H., Hefti, S., Fisk, L.A. et al. (1999, January). The transition between fast and slow solar wind from composition data. *Space Science Reviews 87*: 353–356. https://doi.org/10.1023/A:1005126718714.

2
The Heating of the Solar Corona

Nicholeen Viall[1], Ineke De Moortel[2], Cooper Downs[3], James A. Klimchuk[1], Susanna Parenti[4], and Fabio Reale[5]

ABSTRACT

The solar corona, the outer atmosphere of the Sun, is heated to millions of Kelvin. This is several orders of magnitude hotter than the photosphere, the optical surface of the Sun, below, and a mystery that has baffled scientists for centuries. The answer to the question of how the solar corona is heated lies in the crucial magnetic connection through the atmosphere of the Sun. The magnetic field that threads the corona extends below the solar photosphere, where convective motions drag the magnetic field footpoints, tangling and twisting them. The chromosphere is the atmospheric layer above the photosphere, and the magnetic field provides an important connection between these layers. The exchange of mass and energy between the chromosphere and the corona is an essential piece of this puzzle. However, the connection between the chromosphere and the corona is a challenging piece of the puzzle both observationally and computationally, as it is highly complex in space and time. We describe the history of the observations and the theoretical understanding of the heating of the solar atmosphere, and end with future prospects of solving the coronal heating problem.

2.1. HISTORICAL VIEWPOINT ON THE CORONAL HEATING PROBLEM

The outer atmosphere of the Sun starts at the photosphere, the optical surface of the Sun, followed by the chromosphere, a thin transition region, and ends in a hot corona above. Understanding the heating of the solar atmosphere is important for understanding the fundamental physical processes governing our star, and also for understanding the space weather effects on Earth's atmosphere and ionosphere. Planetary ionospheres in general are created by the ionization of their atmosphere by ultraviolet (UV) and X-ray emissions from their parent star. This makes understanding these emissions important for understanding stellar coronae and exoplanetary systems in general, as well as space weather at other planets in our solar system. Radiation at UV and X-ray wavelengths also heats Earth's mesosphere, impacting the drag on spacecraft orbiting at these altitudes. Solar emission at these wavelengths exhibit large changes over short timescales in the case of flares (see Chapter 5), or more slowly, as the solar atmosphere and its heating evolve over the 11-year sunspot cycle (see Chapter 3). Lastly, the acceleration of the solar wind (see Chapter 1) plasma that fills the solar system is closely tied to the physics of the heating of the solar atmosphere and the hot corona.

[1] *NASA Goddard Space Flight Center, Greenbelt, Maryland, USA*

[2] *School of Mathematics and Statistics, University of St Andrews, St Andrews, United Kingdom; and Rosseland Centre for Solar Physics, University of Oslo, Norway*

[3] *Predictive Science Inc., San Diego, California, USA*

[4] *Université Paris-Saclay, CNRS, Institut d'Astrophysique Spatiale, Orsay, France*

[5] *Dipartimento di Fisica & Chimica, Universitá di Palermo, Palermo, Italy*

Space Physics and Aeronomy Collection Volume 1: Solar Physics and Solar Wind, Geophysical Monograph 258, First Edition.
Edited by Nour E. Raouafi and Angelos Vourlidas.

DOI: 10.1002/9781119815600.ch2

Figure 2.1 A total solar eclipse viewed in white light from Earth. (*Source*: S. Habbal, M. Druckmüller and P. Aniol, NASA/GSFC, taken on 20 March 2015 from Svalbard, Norway.)

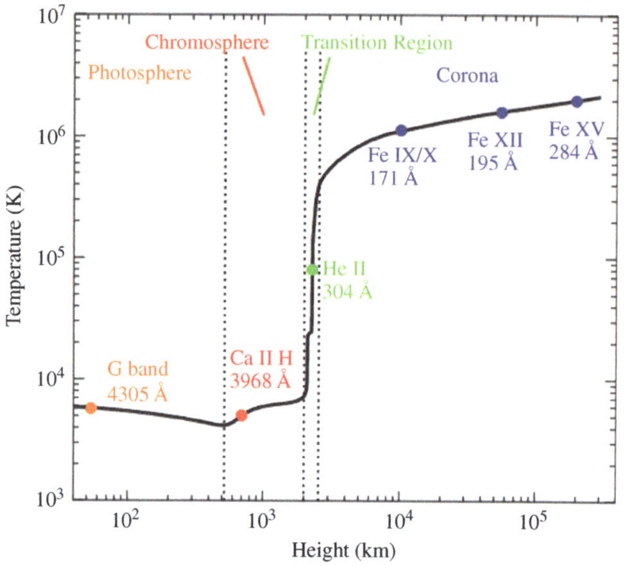

Figure 2.2 Plotted is a typical temperature as a function of height defining the different layers of the solar atmosphere. The mean formation heights of several commonly used spectral bands and lines are shown for reference. (*Source*: Yang et al., 2009. © 2009.)

The corona has been observed for millennia from total solar eclipses, such as the total solar eclipse observed from Svalbard, Norway, on 20 March 2015, shown in Figure 2.1. Total solar eclipses occur when the Moon crosses in between the Sun and Earth, blocking the emission from the Sun's photosphere, which is 10^5–10^6 times brighter in white light emission than the Sun's corona. Amazingly, only in the last centuries was it realized that the corona belongs to the Sun, and not to the Moon. In 1605, Hannes Kepler speculated that the corona surrounded the Sun, and after a total solar eclipse in 1724, Giacomo Filippo Maraldi concluded that the corona was indeed part of the Sun, because the Moon was observed to traverse the Sun.

With the advent of spectroscopy, important aspects of the solar atmosphere were revealed. Work by Grotrian (1939) and Edlén (1943) identified forbidden lines in the emission spectra of the corona, and it was recognized that these lines required highly ionized species. See Peter and Dwivedi (2014) for a review of these measurements and their historical context. Helium was later identified in a yellow spectral line in the optical spectrum in a prominence observation in 1868 (Kochhar, 1991), 10 years before it was identified on Earth (Palmieri, 1881). Also, a strong green emission line emerged in the optical spectrum, identified during the 1869 solar eclipse by Young and Harkness, which was of unknown origin. It was temporarily identified as a new element, coronium, in analogy to helium. It took nearly 60 years for scientists to conclude that the corona is not made in the same way as the underlying chromosphere and photosphere: the green line was ultimately identified as a forbidden line of the highly ionized iron (Fe XIV). This highly ionized state of iron requires extremely high temperatures, and so it was then realized that the corona is extremely hot, at temperatures above 1 million K, more than a hundred times the temperature of the photosphere. Although the energy source must be the flows in the convection layer of the Sun (see Chapter 7; solar interior), the heat cannot be simply conducted from below.

In Figure 2.2, we illustrate this extreme temperature gradient by showing a typical temperature gradient as a function of height (adapted from Yang et al., 2009). Also plotted are several commonly used spectral bands and lines at their mean formation height. The photosphere is the lowest layer, where the gas transitions from optically thick to optically thin; most solar optical light is from this layer. The next layer is the solar chromosphere, followed by a thin transition region and, finally, the solar corona, which is several orders of magnitude hotter than the chromosphere and photosphere below.

Thus, the problem of coronal heating was born. The questions became: What makes the corona so hot? What is the carrier that transports the solar energy to the corona? Once that energy reaches the corona, how is it dissipated? At temperatures above 1 million K most of the energy is radiated in the extreme ultra violet (EUV) and X-ray band, but these wavelengths cannot be observed from the ground, because they are absorbed by the terrestrial atmosphere. Before technology allowed the launch of telescopes with the capabilities of observing these wavelengths in outer space, the first hypotheses for coronal heating were made. The transport and subsequent dissipation of energy by waves was first considered as a potential heating mechanism for the solar atmosphere. Both Biermann (1946) and Schwarzschild (1948) realized that the "noise" (acoustic waves) emitted by the convection zone would propagate into the solar atmosphere and could be a possible energy source to heat the chromosphere and corona (see, e.g., Narain & Ulmnschneider 1990, 1996; Stein & Leibacher, 1974, for a review). The idea was that acoustic waves coming up from the solar surface and generated by the solar granulation are similar to a field of pistons. According to this theory, the waves steepen into shock waves while traveling in the chromosphere and eventually dissipate, releasing their energy in the corona. Detailed steady-state models were developed that treated the whole corona as a uniformly stratified medium, also known as a plane-parallel atmosphere. At this point, it was thought that the magnetic field had only the role of shaping the visible structures because of preferential heat conduction along the field lines (see review by Kuperus, 1969). Although it seems that acoustic waves do play a role in chromospheric heating, their energy is dissipated before they reach the corona, and they do not directly contribute to coronal heating. Alfvén & Lindblad (1947), on the other hand, suggested that magnetohydrodynamic (MHD), or Alfvén, waves could be a prime candidate for supplying energy to the solar atmosphere (see, e.g., Hollweg, 1978, or Wentzel, 1974). MHD waves remain prime candidates for heating throughout the solar atmosphere and into the solar wind.

The breakthrough for the coronal heating problem came from solar X-ray observations during the Space Age, when it first became clear that the active region (AR) and quiet Sun (QS) corona comprise magnetically confined plasma, and that a plane-parallel atmosphere is not an accurate model. The first observations were made from rocket flights above the atmosphere. Pinhole cameras had arcmin (arc minute = 1/60th of a degree) resolution and revealed that intense X-ray emission comes from localized and relatively small regions of activity. The localized regions of X-ray emission were observed to be correlated with H-alpha and Ca-K plages, which themselves are chromospheric bright patches associated with concentrations of magnetic field, and also correlated with radio emission (see review by Peres & Vaiana, 1990). The quantum leap occurred with the development of grazing incidence optics for X-ray astronomy by Riccardo Giacconi's group in the 1960s (e.g., Giacconi & Rossi, 1960). X-ray photons were reflected by grazing metallic coaxial and confocal surfaces at very small angles of a few degrees. Grazing incidence observations improved continuously and in the late 1960s achieved a spatial resolution below the arcmin threshold to a few arcseconds (1 arcsec = 1/60th arcmin). This was sufficient to localize the most intense emission between concentrated regions of opposite magnetic polarity, ARs, showing the crucial role that the magnetic field plays in coronal heating.

Already from the early extensive rocket program, the morphology of the corona was firmly established: bright arch-like structures connecting regions of opposite photospheric magnetic polarity—called coronal loops—are the fundamental features of the corona (Vaiana et al., 1968, 1973). However, coronal magnetic field measurements were not yet possible, and so this alignment could not be established directly. Later it was largely confirmed by models that extrapolate the photospheric field into the corona (Poletto et al., 1975). With the extrapolation, it became clear that the magnetic pressure is much higher than the thermal pressure in the corona, high enough to confine the hot coronal plasma (see Chapter 3).

Further progress was made when the observations from telescopes onboard satellite missions allowed for temporal coverage in the X-ray and EUV beyond a few minutes. This was the case of the Skylab mission in 1973–1974, which allowed the study of the evolution of ARs, coronal holes, and very small bright points (Golub et al., 1975). It was confirmed that the bright corona is entirely made up of magnetic loops of all sizes confining hot plasma. Coronal holes correspond instead to regions with a dominant magnetic polarity in the photosphere, where the magnetic field opens to interplanetary space, and are therefore the natural source of the high-speed solar wind (Krieger et al., 1973, and see Chapter 1).

X-ray observations of the solar corona opened a new scenario for the physics of the solar corona and the problem of coronal heating: modeling the atmosphere as individual loops. Figure 2.3 shows a comparison of an X-ray image taken in 3–60 Å on 7 March 1970 from an Aerobee rocket (VanSpeybroeck et al., 1970) with an X-ray image of the Sun taken with Hinode XRT (X-Ray Telescope Golub et al., 2007) in the Al-mesh filter on 26 February 2014, to illustrate the advancement of X-ray imaging. A first classification of loops was attempted to connect the maximum plasma temperature and the pressure at the base of the corona with the size of the loops, and the underlying photosphere and chromosphere structures. It was determined that coronal loop heights are typically in the range between 1,000 km up to more than 100,000 km. With observations in different filters, it became possible to measure densities in the corona of 10^7–10^{10} cm^{-3}, and temperatures of 1 to 4 million K, exceeding 10 million in flares. The intensity of the observed X-ray emission is directly correlated with the field complexity of the underlying magnetic field. These conditions determined the basis for a new investigation of coronal heating. Although the energy from the mechanical motions in the convection zone below the photosphere is crucial, the magnetically structured corona indicates that the key is the conversion

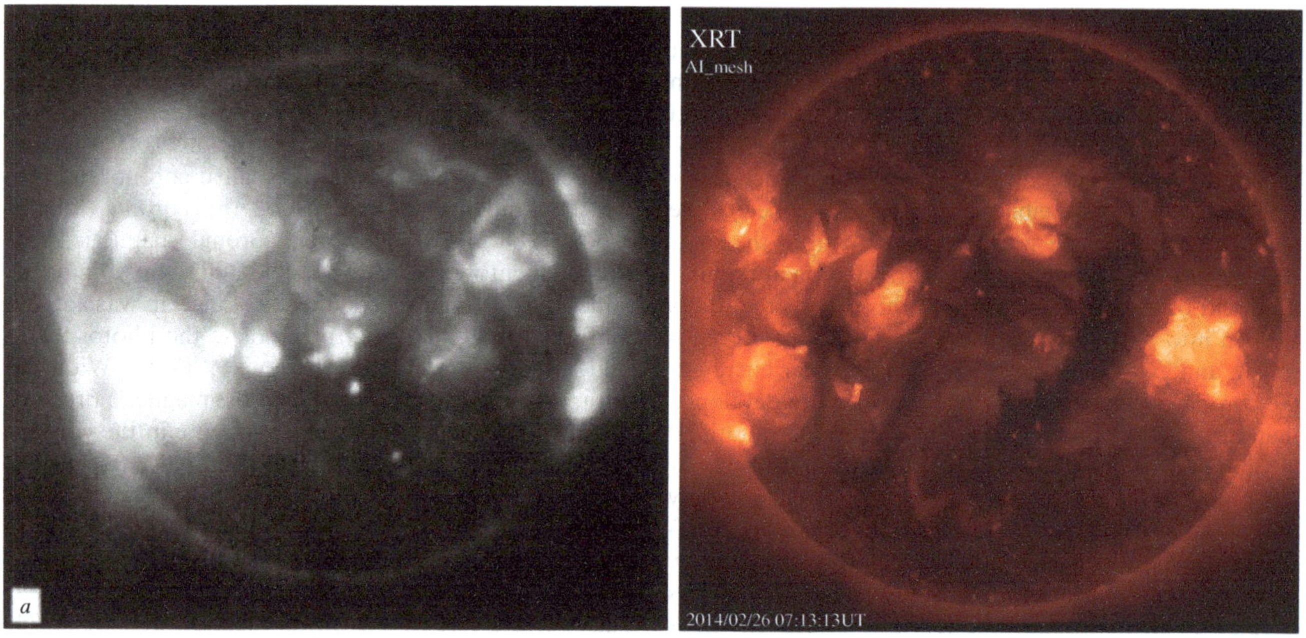

Figure 2.3 Left shows an X-ray image taken on 7 March 1970 from an *Aerobee* rocket; right shows an X-ray image taken on 26 February 2014, with Hinode XRT in the Al-mesh filter. (*Source*: (Left) Image from Van Speybroeck et al., 1970; (Right) Image courtesy of Smithsonian Center for Astrophysics, JAXA, and NASA.)

of magnetic energy into heat. Indeed, estimates of the Poynting flux associated with photospheric magnetic footpoint motions indicate there is more than enough energy available to account for both the chromospheric and coronal energy requirements. For example, Parnell and De Moortel (2012) estimate the Poynting flux to be on the order of 2.5×10^7erg cm^{-2} s^{-1} to 2.5×10^9 erg cm^{-2} s^{-1} in QS and AR corona, respectively (see also, e.g., Klimchuk, 2006).

In coronal conditions, the plasma is highly ionized and magnetized. Plasma particles spiral around the field lines with a typical gyroradius of about 10 cm, and their mean free path at 1 million K and a density of 10^9 cm^{-3} is about 100 km, both shorter than the typical length of coronal structures. The particles are therefore free to move only along the field lines, and the confined plasma can be described as a compressible fluid. At a million K, thermal conduction for a fully ionized gas is strongly guided by the magnetic field; that is, it is very efficient along the field lines, but strongly inhibited across them (Spitzer, 1962). The thermal conduction coefficient has a steep dependence on the temperature ($T^{5/2}$) and becomes extremely efficient above 1 million K. Therefore, if we consider a closed magnetic structure, the plasma confined in it moves and transports energy mostly along the magnetic field lines. Each loop is therefore dynamically and thermally insulated from the others. This was a radical change of view for the solar atmosphere: the corona is no longer described by a single stratified atmosphere, but it is structured into a collection of a myriad of independent atmospheres, each contained in single coronal loop flux tubes.

Recent progress has been more gradual and extensively employed physical models of the loop atmosphere. One difficulty has been the inability to localize the heating. The efficient thermal conduction (possibly combined with mass flows) immediately redistributed energy along the magnetic field, so evidence of localization is rapidly lost. The coronal plasma is optically thin, and its emission is a function of the product of the temperature and the square of the density, so the bright regions are typically the densest regions, not necessarily the hottest, though the two parameters are related. Figure 2.4 shows a composite image of the solar corona taken by the Solar Dynamics Observatory's (SDO) Atmospheric Imaging Assembly (AIA). The images were taken at nearly the same time in the 211 Å (red), 193 Å (green), and 171 Å (blue) filters. The 211 Å channel peak sensitivity is to plasma around 2 MK; the 193 Å channel peak sensitivity is around 1.5 MK; and the 171 Å channel peak sensitivity is around 0.8 MK. Bright, dense ARs are visible, as well as darker, cooler coronal holes at the poles of the Sun.

As a starting point for modeling a coronal loop, long-term X-ray observations showed that the lifetimes of the X-ray bright coronal loops are generally significantly larger than the cooling times for radiation and thermal conduction. The implication is that the heating powering the loops must be rather continuous. As a result,

Figure 2.4 Shown is a composite image from the SDO. Three images of the Sun taken at nearly the same time with the AIA are superimposed. From hotter to cooler plasma, images taken in the 211 Å (red), 193 Å (green), and 171 Å (blue) channels are shown. Bright, dense active regions (ARs) are visible, as are darker, cooler coronal holes at the poles of the Sun. (*Source*: Courtesy of NASA/GSFC.)

coronal loops were systematically described with hydrostatic models, with each loop having unique parameters independent of the others. As a first approximation, gravity can be neglected, so the loop internal structure is essentially dictated by the energy balance between the losses by radiation and thermal conduction and the heating, which can be described as an empirical function of the coordinate along the loop. By solving the equation of energy conservation under the simplifying assumption of a heating distributed uniformly along the loop, it is possible to derive scaling laws connecting important loop parameters, that is, the maximum temperature, the length, the pressure at the loop base, and the heating rate per unit volume (Rosner et al., 1978). These so-called RTV scaling laws were compared with observations and have been a reference point for loop physics for decades.

The first challenge to the RTV scaling laws was that they generally do not apply to the most highly impulsive coronal events: coronal flares. Flares are highly localized explosions where the plasma is heated above 10 million K for a few minutes and often reaches and even overtakes the brightness of the whole remaining corona (see Chapter 5). The observed evolution generally consists of a fast brightening followed by slower decay. Flares are often observed on many other stars, where they can be orders of magnitude more intense that on the Sun. Because their duration is generally small, and the physics is very complex, their investigation required the development of time-dependent hydrodynamic models of plasma confined in a single flux tube. This is the basis for one-dimensional (1D) nanoflare models of coronal heating, described further in section 2.3.2.

From theoretical considerations, Parker hypothesized that the same mechanism that produces solar flares, but with a factor-of-10^9 less energy, could be responsible for heating the corona. In these nanoflares, as they were named, the X-ray corona is heated by dissipation at the many small current sheets forming as a consequence of the continuous shuffling of the footpoints of the magnetic field in the photospheric convection (Parker, 1988). Rapid small-scale reconnection of the magnetic field across the discontinuities destroys them as fast as they are created by the motions of the footpoints. The critical angle between the magnetic fields across the current sheet when reconnection occurs can be estimated by relating the Poynting flux associated with the photospheric footpoint motion of the magnetic fields with the energy budget of the corona (Klimchuk, 2006). If the reconnection occurs before this critical angle, then the corona would be cooler than observed, and if the reconnection occurs after this critical angle, then the corona would be hotter than observed. Much attention has then been devoted to studying coronal magnetic reconnection (Priest, 1996) through modeling with detailed magnetic treatment and simplified coronal atmosphere. After first attempts in 2D geometry, it was soon realized that reconnection must be modeled in a 3D geometry to be realistic (Priest, 1999).

The other leading theory of coronal heating, as mentioned above, is through the dissipation of MHD waves. The presence of waves and oscillations of magnetic structures throughout the solar atmosphere has been firmly established by observational evidence. Indeed, recent high-cadence and high-resolution observations have revealed an abundance of waves, oscillations, and other quasi-periodic disturbances present in the solar atmosphere (see, e.g., reviews by Arregui et al., 2012; Arregui, 2015; Banerjee et al., 2007; De Moortel & Nakariakov, 2012; Jess et al., 2015; Mathioudakis et al., 2013; Nakariakov & Verwichte, 2005; Zaqarashvili & Erdélyi, 2009). In most instances, the perturbations have been interpreted in terms of MHD modes, and in many cases, these observed waves are reported to contain a substantial amount of energy.

In the next section, we describe the evidence for different heating mechanisms in the chromosphere and corona and describe the observational diagnostics used.

2.2. OBSERVATIONAL CONSTRAINTS ON CHROMOSPHERIC AND CORONAL HEATING

Currently, observations of the chromosphere and closed-field corona are limited to remote measurements of the Sun. No current or planned mission will reach the predicted height of the closed corona (below ~2.5 solar radii) with in situ measurements. As described above, spectroscopic measurements and images taken in different wavelengths are the primary way that information on the chromosphere and corona is obtained. See Del Zanna and Mason (2018) for a thorough review on spectral diagnostics, especially in combination with imaging techniques, in the UV and X-ray of the solar atmosphere. Magnetic field measurements in the photosphere are regularly obtained; however, chromospheric and coronal magnetic fields are more difficult, though radio observations provide a measure (Shibasaki et al., 2011). The standard practice is to extrapolate the magnetic field from the photospheric measurements into the solar atmosphere, either assuming a potential field source surface (PFSS), or a nonlinear force-free (NLFF) model of the magnetic field (see Chapter 3 and the review by Wiegelmann & Sakurai, 2012). Recently, the Coronal Multi-Channel Polarimeter (CoMP) instrument (see Tomczyk et al., 2007) has been measuring the coronal magnetic field with a full latitudinal field of view (FOV) in the low corona (~1.03 to 1.5 Rsun), providing important constraints on the magnetic field. Modeling is still required to make a direct comparison between the 3D magnetic field structure and the plasma properties observed in the chromosphere and corona.

From observations, it is clear that the chromosphere is highly complex, ultimately set in motion from the convection below. Figure 2.5, from Judge (2006), Wedemeyer-Böhm et al. (2009), and Wiegelmann et al. (2014), shows the complexity in the atmosphere of a quiet region of the Sun.

The convection zone drives granulation and supergranulation flows that concentrate the magnetic field into network lanes. The plasma beta (the ratio of the plasma pressure to the magnetic pressure) shows that the photosphere and chromosphere are dominated by plasma pressure, whereas the corona is dominated by magnetic pressure. Finally, much of the magnetic field in the chromosphere closes in low-lying loops, and does not connect up into the corona. So, although the connection to the chromosphere is crucial for understanding the corona, there are regions of the chromosphere where the opposite is not always true.

Withbroe and Noyes (1977) estimated the energy required to sustain the chromosphere and corona in the QS, ARs, and coronal holes, based on their typical observed emissions. These are the baseline requirements for the solar atmosphere. It is what observers and theorists have been working to explain for decades. In Table 2.1, we show these estimates of energy in erg cm^{-2} s^{-1} from the chromosphere and the corona.

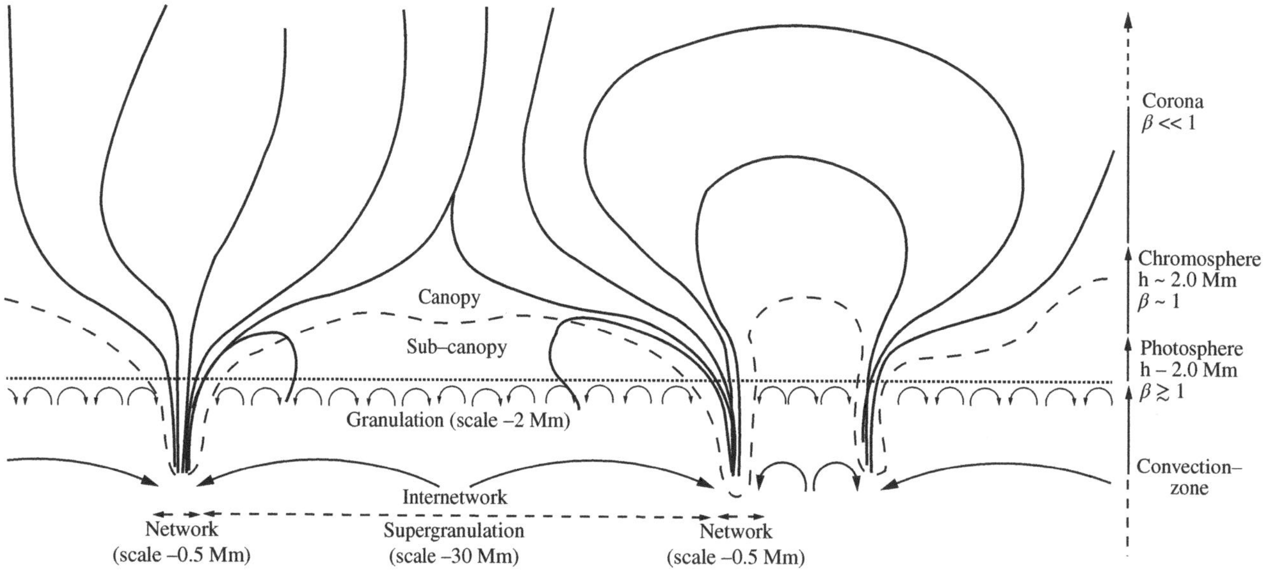

Figure 2.5 This cartoon shows the atmospheric layers of the Sun. Large-scale, supergranular cells concentrate magnetic field in the network, and smaller granulation cells stirs up the internetwork fields. The strongest concentrations of magnetic field in the network expand outward into the corona, whereas the weaker fields close below the corona. Plasma betas at different heights are shown on the right. (*Source*: Wiegelmann et al. 2014. Licensed under CCBY.)

In much of the chapter, we differentiate "wave" heating from "reconnection" heating. The distinction between the two is determined by the timescale of driving and whether the flux tube experiences that driving as a wave or as a slow buildup of energy. The relevant time scale for the flux tube is the length of the flux tube divided by the Alfvén speed in the loop. Energy buildup and dissipation that is shorter than this timescale is an "AC" mechanism, whereas energy buildup and dissipation on longer timescales is "DC" (e.g., Cranmer, 2009; Zirker, 1993). This is a simplified version of the complex physics that occurs in the actual Sun, as we will see in much of the discussion that follows. That is, the two concepts are not mutually exclusive, as magnetic reconnection launches waves, and

Table 2.1 Coronal Temperature and Energy Requirements of the Solar Atmosphere

Location	Quiet Sun (QS)	Coronal Hole	Active Region (AR)
Coronal temperature (K)	1.1–1.6 × 10^6	10^6	2.5 × 10^6
Coronal energy loss (erg/cm^{2}s)	3×10^5	8×10^5	10^7
Chromospheric energy loss (erg/cm^{2}s)	4×10^6	4×10^6	2×10^7

Source: Withbroe and Noyes, 1977.

waves can bring field lines together, forcing the onset of reconnection. Many of their observational signatures are not easily distinguishable, which is one reason why progress has been difficult. Nevertheless, it is still a useful conceptual tool to separate waves from reconnection, and so, where useful, we have done so in what follows.

2.2.1. Observations of Chromospheric Heating

The first thing that is immediately obvious from the Withbroe and Noyes (1977) result is that the chromosphere has a much greater energy requirement than the corona, even though the corona is strikingly hotter. The chromosphere is partially ionized, making ion-neutral effects important, and radiates strongly in Lyman alpha, CaII, MgII, and the H$^-$ continuum.

In a more complex version of Figure 2.5, in Figure 2.6, we show a figure adapted from Wedemeyer-Böhm and Wöger (2008) and Wedemeyer-Böhm et al. (2009). This version includes many of the complex observational aspects of the chromosphere. See Wedemeyer-Böhm et al. (2009) for an excellent summary of these phenomena. An important takeaway is that the zoo of phenomena and dynamics that occur in the chromosphere is largely a function of the magnetic domain. For example, fibrils and spicules are associated with the concentrated network fields and intergranular lanes, whereas the wave dynamics and small, weak magnetic loops are associated with the internetwork. Wave propagation connects the

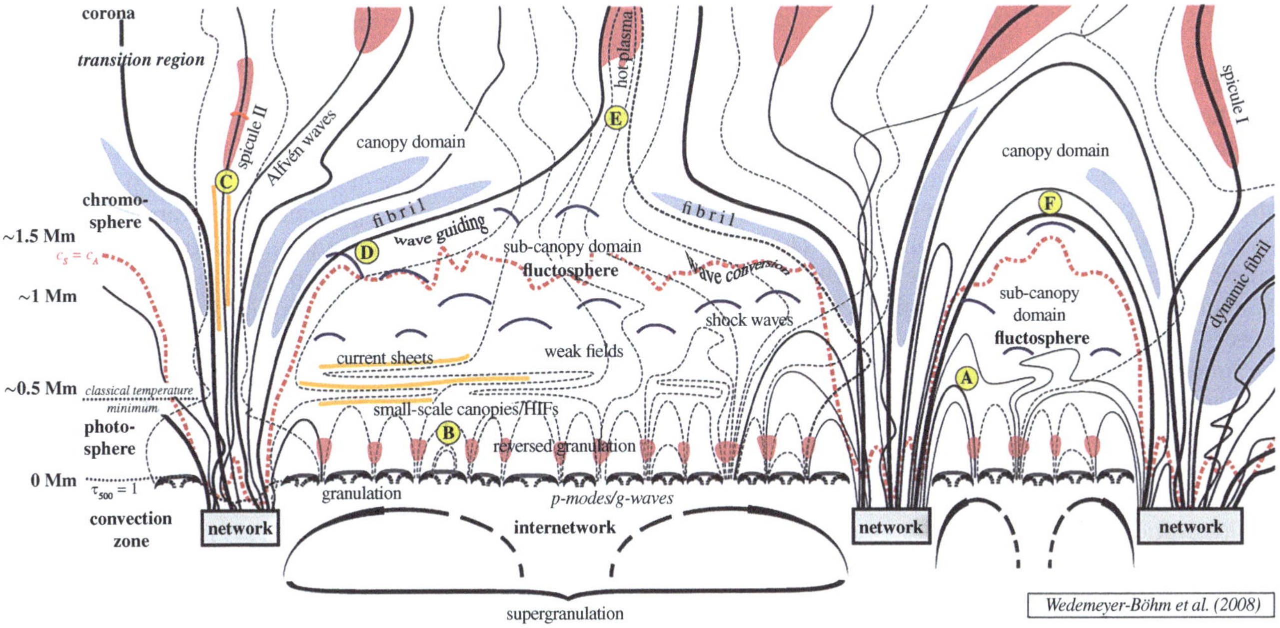

Figure 2.6 Similar to Figure 2.5, with additional complexity of the observations added. (*Source*: Wedemeyer-Böhm et al., 2009. © 2009 Springer Nature.)

different regions in ways that are not immediately clear when looking at the magnetic field connections of Figure 2.5 alone.

Observations of Chromospheric Waves

In the chromosphere, magneto-acoustic waves arising from convective flows propagating upward into the solar atmosphere are expected to shock readily, which would lead to strong brightenings and heating as demonstrated, for example, by Carlsson and Stein (1992, 1997). Using Interface Region Imaging Spectrograph (IRIS) and Swedish Solar Telescope (SST) observations of bright grains in internetwork regions in the quiet chromosphere, Martinez-Sykora et al. (2015) suggested that such upward propagating acoustic waves can even reach the transition region as, on occasion, the characteristic sawtooth shock signature can be found in the IRIS Si IV spectra.

In addition to the acoustic mode shocks, observations have revealed a multitude of both compressive and incompressive waves in photospheric and chromospheric structures (see, e.g., Carlsson et al., 1997; Judge et al., 2001; Wikstøl et al., 2000 for early examples using Transition Region and Coronal Explorer (TRACE) and Solar Ultraviolet Measurements of Emitted Radiation (SUMER)). Here, we will focus on examples that provide direct evidence relevant to heating rather than an exhaustive review of wave observations.

Chromospheric spicules and dynamic fibrils have been widely reported to support a variety of different and complex wave modes. For example, using Hinode data, De Pontieu et al. (2007) reported transverse oscillating ("swaying") spicules with amplitudes of around 10–25 km/s and periods on the order of 100–500s. These "Alfvénic" waves contain a substantial amount of energy flux, potentially sufficient to heat the QS corona and/or accelerate the solar wind (assuming a transmission coefficient of energy from the chromosphere to the corona of about 3%). Numerical simulations by Martinez-Sykora et al. (2017) show a natural association between (type II) spicules and transverse waves from the rapid release of magnetic tension. Analyzing transverse motions along a large number of spicules, Okamoto and De Pontieu (2011) found a complex interplay between upward and downward propagating waves as well as standing waves, where it is suggested that the upward propagating waves are produced near the solar surface and the downward propagating waves are the result of reflection off the transition region. Analyzing Rapid Oscillations in the Solar Atmosphere (ROSA) data of chromospheric mottles, Kuridze et al. (2013) also report a combination of propagating and standing transverse oscillations, where they suggest that the sudden decrease in phase speed and transverse velocity could be due to the waves becoming nonlinear near the canopy. Morton et al. (2014) analyzed the evolution of the transverse wave properties for a spicule observed with Hinode/Solar Optical Telescope (SOT) and found a rapid initial increase in the wave amplitude due to the large changes in density and magnetic field strength, followed by a decrease in amplitude at larger heights, which the authors attributed to wave damping.

Figure 2.7 shows an example of transverse waves observed in Hinode SOT Ca II H 3968 Å, adapted from De Pontieu et al. (2007). Panel A is an image of the limb of the Sun showing that the spicules dominate the chromosphere. Panels B and D show space-time plots of the Ca intensity at the slits marked "1," and "2," respectively, in Panel A. Swaying back-and-forth motions are clear in the plots. Panels C and E show the results of a Monte Carlo simulation of spicules carrying Alfvén waves.

As well as the incompressive (transverse) waves and oscillations described so far, compressive modes have been observed in photospheric and chromospheric structures, usually by combining observations of modulations in intensity and variations in the width of the observed structures. For example, using Rapid Oscillations in the Solar Atmosphere (ROSA) observations of magnetic pores in the photosphere, Morton et al. (2011) interpreted out-of-phase periodic variations in pore size and intensity in terms of compressive MHD sausage modes, with periods ranging from 30 s to 450 s. Due to the similarity in periods, these authors suggested that the observed sausage modes could be driven by the global acoustic modes of the solar interior. Analyzing Dunn Solar Telescope (DST) observations in wavelengths covering the photosphere to the base of the Transition Region, Grant et al. (2015) found evidence of damping of sausage modes in a magnetic pore. Keys et al. (2018) again analyzed sausage modes in magnetic pores and found that surface modes were present more frequently in the data and that the sausage modes appeared to carry more energy than body modes. Frequencies were found to be in the range of 2–12 mHz (although the authors find no evidence of surface modes above 10 mHz).

Morton et al. (2012) provided observational evidence of concurrent compressive (sausage) and incompressive (kink) modes in chromospheric fibrilar structures. Estimating that 4%–5% of chromospheric magnetic flux tubes penetrate into the corona, the total energy flux able to reach the corona is estimated to be on the order of $1.7 \times 10^5 \pm 1.1 \times 10^5$ erg cm^{-2} s^{-1}for incompressible transverse waves and about $4.6 \times 10^5 \pm 1.5 \times 10^5$ erg cm^{-2} s^{-1} for compressive waves. Jafarzadeh et al. (2017a) and Gafeira et al. (2017) similarly report on ubiquitous transverse and compressive waves in slender Ca II H fibrils in SUNRISE data. Jafarzadeh et al. (2017b) also find compressible and incompressible (transverse) waves in small-scale magnetic features observed with SUNRISE.

Although harder to observe, torsional perturbations have also been reported in the lower solar atmosphere

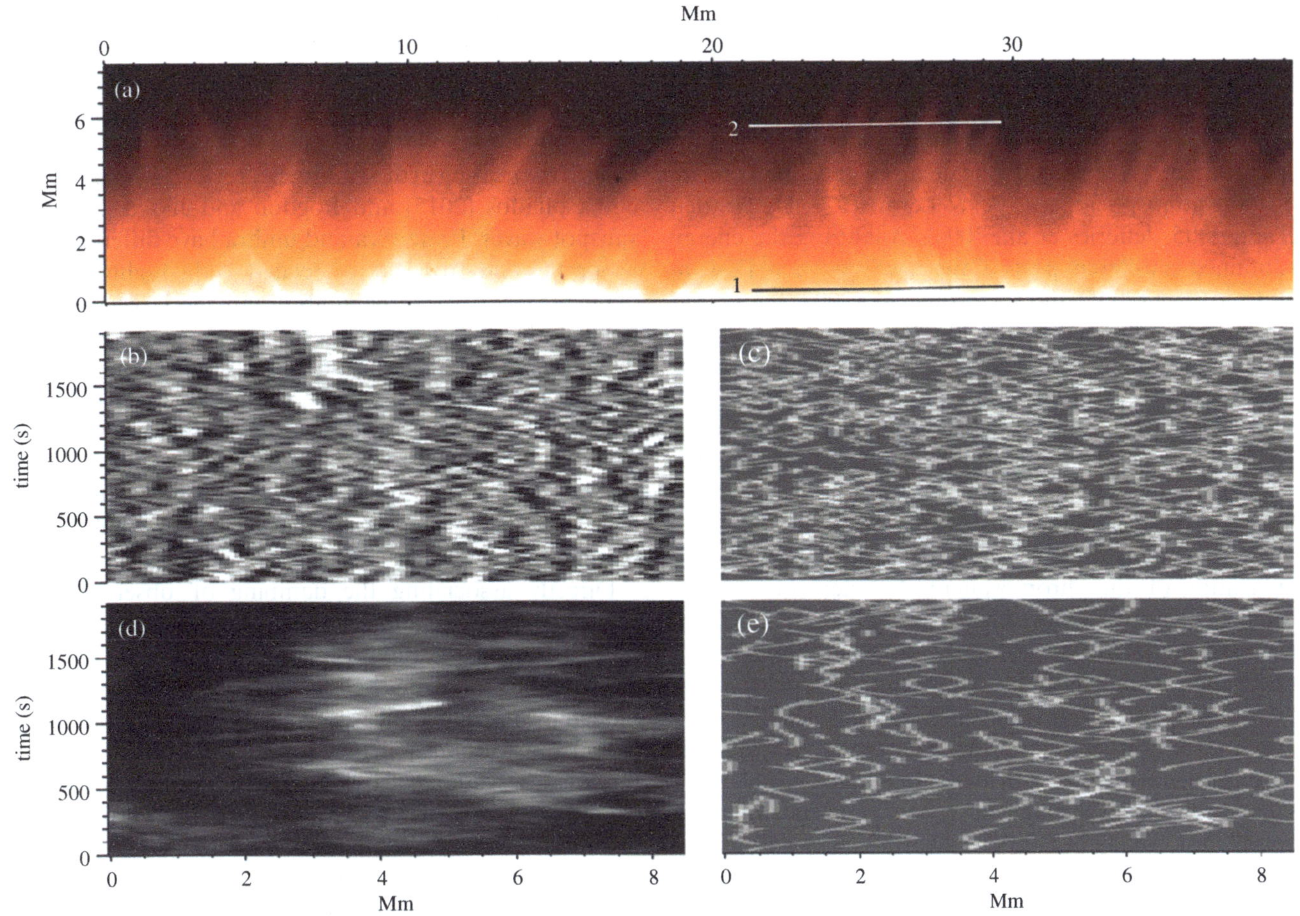

Figure 2.7 Alfvén waves observed in the chromosphere. (A) Spicules in Hinode SOT Ca II H 3968 Å image. (B) A space-time plot of the Ca intensity at the slit marked "1" in Panel A. (D) Same as B, but at slit location "2." C and E show Monte Carlo simulations in which spicules carry Alfvén waves. (*Source*: From De Pontieu et al., 2007).

from analyzing line width and Doppler velocities. For example, based on variations in line width in H_a, Jess et al. (2009) suggested the presence of torsional Alfvén waves in a large photospheric bright point group. The global average energy associated with these oscillations was estimated to be on the order of 2.4×10^5 erg cm^{-2} s^{-1}. Such torsional motions could be associated with surface vortex motions (e.g., Bonet et al., 2008; Wedemeyer-Böhm et al., 2009), propagating up into the solar atmosphere. Spicules (and dynamic fibrils) have also been found to undergo rotational motions (e.g., De Pontieu et al., 2012, 2014; Rouppe van der Voort et al., 2015; Skogsrud et al., 2015), further increasing the wave energy budget associated with spicules. Detailed analysis of the spectra revealed evidence of heating, potentially associated with the rapid torsional motions. Analyzing SST observations, Srivastava et al. (2017) found evidence for high-frequency (~12–42 mHz) torsional motions in thin spicular structures in the chromosphere. Combining the observations with numerical modeling, they suggest that even after partial reflection off the transition region, a substantial amount of energy (~10^5 erg cm^{-2} s^{-1}) is transferred to the overlying corona, providing sufficient Poynting flux to both heat the overlying corona and accelerate the solar wind. Again using SST observations, Stangalini et al. (2017) found evidence for the presence of helical perturbations in chromospheric small-scale magnetic elements, which the authors interpret as kink-like oscillations with an elliptic polarization.

The (selective) examples described above provide a flavor of the many types of observations of waves and oscillations in the lower solar atmosphere. For a more comprehensive overview, we refer the interested reader to Jess et al. (2015) and the excellent introductions in many of the papers cited above. However, even from the small selection of observational examples presented here, it is clear that many of the observed waves and oscillations are reported to contain substantial amounts of energy. Given the ubiquitous nature of some of the described waves and oscillations, observations have now established that the chromosphere is a reservoir of wave energy contained in compressive, incompressive, and torsional motions. Although gravitational stratification is expected to lead to a "cutoff" frequency, the inclination of the magnetic field reduces the effect of gravity and hence allows for a broader spectrum of waves to propagate from the photosphere into the atmosphere (e.g., Bel & Leroy, 1977; De Pontieu et al., 2005; Jefferies et al., 2006). Analyzing the energy contained in different parts of the wave spectrum, Fossum and Carlsson (2005) and Carlsson et al. (2007) found that the energy contained in the low-frequency part of the spectrum (<5 mHz) was substantially larger than in high-frequency waves (see also more recent work by Rajaguru et al., 2019, who estimate the acoustic energy flux in the 2–5 mHz frequency range to be on the order of 2.25–2.6×10^{6} erg cm^{-2} s^{-1} between the upper photosphere and the lower chromosphere).

There is plenty of wave energy to play a significant role in heating of the chromosphere. The next question is whether or not that wave energy is dissipated, and if so, where. A number of authors have looked at the change in wave energy in height by comparing observations of waves and oscillations in different wavelengths. For example, Grant et al. (2015) interpreted the gradual decrease in energy flux with height in upward propagating sausage modes in a magnetic pore as evidence of wave damping. Looking at Hinode and IRIS observations in a sunspot umbra, Kanoh et al. (2016) estimate upward energy fluxes associated with slow magneto-acoustic waves on the order of 2×10^{7} erg cm^{-2} s^{-1} at the photosphere, decreasing to 8.3×10^{4} erg cm^{-2} s^{-1} at the level of the lower transition region. If all of this "missing" wave energy is dissipated, it would be more than what is required to maintain chromospheric temperatures in the sunspot umbra. The remaining upward energy flux in the transition region, however, is too small to make a significant contribution to coronal heating. Similarly, Krishna Prasad et al. (2017) study the damping characteristics of slow magneto-acoustic waves at multiple heights in a sunspot umbra and find an energy flux ~1.3×10^{7} erg cm^{-2} s^{-1} at the photosphere, decreasing to about 40 erg cm^{-2} s^{-1} at a height of about 2 Mm (observed in IRIS 1330 Å), again indicating significant damping. These authors also find a frequency dependence in the energy decrease, with shorter damping lengths observed for oscillations with higher frequencies. A similar dependence on frequency was found by Morton et al. (2014) for the damping of chromospheric kink waves when comparing ROSA and CoMP data. Complexities and uncertainties such as line formation heights, opacity, and wave mode identification, among others, make estimating the changes in wave energy fluxes with height less than straightforward. Despite this, these examples highlighted here clearly show a significant decrease in wave energy flux with height where the "lost" energy could make a substantial contribution to heating the local plasma.

Sunspots are known to exhibit myriads of waves and oscillations. The above were instances where energy estimates show compelling evidence that the waves contributed to the heating in the umbra. See reviews by Khomenko and Collados (2015) and Bogdon and Judge (2006) on sunspot waves. Penumbra and umbra have different wave properties and propagation, related to the highly inclined magnetic field of the penumbra and the largely vertical field of the umbra. There is also a connection between dynamics in each, with a possible connection between umbral flashes and running penumbral waves (van der Voort et al., 2003). There are global oscillations with periods as long as hours to days, but the dominant modes are several minutes. The atmospheric layers above the umbra, from the chromosphere up to the corona, all exhibit 3-min oscillations.

Directly associating the damping of observed wave energy with dissipation and heating remains elusive, though. The observations of wave modes in spicules cited above come tantalizingly close, as do the observations and modeling of prominence oscillations by Okamoto et al. (2015) and Antolin et al. (2015). There is a very rich literature of observations of oscillations in prominences with a comprehensive review provided by Arregui et al. (2012). Okamoto et al. (2007) reported on transverse waves in Hinode observations of prominences with a significant Poynting flux but with no direct evidence of heating. Using IRIS observations in multiple wavelengths and numerical modeling, Okamoto et al. (2015) and Antolin et al. (2015) combined the measurements of transverse (plane of the sky) displacements with line-of-sight (Doppler) measurements in multiple wavelengths to show evidence of significant heating from chromospheric to transition region temperatures. The authors suggest that the observed phase differences of about 180° between transverse motions in the plane-of-sky and line-of-sight velocities of the oscillating prominence threads are consistent with resonant absorption and the triggering of the Kelvin–Helmholtz instability (KHI) at the boundaries of the oscillating threads, where the instability creates thin current sheets, allowing the wave energy to be dissipated. Further evidence of wave-based heating is presented by Grant et al. (2018), who interpret DST Interferometric Bidimensional Spectrometer (IBIS) observations in terms of Alfvén wave dissipation in the form of shock fronts close to chromospheric umbra boundaries, consistent with observed local temperature enhancement of about 5%.

Observations of Reconnection Heating in the Chromosphere

As seen in the previous section and Figure 2.6, the chromosphere is highly complex with many different types of waves and shocks that heat the chromosphere. In this section, we discuss compelling observations that magnetic reconnection is likely playing a role in the heating of the chromosphere too.

One of the most striking observations of reconnection heating of the chromosphere is the observations of jets. We refer the reader to Chapter 6 for a full discussion of these phenomena, and discuss here the resulting atmospheric heating that is observed. Hinode observations have revealed that almost all parts of the AR chromosphere are covered by numerous tiny jets or jetlike features, which are thought to be caused by magnetic reconnection. Shibata et al. (2007) presented observations of chromospheric anemone jets, which are observed abundantly in Hinode SOT measurements. The average energy flux of anemone jets is 7×10^3 erg cm^{-2} s^{-1}. Observations suggest that anemone jets contribute to heating the chromosphere, but that not much of the associated energy heats the corona. Other types of jets observed to heat the atmosphere that are thought to be associated with magnetic reconnection include penumbral microjets (Katsukawa et al., 2007) and jets in the network regions observed with IRIS (Tian et al., 2014). Many jets are observed to heat up to at least 10^5 K in IRIS, confirming that they are involved in heating the chromosphere and transition region, and possibly even contribute to the solar wind.

Another striking observation that is highly suggestive of magnetic reconnection is the observation of flux cancellation in photospheric magnetic field measurements. The very high-resolution Imaging Magnetograph eXperiment (IMaX) instrument on two flights of the SUNRISE balloon mission (Solanki et al., 2010, 2017) show that photospheric flux cancellation, and therefore reconnection, is happening at a much higher rate than previously thought. IMaX has a factor-of-six higher resolution compared to the resolution of the Helioseismic and Magnetic Imager (HMI) measurements on SDO. The QS flux emergence and cancellation rate was recently shown by Smitha et al. (2017) to be 1100 Mx cm^{-2} day^{-1} which is an order of magnitude larger than previously thought, and parasitic polarities in nearby AR footpoints have a flux cancellation rate of 10^{15} Mx cm^{-2} s^{-1} (Chitta et al., 2017b). Following these observations, Priest et al. (2018) showed analytically that magnetic reconnection driven by such photospheric flux cancellation releases heat in the chromosphere and that the reconnection heights they model are consistent with those from magnetic field extrapolations for the observed chromospheric bursts (e.g., Chitta et al., 2017a; Tian et al., 2018). Chitta et al. (2019) studied plages in the vicinity of ARs with SST and AIA. Using their 3D radiation MHD simulation of magneto-convection, they estimate that the observed flux emergence and cancellation events are associated with a transfer of energy in excess of 1×10^9 erg cm^{-2} s^{-1} through the photosphere.

Another place where reconnection occurs is in the photospheric network and internetwork magnetic fields. Gošić et al. (2018) provides a nice discussion of recent attempts to determine whether magnetic flux cancellation events in the photospheric network and internetwork fields provide enough energy to heat the chromosphere. Determining whether there is enough energy is only the first step; estimates of how much energy is transferred to chromosphere and corona are extremely difficult and typically involve data-model comparisons. Wiegelmann et al. (2013) and Chitta et al. (2014) used temporal evolution of the magnetic connectivity extrapolated from IMaX observations and concluded that there is not enough energy released in the chromosphere from photospheric network and internetwork fields of the QS. Gošić et al. (2018) used coordinated IRIS and SST observations of cancellations of internetwork magnetic fields to make another estimate of chromospheric heating from reconnection. They found that the cancellation events are associated with clear evidence of local heating in the atmosphere, with chromospheric temperature increases of more than 10^3 K. However, the global estimate based on the cancellation events averaged over time and area is $\approx 1 \times 10^{-2}$ erg cm^{-3} s^{-1}, which is an order of magnitude too low to power the global chromosphere.

It is now believed that chromospheric "bombs" and "bursts" are due to magnetic reconnection as well. Ellerman bombs were the first of such events observed. They are brightenings at temperatures of ~10,000 K that seem to be due to reconnection events from emerging flux. Ellerman bombs are estimated to release between 10^{23} and 10^{26} ergs of radiative energy per event (Reid et al., 2016). Georgoulis et al. (2002) studied the statistical properties of Ellerman bombs and found that they generally exhibit power-law distribution functions. They occur preferentially near magnetic neutral lines and are associated with photospheric downflows. Georgoulis et al. (2002) estimated Ellerman bombs to have total energies (not just the radiated energy) in the range (10^{27}, 10^{28}) ergs, and temperature enhancements of $\sim 2 \times 10^3$ K.

With IRIS, bomb and burst events are now observed low in the solar atmosphere in the UV with temperatures up to 8×10^4 K. Peter et al. (2014) present an event observed with IRIS that they compare to an Ellerman bomb. These bombs are also believed to be due to magnetic reconnection; however, the IRIS bombs are much hotter than Ellerman bombs, and are observed to heat to at least chromospheric, and even transition region temperatures. The plasma in the photosphere is observed to be heated to almost 100,000 K, and there is a bidirectional flow channeled by the magnetic field. The normal temperature stratification (a very hot corona above a hot transition region gas above the chromosphere) is turned upside down. Figure 2.8 shows a cartoon

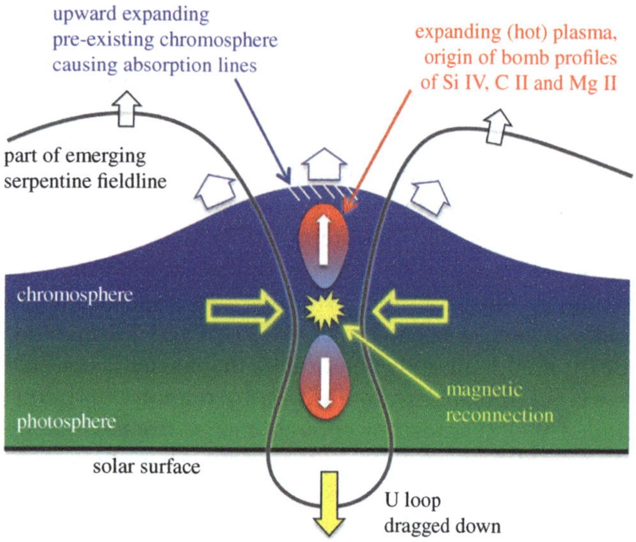

Figure 2.8 A cartoon representation of a magnetic reconnection event generating hot plasma embedded in cooler material in the chromosphere. (*Source*: Peter et al., 2014. © 2014 The American Association for the Advancement of Science [AAAS].)

representation of the reconnection event and plasma heating, adapted from Peter et al. (2014).

In addition to the similarity to Ellerman bombs, chromospheric bombs and bursts in UV show additional evidence for intermittent magnetic reconnection. Rouppe van der Voort et al. (2017) used the CHROMIS instrument Ca II K line at the SST and found blob-like features at Alfvénic speeds followed by the heating of plasma to transition region temperatures in IRIS, which is consistent with predicted signatures of reconnection (Innes et al., 2015). A unifying picture is coming together. Hansteen et al. (2017) showed with a 3D radiative MHD simulations of magnetic flux emergence that reconnection between emerging bipoles triggers Ellerman bombs in the photosphere, UV bursts in the middle and low chromosphere and small nanoflares and microflares that occur in the upper chromosphere but produce 10^6 K plasma. Depending on where the reconnection occurs, the energy released through flux cancellation can produce a range of "bomb" and "burst" signatures, such as the Ellerman bomb in the wings of Hα in the QS and around sunspots (Hansteen et al., 2017; Rouppe van der Voort et al., 2016), ultraviolet (UV) bursts in the chromosphere (Peter et al., 2014) and explosive events in the transition region (Brueckner & Bartoe, 1983; Innes et al., 2011).

The transition region is a layer where thermal conduction is a heating term, in contrast to the corona, where it is a cooling term. Although the transition region is geometrically thin, it spans temperatures from chromospheric (~10^4 K) to coronal (~10^6 K). A long-standing problem has been that the observed emission at temperatures below about 10^5 K is much greater than that predicted through thermal conduction from the hot corona alone (Gabriel, 1976). Feldman (1983) proposed that this excess "transition region" emission could be due to structures that were not resolvable at that time, "unresolved fine structures" or UFS. This emission is not "transition region" emission in the thermal conduction definition of transition region; rather, these would be structures whose peak temperatures were on the order of 100,000 K. Figures 2.5 and 2.6 show that, in the QS, a large amount of magnetic flux is expected to close back to the surface of the Sun before reaching coronal heights. Trujillo et al. (2018) showed that modeling of observations taken with the Chromospheric Lyman-Alpha Spectro-Polarimeter (CLASP) are indeed consistent with a highly corrugated magnetic field structure in the transition region. Antiochos and Noci (1986) showed that loops with heights above the chromosphere that are lower than 5000 km have static solutions with a temperature maximum of less than 100,000 K. An alternate explanation is required for the large unipolar areas (plage) in ARs, where the small UFS loops do not exist. Type II spicules are another possibility (Klimchuk, 2012).

Though there has been observational evidence that these low-lying structures contribute to the emission at these temperatures for years (e.g., Dowdy et al., 1986; Feldman et al., 1999; Vourlidas et al., 2010), it was the high-resolution IRIS observations that confirmed low-lying structures that are likely the UFS. Magnetic reconnection is likely involved in their heating (Hansteen et al., 2014). These IRIS low-lying loops are highly dynamic; both heating and cooling have been observed. They are heated to hundreds of thousands of K, but never reach coronal temperatures, and have scales of only about 2–6 Mm. Hansteen et al. (2014) demonstrated that their high densities cause very rapid radiative cooling, and hence these loops cool before they reach coronal temperatures. Interestingly, Pereira et al. (2018) used SST and showed a lack of chromospheric heating associated with the IRIS low-lying loops. They argued that magnetic reconnection created the loops based on the observations of Y-shaped jets above the loop.

As described earlier, waves and reconnection are not mutually exclusive, and in fact there seem to be very compelling examples where magnetic reconnection results in waves, and where waves drive magnetic reconnection. Song et al. (2017) concluded that magnetic reconnection events in a light bridge triggered oscillations in the nearby sunspot umbra. Tian et al. (2014) presented IRIS observations of magnetic reconnection jets that launched upward propagating waves. With causality going the other direction, many explosive events are observed to occur at intervals of 3–5 min, close to the periodicities produced by solar granulation (Chae et al., 1998; Doyle et al., 2006; Ning et al., 2004). Heggland et al. (2009) simulated wave-induced magnetic reconnection in the solar atmosphere and reproduced many aspects of the observations of periodic reconnection in chromospheric jets.

2.2.2. Observations of Coronal Heating

Coronal heating cannot be understood without understanding of the coupling from the photosphere through the chromosphere and transition region into the corona (see review by Wedemeyer-Böhme et al., 2009). The complexity of this connection is illustrated in Figure 2.6. The atmosphere is dynamic on many spatial and temporal scales, and therefore the coupling is dynamic too. The exchange of mass between the chromosphere and corona through chromospheric evaporation and condensation is crucial to understanding the coronal observations. Most coronal heating occurs on spatial scales that are not resolved by current instrumentation. The corona is optically thin, so even with perfect resolution, there would still be line-of-sight effects that make heating events that are physically separate appear to overlap. As with the chromospheric studies described above, images and spectroscopy in the X-ray and UV are the primary observational tools for understanding coronal heating.

Both waves and reconnection are predicted to produce impulsive heating. As reviewed in Klimchuk and Hinode Review Team (2019), there are several independent lines of evidence that point to the existence of impulsive heating events. TRACE data combined with the extensive X-ray observations with the Yohkoh mission were among the first to show that many coronal loops are not in a steady state. In particular, many coronal loops are diagnosed in the EUV to be denser than predicted by scaling laws of hydrostatic equilibrium conditions for their temperature. Presumably, this means that they are observed while they are cooling, and therefore the heating of each magnetic strand is not continuous, but rather irregular, or impulsive (Klimchuk, 2006). Other evidence for impulsive heating includes the fact that loop intensities as a function of height are not in agreement with the hydrostatic scale height, emission measure distributions with shallow slopes coolward of their peak, time lags between emissions formed at different temperatures, asymmetries in the distribution of intensity fluctuations, footpoints (transition region) not sufficiently bright, and the existence of very hot (>5 MK) plasma. A review of these observational aspects is found in Reale (2014) and Klimchuk (2006, 2015).

However, as we will show in the descriptions of observational coronal heating constrains that follow, many observations are consistent with steady heating too. Because impulsive and steady heating both seem to be consistent with observations at certain times and locations, current research focuses on the frequency distribution of the heating events and requires a range of frequencies, rather than either steady heating or impulsive heating (e.g., Bradshaw & Viall, 2016; Cargill, 2014; Reep et al., 2013). Low-frequency heating events occur at cadences that allow plasma cooling and evolution between them, and high-frequency heating is effectively steady heating in terms of the plasma evolution (see discussion in section 2.3 on modeling), and produces observations consistent with steady heating, though they too are impulsive. Note that much of the current literature uses "impulsive heating events" and "nanoflares" interchangeably, even in situations where the mechanism is not necessarily magnetic reconnection, though we only use "nanoflare" in the reconnection sense in this chapter.

Direct evidence of wave heating or reconnection heating is the exception, not the rule. Due to the unresolved nature of the events, most observations of coronal heating do not conclusively or immediately point to either waves or reconnection. Rather, they put constraints on the characteristics of the mechanisms' timing and spatial distribution, which then must be compared to forward-modeled simulations of coronal heating (section 2.3). Also, as described for the chromosphere, magnetic reconnection and waves in the corona are observed and predicted to occur in concert, complicating the observations further. Due to these limitations, the current state of the art is to use observations to determine the spatial and temporal distribution of coronal heating. We next describe such observational constraints that do not yet point conclusively to waves or reconnection.

Individual Brightenings

One observational constraint is transient brightenings observed in the UV and X-ray. The nanoflare concept introduced first by Parker (1988) was the result of these observed properties of the X-ray, UV corona, and transition region, which showed a space and time variability at small scales. Typical events that Parker considered are those with an X-ray illumination variation between 20% and 100% over characteristic timescales of 20–60 s. Such observations suggested a "heat input composed of many localized impulsive bursts of energy." By this time there was already observational evidence that brightenings from larger events were less frequent than from the smaller ones. Parker proposed the name *nanoflare* for any of these events having an associated inferred energy below the microflare value ($\leq 10^{27}$ ergs). Already in his paper Parker had introduced concepts such as impulsive, unresolved events, clusters of nanoflares, and statistics of nanoflares with the idea that the corona is the result of a large number of nanoflares happening all the time. He showed that quantitatively they could roughly explain the existence of the corona. However, it is still unclear how many of these brightenings are indeed reconnection events.

Observations of individual brightenings show that they have power-law distributions. This has been measured in the Fourier spectra of the radiance (Ireland et al., 2015), the spatial and temporal scales of the brightenings, and the inferred thermal energy. In particular, the index of the slope in the energy distribution remains close to −2. Table 1 of Benz and Krucker (2002) summarizes estimated slopes from several studies, which range from −1.35 to −2.59. One important aspect to keep in mind is that soft X-ray and EUV observations infer the amount of energy of each event from the observed radiance. This is the plasma response to the injected energy. The hypothesis behind studying the plasma radiation is that all the energy injected in the loop in the form of heating by the magnetic field is radiated away to cool the system and that the plasma response to this heating preserves the heating statistical properties. However, Parenti et al. (2006, 2008) have tested these two hypotheses, and they showed that in general they do not hold: only the very high-temperature emission (~10 MK) retains memory of the statistical properties of the energy heating the system, whereas at lower temperatures, including the average 1–3 MK corona, such information is lost. See also Benz and Krucker (2002), Buchlin et al. (2005), and Jess et al. (2019) for a description of the difficulties in estimating the geometry and energies associated with brightening, and therefore estimating power-law slopes from them. It is clear that the direct relation made from radiance to heating properties is not straightforward, and other diagnostic methods should be used to investigate small-scale heating.

Loop Fine Structure

When subarcsec monitoring of the solar corona in the extreme UV (EUV) was made with the TRACE mission, it was clear that the coronal loops observed with prior instrumentation consist of bundles of even thinner coronal loops, or strands (Lenz et al., 1999). The data arising from high-resolution instruments in the soft-X-ray (HINODE/XRT) and EUV-UV (TRACE, AIA/SDO and IRIS), as well as the few observations from sounding rockets (Hi-C, pixels of about 75 km), have confirmed that loops have substructure at least down to less than 1000 km (e.g., Reale et al., 2007; see discussion in Klimchuk, 2015). Such detected strands may contain even finer structure.

There is a long history of literature concerning observational results of the thermal structure of loops, which is an indication of the elemental cross-scale heating scale. If an observed coronal loop shows isothermal plasma, then this could be evidence that the heating mechanism deposited energy over the entire width of the loop at the same time, whereas multithermal plasma indicates heating events occurring on substructures within the loop. The results are controversial, particularly concerning warm loops (see Del Zanna and Mason, 2003; Cirtain et al., 2007; Reale, 2014; and Tripathi et al., 2009 for a full review). Some observations indicate that even a single observed coronal loop is typically a compilation of many subresolution magnetic field strands, each independently undergoing heating and cooling cycles and are multithermal (Schmelz et al., 2001), whereas other observations indicate monolithic, resolved, or almost-resolved loops (Aschwanden & Nightingale, 2005; Aschwanden & Peter, 2017; Brooks et al., 2012). Some of these differences come from the different diagnostic techniques used, instrumental limitations, and the uncertainties in the data. There is also a possibility that there exist different classes of loops subject to different heating mechanisms and/or different frequency of heating events. Warm, ~1 MK loops have been found to be either isothermal or multithermal at a given time, and quite time variable. Hotter, ~3 MK loops in AR cores are multithermal but more stable in time.

Spectroscopic filling factors measured in the EUV are another indication of substructures. They are a measure of how much the observed structure is filled with plasma at the temperature of the emission, and are another indication of substructure. Warren et al. (2008) measured filling factors in AR coronal loops, and Hahn and Savin (2016) measured QS and coronal hole filling factors; both papers found filling factors of around 10%, a strong indication of substructure.

Differential Emission Measure (DEM)

The DEM is a measure of the amount of plasma at a given temperature. In ARs in general, the DEM has a triangular shape (in the log-log scale), with the two sides of the peak being fitted with power-law functions that can change depending on the AR. Figure 2.9 shows one example of the DEM in the core of an AR derived from Hinode/EIS. The power-law index (α) computed over the lower temperatures was found to vary in the range $1.6 \leq \alpha \leq 5.2$ (e.g., see review of the observations in Bradshaw et al., 2012). This range is likely due to physical variations between different ARs, with some AR cores having slopes more consistent with high-frequency heating (Warren et al., 2011), and a possible dependence on the AR age (Ugarte-Urra & Warren, 2012). Warren et al. (2012) surveyed 15 ARs, and found that steeper slopes are correlated with the total unsigned photospheric magnetic flux as well as the relative amount of hot plasma along the line of sight. Note, however, that there are also instrumental uncertainties in the calculation of DEMs (Guennou et al., 2013).

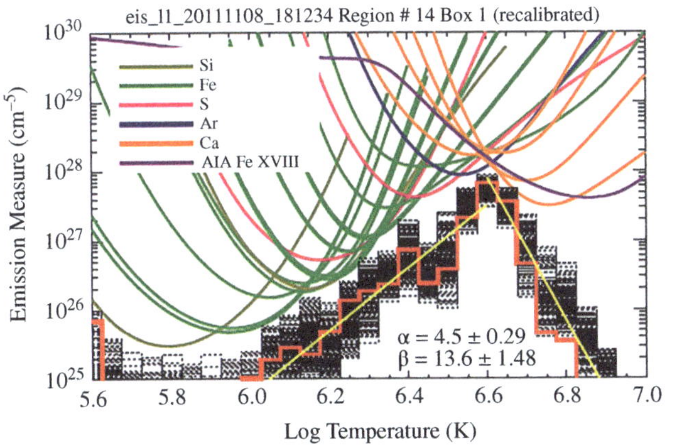

Figure 2.9 Active region (AR) differential emission measure curve. The emission measure distribution is the red curve. The black curves are solutions that use perturbed intensities. The parameters α and β are the unsigned slopes of the emission measure below and above log T = 6.6. (*Source*: Warren et al, 2014. © 2014 IOP Publishing.)

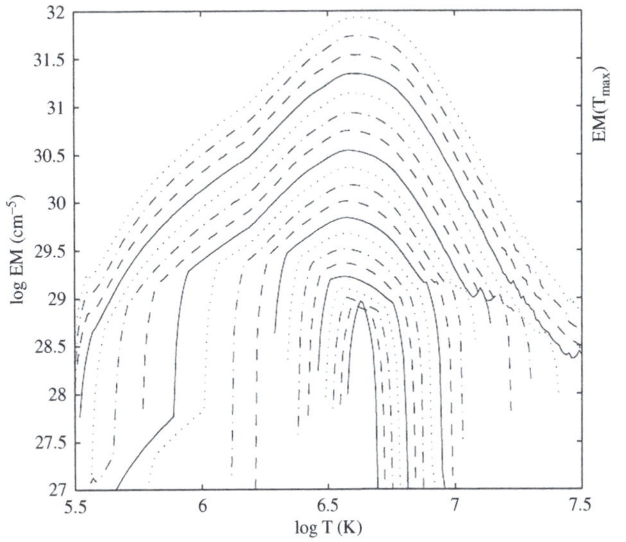

Figure 2.10 Emission measure as a function of temperature for nanoflare trains with constant energy nanoflares. The 20 curves are associated with different delay times between the nanoflares (between 250 and 5000 s). The lowest curve corresponds to a delay time of 250 s and the highest to 5000 s. Each curve is shifted vertically by 0.2 on a log scale with respect to the previous one as the delay time increases. The four line styles break the delay time up into groups of 1000 s. (*Source*: Cargill, 2014. © 2014 IOP Publishing.)

Models of impulsive heating (i.e., where the heating as a function of space and time is assumed, but a particular mechanism is not; see section 2.3.2) have shown that the cross-field thermal structure (represented by a DEM, for instance) carries important information about how the heating energy is distributed and released within a flux tube. Figure 2.10, shows an example where many flux tubes are in the simulated observation, which would result due to either line-of-sight effects and/or because the flux tube diameters are smaller than the resolution of the instrument. The plasma temperature of the many flux tubes changes depending on the frequency of the heating events, assuming a constant total injected energy. High-frequency events produce an almost isothermal plasma along the line of sight. As the heating frequency decreases, the plasma becomes more multithermal. It is important to note that in simulations of impulsive heating, such as this one, the coronal emission is dominated by plasma in its cooling phase (Klimchuk, 2006).

Light Curves

Another way to understand the properties of the heating in loops is to characterize their life cycle through their light curves, or emission at different wavelengths, as a function of time. These can give an estimate of the temporal distribution of the heating. The temporal stability over a few hours of the emission and the temperature of hot loops in the core of the AR are consistent with high-frequency heating which prevents the plasma from cooling (e.g., Porter & Klimchuk, 1995; Schrijver et al., 2004; Warren & Winebarger, 2006; Warren et al., 2011).

In contrast, the cooling of loops has been detected through light curves since, at least, the Yohkoh and Skylab era (see Reale, 2014, for a full review), but only recently was it possible to measure the full lifecycle temperature range (from a few megakelvins to chromospheric temperatures), which provides a better constraint on the theories. Some of the coronal loops are seen to be cooling from hotter loops (e.g., Ugarte-Urra et al., 2009; Viall & Klimchuk, 2011; Winebarger & Warren, 2005). The observation of a stable emission over several hours together with the associated small, but detectable fluctuations (~15%), are compatible with a loop made of a substructures where each substructure is heated and cools fully, but the loop in aggregate is continuously heated (e.g., Sakamoto et al., 2008; Terzo et al., 2011). In support of this general picture has been the arrival of the SDO/AIA high spatial and temporal resolutions EUV data. Viall and Klimchuk (2012, 2015, 2016) have developed and used a time lag method (a delay in the peak of emission seen in the different AIA channels) to highlight the full cycle of plasma cooling, both in observable loops and in the diffuse surrounding unstructured areas. Analyzing the same 15 ARs dataset used by Warren et al. (2012), Viall and Klimchuk (2017) measured different classes of cooling cycles that could be explained using trains of impulsive heating events with a range of different frequencies, high and low. Combining forward modeling with the observations, (Bradshaw & Viall, 2016) suggested that the majority of the plasma was heated by impulsive heating events at a range of frequencies, with a typical frequency of several thousand seconds.

Non-Thermal Broadening

Another observational signature expected from impulsive heating is spectral line broadening due to the high-speed plasma evaporating from the chromosphere into the coronal loop. High velocities, on the order of 100 km/s or more, are predicted in the early phase of the cooling cycle, when the strand is still weakly filled by plasma. Patsourakos and Klimchuk (2006) showed that this signature could be measured by observing the profiles of hot lines, such as Fe XVII, whereas cooler lines are less affected by the heating. However, when looking at the observations, no clear evidence has been found (e.g., Brooks and Warren 2016). It is possible that the present instrumentation does not have a high enough spectral resolution and sensitivity to isolate such weak signatures from the bulk coronal emission. Wave motion, even in the absence of heating, will also contribute to non-thermal line widths, with broadening of the non-thermal line width attributed to unresolved wave motions instead of flows. Brooks and Warren (2016) investigated non-thermal line broadening in Hinode/EIS observations in the core of 15 non-flaring ARs and find that their results are inconsistent with all current models of heating: Alfvén wave turbulence, reconnection jets in the corona, and chromospheric evaporation induced by coronal nanoflares. A straightforward conclusion of non-thermal broadening cannot be inferred from the current model predictions.

Coronal Rain

Another observational constraint on coronal heating is coronal rain. Coronal rain consists of cool, chromospheric temperature plasma observed to fall down from the corona along magnetic field lines (Antolin & Rouppe van der Voort, 2012; Auchere et al., 2014; Froment et al., 2015, 2017; Müller et al., 2005). When it is observed outside of flaring regions, coronal rain seems to require a condition called thermal non-equilibrium, which requires high-frequency heating that is concentrated toward the footpoints (see section 2.3.2 for a discussion of the theory of thermal non-equilibrium and how it is constrained by observations). Theoretically, thermal non-equilibrium is described with steady heating; however, the heating that produces coronal rain can also be impulsive, provided that the heating is at a high enough frequency relative to the plasma cooling time (Karpen & Antiochos, 2008). Antolin and Rouppe van der Voort (2012) measured the statistical properties of coronal rain outside of flares using the CRisp Imaging Spectro Polarimeter (CRISP) instrument on the SST. They found that, on average, they are 310 km wide, 710 km long, and at temperatures of below 7000 K. It is estimated that the fraction of the coronal volume with coronal rain is between 7% and 30%. An additional observational constraint on their formation is that sometimes coronal rain occurs periodically and in conjunction with periodic intensity enhancements (Auchère et al., 2018).

Doppler Shifts

Observed Doppler shift patterns generally show that lower-temperature plasma is red-shifted, whereas higher-temperature plasma is blue-shifted. The Doppler shifts show center-to-limb variations consistent with a cosine (theta) variation (Peter & Judge, 1999), and the general interpretation of the red shifts is that they are downflows from a hotter corona on the flux tube (Chae et al., 1998). The transition between up- and down-flowing plasma is measured to be somewhere between 0.5 and 1.0 MK (e.g., Chae et al., 1998; Dadashi et al., 2011; Peter & Judge, 1999).

Elemental Fractionation

Spectral observations show that elements with a low first ionization potential (FIP)—below 10 eV—are enhanced in abundance in the closed-field corona relative to their photospheric abundances. The chromosphere, where these low-FIP elements would be ionized whereas higher-FIP elements remain neutral, is thought to be the location where this difference in relative abundance is set, due to forces acting on ions but not on neutral elements. FIP abundance is an important tracer for solar wind formation because it is "frozen" into the solar wind plasma, and does not evolve in the solar wind as it travels away from the Sun. See the review by Laming (2015) for a thorough survey of the topic, and Chapter 1 for the use of in situ measurements of the relative abundance of low FIP elements in the solar wind to make a connection with the solar atmosphere.

Footpoints and Moss

Due to the crucial chromosphere–transition region–corona connection, the footpoints of coronal field lines can be an important diagnostic of coronal heating. Diagnostic techniques include comparing emissions formed at different temperatures, and therefore connections between different heights (e.g., Vourlidas et al., 2001), as well as its time variability, which is an indication of the heating frequency (Antiochos et al., 2003; Testa et al., 2014). Moss, an example of footpoint emission, is a low-height (3000 km), 1 MK plasma that was originally identified with the TRACE imager. Moss is now known to be the footpoints of hotter, 3–10 MK loops, with a brightness that scales with loop pressure and filling factor (e.g., Berger et al., 1999; De Pontieu et al., 1999; Fletcher & De Pontieu, 1999; Martens et al., 2000; Schrijver et al., 1999).

Observations of Coronal Wave Heating

Although periodic variations were detected in early observations (see, e.g., Antonucci et al., 1984; Aschwanden, 1987; Deubner & Fleck, 1989; Rosenberg, 1970; Trottet et al., 1979; or Tsubaki, 1988; Vernazza et al., 1975, for an overview), it was the increased spatial resolution and wavelength coverage of Solar and Heliospheric Observatory (SOHO) and TRACE that provided spatially resolved observations of waves and oscillations in the solar atmosphere. Observations of standing and propagating waves were discovered in a wide range of coronal structures such as, for example, intensity (density) variations in coronal fan loops (see, e.g., De Moortel, 2009, for an overview), transverse loop oscillations (e.g., Aschwanden et al., 1999; Nakariakov et al., 1999), and perturbations in coronal plumes (e.g., DeForest & Gurman, 1998; Ofman et al., 1997).

Present-day observations by a range of imaging and spectroscopic instruments have allowed us to establish the presence of waves and oscillations in almost all parts of the solar corona and out into the solar wind. We have been able to directly measure the oscillatory displacement of coronal loops, intensity variations in magnetic structures, and variations in Doppler velocities, line widths, and periodic radio signatures. The study of MHD waves has two major applications within solar physics, namely, coronal (or magneto) seismology (Uchida, 1970; Roberts et al., 1984—see, e.g., Nakariakov & Verwichte, 2005, or De Moortel & Nakariakov, 2012, for a review on coronal seismology) and as a potential coronal heating mechanism (see reviews by, e.g., Arregui, 2015; De Moortel & Browning, 2015; Klimchuk, 2006, 2015; Parnell & De Moortel, 2012; Walsh & Ireland, 2003), and for comprehensive overviews of observations of waves and oscillations, we refer the reader to the reviews listed above. We will restrict ourselves here to looking at a few particular examples and their potential relation to coronal heating.

Early observations by SOHO and TRACE revealed the presence of quasi-periodic intensity perturbations propagating along coronal fan loops, mostly at the edges of ARs (see, e.g., Berghmans & Clettea, 1999; De Moortel et al., 2000; Robbrecht et al., 2001; Schrijver et al., 1999). These propagating coronal disturbances (PCDs) were found to have amplitudes on the order of a few percent compared to the background intensity, periods of a few minutes (typically 2–10 min although periods on the order of 15 min have been reported in polar plumes) and propagation speeds ~100 km/s (when observed in wavelengths such as 171 A, i.e., around 1 MK). The combination of these properties, and in particular the propagation speed close to the local sound speed, gave rise to the interpretation of PCDs as propagating slow magneto-acoustic waves (De Moortel et al., 2002a, 2002b), although others suggested an alternative interpretation in terms of flows. Combining the imaging observations with spectroscopic measurements (e.g., from Hinode/EIS) led to a more complicated picture than was first apparent. The spectroscopic observations showed that the low-amplitude, periodic PCDs were present not only in the intensity but also in the Doppler velocity measurements, and both the wave (e.g., Banerjee et al., 2009; Kitagawa et al., 2010; Krishna Prasad et al., 2011; Mariska & Muglach, 2010; Marsh et al., 2011) and flow interpretation (e.g., Bryans et al., 2010; Doschek et al., 2007; Guo et al., 2010; Hara et al., 2008; He et al., 2010; Peter, 2010; Sakao et al., 2007; Ugarte-Urra & Warren, 2011) became abundant in the literature. The ambiguous interpretation was highlighted by two papers analyzing the same dataset but putting forward a different interpretation (De Pontieu & McIntosh, 2010; Wang et al., 2009). Wang et al. (2009) compare the intensity and Doppler shift perturbations and, finding that these are approximately in phase, interpret the PCDs as slow magneto-acoustic waves. However, De Pontieu and McIntosh (2010) also look at the line widths and line asymmetries and, after comparing with model results suggest that the in-phase oscillations in intensity, Doppler velocity, line widths, and line asymmetries are consistent with an interpretation in terms of quasi-periodic upflows.

Distinguishing between the interpretations in terms of the propagation of slow magneto-acoustic waves and quasi-periodic upflows is less straightforward than it perhaps might seem. When looking at the asymmetries in the spectral line profiles, PCDs are often accompanied by quasi-periodically occurring enhancements of the same order of magnitude in the blue wing of the spectral line. De Pontieu and McIntosh (2010) and Tian et al. (2011) show that fitting line profiles that exhibit such an enhancement in the blue wing with a single Gaussian (rather than with a double Gaussian) mimics the properties of the intensity perturbations in the imaging data. However, Verwichte et al. (2010) argue that the periodically varying double-Gaussian fit could be composed of an oscillating dominant (core) component and an additional small, stationary blue-wing component, consistent with an interpretation in terms of slow magneto-acoustic waves. De Moortel et al. (2015) used forward modeling to derive observational signatures associated with simple models of slow magneto-acoustic wave or periodic flow propagation. Although there were a number of observational differences between the flow and wave scenarios, it was not possible to establish clear, robust observational characteristics that can be used in isolation (i.e., without needing a like-for-like comparison between the models). The identification problem is further exacerbated by the fact that any distinguishing characteristics require extensive analysis of the spectroscopic data, which is nontrivial with the low signal-to-noise levels of these small-amplitude perturbations.

Recent work by De Pontieu et al. (2017), based on the numerical simulations by Martinez-Sykora et al. (2017), demonstrated that the interpretation of PCDs as either propagating slow waves or episodic flows might be too simplistic. Similarities in properties have previously hinted at a close relationship between PCDs and type II spicules (De Pontieu et al., 2005; De Pontieu & Erdélyi, 2006), which was firmly established by De Pontieu et al. (2017) by combining IRIS and AIA observations with numerical modeling. It was found that PCDs in plage region loops are not simply a signature of slow waves or flows but are part of a complex set of events linking the generation of spicular flows with shock waves that propagate into the corona as well as plasma heating through the dissipation of electrical currents and magnetic waves. Here, PCDs are not just a small-amplitude perturbation of the background intensity but the result of the formation of new coronal strands (an aspect missed by the often-used running difference analysis of PCDs). The complexity naturally explains the wide range of PCD propagation speeds reported in the literature, as the observed speeds are the result of a mixture of real flows, remnants of shock waves, and dissipation of currents in the coronal volume (i.e., local heating followed by thermal conduction). In addition, this model also resolves the ambiguous interpretations in terms of periodic flows (usually observed lower down in the loops) and slow magneto-acoustic waves (observed along a greater extent of the loops). Despite their small energy budget (~300 erg cm^{-2} s^{-1}—see, e.g., De Moortel et al., 2002a), their omnipresence in the solar atmosphere and their close link with spicule-driven heating imply that PCDs potentially play an important role in the mass and energy flow in the solar corona.

The increased spatial and temporal resolution of TRACE also gave us the first imaging observations of standing transverse loop oscillations induced by a nearby impulsive event (Aschwanden et al., 1999; Nakariakov et al., 1999). Since then, transverse oscillations have been reported in many of the solar atmosphere's magnetic structures such as prominences (e.g., Okamoto et al., 2007), magnetic bright points and intergranular lanes (Jess et al., 2012), chromospheric spicules (e.g., De Pontieu et al., 2007) and mottles (e.g., Kuridze et al., 2012), coronal loops (e.g., McIntosh et al., 2011), coronal rain (e.g., Antolin and Verwichte, 2011), and jets (e.g., Cirtain et al., 2007). Similar transverse perturbations were reported in Doppler shift observations of large, off-limb, coronal loops by Tomczyk et al. (2007) and Tomczyk and McIntosh (2009). Many other examples can be found in the literature. We will focus on the potential relevance of these perturbations for coronal heating.

Estimates of the energy associated with the observed propagating transverse waves vary substantially. For example, McIntosh et al. (2011) analyze SDO/AIA and report on small-amplitude transverse displacements in a coronal hole region, QS and AR loops. They suggest that these ubiquitous low-amplitude oscillatory displacements are footpoint driven and contain sufficient energy to account for the heating requirements of the QS and coronal holes (~1–2 × 10^5 erg cm^{-2} s^{-1}). In the AR loops, however, the energy is estimated to be about 10^5 erg cm^{-2} s^{-1}, which is at least two orders of magnitude too small (where Withbroe & Noyes estimate that about 2 × 10^7 erg cm^{-2} s^{-1} is needed to account for the heating of ARs). In contrast to the energy budget reported by McIntosh et al. (2011), Tomczyk et al. (2007) report a much smaller energy budget estimated from CoMP observations of propagating disturbances in off-limb coronal loops. However, McIntosh and De Pontieu (2012) find that the off-limb line-of-sight superposition of Doppler velocities "hides" most of the wave energy flux, which is instead present in the large, observed non-thermal line broadening. Using 3D numerical simulations of oscillations in cylindrical flux tubes, De Moortel and Pascoe (2012) demonstrated that the superposition of randomly directed oscillating transverse displacements does indeed result in "observed" line-of-sight integrated Doppler velocities with an energy budget that is at least an order of magnitude smaller than the energy present in the three-dimensional coronal (numerical) volume. Therefore, the energy budget estimated by McIntosh et al. (2011) in the AR loops is likely to be a lower limit due to the superposition of a large number of oscillating loop strands. Off-limb, one would also expect a large number of loops with randomly directed transverse displacements along the line of sight, which could account for the low-energy budget reported by Tomczyk et al. (2007). In addition, mode identification further complicates the derivation of an estimated energy budget from observed waves and oscillations (Goossens et al., 2013).

To be relevant in the context of coronal heating, observed waves and oscillations not only need to contain a substantial amount of energy, but this energy needs to be dissipated on the appropriate time scales and in the right places. It is important to note here that damping (of the perturbation amplitudes) and dissipation are not necessarily the same. For example, waves can undergo "damping" due to a change in the background medium or by transferring energy to a different wave mode. The standing transverse loop oscillations induced by impulsive events (see, e.g., Aschwanden et al., 1999, and Schrijver et al., 1999, for an early overview) are observed to undergo very rapid damping, within a few oscillation periods, but their contribution to (local) heating, if any, remains unclear. They do not occur frequently enough to really be relevant for coronal heating and in addition, their rapid damping is likely due to resonant absorption (see, e.g., Goossens et al., 2006; Goossens et al., 2011; Ruderman & Erdélyi, 2009; Ruderman & Roberts, 2002), which transfers energy from the observed transverse loop oscillation to azimuthal Alfvén waves in the loop boundary. Although subsequent phase mixing of the Alfvén waves in the loop boundary region could lead to local heating, this is unlikely to sustain the loop at coronal temperatures (Cargill et al., 2016). Aschwanden and Terradas (2008) find that the timescales on which the oscillating loops cool through the EUV passbands are compatible with heating and subsequent cooling, but this heating cannot necessarily be attributed to wave heating from the transverse loop oscillation.

The situation is different, however, for the propagating transverse (Alfvénic) waves observed by, for example, Tomczyk et al. (2007) and McIntosh et al. (2011). The CoMP observations not only show that these perturbations are ubiquitous through the solar corona, Tomczyk and McIntosh (2009) suggest that the waves must undergo rapid, in situ damping to explain the observed disparity between outward (propagating upward toward the loop apex) and inward (propagating down from the loop apex) wave power. The rapid damping of the transverse perturbations can be explained by mode coupling, again transferring energy from the observed oscillating displacements to small-scale azimuthal Alfvén waves in the tube boundary (see, e.g., De Moortel et al., 2016, for a review) but the contribution to local heating from the subsequent phase mixing in the loop boundary has so far not been found to be viable (Pagano & De Moortel, 2017, 2019). However, Morton et al. (2015) reported on the observation of counter-propagating Alfvénic waves which could lead to the development of turbulence and enhanced wave dissipation, which was not included in the model by Pagano et al. (2018) . Morton et al. (2016) suggested that regardless of local topology, the CoMP spectra reveal the presence of enhanced power around 3 mHz, suggesting a global link between the observed coronal transverse perturbations and the global p-modes. In addition, analyzing CoMP data from a number of different dates during 2012–2015, Morton et al. (2019) recently suggested the possible presence of a basal Alfvénic flux throughout the solar cycle. On the other hand, Morton and McLaughlin (2013) exploit the high resolution provided by Hi-C to try and confirm the ubiquitous nature of wave activity in the solar corona but find that low-frequency (50–200 s) wave activity in the coronal loops is generally of low energy and that some structures do not show any periodic transverse displacements at all.

The observations of waves and oscillations in the solar corona described so far are all direct measurements of periodic variations in observables such as intensity or Doppler velocity. Indirect evidence of the presence of waves can also be derived from observations of non-thermal line widths, with broadening of the non-thermal line width attributed to unresolved wave motions. For example, McClements et al. (1991) demonstrated that if Alfvén waves are present, there should be a center-to-limb broadening of the observed line widths, a concept that was used to suggest the presence of Alfvén waves based on observed changes in the line broadening by, for example, Hassler et al. (1990) or Erdélyi et al. (1998). As pointed out by McIntosh and De Pontieu (2012), a considerable amount of wave energy could be "hiding" in the non-thermal line widths. Changes in observed line widths could therefore be indicative of changes in the wave amplitude. Such variations have been analyzed by, for example, Harrison et al. (2002), who found a decrease in EUV line width with height above the limb, which they suggested could be due to wave damping (see also, e.g., O'Shea et al., 2003; Sakurai et al., 2002; Zaqarashvili, 2003; Wilhelm et al., 2004). A similar decrease in the non-thermal line widths was found by, for example, Hahn et al. (2012), Bemporad and Abbo (2012), and Hahn and Savin (2013) in polar coronal holes, which these authors also attribute to damping of Alfvén waves (see also, e.g., Banerjee et al., 1998, 2009b). In coronal holes, dissipation of wave energy is not only relevant for coronal heating but also for the acceleration of the fast solar wind (see review by, e.g., Banerjee et al., 2011; Cranmer, 2009). Hahn et al. (2012) suggest that the observed rapid decrease in line width could account for up to 70% of the energy required to heat the polar coronal hole and drive the solar wind acceleration (where Hahn & Savin, 2013, estimate the initial energy flux density present as $6.7 \pm 0.7 \times 10^5$ erg cm^{-2} s^{-1}). Hahn and Savin (2014) find that the spatial profile of the wave damping is correlated with the loop length, suggesting that the mechanism responsible for the damping of the waves depends on the global loop properties. Oscillating transverse displacements have also been measured directly in coronal plumes from SDO/AIA observations by, for example, Thurgood et al. (2014), who find that the energy budget contained within these resolved oscillations is insufficient to account for the acceleration of the solar wind.

Observations of Coronal Heating from Magnetic Reconnection

Parker proposed that the origin of nanoflares is the dissipation of magnetic tangential discontinuities (reconnections) that are created by the photospheric random motion of a bundle of magnetic field footpoints. Hard X-ray observations show that the detected events in the flare-microflare energy range ($\geq 10^{27}$ ergs, over about three decades) are distributed in energy following a power law with an index greater or close to, −2 (see, e.g., Crosby et al., 1993; Lin et al., 1984, and references therein). Similar studies have also been addressed observing small impulsive transients in the soft X-ray (using Yohkoh/SXT) and EUV (SOHO/EIT, SUMER, and TRACE data) both in QS and ARs (Aletti et al. 2000; Aschwanden & Parnell, 2002; Bazarghan et al., 2008; Benz et al. 2005; Berghmans et al., 1998; Buchlin et al., 2006; Krucker & Benz, 1998; Parnell & Jupp, 2000; Pauluhn & Solanki, 2007). Hudson (1991) pointed out that a power law steeper than −2 is required for nanoflares to be energetically important (compared to larger flares). Since nanoflares are not directly observable, this requires extrapolation from the observed events, which are much more energetic. There are two problems with this extrapolation. First, as mentioned previously, there are large uncertainties in the measured power-law slope. Second, an extrapolation from flare energies to nanoflare energies may not be appropriate. Yashiro et al. (2006) have found that the slopes for flares that are and are not associated with CMEs are different, with the confined flare slope being steeper than −2. Nanoflares are more likely related to confined flares than to eruptive flares.

Even in resolved flares, it is the consequences of reconnection on the plasma that is observed, such as bidirectional jets (Innes et al., 1997). Such signatures are expected to be especially difficult to detect in nanoflare-associated jets. For example, reasons that it may be difficult to detect nanoflare jets are if jet energy is converted to thermal energy rapidly, if the jet interacts with the surrounding plasma, if the filling factor is $<<1$, 3D effects, tearing islands, and if thermal conduction cools the associated shocks rapidly (e.g., see Cargill, 1996; Klimchuk, 1998, for a discussion of these effects).

As with chromospheric heating, one way to estimate the amount of reconnection is through the change in the magnetic field and estimated flux cancellation. Using a potential field model calculated from SUNRISE/IMaX data, Wiegelmann et al. (2013) found that the connectivity in the upper solar atmosphere changes rapidly, suggesting magnetic reconnection. However, both Wiegelman et al. (2013) and Chitta et al. (2014) estimated an upper limit of the free magnetic energy that would be available to magnetic reconnection, and found that it was too small to explain the coronal heating in the QS. We note that the above estimates do not take into account the free magnetic energy associated with tangling and twisting of the field on small scales. Such energy could easily be enough to power the QS (Klimchuk, 2006, 2015).

Evidence and theory suggest that X-ray bright points are produced by flux cancellation (e.g., Archontis & Hansteen, 2014; Falconer et al., 1999; Longcope, 1998; Martin et al., 1985; Parnell & Galsgaard, 2004; Parnell & Priest, 1995; Priest et al., 1994, and review by Priest et al., 2018). Flux cancellation has been associated with coronal loop brightenings, also implicating magnetic reconnection (Tiwari et al. 2014; Huang et al. 2018). Chitta et al. (2018) observed fluctuations in the footpoint of a coronal loop observed in the171 Å channel of SDO/AIA, concurrent with flux cancellation. They concluded that the flux cancellation could provide energy to heat the corona to temperatures over 5 MK. Braiding followed by reconnection has been observed recently in the Hi-C, high-resolution observations of an AR in the EUV (Cirtain et al., 2007). The event observed is shown in Figure 2.11. Panel e shows an image observed with Hi-C, while c shows an unsharp mask version of the image. Panels a, b, d, and f show concurrent SDO/AIA images of the same region. The event was a C1.7 flare, which might share similar physics to smaller nanoflares.

One important observation that is unique to magnetic reconnection is very hot, >5MK plasma. The DEM of ARs generally decreases quite sharply above the peak at 3MK, and it is difficult to quantify the highest detectable temperature emission, though attempts have been made. The very hot plasma is emitted in the very early phase of the cooling (e.g., Cargill, 1994; Cargill et al., 1995). For this reason, it can possibly preserve more information on the heating properties (Parenti et al., 2006). In the recent decade, several efforts have been made to detect such plasma. It has been identified in on-disk AR using EUV, soft X-ray, and hard X-ray (e.g., Brosius et al., 2014; Hannah et al., 2016; Marsh et al., 2018; Patsourakos & Klimchuk, 2009; Reale & Ciaravella, 2006; Reale et al., 2009a, 2009b; Schmelz et al., 2016; Testa et al., 2011; Warren et al., 2012; Figure 2.12), even though the quantification with respect to the background, 3 MK plasma has been a very difficult task. EUV data rely in the detection of Fe XVIII-XIX emission (formed around $\log T=6.9$) in deep exposures. The general agreement is that it is present in an amount consistent with nanoflares. Most recently (Parenti et al., 2017) quantified the hot plasma in off-limb AR data using very deep exposures of Fe XIX: up to ~0.1% of the plasma from the bulk corona could reach ~10 MK. For the first time, it was shown that this hot plasma was persistently present almost everywhere in the core of the AR up to about 9 Mm above the limb (Figure 2.13). A smaller amount of hot plasma was detected on the disk by the rocket Focusing Optics X-ray Solar Imager (FOXSI-2) in hard X-ray (Ishikawa et al., 2017).

Table 2.5 (from Parenti et al., 2017) summarizes from the literature the index (β) of the inferred power-law slope above 5 MK, which shows an important variability

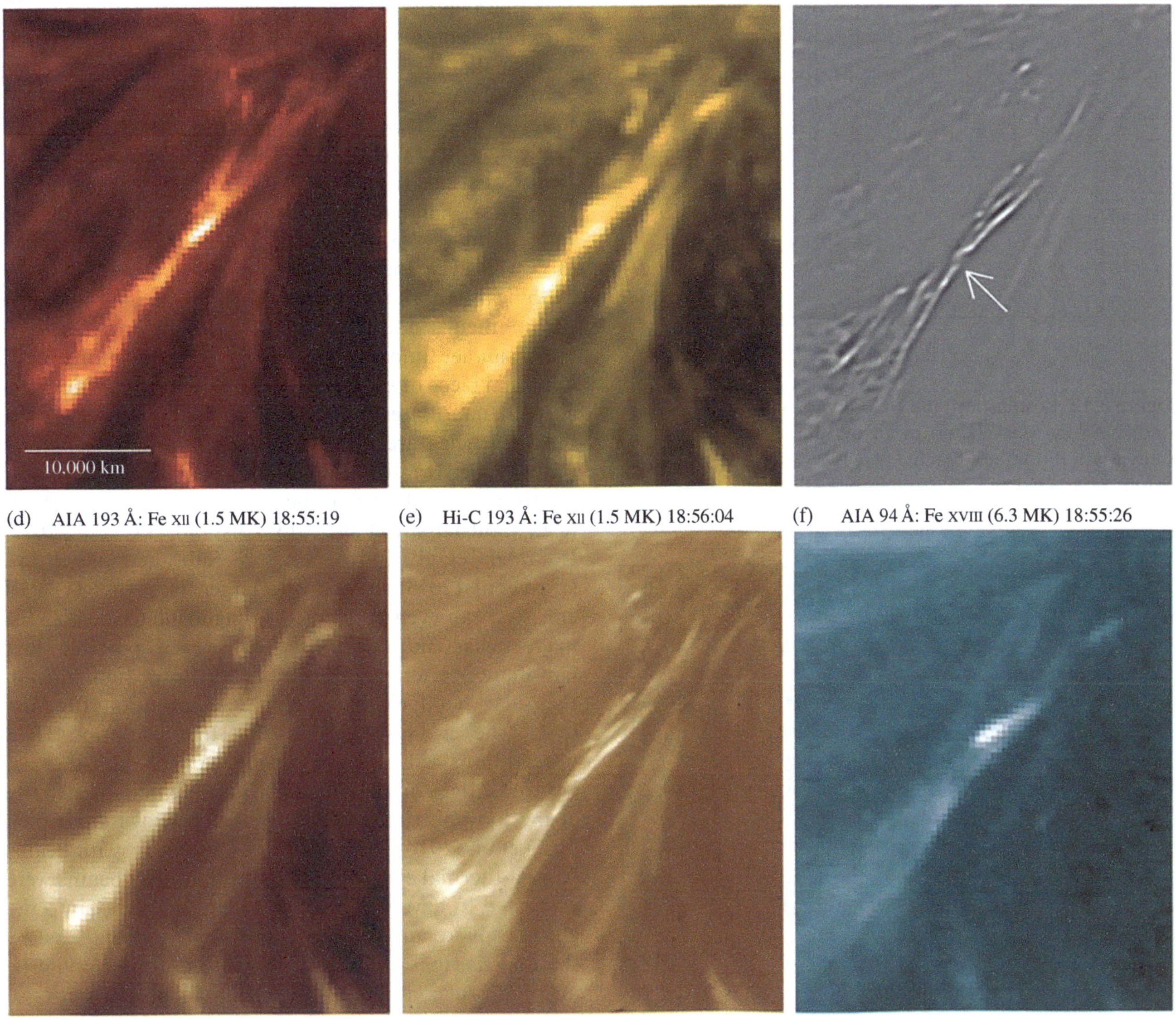

Figure 2.11 Magnetic reconnection observed after braiding in Hi-C for a C1.7 flare. A, b, d, and f show the event viewed in different AIA passbands. E shows the Hi-C image, and c shows an unsharp mask version of the Hi-C. (*Source*: Adapted from Cirtain et al., 2007.)

($-14 < \beta < -4.4$). At high temperatures, the EUV and X-ray emissions become weak, inferring the slope becomes very difficult, and the uncertainties in the measure increase. In some cases, the power-law shape has not been confirmed (Barnes et al., 2016). It is also possible that changes in the DEM shape (and so in the power-law index) may occur during the AR lifetime (e.g., Ko et al. 2016). More work has to be done in this respect. Indeed, the properties of this very high-temperature plasma are not yet completely understood. The physics involved may be more complex as at the very high temperature this tenuous plasma can be in non-equilibrium conditions (Bradshaw & Cargill 2006; Bradshaw, 2009). Ionization non-equilibrium, as well as departures from Maxwellian distributions in general, are important to consider in the interpretation of observations (Dudík et al., 2017).

More generally, the presence of hard X-rays (HXR) indicate hot >5MK plasma and can be used to identify nanoflares. HXR emission can also be used to test whether nanoflares have similar physical processes as larger flares, namely, magnetic reconnection that results in particle acceleration. Microflares are observed to have HXR emission and hot plasma. RHESSI observations (review by Hannah et al., 2011) show that microflares resemble more energetic flares in many respects, including that they occur only in AR, have high temperatures and

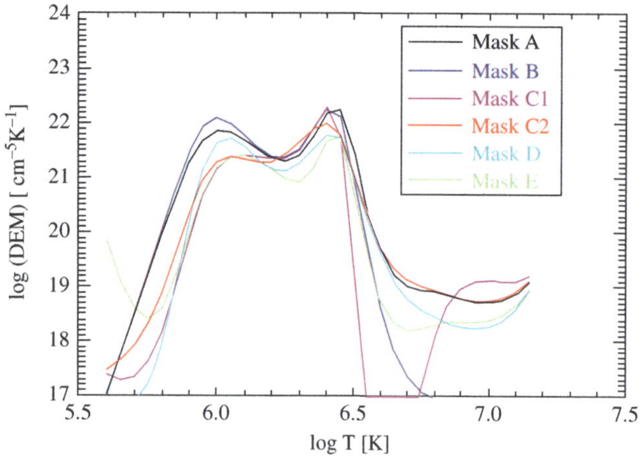

Figure 2.12 Results from the DEM inversion for different areas (indicated by Mask letter) in an off-limb active region (AR). Only Mask B does not reach a very high temperature. (*Source*: Parenti et al. (2017). © 2017 IOP Publishing.)

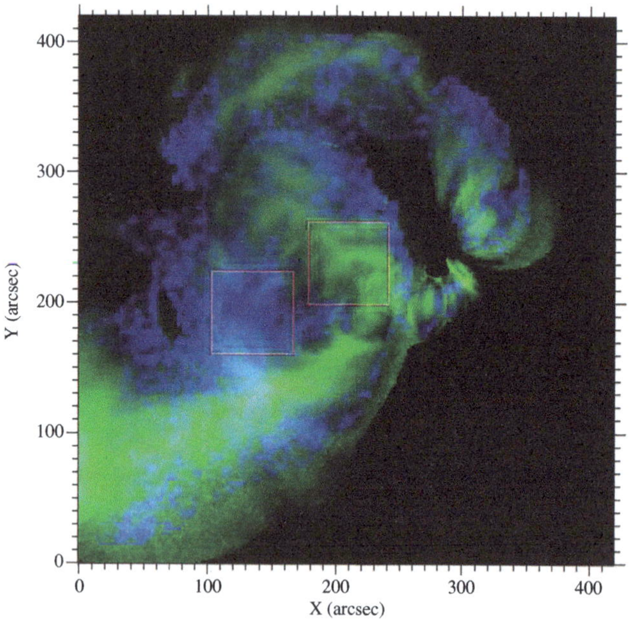

Figure 2.13 Map of temperature in an active region (AR). The Green scale refers to an average of 3 MK, and the blue scale localizes the plasma above 5 MK. (*Source*: From Reale et al., 2009).

non-thermal emission, and non-thermal energy deposition for microflares was confirmed with Nuclear Spectroscopic Telescope Array (NuSTAR; Glesener et al., 2017; Wright et al., 2017). Although the amount of energy in microflares does not account for the global coronal heating budget, the observations show that they supply substantial coronal heating in the locations where they occur.

Because microflares are part of a distribution of flares, it is possible that flares smaller than microflares could supply the energy for the global coronal. In order to see if even smaller events, which could explain coronal heating, exist, data from the (FOXSI) rocket and the NuSTAR have been used. Marsh et al. (2018) showed that these HXR data were consistent with nanoflares and ruled out steady heating. Kuhar et al. (2018) showed evidence of three "flares" in NuSTAR that are eight orders of magnitude fainter than flares—smaller than microflares. However, even these smaller flares are larger than the canonical nanoflare. Importantly, they had substructure within them, indicating that they were comprised of many smaller fundamental magnetic reconnection events.

A final piece of evidence of magnetic reconnection and coronal nanoflares is beams of non-thermal electrons. Such particles are observed in association with larger flares. Type III radio emission is a possible diagnostic of electron beams that may be associated with nanoflares (e.g., Saint-Hilaire et al., 2013; Alissandrakis et al., 2015; Bouratzis et al., 2016; James & Subramanian, 2018). Chen et al. (2013) showed with larger flares that Type III emission can even be used as a diagnostic to localize the magnetic reconnection site. It is even possible that nanoflare-accelerated electron beams leave signatures in the solar wind electron velocity distribution function (e.g., Che & Goldstein, 2014). High-resolution IRIS observations also show evidence of non-thermal electrons. IRIS observed intensity and velocity variability at the footpoint of a hot, dynamic coronal loop, which Testa et al. (2014) concluded were consistent with beams of non-thermal electrons in a nanoflare. The beams were estimated to have deposited $<10^{25}$ erg of energy into the chromosphere and TR. Indeed, the energy deposited by the non-thermal particles in the chromosphere could be an important contribution to the coronal plasma response in a nanoflare (Brown et al., 2000).

2.3. MODELING OF THE SOLAR ATMOSPHERE

The ultimate goal of theory and modeling is to understand the fundamental physics well enough to quantitatively predict the observables described above. Modeling the relevant physics for the entire solar atmosphere is not possible with current computational resources. For example, it is not possible to resolve the thin current sheets and their dissipation regions up through the global corona scales or even active region scales simultaneously. Within these computational limitations and the available observations, there are three main modeling strategies for studying coronal heating, each with its strengths and weaknesses. The important point is that they all must approximate or represent some aspects of the physics. In some, it is immediately apparent, and in some, the approximations are more subtle. In all cases, there are situations and interpretations that are appropriate, and conclusions and applications that are not.

The first modeling approach is to perform detailed simulations of the energy release process (broadly categorized as magnetic reconnection or waves, though as we have outlined above, they are not mutually exclusive). To do this accurately requires zooming in on a very small region of the solar atmosphere, such as an individual current sheet. Resistive MHD is generally adequate for coronal conditions, though kinetic effects may become important as smaller and smaller structures develop. A key requirement is achieving a high enough resolution, because the numerical resolution affects the dissipation and heating (e.g., Galsgaard & Nordlund, 1996). The disadvantage of this approach is that coupling to larger scales and to the lower atmosphere cannot be easily included. Such coupling can affect the energy release itself, because there is often a feedback between large-scale stresses and the reconnection that allows the field to evolve. Equally important is the energetic and dynamic coupling between the corona and transition region/chromosphere, which dramatically impacts the plasma properties and radiative output of the corona. This can cause serious problems when comparing predicted observations with actual observations to evaluate the models.

The second approach to studying coronal heating is to treat the heating in an ad hoc manner by simply specifying its spatial and temporal dependencies. Then, the response of the plasma, including the crucial coupling with the lower atmosphere, can be simulated with great accuracy. These are the so-called hydrodynamic loop models discussed in section 2.1. The advantage here is that meaningful comparisons with observations can be made. The spatial and temporal dependencies of the heating can be changed to determine which ones give the best agreement. Because they are computationally less expensive than global models (model type three), the massive parameter space of the possible heating mechanisms and their associated spatial and temporal dependencies can be investigated efficiently. The disadvantage is that the properties of the heating are assumed and not necessarily motivated by a particular physical mechanism. Nevertheless, the observationally derived properties place valuable constraints on any theories for the mechanism, current or future.

The third approach attempts to combine the first two by treating both the energy release process and the response of the plasma, including the production of observational signatures. This is sometimes called *ab initio*, or first principles, modeling. Impressive models of active regions and even the global corona have been obtained. However, present-day computers are not capable of accurately simulating simultaneously the energy release process and the coupling to the lower atmosphere and the global scales. The dominant heating mechanism in these simulations is ohmic dissipation of relatively large-scale currents, and it is not yet understood if this is a reasonable proxy for what is heating the solar atmosphere. Measurements of kG elemental flux tubes in the photosphere lead to the conclusion that an individual AR contains on the order of 100,000 current sheets or more (Klimchuk, 2015). Simulations of the range of physical effects relevant for the reconnection of even a single current sheet is highly challenging (Daughton et al., 2011; Huang & Battacharjee, 2016; Leake et al. 2020), let alone for 100,000. The actual heating mechanism operating on the Sun (fundamentally involving small-scale structures and viscosity, according to models of type 1) has not yet been established. A crucial aspect is the onset of reconnection. As discussed by Klimchuk (2015), reconnection must remain switched off to allow magnetic stresses to build, and switch on to release the stored energy only after a sufficient level of stress has been reached. If reconnection were to happen too soon, the corona would be cooler than observed.

All three approaches to the coronal heating problem provide useful information. Eventually, they will converge on a common solution. But the limitations of each approach must be recognized at all times.

2.3.1. Approach One: Local MHD Models

Wave Heating

As is evident from the observations described above, the key question for wave-based heating of the solar atmosphere is no longer whether waves and oscillations are present but whether they contain sufficient energy and whether wave dissipation occurs on the right timescales and in the right locations. Alfvén(ic) waves, in particular, are seen as a highly effective way to transport energy from below the Sun's surface out into the solar atmosphere, and the suggestion of Alfvén-wave-based heating of the solar atmosphere was already made by Alfvén (1947) (see also, e.g., Wentzel, 1974; Wentzel, 1976; or Hollweg, 1978). However, classical resistive or viscous dissipation of the Alfvén wave energy is slow, as it is proportional to the viscous and magnetic Reynolds numbers (both $>>1$). For example, Porter et al. (1994) pointed out that even though observed non-thermal line broadenings might imply sufficient wave fluxes, this does not necessarily mean that this wave energy is readily converted into heat. Arregui (2015) argued that given the rapid damping of many of the observed (standing) transverse loop oscillation (generally within a few oscillation periods), it is not at all clear that the time scales involved even allow for heating to occur during the actual oscillations.

To enhance the dissipation rate of Alfvén waves, mechanisms such as resonant absorption (Ionson, 1978; Tataronis & Grossman, 1973) and phase mixing (Heyvaerts & Priest, 1983) were proposed. Both resonant absorption and phase mixing depend on the presence of a variation in the local Alfvén speed and the fact that individual field lines can oscillate with their own Alfvén frequency. Resonant absorption occurs when, on a specific magnetic surface, a "global" mode (reflects the large-scale plasma and magnetic structure) is in resonance with local oscillations. This will lead to the transfer of energy from large-scale perturbations to small length scales at a rate that is independent of the diffusion coefficients, giving effective damping of the large-scale mode (see, e.g., Hollweg & Sterling, 1984; Poedts et al., 1989; Poedts et al., 1990; Ofman et al., 1995, or Goossens et al., 2011, for a review of resonant absorption). Resonant absorption itself is a damping rather than a dissipation mechanism; to provide heating, dissipation of the wave energy must still occur within the resonant layer, which is often implicitly assumed to happen on timescales similar to the damping of the original global perturbation.

Phase mixing occurs when Alfvén waves gradually move out of phase on neighboring field lines with different Alfvén speeds (due to a transverse inhomogeneity in the density and/or the magnetic field strength), generating increasingly large transverse gradients, again leading to more efficient dissipation. This leads to a significant enhancement of the dissipation, which under the assumption of strong phase mixing becomes proportional to the cube root of Rm (the magnetic Reynolds number) and Re (the viscous Reynolds number) (Heyvaerts & Priest, 1983). Although theoretical models of the individual mechanisms are usually presented, there is a natural close interplay between resonant absorption and phase mixing, where the small-scale oscillations in the resonant layer are expected to undergo phase mixing due to the local variation in the Alfvén speed (see, e.g., Ruderman et al., 1997a, 1997b). Apart from resonant absorption and phase mixing, other mechanisms such as mode conversion (to modes that might more readily dissipate) and ion-cyclotron resonance have also been proposed (see reviews by, e.g., Aschwanden, 2004; Erdélyi & Ballai, 2007; Goedbloed & Poedts, 2004; Klimchuk, 2006, and references therein). Ion-cyclotron waves in particular are likely important for heating coronal holes and the acceleration of the fast solar wind (see Cranmer et al., 2017, and Chapter 1 on the solar wind).

There is a very substantial body of literature on the theoretical and numerical development of wave-based heating models, particularly in the 1980s and early 1990s (see, e.g., Aschwanden, 2004; Goedbloed & Poedts, 2004, or Goossens et al., 2011, for more detailed reviews). For resonant absorption and/or phase mixing to occur, a gradient in the Alfvén speed has to be present. Also implicit in the models is the presence of an ignorable coordinate, where Parker (1991) argues that this might not readily be the case in the solar atmosphere, leading to coupling to other modes. The most often-used equilibrium configuration for wave-heating models is a straight (and uniform) magnetic field with a transverse density gradient (balanced by a change in temperature to maintain cross-field pressure balance), although see, for example, Poedts et al. (1990) for resonant absorption in a sheared magnetic field. Most modeling then injects waves into this equilibrium (either through imposing a perturbation on the equilibrium or through boundary driving) and investigates the heating through dissipation of the wave energy. However, this is where wave-based heating mechanisms face a fundamental problem. In magnetically closed loops, the density is related to the magnitude of the heating (see, e.g., Klimchuk, 2006, or Reale, 2014), which immediately begs the question of whether wave heating can self-consistently sustain the assumed (equilibrium) density profile or if there is an additional (alternative) heating mechanism needed to sustain the high-density region and prevent draining (e.g., Bradshaw & Cargill, 2010). If the former, then the imposed density structure must be compatible with the spatial profile of the wave heating. If the latter, then there are new questions to answer, such as: what is the required (background) heating mechanism, how does it affect the wave-based heating, and does this automatically imply that the contribution from wave-based heating must be small?

Cargill et al. (2016) assessed the self-consistency of assuming a fixed (preexisting) density gradient by examining the evolution of the coronal density when wave heating takes place. They found that heating due to phase mixing of Alfvén waves cannot sustain the assumed density. Conceptually, this is relatively easy to understand. Phase mixing leads to enhanced dissipation (local heating) where the gradient in the Alfvén speed, that is, in the density, is largest. However, cooling and draining are proportional to the density (not to the gradient of the density) and therefore will occur faster where the density is highest. In other words, the imposed density profile (which leads to phase mixing) cannot be sustained self-consistently by the wave heating. The same basic conclusion holds for resonant absorption (Ofman et al., 1998). Cargill et al. (2016) also look at the related question of whether density changes induced by the wave heating affect the imposed (equilibrium) density profile sufficiently to alter the wave propagation and dissipation (see also Ofman et al., 1998). It is found that including feedback of the heating on the density gradient can lead to significant local structuring but that this occurs only on timescales longer than the thermal evolution of the loop (i.e., the changes to the density profile due to cooling and draining are more significant than the local structuring due to the wave heating). In addition, the authors point out again that transport coefficients need to be substantially enhanced to obtain effective heating. Pagano and De Moortel (2017) study heating due to phase mixing in 3D MHD simulations where transverse footpoint displacements trigger kink modes that mode couple to torsional Alfvén modes in the boundary shell of the cylindrical flux tube. As expected, the Alfvén waves phase mix in the boundary layer of the loop but even with the use of (excessively) high values of magnetic resistivity and strong footpoint drivers, the heating due to phase mixing is insufficient to be relevant for coronal heating. Further work by Pagano et al. (2018) and Pagano and De Moortel (2019) extended this study to include multiharmonic loop oscillations and an observed power spectrum, but the basic conclusion remains unchanged.

So, can the slow dissipation rate of the Alfvén wave energy that seems to be inherent to the basic phase mixing and resonant absorption models be increased? One possibility is the development of the Kelvin–Helmholtz instability and/or the onset of turbulence (see, e.g., Browning & Priest, 1984; Lee & Roberts, 1986; Hollweg, 1986; Hollweg & Yang, 1988; or Hollweg & Johnson, 1988, for early discussions). A turbulent cascade of wave energy down to smaller length scales where dissipation becomes efficient might address the problem of the slow dissipation rate of Alfvén waves. A frequently studied model is turbulence induced by the (internal) reflection of waves (e.g., Matthaeus et al., 1999; Suzuki & Inutsuka, 2005; Verdini & Velli, 2007). Numerical modeling by van Ballegooijen et al. (2011) showed that reflections at the transition region could lead to such a turbulent cascade of Alfvén waves in the chromosphere down to dissipative length scales and could account for the required chromospheric heating rate (see also Asgari-Targhi et al., 2013, 2014; van Ballegooijen et al., 2014), though the present Alfvén wave turbulence modeling currently does not satisfy observational constraints (van Ballegooijen et al., 2017). However, the model uses reduced MHD and hence compressive effects such as the coupling between Alfvén waves and slow modes or the formation of (compressive) shocks cannot be evaluated (see also, e.g., Tu & Song, 2013). In the chromosphere, compressive effects are expected to be important, with several authors showing that shock heating might in fact be the dominant mechanism (e.g., Hollweg et al., 1982; Kudoh & Shibata, 1999; Matsumoto & Suzuki, 2012, 2014; Arber et al., 2016). In the corona however, the compressive effects might be less significant. A cascade of Alfvénic turbulence is also present in the global models of coronal heating and solar wind by, for example, Sokolov et al. (2013) and van der Holst et al. (2014), though it is parameterized (see section 2.3.3). A recent paper by Magyar et al. (2017) looked at the development of turbulence-like behavior for unidirectionally propagating Alfvénic waves, that is, without internal reflections, in a perpendicularly inhomogeneous plasma. The authors find that this generalized phase mixing model can lead to "uniturbulence" with complex, ribbon-like current structures throughout the 3D volume. Enhanced dissipation can also be associated with the presence of an instability such as the Kelvin–Helmholtz instability (KHI), and the strong shear motions at the edges of transverse oscillating loops have been shown to lead to the rapid development of the KHI (e.g., Antolin et al., 2014; Browning & Priest, 1984; Ofman et al., 1994; Terradas et al., 2008; Uchimoto et al., 1991). However, whether this is sufficient to make wave-based heating significant is still under debate (e.g., Howson et al., 2017a, 2017b; Guo et al., 2019; Karampelas et al., 2017).

If wave-induced turbulence (or unresolved Alfvénic motions) makes a viable contribution to heating of the solar atmosphere, this contribution should be evident in observed line broadening (see also, e.g., De Moortel et al., 2014; Liu et al., 2014; Morton et al., 2015 for indirect evidence for the possible presence of Alfvénic turbulence). For example, Carlsson et al. (2015) find that the width of the chromospheric O I 135.6 nm line implies non-thermal velocities less than 10 km/s, whereas De Pontieu et al. (2015) find that non-thermal line broadening at the transition region (IRIS Si IV) footpoints of coronal loops is limited to about 20 km/s. Such observations place constraints on the amount of unresolved Alfvénic turbulence in the (lower) solar atmosphere, and, so far it is not clear that sufficient power is present to play a significant role in coronal heating.

We note here that we have so far only described wave heating based on dissipation in a fully ionized plasma. Whereas this assumption is most likely valid for the solar corona, in the chromosphere, dissipation due to ion-neutral collisions becomes important (e.g., Khodachenko et al., 2004; Khomenko & Collados, 2012), and several authors have shown this can lead to efficient chromospheric heating (e.g., De Pontieu et al., 2001; Goodman, 2011; Leake et al., 2005; Song & Vasylunas, 2011; Tu & Song, 2013).

Modeling can establish whether wave heating can be efficient enough to obtain viable heating rates, and observations can place limits on the available wave energy flux. However, to assess whether any particular (wave) heating model is indeed operating in the solar atmosphere, observational signatures need to be established through forward modeling. For such observational characteristics to be meaningful, the full thermodynamic response of the plasma (including thermal conduction and radiation) needs to be included in the modeling.

So far, it has not proved easy to establish unique observational signatures associated with wave-based heating mechanisms (see Taroyan & Erdélyi, 2009, for a review of heating diagnostics with MHD waves). For example, it is not clear whether wave-based heating would appear near-continuous or episodic ("bursty") and hence, difficult to distinguish from reconnection-based heating models (e.g., Antolin & Shibata, 2010; De Groof & Goossens, 2002; Klimchuk, 2006; Mendoza-Briceño et al., 2005; Mendoza-Briceño & Erdélyi, 2006; Moriyasu et al., 2004; Ofman et al., 1998). If resonant absorption and/or phase mixing leads to drifting of the heating layers (Cargill et al., 2016; Ofman et al., 1998), heating may appear impulsive (depending on the resolution) and could resemble a nanoflare storm. Indeed, based on 3D reduced-MHD numerical models of Alfvén wave propagation, van Ballegooijen and collaborators (Asgari-Targhi et al., 2013, 2014; van Ballegooien et al., 2011, 2014) found that the spatial and temporal distribution of heating events was similar to the profile of a nanoflare storm. Although these authors pointed out that the wave-based coronal heating rate was dependent on the loop parameters (with an increase in heating rate with coronal field strength and a decrease with loop length), this does not provide robust, observational characteristics that can be used in isolation as it relies on being able to make a direct comparison between different loops (as well as being able to reliably measure the field strength)—a limitation that is likely to be true for any parameterization of heating mechanisms in terms of loop parameters such as field strength or length. On the other hand, investigating the different observational signatures of heating by nanoflares or Alfvén wave dissipation, Antolin et al. (2008, 2010) suggested that coronal rain could provide insight into coronal heating mechanisms. Comparing observed estimated heating profiles with analytically and numerically obtained profiles, Van Doorsselaere et al. (2007) suggest that any wave-heating mechanism should be dominated by a resistive rather than a viscous phenomenon. Taroyan et al. (2007) demonstrated that the power spectrum of Doppler shift oscillations can be used to distinguish between uniform and footpoint heating of coronal loops and to estimate the average energy of a single heating event.

Magnetic Reconnection

The other types of simulations that fall under the heading of this first modeling approach are those that study dissipation during magnetic reconnection. The time-dependent nature of reconnection was hypothesized from the current sheet tearing mode instability (Furth et al., 1963), well before detailed numerical simulations demonstrated their existence. The Parker idea of nanoflares is that photospheric motions that twist and tangle the fields result in the formation of tangential discontinuities, which correspond to currents. Though braided fields are not often observed, elementary magnetic fields are not resolved by current instruments, and, given their aspect ratios, misalignments are not expected to be obvious in existing instrumentation (Wilmot-Smith, 2015). The process by which these field lines relax back into a force-free equilibrium has been the focus of many recent experiments. See Wilmot-Smith (2015) and Cargill et al. (2010) for excellent reviews on reconnection experiments. A general result is that thin current layers form, and a highly dynamic, but statistical steady state is achieved where the Poynting flux and dissipation are coupled on long timescales, but decoupled on short timescales. In the numerical models, dissipation can be both ohmic and viscous, and is bursty and spatially localized.

Moving to 3D theory and numerical simulations of magnetic reconnection provided important new insight. In particular, Hornig and Priest (2003), Priest et al. (2003), and Pontin et al. (2005) demonstrated that it is not possible to follow individual field lines through the diffusion region. Instead, the reconnection in 3D takes a finite volume and amount of time and involves many field lines at any one instant (not just two). The coronal tectonics framework with driving from the magnetic carpet was advanced by Priest et al. (2002), with dissipation at the separatrix surfaces. Currents can form and reconnection can take place both at the central reversal and current sheets, but also separatrices (Priest & Forbes, 2000). These separatrix surfaces occur because of the motion and localization of the magnetic field (Schrijver et al., 1999). Related, quasi-separatrix layers (QSLs; Démoulin et al., 1997; Priest & Démoulin, 1995; Titov et al., 2002) are locations where the magnetic connectivity has steep gradients, rather than being truly discontinuous. Current sheets can also form at QSLs (Titov, 2007), and electric currents may dissipate there as well (Aulanier et al., 2005; Bellan, 2006; Pariat et al., 2009). Simulations have advanced to the point of applying the tectonics framework to loop models of magnetic reconnection. Such simulations model the magnetic field line as straight and occurring between two parallel plates, with velocities applied at the boundaries. There is no flux emergence or cancellation in these models; instead, they investigate how footpoint motions of different types (rotation, shears) result in reconnection and dissipation (De Moortel & Glasgaard, 2006a,b; Mellow et al., 2005; Wilmot-Smith & De Moortel, 2007; Knizhnik et al. 2018, 2019; Leake et al., 2020;).

Boundary shears, or flows oppositely directed across a neutral flow line, were applied first in a loop simulation by van Ballegooijen (1988). They found that shear driving produced QSLs (Titov et al., 2002), locations where magnetic reconnection is likely to occur. Resistive 3D MHD models show that shearing generates electric currents, and the strength of the currents grows exponentially (Galsgaard & Nordlund, 1996). The current sheet fragments, the magnetic structure is complex, and the dissipation is bursty.

Moving on from simple shear motions, others have approximated solar granulation with rotational motions (e.g., Gomez et al., 2000; Longcope et al., 1993; Rapazzo et al., 2007), based on the principle that twisting the photospheric footpoints should produce dissipation at many current sheets (Tucker, 1973). Random circulation patterns at small scales can be linked to energetic heating events occurring with power-law frequency distributions (Knizhnik et al., 2018), and such experiments yield insights into the possible relationship of the coronal heating problem to the helicity injection and condensation processes that shape the large-scale corona (Antiochos, 2013). The kink instability (Hood & Priest, 1979) can also result in line-tied coronal field lines if they are twisted past a certain threshold. Once this threshold is reached and the kink instability sets in, strong current sheets develop, and energy is dissipated by magnetic reconnection (Browning et al., 2008; Hood et al., 2009).

In contrast to these footpoint-driving motions, loop simulation can begin with a braided loop (Longbottom et al. 1998). The conclusion of the braiding experiments is that the nature of the photospheric tangling and flows will impact the resulting heating (Wilmot-Smith et al., 2011). Recent advancements have included braiding with footpoint motions, and have produced models that account for the connection of the corona with the chromosphere and the important mass and energy exchanged between the two, through the transition region. These have resulted in predictions of realistic coronal temperatures that result from the energy dissipation of magnetic field stresses from rotating the magnetic footpoints (Reale et al., 2016) and from random footpoint motions (Dahlburg et al., 2016). Both cases produce EUV emission from hot, multi-million temperature plasma, which is crucial for testing theories with observations, and both find the heating rate in the loops to be highly structured in space and time.

Flux emergence and cancellation is another way for footpoint motions to generate magnetic reconnection (e.g., Parnell & Priest, 1995; Welsch, 2006). Priest et al. (2018) recently expanded the flux cancellation concept analytically in 3D for the case of coronal heating, though flux emergence and cancellation had been studied extensively for larger flares. Syntelis et al. (2019) modeled flux cancellation and showed that hot ejections with temperatures of coronal loops and/or cool ejections can be formed depending on the height of the reconnection. Chen et al. (2014) modeled flux emergence in the formation of a solar AR, and showed that footpoint braiding can be an inherent feature of flux emergence. The simulated emergence-braiding generates currents that dissipate and can heat the corona to temperatures of 1 MK. In a QS example, Meyer et al. (2013) estimated flux cancellation rates using the HMI/ SDO and showed that there is enough energy to maintain the radiative losses of the QS corona, with a lower limit of 8.7×10^4 erg cm^{-2} s^{-1}.

Under wave models, we described the onset of turbulence from waves. MHD turbulence can also be set in motion by slow footpoint motions under the same "DC" regime as magnetic reconnection (e.g., Mikić et al., 1989). The turbulence converts energy to ohmic heating; that is, the dissipation of electric currents (Hendrix & van Hoven, 1996), and again, the results show an intermittent distribution of events in space and time (e.g., Dmitruk et al., 1998; Einaud et al., 1996; Georgoulis et al., 1998; Rappazzo et al., 2008), with well-developed power laws in energy spectra found in the simulations (Georgoulis et al., 1998; Rapazzo et al., 2010), even when driven by simple shear motion (Dahlburg et al., 2009). For a general background on MHD instabilities, turbulence, and magnetic reconnection, see Biskamp (1997); for a review of the impact of turbulence in the solar environment on reconnection and heating, see Matthaeus and Velli (2011); and for a review on MHD of magnetic reconnection in a turbulent media, see Browning and Lazarian (2013). Because particle acceleration may occur with nanoflare reconnection, it is an important aspect to capture in reconnection simulations as well. Particle acceleration is highly dependent on the nature of the current sheet, or current sheet fragments (Cargill et al., 2012), so turbulence may be an important path to accelerating non-thermal particles (Browning & Lazarian, 2013; Vlahos et al., 2016). Baumann and Nordlund (2012) discuss results from particle-in-cell simulations that couple the kinetic scales with MHD scales, which is an important advancement in linking particle acceleration with coronal heating.

Theories of avalanches and self-organized-criticality may also prove important for explaining nanoflare storms. They provide a useful description of larger solar flares (Lu et al., 1993; Vlahos et al., 1995), and extrapolating the application to nanoflares shows promise (Georgoulis et al., 1998). See Charbonneau et al. (2001) for a nice review on avalanche models. Detailed 3D MHD simulations of avalanches show that an individual unstable magnetic strand (elemental field line) can start an avalanche, triggering other strands to go unstable, even when they are below marginal stability (Hood et al., 2016; Tam et al., 2015).

2.3.2. Approach Two: Field-Aligned Hydrodynamic Modeling

We now discuss the second approach in more depth. It has received the most attention through the years.

Our discussion concerns the atmosphere in an individual magnetic flux tube for which the plasma is approximately uniform over the cross section. We refer to this as a strand. It is the fundamental building block of the magnetically closed corona. Observationally distinguishable loops in coronal images are bundles of unresolved strands. The diffuse component of the corona, which contains more plasma and produces more emission than distinct loops (Viall & Klimchuk, 2011), is also made up of strands. Because the plasma beta is large and thermal conduction is highly efficient along the magnetic field but inhibited across it, we can treat each strand as a rigid thermally insulated pipe. Strands are rooted in the photosphere at both ends and are often treated as semicircular. The shape has little influence as long as the maximum height is less than the gravitational scale height, which is usually the case in ARs.

We first consider steady (time-independent) heating. If both the heating and the cross-sectional area are symmetric about the loop midpoint, the loop atmosphere will be in static equilibrium. (An important exception is when the heating is highly concentrated at low coronal altitudes, which we discuss below.) Roughly one third of the energy deposited in the coronal portion is radiated directly to space. The remaining two thirds is thermally conducted down to the transition region at the footpoints and radiated to space from there. Although the transition region is very thin, its emissivity (ergs cm^{-3} s^{-1}) is far stronger than in the corona, so it can accommodate the downward energy flux. The equilibrium corona is thus characterized by a balance between the energy input from coronal heating and cooling from radiation and thermal conduction. The energy balance in the transition region is primarily between heating from thermal conduction and cooling from radiation. There may also be some direct heating that occurs locally, but it is usually thought to be small owing to the thin nature of the transition region. If we define the top of the transition region to be the place where thermal conduction switches from a cooling term above to a heating term below (Klimchuk et al., 2008), then it occurs at a temperature of roughly 60% of the maximum temperature in the strand. If the maximum is 3 MK (where the emission measure distribution peaks in ARs), the transition region will extend to 1.8 MK. Thus, much of the emission from ARs that is normally thought of as "coronal" actually comes from the transition region of the hot coronal loops.

Because both wave and magnetic reconnection heating are unlikely to be steady (Klimchuk, 2006), an important question concerns the frequency with which nanoflares repeat in a given strand. If they repeat on a time scale that is short compared to a cooling time, the state of the plasma is similar to what it would be if the heating were truly steady. This is called high-frequency heating. We now discuss the opposite case of low-frequency heating, where the plasma has time to cool fully between successive events.

When a heating event occurs in a strand that is otherwise experiencing minimal heating, the temperature rises suddenly. It can reach values approaching or even exceeding 10 MK. Thermal conduction is very strong at these elevated temperatures, and the coronal plasma cools quickly. The intense downward conduction flux heats the plasma in the transition region and top of the chromosphere, raising its pressure and causing it to expand rapidly into the corona. This is the well-known process of chromospheric evaporation.

As the coronal temperature falls and its density increases from evaporation, thermal conduction becomes less effective, and radiation takes over as the dominant cooling mechanism. The plasma continues to cool after evaporation ceases completely, and it begins to drain back down to the surface as the pressure support declines. This is known as chromospheric condensation, the inverse process of evaporation. Eventually the strand reaches a final state characterized by low temperature and density.

Because the thermodynamic variables (temperature, pressure, and density) are reasonably uniform along the coronal portion of the strand at any given time, it is meaningful to consider their field-line averages. The evolution of the average values can be well described by considering the effect of enthalpy (the flux of internal energy carried by flowing plasma). During evaporation, the excess of the downward thermal conduction flux compared to transition region radiation drives an upward enthalpy flux. During condensation, the radiation from the transition region exceeds the downward thermal conduction flux, and a downward enthalpy flux makes up the difference. This is the basis of the "0D" hydrodynamics code EBTEL (Cargill, Bradshaw, & Klimchuk, 2012; Klimchuk et al., 2008).

Observations seem most consistent with heating events that occur with a distribution of frequencies and for which the distribution is centered roughly on a frequency corresponding to an inverse cooling time (~1000 s) (Cargill et al. 2015). Some heating events have higher frequency, giving rise to quasi-steady conditions, and others have lower frequency, giving rise to the very hot, though faint, plasma that is often called the smoking gun of nanoflares (Cargill, 1994).

As mentioned above, symmetric steady heating does not always produce an equilibrium state. If it is concentrated at low coronal altitudes, a fascinating situation called thermal non-equilibrium (TNE) can occur (Antiochos & Klimchuk, 1991). With TNE, no solution to the steady-state fluid equations exists. The strand atmosphere is constantly evolving even though the heating is not. The strand is essentially searching for a nonexistent equilibrium, and as it does so it periodically undergoes a thermal runaway, related to, but different from a thermal instability (Klimchuk, 2019). This usually results in the formation of a cold condensation of chromospheric temperature high in the corona. The condensation slides down the loop leg to the solar surface, and the cycle repeats. This is the generally accepted explanation of coronal rain (Antolin et al., 2010; Muller et al., 2004; Schrijver, 2001). If there is a dip in the magnetic field, that is, a region of upward concavity, the condensation may settle in the dip and grow. This is believed to be the process by which prominences form (Antiochos et al., 1999; Karpen et al., 2006; Luna et al., 2012). Long-period (multi-hour) loop pulsations (Auchere et al., 2014; Froment et al., 2015, 2017) and periodic reconnection events at null point topologies (Mason et al., 2019) have also been attributed to TNE. It is possible that TNE is even more widespread than just these relatively isolated phenomena, perhaps playing an important role in the diffuse component of the corona (Downs et al., 2016; Winebarger et al., 2016; Mok et al., 2016).

It has recently been recognized that asymmetries in the heating and/or cross-sectional area affect TNE (Froment et al., 2018; Mikić et al., 2013). If the asymmetries are modest, a cold condensation may not form. A thermal runaway begins, but the plasma reheats before it drops significantly below 1 MK. This is sometimes called incomplete condensation. However, if the asymmetries are larger than roughly a factor of 3, TNE is prevented from occurring altogether (Klimchuk & Luna, 2019). A steady-state end-to-end flow develops instead. The essence of TNE is that heating is too weak at high altitudes to balance the radiation, which is set by the strong heating at low altitudes. If the asymmetry-driven flow is fast enough, a gradient in the enthalpy flux is able to balance the excesses radiation, preventing TNE from occurring.

Before moving on to discuss models of the third type, we point out the existence of what might be called hybrid models. These take information on heating from "MHD-based" models that cannot accurately predict the plasma response and radiation signature, and feed that heating information into the hydrodynamic models that can. Examples include (e.g., Buchlin et al., 2007; Gontikakis et al., 2013; López Fuentes & Klimchuk, 2015; Moraitis, et al., 2016; Parenti et al., 2006; Reale et al., 2005; Walsh & Galtier, 2000).

2.3.3. Approach Three: Global Models of the Solar Atmosphere

One of the more challenging aspects of studying the coronal heating problem stems from the inherently multi-dimensional nature of the corona itself. This is somewhat paradoxical, considering how the dominance of the magnetic field and the great efficiency of field-aligned electron heat conduction has traditionally led to the corona being thought of as a collection of independent, stratified atmospheres collimated along flux tubes. It is precisely this character, however, that ensures a litany of magnetic conditions are simultaneously present in the coronal volume and baked into our remote-sensing diagnostics. From the hot plasma threaded by strong magnetic fields in ARs, to the warm, diffuse plasma in the meandering weak-field regions of the QS, to the colder, outflowing plasma that becomes the solar wind along open flux tubes, each system presents a different set of heating conditions and field geometries, each defining its own unique piece of the coronal heating puzzle.

As mentioned previously, we typically observe coronal plasma via optically thin emission lines. In this case, any intensities measured by a detector represent the integrated emission along the line of sight and thus include contributions from the entire coronal (and transition region) volume along it. Because the governing equations are strongly linked to the magnetic field geometry (loop length, areal expansion, and inclination), changes in the hydrodynamic character of coronal plasmas become particularly important as the magnetic field properties change along a given line of sight or from region to region. In this light, extending our intuition gleaned from analytic theory and single flux tube modeling to the myriad complex and inherently 3D conditions present in the corona is a nontrivial task.

One way to tackle this problem in models is to consider the "realistic," 3D nature of the coronal magnetic field by incorporating information from observations. Typically, this falls into two categories: employing 3D coronal magnetic field extrapolations, or using full 3D magnetohydrodynamic (MHD) computations, the latter of which capture the hydrodynamic and magnetic evolution simultaneously. In either case, such models typically incorporate measurements of the magnetic field in the photosphere as the inner boundary condition, enabling one to study the entire corona (or a subregion) at a particular point in time.

A popular use of magnetic field extrapolations in coronal heating studies has been to test empirical scaling laws of coronal heating, in particular, the relationship between the coronal heating rate and the magnetic field strength, field stratification, and loop length. As illustrated by Mandrini et al. (2000), one can relate the scaling of loop properties in a 3D magnetic field model with the predicted scaling of various coronal heating theories, narrowing the plausible range of theories. Furthermore, simple hydrostatic loop modeling along selected loop bundles can be combined with forward modeling of EUV and soft X-ray emission to test which (if any) heating formulation may be most consistent with observations (e.g., Lundquist et al., 2008; Schrijver et al., 2004). A modern take on this technique is to simulate the hydrodynamic evolution of hundreds to thousands of loop bundles within a given field model using either 1D hydrodynamic models or a more efficient "0D" approach that captures the average evolution of loop plasma in response to a specified heat input (i.e., the EBTEL model; Klimchuk et al., 2008). In this way, the time-dependent evolution of a given subregion of the corona can be studied in the context of realistic geometries and observables, greatly extending our ability to characterize and contextualize coronal heating models (e.g., Bradshaw & Viall, 2016; Nita et al., 2018; Ugarte-Urra et al., 2017).

That being said, solving simplified 0D or 1D hydrodynamic loop models along realistic magnetic loop bundles is not without its disadvantages. Lines of sight passing through the corona will include emissivity contributions over a range of heights and loop bundles; therefore, selecting loop bundles in such a way that they sufficiently sample the variety of loop properties in the 3D volume that contribute to the emissivity is a matter of art. Often, the method used to fill the loop bundles results in empty voxels between filled loops, which is unphysical. At a certain point, if the spatially resolved hydrodynamic evolution is to be simulated accurately for many thousands of loop bundles, it may make more sense to do the calculation in 3D, where the full dynamic range of the magnetic field and volume-filling nature of the corona can be properly accounted for. The utility of doing so is nicely illustrated in Mok et al. (2016), which still uses a realistic magnetic field extrapolation but efficiently solves for the parallel hydrodynamic transport, including gradients and expansion in the magnetic field. In this case, they illustrated how a steady, but stratified coronal heating profile (concentrated at loop footpoints) could lead to rich heating and cooling signatures, exciting TNE, and coronal rain cycles in a significant portion of an AR.

Moving toward increasing sophistication, the standard approach for modeling the coupled hydrodynamic and magnetic evolution of astrophysical plasmas on large, macroscopic scales is to use some variant of the MHD equations in a numerical model. For capturing coronal heating in realistic field contexts, the ideal MHD equations must be extended (at a minimum) to include the relevant thermodynamic terms in the energy equation: coronal heating, optically thin radiative losses, and field-aligned electron heat conduction, a variant sometimes referred to as "thermodynamic" MHD (e.g., Lionello et al., 2009). Solving for these terms allows the plasma density and temperature to be computed with realistic thermodynamic energy balance along various magnetic structures in the corona. Solutions can also be benchmarked by the forward modeling of EUV and X-ray observables (including SDO/AIA bandpasses), allowing the solution to be compared directly with observations. Some more specialized "radiative" MHD models are expressly designed to connect the dynamics of the convection zone directly to the low corona, incorporating prescriptions for radiative transfer in the chromosphere and below (e.g., Abbett, 2007; Gudiksen et al., 2011).

It should be mentioned that solving for electron heat conduction substantially increases the difficulty of a given hydrodynamic or MHD computation. This is because of the stiff, parabolic nature of the term itself and the extremely small conduction length scales that are present in the transition region (dictated by the steep dependence of the conductivity with temperature). Resolving the conduction length scales in the transition region can be essential for describing the hydrodynamics of coronal plasmas (Bradshaw & Cargill, 2013), and this requirement implies strict limitations on the grid size and/or feasibility of multidimensional simulations. Various practical solutions include treating the transition region as an unresolved discontinuity (Johnston et al., 2017), broadening the conduction length scale in a manner that preserves the coronal hydrodynamics (Johnston et al., 2019; Lionello et al., 2009), or using an alternate form of the conduction term itself (Rempel, 2017)

For our purposes, it helps to divide MHD models that study coronal heating further into two broad categories: (a) Local, *ab initio* models that attempt to capture all or some of the physical mechanisms that cause heating on small scales, and (b) global models that capture both the large and small scales of the magnetized corona. The first category includes models that attempt to capture the connection between surface flows in the photosphere (where the plasma β is large) and energy dissipation in the corona. Seminal 3D simulations by Gudiksen and Nordlund (2005) illustrated how footpoint motions of the magnetic field in the photosphere could lead to Joule (resistive) dissipation in the coronal volume, and the formation of plasma structures that resemble the cool loop-like structures observed in EUV. Other simulations have further elucidated the Joule heating mechanism (Bingert & Peter, 2011), as have complementary scenarios involving Alfvénic fluctuations powering the corona and solar wind (De Pontieu et al., 2007). Simulations capturing the complete convection zone to corona connection are also crucial for detailing the way in which the local convective dynamo continually stresses the corona, and how the relevant energies convert and vary as a function of height in the solar atmosphere (Abbett, 2007).

As computational power grows, and methods for approaching this problem become more sophisticated, new details are emerging on exactly how such braiding or

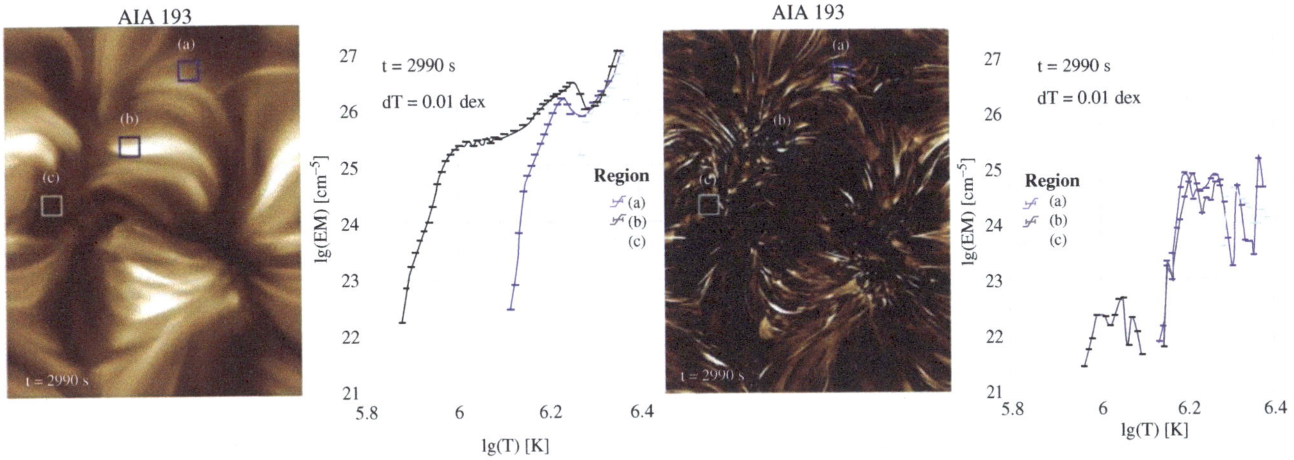

Figure 2.14 High-resolution radiative MHD simulation spanning the convection zone and corona for a 24×24 Mm patch of quiet Sun (from Kanella & Gudiksen, 2019). Here, forward-modeled SDO/AIA 193Å images are computed for the total volume (left) and sub-volumes with clearly identified heating events (right). The emission measure distributions (EM vs. T) are plotted for specific lines of sight. (*Source*: Kanella & Gudiksen, 2019. © 2019.)

"DC" heating mechanisms dissipate energy and how we observe it. For example, recent work by Rempel (2017) shows that when the magnetic Prandtl number of the simulation increases, viscous heating by strong Lorentz force induced flows may dominate direct Joule dissipation. Similarly, forward modeling of line-of-sight integrated plasma emissivities from high-quality, high-resolution simulations (e.g., Kanella & Gudiksen, 2019) are increasingly able to test and challenge our assumptions of how DC heating mechanisms should manifest observationally. An example of results from a high-resolution radiative MHD simulation spanning the convection zone and corona for a 24×24 Mm patch of QS (Kanella & Gudiksen, 2019) is shown in Figure 2.14. The left panel shows the predicted SDO/AIA emission in the 193 Å channel computed over the entire volume, while the right volume shows sub-volmes and identifies individual heating events. The emission measure distributions (EM vs. T) for specific lines-of sight are also shown, illustrating how a background of diffuse multithermal-thermal plasma may conceal individual heating events. In such a setup, Joule heating can be systematically characterized in space and time and connected to the evolution of plasma quantities and observables. In particular, these efforts inform us on how impulsively heated plasma structures in 3D relate to our nominal, simplified notions of loop geometry and cross sections.

The second broad category of MHD models capture the 3D, global volume of the solar corona, and provide a complementary and powerful tool for studying the coronal heating problem. Because it is not yet feasible to simultaneously capture the full volume of the corona ($<1-20\ R_s$) and the scales where dissipation occurs in a single numerical model (<200 km and smaller), global models will generally specify the heating term directly, using either empirical scaling laws (Downs et al., 2010; Lionello et al., 2009) or by solving an additional set of equations that encapsulate the macroscopic effects of an unresolvable heating mechanism (van der Holst et al., 2014; Mikić et al., 2018).

Including the global structure of the corona has a number of key advantages. First, by including the full spherical volume of the solar corona, one naturally captures a large dynamic range of magnetic field strengths, loop geometries, and plasma β regimes. This ensures that a wide parameter space can be explored with a given heating formulation. Assessing the influence of a heating model on all morphological regions in the corona at the same time (ARs, QS, coronal holes) helps build a complete perspective on coronal heating and identify the missing or problematic aspects of a given approach. Second, by including the thermodynamic heating and transport processes, such models naturally provide a realistic background of temperatures and densities in the corona, providing a starting point for studying dynamic, large-scale phenomena such as CMEs, where forward modeling of coronal observables is an essential piece connecting theory to observations (Downs et al., 2012; Jin et al., 2016).

Third, an often-overlooked but important aspect of the coronal heating problem is the way in which the plasma heating and acceleration can directly influence the coronal magnetic field structure and morphology. Although ARs are typically thought of as magnetically dominated, this is not generally true in the extended corona and solar wind. In fact, it is the interplay between

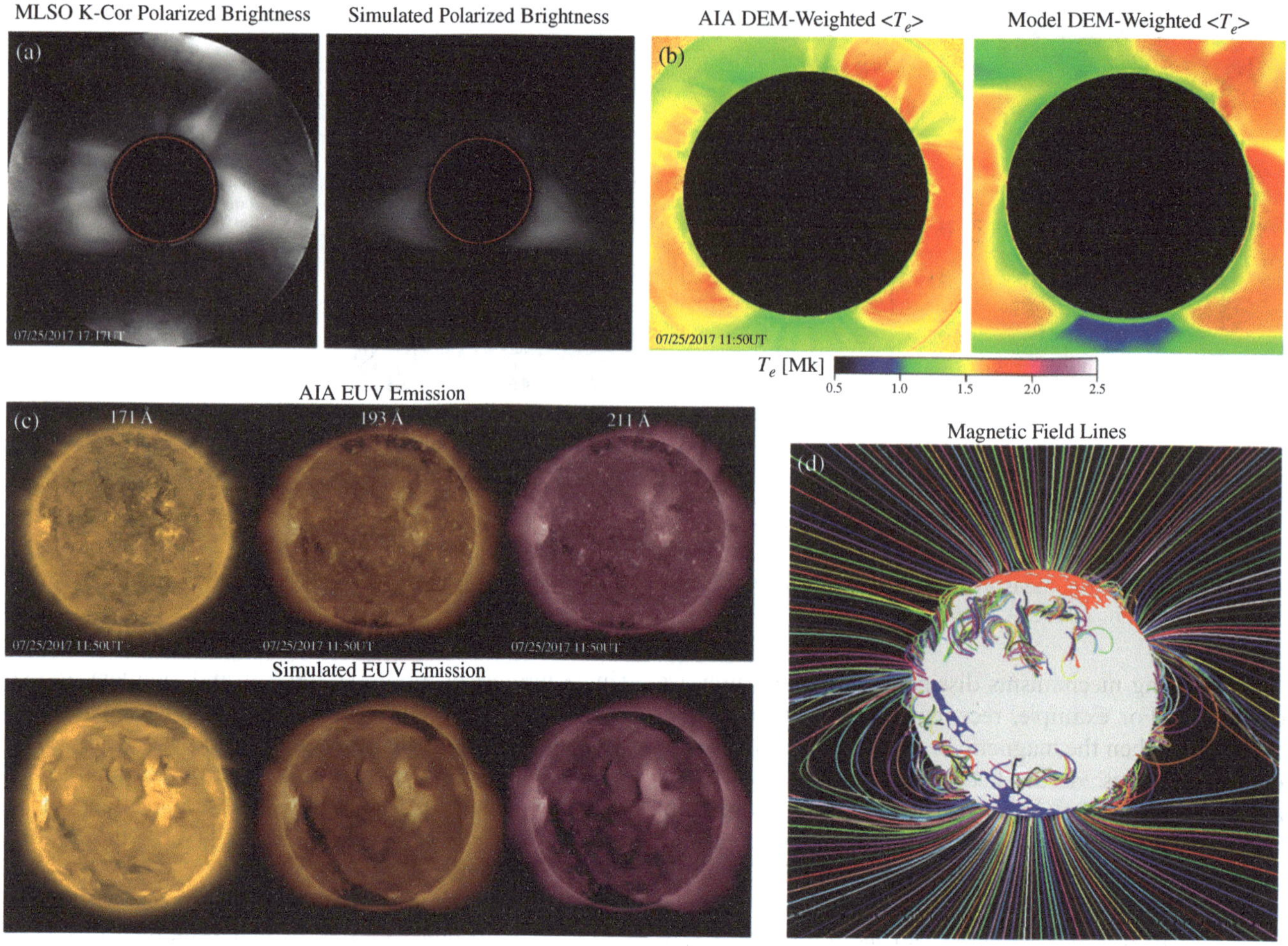

Figure 2.15 Diagnostics from a thermodynamic MHD model of the global corona and their comparison with observations. (a) Radially filtered K-Corona brightness in visible light. (b) Average electron temperatures. (c) EUV emission in three of the SDO/AIA imaging bandpasses. (d) The underlying magnetic field, visualized with traces near the plane of sky and a map of the open (blue/red) and closed (gray) flux at the inner boundary. (*Source*: Mikić et al., 2018.)

the thermal and ram pressures of the heated plasma with the overlying magnetic field that opens the coronal magnetic fields in the first place, creating bipolar helmet streamers and unipolar pseudostreamers (and perhaps continually modulating them; e.g., Mason et al., 2019). Assessing how and where a given heating model can affect the global coronal morphology in EUV (Downs et al., 2010) and white light (Mikić et al., 2018) can provide complementary information and constraints to the overall magnitude and scaling of coronal heating mechanisms. Figure 2.15 shows an example comparing the results of a 3D MHD global model with observations (from Mikić et al., 2018). Panel a shows the predicted radially filtered K-Corona brightness in visible light, which is sensitive to electron density, next to that observed at Mauna Loa Solar Observatory (MLSO). Panel b shows the predicted electron temperatures and those measured by SDO/AIA, determined by fitting EUV intensities off the limb. Panel c shows the observed and predicted EUV emission in three of the SDO/AIA imaging bandpasses. Panel d shows the underlying magnetic field, visualized with traces near the plane of sky and a map of the open (blue/red) and closed (gray) flux at the inner boundary

Along these lines, it is now possible to explore the unification of coronal heating and solar wind acceleration in such models. It has long been postulated that the turbulent fluctuations observed in the solar wind have a low coronal origin (at least in part), and a number of theories exist that link them (see review by Cranmer et al., 2015 and Chapter 1, on the solar wind). There has been a recent trend to develop tractable formalisms that capture the macroscopic propagation, reflection, and dissipation of low-frequency Alfvénic turbulence, which is used to specify coronal heating and solar wind acceleration in global models (Mikić et al., 2018; Usmanov et al., 2018; van der Holst et al., 2014). Although specific implementations vary, the key concept is to leverage the natural scaling of the Poynting flux of waves injected into the corona with the magnitude of the magnetic field at the base of a given flux tube. Nicely articulated by Sokolov et al. (2013), this provides the large dynamic range of heating required for the different morphological regions of the corona, and such formalisms have reasonable scaling properties for both open (Lionello et al., 2014), and closed-field regions of the corona (Downs et al., 2016). Exploring the extent to which phenomenological modeling can capture the basic properties of the corona and solar wind is key for delineating when or where the small-scale or stochastic aspects of coronal heating are (or are not) essential for describing the global state of the corona and inner heliosphere.

Ultimately, whether using extrapolations for hydrodynamic modeling, conducting sophisticated *ab initio* experiments, or simulating the global corona, all of these varied approaches incorporate the realistic multidimensional nature of the coronal magnetic field in some way. Perhaps the most important commonality of these multidimensional approaches is that they enable the forward modeling of coronal observables, and their interpretation in direct observational contexts. This latter aspect is key for characterizing heating mechanisms, and is becoming even more important as our remote-sensing diagnostics become more sensitive and varied.

For example, although static EUV and soft X-ray morphologies have been the primary focus of a number of coronal heating studies, it is now clear that the time dependence of coronal emission may pose an additional set of constraints on how, when, and where coronal heating is deposited. Capturing how certain formalisms may or (may not) explain the predominant heating and cooling cycles observed in line-of-sight integrated emission from ARs (e.g., time lags; Viall & Klimchuk, 2012) is an important next step in vetting competing heating scenarios (Bradshaw & Viall, 2016; Winebarger et al., 2016).

2.4. FUTURE PROSPECTS FOR MODELING AND OBSERVATIONS

The understanding of the heating and energization of the solar atmosphere has come a long way from the initial approximation of a plane parallel atmosphere with three static layers: chromosphere, transition region, and corona. The magnetic field is fundamental to the heating. Spatial and temporal dynamics on multiple scales resulting in impulsive heating seem to be unavoidable. Many improvements in understanding small-scale impulsive heating have been made in the past two decades, but there is plenty of room for further development.

Observations have now clearly established that waves and oscillations are prevalent throughout all structures of the solar atmosphere, and many of the observed oscillations are reported to contain a substantial energy budget, comparable to, for example, the QS heating requirements. However, establishing the actual contribution of the observed waves and oscillations to heating of the solar atmosphere remains an open challenge, as there are still substantial difficulties to be addressed for wave-based heating mechanisms to be viable, particularly in the corona of magnetically closed regions such as AR loops. Likewise, it is clear that magnetic reconnection provides heating to the solar atmosphere, but issues still need to be addressed before definitive conclusions can be made. A more accurate characterization of the very hot emission is also needed, as the hot, >5 MK temperature plasma is a unique indication of magnetic reconnection. The connection between the layers of the atmosphere is crucial, and further quantitative observational constraints on the relation between the chromosphere and corona are needed, as the new high-resolution IRIS results seem to be just the tip of the iceberg. It is clear now that in the highly structured and dynamical solar atmosphere, multiple heating mechanisms will be at work (most likely simultaneously), and the true "coronal heating question" is to determine the relative contribution of different mechanisms in different structures and under different circumstances.

New telescopes and instrumentation that are poised to provide new and/or high-fidelity measurements of the solar atmosphere now or in the near future include Solar Orbiter (February 09, 2020), and the ground-based telescopes Daniel K. Inoue Solar Telescope (DKIST), the European Solar Telescope (EST), and The Atacama Large Millimeter/submillimeter Array (ALMA). The SPICE spectrometer on Solar Orbiter will provide measurements of electron temperatures in the outer corona; DKIST has been designed to make polarimetric measurements of the magnetic field of the off-limb low corona at 0.1 arcsec spatial resolution, which will be an important advancement in the understanding of the AR magnetic configuration; and ALMA primarily probes chromospheric temperatures and heating at high spatial, temporal, and spectral resolution (Wedemeyer et al., 2016).

Coordinating these synergistic observations with each other and with models is crucial for further advancement and understanding. For example, because the DKIST magnetic field diagnostic involves line-of-sight convolutions of the temperature and density as well as the magnetic field strength and/or orientation, interpreting such measurements will increasingly require context provided by models that capture both realistic coronal heating mechanisms and magnetic field geometries.

Solar Orbiter has the unique characteristics of approaching the Sun down to 0.3 AU and moving out of the ecliptic. Measurements of the photospheric magnetic field (instrument PHI), together with the UV-EUV imaging (full sun at medium resolution with EUI/FSI and high resolution with EUI/HRI) and spectroscopic data (SPICE) from a unique point of view will be made. For the first time, ARs will be observed outside of the ecliptic, providing additional information on their topology, magnetic field, and plasma flow directions. Occasionally, depending on the spacecraft–Sun–Earth angle, coordinated observations will be made with other observatories. These include, for instance, DKIST and SDO. Different LOS measurements of the photospheric magnetic field (PHI, DKIST, and SDO/HMI), combined with UV imaging (SDO/AIA and EUI/FSI) will provide important constraints to the 3D geometry of ARs. If still available, STEREO could also contribute with different orbit configurations with Solar Orbiter, and, with Parker Solar Probe, would provide white light images of the corona and solar wind outflow. For the first time, it will be possible to perform stereo-spectroscopy using, if still available, HINODE/EIS and IRIS and DKIST, with SPICE. We could achieve a more precise diagnostics of plasma flow and constrain the line-of-sight effects on the density and temperature measurements.

Looking toward the near future for modeling, the potential unification of models that simultaneously describe heating and solar wind acceleration opens up new ways to study the coronal heating problem. It has been known for some time that the relative charge states, elemental abundances, fluctuation content, and speed distribution of the solar wind (measured in situ by spacecraft) encode key details of the underlying heating mechanism. Ensuring that a given model satisfies not only remote-sensing diagnostics but in situ measurements as well provides tighter constraints on the models (e.g., Oran et al., 2015) and forces us to broaden our perspective beyond the low corona when studying this classic problem. The in situ plasma measurements made with Parker Solar Probe and Solar Orbiter compared with the remote-sensing spectral diagnostics with the SPICE spectrometer on Solar Orbiter will provide important constraints in this way (Del Zanna & Mason, 2018). Likewise, the connection between coronal heating solar wind and particle acceleration could be also probed with spectroscopic observations.

In summary, the heating of the solar atmosphere involves universal, fundamental physical processes such as waves, magnetic reconnection, turbulence, and particle acceleration. In addition, it is important for understanding the solar wind that fills the solar system, and space weather interactions with Earth and other solar system planets, as well as for understanding stellar atmospheres and how they create and modulate the space environment around their exoplanets.

ACKNOWLEDGMENTS

N.M.V and J.A.K. are supported by the NASA Heliophysics Internal Scientist Funding Model. I.D.M. has received support from the European Union Horizon 2020 research and innovation program (grant agreement No. 647214) and the Research Council of Norway through its Centres of Excellence scheme, project number 262622. S.P. acknowledges the funding by CNES through the MEDOC data and operations center. CD was supported by the NASA Heliophysics Supporting Research program, grant 80NSSC18K1129.

REFERENCES

Abbett, W. P. (2007). The magnetic connection between the convection zone and corona in the quiet Sun. *The Astrophysical Journal*, *665*, 1469. doi: 10.1086/519788.

Aletti V., Velli, M., Bocchialini, K., Einaudi, G., Georgoulis, M., & Vial, J.-C. (2000). Microscale structures on the quiet Sun and coronal heating. *The Astrophysical Journal*, *544*, 550.

Alfvén, H. and Lindblad, B. (1947). Granulation, magneto-hydrodynamic waves, and the heating of the solar corona. *Monthly Notices of the Royal Astronomical Society*, *107*, 211–219.

Alissandrakis, C. E., Nindos, A., Patsourakos, S., Kontogeorgos, A., & Tsitsipis, P. (2015). A tiny event producing an interplanetary type III burst. *Astronomy & Astrophysics*, *582*, 52.

Antiochos, S. K. (2013). Helicity condensation as the origin of coronal and solar wind structure. *The Astrophysical Journal*, *772*, 72.

Antiochos, S. K., & Klimchuk, J. A. (1991). A model for the formation of solar prominences. *The Astrophysical Journal*, *378*, 372.

Antiochos, S. K., MacNeice, P. J., Spicer, D. S., & Klimchuk, J. A. (1999). The dynamic formation of prominence condensations. *The Astrophysical Journal*, *512*, 985.

Antiochos, S. K., & Noci, G. (1986). The structure of the static corona and transition region. *The Astrophysical Journal*, *301*, 440.

Antiochos, S. K., Karpen, J. T., DeLuca, E. E., Golub, L., & Hamilton, P. (2003). Constraints on active region coronal heating. *The Astrophysical Journal*, *590*, 547.

Antolin, P., Okamoto, T. J., De Pontieu, B., Uitebroek, H., Van Doorselaere, T., & Yokoyama, T. (2015). Resonant absorption of transverse oscillations and associated heating in a solar prominence. II- Numerical aspects. *The Astrophysical Journal*, *809*, 72. doi: 10.1088/0004-637X/809/1/72.

Antolin, P., & Rouppe van der Voort, L. (2012). Observing the fine structure of loops through high-resolution spectroscopic observations of coronal rain with the CRISP instrument at the Swedish Solar Telescope. *The Astrophysical Journal*, *745*(2), article id. 152, 21 pp.

Antolin, P., & Shibata, K. (2010). The role of torsional Alfvén waves in coronal heating. *The Astrophysical Journal, 712*, 494–510.

Antolin, P., Shibata, K., & Vissers, G. (2010). Coronal rain as a marker for coronal heating mechanisms. *The Astrophysical Journal, 716*, 154.

Antolin, P., & Verwichte, E. (2011). Transverse oscillations of loops with coronal rain observed by Hinode/Solar Optical Telescope. *The Astrophysical Journal, 736*, 121.

Antolin, P., Shibata, K., Kudoh, T., Shiota, D., & Brooks, D. (2008). Predicting observational signatures of coronal heating by Alfvén waves and nanoflares. *The Astrophysical Journal, 688*, 669–682.

Antolin, P., Yokoyama, T., & Van Doorsselaere, T. (2014). Fine strand-like structure in the solar corona from magnetohydrodynamic transverse oscillations. *Astrophysical Journal Letters, 787*, L22.

Antonucci, E., Gabriel, A. H., & Patchett, B. E. (1984). Oscillations in EUV emission lines during a loop brightening. *Solar Physics, 93*, 85–94.

Arber, T. D., Brady, C. S., & Shelyag, S. (2016). Alfvén wave heating of the solar chromosphere: 1.5D models. *The Astrophysical Journal, 817*, 94.

Arregui, I., Oliver, R., & Ballester, J. L. (2012). Prominence oscillations. *Living Reviews in Solar Physics, 9*, 2.

Arregui, I. (2015). Wave heating of the solar atmosphere. *Philosophical Transactions of the Royal Society A, 373*, 20140261.

Aschwanden, M. J. (1987). Theory of radio pulsations in coronal loops. *Solar Physics, 111*, 113–136.

Aschwanden, M. J. (2004). *Physics of the solar corona. An introduction*. Chichester, UK: Praxis Publishing Ltd.

Aschwanden, M. J., Fletcher, L., Schrijver, C. J., & Alexander, D. (1999). Coronal loop oscillations observed with the transition region and coronal explorer. *The Astrophysical Journal, 520*, 880.

Aschwanden, M. J., & Nightingale, R. W. (2005). Elementary loop structures in the solar corona analyzed from TRACE triple-filter images. *The Astrophysical Journal, 633*, 499.

Aschwanden, M. J., & Peter, H. (2017). The width distribution of loops and strands in the solar corona—are we hitting rock bottom? *The Astrophysical Journal, 840*, 4.

Aschwanden, M.J., & Parnell, C.E. (2002). Nanoflare statistics from first principles: Fractal geometry and temperature synthesis. *The Astrophysical Journal, 572*, 1048.

Aschwanden, M. J., & Terradas, J. (2008). The effect of radiative cooling on coronal loop oscillations. *Astrophysical Journal Letters, 686*, L127–L130.

Asgari-Targhi, M., van Ballegooijen, A. A., Cranmer, S. R., & DeLuca, E. E. (2013). The spatial and temporal dependence of coronal heating by Alfvén wave turbulence. *The Astrophysical Journal, 773*, 111.

Asgari-Targhi, M., van Ballegooijen, A. A., & Imada, S. (2014). Comparison of extreme ultraviolet imaging spectrometer observations of solar coronal loops with Alfvén wave turbulence models, *The Astrophysical Journal, 786*, 28.

Auchere, F., Bocchialini, K., Solomon, J., & Tison, E. (2014). Long-period intensity pulsations in the solar corona during activity cycle 23. *Astronomy & Astrophysics, 563*, A8.

Auchère, F., Froment, C., Soubrié, E., Antolin, P., Oliver, R., & Pelouze, G. (2018). The coronal monsoon: Thermal nonequilibrium revealed by periodic coronal rain. *The Astrophysical Journal, 853*, 176.

Banerjee, D., Erdélyi, R., Oliver, R., & O'Shea, E. (2007). Present and future observing trends in atmospheric magnetoseismology. *Solar Physics, 246*, 3.

Banerjee, D., Pérez-Suárez, D., & Doyle, J. G. (2009b). Signatures of Alfvén waves in the polar coronal holes as seen by EIS/Hinode. *Astronomy & Astrophysics, 501*, L15–L18.

Banerjee, D., Gupta, G. R., & Teriaca, L. (2011). Propagating MHD waves in coronal holes. *Space Science Reviews, 158*, 267–288.

Banerjee, D., Teriaca, L., Doyle, J. G., & Wilhelm, K. (1998). Broadening of SI VIII lines observed in the solar polar coronal holes. *Astronomy & Astrophysics, 339*, 208–214.

Banerjee, D., Teriaca, L., Gupta, G. R., Imada, S., Stenborg, G., & Solanki, S. K. (2009a). Propagating waves in polar coronal holes as seen by SUMER & EIS. *Astronomy & Astrophysics, 499*, L29–L32.

Barnes, W.T., Cargill, P.J., & Bradshaw, S.J. (2016). Inference of heating properties from "hot" non-flaring plasmas in active region cores. II. Nanoflare trains. *The Astrophysical Journal, 833*, 217.

Baumann, G., & Nordlund, A. (2012). Particle-in-cell simulation of electron acceleration in solar coronal jets. *The Astrophysical Journal, 759*, 9.

Bel, N., & Leroy, B. (1977). Analytical study of magnetoacoustic gravity waves. *Astronomy & Astrophysics, 55*, 239.

Bemporad, A., & Abbo, L. (2012). Spectroscopic signature of Alfvén waves damping in a polar coronal hole up to 0.4 solar radii. *The Astrophysical Journal, 751*, 110.

Benz, A. O., & Krucker, S. (2002). Energy distribution of microevents in the quiet solar corona. *The Astrophysical Journal, 568*, 413.

Benz, A. O., Grigis, P. C., Csillaghy, A., and Saint-Hilaire, P. (2005) Survey on solar X-ray flares and associated coherent radio emissions. *Solar Physics, 226*, 121. doi: 10.1007/s11207-005-5254-.

Berger, T. E., De Pontieu, B., Schrijver, C. J., & Title, A. M. (1999). High-resolution imaging of the solar chromosphere/corona transition region. *The Astrophysical Journal, 519*, 97.

Berghmans, D., Clette, F., & Moses, D. (1998). Quiet Sun EUV transient brightenings and turbulence. A panoramic view by EIT on board SOHO. *Astronomy and Astrophysics, 336*, 1039.

Berghmans, D., & Clette, F. (1999). Active region EUV transient brightenings–first results by EIT of SOHO JOP 80. *Solar Physics, 186*, 207–229.

Biermann, L. (1946). Zur Deutung der chromosphärischen Turbulenz und des Exzesses der UV-Strahlung der Sonne. *Naturwissenschaften, 33*, 118–119.

Bingert, S., & Peter, H. (2011). Intermittent heating in the solar corona employing a 3D MHD model. *Astronomy & Astrophysics, 530*, A112. doi: 10.1051/0004-6361/201016019.

Bogdan, T. J., & Judge P. G. (2006). Observational aspects of Sunspot oscillations. *Philosophical Transactions of the Royal Society A: Mathematical, Physical and Engineering Sciences, 364*. http://doi.org/10.1098/rsta.2005.1701.

Bonet, J. A., Márquez, I., Sánchez Almeida, J., Cabello, I., & Domingo, V. (2008). Convectively driven vortex flows in the Sun. *The Astrophysical Journal Letters, 687*, L131.

Bouratzis, C., Hillaris, A., Alissandrakis, C. E., Preka-Papadema, P., Moussas, X., Caroubalos, C., et al. (2016). High resolution observations with Artemis-IV and the NRH I. Type IV associated narrow-band bursts. *Astronomy & Astrophysics*, *586*, 29.

Bradshaw, S. J., & Cargill, P.J. (2006). Explosive heating of low-density coronal plasma. *Astronomy & Astrophysics*, *458*, 987.

Bradshaw, S. J. (2009). A numerical tool for the calculation of non-equilibrium ionisation states in the solar corona and other astrophysical plasma environments. *Astronomy & Astrophysics*, *502*, 409.

Bradshaw, S. J., & Cargill, P. J. (2010). The cooling of coronal plasmas. III. Enthalpy transfer as a mechanism for energy loss. *The Astrophysical Journal*, *717*, 163.

Bradshaw, S. J., & Cargill, P. J. (2013). The influence of numerical resolution on coronal density in hydrodynamic models of impulsive heating. *The Astrophysical Journal*, *770*, 12.

Bradshaw S. J., Klimchuk, J. A., & Reep, J.W. (2012). Diagnosing the time-dependence of active region core heating from the emission measure: I. Low-frequency nanoflares. *The Astrophysical Journal 758*, 53.

Bradshaw, S. J., & Viall, N. M. (2016). Patterns of activity in a global model of a solar active region. *The Astrophysical Journal*, *821*, 63. doi: 10.3847/0004-637X/821/1/63.

Brooks, D. H., & Warren, P. (2016). Measurements of non-thermal line widths in solar active regions. *The Astrophysical Journal*, *820*, 63.

Brooks, D. H., Warren, H. P., & Ugarte-Urra, I. (2012). Solar coronal loops resolved by Hinode and the Solar Dynamics Observatory. *Astrophysical Journal Letters*, *755*, 33.

Brosius, J. W., Daw, A. N., & Rabin, D. M. (2014). Pervasive faint Fe XIX emission from a solar active region observed with EUNIS-13: Evidence for nanoflare heating. *The Astrophysical Journal*, *790*, 112.

Brown, J. C., Krucker, S., Güdel, M., & Benz, A. O. (2000). Mechanisms for dynamic coronal mass supply via evaporative solar "micro-events". *Astronomy & Astrophysics*, *359*, 1185.

Browning, P. K., & Priest, E. R. (1984). Kelvin-Helmholtz instability of a phased-mixed Alfven wave. *Astronomy & Astrophysics*, *131*, 283.

Browning, P. K., Gerrard, C., Hood, A. W., Kevis, R., & van der Linden, R. A. M. (2008). Heating the corona by nanoflares: Simulations of energy release triggered by a kink instability. *Astronomy & Astrophysics*, *485*, 837.

Browning, P., & Lazarian, A. (2013). Notes on magnetohydrodynamics of magnetic reconnection in turbulent media. *Space Science Reviews*, *178*, 325.

Bryans, P., Young, P. R., & Doschek, G. A. (2010). Multiple component outflows in an active region observed with the EUV Imaging Spectrometer on Hinode. *The Astrophysical Journal*, *715*, 1012–1020.

Buchlin, E., Vial, J.-C., & Lemaire, P. (2006), A statistical study of SUMER spectral images: events, turbulence, and intermittency. *Astronomy & Astrophysics*, *451*, 1091.

Buchlin, E., Galtier, S., & Velli, M. (2005). Influence of the definition of dissipative events on their statistics. *Astronomy & Astrophysics*, *436*, 355.

Buchlin, E., Cargill, P. J., Bradshaw, S. J., & Velli, M. (2007). Profiles of heating in turbulent coronal magnetic loops. *Astronomy & Astrophysics*, *469*, 347.

Cargill, P. J. (1994). Some implications of the nanoflare concept. *The Astrophysical Journal*, *422*, 381.

Cargill, P. J. (1996). A possible difficulty with a nanoflare-heated corona. *Solar Physics*, *167*, 267.

Cargill, P. J. (2014), Active region emission measure distributions and implications for nanoflare heating. *The Astrophysical Journal*, *784*, 49.

Cargill, P. J., Mariska, J. T., & Antiochos, S.K. (1995), Cooling of solar flares plasmas. 1: Theoretical considerations. *The Astrophysical Journal*, *439*, 1034.

Cargill, P. J., Bradshaw, S. J., & Klimchuk, J. A. (2012). Enthalpy-based thermal evolution of loops. II. Improvements to the model. *The Astrophysical Journal*, *752*, 161.

Cargill, P. J., De Moortel, I., & Kiddie, G. (2016). Coronal density structure and its role in wave damping in loops. *The Astrophysical Journal*, *823*, 31.

Cargill, P., Parnell, C., Browning, P., De Moortel, I., & Hood, A. (2010). Magnetic reconnection in the solar atmosphere: from proposal to paradigm. *Astronomy & Geophysics*, *51*(3), 3.31–3.35. https://doi.org/10.1111/j.1468-4004.2010.51331.x.

Cargill, P. J., Warren, H. P., & Bradshaw, S. J. (2015). Modelling nanoflares in active regions and implications for coronal heating mechanisms. *Philosophical Transactions of the Royal Society A: Mathematical, Physical and Engineering Sciences*, *373*(2042). https://doi.org/10.1098/rsta.2014.0260.

Cargill, P. J., Vlahos, L., Baumann, G., Drake, J. F., & Nordlund, Å. (2012). Current fragmentation and particle acceleration in solar flares. *Space Science Reviews*, *173*, 223.

Carlsson, M., Judge, P. G., & Wilhelm, K. (1997). SUMER observations confirm the dynamic nature of the quiet solar outer atmosphere: The internetwork chromosphere. *The Astrophysical Journal*, *486*, L63–L66.

Carlsson, M., & Stein, R. F. (1992). Non-LTE radiating acoustic shocks and Ca II K2V bright points. *Astrophysical Journal Letters*, *397*, L59.

Carlsson, M., & Stein, R. F. (1997). Formation of solar calcium H and K bright grains. *The Astrophysical Journal*, *481*, 500.

Carlsson, M., Hansteen, V. H., & De Pontieu, B. (2007). Can high frequency acoustic waves heat the quiet Sun chromosphere? *Publications of the Astronomical Society of Japan*, *59*, S663–S668.

Carlsson, M., Leenaarts, J., & De Pontieu, B. (2015). What do IRIS observations of Mg II k tell us about the solar plage chromosphere? *The Astrophysical Journal*, *809*, L30.

Chae, J., Yun, H. S., & Poland, A. I. (1998). Temperature dependence of ultraviolet line average Doppler shifts in the quiet Sun. *The Astrophysical Journal Supplement*, *114*, 151.

Charbonneau, P., McIntosh, S. W., Liu, H.-L., & Bogdan, T. J. (2001). Avalanche models for solar flares. *Solar Physics*, *203*, 321.

Che, H., & Goldstein, M. L. (2014). The origin of non-Maxwellian solar wind electron velocity distribution function: Connection to nanoflares in the solar corona. *The Astrophysical Journal*, *795L*, 38.

Chen, F., Peter, H., Bingert, S., & Cheung, M. C. M. (2014). A model for the formation of the active region corona driven by magnetic flux emergence. *Astronomy & Astrophysics*, *564*, 12.

Chen, B., Bastian, T. S., White, S. M., Gary, D. E., Perley, R., Rupen, M., et al. (2013). Tracing electron beams in the Sun's corona with radio dynamic imaging spectroscopy. *Astrophysical Journal Letters, 763L*, 21.

Chitta, L. P., Kariyappa, R., van Ballegooijen, A. A., DeLuca, E. E., & Solanki, S. K. (2014). Nonlinear force-free field modeling of the solar magnetic carpet and comparison with SDO/HMI and Sunrise/IMaX observations. *The Astrophysical Journal, 793*, 112.

Chitta, L. P, Sukarmadii, A. R. C., Rouppe van der Voort, L., & Peter, H. (2019). Energetics of magnetic transients in a solar active region plage. *Astronomy & Astrophysics, 623*, 176.

Chitta, L. P., Peter, H., Young, P. R. & Huang, Y. -M. (2017a) Compact solar UV burst triggered in a magnetic field with a fan-spine topology. *Astronomy & Astrophysics, 605*, A49, doi :10.1051/0004-6361/201730830.

Chitta, L. P., Peter, H., Solanki, S. K., Barthol, P., Gandorfer, A., Gizon, L., et al. (2017b). Solar coronal loops associated with small-scale mixed polarity surface magnetic fields. *The Astrophysical Journal Supplement Series, 229*, 4. doi:10.3847/1538-4365/229/1/4.

Chitta, L. P., Peter, H., & Solanki, S. K. (2018). Nature of the energy source powering solar coronal loops driven by nano-flares. *Astronomy & Astrophysics, 615*, 9. doi: 10.1051/0004-6361/201833404.

Cirtain, J. W., Golub, L., Lundquist, L., van Ballegooijen, A., Savcheva, A., Shimojo, M., et al. (2007). Evidence for Alfvén waves in solar X-ray jets. *Science, 318*, 1580.

Cranmer, S. R. (2009). Coronal holes. *Living Reviews in Solar Physics, 6*, 3. https://doi.org/10.12942/lrsp-2009-3.

Cranmer, S. R., Asgari-Targhi, M., Miralles, M. P., Raymond, J. C., Strachan, L., Tian, H. et al. (2015). The role of turbulence in coronal heating and solar wind expansion. *Philosophical Transactions of the Royal Society of London Series A, 373*, 20140148. doi: 10.1098/rsta.2014.0148.

Cranmer, S. R., Gibson, S. E., & Riley, P. (2017). Origins of the ambient solar wind: Implications for space weather. *Space Science Reviews, 212*, 1345.

Crosby, N. B., Aschwanden, M. J., & Dennis, B. R. (1993). Frequency distributions and correlations of solar X-ray flare parameters. *Solar Physics, 143*, 275.

Dadashi, N., Teriaca, L., & Solanki, S. K. (2011). The quiet Sun average Doppler shift of coronal lines up to 2 MK. *Astronomy & Astrophysics, 534*, 90.

Dahlburg, R. B., Einaudi, G., Taylor, B. D., Ugarte-Urra, I., Warren, H. P., Rappazzo, A. F., et al. (2016). Observational signatures of coronal loop heating and cooling driven by footpoint shuffling.*The Astrophysical Journal, 817*(1), article id. 47, 15 pp. doi: 10.3847/0004-637X/817/1/47.

Dahlburg, R. B., Liu, J.-H., Klimchuk, J. A., & Nigro, G. (2009). Explosive instability and coronal heating. *The Astrophysical Journal, 704*, 1059.

Daughton, W., Roytershteyn, V., Karimabadi, H., Yin, L., Albright, B. J., Bergen, B., et al. (2011). Role of electron physics in the development of turbulent magnetic reconnection in collisionless plasmas. *Nature Physics, 7*, 539.

DeForest, C. E., & Gurman, J. B. (1998). Observation of quasi-periodic compressive waves in solar polar plumes. *The Astrophysical Journal, 501*, L217.

De Groof, A., & Goossens, M. (2002). Fast and Alfvén waves driven by azimuthal footpoint motions-II. Random driver. *Astronomy & Astrophysics, 386*, 691–698.

Del Zanna, G., & Mason, H. E. (2003). Solar active regions: SOHO/CDS and TRACE observations of quiescent coronal loops. *Astronomy & Astrophysics, 406*, 1089–1103.

Del Zanna, G., & Mason, H. E. (2018). Solar UV and X-ray spectral diagnostics. *Living Reviews in Solar Physics, 15*, 5. https://doi.org/10.1007/s41116-018-0015-3.

De Moortel, I., Antolin, P., & Van Doorsselaere, T. (2015). Observational signatures of waves and flows in the solar corona. *Solar Physics, 290*, 399.

De Moortel, I., & Browning, P. (2015). Recent advances in coronal heating. *Philosophical Transactions of the Royal Society A, 373*, 20140269.

De Moortel, I. (2009). Longitudinal waves in coronal loops. *Space Science Review, 149*, 65–81.

De Moortel, I., & Galsgaard, K. (2006a). Numerical modelling of 3D reconnection due to rotational footpoint motions. *Astronomy & Astrophysics, 451*, 1101–1115. doi:10.1051/0004-6361:20054587.

De Moortel, I., & Galsgaard, K. (2006b). Numerical modelling of 3D reconnection II. Comparison between rotational and spinning footpoint motions. *Astronomy & Astrophysics, 459*, 627–639. doi:10.1051/0004-6361:20065716.

De Moortel, I., Hood, A. W., Ireland, J., & Walsh, R. W. (2002a). Longitudinal intensity oscillations in coronal loops observed with TRACE–II. Discussion of Measured Parameters. *Solar Physics, 209*, 89–108.

De Moortel, I., Ireland, J., & Walsh, R. W. (2000). Observation of oscillations in coronal loops. *Astronomy & Astrophysics, 355*, L23.

De Moortel, I., Ireland, J., Walsh, R. W., & Hood, A. W. (2002b). Longitudinal intensity oscillations in coronal loops observed with TRACE–I. Overview of measured parameters. *Solar Physics, 209*, 61–88.

De Moortel, I., McIntosh, S. W., Threlfall, J., Bethge, C., & Liu J. (2014) Potential evidence for the onset of Alfvénic turbulence in trans-equatorial coronal loops. *The Astrophysical Journal Letters, 782*, L34.

De Moortel, I., & Nakariakov, V. M. (2012). Magnetohydrodynamic waves and coronal seismology: An overview of recent results. *Philosophical Transactions of the Royal Society A, 370*, 3193.

De Moortel, I., & Pascoe, D. (2012). The effects of line-of-sight integration on multistrand coronal loop oscillations. *The Astrophysical Journal Letters, 746*, 31.

De Moortel, I., Pascoe, D. J., Wright, A. N., & Hood, A. W. (2016). Transverse, propagating velocity perturbations in solar coronal loops. *Plasma Physics and Controlled Fusion, 58*, article id 014001.

De Pontieu, B., Berger, T. E., Schrijver, C. J., & Title, A. M. (1999). Dynamics of transition region 'Moss' at high time resolution. *Solar Physics, 190*, 419.

De Pontieu, B., McIntosh, S. W., Carlsson, M., Hansteen, V. H., Tarbell, T. D., Schrijver, C. J. et al. (2007). Chromospheric Alfvénic waves strong enough to power the solar wind. *Science, 318*, 1574–1577.

De Pontieu, B., Carlsson, M., Rouppe van der Voort, L. H. M., Rutten, R. J., Hansteen, V. H., & Watanabe, H. (2012). Ubiquitous torsional motions in type II spicules. *The Astrophysical Journal Letters*, *752*, L12.

De Pontieu, B., De Moortel, I., Martinez-Sykora, J., & McIntosh, S. (2017). Observations and numerical models of solar coronal heating associated with spicules. *Astrophysical Journal Letters 845*, L18.

De Pontieu, B., Martens, P. C. H., & Hudson, H. S. (2001). Chromospheric damping of Alfvén waves. *The Astrophysical Journal*, *558*, 859.

De Pontieu, B., & McIntosh, S. W. (2010). Quasi-periodic propagating signals in the solar corona: The signature of magnetoacoustic waves or high-velocity upflows? *The Astrophysical Journal, 722*, 1013–1029.

De Pontieu, B., McIntosh, S., Martinez-Sykora, J., Peter, H., & Pereira, T. M. D. (2015). Why is non-thermal line broadening of spectral lines in the lower transition region of the Sun independent of spatial resolution? *The Astrophysical Journal*, *799*, L12.

De Pontieu, B., Erdélyi R., & De Moortel, I. (2005). How to channel photospheric oscillations into the corona. *The Astrophysical Journal*, *624*, 61–64.

De Pontieu, B., & Erdélyi, R. (2006). The nature of moss and lower atmospheric seismology. *Philosophical Transactions of the Royal Society A*, *364*, 383.

De Pontieu, B., Title, A., & Carlsson, M. (2014). Probing the solar interface region. *Science*, *346*, D315.

Deubner, F.-L., & Fleck, B. (1989). Dynamics of the solar atmosphere. I - Spatio-temporal analysis of waves in the quiet solar atmosphere. *Astronomy & Astrophysics, 213*, 423–428.

Dmitruk, P., Gómez, D. O., & DeLuca, E. E. (1998). Magnetohydrodynamic turbulence of coronal active regions and the distribution of nanoflares. *The Astrophysical Journal*, *505*, 974.

Doschek, G. A., Mariska, J. T., Warren, H. P., Brown, C. M., Culhane, J. L., Hara, H., et al. (2007). Non-thermal velocities in solar active regions observed with the Extreme-ultraviolet Imaging Spectrometer (EIS) on Hinode. *The Astrophysical Journal, 667*, L109–L112.

Dowdy, J. F., Jr., Rabin, D., & Moore, R. L. (1986). On the magnetic structure of the quiet transition region. *Solar Physics*, *105*, 35.

Downs, C., Lionello, R., Mikic´, Z., Linker, J. A., & Velli, M. (2016). Closed-field coronal heating driven by wave turbulence. *The Astrophysical Journal*, *832*, 180. doi: 10.3847/0004-637X/832/2/180.

Downs, C., Roussev, I. I., van der Holst, B., Lugaz, N., & Sokolov, I. V. (2012). Understanding SDO/AIA observations of the 2010 June 13 EUV wave event: Direct insight from a global thermodynamic MHD simulation. *The Astrophysical Journal*, *750*, 134. doi: 10.1088/0004-637X/750/2/134.

Downs, C., Roussev, I. I., van der Holst, B., et al. (2010). Toward a realistic thermodynamic magnetohydrodynamic model of the global solar corona. *The Astrophysical Journal*, *712*, 1219. doi: 10.1088/0004-637X/712/2/1219.

Dudík, J., Dzifčáková, E., Meyer-Vernet, N., Del Zanna, G., Young, P. R., Giunta, A. et al. (2017). Nonequilibrium processes in the solar corona, transition region, flares, and solar wind. *Solar Physics*, *292*, 100.

Edlén, B. (1943). Die deutung der emissionslinien im spektrum der sonnenkorona. Mit 6 abbildungen. *Zeitschrift für Astrophysik*, *22*, 30–64.

Einaudi, G., Velli, M., Politano, H., & Pouquet, A. (1996). Energy Release in a Turbulent Corona. *The Astrophysical Journal*, *457*, 113.

Erdélyi, R., & Ballai, I. (2007). Heating of the solar and stellar coronae: a review. *Astronomische Nachrichten*, *328*, 726–733.

Erdélyi, R., Doyle, J. G., Perez, M. E., & Wilhelm, K. (1998). Center-to-limb line width measurements of solar chromospheric, transition region and coronal lines. *Astronomy & Astrophysics, 337*, 287.

Fletcher, L. & De Pontieu, B. (1999). Plasma diagnostics of transition region "moss" using SOHO/CDS and TRACE. *The Astrophysical Journal*, *520*, 135.

Fossum, A., & Carlsson, M. (2005). High-frequency acoustic waves are not sufficient to heat the solar chromosphere. *Nature*, *435*, 919.

Froment, C., Auchère, F., Bocchialini, K., Buchlin, E., Guennou, C., & Solomon, J. (2015). Evidence for evaporation-incomplete condensation cycles in warm solar coronal loops. *The Astrophysical Journal*, *807*, 158.

Froment, C., Auchère, F., Aulanier, G., Mikić, Z., Bocchialini, K., Buchlin, E., & Solomon, J. (2017). Long-period intensity pulsations in coronal loops explained by thermal non-equilibrium cycles. *The Astrophysical Journal*, *835*, 272.

Froment, C., Auchère, F., Mikić, Z., Aulanier, G., Bocchialini, K., Buchlin, E., et al. (2018). On the occurrence of thermal nonequilibrium in coronal loops. *The Astrophysical Journal*, *855*, 52.

Furth, H. P., Killeen, J., & Rosenbluth, M. N., (1963). Finite-resistivity instabilities of a sheet pinch. *Physics of Fluids*, *6*, 459.

Gafeira, R., Lagg, A., Solanki, S. K., Jafarzadeh, S., van Noort, M., Barthol, P., et al. (2017). Oscillations on width and intensity of slender Ca II H Fibrils from Sunrise/SuFI. *The Astrophysical Journal Supplement*, *229*, 7.

Galsgaard, K., & Nordlund, Å. (1996). Heating and activity of the solar corona: 1. Boundary shearing of an initially homogeneous magnetic field. *Journal of Geophysical Research*, *101*, 13445.

Georgoulis, M. K., Rust, D. M., Bernasconi, P. N., & Schmieder, B. (2002). Statistics, morphology, and energetics of Ellerman bombs. *The Astrophysical Journal*, *575*, 506.

Georgoulis, M. K., Velli, M., & Einaudi, G. (1998). Statistical properties of magnetic activity in the solar corona. *The Astrophysical Journal*, *497*, 957.

Glesener, L., Krucker, S., Hannah, I. G., Hudson, H., Grefenstette, B. W., White, S. M. et al. (2017). NuSTAR hard X-ray observation of a sub-A class solar flare. *The Astrophysical Journal*, *845*, 2.

Goedbloed, J. P. H., & Poedts, S. (2004). *Principles of Magnetohydrodynamics*. Cambridge, UK: Cambridge University Press.

Gontikakis, C., Patsourakos, S., Efthymiopoulos, C., Anastasiadis, A., & Georgoulis, M. K. (2013). Combining particle acceleration and coronal heating via data-constrained calculations of nanoflares in coronal loops. *The Astrophysical Journal*, *771*, 126.

Goodman, M. L. (2011). Conditions for photospherically driven Alfvénic oscillations to heat the solar chromosphere by Pedersen current dissipation. *The Astrophysical Journal*, *735*, 45.

Goossens M., Andries J., & Arregui I. (2006). Damping of magnetohydrodynamic waves by resonant absorption in the solar atmosphere. *Philosophical Transactions of the Royal Society A*, *364*, 433–446.

Goossens, M., Erdélyi, R., & Ruderman, M. S. (2011). Resonant MHD waves in the solar atmosphere. *Space Science Reviews*, *158*, 289–338.

Goossens, M., Van Doorsselaere, T., Soler, R., & Verth, G. (2013). Energy content and propagation in transverse solar atmospheric waves. *The Astrophysical Journal*, *768*, 191.

Golub, L., Krieger, A.S., & Vaiana, G.S. (1975) Observation of a Non-Uniform Component in the Distribution of Coronal Bright Points. *Solar Physics 42*, 131.

Golub, L., DeLuca, E., Austin, G., Bookbinder, J., Caldwell, D., Cheimets, P., et al. (2007). The X-Ray Telescope (XRT) for the Hinode mission. *Solar Physics*, *243*, 63.

Gošić, M., de la Cruz Rodríguez, J., De Pontieu, B., Bellot Rubio, L. R., Carlsson, M., Esteban Pozuelo, S., et al. (2018). Chromospheric heating due to cancellation of quiet Sun internetwork fields. *The Astrophysical Journal*, *857*, 48.

Grant, S. D. T., Jess, D. B., Moreels, M. G., Morton, R. J., Christian, D. J., Giagkiozis, I., et al. (2015). Wave damping observed in upwardly propagating sausage-mode oscillations contained within a magnetic pore. *The Astrophysical Journal*, *806*, 132.

Grant, S. D. T., Jess, D. B., Zaqarashvili, T. V., Beck, C., Socas-Navarro, H., Aschwanden, M. J., et al. (2018) Alfvén wave dissipation in the solar chromosphere. *Nature Physics*, *14*, 480–483. https://doi.org/10.1038/s41567-018-0058-3.

Grotrian, W. (1939). Zur frage der deutung der linien im spektrum der sonnenkorona. *Naturwissenschaften*, *27*, 214.

Guennou, C., Auchère, F., Klimchuk, J. A., Bocchialini, K., & Parenti, S. (2013), Can the differential emission measure constrain the timescale of energy deposition in the corona?*The Astrophysical Journal* , *774*, 31.

Gudiksen, B. V., Carlsson, M., Hansteen, V. H., Hayek, W., Leenaarts, J., & Martínez-Sykora, J. (2011). The stellar atmosphere simulation code Bifrost. *Astronomy & Astrophysics*, *531*, A154. doi: 10.1051/0004-6361/201116520.

Gudiksen, B. V., & Nordlund, A. (2005). An ab initio approach to the solar coronal heating problem. *The Astrophysical Journal*, *618*, 1020. doi: 10.1086/426063.

Guo, L.-J., Tian, H., & He, J.-S. (2010). Quasi-periodic outflows observed by the X-ray telescope onboard Hinode in the boundary of an active region. *Research in Astronomy and Astrophysics*, *10*, 1307–1314.

Guo, M., Van Doorsselaere, T., Karampelas, K., Li, B., Antolin, P., & De Moortel, I. (2019). Heating effects from driven transverse and Alfvén waves in coronal loops. *The Astrophysical Journal*, *870*, 55.

Hahn, M., Landi, E., & Savin, D. W. (2012). Evidence of wave damping at low heights in a polar coronal hole. *The Astrophysical Journal*, *753*, 36.

Hahn, M., & Savin, D. W. (2013). Observational quantification of the energy dissipated by Alfvén waves in a polar coronal hole: evidence that waves drive the fast solar wind. *The Astrophysical Journal*, *776*, 78.

Hahn, M., & Savin, D. W. (2014). Evidence for wave heating of the quiet-Sun corona. *The Astrophysical Journal*, *795*, 111.

Hahn, M., & Savin, D. W. (2016). Inferring the coronal density irregularity from euv spectra. *The Astrophysical Journal*, *829*, 1.

Hannah, I. G., Hudson, H. S., Battaglia, M., et al. (2011). Microflares and the statistics of X-ray flares. *Space Science Reviews*, *159*, 263. https://doi.org/10.1007/s11214-010-9705-4.

Hannah, I. G., Grefenstette, B. W., Smith, D. M., Glesener, L., Krucker, S., Hudson, H., et al (2016), The first X-ray imaging spectroscopy of quiescent solar active regions with NuSTAR. *The Astrophysical Journal Letters*, *820*, L14.

Hansteen, V. H., De Pontieu, B., Carlsson, M., Lemen, J., Title, A., Boerner, P., et al. (2014). The unresolved fine structure resolved: IRIS observations of the solar transition region. *Science*, *346*, 6207. doi: 10.1126/science.1255757.

Hansteen, V. H., Archontis, V., Pereira, T. M. D., Carlsson, M., Rouppe van der Voort, L., & Leenaarts, J. (2017). Bombs and flares at the surface and lower atmosphere of the Sun. *The Astrophysical Journal*, *839*, 1.

Hara, H., Watanabe, T., Harra, L. K., Culhane, J. L., Young, P. R., Mariska, J. T., et al. (2008). Coronal plasma motions near footpoints of active region loops revealed from spectroscopic observations with Hinode EIS. *The Astrophysical Journal*, *678*, L67–L71.

Harrison, R. A., Hood, A. W., & Pike, C. D. (2002). Off-limb EUV line profiles and the search for wave activity in the low corona. *Astronomy & Astrophysics*, *392*, 319–327.

Hassler, D. M., Rottman, G. J., Shoub, E. C., & Holzer, T. E. (1990). Line broadening of Mg ll609 and 625 coronal emission lines observed above the solar limb. *The Astrophysical Journal*, *348*, L77–L80.

He, J.-S., Marsch, E., Tu, C.-Y., Guo, L.-J., & Tian, H. (2010). Intermittent outflows at the edge of an active region – a possible source of the solar wind? *Astronomy & Astrophysics*, *516*, A14.

Hendrix, D. L., & van Hoven, G. (1996). Magnetohydrodynamic turbulence and implications for solar coronal heating. *The Astrophysical Journal*, *467*, 887.

Heyvaerts, J., & Priest, E. R. (1983). Coronal heating by phase-mixed shear Alfven waves. *Astronomy & Astrophysics*, *117*, 220–234.

Hollweg, J. V. (1978). Alfvén waves in the solar atmosphere. *Solar Physics*, *56*, 305–333.

Hollweg, J., Jackson, S., & Gallaway, D. (1982). Alfvén waves in the solar atmosphere III. Nonlinear waves on open flux tubes. *Solar Physics*, *75*, 35.

Hollweg, J. V., & Sterling, A. C. (1984). Resonant heating - an interpretation of coronal loop data. *Astrophysical Journal Letters*, *282*, L31–L33.

Hollweg, J. V. (1986). Transition region, corona, and solar wind in coronal holes. *Journal of Geophysical Research*, *91*, 4111.

Hollweg, J. V., & Johnson, W. (1988). Transition region, corona, and solar wind in coronal holes: Some two-fluid models. *Journal of Geophysical Research*, *93*, 9547.

Hollweg, J. V., & Yang, G. (1988). Resonance absorption of compressible magnetohydrodynamic waves at thin "surfaces". *Journal of Geophysical Research*, *93*, 5423.

Hood, A. W., Cargill, P. J., Browning, P. K., & Tam, K. V. (2016). An MHD avalanche in a multi-threaded coronal loop. *The Astrophysical Journal*, *817*, 5.

Hood, A. W., & Priest, E. R. (1979). Kink instability of solar coronal loops as the cause of solar flares. *Solar Physics*, *64*, 303.

Hood, A. W., (2009). Coronal heating by magnetic reconnection in loops with zero net current. *Astronomy & Astrophysics*, *506*, 913.

Hornig, G., & Priest, E. R. (2003). Evolution of magnetic flux in an isolated reconnection process. *Physics of Plasmas*, *10*, 2712.

Howson, T. A., De Moortel, I., Antolin, P. (2017b). Energetics of the Kelvin-Helmholtz instability induced by transverse waves in twisted coronal loops. *Astronomy & Astrophysics*, *607*, A77.

Howson, T. A., De Moortel, I., & Antolin, P. (2017a). The effects of resistivity and viscosity on the Kelvin- Helmholtz instability in oscillating coronal loops. *Astronomy & Astrophysics*, *602*, A74.

Huang, Y.-M., & Battacharjee, A. (2016). Turbulent magnetohydrodynamic reconnection mediated by the plasmoid instability. *The Astrophysical Journal*, *818*, 20.

Hudson, H. S. (1991). Solar flares, microflares, nanoflares, and coronal heating. *Solar Physics*, *133*, 357.

Innes, D. E., Cameron, R. H. & Solanki, S. K. (2011). EUV jets, type III radio bursts and sunspot waves investigated using SDO/AIA observations. *Astronomy & Astrophysics*, *531*, L13. doi: 10.1051/0004-6361/201117255.

Innes, D. E., Inhester, B., Axford, W. I., & Wilhelm, K. (1997). Bi-directional plasma jets produced by magnetic reconnection on the Sun. *Nature*, *386*, 811.

Innes, D. E., Guo, L. -J., Huang, Y. -M., & Bhattacharjee, A. (2015). Iris si iv line profiles: An indication for the plasmoid instability during small-scale magnetic reconnection on the Sun. *The Astrophysical Journal*, *813*, 2, DOI 10.1088/0004-637X/813/2/86.

Ionson, J. A. (1978). Resonant absorption of Alfvénic surface waves and the heating of solar coronal loops. *The Astrophysical Journal*, *226*, 650–673.

Ishikawa, S., Glesener, L., Krucker, S., Christe, S., Buitrago-Casas, J. C., Narukage, N., et al. (2017). Detection of nanoflare-heated plasma in the solar corona by the FOXSI-2 sounding rocket. *Nature Astronomy*, *1*, 771.

Ireland, J. R., McAteer, T. J., & Inglis, A. R. (2015). Coronal Fourier power spectra: Implications for coronal seismology and coronal heating. *The Astrophysical Journal*, *798*, 1.

Jafarzadeh, S., Solanki, S. K., Gafeira, R., van Noort, M., Barthol, P., Blanco Rodríguez, J., et al. (2017a). Transverse oscillations in slender Ca II H Fibrils Observed with Sunrise/SuFI. *The Astrophysical Journal Supplement*, *229*, 9.

Jafarzadeh, S. Solanki, S. K., Stangalini, M., Steiner, O., Cameron, R. H., & Danilovic, S. (2017b). High-frequency oscillations in small magnetic elements observed with Sunrise/SuFI. *The Astrophysical Journal Supplement*, *229*, 10.

James, T., & Subramanian, P. (2018). Energetics of small electron acceleration episodes in the solar corona from radio noise storm observations. *Monthly Notices of the Royal Astronomical Society*, *479*, 1603.

Jefferies, S. M., McIntosh, S. W., Armstrong, J. D., Bogdan, T. J., Cacciani, A., & Fleck, B. (2006). Magnetoacoustic portals and the basal heating of the solar chromosphere. *Astrophysical Journal Letters*, *648*, 151–155.

Jess, D. B., Dillon, C. J., Kirk, M. S., Reale, F., Mathioudakis, M., Grant, S. D. T., et al. (2019). Statistical signatures of nanoflare activity. I. Monte Carlo simulations and parameter-space exploration. *The Astrophysical Journal*, *871*, 133.

Jess, D. B., Mathioudakis, M., Erdélyi, R., Crockett, P. J., Keenan, F. P., & Christian, D. J. (2009). Alfvén waves in the lower solar atmosphere. *Science*, *323*, 1582–1585.

Jess, D. B., Morton, R. J., Verth, G., Fedun, V., Grant, S. D. T., & Giagkiozis, I. (2015). Multiwavelength studies of MHD waves in the solar chromosphere. *Space Science Reviews*, *190*, 103–161.

Jess, D. B., Shelyag, S., Mathioudakis, M., Keys, P. H., Christian, D. J., & Keenan, F. P. (2012). Propagating wave phenomena detected in observations and simulations of the lower solar atmosphere. *The Astrophysical Journal*, *746*, 183.

Jin, M., Schrijver, C. J., Cheung, M. C. M., DeRosa, M. L., Nitta, N. V., & Title, A. M. (2016). A numerical study of long-range magnetic impacts during coronal mass ejections. *The Astrophysical Journal*, *820*, 16. doi: 10.3847/0004-637X/820/1/16.

Johnston, C. D., & Bradshaw, S. J. (2019). A fast and accurate method to capture the solar corona/transition region Enthalpy exchange. *Astrophysical Journal Letters*, *873*, L22.

Johnston, C. D., Hood, A. W., Cargill, P. J., & De Moortel, I. (2017). A new approach for modelling chromospheric evaporation in response to enhanced coronal heating - I. The method. *Astronomy & Astrophysics*, *597*, A81.

Judge, P., (2006). Observations of the solar chromosphere. In J. Leibacher, R.F. Stein, & H. Uitenbroek (Eds.), Solar MHD theory and observations: A high spatial resolution perspective (Vol. 354, p. 259). Astronomical Society of the Pacific Conference Series, ASP, San Francisco.

Judge, P. G., Tarbell, T. D., & Wilhelm, K. (2001). A study of chromospheric oscillations using the SOHO and TRACE spacecraft. *The Astrophysical Journal*, *554*, 424–444.

Kanella, C., & Gudiksen, B. V. (2019). Identification of coronal heating events in 3D simulations. *Astronomy & Astrophysics*, *621*, A95. doi: 10.1051/0004-6361/201833634.

Kanoh, R., Shimizu, T., & Imada, S. (2016). Hinode and IRIS observations of the magnetohydrodynamic waves propagating from the photosphere to the chromosphere in a sunspot. *The Astrophysical Journal*, *831*, 24.

Karampelas, K., Van Doorsselaere, T., & Antolin, P. (2017). Heating by transverse waves in simulated coronal loops. *Astronomy & Astrophysics*, *604*, A130.

Karpen, J. T., & Antiochos, S. K. (2008). Condensation formation by impulsive heating in prominences. *The Astrophysical Journal*, *676*, 658.

Karpen, J. T., Antiochos, S. K., & Klimchuk, J. A. (2006). The origin of high-speed motions and threads in prominences. *The Astrophysical Journal*, *637*, 531.

Katsukawa, Y., Berger, T. E., Ichimoto, K., Lites, B. W., Nagata, S., Shimizu, T., et al. (2007). Small-scale jetlike features in penumbral chromospheres. *Science*, *318*, 1594.

Keys, P. H., Morton, R. J., Jess, D. B., Verth, G., Grant, S. D. T., Mathioudakis, M., et al. (2018). Photospheric observations of surface and body modes in solar magnetic pores. *The Astrophysical Journal*, *857*, 28.

Khodachenko, M. L., Arber, T. D., Rucker, H. O., & Hanslmeier, A. (2004). Collisional and viscous damping of MHD waves in partially ionized plasmas of the solar atmosphere. *Astronomy & Astrophysics*, *422*, 1073.

Khomenko, E., & Collados, M. (2012). Heating of the magnetized solar chromosphere by partial ionization effects. *The Astrophysical Journal*, *747*, 87.

Khomenko, E., & Collados, M. (2015). Oscillations and waves in sunspots. *Living Reviews in Solar Physics*, *12*(1), 6. doi10.1007/lrsp-2015-6.

Kitagawa, N., Yokoyama, T., Imada, S., & Hara, H. (2010). Mode identification of MHD waves in an active region observed with Hinode/EIS. *The Astrophysical Journal*, *721*, 744–749.

Klimchuk, J. A. (2000). Cross-sectional properties of coronal loops. *Solar Physics*, *193*, 53.

Klimchuk, J. A. (2006). On solving the coronal heating problem. *Solar Physics*, *234*, 41–77.

Klimchuk, J. A. (1998). Solar jets and coronal plumes. *Proceedings of an International meeting, Guadeloupe*, France, 23–26 February 1998, Publisher: Paris: European Space Agency (ESA), 1998, ESA SP-421. ISBN: 9290926848, p. 233.

Klimchuk, J. A. (2012). The role of type II spicules in the upper solar atmosphere. *Journal of Geophysical Research*, *117*, A12102. doi: 10.1029/2012JA08170.

Klimchuk, J. A. (2015). Key aspects of coronal heating. *Philosophical Transactions of the Royal Society A*, *373*, 20140256. http://dx.doi.org/10.1098/rsta.2014.0256.

Klimchuk, J. A. (2019). The distinction between thermal nonequilibrium and thermal instability. *Solar Physics*, *294*, 173. doi: 10.1007/s11207-019-1562-z.

Klimchuk, J. A., & Hinode Review Team (2019). "Achievements of Hinode in the First Eleven Years (Nanoflare Heating: Observations and Theory)," *Publications of the Astronomical Society of Japan*, *71*(5), R1. doi: 10.1093/pasj/psz084.

Klimchuk, J. A., & Luna, M. (2019). The role of asymmetries in thermal nonequilibrium. *The Astrophysical Journal*, *884*, 68.

Klimchuk, J. A., Patsourakos, S., & Cargill, P. J. (2008). Highly efficient modeling of dynamic coronal loops. *The Astrophysical Journal*, *682*, 1351. doi: 10.1086/589426.

Knizhnik, K. J., Uritsky, V. M., Klimchuk, J. A., & DeVore, C. R. (2018). Power-law statistics of driven reconnection in the magnetically closed corona. *The Astrophysical Journal*, *853*, 82.

Knizhnik, K. J., Antiochos, S. K., Klimchuk, J. A., & DeVore, C. R. (2019). The role of magnetic helicity in coronal heating. *The Astrophysical Journal*, *883*, 26.

Ko, Y.-K., Young, P. R., Muglach, K., Warren, H. P., & Ugarte-Urra, I. (2016). Correlation of coronal plasma properties and solar magnetic field in a decaying active region. *The Astrophysical Journal*, *826*, 126.

Kochhar, R. K. (1991). French astronomers in India during the 17th - 19th centuries. *Journal of the British Astronomical Association*, *101*(2), 95–100.

Krieger, A. S., Timothy, A. F., & Roelof, E. C. (1973). A coronal hole and its identification as the source of a high velocity solar wind stream. *Solar Physics*, *29*, 505.

Krishna Prasad, S., Banerjee, D., & Gupta, G. R. (2011). Propagating intensity disturbances in polar corona as seen from AIA/SDO. *Astronomy & Astrophysics*, *528*, L4.

Krishna Prasad, S., Jess, D. B., Van Doorsselaere, T., Verth, G., Morton, R. J., Fedun, V. et al. (2017). The frequency-dependent damping of slow magnetoacoustic waves in a sunspot umbral atmosphere. *The Astrophysical Journal*, *847*, 5.

Krucker, S., & Benz, A.O. (1998). Energy distribution of heating processes in the quiet solar corona. *Astrophysical Journal Letters*, *501*, L213.

Kudoh, T., & Shibata, K. (1999). Alfvén wave model of spicules and coronal heating. *The Astrophysical Journal*, *514*, 493.

Kuhar, M., Krucker, S., Glesener, L., Hannah, I. G., Grefenstette, B. W., Smith, D. M. et al. (2018). Nustar detection of X-ray heating events in the quiet Sun. *Astrophysical Journal Letters*, *856*, 2.

Kuperus, M. (1969). The Heating of the Solar Corona. *Space Science Reviews*, *9*, 713.

Kuridze, D., Verth, G., Mathioudakis, M., Erdélyi, R., Jess, D. B., Morton, R. J., et al. (2013). Characteristics of transverse waves in chromospheric mottles. *The Astrophysical Journal*, *779*, 82.

Kuridze, D., Morton, R. J., Erdélyi, R., Dorrian, G. D., Mathioudakis, M., Jess, D. B., et al. (2012). Transverse oscillations in chromospheric mottles. *The Astrophysical Journal*, *750*, 51.

Laming, M. J. (2015). The FIP and inverse FIP effects in solar and stellar coronae. *Living Reviews in Solar Physics*, 12(2). doi 10.1007/lrsp-2015-2.

Leake, J. E., Arber, T. D., & Khodachenko, M. L. (2005). Collisional dissipation of Alfvén waves in a partially ionised solar chromosphere. *Astronomy & Astrophysics*, *442*, 1091.

Leake, J. E., Daldorff, L. K. S., & Klimchuk, J. A., (2020), The onset of 3D magnetic reconnection and heating in the solar corona. *The Astrophysical Journal*, *891*, 62.

Lee, M. A., & Roberts, B. (1986). On the behavior of hydromagnetic surface waves. *The Astrophysical Journal*, *301*, 430.

Lenz, D. D., DeLuca, E. E., Golub, L., Rosner, R., & Bookbinder, J. A. (1999). Temperature and emission-measure profiles along long-lived solar coronal loops observed with the transition region and coronal explorer. *The Astrophysical Journal*, *517*, L155.

Lin, R. P., Schwartz, R. A., Kane, S. R., Pelling, R. M., & Hurley, K. C. (1984). Solar hard X-ray microflares. *The Astrophysical Journal*, *283*, 421.

Lionello, R., Linker, J. A., & Mikic´, Z. (2009). Multispectral emission of the Sun during the first whole Sun month: Magnetohydrodynamic simulations. *The Astrophysical Journal*, *690*, 902. doi: 10.1088/0004-637X/690/1/902.

Lionello, R., Velli, M., Downs, C., Linker, J. A., Mikić, Z., & Verdini, A. (2014). Validating a time-dependent turbulence-driven model of the solar wind. *The Astrophysical Journal*, *784*, 120. doi: 10.1088/0004-637X/784/2/120.

Liu, J., McIntosh, S. W., De Moortel, I., Threlfall, J., & Bethge, C. (2014). Statistical evidence for the existence of Alfvénic turbulence in solar coronal loops. *The Astrophysical Journal*, *797*, 7.

Longbottom, A. W., Rickard, G. J., Craig, I. J. D., & Sneyd, A. D. (1998). Magnetic flux braiding: Force-free equilibria and current sheets. *The Astrophysical Journal*, *500*, 471–482. doi:10.1086/305694.

López Fuentes, M., & Klimchuk, J. A. (2015). Two-dimensional cellular automaton model for the evolution of active region coronal plasmas. *The Astrophysical Journal*, *799*, 128.

Lu, E. T., Hamilton, R. J., McTiernan, J. M., & Bromund, K. R. (1993). Solar flares and avalanches in driven dissipative systems. *The Astrophysical Journal*, *412*, 841.

Luna, M., Karpen, J. T., & DeVore, C. R. (2012). Formation and evolution of a multi-threaded solar prominence. *The Astrophysical Journal*, *746*, 30.

Lundquist, L. L., Fisher, G. H., Metcalf, T. R., Leka, K. D., & McTiernan, J. M. (2008). Forward modeling of active region coronal emissions. II. Implications for coronal heating. *The Astrophysical Journal*, *689*, 1388. doi: 10.1086/592760.

Magyar, N., Van Doorsselaere, T., Goossens, M. (2017). Generalized phase mixing: Turbulence-like behaviour from unidirectionally propagating MHD waves. *NatSR*, *7*, 14820.

Mariska, J. T., & Muglach, K. (2010). Doppler-shift, intensity, and density oscillations observed with the extreme ultraviolet imaging spectrometer on Hinode. *The Astrophysical Journal*, *713*, 573–583.

Martens, P. C. H., Kankelborg, C. C., & Berger, T. E. (2000). On the nature of the "Moss" observed by TRACE. *The Astrophysical Journal*, *537*, 471.

Mandrini, C. H., De′moulin, P., & Klimchuk, J. A. (2008). Magnetic field and plasma scaling laws: Their implications for coronal heating models. *The Astrophysical Journal*, *530*, 999. doi: 10.1086/308398.

Marsh, M. S., De Moortel, I., & Walsh, R. W. (2011). Observed damping of the slow magnetoacoustic mode. *The Astrophysical Journal*, *734*, 81.

Marsh, A. J., Smith, D. M., Glesener, L., Klimchuk, J. A., Bradshaw, S. J., Vievering, J., et al. (2018). Hard X-ray constraints on small-scale coronal heating events. *The Astrophysical Journal*, *864*, 5. DOI https://doi.org/10.3847/1538-4357/aad380.

Martinez-Sykora, J., De Pontieu, B., Hansteen. V. H., Rouppe van der Voort, L., Carlsson, M., & Pereira, T. M. D. (2017). On the generation of solar spicules and Alfvénic waves. *Science*, *356*, 1269.

Martinez-Sykora, J., Rouppe van der Voort, L., Carlsson, M., De Pontieu, B., Pereira, T. M. D., Boerner, P., et al. (2015). Internetwork chromospheric bright grains observed with IRIS and SST. *The Astrophysical Journal*, *803*, 44.

Mason, E. I., Antiochos, S. K., and Viall, N. M. (2019). Observations of solar coronal rain in null point topologies. *Astrophysical Journal Letters*, *874*, L33.

Moraitis, K., Toutountzi, A., Isliker, H., Georgoulis, M., Vlahos, L., & Chintzoglou, G. (2016). An observationally-driven kinetic approach to coronal heating. *Astronomy & Astrophysics*, *596*, 56.

Matthaeus, W. H., & Velli, M. (2011). Who needs turbulence? A review of turbulence effects in the heliosphere and on the fundamental process of reconnection. *Space Science Reviews*, *160*, 145.

Matthaeus, W. H., Zank, G. P., Oughton, S., Mullan, D. J., & Dmitruk, P. (1999). Coronal heating by magnetohydrodynamic turbulence driven by reflected low-frequency waves. *The Astrophysical Journal*, *523*, L93–L96.

Mathioudakis, M., Jess, D. B., & Erdélyi, R. (2013). Alfvén waves in the solar atmosphere. *Space Science Reviews*, *175*, 1–27.

Matsumoto, T., & Suzuki, T. K. (2012). Connecting the Sun and the solar wind: The first 2.5-dimensional self-consistent MHD simulation under the Alfvén wave scenario. *The Astrophysical Journal*, *749*, 8.

Matsumoto, T., & Suzuki, T. K. (2014). Connecting the Sun and the solar wind: the self-consistent transition of heating mechanisms. *Monthly Notices of the Royal Astronomical Society*, *440*, 971.

McClements, K. G., Harrison, R. A., & Alexander, D. (1991). The detection of wave activity in the solar corona using UV line spectra. *Solar Physics*, *131*, 41.

McIntosh, S. W., De Pontieu, B., Carlsson, M., Hansteen, V., Boerner, P., & Goossens, M. (2011). Alfvénic waves with sufficient energy to power the quiet solar corona and fast solar wind. *Nature*, *475*, 477–480.

McIntosh, S. W., & De Pontieu, B. (2012). Estimating the "dark" energy content of the solar corona. *The Astrophysical Journal*, *761*, 138.

Mellow, C., Gerrard, C. L., Galsgaard, K., Hood, A. W., & Priest, E. R. (2005). Numerical simulations of the flux tube tectonics model for coronal heating. *Solar Physics*, *227*, 39–60. doi:10.1007/s11207-005-1713-2.

Mendoza-Briceño, C. A., Sigalotti, L. D. G., & Erdélyi, R. (2005). Catastrophic cooling of impulsively heated coronal loops. *The Astrophysical Journal*, *624*, 1080–1092.

Mendoza-Briceño, C. A., & Erdélyi, R. (2006). Intermittent coronal loop oscillations by random energy releases. *The Astrophysical Journal*, *648*, 722–731.

Meyer, K. A., Sabol, J., Mackay, D. H., & van Ballegooijen, A. A. (2013). The storage and dissipation of magnetic energy in the quiet Sun corona determined from SDO/HMI magnetograms. *The Astrophysical Journal*, *770*, 18.

Mikić, Z., Schnack, D. D., & van Hoven, G. (1989). Creation of current filaments in the solar corona. *The Astrophysical Journal*, *338*, 1148.

Mikić, Z., Lionello, R., Mok, Y., Linker, J. A., & Winebarger, A. R. (2013). The importance of geometric effects in coronal loop models. *The Astrophysical Journal*, *773*, 94.

Mikić, Z., Downs, C., Linker, J. A., Caplan, R. M., Mackay, D. H., Upton, L. A., et al. (2018). Predicting the corona for the 21 August 2017 total solar eclipse. *Nature Astronomy*, *2*, 913. doi: 10.1038/s41550-018-0562-5.

Mok, Y., Mikić, Z., Lionello, R., Downs, C., & Linker, J. A. (2016). A three-dimensional model of active region 7986: Comparison of simulations with observations. *The Astrophysical Journal*, *817*, 15. doi: 10.3847/0004-637X/817/1/15.

Moriyasu, S., Kudoh, T., Yokoyama, T., & Shibata, K. (2004). The nonlinear Alfvén wave model for solar coronal heating and nanoflares. *The Astrophysical Journal*, *601*, L107–L110.

Morton, R. J., Erdélyi, R., Jess, D. B., & Mathioudakis, M. (2011). Observations of sausage modes in magnetic pores. *The Astrophysical Journal, 729*, L18.

Morton, R. J., & McLaughlin, J. A. (2013). Hi-C and AIA observations of transverse magnetohydrodynamic waves in active regions. *Astronomy & Astrophysics, 553*, L10.

Morton, R. J., Tomczyk, S., & Pinto, R. F. (2015). Investigating Alfvénic wave propagation in coronal open-field regions. *Nature Communications, 6*, 7813.

Morton, R. J., Tomczyk, S., & Pinto, R. F. (2016). A global view of velocity fluctuations in the corona below 1.3 R ⊙ with CoMP. *The Astrophysical Journal, 828*, 89.

Morton, R. J., Weberg, M. J., & McLaughlin, J. A. (2019). A basal contribution from p-modes to the Alfvénic wave flux in the Sun's corona. *Nature Astronomy, 3*, 223–229.

Morton, R. J., Verth, G., Jess, D. B., Kuridze, D., Ruderman, M. S., Mathioudakis, M. et al. (2012). Observations of ubiquitous compressive waves in the Sun's chromosphere. *Nature Communications, 3*, 1315.

Morton, R. J., Verth, G., Hillier, A., & Erdélyi, R. (2014). The generation and damping of propagating MHD kink waves in the solar atmosphere. *The Astrophysical Journal, 784*, 29.

Müller, D. A. N., Peter, H., & Hansteen, V. H. (2004). Catastrophic cooling and high-speed downflows in solar coronal loops. In A. K. Dupree & A. O. Benz (Eds.), *Stars as Suns: Activity, evolution and planets: Proceedings of the 219th symposium of the International Astronomical Union held during the IAU General Assembly XXV*, Sydney, Australia, 21–25 July 2003. San Francisco, CA: Astronomical Society of the Pacific (ASP).

Müller, D. A. N., De Groof, A., Hansteen, V. H., & Peter, H. (2005). High-speed coronal rain. *Astronomy & Astrophysics, 436*, 1067.

Narain, U., & Ulmschneider, P. (1990). Chromospheric and coronal heating mechanisms. *Space Science Reviews, 54*, 377.

Narain, U., & Ulmschneider, P. (1996). Chromospheric and coronal heating mechanisms II. *Space Science Reviews, 75*, 453.

Nakariakov, V. M., Ofman, L., DeLuca, E. E., Roberts, B., & Davila, J. M. (1999). TRACE observation of damped coronal loop oscillations: Implications for coronal heating. *Science, 285*, 862–864.

Nakariakov, V. M., & Verwichte, E. (2005). Coronal waves and oscillations. *Living Reviews in Solar Physics*, 2(3).

Nita, G. M., Viall, N. M., Klimchuk, J. A., Loukitcheva, M. A., Gary, D. E., Kuznetsov, A. A., et al. (2018). Dressing the coronal magnetic extrapolations of active regions with a parameterized thermal structure. *The Astrophysical Journal, 853*, 66. doi: 10.3847/1538-4357/aaa4bf.

Ofman, L., Davila, J. M., & Steinolfson, R. S. (1994). Nonlinear studies of coronal heating by the resonant absorption of Alfvén waves. *GeoRL, 21*, 2259.

Ofman, L., Davila, J. M., & Steinolfson, R. S. (1995). Coronal heating by the resonant absorption of Alfven waves: Wavenumber scaling laws. *The Astrophysical Journal, 444*, 471–477.

Ofman, L., Klimchuk, J. A., & Davila, J. M. (1998). A self-consistent model for the resonant heating of coronal loops: the effects of coupling with the chromosphere. *The Astrophysical Journal, 493*, 474.

Ofman, L., Romoli, M., Poletto, G., Noci, G., & Kohl, J. L. (1997). Ultraviolet coronagraph spectrometer observations of density fluctuations in the solar wind. *The Astrophysical Journal, 491*, L111.

Okamoto, T. J., Tsuneta, S., Berger, T. E., Ichimoto, K., Katsukawa, Y., Lites, B. W., et al. (2007). Coronal transverse magnetohydrodynamic waves in a solar prominence. *Science, 318*, 1577.

Okamoto, T.J., De Pontieu, B. (2011), Propagating waves along spicules. *The Astrophysical Journal, 736*, 24, doi:10.1088/2041-8205/736/2/L24.

Okamoto, T. J., Antolin, P., De Pontieu, B. Uitenbroek, H., Van Doorsselaere, T., & Yokoyama, T. (2015) Resonant absorption of transverse oscillations and associated heating in a solar prominence. II-Observational aspects. *The Astrophysical Journal, 809*, 71, doi 10.1088/0004-637X/809/1/71.

Oran, R., Landi, E., van der Holst, B., Lepri, S. T., Vásquez, A. M., Nuevo, F. A., et al. (2015). A steady-state picture of solar wind acceleration and charge state composition derived from a global wave-driven MHD model. *The Astrophysical Journal, 806*, 55, doi: 10.1088/0004-637X/806/1/55.

O'Shea, E., Banerjee, D., & Poedts, S. (2003). Variation of coronal line widths on and off the disk. *Astronomy & Astrophysics, 400*, 1065.

Pagano, P., & De Moortel, I. (2017) Contribution of mode-coupling and phase-mixing of Alfvén waves to coronal heating. *Astronomy & Astrophysics, 601*, A107.

Pagano, P., Pascoe, D. J., & De Moortel, I. (2018). Contribution of phase-mixing of Alfvén waves to coronal heating in multi-harmonic loop oscillations. *Astronomy & Astrophysics, 616*, 125.

Pagano, P., & De Moortel, I. (2019). Contribution of observed multi frequency spectrum of Alfvén waves to coronal heating. *Astronomy & Astrophysics, 623*, A37.

Palmieri, L. (1881). La riga dell'Helium apparsa in una recente sublimazione vesuviana [The line of helium appeared in a recently sublimated material [from Mt.] Vesuvius.]. *Rendiconto dell'Accademia delle Scienze Fisiche e Matematiche* (Naples, Italy), *20*, 223, retrieved June 2019.

Parenti, S., del Zanna, G., Petralia, A., Reale, F., Teriaca, L., Testa, P., et al. (2017), Spectroscopy of very hot plasma in non-flaring parts of a solar limb active region: Spatial and temporal properties. *The Astrophysical Journal, 846*, 25.

Parenti, S., Buchlin, E., Cargill, P. J., Galtier, S., & Vial, J.-C. (2006). Modeling the radiative signatures of turbulent heating in coronal loops. *The Astrophysical Journal, 651*, 1219.

Parenti, S., & Young, P. R. (2008), On the ultraviolet signatures of small scale heating in coronal loops. *Astronomy and Astrophysics, 492*, 857. doi: 10.1051/0004-6361:200809928.

Parker, E. N. (1988). Nanoflares and the Solar X-ray corona. *The Astrophysical Journal, 330*, 474.

Parker, E. N. (1991). The phase mixing of Alfven waves, coordinated modes, and coronal heating. *The Astrophysical Journal, 376*, 355.

Parnell, C. E., & De Moortel, I. (2012). A contemporary view of coronal heating. *Philosophical Transactions of the Royal Society A, 370*, 3217–3240.

Parnell, C. E. & Galsgaard, K. (2004) Elementary heating events – magnetic interactions between two flux sources - II. Rates of flux reconnection. *Astronomy and Astrophysics*, *428*, 595. doi: 10.1051/0004-6361:20034350.

Parnell, C. E., & Jupp, P. E. (2000), Statistical analysis of the energy distribution of nanoflares in the quiet Sun. *The Astrophysical Journal*, *529*, 554.

Parnell, C. E. & Priest, E. R. (1995), A converging flux model for the formation of an X-ray bright point above a supergranule cell. *Geophysical and Astrophysical Fluid Dynamics*, *80*, 255 DOI: 10.1080/03091929508228958.

Patsourakos, S., & Klimchuk, J. A.(2006). Nonthermal spectral line broadening and the nanoflare model. *The Astrophysical Journal*, *647*, 1452.

Patsourakos, S., & Klimchuk, J. A. (2009). Spectroscopic observations of hot lines constraining coronal heating in solar active regions. *The Astrophysical Journal*, *696*, 760.

Peter, H. (2010). Asymmetries of solar coronal extreme ultraviolet emission lines. *Astronomy & Astrophysics*, *521*, A51.

Peter, H., & Dwivedi, B. N. (2014). Discovery of the Sun's million-degree hot corona. *Frontiers in Astronomy and Space Sciences*, *1*, 2. doi: 10.3389/fspas.2014.00002.

Peter, H., & Judge, P. G. (1999). On the Doppler shifts of solar ultraviolet emission lines. *The Astrophysical Journal*, *522*, 1148.

Peter, H., Tian, H., Curdt, W., Schmit, D., Innes, D., De Pontieu, B., et al. (2014). Hot explosions in the cool atmosphere of the Sun. *Science*, *346*, 6207. doi: 10.1126/science.1255726.

Peres G., & Vaiana, G. S. (1990) . X-ray observations, scaling laws and magnetic fields. *Memorie della Societa Astronomica Italiana 61*, 401.

Poedts S., Goossens M., & Kerner W. (1989). Numerical simulation of coronal heating by resonant absorption of Alfvén waves. *Solar Physics*, *123*, 83–115.

Poedts, S., Goossens, M., & Kerner, W. (1990). On the efficiency of coronal loop heating by resonant absorption. *The Astrophysical Journal*, *360*, 279–287.

Pontin, D. I., Galsgaard, K., Hornig, G., & Priest, E. R., (2005). A fully magnetohydrodynamic simulation of three-dimensional non-null reconnection. *Physics of Plasmas*, *12*, 052307.

Porter, L. J., Klimchuk, J. A., & Sturrock, P. A. (1994). The possible role of high-frequency waves in heating solar coronal loops. *The Astrophysical Journal*, *435*, 482–501.

Porter, L. J., & Klimchuk, J. A. (1995). Soft X-ray loops and coronal heating. *The Astrophysical Journal*, *454*, 499.

Poletto, G., Vaiana, G. S., Zombeck, M. V., Krieger, A. S., & Timothy, A. F. (1975). a comparison of coronal X-ray structures of ARs with magnetic fields computed from photospheric observations. *Solar Physics, 44*, 83.

Priest, E. R., & De'moulin, P. (1995). Three-dimensional magnetic reconnection without null points: 1. Basic theory of magnetic flipping. *Journal of Geophysical Research*, *100*, 23,443–23,463.

Priest, E. R., Parnell, C. E., & Martin, S. F. (1994). A converging flux model of an X-ray bright point and an associated canceling magnetic feature. *The Astrophysical Journal*, *427*, 459. doi: 10.1086/174157.

Priest, E. R. (1996). Coronal heating by magnetic reconnection. *Astrophysics and Space Science, 237*, 49.

Priest, E. R. (1999). Heating the solar corona by magnetic reconnection. *Astrophysics and Space Science*, *264*, 77.

Priest, E. R., & Forbes, T. G. (2000). *Magnetic Reconnection*. Cambridge University Press.

Priest, E. R., Hornig, G., & Pontin, D. I. (2003). On the nature of three-dimensional magnetic reconnection.*Journal of Geophysical Research*, *108*, 1285.

Priest, E. R., Chitta, L. P., & Syntelis, P. (2018). A cancellation nanoflare model for solar chromospheric and coronal heating. *Astrophysical Journal Letters*, *862*(2), L24.

Priest, E. R., Heyvaerts, J. F., & Title, A. M. (2002). A flux-tube tectonics model for solar coronal heating driven by the magnetic carpet. *The Astrophysical Journal*, *576*, 533–551. doi:10.1086/341539.

Pereira, T. M. D., Rouppe van der Voort, L., Hansteen, V. H., & De Pontieu, B. (2018). Chromospheric counterparts of solar transition region unresolved fine structure loops. *Astronomy & Astrophysics*, *611*, L6, 4.

Rappazzo, A. F., Velli, M., Einaudi, G., & Dahlburg, R. B. (2008). Nonlinear dynamics of the Parker scenario for coronal heating. *The Astrophysical Journal*, *677*, 1348.

Rappazzo, A. F., Velli, M., & Einaudi, G. (2010). Shear photospheric forcing and the origin of turbulence in coronal loops. *The Astrophysical Journal*, *722*, 65.

Rajaguru, S. P., Sangeetha, C. R., & Tripathi, D. (2019). Magnetic fields and the supply of low-frequency acoustic wave energy to the solar chromosphere. *The Astrophysical Journal*, *871*, 155.

Reale, F., & Ciaravella, A. (2006), Analysis of a multi-wavelength time-resolved observation of a coronal loop. *Astronomy and Astrophysics*, *449*, 1177.

Reale, F., Orlando, S., Guarrasi, M., Mignone, A., Peres, G., Hood, A. W., et al. (2016). 3d MHD modeling of twisted coronal loops. *The Astrophysical Journal*, *830*(1), article id. 21, 15 pp.

Reale, F., Parenti, S., Reeves, K. K., Weber, M., Bobra, M. G., Barbera, M., et al. (2007), Fine thermal structure of a coronal active region. *Science*, *318*, 1582.

Reale, F., Testa, P., Klimchuk, J. A., & Parenti, S., (2009b). Evidence of widespread hot plasma in a nonflaring coronal active region from Hinode/X-ray telescope. *The Astrophysical Journal*, *698*, 756.

Reale, F. (2014). Coronal loops: Observations and modeling of confined plasma. *Living Reviews in Solar Physics*, *11*, 4.

Reale, F., McTiernan, J. M., & Testa, P. (2009a). Comparison of Hinode/XRT and RHESSI detection of hot plasma in the non-flaring solar corona. *The Astrophysical Journal Letters*, *704*, L58.

Reale, F., Nigro, G., Malara, F., Peres, G., & Veltri, P. (2005), Modeling a coronal loop heated by magnetohydrodynamic turbulence nanoflares. *The Astrophysical Journal*, *633*, 489.

Reep, J. W., Bradshaw, S. J., & Klimchuk, J. A. (2013). Diagnosing the time dependence of active region core heating from the emission measure. II. nanoflare trains. *The Astrophysical Journal*, *764*, 193.

Reid, A., Mathioudakis, M., Doyle, J. G., Scullion, E., Nelson, C. J., Henriques, V., et al. (2016). Magnetic flux cancellation in Ellerman bombs. *The Astrophysical Journal*, *823*, 110.

Rempel, M. (2017). Extension of the MURaM radiative MHD code for coronal simulations. *The Astrophysical Journal*, *834*, 10. doi: 10.3847/1538-4357/834/1/10.

Robbrecht, E., Verwichte, E., Berghmans, D., Hochedez, J. F., Poedts, S., & Nakariakov, V. M. (2001). Slow magnetoacoustic waves in coronal loops: EIT and TRACE. *Astronomy & Astrophysics*, *370*, 591–601.

Roberts, B., Edwin, P. M., & Benz, A. O. (1984). On coronal oscillations. On coronal oscillations. *The Astrophysical Journal*, *279*, 857–865.

Rosenberg, H. (1970). Evidence for MHD pulsations in the solar corona. *Astronomy & Astrophysics*, *9*, 159.

Rosner, R., Tucker, W. H., & Vaiana, G. S. (1978). Dynamics of the quiescent solar corona. *The Astrophysical Journal*, *220*, 643.

Rouppe van der Voort, L., de Pontieu, B., Pereira, T. M. D., Carlsson, M., & Hansteen, V. (2015). Heating signatures in the disk counterparts of solar spicules in interface region imaging spectrograph observations. *Astrophysical Journal Letters*, *799*, L3.

Rouppe van der Voort, L., De Pontieu, B., Scharmer, G. B., de la Cruz Rodríguez, J., Martínez-Sykora, J., Nóbrega-Siverio, D., et al. (2017). Intermittent reconnection and plasmoids in UV bursts in the low solar atmosphere. *Astrophysical Journal Letters*, *851*, 1.

Rouppe van der Voort, L. H. M., Rutten, R. J. & Vissers, G. J. M. (2016), Reconnection brightenings in the quiet solar photosphere. *Astronomy & Astrophysics*, *592*, 100, doi: 10.1051/0004-6361/201628889.

Ruderman, M. S., & Roberts, B. (2002). The damping of coronal loop oscillations. *The Astrophysical Journal*, *577*, 475.

Ruderman, M. S., & Erdélyi, R. (2009). Transverse oscillations of coronal loops. *Space Science Review*, *149*, 199–228.

Ruderman, M. S., Berghmans, D., Goossens, M., & Poedts, S. (1997a). Direct excitation of resonant torsional Alfven waves by footpoint motions. *Astronomy & Astrophysics*, *320*, 305–318.

Ruderman, M. S., Goossens, M., Ballester, J. L., & Oliver, R. (1997b). Resonant Alfven waves in coronal arcades driven by footpoint motions. *Astronomy & Astrophysics*, *328*, 361–370.

Saint-Hilaire, P., Vilmer, N., & Kerdraon, A. (2013). A decade of solar type III radio bursts observed by the Nançay Radioheliograph 1998-2008. *The Astrophysical Journal*, *762*, 1.

Sakao, T., Kano, R., Narukage, N., Kotoku, J., Bando, T., DeLuca, E. E., et al. (2007). Continuous plasma outflows from the edge of a solar active region as a possible source of solar wind. *Science*, *318*, 1585–1588.

Sakurai, T., Ichimoto, K., Raju, K. P., & Singh, J. (2002). Spectroscopic observation of coronal waves. *Solar Physics*, *209*, 265.

Sakamoto, Y., Tsuneta, S., & Vekstein, G. (2008), Observational appearance of nanoflares with SXT and TRACE. *The Astrophysical Journal*, *689*, 1421.

Schmelz, J. T., Christian, G. M., & Matheny, P. O. (2016). Hot plasma from solar active-region cores: Constraints from the Hinode X-ray telescope. *The Astrophysical Journal*, *833*, 2.

Schmelz, J. T., Scopes, R. T., Cirtain, J. W., Winter, H. D., & Allen, J. D. (2001). Observational constraints on coronal heating models using coronal diagnostics spectrometer and soft X-ray telescope data. *The Astrophysical Journal*, *556*, 896S.

Schrijver, C. J. (2001). Catastrophic cooling and high-speed downflow in quiescent solar coronal loops observed with TRACE. *Solar Physics*, *198*, 325. doi: 10.1023/A:1005211925515.

Schrijver, C. J., Title, A. M., Berger, T. E., Fletcher, L., Hurlburt, N. E., Nightingale, R. W., et al. (1999). A new view of the solar outer atmosphere by the transition region and coronal explorer. *Solar Physics*, *187*, 261.

Schrijver, C. J., Sandman, A. W., Aschwanden, M. J., & De Rosa, M. L. (2004). The coronal heating mechanism as identified by full-sun visualizations. *The Astrophysical Journal*, *615*, 512. doi: 10.1086/424028.

Schwarzschild, M. (1948). On noise arising from the solar granulation. *The Astrophysical Journal*, *107*, 1.

Skogsrud, H., Rouppe van der Voort, L., De Pontieu, B., & Pereira, T. M. D. (2015). On the temporal evolution of spicules observed with IRIS, SDO, and Hinode. *The Astrophysical Journal*, *806*, 170.

Shibata, K., Nakamura, T., Matsumoto, T., Otsuji, K., Okamoto, T. J., Nishizuka, N., et al. (2007). Chromospheric anemone jets as evidence of ubiquitous reconnection. *Science*, *318*, 5856, 1591. https://doi.org/10.1126/science.1146708.

Shibasaki, K., Alissandrakis, C. E., & Pohjolainen, S. (2011). Radio emission of the quiet Sun and active regions. *Solar Physics*, *273*, 309. https://doi.org/10.1007/s11207-011-9788-4.

Smitha, H. N., Anusha, L. S., Solanki, S. K., and Riethmüller, T. L. (2017). Estimation of the Magnetic Flux Emergence Rate in the Quiet Sun from Sunrise Data. *The Astrophysical Journal Supplement Series*, *229*, 17. https://doi.org/10.3847/1538-4365/229/1/17.

Song, P., & Vasyliūnas, V. M. (2011). Heating of the solar atmosphere by strong damping of Alfvén waves. *Journal of Geophysical Research*, *116*, A09104.

Song, D., Chae, J., Kwak, H., Kano, R., Yurchyshyn, V., Moon, Y.-J., et al. (2017). Three-minute sunspot oscillations driven by magnetic reconnection in a light bridge. *Astrophysical Journal Letters*, *850*, 2, L33. doi10.3847/2041-8213/aa9a36.

Sokolov, I. V., van der Holst, B., Oran, R., Downs, C., Roussev, I. I., Jin, M., et al. (2013). Magnetohydrodynamic waves and coronal heating: Unifying empirical and MHD turbulence models. *The Astrophysical Journal*, *764*, 23.

Spitzer, L. (1962). *Physics of Fully Ionized Gases*. New York: Interscience (2nd edition).

Stangalini, M., Giannattasio, F., Erdélyi, R., Jafarzadeh, S., Consolini, G., Criscuoli, S., et al. (2017). Polarized kink waves in magnetic elements: Evidence for chromospheric helical waves. *The Astrophysical Journal*, *840*, 19.

Stein, R. F., & Leibacher, J. (1974). Waves in the solar atmosphere. *Annual Review of Astronomy and Astrophysics*, *12*, 407–435.

Suzuki, T. K., & Inutsuka, S. (2005). Making the corona and the fast solar wind: a self-consistent simulation for the low-frequency Alfvén waves from the photosphere to 0.3 AU. *The Astrophysical Journal, 632*, L49–L52.

Syntelis, P., Priest, E. R., & Chitta, L. P. (2019). A cancellation nanoflare model for solar chromospheric and coronal heating. II. 2D theory and simulations. *The Astrophysical Journal, 872*, 32.

Tam, K. V., Hood, A. W., Browning, P. K., & Cargill, P. J. (2015). Coronal heating in multiple magnetic threads. *Astronomy & Astrophysics, 580*, 122.

Taroyan, Y., Erdélyi, R., Doyle, J. G., & Bradshaw, S. J. (2007). Analysis of power spectra of Doppler shift time series as a diagnostic tool for quiescent coronal loops. *Astronomy & Astrophysics, 462*, 331–340.

Taroyan, Y., & Erdélyi, R. (2009). Heating diagnostics with MHD waves. *Space Science Reviews, 149*, 229–254.

Tataronis, J. A., & Grossman, W. (1973). Decay of MHD waves by phase mixing I. The sheet-pinch in plane geometry. *Zeitschrift für Physik, 261*, 203.

Terradas, J., Andries, J., Goossens, M., Arregui, I., Oliver, R., & Ballester, J. L. (2008). Nonlinear instability of kink oscillations due to shear motions. *Astrophysical Journal Letters, 687*, L115.

Terzo, S., Reale, F., Miceli, M., Klimchuk, J. A., Kano, R., & Tsuneta, S. (2011), Widespread nanoflare variability detected with Hinode/X-ray telescope in a solar active region. *The Astrophysical Journal, 736*, 111.

Testa, P., Reale, F., Landi, E., DeLuca, E. E., & Kashyap, V. (2011), Temperature distribution of a non-flaring active region from simultaneous Hinode XRT and EIS observations. *The Astrophysical Journal, 728*, 30.

Testa, P., De Pontieu, B., Allred, J., Carlsson, M., Reale, F., Daw, A., et al. (2014). Evidence of nonthermal particles in coronal loops heated impulsively by nanoflares. *Science, 346*, 1255724 doi: 10.1126/science.1255724.

Thurgood, J. O., Morton, R. J., & McLaughlin, J. A. (2014). First direct measurements of transverse waves in solar polar plumes using SDO/AIA. *Astrophysical Journal Letters, 790*, L2.

Tian, H., McIntosh, S. W. & De Pontieu, B. (2011). The spectroscopic signature of quasi-periodic upflows in active region timeseries. *The Astrophysical Journal, 727*, L37.

Tian, H., Delica, E. E., Cranmer, S. R., De Pontieu, B., Peter, H. Martinez-Sykora, J., et al. (2014). Prevalence of small-scale jets from the networks of the solar transition region and chromosphere. *Science, 346B*, 315.

Tian, H., Zhu X., Peter, H., Zhao, J., Samanta, T., & Chen, Y. (2018) Magnetic reconnection at the earliest stage of solar flux emergence. *The Astrophysical Journal, 854*, 2, doi10.3847/1538-4357/aaaae6.

Titov V. S. (2007) Generalized squashing factors for covariant description of magnetic connectivity in the solar corona. *The Astrophysical Journal, 660*, 863.

Titov, V. S., Hornig, G., & Démoulin, P. (2002). Theory of magnetic connectivity in the solar corona. *Journal of Geophysical Research, 107*, 1164. doi:10.1029/2001JA000278.

Tomczyk, S., & McIntosh, S. W. (2009). Time-distance seismology of the solar corona with CoMP. *The Astrophysical Journal, 697*, 1384–139.

Tomczyk, S., McIntosh, S. W., Keil, S. L., Judge, P. G., Schad, T., Seeley, D. H., et al. (2007). Alfvén waves in the solar corona. *Science, 317*, 1192.

Tripathi, D., Mason, H. E., Dwivedi, B. N., Del Zanna, G., & Young, P. R. (2009). Active region loops: Hinode/extreme-ultraviolet imaging spectrometer observations. *The Astrophysical Journal, 694*, 1256.

Trottet, G., Pick, M., & Heyvaerts, J. (1979). Gyro-synchrotron modulation in the moving type IV bursts. *Astronomy & Astrophysics, 79*, 164–168.

Trujillo Bueno, J., Štěpán, J., Belluzzi, L., Asensio Ramos, A., Manso Sainz, R., del Pino Alemán, T. et al. (2018). CLASP Constraints on the Magnetization and Geometrical Complexity of the Chromosphere-Corona Transition Region *The Astrophysical Journal Letters, 866*, L15. https://doi.org/10.3847/2041-8213/aae25a.

Tsubaki, T. (1988). Observations of periodic oscillations or waves in the solar corona and prominences. In Solar and stellar coronal structure and dynamics. *Proceedings of the Ninth Sacramento Peak Summer Symposium*, Sunspot, NM, 17–21 August 1987, pp. 140–149. Sunspot, NM: National Solar Observatory.

Tucker, W. H. (1973). Heating of solar active regions by magnetic energy dissipation: The steady-state case. *The Astrophysical Journal, 186*, 285.

Tu, J., & Song, P. (2013). A study of Alfvén wave propagation and heating the chromosphere. *The Astrophysical Journal, 777*, 53.

Uchida, Y. (1970). Diagnosis of coronal magnetic structure by flare-associated hydromagnetic disturbances. *Publications of the Astronomical Society of Japan, 22*, 341–364.

Uchimoto, E., Strauss, H. R., & Lawson, W. S. (1991). Kelvin-Helmholtz instability in an Alfvén resonant layer of a solar coronal loop. *Solar Physics, 134*, 111.

Ugarte-Urra, I., & Warren, H. P. (2012). Is active region core variability age dependent? *The Astrophysical Journal, 761*, 21.

Ugarte-Urra, I., & Warren, H. P. (2011). Temporal variability of active region outflows. *The Astrophysical Journal, 730*, 37.

Ugarte-Urra, I., Warren, H. P., Upton, L. A., & Young, P. R. (2017). Modeling coronal response in decaying active regions with magnetic flux transport and steady heating. *The Astrophysical Journal, 846*, 165. doi: 10.3847/1538-4357/aa8597.

Ugarte-Urra, I., Warren, H. P., & Brooks, D. H. (2009). Active region transition region loop populations and their relationship to the corona. *The Astrophysical Journal, 695*, 642.

Usmanov, A. V., Matthaeus, W. H., Goldstein, M. L., & Chhiber, R. (2018). The steady global corona and solar wind: A three-dimensional MHD simulation with turbulence transport and heating. *The Astrophysical Journal, 865*, 25. doi: 10.3847/1538-4357/aad687.

Vaiana, G. S., Reidy, W. P., Zehnpfennig, T., van Speybroeck, L., Giacconi, R. (1968). X-ray structures of the Sun during the importance 1N Flare of 8 June 1968. *Science 161*, 564.

Vaiana, G. S., Krieger, A. S., Timothy, A. F. (1973). Identification and Analysis of Structures in the Corona from X-Ray Photography. *Solar Physics 32*, 81.

van Ballegooijen, A. A. (1988a). Force free fields and coronal heating Part 1. The formation of current sheets. *Geophysical &*

Astrophysical Fluid Dynamics, *41*, 181–211. doi:10.1080/03091928808208850.

van Ballegooijen, A. A., Asgari-Targhi, M., Cranmer, S. R., & DeLuca, E. E. (2011). Heating of the solar chromosphere and corona by Alfvén wave turbulence. *The Astrophysical Journal*, *736*, 3.

van Ballegooijen A. A., Asgari-Targhi M., & Berger M. A. (2014). On the relationship between photospheric footpoint motions and coronal heating in solar active regions. *The Astrophysical Journal*, *787*, 87.

van Ballegooijen, A. A., Asgari-Targhi, M., & Voss, A. (2017). The heating of solar coronal loops by Alfvén wave turbulence. *The Astrophysical Journal*, *849*, 46.

van der Holst, B., Sokolov, I. V., Meng, X., Jin, M., Manchester, W. B., IV, Tóth, G., & Gombosi, T. I. (2014). Alfvén wave solar model (AWSoM): Coronal heating. *The Astrophysical Journal*, *782*, 81. doi: 10.1088/0004-637X/782/2/81.

Vanspeybroeck, L. P., Krieger, A. S., & Vaiana, G. S. (1970). X-ray photographs of the Sun on March 7, 1970. *Nature*, *227*, 818. https://doi.org/10.1038/227818a0.

Verdini, A., & Velli, M. (2007). Alfvén waves and turbulence in the solar atmosphere and solar wind. *The Astrophysical Journal*, *662*, 669–676.

Vernazza, J. E., Foukal, P. V., Noyes, R. W., Reeves, E. M., Schmahl, E. J., Timothy, J. G., et al. (1975). Time variations in extreme-ultraviolet emission lines and the problem of coronal heating. *The Astrophysical Journal Letters*, *199*, L123–L126.

Verwichte, E., Marsh, M., Foullon, C., Van Doorsselaere, T., De Moortel, I., Hood, A. W., et al. (2010). Periodic spectral line asymmetries in solar coronal structures from slow mangnetoacoustic waves. *The Astrophysical Journal*, *724*, L194–L198.

Viall, N. M., & Klimchuk, J. A. (2011). Patterns of nanoflare storm heating exhibited by an active region observed with Solar Dynamics Observatory/Atmospheric Imaging Assembly. *The Astrophysical Journal*, *738*, 24.

Viall, N. M., & Klimchuk, J. A. (2012). Evidence for widespread cooling in an active region observed with the SDO atmospheric imaging assembly. *The Astrophysical Journal*, *753*, 35. doi: 10.1088/0004-637X/753/1/35.

Viall, N. M., Klimchuk, J. A. (2015). The transition region response to a coronal nanoflare: forward modeling and observations in SDO/AIA. *The Astrophysical Journal*, *799*, 58.

Viall, N. M., & Klimchuk, J. A. (2016). Signatures of steady heating in time lag analysis of coronal emission. *The Astrophysical Journal*, *828*, 76.

Viall, N. M., & Klimchuk, J. A. (2017). A survey of nanoflare properties in active regions observed with the solar dynamics observatory. *The Astrophysical Journal*, *842*, 108.

Vlahos, L., Georgoulis, M., Kluiving, R., & Paschos, P. (1995). The statistical flare. *Astronomy & Astrophysics*, *299*, 897.

Vlahos, L., Pisokas, T., Isliker, H., Tsiolis, V., & Anastasiadis, A. (2016). Particle acceleration and heating by turbulent reconnection. *The Astrophysical Journal*, *827L*, 3.

Vourlidas, A., Klimchuk J. A., Korendyke, C. M., Tarbell, T. D., & Handy, B. N. (2001). On the correlation between coronal and lower transition region structures at arcsecond Scales. *The Astrophysical Journal*, *563*, 374.

Vourlidas, A., Sanchez Andrade-Nuño, B., Landi, E., Patsourakos, S., Teriaca, L., Schühle, U., et al. (2010). The structure and dynamics of the upper chromosphere and lower transition region as revealed by the subarcsecond VAULT observations. *Solar Physics*, *261*, 53. doi: 10.1007/s11207-009-9475-x.

Walsh, R. W., & Ireland, J. (2003). The heating of the solar corona. *The Astronomy and Astrophysics Review*, *12*, 1.

Walsh, R. W., & Galtier, S. (2000). Intermittent heating in a model of solar coronal loops. *Solar Physics*, *197*, 57.

Wang, T. J., Ofman, L., Davila, J. M., & Mariska, J. T. (2009). Hinode/EIS observations of propagating low-frequency slow magnetoacoustic waves in fan-like coronal loops. *Astronomy & Astrophysics*, *503*, L25–L28.

Warren, H. P., & Winebarger, A.R. (2006) Hydrostatic modeling of the integrated soft X-ray and extreme ultraviolet emission in solar active regions. *The Astrophysical Journal*, *645*, 711.

Warren, H. P., Brooks, D. H., & Winebarger, A. R. (2011), Constraints on the heating of high-temperature active region loops: Observations from Hinode and the Solar Dynamics Observatory. *The Astrophysical Journal*, *734*, 90.

Warren, H. P., Ugarte-Urra, I., & Landi, E. (2014), The absolute calibration of the EUV imaging spectrometer on Hinode. *The Astrophysical Journal Supplement Series*, *213*, 11.

Warren, H. P., Winebarger, A. R., & Brooks, D. H. (2012), A systematic survey of high-temperature emission in solar active regions. *The Astrophysical Journal*, *759*, 141.

Warren, H. P., Ugarte-Urra, I., Doschek, G. A., Brooks, D. H., & Williams, D. R. (2008). Observations of active region loops with the EUV imaging spectrometer on Hinode. *The Astrophysical Journal*, *686*, 131.

Wedemeyer, S., Bastian, T., Brajša, R., Hudson, H., Fleishman, G., Loukitcheva, M., et al. (2016). Solar science with the Atacama large millimeter/submillimeter array—A new view of our Sun. *Space Science Reviews*, *200*, 1.

Wedemeyer-Böhm, S., & Rouppe van der Voort, L. (2009). Small-scale swirl events in the quiet Sun chromosphere. *Astronomy & Astrophysics*, *507*, L9.

Wedemeyer-Böhm, S., & Wöger, F. (2008). Small-scale structure and dynamics of the lower solar atmosphere. *In IAU Symposium*, vol. *247*, pp. 66–73.

Wedemeyer-Böhm, S., Lagg, A., & Nordlund, A. (2009). Coupling from the photosphere to the chromosphere and the corona.*Space Science Reviews*, *144*(1), 317. doi10.1007/s11214-008-9447-8.

Wentzel, D. G. (1974). Coronal heating by Alfven waves. *Solar Physics*, *39*, 129–140.

Wentzel, D. G. (1976). Coronal heating by Alfvén waves, II. *Solar Physics*, *50*, 343–360.

Wiegelmann, T., Thalmann, J. K., & Solanki, S. K. (2014). The magnetic field in the solar atmosphere. *The Astronomy and Astrophysics Review*, *22*(1), 78. doi10.1007/s00159-014-0078-7.

Wiegelmann, T., & Sakurai, T. (2012). Solar force-free magnetic fields. *Living Reviews in Solar Physics*, 9(5). https://doi.org/10.12942/lrsp-2012-5.

Wiegelmann, T., Solanki, S. K., Borrero, J. M., Peter, H., Barthol, P., Gandorfer, A., et al. (2013). Evolution of the fine structure of magnetic fields in the quiet Sun: Observations from sunrise/IMaX and extrapolations. *Solar Physics*, *283*, 253.

Wikstøl, Ø., Hansteen, V. H., Carlsson, M., & Judge, P. G. (2000). Chromospheric and transition region internetwork oscillations: A signature of upward-propagating waves. *The Astrophysical Journal*, *531*, 1150–1160.

Wilhelm, K., Dwivedi, B. N., & Teriaca, L. (2004). On the widths of the Mg X lines near 60 nm in the corona. *Astronomy & Astrophysics*, *415*, 1133.

Wilmot-Smith, A. L. (2015). An overview of flux braiding experiments. *Philosophical Transactions of the Royal Society A: Mathematical, Physical and Engineering Sciences. 373*, 20140265, http://doi.org/10.1098/rsta.2014.0265.

Wilmot-Smith, A. L., Pontin, D. I., Yeates, A. R., & Hornig, G. (2011). Heating of braided coronal loops. *Astronomy & Astrophysics*, *536*, A67. doi:10.1051/0004-6361/201117942.

Wilmot-Smith, A. L., & De Moortel, I. (2007). Magnetic reconnection in flux tubes undergoing spinning footpoint motions. *Astronomy & Astrophysics*, *473*, 615–623. doi:10.1051/0004-6361:20077455.

Winebarger, A. R., Lionello, R., Downs, C., Mikić, Z., Linker, J., & Mok, Y. (2016). An investigation of time lag maps using three-dimensional simulations of highly stratified heating. *The Astrophysical Journal*, *831*, 172. doi: 10.3847/0004-637X/831/2/172.

Winebarger, A. R., & Warren, H.P. (2005). Cooling active region loops observed with SXT and TRACE. *The Astrophysical Journal*, *626*, 543.

Withbroe, G. L., & Noyes, R. W. (1977). Mass and energy flow in the solar chromosphere and corona. *Annual Review of Astronomy and Astrophysics*, *15*, 363–387.

Wright, P. J., Hannah, I. G., Grefenstette, B. W., Glesener, L., Krucker, S., Hudson, H. S., et al. (2017). Microflare heating of a solar active region observed with NuSTAR , Hinode / XRT, and SDO /AIA. *The Astrophysical Journal*, *844*, 2.

Yang, S. H., Zhang, J., Jin, C. L., Li, L. P., & Duan, H. Y. (2009). Response of the solar atmosphere to magnetic field evolution in a coronal hole region. *Astronomy & Astrophysics*, *501*(2), 745–753.

Zaqarashvili, T. V. (2003). Observation of coronal loop torsional oscillation. *Astronomy & Astrophysics*, *399*, L15.

Zaqarashvili, T. V., & Erdélyi, R. (2009). Oscillations and waves in solar spicules. *Space Science Reviews*, *149*, 355.

Zirker, J. B. (1993). Coronal heating. *Solar Physics*, *148*, 43.

3
Solar Magnetism and Radiation

Gordon Petrie, Serena Criscuoli, and Luca Bertello

ABSTRACT

The Sun's magnetic field plays a leading role in almost all branches of solar physics, and the release of electromagnetic energy as radiation dominates the solar energy that reaches us on Earth. We have no direct way of measuring the Sun's magnetic field. We make estimates through the analysis of spectral and spectropolarimetric observations typically carried out in absorption lines. We then study the influence of the magnetic field on that radiation at visible, ultraviolet, extreme-ultraviolet, and X-ray wavelengths over different timescales. Here we introduce the Sun's magnetic field and its interrelationship with its radiative output.

3.1. INTRODUCTION

It has often been said that, without its magnetic field, the Sun would be a much less interesting star (e.g., Parker, 1994). It is also true, and essential to life on Earth, that the radiative output of the Sun is what makes our planet habitable. In this chapter, we focus on the connection between the Sun's magnetic field and its radiation. The vast majority of the energy produced in the solar core by nuclear fusion is ultimately transported into space by photons emitted from the surface layers. Without the magnetic field, this energy would be nearly constant. The 11-yr modulation of the solar energy flux is caused by the cyclical pattern of solar magnetic activity.

The solar magnetic field has an approximately 22-year activity cycle, an 11-year sunspot cycle with alternating magnetic polarity, which is believed to be generated by dynamo action of plasma flows in the convection zone. A solar dynamo is a process by which the magnetic field in an electrically conducting solar plasma is maintained against Ohmic dissipation. The magnetic field generated by this dynamo emerges through the photosphere in the form of sunspots and smaller bipolar magnetic regions, and is dispersed and spread across the surface to form an organized structure on a global scale that includes the dipole-like configuration of the polar fields. This large-scale solar field extends to permeate and structure the heliosphere.

Like all astrophysicists, we rely on remote-sensing observations. Magnetic field information is extracted from the solar surface and from the atmospheric layers above it. Information on the physical conditions of the Sun reaches us as light spectra detected by telescopes. To measure the Sun's magnetic field and to study its effects on the Sun's radiative output, we observe light as a function of wavelength and polarization.

The radiative output of the Sun is modulated by the magnetic activity cycle. This variation is particularly strong in the radiation from the upper atmospheric layers, the chromosphere and the corona. In these layers, the magnetic field is dominant in the balance of forces, the dissipation of magnetic energy drives the thermodynamics, and the magnetic field determines the structure of the plasma, which is believed to be entrained by the magnetic field. The plasma structure in these layers is therefore strongly dependent on the activity cycle. Consequently, the intensities at ultraviolet (UV), extreme-ultraviolet

National Solar Observatory, Boulder, Colorado, USA

Space Physics and Aeronomy Collection Volume 1: Solar Physics and Solar Wind, Geophysical Monograph 258, First Edition.
Edited by Nour E. Raouafi and Angelos Vourlidas.

DOI: 10.1002/9781119815600.ch3

(EUV), and radio wavelengths vary by factors in the range 2–20 between solar cycle minimum and maximum, whereas the soft X-ray flux changes by a factor on the order of 100. The frequency-integrated total solar irradiance (TSI) at Earth as measured by bolometric instruments in space also changes in phase with the solar cycle by a factor of order 0.1%.

In this chapter, we review the basic properties of the solar magnetic field and its effect on solar radiation. In Section 3.2, we give a brief history of the measurement of the solar magnetic field, first for sunspots by Hale and later for large-scale weak fields by Babcock, leading to the standard Babcock–Leighton phenomenological model for the solar cycle. We also discuss some remaining challenges of solar magnetic field measurement. In Section 3.3, we describe the key features of the solar photosphere, chromosphere, and corona that contribute to the magnetic variability of the Sun's radiation: sunspots, faculae, filaments, coronal holes, and flares. In Section 3.4, we discuss solar irradiance, the electromagnetic energy reaching 1 AU per unit time and area, and describe observations and modeled reconstructions of its cyclical behavior, as a function of wavelength (spectral solar irradiance) and also integrated over wavelength (TSI). We relate these quantities to the surface magnetism and also discuss other physical mechanisms for their variability. In Section 3.5, we describe a major information source for solar surface magnetism and irradiance variability throughout the 20th century, namely, Ca II K observations (spectroheliograms and disk-integrated spectra). The numerous spectroheliogram archives, in particular those from Mt. Wilson and Kodaikanal, together describe about ten cycles of solar activity.

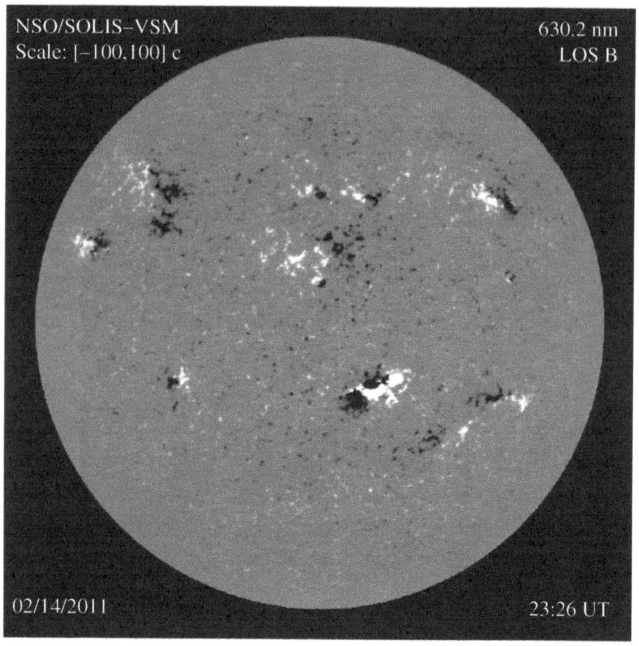

Figure 3.1 Full-disk image of the photospheric longitudinal (line-of-sight) magnetic field taken on 14 February 2011 by the SOLIS/VSM instrument. The gray scale indicates a longitudinal magnetic flux density of positive (light) and negative (dark) polarity, saturated at ±100 G. North is directly upward. A large bipolar active region is located near the disk center, and smaller bipoles are distributed over low latitudes, between +30° and −30°, and decayed active-region magnetic flux is spread over larger areas. (*Source*: Gordon Petrie.)

In the following sections, we describe how these fields are measured and how they are organized in the manner observed.

3.2. THE SOLAR MAGNETIC FIELD

Figure 3.1 shows a full-disk image of the photospheric longitudinal (line-of-sight) magnetic field component on a typical day of moderate solar activity. This image shows the global distribution of the surface field into compact magnetic bipoles approximately aligned with the equator and confined to low latitudes (between about ±30° and sometimes referred to as active latitudes), and large-scale weak fields covering the surface at all latitudes, composed of small-scale concentrations of magnetic flux.

Figure 3.2 shows a high-resolution image of this longitudinal field component, together with the transverse field component and the continuum emission intensity. This high-resolution image shows the weak fields in detail: their mixed magnetic polarity, their arrangement in small flux concentrations of different sizes, and their relationship to the convection patterns of the photosphere as shown in the continuum intensity image.

3.2.1. Measurement of Solar Magnetic Field

The Zeeman (1897) effect enables us to understand the splitting of spectral lines and their polarization by the magnetic field, and to use this effect to measure the magnetic field. Solar magnetograph instruments measure the solar magnetic field via spectropolarimetry, that is, the measure of light polarization as a function of wavelength. An atom, represented by a dipole oscillator, absorbs light near its oscillation frequency μ_0. When the atom is placed in a magnetic field, the motion of the electron along the field is not affected, but in the plane perpendicular to the field the electron precesses at the Larmor frequency ω, counterclockwise as viewed in the direction of the field. The motion of the precessing oscillator is the superposition of two oppositely directed circular motions, a left-handed motion with frequency $\mu_0 + \omega$ and a right-handed motion with frequency $\mu_0 - \omega$, known as σ components. When we observe the atom from the direction of the

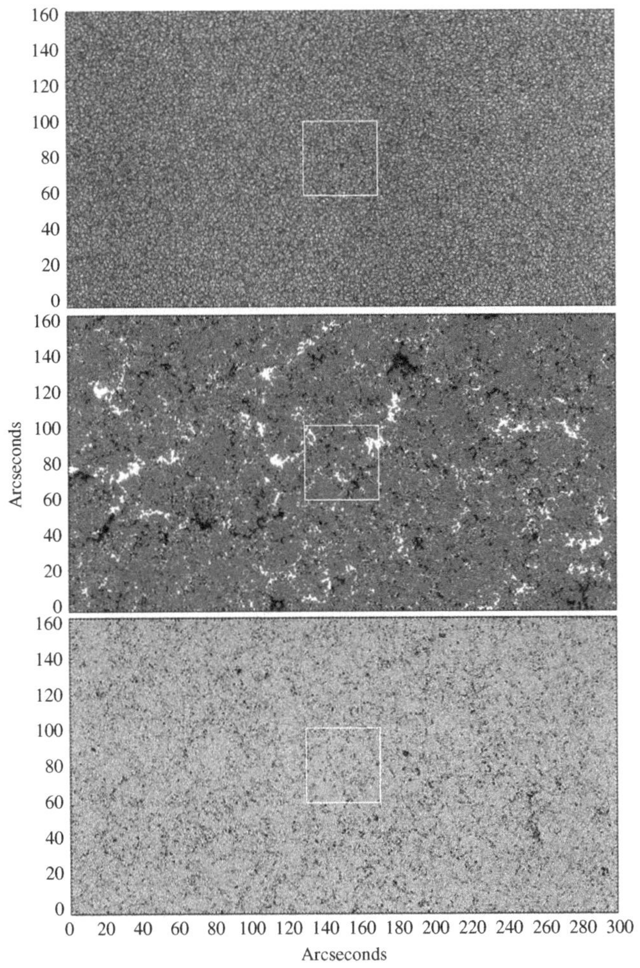

Figure 3.2 Quiet Sun at the disk center as observed by the *Hinode* SOT-SP on 10 March 2007 between 11:37 and 14:37 UT. The top panel shows a continuum intensity map at 630 nm. The other panels display the corresponding maps of (middle) longitudinal magnetic flux density and (bottom) transverse flux density. They were derived from the circular and linear polarization signals observed in the Fe I 630 nm line pair. The grayscales saturate at ±50 G and ±200 G, respectively. (*Source*: Images adapted from Lites et al., 2008; by Bellot Rubio and Orozco Suárez, 2019.)

magnetic field, the left-handed component, which has the shorter wavelength, absorbs the left-handed circularly polarized light selectively. The original unpolarized light therefore becomes right-handed circularly polarized on passing through the plasma. The opposite happens with the longer-wavelength component. Observed from a direction perpendicular to the magnetic field, these two σ components absorb the light whose electric vector is in the plane perpendicular to the magnetic field. Thus, the light also becomes linearly polarized on passing through the solar plasma. Therefore, circular polarization is observed along the field, and linear polarization is observed perpendicular to the field. Magnetic fields in the solar photosphere are measured using this polarizing effect of magnetic fields on the solar radiation field. A magnetograph measures this polarization, which is described in terms of the Stokes parameters I, Q, U, and V. Stokes Q and U represent the linear polarization and V the circular polarization. Stokes I is the total intensity satisfying $I^2 \geq Q^2 + U^2 + V^2$ with equality when the light is fully polarized. Although this principle is common to all magnetographs, there are very great differences in the details of how the polarization measurements are sampled in space, time, and wavelength with different instruments.

Hale (1908) was the first to take advantage of the opportunity presented by the discovery of the Zeeman effect to measure the magnetic field in a sunspot, the first measurement of a magnetic field outside Earth. This application then led to his discovery (Hale et al., 1919) of Hale's sunspot polarity law governing the east–west orientation of the polarities of sunspot pairs, which is opposite in each solar (north/south) hemisphere and alternates with every new 11-year cycle. Hale's law is illustrated by the fields in Figure 3.1, bearing in mind that not all active regions are simple bipoles. The Hale cycle for the solar magnetic field is therefore 22 yr, twice the length of the sunspot cycle.

Another law reported by Hale et al. (1919) is Joy's law for sunspot tilt: the sunspot pairs appear tilted systematically (in a statistical sense) with respect to the equator such that the leading sunspot of the pair, in terms of the direction of solar rotation, is closer than the trailing sunspot to the equator. This pattern is also evident in Figure 3.1. These observational laws serve as fundamental constraints on dynamo theory.

Magnetic fields outside sunspots are too weak to measure with the technology that was available to Hale. Their measurement had to wait until the introduction of the photoelectric magnetograph by Babcock (1953). Instead of manually measuring wavelength positions of spectral lines on photographic plates, photomultipliers served as detectors, and electro-optical polarization modulation was provided by a Pockels cell, an optical crystal where a phase delay can be controlled by applying a variable electric voltage. Light from the blue and red wings of a Zeeman-sensitive spectral line was separated, and a tilted glass plate near the spectral focal plane kept the spectral line centered with respect to the blue and red exit slits. The tilt angle of this plate served as the Doppler shift measurement. The modulated polarization signals from the two exit slits were differenced (Stokes V profiles are anti-symmetric) and then demodulated to give the magnetic field measurement along the line of sight.

Because photomultipliers are 1-pixel detectors, these first magnetograms were constructed as raster scans of the full solar disk. The pixel size was large, on the order of

1 arcmin, but the sensitivity was good enough to detect the ≈1 G weak large-scale magnetic field of the photosphere. Thus, H. D. Babcock could begin the first full-disk synoptic magnetogram program at Mt. Wilson Observatory.

Following the initiation of the Mt. Wilson synoptic program, Howard (1967) began the first daily full-disk synoptic magnetogram observations from which full-surface synoptic magnetograms were constructed (Howard, 1989). Similar programs began at NSO Kitt Peak (Harvey et al., 1980; Livingston et al., 1976) and at WSO (Svalgaard et al., 1978) in the 1970s. The NSO and WSO programs continue to the present day, with instrument upgrades at NSO (Keller et al., 2003). Spaceborne full-disk synoptic magnetogram programs, Solar and Heliospheric Observatory (SOHO)/Michelson Doppler Imager (MDI) (1996–2011; Scherrer et al., 1995) and Solar Dynamics Observatory (SDO)/Helioseismic and Magnetic Imager (HMI) (2010–present; Scherrer et al., 2012) have flown on NASA satellites.

Various methods have been used to calculate the properties of the magnetic field in different subdomains of the solar atmosphere. Most quantitative estimates are derived using the Zeeman polarization effects on atomic spectral lines. The objective is to map the solar magnetic field in heliographic coordinates and height from an observation of the Stokes vector (intensity and polarization of the sunlight). There are trade-offs between the high spatial resolution necessary to resolve small-scale magnetic fields, the high polarimetric sensitivity necessary for the detection of weak magnetic fields, and the full-disk coverage needed for observations of global field structure and its evolution, global dynamo research, and space weather projects.

Even from the Sun there is not enough light available to meet the needs of all projects with a single observation program, and so telescopes are designed to prioritize some requirements over others. For example, in practice, most measurements of solar magnetic fields are made with either spatially scanning slit spectrographs such as the Hinode Solar Optical Telescope Spectro-Polarimeter (SOT-SP) and Synoptic Optical Long-term Investigations of the Sun Vector Spectro-Magnetograph (SOLIS/VSM), where simultaneous measurement of the spectral line with good wavelength resolution is important, or wavelength-scanning narrow-band filtergraphs, such as Global Oscillations Network Group (GONG), or the SDO/HMI, where a high-speed simultaneous snapshot of a spatial field of view is the most important consideration. In the former case, slow and non-simultaneous spatial sampling is the price paid for a simultaneous and high-resolution measurement of the spectral line and polarization. In the latter case, high cadence and simultaneous spatial sampling come at the price of not observing the wavelength and polarization simultaneously.

With different treatment of the Sun's polarized light, we can infer either the full vector field or only its longitudinal (line-of-sight) component. The advantage of measuring only the longitudinal component is simplicity, exploiting the nearly linear dependence of the longitudinal Zeeman effect for circular polarization on the longitudinal component of the observed magnetic field. Complex inversion techniques, often leading to ambiguous and model-dependent answers, can be avoided in the longitudinal case. The first solar magnetograms of Babcock (1953) were of this kind, and the majority of full-disk magnetograms from synoptic observing programs still provide the longitudinal component. Because most of the magnetic field outside active regions is approximately vertical (Petrie & Patrikeeva, 2009; Svalgaard et al., 1978), these longitudinal field measurements are easily converted into data for the radial field component. Longitudinal magnetograms have therefore been extensively used to study the distribution and evolution of magnetic flux over the full solar surface, and for extrapolating models of the coronal magnetic field (Wiegelmann et al., 2017).

Measurements of the full field vector are complicated by difficulties of detection and interpretation of the signal for the transverse field component (Harvey, 1969). The Zeeman effect for linear polarization is proportional to the square of the transverse field strength, and is therefore weaker than the Zeeman effect for circular polarization, which exhibits linear dependence on the longitudinal field strength. It is due to this quadratic dependence that the orientation of the transverse field suffers from the so-called 180° ambiguity in azimuth that must be resolved (by using further assumptions) so that the data can be used for most scientific applications.

Though the longitudinal component can support studies based mostly on large-scale fields outside active regions, the study of solar activity depends heavily on measurements for the full vector field. In active regions, especially near boundaries where the opposite magnetic polarities meet, here referred to as neutral lines, and in sunspot penumbrae, the horizontal field component becomes dominant, and its correct interpretation is mandatory. Often, strong horizontal fields are associated with strong shear and free magnetic energy (i.e., the magnetic energy difference between a non-potential field and its equivalent lowest-energy, potential-field state) capable of fueling major flares and initiating coronal mass ejections. Furthermore, the helicity content of solar magnetic fields gives us critical information on the nature of the solar dynamo (Gosain et al., 2013; Liu et al., 2014; Pevtsov et al., 1994, 1995; Seehafer, 1990); see Section 3.2.3.

Synoptic observing programs for the full photospheric vector over the full solar disk are currently run by SOLIS/VSM and SDO/HMI. Vector synoptic maps are still in their infancy, but they have already become useful for monitoring trends in magnetic helicity (Gosain et al., 2013) and relating them to subsurface flow patterns (Komm & Gosain, 2015).

Full-disk magnetic field measurements based on the Zeeman effect for the deep chromosphere have been taken daily at NSO since the mid-1980s (Jones, 1985) and were carried out at MWO between the mid-1980s and 2013. For the high chromosphere and corona, the Zeeman effect can be applied only to strong fields in active regions and to infrared spectral lines whose long wavelengths enhance sensitivity due to the dependence of the Zeeman effect on the square of the wavelength.

Besides the Zeeman effect, the Hanle effect (Hanle, 1924) provides an alternative means of measuring solar magnetic fields. The Hanle effect is the change in linear polarization of a spectral line by the magnetic field. Whereas the Zeeman effect is polarization of light by a magnetic field, the Hanle effect modifies and depolarizes light that has already been polarized elsewhere by, for example, a scattering process. In the case of the solar atmosphere observed above the limb where scattered light is linearly polarized parallel to the limb, the Hanle effect can be interpreted as a rotation of the direction of linear polarization as well as a weakening of this polarization (Stenflo, 1982). This approach can provide information unavailable from the Zeeman effect that can be applied to weak fields such as in the high chromosphere and corona. Unlike the Zeeman effect, the Hanle effect can be used to diagnose mixed-polarity magnetic structures smaller than the spatial resolution of the instrument via the depolarization of spectral lines by small-scale turbulent flows. The Hanle effect was first used by solar physicists to measure the magnetic fields of prominences (Leroy et al., 1977), and revealed seething micro-turbulent fields of strength 10–100 G (Stenflo, 1982). Raouafi et al. (2016) simulated the Hanle effect signal in the H I Ly-α and He I 10830 Å lines that one can expect from coronal magnetic fields. They concluded that a combination of information from spectral lines with different sensitivities to the Hanle effect may give the best information on the coronal field.

Coronal spectropolarimetry is challenging for several reasons. Off-limb spectral line intensities are 10^5–10^6 times fainter than the typical disk intensity. The amplitude of the linear polarization (Q and U) is on the order of 5% to 10% that of the line signal strength, and the V polarization is about 10^4 times fainter than the line intensity, or about 10^9 times fainter than the disk intensity. Despite these challenges, Firor and Zirin (1962) showed that the infrared forbidden lines of Fe XIII may be used to diagnose the properties of the coronal plasma. The calculations by House (1977) for the line formation and the observations by Querfeld (1977) showed that linear polarization of the 1074.7 nm Fe XIII line is sensitive to the coronal magnetic field orientation, and Kuhn (1995) demonstrated that Zeeman splitting observations in the circular polarization of the 1074.7 nm line could achieve line-of-sight field sensitivity of a few gauss. Judge (1998) modeled the Zeeman sensitivity of the Fe XIII lines, requiring a photometric sensitivity of 10^{-4} for line-of-sight field sensitivity of a few gauss, a measurement demonstrated by Lin et al. (2000). Lin et al. (2004) determined the line-of-sight magnetic flux density and transverse field orientation of a coronal loop system in the Fe XIII line at 1074.7 nm using the 0.46 m off-axis reflecting coronagraph SOLARC (Solar Observatory for Limb Active Regions and Coronae). By integrating for 70 min, they produced a coronal magnetogram with 20 arcsec spatial resolution, field of view 0.1 to 0.45 solar radii above the limb, and sensitivity of about a gauss. The line-of-sight flux density was about 4 G about 100 arcsec above an active region. Being much stronger than the circular polarization Stokes V, the linear polarization signal has been used to constrain the magnetic field direction in twisted coronal magnetic structures called prominence cavities (see Section 3.3.2) in the plane-of-sky (Bak-Stęślicka et al., 2013; Rachmeler et al., 2013) using the HAO Coronal Multi-channel Polarimeter (CoMP) at Mauna Loa Solar Observatory (Tomczyk et al., 2008).

3.2.2. Cyclical Behavior of the Global Solar Field

The first synoptic magnetogram observations performed at Mt. Wilson by H. W. Babcock (1953) led H. D. Babcock (1959) to discover that the Sun's magnetic cycle extends to the global field including the polar fields. H. W. Babcock (1961) could then present a comprehensive empirical picture of global solar magnetism over a Hale (magnetic) cycle. This global solar field was found to exhibit cyclical behavior in the strength of both its low-latitude activity and its polar fields. The low- and high-latitude field strengths oscillate in anti-phase with each other.

The behavior of the Sun's magnetic field is intimately related to distinctive plasma flow patterns. The convection zone, including the photosphere, rotates differentially such that the equator rotates faster than the poles, and there is a thin shear layer near the surface and a larger one at the bottom of the convection zone called the tachocline. The photospheric rotation period is 24.47 days at the equator and almost 38 days at the poles. The shearing motion of the differential rotation is believed to play a critical role in maintaining the Sun's magnetic field from cycle to cycle. The other major large-scale flow pattern is a poleward meridional flow of about 20 m/s in each hemisphere. Conservation of mass implies an equatorward return flow of lower mean speed low in the convection zone, though evidence of multiple meridional

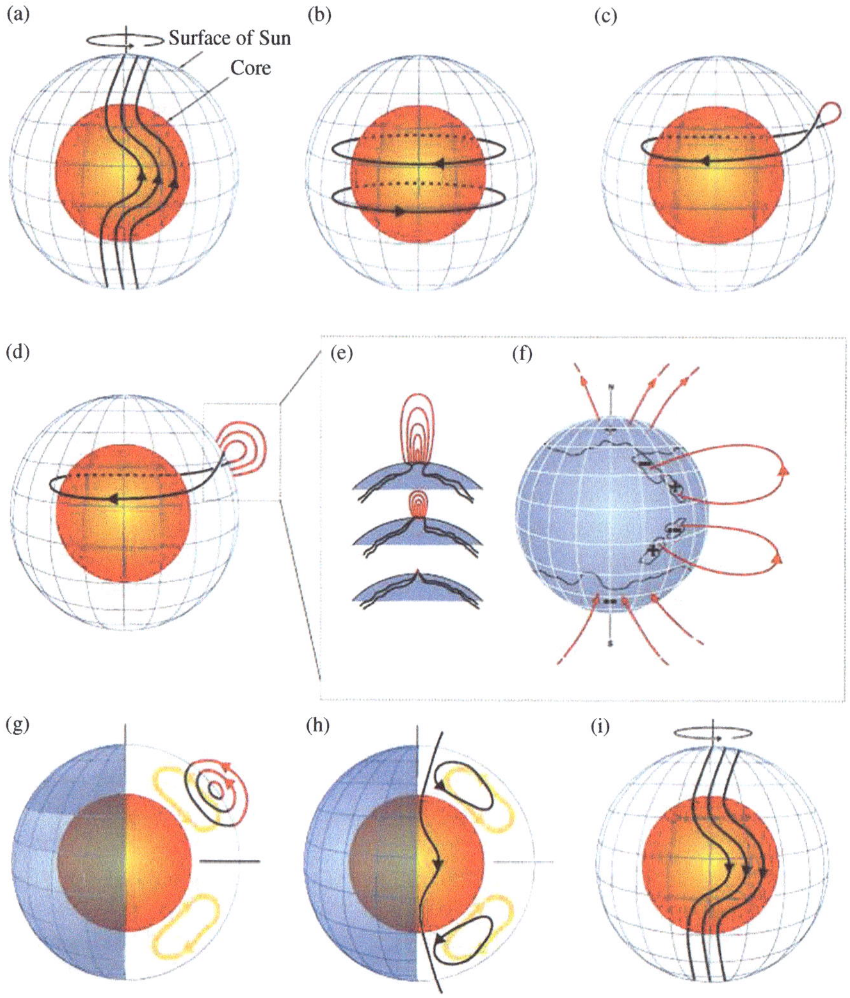

Figure 3.3 Schematic diagram of the major global photospheric flux-transport dynamo processes. The red inner sphere represents the Sun's radiative core and the blue mesh the solar surface. In between is the solar convection zone. (a) Shearing of poloidal field by the Sun's differential rotation near the bottom of the convection zone. (b) The toroidal field resulting from this shearing by differential rotation. (c) When the toroidal field is strong enough, buoyant loops rise to the surface, twisting as they rise due to the Coriolis force. Tilted bipoles are formed from these loops. (d–f) Additional flux emerges (d, e) and spreads (f) in latitude and longitude from decaying spots. (g) Meridional flow (yellow circulation with arrows) carries surface magnetic flux poleward, causing the polar fields to reverse. (h) Some of this flux is transported downward to the bottom and toward the equator. These poloidal fields have sign opposite to those at the beginning of the sequence, in (a). (i) This reversed poloidal flux is sheared by the differential rotation to produce the new toroidal field opposite in sign to that shown in (b). (*Source*: Dikpati & Gilman, 2007. © 2007 IOP Publishing.) Babcock (1961) did not explicitly include meridional circulation in his model, though he speculated about its existence.

flow cells has been reported (Jackiewicz et al., 2015; Kholikov et al., 2014; Zhao et al., 2013). The supergranular flow field (Leighton et al., 1962) concentrates the decayed active-region flux into narrow lanes at the boundaries of convection cells (Parker, 1963). Because the boundaries are non-stationary but move randomly, and the plasma forces dominate the dynamics, the magnetic elements are assumed to move randomly and independently of each other, in patterns often modeled as diffusion (Leighton, 1964, 1969; Sheeley et al., 1985).

Figure 3.3, due to Dikpati and Gilman (2007), shows the different stages of the solar cycle according of the phenomenological model presented by Babcock (1961). In the Sun's convection zone, the plasma rotates differentially such that the low latitudes rotate faster than the high latitudes. At the beginning of a cycle, the global field has a dipole-like structure, mostly confined to meridional planes (referred to as a poloidal field). This field, being frozen into the plasma (Alfvén, 1943), is wound up by the differential rotation in the convection zone, building up a

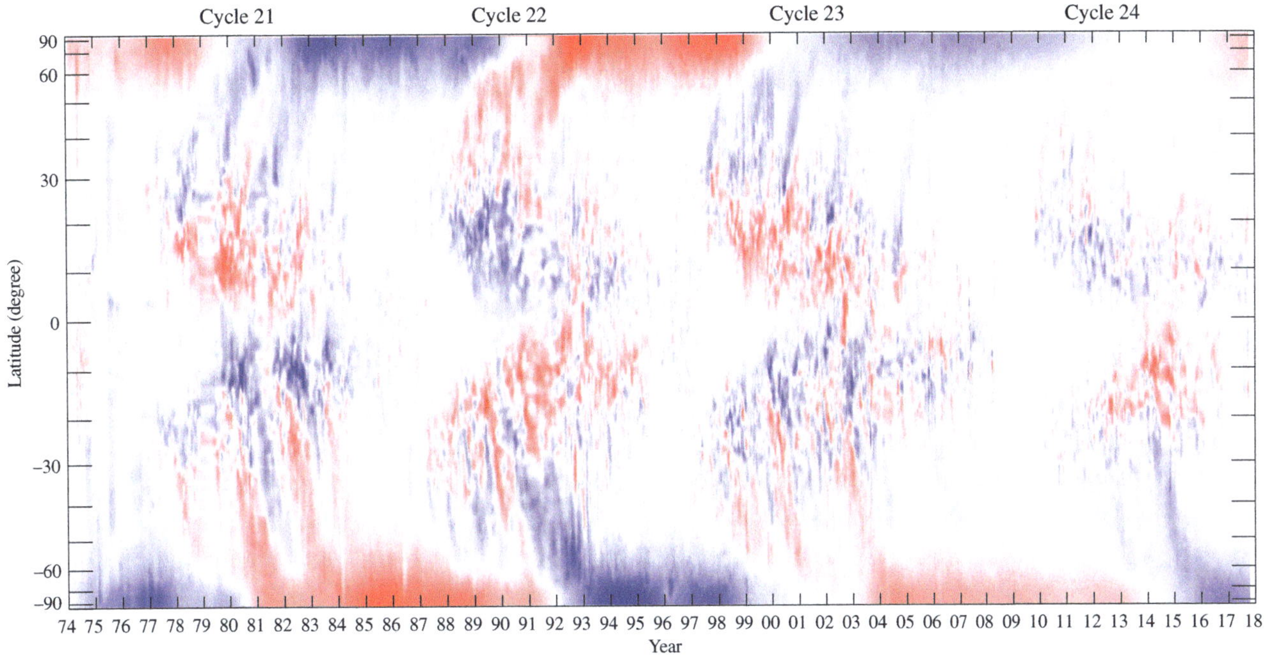

Figure 3.4 Butterfly diagram (latitude–time plot) based on Kitt Peak magnetogram data, summarizing the photospheric radial field distributions derived from the longitudinal photospheric field measurements. Each pixel is colored to represent the average field strength at each time and latitude. Red/blue represents positive/negative flux, with the color scale saturated at 15 G. (*Source*: Petrie, 2013. © 2013 IOP Publishing.)

field component along a latitudinal circle (referred to as the toroidal field component). When this toroidal field becomes strong enough to be susceptible to buoyancy instabilities, some of the field rises and emerges through the photosphere as sunspot pairs with the Hale polarity pattern of its cycle and hemisphere. In the process, the Joy's law tilt pattern discussed in Section 3.2.1 is produced by the Coriolis force acting on the field during its buoyant rise. This Joy's law tilt ensures that the trailing sunspot polarity in each hemisphere is the opposite of the polarity of the polar field in that hemisphere. As sunspots age, they are fragmented and dispersed by diffusion-like near-surface convective flow patterns. The Joy's law bipole tilt causes the polarity of the flux dispersed to each pole to preferentially have an opposing polarity to the polar field in that hemisphere. The dispersal of the sunspot fields throughout a cycle therefore slowly cancels the polar field in each hemisphere and creates a new polar field of reversed polarity in its place. Thereupon a cycle begins with opposite polarity to its predecessor. From this descriptive phenomenological model, Leighton (1964, 1969) gave a numerical model for the solar magnetic cycle. The Babcock–Leighton framework has become and remained standard in the study of global solar flux transport.

The solar magnetic field is believed to be generated by dynamo action of plasma flows in the interior. A solar dynamo is a process by which the magnetic field in an electrically conducting solar plasma is maintained against Ohmic dissipation. Dynamo models are usually described by coupled equations, one for the vector potential of the poloidal field and the other for the toroidal field component (Charbonneau, 2010). The conversion from poloidal to toroidal field is achieved by interior differential rotation as exemplified above by the Babcock–Leighton model. However, the equation for the poloidal field had no equivalent source term (Cowling, 1933) until Parker (1955) introduced one representing the effect of helical turbulent flows on buoyantly rising loops, a theory generalized to a statistical description of poloidal field generation by turbulent flows by Steenbeck and Krause (1969) and Krause and Raedler (1980). The Babcock–Leighton framework summarized above is just one example of a dynamo model in which the poloidal field creation process is represented by the emerging bipoles' Joy's law tilt.

One useful way to summarize the evolution of the global photospheric field over multiple activity cycles is to plot a latitude–time diagram, known as a butterfly diagram because of the distinctive butterfly-wing-like patterns in such plots. Figure 3.4 shows a magnetic butterfly diagram based on NSO Kitt Peak and SOLIS/VSM full-disk magnetograms. This plot and the subsequent plots of this section are discussed in more detail by Petrie (2015). Magnetic butterfly diagrams emphasize the high

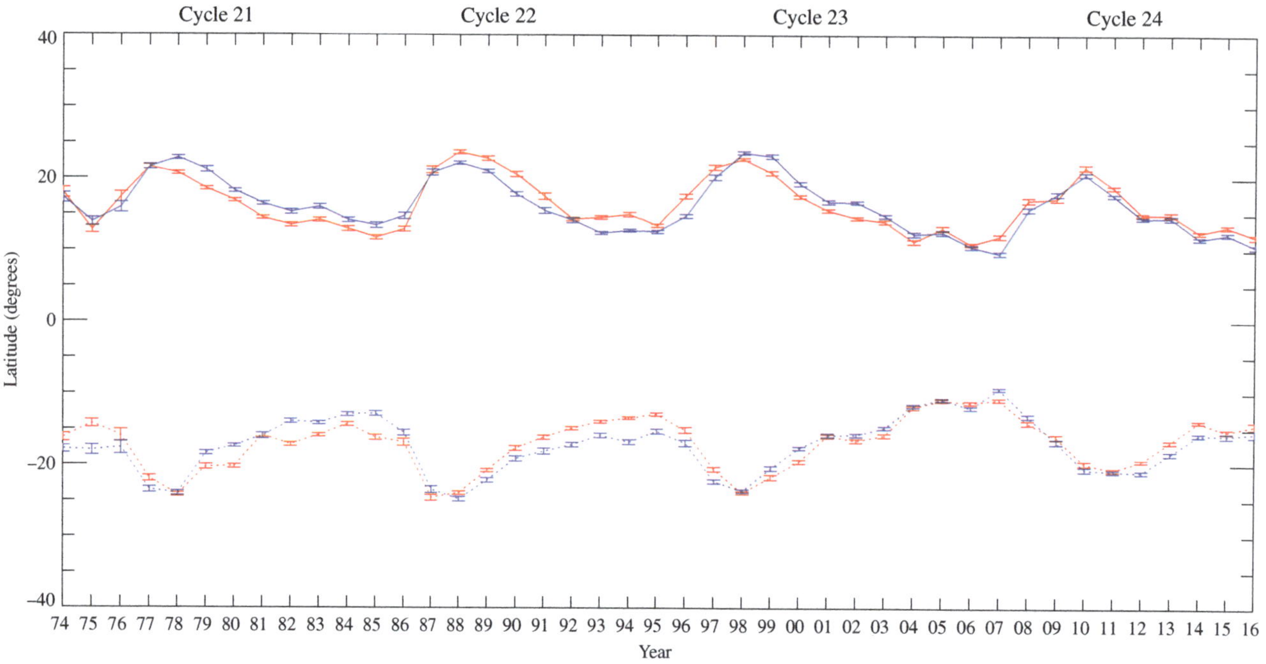

Figure 3.5 The latitudinal centroids of the active-region fields as functions of time for positive (red) and negative (blue) polarities in the northern (solid lines) and southern (dotted lines) hemispheres for the Kitt Peak data summarized in Figure 3.4. The error bars indicate the standard deviations of the annual means. The displacements between the positive and negative centroids are a consequence of and consistent with Joy's law. (*Source*: Petrie, 2012. © 2012 IOP Publishing.)

degree of organization in the long-term behavior of the fields at low and high latitudes, and the systematic magnetic patterns traveling from low to high latitudes. The active fields emerge at low latitudes; they begin each cycle emerging at latitudes around ±30° and subsequently emerge at progressively lower latitudes on average, a phenomenon referred to as Spörer's law, creating the distinctive wings of the butterfly patterns first reported by Maunder (1913). The diagram also shows the change of polarity of the polar fields once each cycle, coinciding with activity maximum but not occurring simultaneously at the two poles. Between the active and polar latitudes, around ±50°, there is clear evidence of poleward flux transport of both polarities in each hemisphere during each cycle, which appears most intense during the most active phases of the cycle. The poleward drift of weak, decayed magnetic flux (Bumba & Howard, 1969) appears in Figure 3.4 as plumes of dominant trailing-polarity flux. These motions are sometimes referred to as poleward surges or "rush to the poles." The polar fields can be seen to respond to these poleward motions of flux, and their largest changes, including those resulting in a change in their polarity, correlate well in time, strength, and polarity with the most intense poleward surges. There patterns are consistent with the Babcock–Leighton phenomenological described above.

A more quantitative representation of the bipole tilt patterns can be produced by plotting the average latitudes of the positive and negative magnetic fields of the active regions in each hemisphere (Petrie, 2012). Figure 3.5 shows such a plot based on the same NSO Kitt Peak and SOLIS/VSM full-disk magnetograms that were used to produce Figure 3.4. Because only strong fields are included, these average latitudes represent the Joy's law tilt of the bipolar active regions and not the transport of the weak, decayed field. As expected from the discussion above, the average latitudes of the positive and negative magnetic fluxes of the active regions in each hemisphere do indeed swap positions each cycle.

Similarly, Figure 3.6 shows plots of the average field strengths at polar latitudes based on measurements from NSO Kitt Peak and WSO. At WSO, the north and south polar line-of-sight field strengths are measured daily in the 3′ apertures nearest the poles, north and south. The NSO/KP estimates are derived from the butterfly map shown in Figure 3.4 and represent the radial field component. This explains some of the difference between the two data sets.

There are also well-documented calibration issues with the WSO data (Svalgaard et al., 1978), causing the fields to be underestimated by a factor of 1.8. Notwithstanding these differences, it is clear that the two data sets are well correlated with each other.

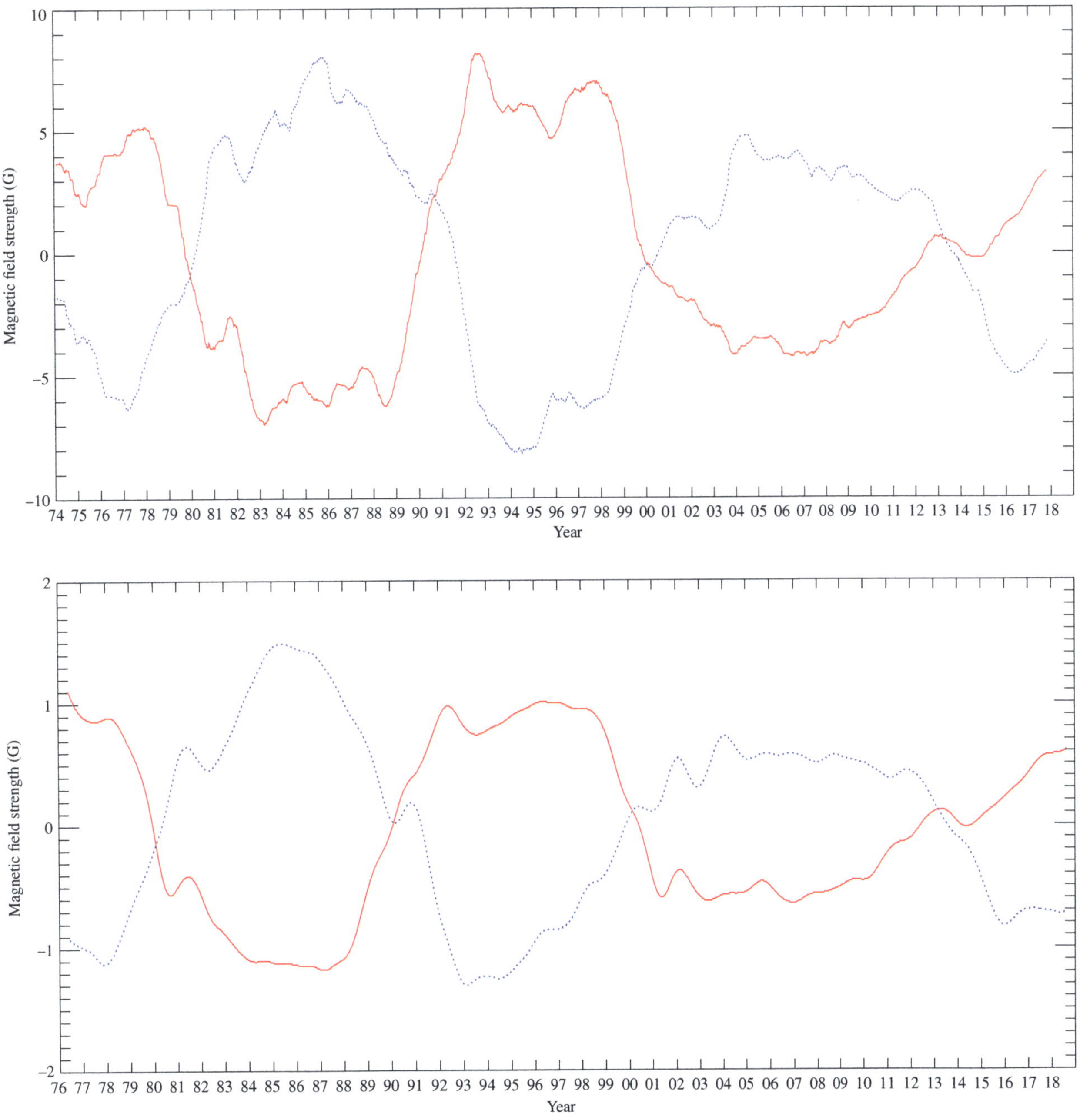

Figure 3.6 Ten-day averages of the north (red solid lines) and south (blue dotted lines) polar fields measured at Kitt Peak (top) and Wilcox (bottom) are plotted against time. The Kitt Peak data are for the radial field component and derive from heliographic latitudes between latitudes ranging from about ±63° to about ±70°. The Wilcox measurements are for the line-of-sight field component and come from the 3′ apertures nearest the poles, which cover between about ±55° and the poles. Changes tend to occur earlier in the Wilcox curves than in the Kitt Peak curves because the Wilcox data derive from lower latitudes on average. (*Source*: Petrie, 2013. © 2013 IOP Publishing.)

The relationship between the bipoles' emergence patterns and the polar field changes also becomes apparent: large polar field changes occur (Figure 3.6) when the activity is strong (Figure 3.4) and the average latitudes of the two magnetic polarities are well separated in latitude in each hemisphere (Figure 3.5). The polar fields became weaker after their polarity reversal during cycle 23 because, though the activity remained quite strong, the positive and negative polarities in each hemisphere appeared with approximately equal latitudes on average,

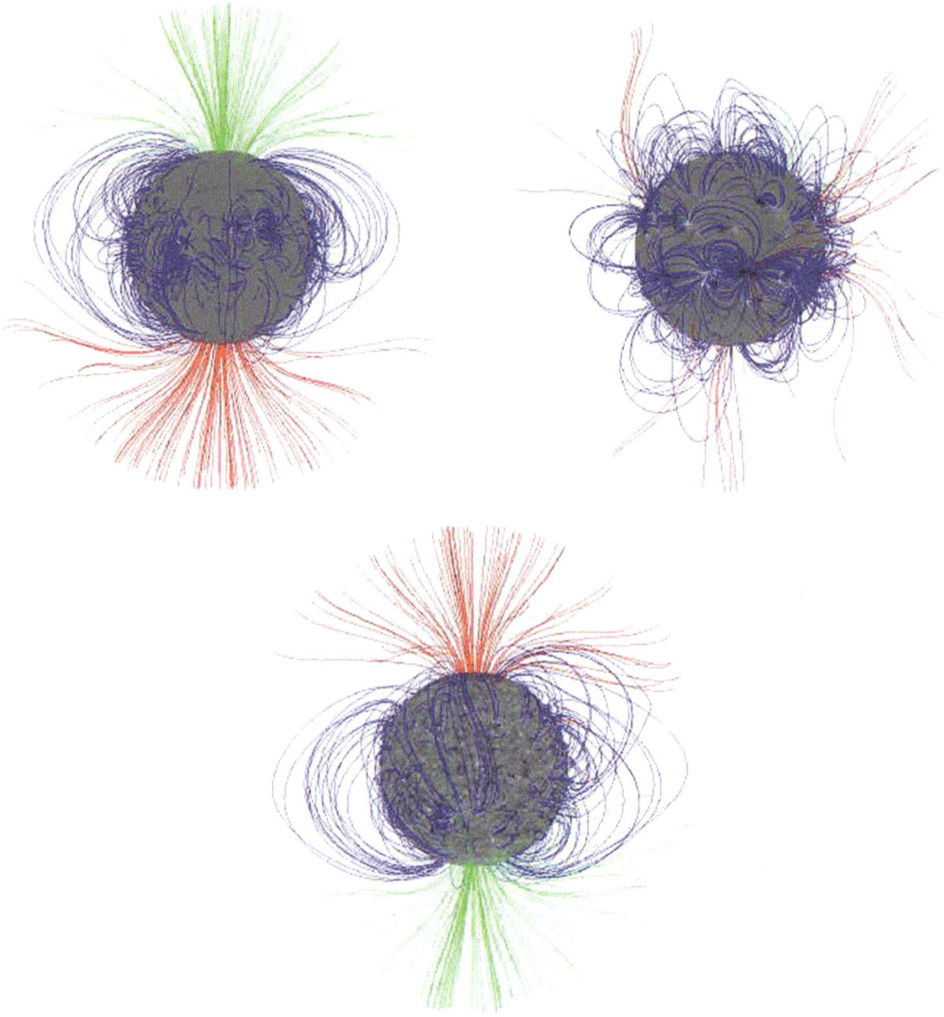

Figure 3.7 Potential-field source-surface models representing the beginnings of years 1996 (top left), 2000 (top right), and 2009 (bottom). The photospheric radial field strength is represented by the gray scale, saturated at 100 G with white/black indicating positive/negative polarity. Green/red field lines represent open fields of positive/negative polarity, and blue lines represent closed fields. Because the solar field cannot have a monopole component, the positive and negative open magnetic flux must balance at all times. In the example from 2000, representing solar maximum, the dipole axis is close to the equatorial plane, and almost all of the positive open flux (green lines) is hidden at the far side of the Sun, and almost all of the visible open lines have negative polarity (red). In 1996 and 2009, the dipole axis was closely aligned with the rotation axis, which is typical of solar minimum. (*Source*: Petrie, 2013. © 2013 IOP Publishing.)

so that their net effect on the development of the polar fields was small.

Figure 3.7 shows three "hairy-ball" plots of potential-field source-surface (PFSS) models (Altschuler & Newkirk, 1969; Schatten et al., 1969) of the coronal magnetic field from the years 1996, 2000, and 2009. The plots therefore represent the cycle 22 minimum, and cycle 23 maximum and minimum, respectively. The differences between the minimum plots and the maximum plot clearly emphasize the much simpler, nearly axisymmetric axial dipole structure associated with dominant polar fields during solar minimum, and the complex three-dimensional structure that is produced by multiple active regions in the absence of strong polar fields.

The polar regions are very important to several of the major branches of global solar physics, from the solar dynamo to the acceleration of the fast solar wind. However, polar field measurements are very challenging. The data plotted in Figure 3.6 are averages based on measurements at relatively low spatial resolution of high-latitude fields, but not of the poles themselves. The tilt angle between the solar rotation axis and our viewpoint

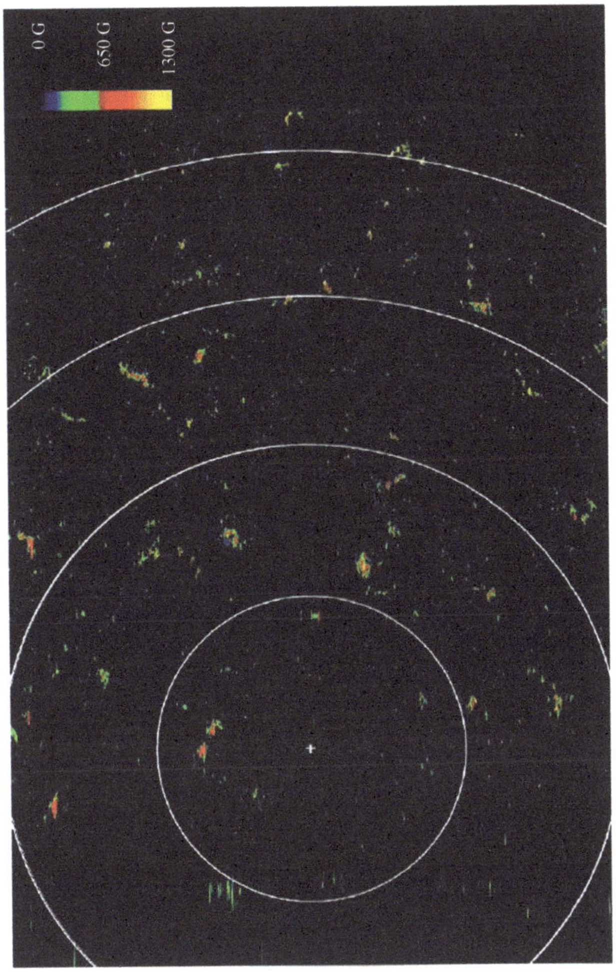

Figure 3.8 South polar view of the magnetic field strength taken at 12:02:19–14:55:48 UT on 16 March 2007 by the Hinode SOT-SP. The original observing field of view was converted to a map representing the view from above the south pole. East is to the left, west is to the right, and the observation was carried out from the top down. Spatial resolution is lost near the extreme limb (i.e., near the bottom of the figure). Latitudinal lines for 85°, 80°, 75°, and 70° are shown as large circles, and the plus sign marks the south pole. (*Source*: From Tsuneta et al., 2008).

on Earth's ecliptic plane is approximately 7.25°, meaning that the viewing angle of the poles is never less than about 83°. Optimal polar viewing angles only occur annually, on 6 March for the south pole and 8 September for the north pole. Over the first/second half of each year, the north/south pole is unobservable from the direction of Earth.

Using the *Hinode* SOT-SP, Tsuneta et al. (2008) observed the south polar field on 16 March 2007. Figure 3.8 shows a spatial map of the measured magnetic field strength. A highly structured, nonuniform distribution of intense, unipolar magnetic features is strikingly evident in the map, with numerous sizable concentrations of kilogauss flux quite evenly distributed across it. The vertical kilogauss patches evolved on timescales around 5–10 hr compared to < 30 min for the horizontal fields, consistent with observations of seething horizontal photospheric fields by Harvey et al. (2007). The average area and magnetic flux tend to be larger for polar concentrations than for quiet-Sun concentrations at low latitudes, and the polar concentrations have a strong polarity bias not generally found at low latitudes (Ito et al., 2010; Shiota et al., 2012). The estimated total polar flux according to these measurements (above 70°) was 5.6×10^{21} Mx (Tsuneta et al., 2008). Thus, the total polar magnetic flux was less than that of a major active region. If the polar fields were evenly distributed, they would have an average strength roughly consistent with the values derived from lower-resolution synoptic line-of-sight measurements. However, the decreasing poleward trend in magnetic flux density that is visible in Figure 3.8 and quantified by Petrie (2017) appears to be an artifact due to foreshortening near the limb. This issue requires further study.

3.2.3. Magnetic Helicity

Magnetic helicity is a measure of the complexity of a magnetic field, including such properties as twist, shear, writhe, and interlinkage. For a more detailed introduction, the reader is referred to Pevtsov et al. (2014). Here, *twist* denotes the internal twisting of a magnetic field about a flux tube axis and *writhe* the twisting or kinking of this flux tube axis. To an excellent approximation, magnetic helicity is conserved in solar plasmas on the scales of the structures that we discuss in this chapter. This is because the high electrical conductivity of the plasma confines resistive phenomena such as magnetic reconnection to small scales so that their effect on the magnetic helicity is very small (Berger, 1984). Magnetic helicity can, however, move across length scales, across structures, and from structure to structure.

Solar magnetic fields and associated structures are observed to have preferentially negative magnetic helicity in the northern hemisphere and positive magnetic helicity in the southern hemisphere. Using photospheric vector magnetograms of active regions, Seehafer (1990), Pevtsov et al. (1994, 1995), Zhang et al. (2010), and Liu et al. (2014) showed that this hemispheric helicity sign rule is consistent but statistically weak, satisfied by about 60–75% of active regions. A summary of the hemispheric helicity rule from the surface magnetic fields in active regions and the quiet Sun to interplanetary magnetic field was given by Pevtsov (2008).

Magnetic helicity is closely associated with the solar dynamo and solar activity. A key component of the dynamo process described in Section 3.2.2 is the interior differential rotation, which winds up the interior field and generates and amplifies its toroidal component. The amount of helicity produced by this process can be estimated from surface field measurements and a model for the interior differential rotation (Berger & Ruzmaikin, 2000), as can the quantity of toroidal magnetic flux (Cameron & Schüssler, 2015).

The effect of differential rotation on individual active regions was first studied by DeVore (2000). Differential rotation rotates each polarity of a bipole around its own axis, and also changes the two polarities' positions relative to each other. Démoulin et al. (2002b) showed that these two effects contribute approximately the equal amounts of magnetic helicity but of opposite sign, so that the net effect of differential rotation on an active region is generally small. Démoulin et al. (2002a) and Green et al. (2002) showed that the long-term effect of differential rotation on active-region helicity is small. However, the helicity generated by interior differential rotation over a 22-yr magnetic cycle has the same sign in each hemisphere as indicated by coronal images and photospheric vector magnetograms, negative in the northern hemisphere and positive in the south hemisphere, in equal quantity in the two hemispheres according to the conservation of magnetic helicity.

Choudhuri (2003) presented an explanation for the hemispheric helicity rule consistent with the Babcock–Leighton model described above. The toroidal field generated in the interior by differential rotation rises buoyantly, meeting the poloidal field on its journey to the surface. This poloidal field is dragged upward with the buoyant toroidal field and wraps around it, giving it magnetic helicity. This model gives magnetic helicity in each hemisphere of sign agreeing with the observed hemispheric rule. A model by Chatterjee et al. (2006) based on this idea also reproduces the observed pattern of a change in helicity sign from the central to the outer parts of sunspots (Gosain et al., 2013; Zhang, 2006).

The Coriolis force is proportional to the sine of solar latitude, but there is no such dependence to be found in the observed proxies of magnetic helicity (Pevtsov et al., 1995). These proxies instead show a weak hemispheric bias with large statistical scatter. Longcope et al. (1998) demonstrated that turbulence in the convection zone could introduce magnetic helicity on small spatial scales of the correct sign bias and with large statistical scatter in each hemisphere, provided that the turbulent flows have nonzero net helicity. They called this effect the Σ-effect. Consistent with this picture, Nandy (2006) showed that the helicity of active regions is independent of latitude, and that larger active regions tend to adhere less to the hemispheric helicity sign rule, and suggested that this is because large, strong magnetic structures respond less to small-scale turbulence. We will briefly describe the role played by magnetic helicity in filament formation in Section 3.3.2 and in solar eruptive phenomena in Section 3.3.5.

3.3. SOLAR MAGNETIC FEATURES WITH RADIATIVE SIGNATURES

In this section, we introduce some key solar magnetic structures that define the radiation patterns of the solar photosphere, chromosphere, and corona. We briefly describe sunspots, faculae, filaments, coronal holes, coronal loops, and solar eruptions: flares and coronal mass ejections (CMEs).

3.3.1. Features of the Photosphere in Visible Light

The most obvious signature of solar magnetic fields in visible light, visible in some cases with the naked eye (with suitable precautions), is the sunspot (Solanki, 2003b; Thomas & Weiss, 2008). Sunspots have been regularly observed with telescopes for centuries, since the time of Galileo, and have been known to be magnetic structures since Hale (1908) measured the magnetic field of a sunspot, the first measurement of a magnetic field outside Earth. The magnetic features of the photosphere range from sunspots, the largest and rarest, through pores, generally smaller structures of sunspot-strength field of a couple of Mm without penumbrae, to the smaller magnetic elements forming faculae and the network, each feature a few hundred km in diameter. Figure 3.9 shows a high-resolution image of a sunspot, from the Big Bear Solar Observatory (BBSO) Goode Solar Telescope. As this figure illustrates, the sunspot is characterized by a dark core, called the umbra, and a less dark, fibril-structured halo called the penumbra.

Sunspots that do not develop penumbrae are referred to as pores, and they are generally smaller than sunspots. A sunspot may lose its penumbra as it decays, in which case it is referred to as a naked sunspot (Rempel, 2015). Umbrae sometimes feature bright filaments called light bridges or bright dots called umbral dots. The penumbral structure is dominated by elongated alternating bright and dark fibrils, and by penumbral grains similar to umbral dots. The wavelength-integrated intensity of light from umbrae is typically about 25% of the quiet-Sun value, and from penumbrae about 75%. Sunspots range in size and lifetime from 3–60 Mm and about 6 hr to several months. Sunspots decay steadily soon after reaching their maximum size. The decay is believed to be caused by turbulent flows of the near-surface plasma diffusing the sunspot's magnetic field. Sunspots have field strength 2500–3500 G in their darkest umbral regions and 700–1000 G in the outer penumbra. At the center of the spot, the field is approximately vertical, whereas it is inclined

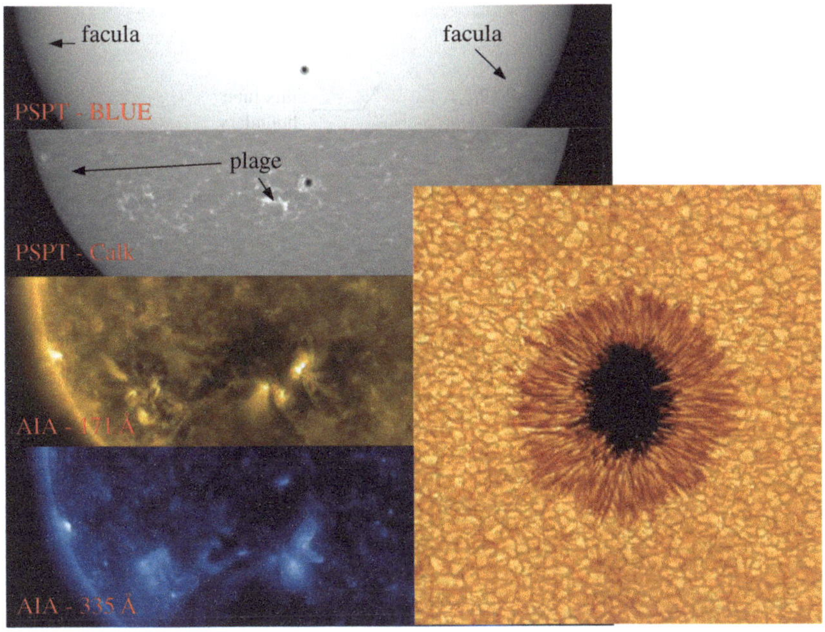

Figure 3.9 A TiO band (706 nm) photospheric image of a sunspot observed on 2 July 2010 by the BBSO Goode Solar Telescope (right). Also shown are slices of four nearly simultaneous full-disk images: blue visible (top left, representing the photosphere) and Ca II K (second-top-left, representing the chromosphere) images from the Precision Solar Photometric Telescope (PSPT), and 171 Å (second-bottom-left, representing the quiet corona around 0.6 MK) and 335 Å (bottom-left, representing the active corona around 2.5 MK) passband images from SDO/AIA. In the PSPT images, the sunspot is near the center, and faculae and plage are indicated. The AIA images show the corresponding coronal emission. (*Source*: Gordon Petrie.)

by 70–80° to the vertical at the edge of the penumbra. Above the solar surface the field lines of a sunspot expand to form a horizontal canopy of loop structures, visible in the SDO/Atmospheric Imaging Assembly (AIA) images in Figure 3.9.

As discussed in Section 3.2, the photospheric magnetic field is concentrated in bipolar active regions which are confined to active latitudes, within 30° of the equator, and in a network of small flux concentrations distributed over the entire solar surface. In the quiet Sun, that is, outside active regions, the small-scale magnetic flux concentrations form a network outlining the borders of supergranular cells with length scale 20–40 Mm. The magnetic elements forming faculae and the network are located in the downflow lanes between granules, associated with intergranular accumulations of magnetic flux. The BBSO image in Figure 3.9 also shows this convection-dominated supergranular cell structure of the quiet-Sun photosphere.

In active regions, before they decay, the field is usually concentrated into sunspots surrounded by small, discrete magnetic features collectively known as faculae or plage (Keller, 1992; Solanki, 1993; de Wijn et al., 2009). Plage and faculae are composed of high-density collections of small-scale magnetic flux concentrations. A plage is defined as an extended bright region observed in the cores of chromospheric spectral lines roughly co-spatial with faculae, which are small features that appear bright when observed near the solar limb in photospheric spectral lines and the continuum. Examples of plage and faculae are shown in the PSPT images in Figure 3.9. In the photosphere, the web-like network of intergranular lanes is observed to be dark, whereas the corresponding network and plage structure in the chromosphere is bright.

The above trends are well explained by the so-called flux tube model (e.g., Fabiani Bendicho et al., 1992; Spruit, 1976, 1981; Steiner, 2005). A magnetic flux tube is a tube-like concentration of magnetic field lines/magnetic flux. Typically, such flux tubes are vertically oriented and distributed across the intergranular lanes as they penetrate the ambient photosphere (which can be approximated as a magnetic-field-free plasma). A collection of vertical flux tubes is manifested as regions of high magnetic flux: faculae. In these high-magnetic-flux regions, convection is inhibited by the magnetic pressure. This reduces both the density and the temperature within the tube, and therefore the opacity. The radiation propagating from the surrounding hotter plasma can then penetrate within the tube, thus heating it. The efficiency of this heating at each geometric height for structures in radiative equilibrium depends on the ratio between the width of the tube and the horizontal optical depth (e.g.,

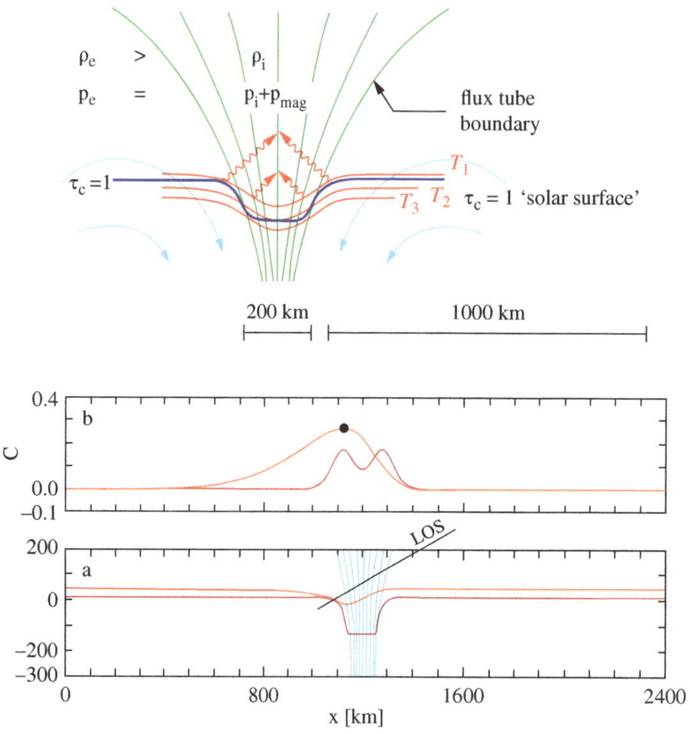

Figure 3.10 Schematic representation of a magnetic flux tube (green lines in top image) and of the emergent intensity. Top: Because of the magnetic pressure, temperature and density are reduced within a flux tube, and the optical depth unity surface (blue thick line) forms deeper in the atmosphere than in the surrounding un-magnetized plasma. At layers slightly above the base of the photosphere, the reduced density allows radiation from below and around the tube to penetrate within it, thus inducing an increase of temperature. Bottom: The emergent intensity (panel b) is a function of the temperatures sampled by the optical depth unity surface (panel a). In both panels, red denotes the vertical line of sight, whereas orange denotes an inclined line of sight. Because of the reduced opacity within the tube, along a tilted angle the optical depth unity surface samples the hotter flanks (hot walls) of the tube, so that the structure appears brighter than the surrounding plasma. (*Source*: Steiner, 2007. © 2007 AIP Publishing.)

Criscuoli & Rast, 2009; Pizzo et al., 1993): if the tube has dimensions smaller than or comparable with the horizontal optical depth, then the radiation can penetrate within the whole tube and increase the temperature within it; if the tube is larger than the horizontal optical depth, the radiation cannot penetrate the whole structure, which then becomes cooler than the surrounding plasma. In the solar photosphere, the transition between these two regimes occurs approximately at 500 km (Knoelker & Schuessler, 1988; Spruit, 1976). In the upper photosphere and chromosphere, the tube is heated by radiation coming from below (e.g., Knoelker & Schuessler, 1988; Pizzo et al., 1993), and by other mechanisms such as dissipation of electric currents, waves, and shocks (Goodman, 2000; Khomenko et al., 2018; Song, 2017), so that the temperature within magnetic structures is higher than the surrounding plasma.

Figure 3.10 shows a schematic representation of a flux tube. The structure, represented by the green lines, is mostly vertical in the photosphere while it expands with height. The fanning results from the pressure of the plasma surrounding the flux tube decreasing more rapidly with height than the gas pressure within the flux tube (Spruit, 1981; Steiner, 2007). As shown in the figure, the optical depth unity along the axis of the tube forms deeper than the surrounding plasma so that the radiation intensity escaping from the tube will in general be different from the intensity emitted from the quiet plasma. More specifically, at wavelengths forming in the deepest layers, magnetic structures have negative or null photometric contrast, whereas at wavelengths forming in the higher layers, magnetic structures appear bright. At slanted viewing angles, again due to the reduced opacity within the tube, the optical depth unity surface forms on the side (or "hot wall") of the tube opposite the observer, thus sampling the hotter granulation. Consequently, magnetic structures appear brighter and brighter the closer they are to the solar limb. Ultimately, according to magnetic flux tube models, the radiative emission of a magnetic structure depends on the size, the associated magnetic field strength, position over the disk, and the spectral range. These findings are supported by observations (e.g., Berger et al., 2007; Criscuoli et al., 2017; Ermolli et al., 2007; Ortiz et al., 2002; Viticchié et al., 2010) and by three-dimensional magneto hydrodynamic simulations of the solar atmosphere (e.g., Criscuoli, 2013; Peck et al., 2019; Tritschler & Uitenbroek, 2006; Vögler, 2005).

3.3.2. Chromospheric Filaments

Solar filaments (called prominences when observed at the limb) are long-lived, relatively cool, and dense structures suspended by the magnetic field at heights on the order of 50 Mm above the solar surface (e.g., Mackay et al., 2010). Chromospheric filaments are observed in, for example, Hα or He I 10830 Å images. Observations of filaments at high latitudes are, along with polar faculae, the longest available data set providing information on solar magnetism at high latitudes. Filaments and filament channels delineate neutral lines, so the overall structure of the magnetic field can be reconstructed from such data (Durrant & Wilson, 2003; McIntosh, 1972).

Filaments are always found straddling photospheric magnetic neutral lines. They can appear within mature active regions, within activity complexes composed of multiple bipolar active regions, at boundaries of active regions, and in the quiet Sun, particularly at high latitudes, where the poleward surges of decayed active-region flux meet the polar field. Filaments within active regions or activity complexes are known as active-region filaments, and are found only at low latitudes. Filaments outside active regions, known as quiescent filaments, are found at all latitudes. Quiescent filaments are aligned along neutral lines between large patches of magnetic elements with opposite magnetic polarity bias on each side. In active-region filaments, the fields tend to be stronger and the neutral lines better defined.

A filament consists of a spine running horizontally along the top of the filament and barbs protruding from the sides of the filament, extending down from the spine to the chromosphere below. High-resolution observations show that spines and barbs consist of thin threads about 200 km across. Although filaments can last many days, the threads have lifetimes of only a few minutes. Short threads are often seen in quiescent filaments, whereas the threads in active-region filaments tend to be longer (Lin et al., 2008). The magnetic structures of the two classes of filament appear to be qualitatively different. When observed above the limb, filaments are referred to as prominences. Prominences' appearances vary greatly, some having predominantly horizontal threads, apparently aligned with the magnetic field, whereas others have nearly vertical threads clearly not aligned with the magnetic field, separated by dark regions. Figure 3.12 shows an example of each case. Berger et al. (2008) showed an example with bright threads and blobs flowing downward at about 10 km s^{-1} and dark regions moving upward at about 20 km s^{-1}.

Filaments form in regions around magnetic neutral lines in the chromosphere where fibrils align parallel to the neutral line. Such regions are known as filament channels. As observed above the limb in white light, quiescent prominences are located in dark cavities at the bases of streamers, tall closed coronal structures that we will describe briefly in Section 3.3.3. It is believed that filament channels and prominence cavities are the same magnetic structure observed under different conditions. The prominence cavity is believed to be a magnetic flux rope, forming a three-part structure with the prominence and the streamer (Low & Hundhausen, 1995). Prominences are therefore embedded within highly non-potential magnetic fields with dominant axial magnetic field.

In filament channels, the fibrils emanate from magnetic elements in the channel with oppositely directed streaming patterns on the two sides of the channel. These patterns indicate that the horizontal axial field must point in the same direction on the two sides of the filament channel. The magnetic field structure has a handedness or "chirality" according to the direction of the axial field as seen from the positive side of the filament channel (Martin, 1992). The axial field direction is often deduced from the fibril structure of the filament as seen in Figure 3.11. The hemispheric helicity sign rule described in Section 3.2.2 is also satisfied by filaments (Martin, 1998): a significant preference for dextral/sinistral filaments to form in the northern/southern hemisphere has been observed at active and high latitudes (Leroy et al., 1983; Martin et al., 1994). In active-region filaments, the handedness is consistent with the overall helicity of the region (Rust & Martin, 1994; Wang, 2013). The filament handedness patterns are believed to be associated with magnetic flux that traps magnetic helicity (Berger, 1984) undergoing cancellation of its vertical field component at neutral lines, concentrating the helicity there (Wang, 2013; Yeates & Mackay, 2012) and setting the scene for filament formation. Many authors have found that the hemisphere helicity rule is stronger for filaments than for active regions (e.g., Pevtsov et al., 2003). Wang (2013) argued that this is due to filaments developing at magnetic neutral lines where much vertical field has cancelled, leaving a dominant horizontal field component near the neutral line and making the helicity sign more obvious and easier to detect than in newly emerged active regions.

In some models of filament formation, a horizontal twisted flux tube emerges from the convection zone through the photosphere and into the chromosphere and

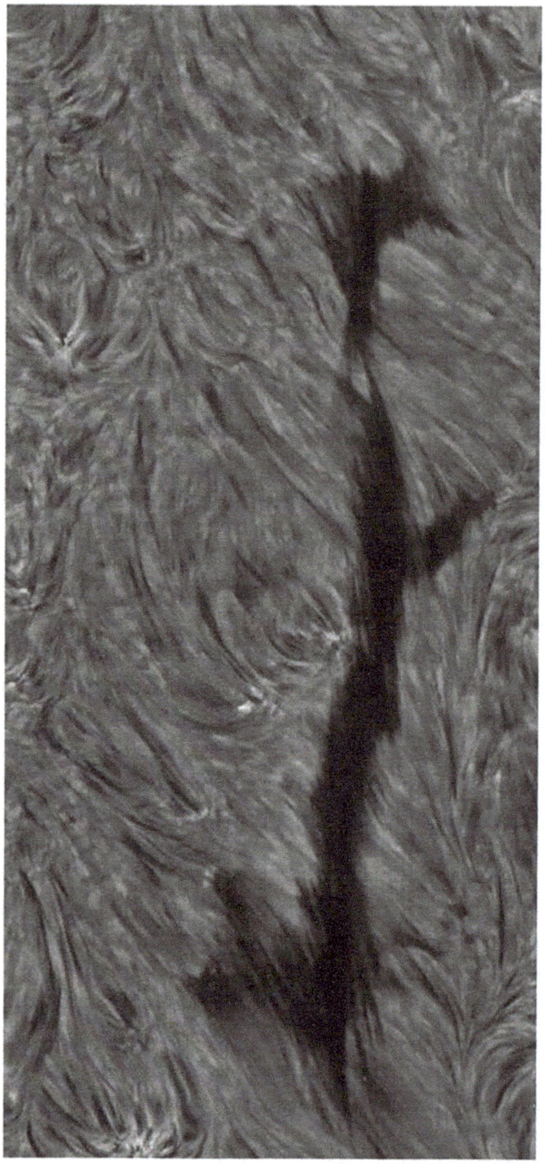

Figure 3.11 High-resolution Hα image of a filament on the disk. A main filament spine, barbs protruding from the sides of the spine, and fine-structure fibrils are all evident. This image was taken by the Dutch Open Telescope on La Palma. From Heinzel and Anzer (2006). (*Source*: Courtesy of R. J. Rutten.)

corona (Low, 1994; Rust & Kumar, 1994). This picture has been supported by multi-height observations of coherent flux emergence indicating a macroscopic horizontal flux tube structure, resulting in active-region filaments (López Ariste et al., 2006). Figure 3.13 shows an observed concave-upward surface magnetic field pattern associated with the emergence of a flux-rope structure. As Figure 3.14 illustrates, other models invoke surface flux-transport patterns to turn an arcade of coronal magnetic loops into a filament (van Ballegooijen & Martens, 1989). In such models, the loop footpoints are advected

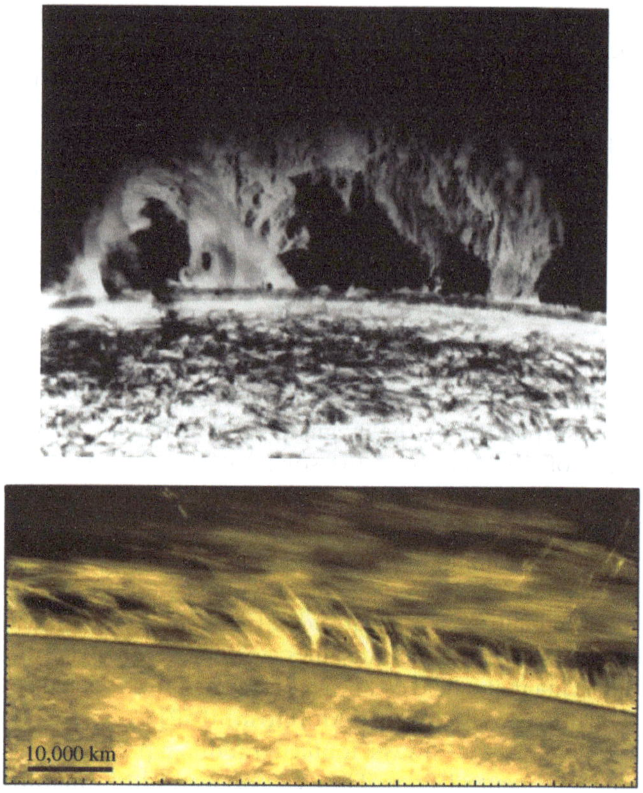

Figure 3.12 Two contrasting examples of solar prominences observed above the limb in different wavelengths. (*Source*: Big Bear Observatory, Caltech.) Top: A Ca II H Hinode Solar Optical Telescope (SOT) (*Source*: image from 9 November 2006 from Okamoto et al. 2007); Bottom: A BBSO Hα image observed in 1970.

by counter-streaming flows in opposite directions parallel to the neutral line, shearing the loop arcade, and the footpoints are moved toward the neutral line by inflows, causing them to pinch together and reconnect with each other to form a long axial filament field along the neutral line, accompanied by a short loop that can submerge. Whether the magnetic field around a prominence has a normal (filament field points from the negative to the positive side of the photospheric neutral line; Kippenhahn & Schlüter, 1957) or inverse (filament field points from the positive to the negative side of the photospheric neutral line; Kuperus & Raadu, 1974) magnetic configuration also has implications for the formation process of the filament (Low & Hundhausen, 1995).

3.3.3. Coronal Loops

The corona is the part of the solar atmosphere that appears in visible light during eclipses at the faint "crown" of emission around the occulted solar disk. The coronal plasma temperature of ≈1 Mk and higher is much hotter than in the photosphere or chromosphere. The coronal radiation features spectral lines that were initially believed to be due to a new element, named coronium, before it was discovered that highly ionized iron is responsible (Grotrian, 1939).

A coronal loop is a magnetic flux tube structure extending into the corona with both ends anchored to the photosphere (Reale, 2014). The plasma inside a coronal loop, in contrast to a coronal hole discussed in Section 3.3.4, is confined within the coronal loop. When this plasma is heated, its pressure and density increase, and it becomes brighter than its surroundings as seen in the bright thread-like structures in Figure 3.15.

Coronal loop structures were first identified in images from an early rocket mission (Vaiana et al., 1968) of a bright loop structure associated with an X-ray flare. The Orbiting Solar Observatory IV (OSO IV) (Krieger et al., 1972) offered low-resolution images of the X-ray coronal structure, but it was the *Skylab* X-ray telescope that allowed Golub et al. (1982) to study coronal loop evolution from the emergence of short, bright loops to their progressive expansion and decreasing brightness as their active-region magnetic field decayed. Recall that expanding active-region loops play a role in the Babcock–Leighton model for the activity cycle.

All of the corona visible in X-rays consists of magnetic loops whose lifetimes are much longer than their expected cooling times (Rosner et al., 1978). *Skylab* also revealed long-lived loops at EUV and UV temperatures. However, coronal active-region emission peaks at X-ray temperatures above 2 MK (Peres et al., 2000; Reale et al., 2009; Warren et al., 2011). Extended loops are well observed in EUV—see Figure 3.15 for an example from the Transition Region and Coronal Explorer (TRACE) 171 Å passband.

The tallest loop structures in the corona are called streamers. These are bright in visible light and are stretched outward to cusp-like apexes by the expanding solar wind plasma. It is streamers that form the distinctive crown-like patterns observed during eclipses.

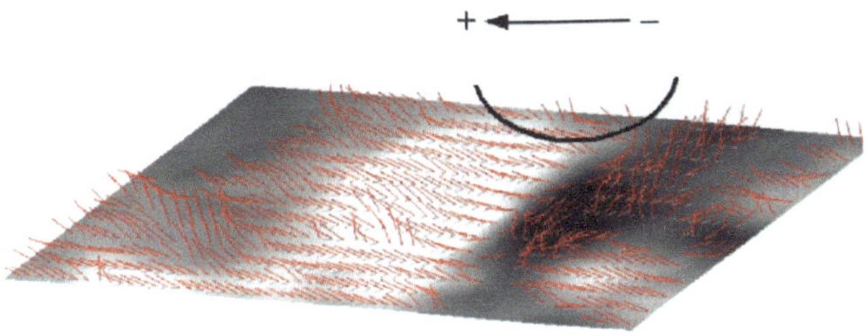

Figure 3.13 Three-dimensional representation of the photospheric vector magnetic field measured at a magnetic neutral line beneath a filament. Where the field vector in the negative polarity (black) points toward the neutral line and in the positive polarity (white) points away from it, the magnetic field is curved concave-upward like the bottom of a flux rope, as illustrated. (*Source*: López Ariste et al., 2006. © 2006.)

3.3.4. Coronal Holes and Solar Wind Structure

Coronal holes are regions of the solar atmosphere whose magnetic field is open to the heliosphere and whose plasma density is low compared to closed-field regions (see the reviews by Cranmer, 2009; Harvey, 2013). They are observed as dark patches on the disk in X-ray

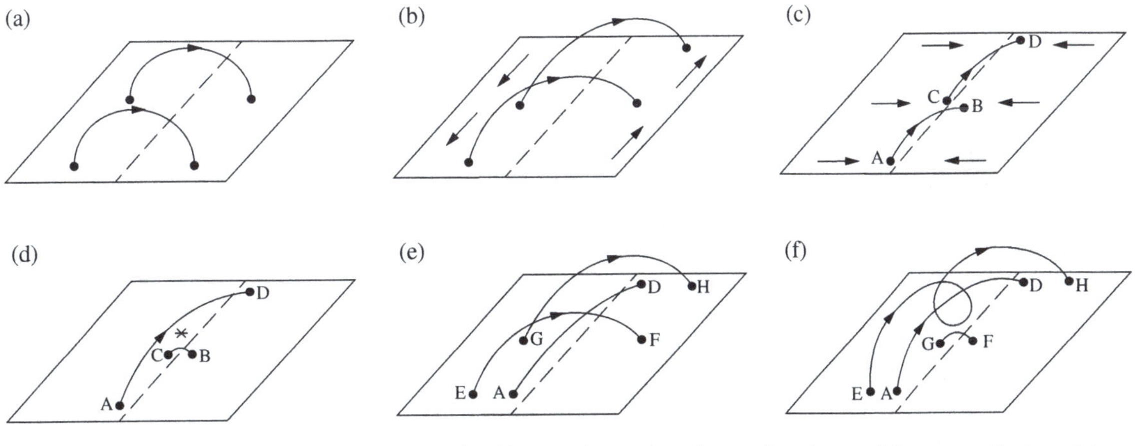

Figure 3.14 A diagram representing a model for filament formation due to shearing and flux cancellation. A loop arcade with footpoints in the photosphere extends into the chromosphere or low corona. The field is initially (a) an unsheared potential field, which is (b) sheared by photospheric footpoint motions. (c) Photospheric inflows toward the neutral line further increase the shear. (d) Magnetic reconnection produces along axial loop AD, and also a short loop CD that submerges. (e) Other loops EF and GH are pushed toward the neutral line, and (f) further reconnection produces the helical filament loop EH and a short loop GF that is submerged. (*Source*: van Ballegooijen and Martens, 1989. © 1989 IOP Publishing.)

Figure 3.15 A TRACE 171 Å image of the coronal loop structure above an active region. (*Source*: M. Aschwanden et al. [LMSAL]/TRACE/National Aeronautics and Space Administration, NASA.)

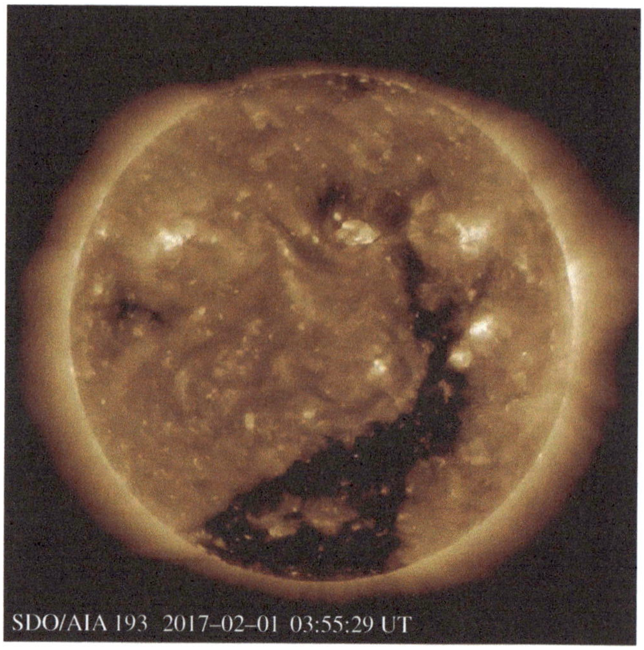

Figure 3.16 An SDO/AIA 193 Å image of the full solar disk showing a prominent south polar coronal hole extension to low latitudes. (*Source*: GSFC/Solar Dynamics Observatory/National Aeronautics and Space Administration, NASA.)

and EUV images, and as low-intensity regions above the limb in coronagraph and eclipse images. Figure 3.16 shows an example of an EUV image from the SDO/AIA in the 193 Å passband with a large extension of the south polar coronal hole to low latitudes. In models for the coronal magnetic field, all coronal fields open to the heliosphere may be interpreted as coronal holes. However, the agreement between observational and theoretical definitions of coronal holes is not perfect.

Low-intensity regions above the limb were first observed by Waldmeier (1956, 1957) in the green spectral line at 5303 Å. EUV images of coronal holes on the disk (Newkirk, 1967) and over the poles (Tousey et al., 1968) began to appear from early rocket missions. OSO measurements showed that coronal hole plasma is cooler and less dense than neighboring plasma (Munro & Withbroe, 1972), results that were confirmed by *Skylab* measurements (Huber et al., 1974). Coronal holes have been studied using limb observations of the Fe xiv 5303 Å coronal green line (Waldmeier, 1957), X-ray, and EUV observations, He I 10830 Å spectroheliograms (Harvey et al., 1975), and models such as the PFSS (Altschuler & Newkirk, 1969; Schatten et al., 1969), magnetohydrodynamic (e.g., Linker et al., 2011) and magnetofrictional (Mackay & Yeates, 2012) models. Harvey and Recely (2002) identified three classes of coronal hole: polar coronal holes confined to high latitudes, isolated coronal holes at active latitudes associated with remnants of decaying active regions, and transient coronal holes that briefly appear after CMEs. During activity minima, the two polar coronal holes are large and dominant, as Figure 3.7 illustrates.

The contrast between solar wind structure during solar minimum and maximum has been emphasized by in situ measurements from outside the confines of the ecliptic plane by the *Ulysses* spacecraft. *Ulysses* observations of the solar wind from both polar coronal holes have shown clear long-term variations in the solar wind parameters (McComas et al., 2008) and in the interplanetary magnetic field (Smith & Balogh, 2008). *Ulysses* completed orbits over both poles roughly every six years between 1992 and 2008, nearly three full orbits. Figure 3.17 summarizes the solar wind structure during the three orbits. The first orbit revealed a simple bimodal structure in the solar wind around the solar minimum of cycle 22, with two distinct kinds of wind: (a) a fast, tenuous and relatively homogeneous wind at high latitudes originating from the polar coronal holes, and (b) slower, denser, and highly variable wind at lower latitudes. Around solar minimum the band of solar wind variability was narrow, confined to a few tens of degrees around the equator, consistent with a small dipole tilt angle.

The second orbit caught the rise and maximum of cycle 23 and showed a complex solar wind structure at all latitudes, including streamers and low-latitude coronal holes. The decline and minimum of cycle 23 were covered by the end of the second orbit and third orbit of

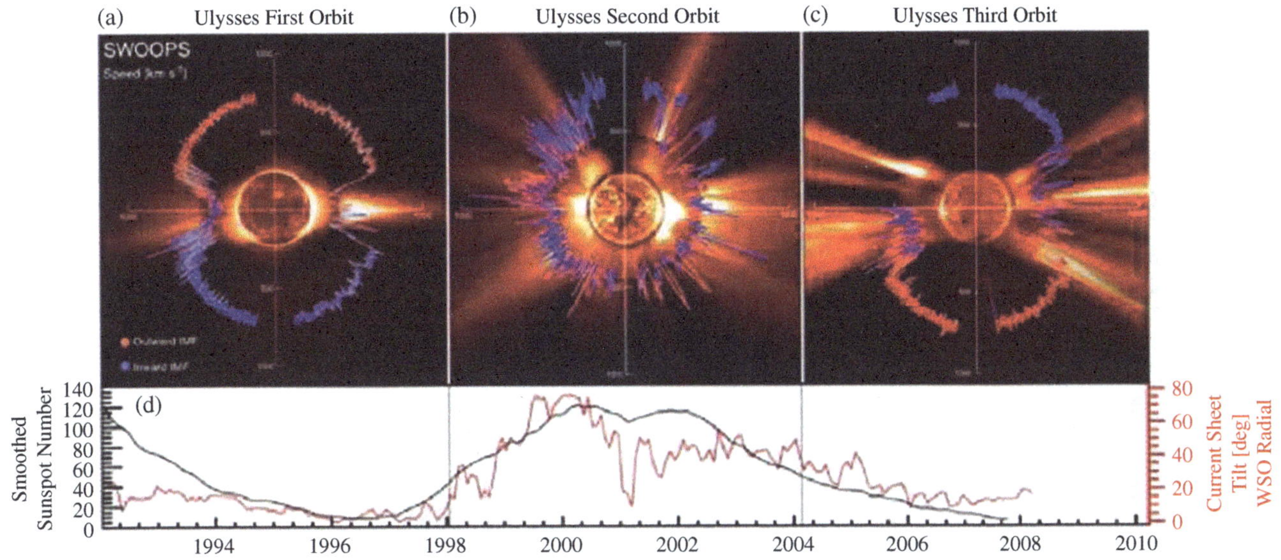

Figure 3.17 (a–c) Polar plots of the solar wind speed for *Ulysses*'s three polar orbits colored to indicate the measured magnetic polarity. In each, the earliest times are at the nine o'clock position and increase counterclockwise. (d) Contemporaneous values for the smoothed sunspot number (black) and heliospheric current sheet tilt (red), lined up to match panels a–c. In panels a–c, the solar wind speed is plotted around characteristic solar images for solar minimum for cycle 22 (17 August 1996), solar maximum for cycle 23 (7 December 2000), and solar minimum for cycle 23 (28 March 2006). From the center out, images from the SOHO Extreme ultraviolet Imaging Telescope (Fe XII at 1950 nm), the Mauna Loa K coronagraph (700–950 nm), and the *SOHO* C2 white-light coronagraph are blended. (*Source*: McComas et al. 2008. © 2008 John Wiley & Sons.)

Ulysses. At this time, the heliosphere in a simpler state but with a significantly more complex structure than observed during the first orbit. There was a more complicated streamer structure, with greater tilt with respect to the equator and a less planar belt of low-speed flow. The simple bimodal solar wind structure of the left and right plots in Figure 3.17 corresponds to the simple dipolar structure in the top left and bottom plots of Figure 3.7.

3.3.5. Flares and CMEs

Flares are localized brightening on the Sun that can span all wavelengths from radio to gamma rays (Benz, 2017). They are not to be confused with the CMEs that often accompany them. Flares brighten across the electromagnetic spectrum over timescales of minutes to hours as part of a conversion of magnetic energy into, besides electromagnetic energy, heat and motion. Figure 3.18 shows X-ray image reconstructions based on data from the Reuven Ramaty High Energy Solar Spectroscopic Imager (RHESSI) satellite overplotted on Hα images for two examples. Flares are powered by magnetic energy stored in the coronal magnetic field, the energy excess over the minimum, current-free energy state of the field, called the free magnetic energy. Some of this energy is released during the flare by magnetic reconnection, a process whereby the magnetic field reconfigures without violating the solenoidal condition. In the reconnection region, the plasma is locally heated to tens of millions of Kelvin, and electrons are accelerated to superthermal energies. Magnetic reconnection is not unique to flares. We have already seen it invoked as part of the Babcock–Leighton model for the solar cycle and in models of filament formation. What sets flares apart is the impulsiveness of the process. A slow buildup of magnetic energy in the corona is followed by a brief, violent event.

This localized energy release may be a part of a much larger instability of the coronal magnetic structure, which may expand and be expelled from the corona in a CME. CMEs consist of large structures that can take a variety of forms: some have the classical "three-part" structure of a compressed plasma envelope ahead of a flux-rope cavity containing a bright filament, whereas others display a more complex geometry or appear as narrow jets (Webb & Howard, 2012). They can drive interplanetary shocks, which are a key source of solar energetic particles, and they are a major contributor to severe space weather at Earth. A CME is a separate magnetic process from a flare, but CMEs often lead to reconnection and are often accompanied by flares. Although flares and CMEs often accompany each other, each can occur without the other. However, they often occur together,

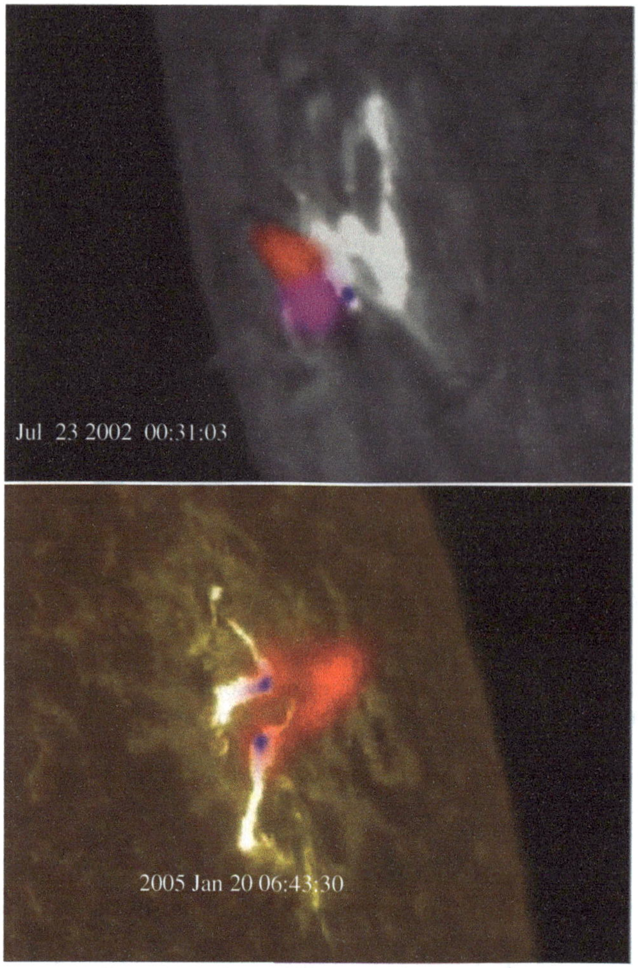

Figure 3.18 Top: Soft X-ray (red), hard X-ray (blue), and gamma-ray (purple) image reconstructions based on observations by the *RHESSI* satellite are overplotted on an Hα image. Bottom: *RHESSI* flare image reconstructions of soft X-rays (red 812 keV) and hard X-rays (blue 2050 keV) overlaid on an Hα image. (*Source*: Benz, 2017. Flare Observations, Springer Nature.)

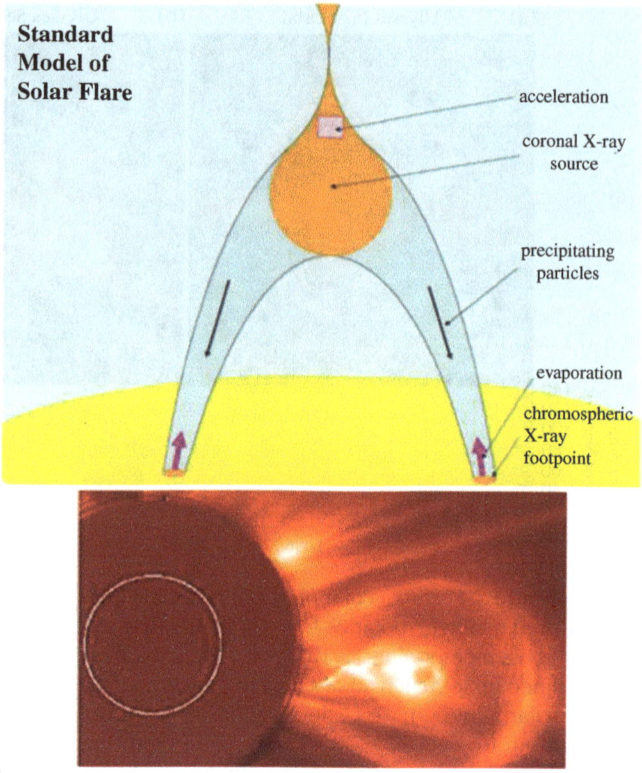

Figure 3.19 Top: Schematic drawing of the standard CSHKP (Carmichael Sturrock Hirayama Kopp Pneuman) flare scenario for energy release at magnetic reconnection sites where loop legs come into contact. Bottom: Schematic drawing of the one-loop flare model. (*Source*: Benz, 2017. Licensed under CCBY 4.0.) Right: White-light *SOHO*/LASCO coronagraph observation of an apparent X-point behind a coronal mass ejection. (*Source*: NASA.)

and it seems likely that any flare and CME accompanying each other are two parts of a common magnetic process.

Small flares can occur anywhere on the Sun, from sunspots to the quiet Sun. Large flares, however, tend to occur in active regions with a complex magnetic structure, particularly around strong, sheared magnetic neutral lines and in sunspots with mixed-polarity umbrae known as δ-spots. Large flares are often prompted by the emergence of strong magnetic flux from the interior and its subsequent interaction with preexisting active-region flux.

The coronal magnetic field can acquire large-scale shear through the interaction of different flux systems due to, for example, flux emergence, or through motion of photospheric footpoints. The characteristic timescale of the corona is fast (1000 km s^{-1}), and pre-flare coronal magnetic evolution occurs on longer timescales of hours to days, passing through a series of quasi-static equilibrium configurations. In this fashion, the coronal field can quietly store vast quantities of free magnetic energy. When magnetic reconnection takes place, prompted by a loss of equilibrium or an instability, some of this free magnetic energy is released in the chromosphere or corona. An increase in photospheric field strength is often observed during flares, which is consistent with the loss of magnetic energy and magnetic pressure called a magnetic implosion by Hudson (2000).

A standard, two-dimensional, much-simplified picture of the coronal magnetic field structure associated with a flare is provided by the CSHKP model, which is represented in Figure 3.19 (top). Named after Carmichael (1964), Sturrock (1966), Hirayama (1974), and Kopp and Pneuman (1976), this structure comprises a coronal loop system whose legs are pinched together to form an electric current sheet. As a result of reconnection at this current sheet, the top portion of the loop system is ejected

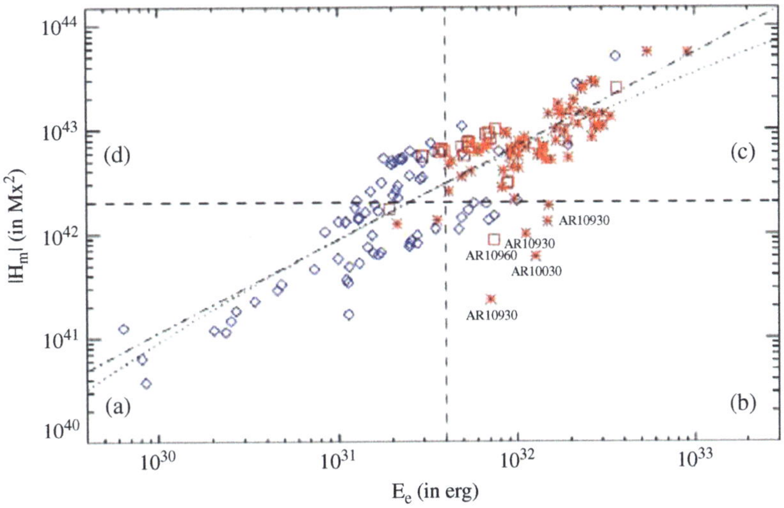

Figure 3.20 Scatter plot of free magnetic energy against relative magnetic helicity for 42 active regions. Blue diamonds represent non-flaring active regions, and red squares and asterisks represent M- and X-class flaring active regions, respectively. Dashed lines indicate the estimated thresholds for relative magnetic helicity ($\approx 2 \times 10^{42}$ Mx^2) and free magnetic energy ($\approx 4 \times 10^{31}$ erg) above which active regions are likely to produce major flares. The dotted and dash-dotted lines show the least-squares best fit and the least-squares best logarithmic fit, respectively. Some active regions are identified by their NOAA active region numbers. (*Source*: Tziotziou et al., 2012. © 2012 IOP Publishing.)

as a plasmid or a CME. A coronagraph image of a CME is shown in Figure 3.19 (bottom) with an apparent X-point at the apex of the loop system beneath it, consistent with the CSHKP model.

Because magnetic helicity is, to an excellent approximation, conserved in the solar atmosphere, and magnetic helicity is steadily being created in the solar interior and introduced into the solar atmosphere in each hemisphere, CMEs are believed to be the Sun's primary means of expelling excess magnetic helicity (Low, 1996, 2001; Rust, 1994). Magnetic helicity cannot be destroyed in bulk, and so, without CMEs, magnetic helicity of opposite sign would build up without bound in each hemisphere unless bodily removed at approximately the same rate that it is created.

Correlations between magnetic helicity and flare and CME activity have been found. Nindos and Andrews (2004) modeled the coronal magnetic fields of 78 active regions. They used linear force-free field models where the magnetic twist parameter α, the ratio between the vertical magnetic field and vertical electric current, is constant. They fitted these models to the vertical photospheric magnetic field data and adopted the value of α that minimized the difference between the modeled and observed horizontal field at the photosphere. They found that the resulting values of the parameter α and of the coronal magnetic helicity were statistically larger for CME-producing active regions than for non-CME-producing regions. LaBonte et al. (2007) found similar results for X-class flares. Tziotziou et al. (2012) estimated the free magnetic energy and magnetic helicity for 42 active regions and found that the modulus of the magnetic helicity is well correlated with the free magnetic energy, and that eruptive and non-eruptive active regions are well separated in their plot of the modulus of magnetic helicity against free magnetic energy.

3.4. THE SUN AS A STAR: SOLAR IRRADIANCE VARIABILITY

The interaction of the solar magnetic field with the dynamic plasma generates the large variety of structures and phenomena illustrated in the previous section. Each of these features is characterized by very different physical properties, which, in turn, result in characteristic radiative emissions. In the following chapter, we describe how the appearance in the solar atmosphere of magnetic structures affects the disk-integrated solar radiation, or sun-as-a-star measurements, focusing on the modulation occurring at the 11-year solar cycle, which is known to affect Earth's atmosphere physical properties and its climate.

Solar irradiance is the electromagnetic energy emitted by the Sun in the unit of time and area at a distance of one Astronomical Unit received at the top of Earth's atmosphere. The TSI is the irradiance integrated over the whole energy spectrum, whereas the term spectral solar irradiance (SSI) refers to the irradiance integrated over

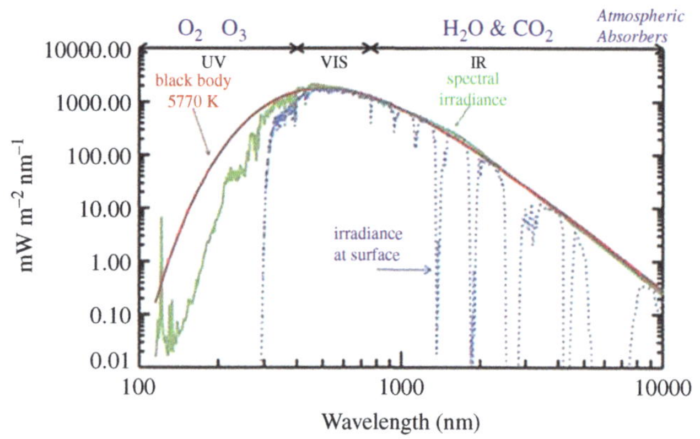

Figure 3.21 Solar spectral irradiance received at the top of Earth's atmosphere (green) and at Earth's surface (blue). For comparison, the red line shows the spectral dependence of the radiation of a black body at 5570 K. (*Source*: Lean, 2017. © 2017 Oxford University Press.)

specific spectral regions. The Sun is a G2-type star, whose spectrum is approximated by that of a black body at 5770 K. About 30% of the solar radiation is reflected back into space by clouds, ice, aerosols, and so on, and, as illustrated in Figure 3.21, approximately 20–25% is absorbed by Earth's atmosphere (mainly by molecular ozone and water vapor) so that approximately 50% of the total radiative energy reaches Earth's surface.

Measurements of the energy emitted by the Sun started at the beginning of the 1800s, when Jean Baptist Joseph Fourier proposed that the energy emitted by the Sun plays an important role in determining the terrestrial surface temperature. The first to measure the solar radiative energy were John Herschel and Claude Pouillet, who performed independent measurements from the Cape of Good Hope and Paris, respectively. By correcting his measurements for the absorption of the atmosphere above Paris, Pouillet estimated the solar energy to be 1228 W · m^{-2} (Dufresne, 2008), which, remarkably, is only about 10% less than the currently accepted value of 1361 W · m^{-2} (see below). His results, published in 1839, also indicated a seasonal variation of about 6%, which is also remarkably close to modern estimates (about 3%). Monitoring of the solar radiation through a variety of instruments started at the end of the 19th century, after Langley and Abbot suggested in the 1880s that variations in solar radiation might influence Earth's climate (Angione, 1981; Foukal, 1994). Because variations in the atmospheric conditions may significantly affect such measurements, solar radiation variability long remained an open issue, which was finally addressed approximately 50 years ago, through measurements carried out by spaceborne instrumentation. Systematic space-based measurements of the TSI started in the late 1970s, with the launch of the HickeyFreidan radiometer aboard the *Nimbus-7* satellite, which has been followed so far by about a dozen instruments (Kopp & Lean, 2011; Kopp, 2014). Each instrument provided different TSI values, not always within the estimated uncertainties.[1] The first measurements reported an average TSI of 1370.56 W · m^{-2}, a value that has been lowered by subsequent measurements (Kopp & Lean, 2011; Meftah et al., 2014). Currently, the International Astronomical Union recommends a TSI average value of 1361 W · m^{-2} (Prša et al., 2016), which is the value first provided by the Total Irradiance Monitor (TIM; Kopp & Lawrence, 2005) aboard the *Solar Radiation and Climate Experiment* (*SORCE*), and confirmed by subsequent missions (Kopp, 2016). Since the first measurements obtained with *ACRIM*, it has been clear that the TSI is not constant, but that it changes over several temporal scales (Kopp, 2016; Shapiro et al., 2017). Variations measured at the 11-year cycle are relevant for Earth climatological studies and are on the order of 0.1%.

Systematic measurements of spectral irradiance started in the 1960s with the launch of the OSO (Neupert, 1964) and have continued since, albeit discontinuously, with various missions. Unfortunately, different instruments monitored in time different part of the spectrum, and, until relatively recently, measurements focused on wavelengths shorter than 400 nm (Ermolli et al., 2013; Schmidtke, 2015). Regular observations at longer wavelengths started in 1996 with the Variability of IRradiance and Gravity Oscillations Sun PhotoMeter (Fröhlich et al., 1997) and even in this case have been continued up to the present with different instruments (Ermolli et al., 2013). In December 2017, the Total and Spectral Solar Irradiance Sensor (TSIS, Richard et al., 2011) was taken aboard the International Space Station. This experiment is scheduled to monitor solar irradiance at least for five years, and it is the continuation of the TIM and Spectral Irradiance Monitor (SIM, Harder et al., 2005) experiments, which have been flying since 2003 aboard the *SORCE* satellite. As with the TSI, it has been clear since the first systematic observations that the SSI varies on different temporal scales and amplitudes that are strongly dependent on the spectral range considered.

The continuous exposure to solar radiation and the general harsh environment of space pose an extreme challenge to the calibration of this type of instrumentation (e.g., Ball et al., 2016). The issue is complicated by the fact that measurements from a single instrument do not usually extend much longer than one cycle, so that records extending over several cycles, necessary to model the impact of solar irradiance on Earth's climate, are produced by "stitching together" measurements obtained by different missions. This, in turn, requires inter-calibration of instruments that are often very different in terms of optical setup, technology, and observational

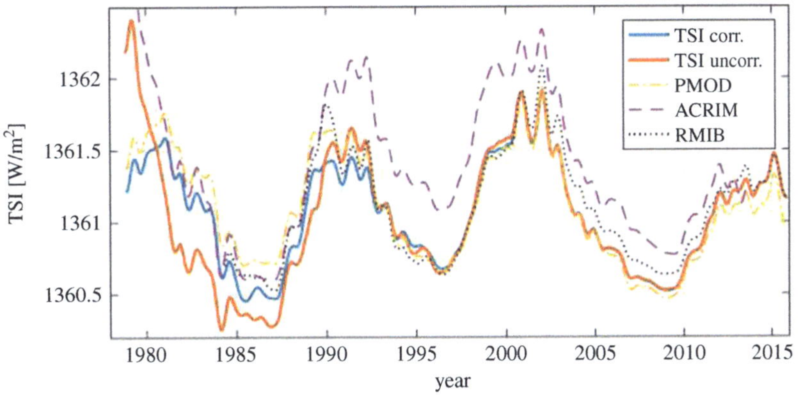

Figure 3.22 Comparison of five different TSI composites. (*Source*: Dudok de Wit et al., 2017. © 2017 John Wiley & Sons.)

strategy, with limited overlapping measurements, all of which make the creation of long time series an extremely challenging task. For the TSI, a few such "composites" have been compiled over the years (e.g., Dudok de Wit et al., 2017; Kopp, 2014), differing in the techniques and assumptions adopted to inter-calibrate the measurements and fill in the gaps. A comparison of various TSI composites presented in the literature is shown in Figure 3.22. Similar efforts for the SSI are also strongly hampered by the intermittency of measurements and by the nonuniform wavelength coverage. The most complete SSI composite, in terms of temporal and spectral coverage, is the one recently created in the framework of the First European Comprehensive Solar Irradiance Data Exploitation (SOLID; Haberreiter et al., 2017), an example of which is shown in Figure 3.23. As a result, irradiance variations at scales longer than one decade are still uncertain.

3.4.1. Irradiance Variability and Surface Magnetism

As mentioned in previous section, both the TSI and the SSI vary on different temporal scales. Radiometric measurements carried out in the late 1970s to early 1980s by instruments aboard *Nimbus-7* and *Solar Space Mission* provided clear evidence that such variations are modulated by the solar surface magnetism (e.g., Hudson, 1988; Willson et al., 1981). In particular, a recent analysis by Shapiro et al. (2017) showed that TSI variations from minutes to hours are dominated by granulation, with amplitudes of approximately 200 ppm. Longer time scales ranging from days to the solar rotation period (27 days) are dominated by the presence of sunspots over the solar disk, with a strong component from faculae, the latter dominating at scales longer than one year. In particular, the facular component dominates at the 11-year cycle scale, thus producing TSI variations in phase with the magnetic activity (as illustrated in

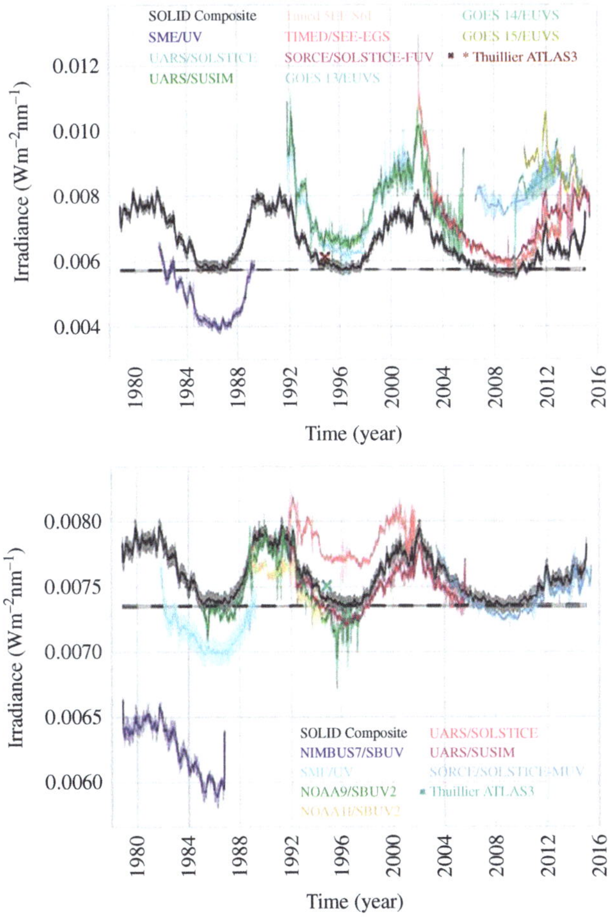

Figure 3.23 Comparison of SSI measurements obtained by various instruments with the SOLID composite. Top: 121.5 nm. Bottom: 200.5 nm. (*Source*: Haberreiter et al., 2017. © 2017 John Wiley & Sons.)

Figure 3.24). Occasional TSI variability is caused by highly energetic flares (Woods et al., 2004) and planetary transits (Kopp et al., 2005).

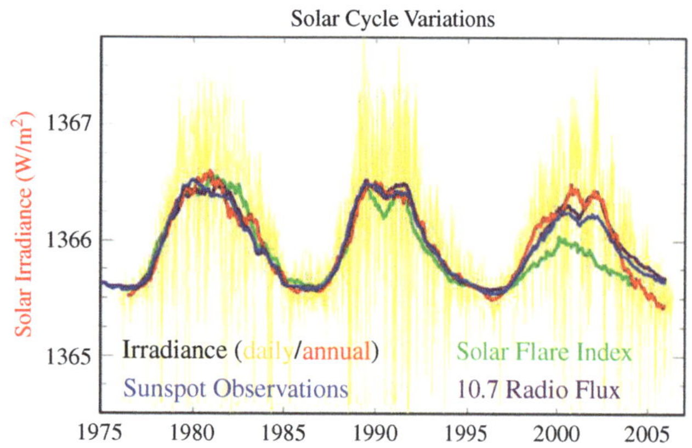

Figure 3.24 Variation of measured total solar irradiance (TSI) compared to the variability of various solar magnetic activity indices. The vertical scale of each index has been adapted to match that of the TSI.

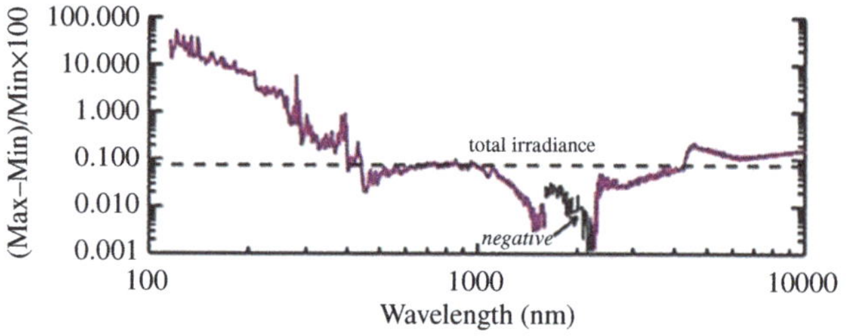

Figure 3.25 Solar spectral irradiance variability. (*Source*: Lean, 2017. © 2017 Oxford University Press.)

SSI variations are strongly dependent on the spectral region considered (Ermolli et al., 2013; Marchenko & DeLand, 2014). Variations in the SSI between a period of maximum of activity and a period of minimum of activity are illustrated in Figure 3.25. The UV undoubtedly varies in phase with the activity cycle, with amplitudes that overall increase with decreasing wavelength, ranging from several percent to less than a percent in the range 300–400 nm. Large variations are observed in strong lines forming in the chromosphere and transition region. The most prominent ones are Mg II h&k, Ca II H & K, and Ly_a, whose core intensities are positively correlated with the activity cycle. The UV spectral region is also dominated by the presence of lines, especially absorption lines of Fe I and molecules as CH, CN, and NH, which become gradually shallower with the increase of activity (e.g., Marchenko & DeLand, 2014, and references therein). Strong energetic events such as flares are known to produce large variations, and are intermittent in nature. The assessment of the SSI variations over long temporal scales in the visible and infrared (IR) is strongly hampered by the fact that it has been monitored for a relatively short temporal range with only a few instruments and that such variations are indeed on the order of only 0.1%. SSI variations in the visible have been reported to be in phase with the cycle, with the exception of data provided by the SORCE/SIM radiometers, which produced a counter-phase signal during the descending phase of cycle 23 (Harder et al., 2009), although there is increasing evidence that the first releases of SORCE/SIM data might suffer from improperly compensated instrumental effects (Mauceri et al., 2018). Measurements obtained by the recently launched TSIS will hopefully shed some light on this important issue. In the IR, at wavelengths longer than approximately 1 μm, the variability is in counter phase with the cycle.

As with the TSI, the SSI also presents variations over time periods related to the 27-day solar rotation. As mentioned above, there is strong empirical evidence that such variations are dominated by the appearance over the solar disk of magnetic structures, such as faculae and sunspots (Section 3.3.1). As illustrated in Figure 3.26, when a group of sunspots rotates into the earthward hemisphere, a brightening is observed in both the SSI and TSI. As the group of sunspots moves toward the central meridian, a gradual dimming of the TSI is observed, whereas the response of the SSI is strongly dependent on the wavelength range, with an increase observed at short wavelength ranges, and a decrease observed in the visible and infrared ranges.

The main physical mechanism by which the surface magnetism modulates the solar irradiance is the different way in which radiation propagates through magnetic structures. Because of the "hot wall" effect explained in Section 3.3, faculae appear bright when observed at slanted angles, and therefore the presence of facular regions toward the limb in general corresponds to an increase in the irradiance, both total and spectral. The contribution of faculae located close to the disk center is instead mostly negligible from the standpoint of the TSI, and is strongly dependent on the wavelength for the SSI. Specifically, the contribution is negligible or negative at wavelengths forming in the photosphere, that is, in the visible and IR ranges, and positive at wavelengths that sample higher layers of the solar atmosphere, that is, in the UV and shorter wavelength ranges. It should be noted that the detailed contribution of faculae strongly depends, as explained in Section 3.3.1, on the physical properties of the magnetic structures, for example, size, magnetic flux, and degree of aggregation (Criscuoli et al., 2017; Criscuoli and Rast, 2009; Ermolli et al., 2007; Ortiz et al., 2002). The contribution of sunspots is instead always negative for the TSI, and for the SSI in the visible and IR ranges, with a maximum negative contribution when these structures are located toward the disk center. In the UV and shorter wavelengths, the contribution is instead always positive. Note that, in spite of the negative contri-

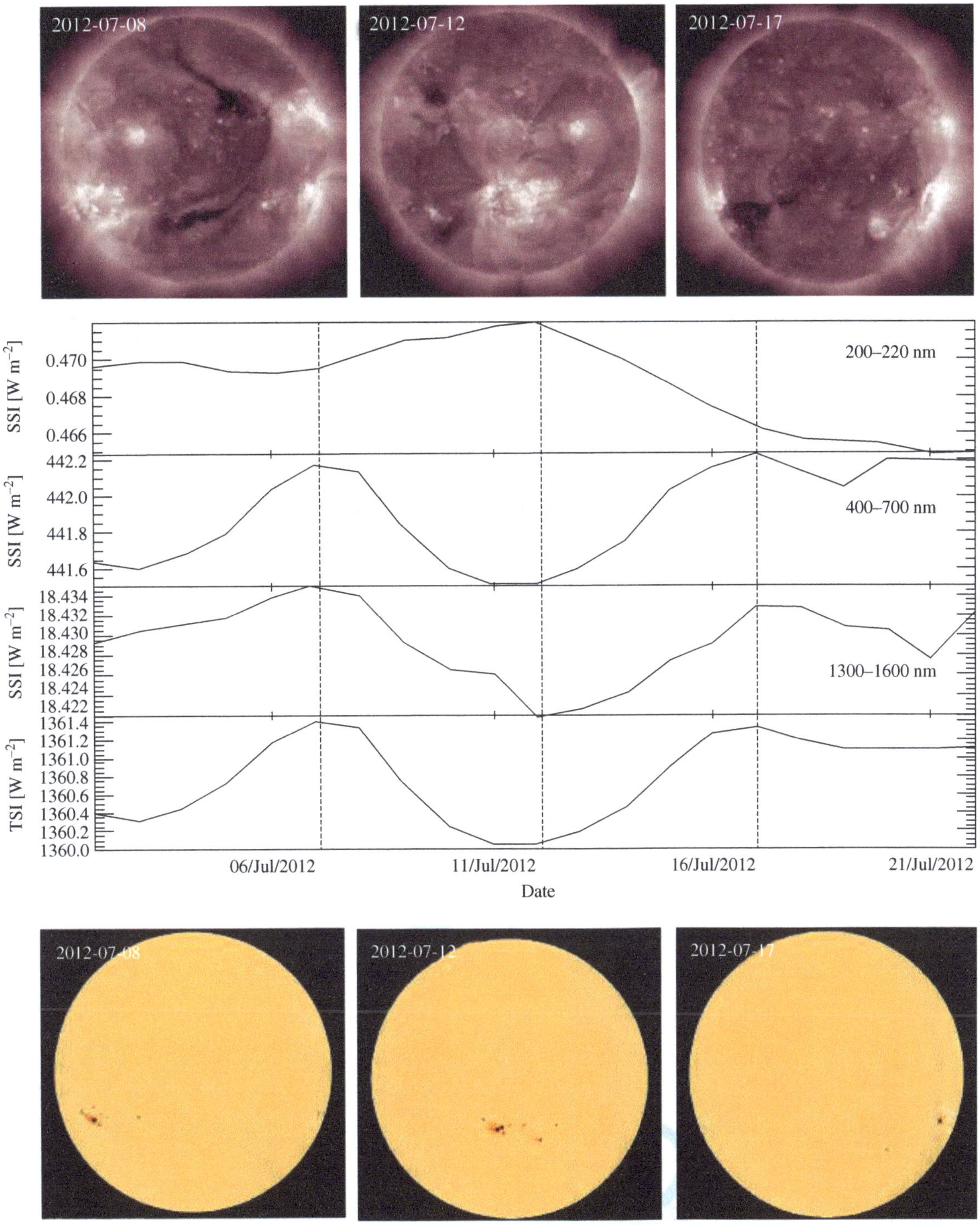

Figure 3.26 Top: Solar images acquired with AIA at 211 nm. Middle: Total solar irradiance (TSI) and solar spectral irradiance (SSI) variations measured with the TIM and SIM radiometers, respectively. Bottom: Intensitygrams acquired with HMI at 617 nm. The dashed vertical lines in the middle panel mark the dates in which data in top and bottom panels were acquired. (*Source*: Gordon Petrie.)

bution from sunspots, variations of the TSI over the 11-year solar cycles are in phase with the magnetic activity, as the positive facular contribution dominates.

3.4.2. Models for Solar Irradiance

The empirical evidence that irradiance variations are modulated by surface magnetism has led to the development of models that allow us to reproduce solar irradiance making use of different types of solar observations.

Semi-empirical models combine full-disk observations and synthetic spectra computed using models of the solar atmosphere to reproduce solar irradiance variations. More specifically, the variation of the area coverage of "archetypal magnetic features" is derived by segmenting full-disk observations. With each of these features is associated an atmosphere model that is used to synthesize the spectrum. The solar spectrum at any point in time is then reconstructed by weighing the synthetic spectra with the area coverage of the different features derived from the analysis of full-disk data. Typically, the solar atmosphere models used for irradiance reconstructions are the one-dimensional, static models derived by Juan Fontenla and collaborators over the years (FAL models, Fontenla et al., 1993; Fontenla et al., 1999; Fontenla et al., 2007; Fontenla et al., 2011; Fontenla et al., 2015), which describe the atmospheric stratification of quiet, network, faculae and sunspot umbra and penumbra (with certain sets of models providing subclasses of these features). Reconstructions based on semi-empirical approaches usually differ according to the observations employed, the number of archetypal features in which images are segmented, the set of atmosphere models, and the radiative transfer codes employed for the spectral synthesis. A schematic representation of the workflow of semi-empirical reconstructions is illustrated in Figure 3.27. The figure shows, in particular, reconstructions making use of full-disk observations acquired with the PSPT (Ermolli et al., 1998; Rast et al., 1999) in the Ca II K and red continuum, segmented into different classes of quiet and magnetic

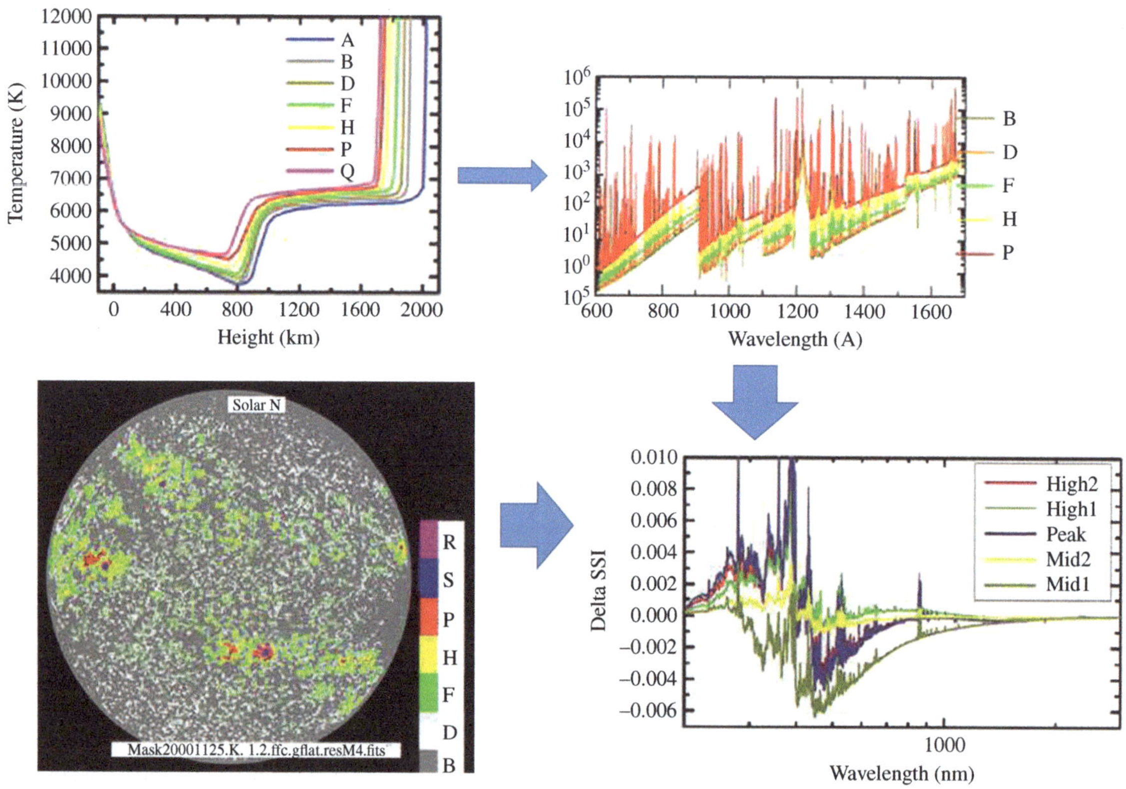

Figure 3.27 Schematic representation of the main steps involved in a semi-empirical irradiance reconstruction. Atmosphere models representing quiet and magnetic features (top left) are used to synthesize spectra (top right). The area and position over the disk at any given point in time of magnetic and quiet features are estimated making use of time series of full-disk images (bottom left). The synthesized spectra are combined with the area and position of the features to reconstruct irradiance variations (bottom right). The images represent in particular the workflow of the Solar Radiation Physical Modeling system, making use of PSPT full-disk observations and syntheses based in this case on atmosphere models published in Fontenla et al. (2011). (*Source*: Modified from Fontenla et al., 2011. © 2011 John Wiley & Sons.)

features with the algorithm described in Fontenla and Harder (2005). The area coverage of such features is then combined with synthetic spectra obtained under non-local thermodynamic equilibrium (NLTE) using FAL models to reconstruct spectral (e.g., Criscuoli et al., 2018; Criscuoli, 2019; Fontenla et al., 2007; Fontenla et al., 2011; Fontenla et al., 2015) and total (Ermolli et al., 2011; Fontenla et al., 2011) irradiance variations. Alternatively, magnetograms in combination with continuum images are also often employed. For instance, Thuillier et al. (2014) and Shapiro et al. (2015) made use of MDI magnetograms and filtergrams, and synthetic spectra obtained in NLTE to reproduce irradiance variations. LTE synthesis is also often employed for reconstructions, but in this case the solar model atmospheres are modified in order to avoid unrealistic emission of spectral lines, especially in the UV, resulting from the chromospheric temperature rise (Unruh et al., 1999). This type of synthesis has been employed, for instance, by Penza et al. (2003), Penza et al. (2006), and Criscuoli et al. (2018) in combination with PSPT observations. The Spectral And Total Irradiance REconstructions (SATIRE-S; e.g., Ball et al., 2014; Krivova et al., 2003; Yeo et al., 2014) also make use of LTE synthesis, but typically the different types of magnetic and quiet features are identified on full-disk magnetograms and filtergrams. More recently, Yeo et al. (2017) reconstructed the TSI employing magnetograms and filtergrams acquired with the HMI and synthetic spectra obtained under LTE assumption using magnetohydrodynamic simulations of the solar photosphere produced with the MURaM code (Rempel, 2014; Vögler, 2005). Finally, reconstructions making use of other types of imagery have also been developed. The EUV variability was reproduced by Lean et al. (2003) following the method described in Warren et al. (2001) making use of full-disk Ca II K and soft X-ray observations and spectra empirically derived from rocket observations. More recently, Haberreiter et al. (2014) and Fontenla et al. (2017) reconstructed SSI variability by employing coronal observations in combination with spectra computed using NLTE synthesis.

Proxy models make use of regression analysis of proxies of magnetic activity (e.g., Ca-index, sunspot number) against solar irradiance measured over particular temporal scales. Several irradiance reconstructions based on this approach have been proposed in the literature, but generally they all make use of at least two types of proxies, one to take into account the contribution of sunspots, and another for faculae and network. The Naval Research Laboratory reconstructions (NRL, (e.g., Lean, 2000; Lean & DeLand, 2012) are most frequently employed for Earth climatological studies. The most recent versions, NRL-TSIv2 and NRL-SSIv2 (Coddington et al., 2016), make use of measurements obtained over the solar-rotation temporal scale with radiometers aboard *SORCE*, regressed against sunspot area and the Mg II index (Viereck et al., 2001), which are proxies for the sunspot and facular contributions, respectively. The recently developed EMPirical Irradiance REconstruction (EMPIRE, Yeo et al., 2017) employs a similar approach and dataset, although it adopts a different mathematical method for the multivariate analysis (which, as discussed below, is the main cause of the discrepancies found with the NRL reconstructions, especially in the reconstructions of the UV). The San Fernando reconstructions (SFO) make use of facular and sunspot photometric indices derived from the analysis of full-disk solar observations acquired in different spectral ranges (e.g., Chapman et al., 2013; Preminger et al., 2002). Other proxies for the facular and network contributions are the flux at 10.7 cm (Tapping, 2013), the Ca II index (e.g., White et al., 1992), and the Ca facular index (e.g., Lean et al., 1998). Because the UV variability is dominated by the facular/network contribution, methods have been developed to reproduce the irradiance making use of these indices. The Mg II index is, in particular, a good proxy for the UV (e.g., Lovric et al., 2017; Thuillier et al., 2012), whereas the flux at 10.7 cm seems to be one of the best proxies for the EUV [10–120 nm] (e.g., Dudok de Wit et al., 2009; Tobiska et al., 2000). Ca II and 10.7 cm flux measurements are available since 1904 and 1945, respectively, so they are also employed instead of the Mg II index as proxies of the facular component for both TSI and SSI reconstructions, at times when Mg II index measurements were not available (e.g., Lean et al., 2001; Yeo et al., 2017).

Models for historical reconstructions. The models described above all make use of data and observations that date back at the earliest to the end of the 19th century. To extend estimates to previous times, other proxies and techniques have been developed. Several reconstructions make use of the sunspot number only, which is available since 1610. In this case, the variation in the distribution of other magnetic features is often estimated by using models (e.g., Bolduc et al., 2012; Dasi-Espuig et al., 2016; Wang et al., 2005), or by empirical relations derived from modern observations (e.g., Egorova et al., 2018) and then combining them with synthetic spectra. Other approaches, necessarily employed for reconstructions prior to 1610, make use of historical measurements of cosmogenic radioisotopes (McCracken & Beer, 2007), whose concentration on Earth is modulated by the properties of the solar magnetic field and therefore by the solar magnetic activity (e.g., Arlt & Weiss, 2014; McCracken et al., 2013; Steinhilber et al., 2008; Usoskin, 2017). Even for long-term or historical reconstructions, irradiance variability is estimated by either a multivariate analysis (e.g., Lean, 2018; Roth & Joos, 2013; Steinhilber

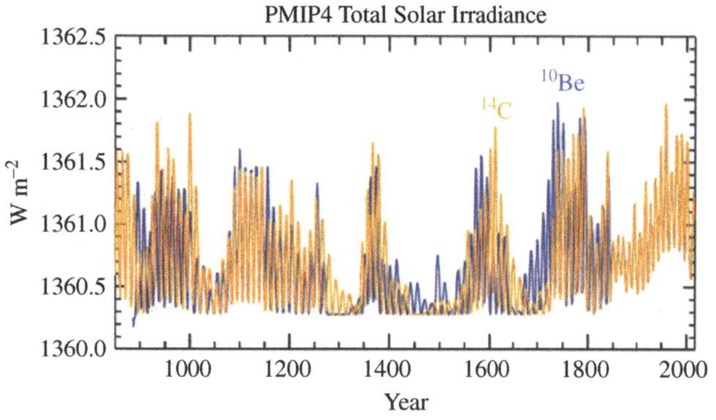

Figure 3.28 TSI variability obtained from two cosmogenic isotopes records using PMIP4 reconstruction. (*Source*: Lean, 2018. © 2018 John Wiley & Sons.)

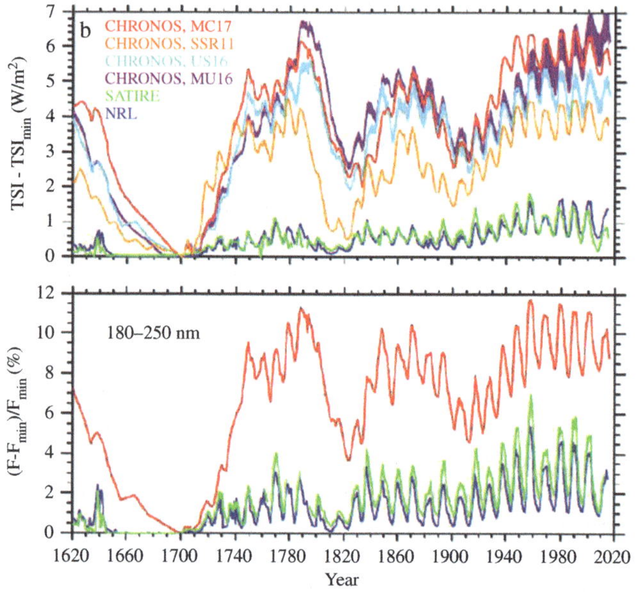

Figure 3.29 Reconstructions of irradiance variability from the Maunder minimum to the present day obtained with the CHRONOS model. The different curves represent reconstructions obtained using as input different time series of the solar modulation potential proxy. The variability obtained with the NRL and SATIRE models is shown for comparison. (*Source*: Egorova et al., 2018. © 2018.)

et al., 2009) or making use of more complex techniques. In the latter case, models are employed to estimate the *open* and *closed* solar magnetic field components (Section 3.3.4) from cosmogenic isotopes, and irradiance is then estimated making use of a semi-empirical approach (e.g., Vieira et al., 2011; Wu et al., 2018). Figure 3.28 shows TSI variability reconstructed using the Paleoclimate Model Intercomparison Project-4 (PMIP4) reconstruction (Jungclaus et al., 2017) obtained with the most rcccnt ^{14}C and ^{10}Bc time series. In spite of the discrepancies obtained using the two proxies, the reconstructions show that the TSI varies on several temporal scales. Besides the 11-year modulation discussed above, periods of grand-maxima alternate with periods of prolonged grand-minima, during which the variability is almost suppressed. These irregularities suggest that the dynamo processes driving the solar cycle (Section 3.2.2) regulate the variability of the solar magnetic field in an intermittent manner (Charbonneau, 2010).

3.4.3. Uncertainties in Irradiance Reconstructions

As with irradiance measurements, discrepancies exist between SSI and TSI reconstructions obtained with different models. In general, models and measurements of the TSI variability show an agreement better than 90% over the recent solar cycles. TSI estimates performed over temporal scales longer than one solar cycle generally show larger disagreement, whereas SSI estimates, especially in the UV, show discrepancies of several tens of percent even for reconstructions of recent cycles. The amplitude of such discrepancies is shown, for instance, in Figure 3.29, which compares reconstructions performed with the Code for the High spectral ResolutiOn recoNstructiOn of Solar irradiance (CHRONOS; Egorova et al., 2018) with NRL and SATIRE reconstructions.

The origin of such discrepancies is diverse. Different models make use of different assumptions, data sets, and methodology. For instance, the EMPIRE and NRL models are both based on proxies regressed on *SORCE* data. Nevertheless, the codes produce different SSI variability and different TSI variability over long temporal scales. Yeo et al. (2017) showed that discrepancies in the SSI variability should be ascribed to differences in the numerical approach used to perform the multivariate analysis, whereas discrepancies in TSI reconstructions most likely result from the different record of sunspot number employed. The effects of using different sunspot number records have also been discussed in detail in Kopp et al. (2016) and more recently by Wu et al. (2018). It is also important to take into account the fact that relations between proxies may vary from cycle to cycle (e.g., Brown & Evans, 1980; Bruevich et al., 2014; Tapping & Morgan, 2017) and show different dependencies during the same cycle (e.g., Bachmann & White, 1994; Criscuoli, 2016; Salabert et al., 2017), which, of course, introduces uncertainties in the reconstructed variability and contributes to the discrepancies obtained with different models. For instance, Foukal (1993) argued that the deviation from linearity between white-light faculae and the Ca II-index suggests that the TSI might be overestimated in models making use of the Ca II-index as proxy for the facular contribution. More recently, Vieira and Solanki

(2010) showed that the relation between the TSI and proxies used for historical reconstructions vary in a manner that is more complex than linear. Another major source of uncertainty is the fact that the estimate of the heliospheric magnetic field through the concentration of cosmogenic isotopes is strongly model-dependent (Usoskin, 2017). By modeling the physical and chemical processes that determine the production, transport, and deposition of cosmogenic isotopes, it is possible to reconstruct the solar modulation potential, which is then converted into properties of the solar photospheric magnetic field. The effects of using different solar modulation potentials on historical irradiance reconstructions are illustrated in Figure 3.29.

Another source of uncertainty is the contribution of the quiet Sun, which is assumed as the reference in several irradiance reconstruction models (Criscuoli & Uitenbroek, 2014). Misidentification of quiet Sun pixels, due, for instance, to data quality or to the (arbitrary) definition of quiet areas on full-disk observations, may not take into account of a variable component. Peck and Rast (2015) showed, for instance, that spectral irradiance variability in the visible spectrum reconstructed utilizing SFO proxies is affected by the criteria adopted to define the quiet Sun. On the other hand, the nature of the variation of the physical properties of the quiet Sun, especially at scales longer than one cycle, is still debated, and it is critical especially for the assessment of irradiance variability from the Maunder minimum (1645–1715, a period in which records indicate an unusually small number of sunspots over the solar disk) to the present day (e.g., Egorova et al., 2018).

3.4.4. Other Physical Mechanisms for Solar Irradiance Variability

Although irradiance reconstruction techniques based on the assumption that irradiance variations are modulated by surface magnetism allow us to reconstruct most of the observed TSI variability, other physical mechanisms have been invoked, especially to take into account possible secular variations. Internal structural changes, resulting from the interaction of magnetic field and plasma convective motions, have been suggested (Sofia & Li, 2005). Such a possibility has been investigated by the use of 1D and 2D stellar evolutionary codes (e.g., Li et al., 2003; Passos et al., 2007; Ventura et al., 2010). Recent estimates obtained making use of state-of-the-art 3D global Radiative Magneto Hydrodynamic simulations (Cossette et al., 2013; Cossette et al., 2017) found an upper limit of 0.01% to the contribution of TSI variations resulting from internal structure changes. Independent observational evidence of such internal changes has been sought in variations of the solar radius and helioseismological measurements. The former has led so far to rather controversial results (Rozelot & Damiani, 2012), with most recent studies pointing to a constant solar diameter (e.g., Meftah et al., 2015; Vaquero et al., 2016). Stronger evidence of internal changes comes instead from helioseismological studies. Independent recent investigations by Kosovichev & Rozelot (2018) and Jain et al. (2018) of the temporal variations of fundamental modes (f-modes) suggest a cyclic variation of the helioseismological radius. Such variations are in modest anti-phase with the magnetic cycle, have amplitudes of a few kilometers, and occur in the sub-surface layers (up to 15 Mm).

3.4.5. Effects of Solar and Stellar Irradiance on Planet and Exoplanet Atmospheres

Solar radiation is unquestionably Earth's primary source of energy, as it is approximately three thousand times larger than the sum of all other sources of energies (Kren et al., 2017). Among such sources, it is the one that most affects Earth's atmosphere and the sustainability of life. Indeed, variations in the TSI and SSI measured at different temporal scales affect the chemistry and dynamics of Earth's atmosphere (Ermolli et al., 2013; Lockwood, 2011; Matthes et al., 2006; Matthes et al., 2017; Seppälä et al., 2014).

Irradiance variations in the extreme-UV affect mostly Earth's ionosphere, and may create disturbances in radio wave propagations and satellite orbits due to air-drag increase (Floyd et al., 2002). The Ly-α (121.6 nm) and the spectral region between 220 and 300 nm are absorbed in the stratosphere, heating the atmosphere and affecting the abundance of ozone, which changes the atmospheric dynamics, thus affecting the tropospheric circulation patterns (top-down mechanism; e.g., Matthes et al., 2006; Seppälä et al., 2014). Radiation at longer wavelengths reaches the troposphere, heating oceans and affecting global circulation and precipitation patterns (bottom-up mechanism; Gray et al., 2010). TSI variations, together with volcanic eruptions, are the major natural contributors to the global changes of temperatures observed since the pre-industrial era. More specifically, the increase in solar magnetic activity from the Maunder minimum throughout the 19th century and volcanic eruptions most likely have antagonistic effects, with the increase in solar activity contributing to the increase in global temperature, and particles ejected into the atmosphere during volcanic eruptions shielding solar radiation and causing temperature decrease. It must be noted that although the importance of these mechanisms has been well established, different models describing the chemical and physical interaction of solar radiation with Earth's atmosphere still lead to discrepant results (Matthes et al., 2017). The challenges described above in providing long-term (e.g., decades and longer), uninterrupted and consistent time series of TSI and SSI measurements also hinder the correct assessment of solar variability on Earth's atmosphere. Moreover, space-based irradiance measurements started about fifty years ago, and have been mostly discontinuous, especially for what concerns the SSI. Reconstructions of solar irradiance obtained from models are in this context extremely useful for providing continuous estimates, and extend them into the past, when measurements were not available. Nevertheless, as with the measurements, the discrepancies described in the previous section between TSI and SSI variations obtained with different irradiance reconstruction models are so large that we cannot accurately assess the effects of solar irradiance variations on the photochemistry of Earth's atmosphere (e.g., Ball et al., 2016; Ermolli et al., 2013; Kopp, 2016).

Studies of solar variability are also fundamental for improving our understanding of stellar variability (see, for instance, the recent reviews by Fabbian et al., 2017; Faurobert, 2019), which, in turn, helps determine the habitable zones of stars (e.g., Gallet et al., 2017). As for our planet, modeling of exoplanet atmospheres requires as fundamental input the radiation emitted by the host star, especially in the UV and shorter wavelengths (e.g., Rugheimer & Kaltenegger, 2018; Shields et al., 2016; Tian et al., 2014). For this reason, great efforts have been dedicated to measuring stellar UV spectra (see Linsky, 2017, for a recent review), with recent particular interest in the spectra of K and M dwarfs (France et al., 2016; Youngblood et al., 2017), their planets being suitable for spectroscopic biomarker searches (Cowan et al., 2015; Kaltenegger, 2017; Scalo et al., 2007; Rugheimer & Kaltenegger, 2018). Stellar irradiance variability induced by magnetic activity may also affect the detectability of exoplanets (Oshagh, 2018). More specifically, the passage over the stellar disk of spots and faculae may induce photometric variations of amplitudes similar to or larger than the photometric variations induced by planetary transits. Similarly, the presence of active regions may alter spectral line profiles, thus hindering exoplanet detections performed through radial velocity measurements. Finally, spectroscopic techniques that allow the physical properties of exoplanet atmospheres to be estimated (e.g., Kreidberg, 2018) require as fundamental input the spectra synthesized through models representing quiet and active regions (faculae and sunspots).

3.5. CALCIUM II K OBSERVATIONS

In previous sections, it was mentioned that proxies, or indices, of solar activity are used in studies of the magnetic field in the past, when measurements were not available, and in studies of solar irradiance variability. Over the years, several proxies have been proposed and monitored to investigate the activity of the Sun and, similarly, of other stars. Proxies are classified as *direct* and *indirect. Direct indices* are measurements typically performed in lines whose intensity is particularly sensitive to magnetic activity, such as the MgII, Ca II, and 10.7 cm radio flux. In the case of the Sun, whose magnetic structures are spatially resolved, sunspot number and area, sunspot group, filaments, and so on, are also widely used proxies. Indirect indices are measurements that can be indirectly related to the magnetic activity, such as the abovementioned ^{10}Be and ^{14}Ca, together with the neutron flux, which is a by-product of the interaction of cosmic rays with Earth's atmosphere. Other widely used proxies are geomagnetic indices, which are direct measurements of the geomagnetic field, whose properties are modulated by the solar magnetic activity. Comprehensive reviews of solar indices are provided in Ermolli et al. (2014), Lockwood (2013), and Usoskin (2017).

Among the various proxies proposed in the literature, in the following we focus on the description of observations carried out at the Ca II K spectral range. Ca II lines are among the most important diagnostics of the solar chromosphere. Ca is one of the most abundant species in the solar atmosphere, and it is mostly ionized at temperatures typical of the photosphere and chromosphere. Therefore, Ca II lines are deep and wide, thus making their observation relatively easy. The cores of the resonant doublet Ca II H&K lines (at 396.8 and 393.3 nm) probe the lower chromosphere, a region spanning approximately 400 to 2500 km. The Ca II infrared lines also probe the chromosphere, but they are formed lower down, so that the H & K lines are by far the most important chromospheric diagnostic in the visible region of the spectrum. Within magnetic structures, the core of Ca II H&K goes into strong emission, thus making these lines an excellent proxy of magnetic activity. The correlation between chromospheric emission observed in Ca lines and magnetic activity has been known since the early measurements of the solar magnetic fields by Babcock, who reported "There appears to be virtually a one-to-one correspondence between prominent regions showing calcium flocculi on the spectroheliograms and strong Bipolar-Magnetic or Multipolar-Magnetic regions on the magnetograms" (Babcock & Babcock, 1955). Such correlation has been confirmed by several subsequent studies carried out with different types of instruments (e.g., Kahil et al., 2017; Leighton, 1959; Narang et al., 2019; Nindos & Zirin, 1998; Ortiz & Rast, 2005; Skumanich et al., 1975). Recently, Pevtsov et al. (2016) showed that Ca II K observations can be employed to produce synthetic magnetograms, or pseudo-magnetograms, an example of which is illustrated in Figure 3.30. Because Ca II K full-disk observations have continued, through

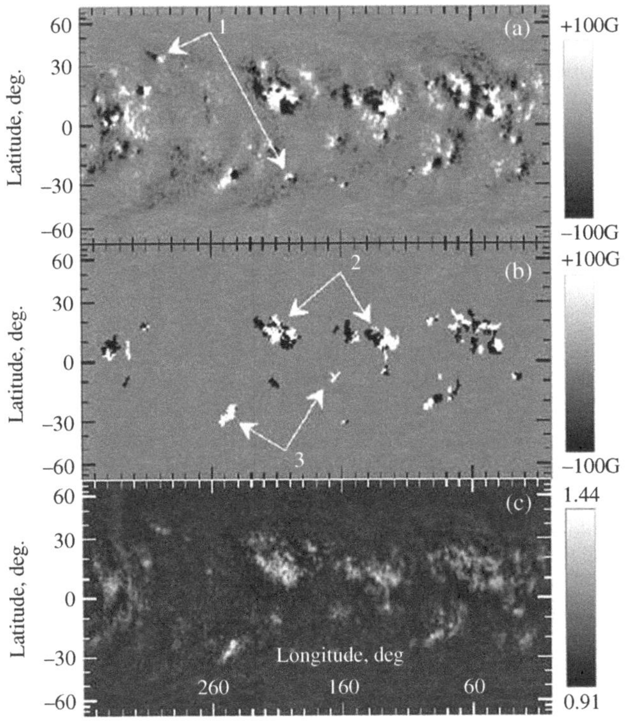

Figure 3.30 (Panel a) Synoptic maps of Carrington Rotation 1708 of pseudo-magnetograms (Panel b) derived by making use of Mount Wilson Ca II K spectroheliograms. (Panel c) The pseudo-magnetogram synoptic map well compares with the one observed at the Mount Wilson Observatory. The arrows represent different types of magnetic regions. (*Source*: Adapted from Pevtsov et al., 2016.)

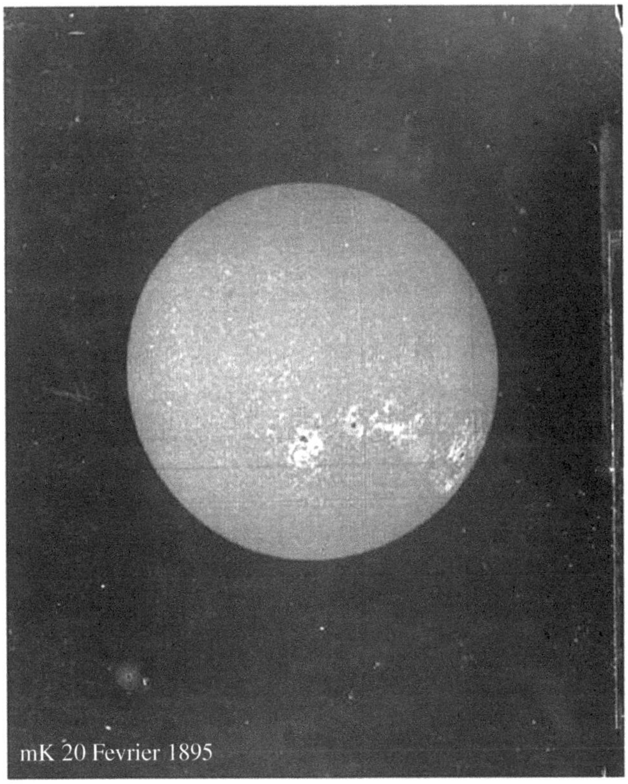

Figure 3.31 An early observation of the solar chromosphere in the Ca II K line. This image was taken on 20 February 1895, and it is part of the Meudon archive of historical spectroheliograms available at http://bass2000.obspm.fr/piwigo/index.php?/category/157. The bright patches visible on the disk are known as plages. They are associated with concentrations of magnetic fields and form a part of the network of bright emissions that characterize the chromosphere. (*Source*: Gordon Petrie.)

various programs, since the beginning on the 1900s, Ca images are the most important proxy to reconstruct the solar magnetic field over the past century. Tziotziou et al. (2012) estimated the free magnetic energy and magnetic helicity for 42 active regions, and found that the modulus of the magnetic helicity is well correlated with the free magnetic energy as shown in Figure 3.20. They also noted that eruptive and non-eruptive active regions are well separated in this figure.

3.5.1. Calcium II K Spectroheliograms

The first spectroheliograph was developed by George Ellery Hale in 1889, and within a few years the first images of the Sun in the Ca II K line became available. Figure 3.31 shows an example of a very early observation taken with a spectroheliograph. The image is from the Meudon archive of historical spectroheliograms, available at http://bass2000.obspm.fr/piwigo/index.php?/category/157. Clearly visible in this image is the presence of plage (Section 3.3.1). However, it was only after the turn of the 20th century that regular observations in the Ca II K line began at the Kodaikanal Observatory (India) and later, in 1915, at the Mount Wilson Observatory (United States). Within the next decade, analogous programs began at several other observatories. Full-disk Ca II K observations started in 1917 at the National Solar Observatory of Japan (Hanaoka, 2013), in 1919 at the Paris-Meudon Observatory in France (Mein & Ribes, 1990), in 1926 at the "Donati" solar tower telescope of the Arcetri Astrophysical Observatory in Italy (Ermolli et al., 2009b), and in 1926 at the Astronomical Observatory of the Coimbra University in Portugal (Garcia et al., 2011). In 1932, Robert Raynolds McMath (1891–1962) extended the spectroheliograph functionality in order to take motions pictures of the Sun.

Although not all these programs are still in operation today, they have provided a unique archive of solar images covering most of the Sun's activity during the past century (see Table 3.1).

The importance of Ca II K spectroheliograms for the study of solar magnetism and irradiance variability is well established. Observations of the solar surface reveal

Table 3.1 Relevant Ca II K Digitized Image Archives. Shown are: name of the observatory, years of operation, number of available images (unknown for some programs), and reference to the subsection in this chapter where the program is briefly discussed

Observatory	Years of Observations	Number of Images	Subsection
Mount Wilson	1915–1985	36,340	5.1
Kodaikanal	1907–1999	22,158	5.2
Meudon	1893–2018	20,986	5.3
Arcetri	1931–1974	5,133	5.4
Coimbra	1926–2007	~16,000	5.5
Mitaka	1917–1974	8,585	
Big Bear	1942–1987		
Kislovodsk	1957–present		
Rome	1964–present		

Information about the instrumentation and the digitized images for some of these historical observations are available from Table 1 in Ermolli et al. (2009a). Access to data archive: Coimbra: http://www.astro.mat.uc.pt/novo/observatorio/site/arquivo.html, Kodaikanal: https://www.iiap.res.in/?q=Kodaikanal_Solar_Observatory, Mitaka: https://solarwww.mtk.nao.ac.jp/en/database.html, Mount Wilson: ftp://howard.astro.ucla.edu/pub/obs/CaK/, Arcetri: https://www.oa-roma.inaf.it/arcetri-plates-archive/, Meudon: http://bass2000.obspm.fr/data_guide.php?what=all, Big Bear: Currently unknown, Kislovodsk: http://en.solarstation.ru/archive, Rome: Currently unknown.
Source: Modified from Ermolli et al. (2009a).

magnetic fields with complex hierarchical structures, evolving on a wide range of different spatial and temporal scales. The most prominent aspect of this variability is the solar cycle of activity, with a period of approximately 11 years for the sunspot numbers and a period of about 22 years for the magnetic polarity (Section 3.2.2). As discussed in Section 3.4, thirty years of satellite measurements of the Sun's energy output have revealed that the solar irradiance changes over the full range of time scales from minutes to decades (e.g., Fröhlich & Lean, 2004)), and this variability is wavelength dependent (Lean, 2001; Solanki & Unruh, 1998). Empirical models have shown that the variability in solar irradiance is indeed modulated by the area variations of the solar surface magnetic features, to a high degree of correlation (Foukal et al., 2004; Foukal & Lean, 1988; Solanki, 2003a). Two of these features, plages and chromospheric magnetic network, account for a significant portion of the Sun's total magnetic flux, UV and EUV variability, which play a critical role in determining the conditions in the heliosphere that directly influence Earth's magnetosphere (see the references in the paper Foukal et al., 2009, for a discussion on this subject). The Mount Wilson and Kodaikanal archives of ionized Ca II K line spectroheliograms provide a fundamental resource for these studies, and have been widely used as proxies to reconstruct the history of the solar magnetism and solar irradiance over the last 100 years and beyond (Foukal, 2003; Foukal et al., 2004).

3.5.2. Mount Wilson Calcium II K Spectroheliograms

Since the start of the 20th century, the monitoring program of the Mount Wilson Observatory has made available to the scientific community a huge number of solar images. This archive contains over 150,000 images of the Sun, which were acquired over a time span in excess of 100 years. The archive includes broadband images called White Light Directs, ionized Ca II K line spectroheliograms, and Hydrogen Balmer alpha spectroheliograms including both images of the solar disk and images of prominences above the solar limb. In 2003, the solar physics group at UCLA began a project to digitize essentially all of the Ca II K and broadband direct images of the archive, with 12 bits of significant precision and up to 3000 by 3000 spatial pixels. This project has completed the digitization of the Ca II K spectroheliogram series that began in 1915 and ended in 1985. Almost 40,000 solar images and step wedge images have been extracted and identified with original log-book parameters of observation time and scan format. Step wedge exposures on the Ca II K spectroheliograph plates began in October 1961 and were continued until the end of the program. The step wedges are rectangular in shape and have been imprinted by exposing the plate to a controlled lightbox after the solar image is obtained. Figure 3.32 shows the distribution of Mount Wilson Ca II K spectroheliograms per year.

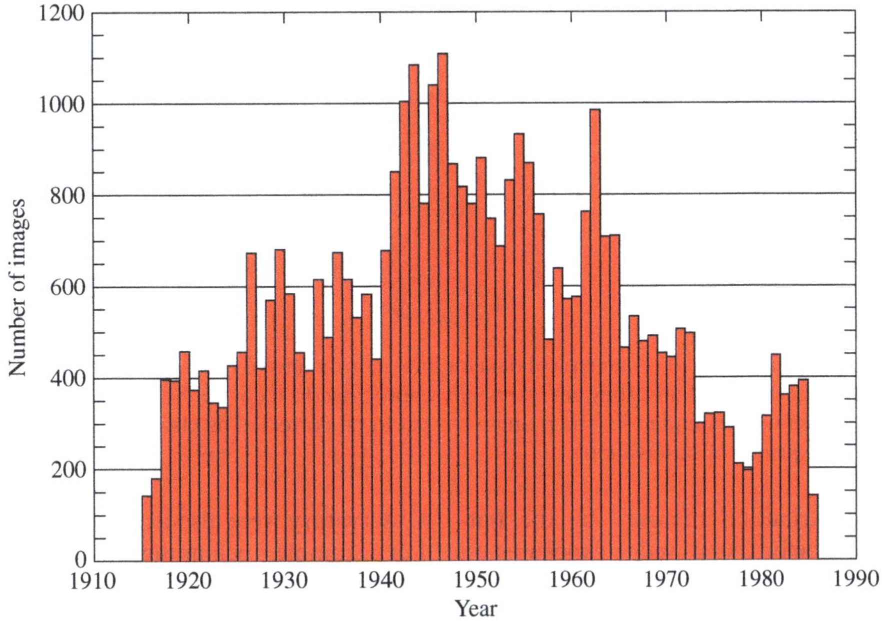

Figure 3.32 Number of available Mount Wilson solar Ca II K spectroheliograms per year. (*Source*: Bertello et al., 2010. © 2010, Springer Nature.)

The analysis of this data set permits a variety of retrospective analyses of the state of solar magnetism and provided a temporal baseline of about 100 years for many solar properties. The chromospheric network is clearly visible on a substantial fraction of the archive images from the Ca II K spectroheliogram sequence. A previous version of this database was used by Foukal and Milano (2001) to study the quiet network contribution to solar irradiance variations. These variations are important because they can be used to investigate the influence of the solar luminosity change on Earth's climate (Foukal et al., 2006). More recently, the digitized images obtained from this archive were independently calibrated by Tlatov et al. (2009) along with Ca II K data from Kodaikanal Observatory and from the National Solar Observatory at Sacramento Peak, for the purpose of comparing the signature of plages and enhanced network. The comparison shows a good agreement with respect to the temporal behavior of the annual and monthly mean plage indices calculated from the three data sets. A more extended comparison, which includes six time series of annual mean plages indices, shows a similar result (Foukal et al., 2009). The image quality and contents of the Ca II K spectroheliogram time series obtained by the digitization of the Arcetri, Kodaikanal, and Mt Wilson photographic archives have been also evaluated by Ermolli et al. (2009a) to estimate their value for studies focusing on timescales longer than the solar cycle.

The calibration of the Mount Wilson CaK images requires several important steps to account for various artifacts, as described in Bertello et al. (2010). First, the images notably present some dust and pit, which it is important to reduce. A Laplacian filter was used for this purpose. The size of the images was then reduced from its original scanned resolution of ~3000 × 3000 to ~866 × 866 spatial pixels, by averaging the pixel values within each 4 × 4 portion of the image. Second, the radii and center of the image were determined using a fairly complex iterative procedure that takes into account the distortion of the image. Because the average distortion of each image amounts to less than ~0.9%, with the radius along the scan/dispersion direction being typically smaller than the radius along the cross-dispersion/slit direction, an average value of the radius was used. This value ranges from about 336 pixels to 340 pixels for all the images investigated, with no significant long-term trends over the 70-year period covered by these observations. A sample of images is shown in Figure 3.33.

A major problem affecting these images is the presence of significant spatial variations in the background intensity level (vignetting). This is due to the relative position between the pupil and the grating, which depends on the coelostat mirror positions and shifts during each exposure due to the scanning of the spectrograph across the solar image. As a result of this effect, the intensity and its gradient are highly variable from one image to another.

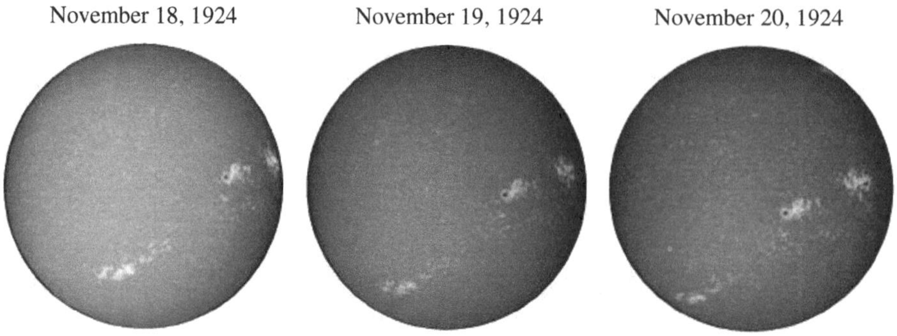

Figure 3.33 These solar images illustrate typical data for three consecutive days extracted from the Mount Wilson archive of Ca II K spectroheliograms. They are essentially unprocessed except for the removal of dust and pits and determination of their centers and radii. The p-angle has yet to be corrected as clearly indicated by the rotation of features in this sequence of observations.

This is the major contribution removed by the flat-fielding. As mentioned before, step wedge exposures on the Ca II K spectroheliograph sequence are available only since 9 October 1961. For those images, it is possible to use, following the ideas presented by de Vaucouleurs (1968), the characteristic curve (H&D) calibration approach to obtain well-calibrated intensity images. Unfortunately, the bulk of the sequence does not offer this possibility, and a different approach is required. In Bertello et al. (2010), a running median filter was applied to determine the large-scale spatial intensity distribution of the image, that is, its background. The image was then divided by this background to obtain the final flat-fielded image.

3.5.3. Kodaikanal Calcium II K Spectroheliograms

The archive at the Kodaikanal Observatory of the Indian Institute of Astrophysics in Bangalore hosts the longest record of spectroheliograms currently available. After the Cambridge spectroheliograph was put into operation, in late 1904, there was some problem with the settings of the second slit at the correct wavelength. Generally, the photographic plate remains fixed, and the spectroheliograph along with the second slit moves. Although spectroheliograms were obtained in 1905 and 1906, the instrument did not perform to its potential until 1907, when regular observations began (Hasan et al., 2010). The 70-micron exit slit of the spectroheliograph corresponds to a 0.5 Å bandpass centered on the core of K3 at λ393.37 nm. Spectroheliograms in the Ca II K line were obtained using the photographic emulsion until 2007. In particular, from the 1970s to late 2007, Scientia EM 23D56 emulsions from the AGFA company were used. After this stock was exhausted, in 2007, images in Ca II K were taken using Kodak film, but the results were not satisfactory, and the program was terminated the same year. Observations in Ca II K at Kodaikanal Observatory were also acquired using other instruments. For example, beginning in 1997, observations were obtained using a filter-based instrument (Daystar), with a 1.2Å bandpass and a Photometrics CCD camera. After 2008, Ca II K measurements, along with broadband images of the Sun, were recorded on two ANDOR CCD cameras simultaneously using the TWIN telescope (Singh & Ravindra, 2012). Although the TWIN telescope is still operational, due to problems with the CCD cameras, observations in the Ca K line have not been obtained since October 2013. Overall, the number of days available at Kodaikanal for making observations of the Sun had become less in the later part of the 20th century as compared to the first half of 20th century. Figure 3.34 shows the distribution of the Kodaikanal solar Ca II K spectroheliograms per year. A quick comparison with Figure 3.32 indicates that the number of daily observations taken at the Kodaikanal Observatory were typically fewer than those at the Mount Wilson Observatory.

At least two major projects have been undertaken in the past to digitize this archive of observations. Makarov et al. (2004) used a commercial scanner, with the setting 1200 × 1200 dpi, to produce 1800 × 1800 8-bit images. More recently, Priyal et al. (2014) have used a 4k × 4k CCD camera, a pixel size of 15μ square, and a 16-bit readout to digitize the images. Images from the first of these two projects were calibrated by Tlatov et al. (2009) to compute a Ca II K plage index and plage contrast using the same approach adopted for the Mount Wilson data. Their analysis shows that the mean plage contrast in the Kodaikanal data is slightly lower than the corresponding value determined from the Mount Wilson images. This is consistent with the fact that the Mount Wilson observations were taken with a moderately narrower spectral bandpass, meaning those two data sets sample different layers in the solar atmosphere. On the other hand, the two monthly averaged plage index time series show a fairly good agreement over the seven solar

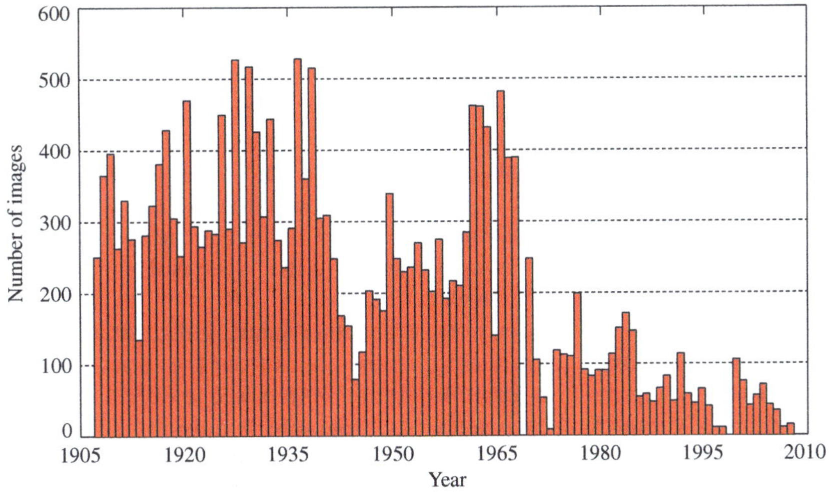

Figure 3.34 Number of calibrated solar Ca II K spectroheliograms per year. (*Source*: Modified from the Kodaikanal Archive.)

cycles (cycle 15 to cycle 21) covered by these observations. The relationship is linear, as shown in Figure 3.35.

Although both Mount Wilson and Kodaikanal images show similar features, there are some systematic differences that are worth noting. Due to the narrower spectral band used in the Mount Wilson observations, 0.35 Å versus 0.50 Å for Kodaikanal observations, off-limb prominences are quite visible in many of the Mount Wilson images but not in the Kodaikanal images. On the other hand, Kodaikanal images show more clearly the presence of umbrae. The two data sets also show a quite different pattern in the large-scale intensity distribution across the solar disk (vignetting), which can be removed by flat-fielding the images.

3.5.4. The Meudon Archive

The series of spectroheliograms of the Paris-Meudon Observatory dates back to 1919, with daily measurements acquired in Hα, Ca II K_{1v}, and Ca II K_3 (core). Annual coverage of the Meudon Ca II K observations is shown in Chatzistergos et al. (2019), figure 2. Except for the period between 1930 and 1964, the annual coverage is in general above or near 0.4. Two spectrographs are operated behind the entrance slit: a prism-spectrograph (2.1 Å mm^{-1}) for K_{1v} and K_3, and a grating spectrograph (3.22 Å mm^{-1}) for Hα (Mein & Ribes, 1990). The output slits are moved in front of the photographic plates so that the solar images are restored automatically. Both input and output slits are curved in order to reduce the geometrical distortions. Despite many improvements being made to the instrumentation between 1919 and 1988, the data collection exhibits a remarkable homogeneity. In its current configuration,

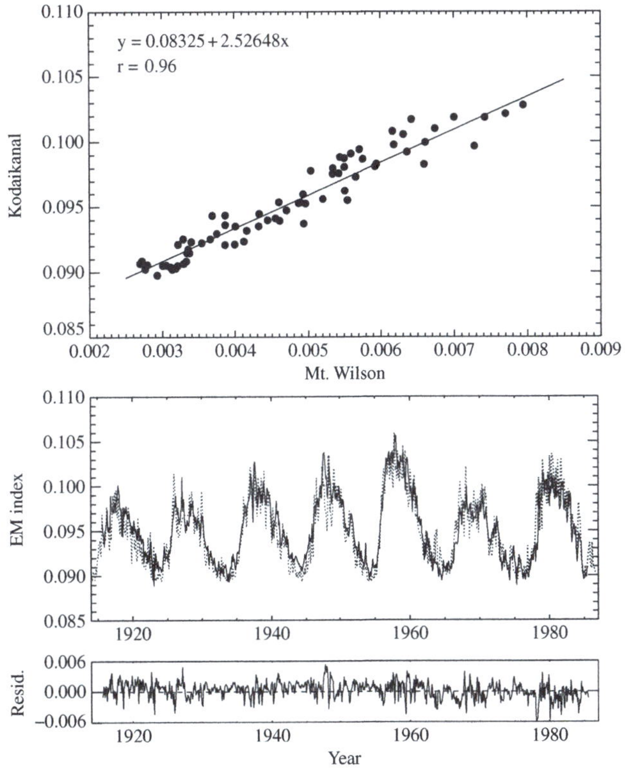

Figure 3.35 Left: Relationship between the annual mean values of the plage index derived from the Mount Wilson and Kodaikanal archive of Ca II K observations taken between 1916 and 1985. The statistical significance (*t* test) of this linear correlation (*r*) is greater than 99.9%. Right: Monthly averages of Kodaikanal (dotted line) and Mount Wilson (solid line) full-disk plage index measurements. Both time series were rescaled into a K line 1-Å emission index equivalent. The lower panel shows the residuals. (*Source*: Bertello et al., 2016.)

images are taken with a 1300 × 1300 14-bit CCD camera, with an image scale of about 1.5 arcsec/pixel. The spectral passband is 0.15Å centered on either K_{1v} or K_3. These images have been used to construct butterfly diagrams and synoptic chromospheric maps to study solar activity and, more recently, to locate radio bursts in connection with active regions.

A subset of images from this database was investigated by Ermolli et al. (2009a) and compared to similar images from other archives, including those previously discussed here. The investigation showed significant differences between observations taken on the same day available from these historical archives (see figure 1 in Ermolli et al., 2009a). For example, the Mount Wilson images show filaments over the solar disk, which are not found in some of the other data. The Kodaikanal images show sunspot regions which are not visible in the Mount Wilson spectroheliograms. On the other hand, the Meudon data show both of these features. However, due mainly to differences in the spectral bandpass, the sizes of the visible bright features (plages) are quite different in all of these images. Geometrical distortions, as revealed by the parameters of the ellipse that best fits the solar limb, are a common problem with those images. The average disk eccentricity for both the Kodaikanal and Mount Wilson images is about 0.12 ± 0.04, whereas for the Meudon images is 0.14 ± 0.02 (Ermolli et al., 2009a). The same study found that, on average, the Kodaikanal images appear to have a slightly higher spatial resolution than the Mount Wilson images, which spatial resolution is close to the one of current Meudon images. Large-scale image inhomogeneities introduced by variations in sky transparency during observations and by instrumental problems vary significantly between the different observing sites. On average, the Kodaikanal images show larger variations in intensity across the disk, followed by the Mount Wilson and Meudon images.

3.5.5. The Arcetri Archive

The digitized archive of the Arcetri spectroheliograms has been described in detail in Ermolli et al. (2009b). We provide here some basic information about the properties of those images. The archive contains almost 13,000 photographic plates of full-disk Ca II K and Hα spectroheliograms acquired from July 1926 to September 1974. Almost half of those images (5976) consist of Ca II K measurements. Between 1926 to 1955, an average of about 100 Ca II K observations per year were taken. This number increased significantly after 1955, reaching almost 400 around 1959–1960. During the last 14 years of the program, the number of observations has continuously and systematically decreased, as clearly indicated by figure 2 in Ermolli et al. (2009a). Observations were taken with a 600 lines/mm grating, and a dispersion of 0.33 mm/Å at 3934Å. Several instrumental changes were made, and different photographic products were used during the lifetime of this program. In particular, original plates stored in the archive have different dimensions and content as described in Ermolli et al. (2009b). One of the major issues related to these observations is the limitation for the study of details of the solar surface imposed by the local seeing at the Arcetri solar tower, which on average is about 5 arcseconds (Righini, 2003; Gasperini et al., 2004). On the other hand, unlike other similar historical datasets, most plates from the series of Arcetri observations contain the step wedge exposures for the calibration of the photographic emulsion to the flux of the incident radiation. This makes this series very suitable for cross-calibration purposes with other archives.

The digitization of this archive was performed by Ermolli et al. (2009b) using a commercial scanner, at 1200 dpi, and a system equipped with a 1K×1K CCD camera. Images were digitized with 16-bit gray-scale dynamic range, and the entire database can be accessed from http://www.oa-roma.inaf.it/solare/index.html.

3.5.6. The Coimbra Archive

In 1907, the Astronomical Observatory of the Coimbra University (Portugal), in collaboration with the Paris-Meudon Observatory, started assembling its first spectroheliograph. It was designed to be a replica of the instrument operating in Meudon, having identical and interchangeable components. A significant advantage of this close collaboration between the two observatories is that in general observations at one site cannot always be obtained every day, mostly due to meteorological conditions. However, the two observatories are very close in longitude to have almost the same window of observations but quite separated from each other to be affected by different weather conditions. Consequently, missing daily observations in one site can be filled with observations from the other site, resulting in a higher duty cycle.

However, it was not until 1926 that the Coimbra instrument began taking regular observations of the full solar disk in the Ca II K spectral line. In the late 1960s, the instrument was relocated from the old location in the city to a new observatory in Santa Clara, near Coimbra. The spectroheliograph was completely renovated in the early 1990s, and equipped with a new objective lens and a new optical branch with diffraction grating to allow for both Ca II K and Hα observations. In 2007, the use of photographic plates was terminated, and since then observations were acquired with a 12-bit CCD camera. Several parameters of solar activity are recorded and tabulated from the daily spectroheliograms including position, dimension, type, and evolution of sunspots, plages, filaments, and prominences. Between 1926 and 2007, photographic spectroheliograms were taken at three different wavelengths: 393.37 nm (Ca II K3), 393.23 nm (Ca II KV1), and 656.28 nm (Hα). The spectral bandwidth for the observations in the Ca II K line was set at 0.015 nm, significantly narrower than what was used for the Kodaikanal measurements. In 2007, after the installation of the CCD camera, changes have been made to the original program, and the bandwidth of the Ca II K observations has been slightly increased to 0.016 nm. Typical images have a size of 1000 × 1000 pixels with a plate scale of 2.2 arcseconds/pixel. Between 1 January 1926 and 31 July 2010, as many as 17,770 observations were taken in the core of the Ca II K line, and 8871 observations in Ca II K1. As shown in figure 2 of Garcia et al. (2011), from 1926 to 1964 the number of observing days for most years (roughly the number of Ca II K observations per year) varies between 170 and 230. This number then dropped significantly due in part to the relocation of the instrument to a new observatory, but started to increase above 200 since the late 1970s, reaching a value of about 300 observations per year at the end of 2010. All photographic material has been scanned and stored as 16-bit FITS and 8-bit JPEG files at http://tangerine.mat.uc.pt/novo/observatorio/site/index.html and at http://bass2000.obspm.fr/home.php.

3.5.7. Other Ca K Line Image Archives

In the mid-20th century, several other observatories conducted synoptic observations of the Sun in Ca II K line. Thus, for example, beginning early 1960, daily spectroheliograms were taken at Evans Solar Facility at NSO/Sacramento Peak observatory in New Mexico, United States. On- and off-band spectroheliograms were observed with a 0.0514 nm spectral bandpass centered at Ca II K line (λ393.367 nm). In addition, a smaller number of off-band images were taken with an offset of 0.038 nm to the red wing of the Ca II K line. The Ca II K line core spectroheliograms from the 1985–1999 period were digitized at 512-pixel, 8-bit resolution. Initially, four images per month (spaced as evenly as the data enabled) in the period 1985–1999 were digitized at 512-pixel, 8-bit resolution. A complete digitization of the entire collection of 7171 plates (6033 plates in the line core, and 1138 with red-wing offset) was carried out later with 1500-pixel, 16-bit resolution (see Tlatov et al., 2009). The spatial resolution of NSO/SP data from this (second) digitization is about 1.2 arcseconds per pixel, so the radius of the solar disk on NSO/SP images is about 750 pixels. The digitized Ca II K line spectroheliograms can be accessed via NSO Digital Library at http://diglib.nso.edu/ftp.html.

Observations in the chromospheric Ca II K line have been taken also at the Kislovodsk Mountain Astronomical Station of the Pulkovo observatory since July 1952. During the first four years of this program, measurements were recorded on photographic plates and then on film until August 2002, when they were replaced by a CCD array. Data from this archive can be retrieved from http://en.solarstation.ru/sun-service/chromosphere.

In 1917, the National Astronomical Observatory of Japan (NAOJ) began taking observations in Ca II K with a Grubb 30-centimeter siderostat and Cooke 18-centimeter objective lens telescope, and a Toepfer double-prism spectroheliograph. Daily observations were taken until the program was terminated in 1974. In total, more than 8585 images were recorded and digitized.

Daily full-disk, photometric images of the Sun in Ca II K line (393.4 nm) wavelengths have also been taken at the San Fernando Observatory (California, United States) since 1988 (Chapman et al., 1992) and at the Baikal Astrophysical Observatory of the Institute of Solar-Terrestrial Physics, Russian Federation (1995–present, http://en.iszf.irk.ru/KCa II-telescope), Mees Observatory of University of Hawaii (1995–1998, http,//www.solar.ifa.hawaii.edu/KLine/index.shtml).

3.5.8. Calcium II K Sun-as-a-Star Measurements

In addition to the disk-resolved spectroheliograms described in the previous section, several long-term Ca II K programs have been monitoring the properties of the Sun-as-a-star. Beginning in November 1976 and until September 2015, observations of the disk-integrated solar Ca II K line were made using the Evans Coronal Facility at the National Solar Observatory/Sacramento Peak, United States (Keil & Worden, 1984; Oranje, 1983). The light from a coelostat was fed to the horizontal Littrow spectrograph, and the spectrum from 3898 to 3954 Å was recorded by rotating the spectrograph grating. The optical setup employed cylindrical lenses to compress the image of the entire solar disk prior to the spectrograph slit. A photomultiplier was used as a light detector. On a typical observing day, the observer would take between 50 and 150 spectral scans. These were averaged, after removing scattered light and dark current, to form a daily mean profile. The Poisson noise level of individual scans is about 10% of the observed signal in the core of the Ca II K line. The random noise of daily averaged profiles estimated from the standard deviation of mean (50–100 individual profiles) is about 3–5% in the 1-Å emission index derived from these data (Keil, 2016). Because the spectral scanning did not extend far into the continuum, the flux calibration was done using the observed intensity at a fixed wavelength offset relative to K-line code (i.e., the average intensity in a 0.528 Å window centered at 1.187 Å in the red wing of the Ca II K line was set to 0.162 of the continuum intensity). The

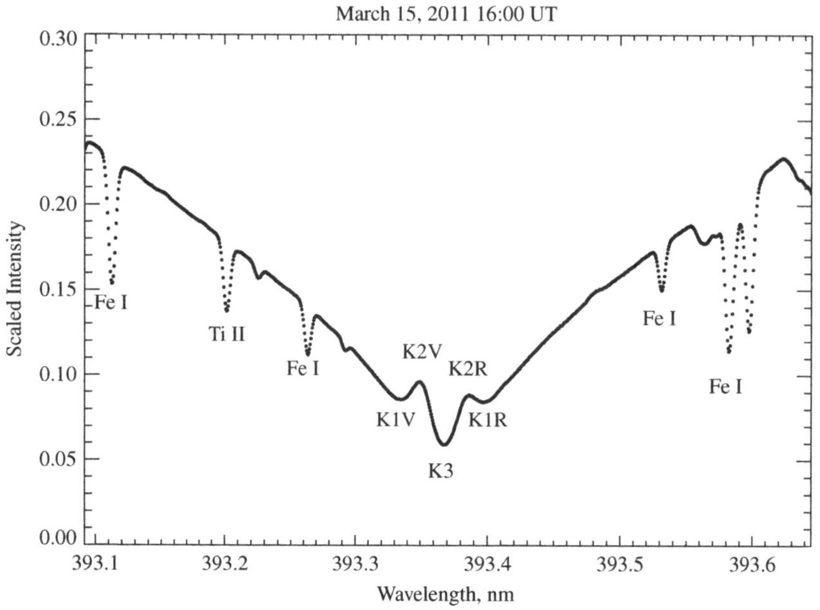

Figure 3.36 Typical Ca II K Sun-as-a-star spectrum obtained by the ISS instrument. (*Source*: Bertello et al., 2016.)

normalized line profiles were used to calculate several parameters such as the 1-Å emission index (the equivalent width of a 1-Å band centered on the K line core); the K3 (the core) intensity; the relative strength of the blue K2 emission peak with respect to the K3 intensity (K2V/K3); the separation of the two emission peaks (K2V–K2R); the line asymmetry, which is the ratio of the blue and red K2 emission maxima (K2V/K2R); the separation of the blue and red K1 minima (K1V–K2R); and the Wilson–Bappu parameter, which is the line width measured between the outer edges of the K2 emission peaks. These parameters are available from the NSO web site at http://nsosp.nso.edu/node/15. The Ca II K-Line Monitoring Program at NSO/Sac Peak had a duty cycle of 10% during 1976–1983 and about 45% between 1984 and 2015. On 1 October 2015, the NSO/Sac Peak Ca II K-Line Monitoring Program was stopped. The Ca K-line Sun-as-a-star time series is now continued using the Integrated Sunlight Spectrometer (ISS), one of the three instruments of SOLIS program Bertello et al. (2011). (We discussed the VSM in Section 3.2.) Figure 3.36 shows a typical Ca II K observation obtained by the ISS instrument. The spectral band is centered at 393.37 nm, with a bandwidth of 0.55 nm and a spectral resolution of about 300,000. Nine parameters are routinely derived from each spectrum, as described in Bertello et al. (2011).

The sun-as-a-star spectra in several photospheric and chromospheric wavelength spectral bands were also recorded at the Kitt Peak Vacuum Telescope (KPVT White and Livingston, 1978). In particular, the Ca II K line profiles were recorded from late 1974 to early 2013. The details of the initial instrument setup can be found in Livingston et al. (2007), but observations during different periods were taken with different instrument setups. Prior to 1992, the observations were taken with a spectrograph grating that did not sample the full solar disk completely (see Livingston et al., 2007, for a discussion of possible effects on data). The cadence varies between 2 and 13 observations per year (1974–1976) to a maximum of 87 observations per year in 2009. The 1-Å emission index derived from these data can be accessed via ftp://solis.nso.edu/kpvt/CaK_Indicies/K-line_stats.dat.

In 1969, M. K. V. Bappu started the monitor program of the Sun-as-a-star Ca II K line at the Kodaikanal Observatory (KKL) to study the Sun's long-term variations (Sivaraman et al., 1987). The observations made at the KPVT and the KKL have similar spectral resolutions, but data from the KPVT has better photometric accuracy because of the electronic and photographic film adopted. The data obtained at Sac Peak has less spectral resolution as compared to KPVT and KKL, but has more coverage around the year. Their initial investigation found that several parameters of the Ca II K line profile varied with solar activity and the Ca II K index varied by about 20% over the solar cycle phase, whereas the central intensity at K3 increased by about 30% during the maximum phase as compared to that at the minimum phase of the solar cycle. Following these observations, Skumanich et al. (1984) proposed a three-component model of cell, network, and plage to explain the observed changes in the Ca-K line profile, using the extant laws of limb darkening. They were able to fit the model profile with the observed one by considering the contribution of two components—cell and network—during the minimum phase. But during the maximum phase, the contribution due to the plage component alone was found to be insufficient to explain the observed profile. Therefore, they suggested an additional contribution due to the "active network component" to achieve good agreement between the model and observed Ca II K line profiles. This assumption explained the existence of extra emission during the maximum phase. However, measurements made at the center of the solar disk over a region of 1–3° showed no variation with the phase of the solar cycle (Livingston et al., 2007; White & Livingston, 1981), suggesting that the extra emission originates from higher latitudes.

To further investigate the findings by Skumanich et al. (1984), in 1986 a new program of solar observations was started to study the dependency of the Ca II K emission with latitude (Singh, 1988). Spectra are taken by using the following methodology: (a) A portion of the Sun's image is allowed to pass through the slit of the spectrograph. (b) While the Sun's image is moved with uniform speed from the east end to the west end on the slit of the spectrograph at a particular latitude, the spectra in the Ca II K wavelength are obtained. (c) The procedure is repeated for each 10° latitude belt, but the spectra are integrated over the longitude belt.

The analysis of Ca-K line spectra as a function of solar latitude and integrated over longitudes taken during the period 1989–2011 indicates that the K1 width reaches its maximum amplitude at various latitude belts at different phases of the solar cycle (Sindhuja et al., 2014). The FWHM of the K1 distribution at different latitudes shows that its width varies by about 30% for the equatorial belt (<30°) and 11% for the polar region (>70°) latitudes. Interestingly, the K1 width varies by about 6% around the 60° latitude during the solar cycle. Analysis of the cross-correlation coefficients of the K1 width between the 35° latitude and other latitude belts as a function of the phase differences indicates that the activity representing the toroidal field (Section 3.2.2) shifted at a uniform rate of about 5.1 m/s in the northern hemisphere from the mid-latitudes toward the equator. In the southern hemisphere, activity shifted at a faster rate, about 14 m/s, in the beginning of the cycle, and the speed decreased with time, yielding an average speed of 7.5 m/s toward the equator. The shift of activity in the higher-latitude belts showed complex behavior, indicating poleward and equatorward migration. These findings, especially the fewer variations in the mid-latitude belts as compared to the polar regions; asymmetry in the speed of the shift in the activity in both hemispheres; and complex variation in the direction of the shift in the activity representing poloidal fields in the mid-latitude belts can guide and constrain solar dynamo models.

In 1994, a Solar-Stellar Spectrograph (SSS) project started at the Lowell Observatory. The SSS project provides regular observations of solar spectral lines in a broad wavelength range including the Ca II K and H lines (Hall & Lockwood, 1998; Hall et al., 2007). Additional details about the SSS project can be found in Pevtsov et al. (2014) and the references therein. Stellar observations with SSS are made with a 1.1-m telescope, and observing runs last typically up to 10 nights per month. Between 20 and 50 stars are observed each night with an exposure time of about 10 min. The main goal of this program of stellar observations is to understand the long-term activity of Sun-like stars with a particular emphasis on characterizing their cycle properties (see, e.g., Hall et al., 2009).

Solar observations are also used to interpret and calibrate stellar measurements of Ca II K and H spectra in terms of stellar activity and cycles (e.g., Buccino et al., 2011; Giampapa et al., 2006; Grunhut et al., 2010; Rutten, 1984; Vieytes et al., 2009). As shown in some studies (e.g., Bertello et al., 2012; Donahue & Keil, 1995; Hasler et al., 2002), the connection between the modulation of the disk-integrated Ca K-line flux and solar surface differential rotation may open the possibility of using these measurements to determine the differential rotation of other main-sequence stars.

In a broader context, Ca II K observations have been used to estimate distances to early-type stars (Megier et al., 2009), and as diagnostics of velocity fields in red giant stars (Vieytes et al., 2011). It has also been suggested to use recurring chromospheric emission visible in Ca II K as a tool to detect massive extra-solar planets orbiting very close to their central stars (Lenz et al., 2011).

ACKNOWLEDGMENTS

The authors thank the referees for helpful comments, and the editors for patience during a difficult time.

NOTE

1. TSI measurements collected by different radiometers are available at https://www.ngdc.noaa.gov/stp/solar/solarirrad.html

REFERENCES

Alfvén, H. (1943). On the existence of electromagnetic-hydrodynamic waves. *Arkiv for Matematik, Astronomi och Fysik, 29B*, 1. ADS.

Altschuler, M. D., & Newkirk, G. (1969). Magnetic fields and the structure of the solar corona. I: Methods of calculating coronal fields. *Solar Physics, 9*, 131. DOI. ADS.

Angione, R. J. (1981). Review of ground-based measurements. In *NASA Conference Publication, 2191*, 11. ADS.

Arlt, R., & Weiss, N. (2014). Solar activity in the past and the chaotic behaviour of the dynamo. *Space Sci. Rev., 186*, 525. DOI. ADS.

Babcock, H. D. (1959). The sun's polar magnetic field. *The Astrophysical Journal, 130*, 364. DOI. ADS.

Babcock, H. W. (1953). The solar magnetograph. *The Astrophysical Journal, 118*, 387. DOI. ADS.

Babcock, H. W. (1961). The topology of the sun's magnetic field and the 22-year cycle. *The Astrophysical Journal*, *133*, 572. DOI. ADS.

Babcock, H. W., & Babcock, H. D. (1955). The sun's magnetic field, 1952-1954. *The Astrophysical Journal*, *121*, 349. DOI. ADS.

Bachmann, K. T., & White, O. R. (1994). Observations of hysteresis in solar cycle variations among seven solar activity indicators. *Solar Physics*, *150*, 347. DOI. ADS.

Bak-Stęślicka, U., Gibson, S. E., Fan, Y., Bethge, C., Forland, B., & Rachmeler, L. A. (2013). The magnetic structure of solar prominence cavities: New observational signature revealed by coronal magnetometry. *The Astrophysical Journal*, *770*(2), L28. DOI. ADS.

Ball, W. T., Krivova, N. A., Unruh, Y. C., Haigh, J. D., & Solanki, S. K. (2014). A new SATIRE-S spectral solar irradiance reconstruction for solar cycles 21-23 and its implications for stratospheric ozone*. *Journal of Atmospheric Sciences*, *71*, 4086. DOI. ADS.

Ball, W. T., Schmutz, W., Fehlmann, A., Finsterle, W., & Walter, B. (2016). Assessing the beginning to end-of-mission sensitivity change of the PREcision MOnitor Sensor total solar irradiance radiometer (PREMOS/PICARD). *Journal of Space Weather and Space Climate*, *6*(27), A32. DOI. ADS.

Bellot Rubio, L., & Orozco Suárez, D. (2019). Quiet Sun magnetic fields: An observational view. *Living Reviews in Solar Physics*, *16*, 1. DOI. ADS.

Benz, A. O. (2017). Flare observations. *Living Reviews in Solar Physics*, *14*(1), 2. DOI. ADS.

Berger, M. A. (1984). Rigorous new limits on magnetic helicity dissipation in the solar corona. *Geophysical and Astrophysical Fluid Dynamics*, *30*, 79. DOI. ADS.

Berger, M. A., & Ruzmaikin, A. (2000). Rate of helicity production by solar rotation. *Journal of Geophysical Research*, *105*(A5), 10481. DOI. ADS.

Berger, T. E., Rouppe van der Voort, L., & Löfdahl, M. (2007). Contrast analysis of solar faculae and magnetic bright points. *The Astrophysical Journal*, *661*, 1272. DOI. ADS.

Berger, T. E., Shine, R. A., Slater, G. L., Tarbell, T. D., Title, A. M., Okamoto, T. J., et al. (2008). Hinode SOT observations of solar quiescent prominence dynamics. *The Astrophysical Journal*, *676*(1), L89. DOI. ADS.

Bertello, L., Pevtsov, A. A., & Pietarila, A. (2012). Signature of differential rotation in sun-as-a-star Ca II K measurements. *The Astrophysical Journal*, *761*, 11. DOI. ADS.

Bertello, L., Ulrich, R. K., & Boyden, J. E. (2010). The Mount Wilson Ca ii K plage index time series. *Solar Physics*, *264*, 31. DOI. ADS.

Bertello, L., Pevtsov, A. A., Harvey, J. W., & Toussaint, R. M. (2011). Improvements in the determination of ISS Ca ii K parameters. *Solar Physics*, *272*, 229. DOI. ADS.

Bertello, L., Pevtsov, A. A., Tlatov, A., & Singh, J. (2016). Solar Ca II K observations. *Asian Journal of Physics*, *25*(3), 295–310.

Bolduc, C., Charbonneau, P., Dumoulin, V., Bourqui, M. S., & Crouch, A. D. (2012). A fast model for the reconstruction of spectral solar irradiance in the near- and mid-ultraviolet. *Solar Physics, 279, 383*. DOI. ADS.

Brown, G. M., & Evans, D. R. (1980). The use of solar faculae in studies of the sunspot cycle. *Solar Physics*, *66*, 233. DOI. ADS.

Bruevich, E. A., Bruevich, V. V., & Yakunina, G. V. (2014). Changed relation between solar 10.7-cm radio flux and some activity indices which describe the radiation at different altitudes of atmosphere during cycles 21-23. *Journal of Astrophysics and Astronomy*, 35, 1. DOI. ADS.

Buccino, A. P., Díaz, R. F., Luoni, M. L., Abrevaya, X. C., & Mauas, P. J. D. (2011). Long-term chromospheric activity in Southern M Dwarfs: Gl 229 a and Gl 752 A. *The Astrophysical Journal*, *141*, 34. DOI. ADS.

Bumba, V., & Howard, R. (1969). Solar activity and recurrences in magnetic-field distribution. *Solar Physics,* 7, *28*. DOI. ADS.

Cameron, R., & Schüssler, M. (2015). The crucial role of surface magnetic fields for the solar dynamo. *Science*, *347*(6228), 1333. DOI. ADS.

Carmichael, H. (1964). *A Process for Flares*, *50*, 451. ADS.

Chapman, G. A., Cookson, A. M., & Preminger, D. G. (2013). Modeling total solar irradiance with San Fernando observatory ground-based photometry: Comparison with ACRIM, PMOD, and RMIB composites. *Solar Physics*, *283*, 295. DOI. ADS.

Chapman, G. A., Herzog, A. D., Lawrence, J. K., Walton, S. R., Hudson, H. S., & Fisher, B. M. (1992). Precise ground-based solar photometry and variations of total irradiance. *Journal of Geophysical Research, 97, 8211*. DOI. ADS.

Charbonneau, P. (2010). Dynamo models of the solar cycle. *Living Reviews in Solar Physics*, 7, 3. DOI. ADS.

Chatterjee, P., Choudhuri, A. R., & Petrovay, K. (2006). Development of twist in an emerging magnetic flux tube by poloidal field accretion. *Astronomy & Astrophysics*, *449*(2), 781. DOI. ADS.

Chatzistergos, T., Ermolli, I., Krivova, N. A., & Solanki, S. K. (2019). Analysis of full disc Ca II K spectroheliograms. II. Towards an accurate assessment of long-term variations in plage areas. *Astronomy & Astrophysics*, 625, A69. DOI. ADS.

Choudhuri, A. R. (2003). On the connection between mean field dynamo theory and flux tubes. *Solar Physics, 215(1), 31*. DOI. ADS.

Coddington, O., Lean, J. L., Pilewskie, P., Snow, M., & Lindholm, D. (2016). A solar irradiance climate data record. *Bulletin of the American Meteorological Society*, *97*, 1265. DOI. ADS.

Cossette, J.-F., Charbonneau, P., & Smolarkiewicz, P. K. (2013). Cyclic thermal signature in a global MHD simulation of solar convection. *The Astrophysical Journal Letters*, 777, L29. DOI. ADS.

Cossette, J.-F., Charbonneau, P., Smolarkiewicz, P. K., & Rast, M. P. (2017). Magnetically modulated heat transport in a global simulation of solar magneto-convection. *The Astrophysical Journal*, *841*, 65. DOI. ADS.

Cowan, N. B., Greene, T., Angerhausen, D., Batalha, N. E., Clampin, M., Colón, K., et al. (2015). Characterizing transiting planet atmospheres through 2025. *Publications of the Astronomical Society of the Pacific, 127, 311*. DOI. ADS.

Cowling, T. G. (1933). The magnetic field of sunspots. *Monthly Notices of the Royal Astronomical Society*, 94, 39. ADS.

Cranmer, S. R. (2009). Coronal holes. *Living Reviews in Solar Physics*, *6*(1), 3. DOI. ADS.

Criscuoli, S. (2013). Comparison of physical properties of quiet and active regions through the analysis of magnetohydrodynamic simulations of the solar photosphere. *The Astrophysical Journal*, *778*, 27. DOI. ADS.

Criscuoli, S. (2016). Angular dependence of the facular-sunspot coverage relation as derived by MDI magnetograms. *Solar Physics*, *291*, 1957. DOI. ADS.

Criscuoli, S. (2019). Effects of continuum fudging on non-LTE synthesis of stellar spectra. I. Effects on estimates of UV continua and solar spectral irradiance variability. *The Astrophysical Journal*, 872, 52. DOI. ADS.

Criscuoli, S., & Rast, M. P. (2009). Photometric properties of resolved and unresolved magnetic elements. *Astrononomy & Astrophysics, 495, 621*. DOI. ADS.

Criscuoli, S., & Uitenbroek, H. (2014). Interpretation of solar irradiance monitor measurements through analysis of 3D MHD simulations. *The Astrophysical Journal*, *788*(2), 151. DOI. ADS.

Criscuoli, S., Norton, A., & Whitney, T. (2017). Photometric properties of network and faculae derived from HMI data compensated for scattered light. *The Astrophysical Journal*, *847*, 93. DOI. ADS.

Criscuoli, S., Penza, V., Lovric, M., & Berrilli, F. (2018). The correlation of synthetic UV color versus Mg II index along the solar cycle. *The Astrophysical Journal*, *865*, 22. DOI. ADS.

Dasi-Espuig, M., Jiang, J., Krivova, N. A., Solanki, S. K., Unruh, Y. C., & Yeo, K. L. (2016). Reconstruction of spectral solar irradiance since 1700 from simulated magnetograms. *Astronomy & Astrophysics*, *590*, A63. DOI. ADS.

de Vaucouleurs, G. (1968). Linearization of characteristic curves in photographic photometry. *Applied Optics* 7, *1513*. DOI. ADS.

de Wijn, A. G., Stenflo, J. O., Solanki, S. K., & Tsuneta, S. (2009). Small-scale solar magnetic fields. *Space Science Reviews*, *144*, 275. DOI. ADS.

Démoulin, P., Mandrini, C. H., Van Driel-Gesztelyi, L., Lopez Fuentes, M. C., & Aulanier, G. (2002a). The magnetic helicity injected by shearing motions. *Solar Physics*, *207*(1), 87. DOI. ADS.

Démoulin, P., Mandrini, C. H., van Driel-Gesztelyi, L., Thompson, B. J., Plunkett, S., Kóvari, Z., et al. (2002b). What is the source of the magnetic helicity shed by CMEs? The long-term helicity budget of AR 7978. *Astronomy & Astrophysics, 382, 650*. DOI. ADS.

DeVore, C. R. (2000). Magnetic helicity generation by solar differential rotation. *The Astrophysical Journal*, *539*(2), 944. DOI. ADS.

Dikpati, M., & Gilman, P. A. (2007). Global solar dynamo models: simulations and predictions of cyclic photospheric fields and long-term non-reversing interior fields. *New Journal of Physics*, *9*, 297. DOI. ADS.

Donahue, R. A., & Keil, S. L. (1995). The solar surface differential rotation from disk-integrated chromospheric fluxes. *Solar Physics, 159, 53*. DOI. ADS.

Dudok de Wit, T., Kretzschmar, M., Lilensten, J., & Woods, T. (2009). Finding the best proxies for the solar UV irradiance. *Geophysical Research Letters*, *36*, L10107. DOI. ADS.

Dudok de Wit, T., Kopp, G., Fröhlich, C., & Schöll, M. (2017). Methodology to create a new total solar irradiance record: Making a composite out of multiple data records. *Geophysical Research Letters*, *44*, 1196. DOI. ADS.

Dufresne, J. L. (2008). La Determination De La Constante Solair Par Claude Pouillet. *La Météorologie*, *30*, 36.

Durrant, C. J., & Wilson, P. R. (2003). Observations and simulations of the polar field reversals in cycle 23. *Solar Physics*, *214*, 23. DOI. ADS.

Egorova, T., Schmutz, W., Rozanov, E., Shapiro, A. I., Usoskin, I., Beer, J., et al. (2018). Revised historical solar irradiance forcing. *Astronomy & Astrophysics, 615, A85*. DOI. ADS.

Ermolli, I., Criscuoli, S., & Giorgi, F. (2011). Recent results from optical synoptic observations of the solar atmosphere with ground-based instruments. *Contributions of the Astronomical Observatory Skalnate Pleso*, 41, 73. ADS.

Ermolli, I., Fofi, M., Bernacchia, C., Berrilli, F., Caccin, B., Egidi, A., et al. (1998). The prototype rise-Pspt instrument operating in Rome. *Solar Physics*, 177, 1. ADS.

Ermolli, I., Criscuoli, S., Centrone, M., Giorgi, F., & Penza, V. (2007). Photometric properties of facular features over the activity cycle. *Astronomy & Astrophysics, 465(1), 305*. DOI. ADS.

Ermolli, I., Solanki, S. K., Tlatov, A. G., Krivova, N. A., Ulrich, R. K., & Singh, J. (2009a). Comparison among Ca II K spectroheliogram time series with an application to solar activity studies. *The Astrophysical Journal*, *698*, 1000. DOI. ADS.

Ermolli, I., Marchei, E., Centrone, M., Criscuoli, S., Giorgi, F., & Perna, C. (2009b). The digitized archive of the Arcetri spectroheliograms. Preliminary results from the analysis of Ca II K images. *Astronomy & Astrophysics*, 499, 627. DOI. ADS.

Ermolli, I., Matthes, K., Dudok de Wit, T., Krivova, N. A., Tourpali, K., Weber, M., et al. (2013). Recent variability of the solar spectral irradiance and its impact on climate modelling. *Atmospheric Chemistry & Physics 13, 3945*. DOI. ADS.

Ermolli, I., Shibasaki, K., Tlatov, A., & van Driel-Gesztelyi, L. (2014). Solar cycle indices from the photosphere to the corona: measurements and underlying physics. *Space Science Reviews, 186, 105*. DOI. ADS.

Fabbian, D., Simoniello, R., Collet, R., Criscuoli, S., Korhonen, H., Krivova, N. A., et al. (2017). The variability of magnetic activity in solar-type stars. *Astronomische Nachrichten*, *338*, 753. DOI. ADS.

Fabiani Bendicho, P., Kneer, F., & Trujillo Bueno, J. (1992). On the photospheric temperature in small-scale magnetic flux concentrations. *Astronomy & Astrophysics*, 264, 229. ADS.

Faurobert, M. (2019). In O. Engvold, J.-C. Vial, & A. Skumanich(Eds.), *Chapter 8 – Solar and stellar variability* (p. 267). DOI. ADS.

Firor, J., & Zirin, H. (1962). Observations of five ionization stages of iron in the solar corona. *The Astrophysical Journal*, *135*, 122. DOI. ADS.

Floyd, L., Tobiska, W. K., & Cebula, R. P. (2002). Solar uv irradiance, its variation, and its relevance to the earth. *Advances in Space Research*, *29*, 1427. DOI. ADS.

Fontenla, J., & Harder, G. (2005). Physical modeling of spectral irradiance variations. *Memorie della Societa Astronomica Italiana*, 76, 826. ADS.

Fontenla, J. M., Avrett, E. H., & Loeser, R. (1993). Energy balance in the solar transition region. III – Helium emission in hydrostatic, constant-abundance models with diffusion. *The Astrophysical Journal*, 406, 319. DOI. ADS.

Fontenla, J. M., Balasubramaniam, K. S., & Harder, J. (2007). Semiempirical models of the solar atmosphere. II. The quiet-sun low chromosphere at moderate resolution. *The Astrophysical Journal*, 667, 1243. DOI. ADS.

Fontenla, J. M., Stancil, P. C., & Landi, E. (2015). Solar spectral irradiance, solar activity, and the near-ultra-violet. *The Astrophysical Journal, 809, 157*. DOI. ADS.

Fontenla, J. M., Harder, J., Livingston, W., Snow, M., & Woods, T. (2011). High-resolution solar spectral irradiance from extreme ultraviolet to far infrared. *Journal of Geophysical Research (Atmospheres)*, *116*, D20108. DOI. ADS.

Fontenla, J. M., Codrescu, M., Fedrizzi, M., Fuller-Rowell, T., Hill, F., Landi, E., et al. (2017). Five years of synthesis of solar spectral irradiance from SDID/SISA and SDO/AIA images. *The Astrophysical Journal, 834, 54*. DOI. ADS.

Fontenla, J., White, O. R., Fox, P. A., Avrett, E. H., & Kurucz, R. L. (1999). Calculation of solar irradiances. I. Synthesis of the solar spectrum. *The Astrophysical Journal*, 518, 480. DOI. ADS.

Foukal, P. (1993). The Curious Case of the Greenwich Faculae. *Solar Physics*, *148*, 219. DOI. ADS.

Foukal, P. (1994). Study of solar irradiance variations holds key to climate questions. *EOS Transactions*, *75*(33), 377. DOI. ADS.

Foukal, P. (2003). Can slow variations in solar luminosity provide missing link between the sun and climate? *EOS Transactions*, *84*, 205. DOI. ADS.

Foukal, P., & Lean, J. (1988). Magnetic modulation of solar luminosity by photospheric activity. *The Astrophysical Journal, 328, 347*. DOI. ADS.

Foukal, P., & Milano, L. (2001). A measurement of the quiet network contribution to solar irradiance variation. *Geophysical Reserach Letters*, *28*, 883. DOI. ADS.

Foukal, P., Bernasconi, P., Eaton, H., & Rust, D. (2004). Broadband measurements of facular photometric contrast using the solar bolometric imager. *The Astrophysical Journal Letters*, *611*, L57. DOI. ADS.

Foukal, P., Fröhlich, C., Spruit, H., & Wigley, T. M. L. (2006). Variations in solar luminosity and their effect on the Earth's climate. *Nature*, *443*, 161. DOI. ADS.

Foukal, P., Bertello, L., Livingston, W. C., Pevtsov, A. A., Singh, J., Tlatov, A. G., et al. (2009). A century of solar Ca ii measurements and their implication for solar UV driving of climate. *Solar Physics, 255, 229*. DOI. ADS.

France, K., Parke Loyd, R. O., Youngblood, A., Brown, A., Schneider, P. C., Hawley, S. L., et al. (2016). The MUSCLES treasury survey. I. Motivation and overview. *The Astrophysical Journal*, 820, 89. DOI. ADS.

Fröhlich, C., & Lean, J. (2004). Solar radiative output and its variability: Evidence and mechanisms. *The Astronomy and Astrophysics Review*, *12*, 273. DOI. ADS.

Fröhlich, C., Crommelynck, D. A., Wehrli, C., Anklin, M., Dewitte, S., Fichot, A., et al. (1997). In-Flight Performance of the Virgo Solar Irradiance Instruments on SOHO. *Solar Physics, 175(2), 267*. DOI. ADS.

Gallet, F., Charbonnel, C., Amard, L., Brun, S., Palacios, A., & Mathis, S. (2017). Impacts of stellar evolution and dynamics on the habitable zone: The role of rotation and magnetic activity. *Astronomy & Astrophysics*, *597*, A14. DOI. ADS.

Garcia, A., Sobotka, M., Klvaňa, M., & Bumba, V. (2011). Synoptic observations with the Coimbra spectroheliograph. *Contributions of the Astronomical Observatory Skalnate Pleso*, 41, 69. ADS.

Gasperini, A., Mazzoni, M., & Righini, A. (2004). La costruzione della Torre Solare di Arcetri nel carteggio Hale-Abetti. *Giornale di Astronomia*, 30(3), 23. ADS.

Giampapa, M. S., Hall, J. C., Radick, R. R., & Baliunas, S. L. (2006). A survey of chromospheric activity in the solar-type stars in the open cluster M67. *The Astrophysical Journal, 651, 444*. DOI. ADS.

Golub, L., Noci, G., Poletto, G., & Vaiana, G. S. (1982). Active region coronal evolution. *The Astrophysical Journal, 259, 359*. DOI. ADS.

Goodman, M. L. (2000). On the mechanism of chromospheric network heating and the condition for its onset in the sun and other solar-type stars. *The Astrophysical Journal, 533, 501*. DOI. ADS.

Gosain, S., Pevtsov, A. A., Rudenko, G. V., & Anfinogentov, S. A. (2013). First synoptic maps of photospheric vector magnetic field from SOLIS/VSM: Non-radial magnetic fields and hemispheric pattern of helicity. *The Astrophysical Journal, 772, 52*. DOI. ADS.

Gray, L. J., Beer, J., Geller, M., Haigh, J. D., Lockwood, M., Matthes, K., et al. (2010). Solar influences on climate. *Reviews of Geophysics*, *48*, RG4001. DOI. ADS.

Green, L. M., López fuentes, M. C., Mandrini, C. H., Démoulin, P., Van Driel-Gesztelyi, L., & Culhane, J. L. (2002). The magnetic helicity budget of a CME-prolific active region. *Solar Physics, 208(1), 43*. DOI. ADS.

Grotrian, W. (1939). Zur Frage der Deutung der Linien im Spektrum der Sonnenkorona. *Naturwissenschaften*, *27*, 214. DOI. ADS.

Grunhut, J. H., Wade, G. A., Hanes, D. A., & Alecian, E. (2010). Systematic detection of magnetic fields in massive, late-type supergiants. *Monthly Notices of the Royal Astronomical Society, 408, 2290*. DOI. ADS.

Haberreiter, M., Delouille, V., Mampaey, B., Verbeeck, C., Del Zanna, G., & Wieman, S. (2014). Reconstruction of the solar EUV irradiance from 1996 to 2010 based on SOHO/EIT images. *Journal of Space Weather and Space Climate*, *4*(27), A30. DOI. ADS.

Haberreiter, M., Schöll, M., Dudok de Wit, T., Kretzschmar, M., Misios, S., Tourpali, K., et al. (2017). A new observational solar irradiance composite. *Journal of Geophysical Research (Space Physics)*, *122*, 5910. DOI. ADS.

Hale, G. E. (1908). On the probable existence of a magnetic field in sun-spots. *The Astrophysical Journal, 28, 315*. DOI. ADS.

Hale, G. E., Ellerman, F., Nicholson, S. B., & Joy, A. H. (1919). The magnetic polarity of sun-spots. *The Astrophysical Journal, 49, 153*. DOI. ADS.

Hall, J. C., & Lockwood, G. W. (1998). The solar activity cycle. I. Observations of the end of cycle 22, 1993 September–1997 February. *The Astrophysical Journal*, 493, 494. DOI. ADS.

Hall, J. C., Lockwood, G. W., & Skiff, B. A. (2007). The activity and variability of the sun and sun-like stars. I. Synoptic Ca II H and K observations. *The Astrophysical Journal*, 133, 862. DOI. ADS.

Hall, J. C., Henry, G. W., Lockwood, G. W., Skiff, B. A., & Saar, S. H. (2009). The activity and variability of the sun and sun-like stars. II. Contemporaneous photometry and spectroscopy of bright solar analogs. *The Astrophysical Journal*, 138(1), 312. DOI. ADS.

Hanaoka, Y. (2013). Long-term synoptic observations of the Sun at the National Astronomical Observatory of Japan. *Journal of Physics Conference Series, Journal of Physics Conference Series*, 440, 012041. DOI. ADS.

Hanle, W. (1924). Über magnetische Beeinflussung der Polarisation der Resonanzfluoreszenz. *Zeitschrift fur Physik*, *30*, 93. DOI. ADS.

Harder, J. W., Fontenla, J. M., Pilewskie, P., Richard, E. C., & Woods, T. N. (2009). Trends in solar spectral irradiance variability in the visible and infrared. *Geophysical Reserach Letters*, *36*, L07801. DOI. ADS.

Harder, J., Lawrence, G., Fontenla, J., Rottman, G., & Woods, T. (2005). The spectral irradiance monitor: Scientific requirements, instrument design, and operation modes. *Solar Physics, 230, 141*. DOI. ADS.

Harvey, J. W. (1969). Magnetic fields associated with solar active-region prominences. PhD thesis, National Solar Observatory.

Harvey, J. W. (2013). The sun in time. *Space Science Reviews*, *176*, 47. DOI. ADS.

Harvey, J. W., Krieger, A. S., Davis, J. M., Timothy, A. F., & Vaiana, G. S. (1975). Comparison of Skylab X-Ray and ground-based Helium observations. In *Bulletin of the American Astronomical Society, Bulletin of the American Astronomical Society*, 7, 358. ADS.

Harvey, J. W., Branston, D., Henney, C. J., Keller, C. U., SOLIS & GONG Teams (2007). Seething horizontal magnetic fields in the quiet solar photosphere. *The Astrophysical Journal Letters*, *659*, L177. DOI. ADS.

Harvey, J., Gillespie, B., Miedaner, P., & Slaughter, C. (1980). Synoptic solar magnetic field maps for the interval including Carrington Rotation 1601-1680, May 5, 1973 – April 26, 1979. *NASA STI/Recon Technical Report N*, 81, 21003. ADS.

Harvey, K. L., & Recely, F. (2002). Polar coronal holes during cycles 22 and 23. *Solar Physics, 211, 31*. DOI. ADS.

Hasan, S. S., Mallik, D. C. V., Bagare, S. P., & Rajaguru, S. P. (2010). Solar Physics at the Kodaikanal observatory: A historical perspective. *Astrophysics and Space Science Proceedings*, *19*, 12. DOI. ADS.

Hasler, K.-H., Rüdiger, G., & Staude, J. (2002). Signature of differential rotation in solar disk-integrated chromospheric line emission. *Astronomische Nachrichten*, *323*, 123. DOI. ADS.

Heinzel, P., & Anzer, U. (2006). On the fine structure of solar filaments. *The Astrophysical Journal, 643(1), L65*. DOI. ADS.

Hirayama, T. (1974). Theoretical model of flares and prominences. I: Evaporating flare model. *Solar Physics*, 34(2), 323. DOI. ADS.

House, L. L. (1977). Coronal emission-line polarization from the statistical equilibrium of magnetic sublevels. I. Fe XIII. *The Astrophysical Journal*, 214, 632. DOI. ADS.

Howard, R. (1967). Magnetic field of the sun (observational). *Annual Review of Astronomy and Astrophysics, 5, 1*. DOI. ADS.

Howard, R. F. (1989). The magnetic fields of active regions. I – Data and first results. *Solar Physics*, 123, 271. DOI. ADS.

Huber, M. C. E., Foukal, P. V., Noyes, R. W., Reeves, E. M., Schmahl, E. J., Timothy, J. G., et al. (1974). Extreme-ultraviolet observations of coronal holes: Initial results from SKYLAB. *The Astrophysical Journal, 194, L115*. DOI. ADS.

Hudson, H. S. (1988). Observed variability of the solar luminosity. *Annual Review of Astronomy and Astrophysics*, *26*, 473. DOI. ADS.

Hudson, H. S. (2000). Implosions in coronal transients. *The Astrophysical Journal, 531(1), L75*. DOI. ADS.

Ito, H., Tsuneta, S., Shiota, D., Tokumaru, M., & Fujiki, K. (2010). *Is the polar region different from the quiet region of the Sun?* The Astrophysical Journal, *719, 131*. DOI. ADS.

Jackiewicz, J., Serebryanskiy, A., & Kholikov, S. (2015). Meridional flow in the solar convection zone. II. Helioseismic inversions of GONG data. *The Astrophysical Journal*, 805(2), 133. DOI. ADS.

Jain, K., Tripathy, S. C., & Hill, F. (2018). The sun's seismic radius as measured from the fundamental modes of oscillations and its implications for the TSI variations. *The Astrophysical Journal Letters*, *859*, L9. DOI. ADS.

Jones, H. P. (1985). Recent studies of magnetic canopies. *Australian Journal of Physics*, 38, 919. ADS.

Judge, P. G. (1998). Spectral lines for polarization measurements of the coronal magnetic field. I. Theoretical intensities. The Astrophysical Journal, 500(2), 1009. DOI. ADS.

Jungclaus, J. H., Bard, E., Baroni, M., Braconnot, P., Cao, J., Chini, L. P., et al. (2017). The PMIP4 contribution to CMIP6 - Part 3: The last millennium, scientific objective, and experimental design for the PMIP4 past1000 simulations. *Geoscientific Model Development*, *10*, 4005. DOI. ADS.

Kahil, F., Riethmüller, T. L., & Solanki, S. K. (2017). Brightness of solar magnetic elements as a function of magnetic flux at high spatial resolution. *The Astrophysical Journal Supplement, 229(1), 12*. DOI. ADS.

Kaltenegger, L. (2017). How to characterize habitable worlds and signs of life. *Annual Review of Astronomy and Astrophysics, 55, 433*. DOI. ADS.

Keil, S. (2016). Personal communication.

Keil, S. L., & Worden, S. P. (1984). Variations in the solar calcium K line 1976-1982. *The Astrophysical Journal*, *276*, 766. DOI. ADS.

Keller, C. U. (1992). Resolution of magnetic flux tubes on the sun. *Nature*, *359*, 307. DOI. ADS.

Keller, C. U., Harvey, J. W., & Solis Team (2003). The SOLIS vector-spectromagnetograph. In J. Trujillo-Bueno & J. Sanchez Almeida (Eds.), *Solar Polarization, Astronomical Society of the Pacific Conference Series* 307, 13. ADS.

Kholikov, S., Serebryanskiy, A., & Jackiewicz, J. (2014). Meridional flow in the solar convection zone. I. Measurements from GONG data. *The Astrophysical Journal*, 784(2), 145. DOI. ADS.

Khomenko, E., Vitas, N., Collados, M., & de Vicente, A. (2018). Three-dimensional simulations of solar magneto-convection including effects of partial ionization. *Astronomy & Astrophysics*, *618*, A87. DOI. ADS.

Kippenhahn, R., & Schlüter, A. (1957). Eine Theorie der solaren Filamente. Mit 7 Textabbildungen. *Zeitschrift fur Astrophysik*, 43, 36. ADS.

Knoelker, M., & Schuessler, M. (1988). Model calculations of magnetic flux tubes. IV – Convective energy transport and the nature of intermediate size flux concentrations. *Astronomy & Astrophysics*, 202, 275. ADS.

Komm, R., & Gosain, S. (2015). Current and kinetic helicity of long-lived activity complexes. *The Astrophysical Journal, 798, 20*. DOI. ADS.

Kopp, G. (2014). An assessment of the solar irradiance record for climate studies. *Journal of Space Weather and Space Climate*, *4*(27), A14. DOI. ADS.

Kopp, G. (2016). Magnitudes and timescales of total solar irradiance variability. *Journal of Space Weather and Space Climate*, *6*(27), A30. DOI. ADS.

Kopp, G., & Lawrence, G. (2005). The Total Irradiance Monitor (TIM): Instrument design. *Solar Physics, 230, 91*. DOI. ADS.

Kopp, G., & Lean, J. L. (2011). A new, lower value of total solar irradiance: Evidence and climate significance. *Geophysical Reserach Letters*, *38*, L01706. DOI. ADS.

Kopp, G., Lawrence, G., & Rottman, G. (2005). The Total Irradiance Monitor (TIM): Science results. *Solar Physics, 230, 129*. DOI. ADS.

Kopp, G., Krivova, N., Wu, C. J., & Lean, J. (2016). The impact of the revised sunspot record on solar irradiance reconstructions. *Solar Physics, 291, 2951*. DOI. ADS.

Kopp, R. A., & Pneuman, G. W. (1976). Magnetic reconnection in the corona and the loop prominence phenomenon. *Solar Physics, 50(1), 85*. DOI. ADS.

Kosovichev, A., & Rozelot, J.-P. (2018). Cyclic changes of the sun's seismic radius. *The Astrophysical Journal, 861, 90*. DOI. ADS.

Krause, F., & Raedler, K.-H. (1980). *Mean-field magnetohydrodynamics and dynamo theory*. ADS.

Kreidberg, L.(2018). *Exoplanet atmosphere measurements from transmission spectroscopy and other planet star combined light observations, 100*. DOI. ADS.

Kren, A. C., Pilewskie, P., & Coddington, O. (2017). Where does Earth's atmosphere get its energy? *Journal of Space Weather and Space Climate*, *7*(27), A10. DOI. ADS.

Krieger, A., Paolini, F., Vaiana, G. S., & Webb, D. (1972). Results from OSO-IV: The long term behavior of X-Ray emitting regions. *Solar Physics, 22(1), 150*. DOI. ADS.

Krivova, N. A., Solanki, S. K., Fligge, M., & Unruh, Y. C. (2003). Reconstruction of solar irradiance variations in cycle 23: Is solar surface magnetism the cause? *Astronomy & Astrophysics*, *399*, L1. DOI. ADS.

Kuhn, J. R. (1995). Infrared coronal magnetic field measurements. In J. R. Kuhn & M. J. Penn (Eds.), *Infrared tools for solar astrophysics: What's next?, 89*. ADS.

Kuperus, M., & Raadu, M. A. (1974). The support of prominences formed in neutral sheets. *Astronomy & Astrophysics*, 31, 189. ADS.

LaBonte, B. J., Georgoulis, M. K., & Rust, D. M. (2007). Survey of magnetic helicity injection in regions producing X-class flares. *The Astrophysical Journal, 671(1), 955*. DOI. ADS.

Lean, J. (2000). Evolution of the sun's spectral irradiance since the maunder minimum. *Geophysical Reserach Letters*, *27*, 2425. DOI. ADS.

Lean, J. (2017). Sun-climate connections. *Oxford Research Encyclopedia of Climate Science*. DOI.

Lean, J. L. (2001). Solar irradiance and climate forcing in the near future. *Geophysical Reserach Letters*, *28*, 4119. DOI. ADS.

Lean, J. L. (2018). Estimating solar irradiance since 850 CE. *Earth and Space Science*, *5*, 133. DOI. ADS.

Lean, J. L., & DeLand, M. T. (2012). How does the sun's spectrum vary? *Journal of Climate*, *25*, 2555. DOI. ADS.

Lean, J. L., Cook, J., Marquette, W., & Johannesson, A. (1998). Magnetic sources of the solar irradiance cycle. *The Astrophysical Journal, 492, 390*. DOI. ADS.

Lean, J. L., White, O. R., Livingston, W. C., & Picone, J. M. (2001). Variability of a composite chromospheric irradiance index during the 11-year activity cycle and over longer time periods. *Journal of Geophysical Research, 106, 10645*. DOI. ADS.

Lean, J. L., Warren, H. P., Mariska, J. T., & Bishop, J. (2003). A new model of solar EUV irradiance variability 2. Comparisons with empirical models and observations and implications for space weather. *Journal of Geophysical Research (Space Physics)*, 108, 1059. DOI. ADS.

Leighton, R. B. (1959). Observations of solar magnetic fields in plage regions. *The Astrophysical Journal, 130, 366*. DOI. ADS.

Leighton, R. B. (1964). Transport of magnetic fields on the sun. *The Astrophysical Journal, 140, 1547*. DOI. ADS.

Leighton, R. B. (1969). A magneto-kinematic model of the solar cycle. *The Astrophysical Journal, 156, 1*. DOI. ADS.

Leighton, R. B., Noyes, R. W., & Simon, G. W. (1962). Velocity fields in the solar atmosphere. I. Preliminary report. *The Astrophysical Journal*, 135, 474. DOI. ADS.

Lenz, L. F., Reiners, A., & Kürster, M. (2011). A search for star-planet interactions in chromospheric lines. In: C. Johns-Krull, M. K. Browning, & A. A. West, (Eds.), *16th Cambridge Workshop on Cool Stars, Stellar Systems, and the Sun, Astronomical Society of the Pacific Conference Series*, *448*, 1173. ADS.

Leroy, J. L., Bommier, V., & Sahal-Brechot, S. (1983). The magnetic field in the prominences of the polar crown. *Solar Physics, 83, 135*. DOI. ADS.

Leroy, J. L., Ratier, G., & Bommier, V. (1977). The polarization of the D3 emission line in prominences. *Astronomy & Astrophysics*, 54, 811. ADS.

Li, L. H., Basu, S., Sofia, S., Robinson, F.J., Demarque, P., & Guenther, D. B. (2003). Global parameter and helioseismic tests of solar variability models. *The Astrophysical Journal, 591, 1267*. DOI. ADS.

Lin, H., Kuhn, J. R., & Coulter, R. (2004). Coronal magnetic field measurements. *The Astrophysical Journal, 613(2), L177*. DOI. ADS.

Lin, H., Penn, M. J., & Tomczyk, S. (2000). A new precise measurement of the coronal magnetic field strength. *The Astrophysical Journal, 541(2), L83*. DOI. ADS.

Lin, Y., Martin, S. F., & Engvold, O. (2008). Filament substructures and their interrelation. In Howe, R., Komm, R.W., Balasubramaniam, K.S., & Petrie, G.J.D. (Eds.), *Subsurface and atmospheric influences on solar activity, astronomical society of the pacific conference series*, *383*, 235. ADS.

Linker, J. A., Lionello, R., Mikić, Z., Titov, V. S., & Antiochos, S. K. (2011). The Evolution of Open Magnetic Flux Driven by Photospheric Dynamics. *The Astrophysical Journal, 731, 110*. DOI. ADS.

Linsky, J. L. (2017). Stellar model chromospheres and spectroscopic diagnostics. *Annual Review of Astronomy & Astrophysics, 55, 159*. DOI. ADS.

Lites, B. W., Kubo, M., Socas-Navarro, H., Berger, T., Frank, Z., Shine, R., et al. (2008). The horizontal magnetic flux of the quiet-sun internetwork as observed with the Hinode spectro-polarimeter. *The Astrophysical Journal, 672, 1237*. DOI. ADS.

Liu, Y., Hoeksema, J. T., & Sun, X. (2014). Test of the hemispheric rule of magnetic helicity in the sun using the helioseismic and magnetic imager (HMI) Data. *The Astrophysical Journal Letters*, *783*, L1. DOI. ADS.

Livingston, W. C., Harvey, J., Pierce, A.K., Schrage, D., Gillespie, B., Simmons, J., & Slaughter, C. (1976). Kitt Peak 60-cm vacuum telescope. *Applied Optics*, *15*, 33. DOI. ADS.

Livingston, W., Wallace, L., White, O. R., & Giampapa, M. S. (2007). Sun-as-a-star spectrum variations 1974-2006. *The Astrophysical Journal, 657, 1137*. DOI. ADS.

Lockwood, M. (2011). Solar physics: Shining a light on solar impacts. *Nature Climate Change*, *1*, 98. DOI. ADS.

Lockwood, M. (2013). Reconstruction and prediction of variations in the open solar magnetic flux and interplanetary conditions. *Living Reviews in Solar Physics*, *10*, 4. DOI. ADS.

Longcope, D. W., Fisher, G. H., & Pevtsov, A. A. (1998). Flux-tube twist resulting from helical turbulence: The Σ-effect. *The Astrophysical Journal, 507(1), 417*. DOI. ADS.

López Ariste, A., Aulanier, G., Schmieder, B., & Sainz Dalda, A. (2006). First observation of bald patches in a filament channel and at a barb endpoint. *Astronomy & Astrophysics, 456(2), 725*. DOI. ADS.

Lovric, M., Tosone, F., Pietropaolo, E., Del Moro, D., Giovannelli, L., Cagnazzo, C., et al. (2017). The dependence of the [FUV-MUV] colour on solar cycle. *Journal of Space Weather and Space Climate*, 7(27), A6. DOI. ADS.

Low, B. C. (1994). Magnetohydrodynamic processes in the solar corona: Flares, coronal mass ejections, and magnetic helicity. *Physics of Plasmas*, *1*(5), 1684. DOI. ADS.

Low, B. C. (1996). Solar activity and the corona. *Solar Physics, 167(1-2), 217*. DOI. ADS.

Low, B. C. (2001). Coronal mass ejections, magnetic flux ropes, and solar magnetism. *Journal of Geophysical Research, 106(A11), 25141*. DOI. ADS.

Low, B. C., & Hundhausen, J. R. (1995). Magnetostatic structures of the solar corona. II. The magnetic topology of quiescent prominences. The Astrophysical Journal, 443, 818. DOI. ADS.

Mackay, D., & Yeates, A. (2012). The sun's global photospheric and coronal magnetic fields: Observations and models. *Living Reviews in Solar Physics*, *9*, 6. DOI. ADS.

Mackay, D. H., Karpen, J. T., Ballester, J. L., Schmieder, B., & Aulanier, G. (2010). Physics of solar prominences: II – Magnetic structure and dynamics. *Space Science Reviews*, *151*, 333. DOI. ADS.

Makarov, V. I., Tlatov, A. G., Singh, J., & Gupta, S. S. (2004). 22-years magnetic cycle in polar activity of the Sun. In A.V. Stepanov, E.E. Benevolenskaya, & A.G. Kosovichev (Eds.), *Multi-wavelength investigations of solar activity, IAU symposium*, *223*, 125. DOI. ADS.

Marchenko, S. V., & DeLand, M. T. (2014). Solar spectral irradiance changes during cycle 24. *The Astrophysical Journal, 789, 117*. DOI. ADS.

Martin, S. F. (1992). In Z. Svestka, B.V. Jackson, & M.E. Machado(Eds.), *An observational conceptual model of the formation of filaments*, 399, 331. ADS.

Martin, S. F. (1998). Conditions for the formation and maintenance of filaments (invited review). *Solar Physics, 182(1), 107*. DOI. ADS.

Martin, S. F., Bilimoria, R., & Tracadas, P. W. (1994). Magnetic field configurations basic to filament channels and filaments. In R.J. Rutten & C.J. Schrijver(Eds.), *Solar Surface Magnetism* (p. 303). ADS.

Matthes, K., Kuroda, Y., Kodera, K., & Langematz, U. (2006). Transfer of the solar signal from the stratosphere to the troposphere: Northern winter. *Journal of Geophysical Research (Atmospheres)*, *111*, D06108. DOI. ADS.

Matthes, K., Funke, B., Andersson, M. E., Barnard, L., Beer, J., Charbonneau, P., et al. (2017). Solar forcing for CMIP6 (v3.2). *Geoscientific Model Development*, 10, 2247. DOI. ADS.

Mauceri, S., Pilewskie, P., Richard, E., Coddington, O., Harder, J., & Woods, T. (2018). Revision of the sun's spectral irradiance as measured by SORCE SIM. *Solar Physics, 293, 161*. DOI. ADS.

Maunder, E. W. (1913). Sun, Place of the, Distribution of sunspots in heliographic latitude, 1874-1913. *Monthly Notices of the Royal Astronomical Society, 74, 112*. ADS.

McComas, D. J., Ebert, R. W., Elliott, H. A., Goldstein, B. E., Gosling, J. T., Schwadron, N. A., et al. (2008). Weaker solar wind from the polar coronal holes and the whole Sun. *Geophysical Reserach Letters*, *35*, 18103. DOI. ADS.

McCracken, K. G., & Beer, J. (2007). Long-term changes in the cosmic ray intensity at Earth, 1428-2005. *Journal of Geophysical Research (Space Physics)*, *112*, A10101. DOI. ADS.

McCracken, K. G., Beer, J., Steinhilber, F., & Abreu, J. (2013). A phenomenological study of the cosmic ray variations over the past 9400 years, and their implications regarding solar activity and the solar dynamo. *Solar Physics, 286, 609.* DOI. ADS.

McIntosh, P. S. (1972). Solar magnetic fields derived from hydrogen alpha filtergrams. *Reviews of Geophysics and Space Physics*, *10*, 837. DOI. ADS.

Meftah, M., Dewitte, S., Irbah, A., Chevalier, A., Conscience, C., Crommelynck, D., et al. (2014). SOVAP/ Picard, a spaceborne radiometer to measure the Total Solar Irradiance. *Solar Physics*, *289*, 1885. DOI. ADS.

Meftah, M., Hauchecorne, A., Irbah, A., Corbard, T., Ikhlef, R., Morand, F., et al. (2015). On the constancy of the diameter of the Sun during the rising phase of solar cycle 24. *The Astrophysical Journal, 808, 4.* DOI. ADS.

Megier, A., Strobel, A., Galazutdinov, G. A., & Krełowski, J. (2009). The interstellar Ca II distance scale. *Astronomy & Astrophysics, 507, 833.* DOI. ADS.

Mein, P., & Ribes, E. (1990). Spectroheliograms and motions of magnetic tracers. Astronomy & Astrophysics, 227, 577. ADS.

Munro, R. H., & Withbroe, G. L. (1972). Properties of a coronal "hole" derived from extreme-ultraviolet observations. *The Astrophysical Journal, 176, 511.* DOI. ADS.

Nandy, D. (2006). Magnetic helicity and flux tube dynamics in the solar convection zone: Comparisons between observation and theory. *Journal of Geophysical Research (Space Physics)*, *111*(A12), A12S01. DOI. ADS.

Narang, N., Banerjee, D., Chandrashekhar, K., & Pant, V. (2019). Association of calcium network bright points with underneath photospheric magnetic patches. *Solar Physics, 294(4), 40.* DOI. ADS.

Neupert, W. M. (1964). Time variations in the solar extreme ultraviolet spectrum observed by OSO I. NASA Special Publication, 50, 49. ADS.

Newkirk, J. G. (1967). Structure of the solar corona. *Annual Review of Astronomy and Astrophysics*, *5*, 213. DOI. ADS.

Nindos, A., & Andrews, M. D. (2004). The association of big flares and coronal mass ejections: What is the role of magnetic helicity? *The Astrophysical Journal, 616(2), L175.* DOI. ADS.

Nindos, A., & Zirin, H. (1998). The relation of CA II K features to magnetic field. *Solar Physics, 179(2), 253.* DOI. ADS.

Okamoto, T. J., Tsuneta, S., Berger, T. E., Ichimoto, K., Katsukawa, Y., Lites, B. W., et al. (2007). Coronal transverse magnetohydrodynamic waves in a solar prominence. *Science*, *318*(5856), 1577. DOI. ADS.

Oranje, B. J. (1983). The CA II K emission from the sun as a star. II – The plage emission profile. Astronomy & Astrophysics, 124, 43. ADS.

Ortiz, A., & Rast, M. (2005) *How good is the Ca II K as a proxy for the magnetic flux? Memorie della Societa Astronomica Italiana, 76, 1018.* ADS.

Ortiz, A., Solanki, S. K., Domingo, V., Fligge, M., & Sanahuja, B. (2002). On the intensity contrast of solar photospheric faculae and network elements. *Astronomy & Astrophysics, 388, 1036.* DOI. ADS.

Oshagh, M. (2018). Noise sources in photometry and radial velocities. In Asteroseismology and exoplanets: Listening to the stars and searching for new worlds (Vol. 49, p. 239). DOI. ADS.

Parker, E. N. (1955). The formation of sunspots from the solar toroidal field. *The Astrophysical Journal, 121, 491.* DOI. ADS.

Parker, E. N. (1963). Kinematical hydromagnetic theory and its application to the low solar photosphere. *The Astrophysical Journal, 138, 552.* DOI. ADS.

Parker, E. N. (1994). *Spontaneous current sheets in magnetic fields: with applications to stellar x-rays* (Vol. 1). International Series in Astronomy and Astrophysics, New York: Oxford University Press, 1994. *1.* ADS.

Passos, D., Brandão, S., & Lopes, I. (2007). On the luminosity evolution of the Sun during the last 7 millennia. *Advances in Space Research*, *40*, 990. DOI. ADS.

Peck, C. L., & Rast, M. P. (2015). Photometric trends in the visible solar continuum and their sensitivity to the center-to-limb profile. *The Astrophysical Journal, 808, 192.* DOI. ADS.

Peck, C. L., Rast, M. P., Criscuoli, S., & Rempel, M. (2019). The solar photospheric continuum brightness as a function of mean magnetic flux density. I. The role of the magnetic structure size distribution. The Astrophysical Journal, 870, 89. DOI. ADS.

Penza, V., Pietropaolo, E., & Livingston, W. (2006). Modeling the cyclic modulation of photospheric lines. *Astronomy & Astrophysics, 454, 349.* DOI. ADS.

Penza, V., Caccin, B., Ermolli, I., Centrone, M., & Gomez, M. T. (2003). Modeling solar irradiance variations through PSPT images and semiempirical models. In: Wilson, A. (Ed.), *Solar variability as an input to the Earth's environment*(p. 299). ESA Special Publication 535. ADS.

Peres, G., Orlando, S., Reale, F., Rosner, R., & Hudson, H. (2000). The Sun as an X-ray star. II. Using the Yohkoh/Soft X-ray telescope-derived solar emission measure versus temperature to interpret stellar X-ray observations. The Astrophysical Journal, 528, 537. DOI. ADS.

Petrie, G. (2017). High-resolution vector magnetograms of the Sun's poles from Hinode: Flux distributions and global coronal modeling. *Solar Physics, 292, 13.* DOI. ADS.

Petrie, G. J. D. (2012). Evolution of active and polar photospheric magnetic fields during the rise of cycle 24 compared to previous cycles. *Solar Physics, 281, 577.* DOI. ADS.

Petrie, G. J. D. (2013). Solar magnetic activity cycles, coronal potential field models and eruption rates. *The Astrophysical Journal, 768, 162.* DOI. ADS.

Petrie, G. J. D. (2015). Solar magnetism in the polar regions. *Living Reviews in Solar Physics 12(1), 5.* DOI. ADS.

Petrie, G. J. D., & Patrikeeva, I. (2009). A comparative study of magnetic fields in the solar photosphere and chromosphere at equatorial and polar latitudes. *The Astrophysical Journal, 699, 871.* DOI. ADS.

Pevtsov, A. A. (2008). What helicity can tell us about solar magnetic fields. *Journal of Astrophysics and Astronomy*, *29*(1-2), 49. DOI. ADS.

Pevtsov, A. A., Balasubramaniam, K. S., & Rogers, J. W. (2003). Chirality of chromospheric filaments. *The Astrophysical Journal, 595(1), 500.* DOI. ADS.

Pevtsov, A. A., Bertello, L., & Marble, A. R. (2014). The sun-as-a-star solar spectrum. *Astronomische Nachrichten*, *335*, 21. DOI. ADS.

Pevtsov, A. A., Canfield, R. C., & Metcalf, T. R. (1994). Patterns of helicity in solar active regions. *The Astrophysical Journal Letters,*. 425, L117. DOI. ADS.

Pevtsov, A. A., Canfield, R. C., & Metcalf, T. R. (1995). Latitudinal variation of helicity of photospheric magnetic fields. *The Astrophysical Journal Letters, 440*, L109. DOI. ADS.

Pevtsov, A. A., Berger, M. A., Nindos, A., Norton, A. A., & van Driel-Gesztelyi, L. (2014). Magnetic helicity, tilt, and twist. *Space Science Reviews, 186*(1-4), 285. DOI. ADS.

Pevtsov, A. A., Virtanen, I., Mursula, K., Tlatov, A., & Bertello, L. (2016). Reconstructing solar magnetic fields from historical observations. I. Renormalized Ca K spectroheliograms and pseudo-magnetograms. Astronomy & Astrophysics, 585, A40. DOI. ADS.

Pizzo, V. J., MacGregor, K. B., & Kunasz, P. B. (1993). A numerical simulation of two-dimensional radiative equilibrium in magnetostatic flux tubes. II – Computational results. The Astrophysical Journal, 413, 764. DOI. ADS.

Preminger, D. G., Walton, S. R., & Chapman, G. A. (2002). Photometric quantities for solar irradiance modeling. *Journal of Geophysical Research (Space Physics), 107*, 1354. DOI. ADS.

Priyal, M., Banerjee, D., Karak, B. B., Muñoz-Jaramillo, A., Ravindra, B., Choudhuri, A. R., et al. (2014). Polar network index as a magnetic proxy for the solar cycle studies. *The Astrophysical Journal Letters, 793, L4*. DOI. ADS.

Prša, A., Harmanec, P., Torres, G., Mamajek, E., Asplund, M., Capitaine, N., et al. (2016). Nominal values for selected solar and planetary quantities: IAU 2015 resolution B3. *The Astrophysical Journal, 152, 41*. DOI. ADS.

Querfeld, C. W. (1977). Observations of Fe XIII 10747 coronal emission-line polarization. Reports of the Lund Observatory, 12, 109. ADS.

Rachmeler, L. A., Gibson, S. E., Dove, J. B., DeVore, C. R., & Fan, Y. (2013). Polarimetric properties of flux ropes and sheared arcades in coronal prominence cavities. *Solar Physics, 288(2), 617*. DOI. ADS.

Raouafi, N. E., Riley, P., Gibson, S., Fineschi, S., & Solanki, S. K. (2016). Diagnostics of coronal magnetic fields through the hanle effect in UV and IR lines. *Frontiers in Astronomy and Space Sciences, 3*, 20. DOI. ADS.

Rast, M. P., Fox, P. A., Lin, H., Lites, B. W., Meisner, R. W., & White, O. R. (1999). Bright rings around sunspots. *Nature, 401*, 678. DOI. ADS.

Reale, F. (2014). Coronal loops: Observations and modeling of confined plasma. *Living Reviews in Solar Physics, 11*(1), 4. DOI. ADS.

Reale, F., Testa, P., Klimchuk, J. A., & Parenti, S. (2009). Evidence of widespread hot plasma in a nonflaring coronal active region from Hinode/X-ray telescope. *The Astrophysical Journal, 698(1), 756*. DOI. ADS.

Rempel, M. (2014). Numerical simulations of quiet sun magnetism: On the contribution from a small-scale dynamo. *The Astrophysical Journal, 789, 132*. DOI. ADS.

Rempel, M. (2015). Numerical simulations of sunspot decay: On the Penumbra-Evershed Flow-Moat flow connection. *The Astrophysical Journal, 814*(2), 125. DOI. ADS.

Richard, E. C., Harber, D., Harder, J. W., Pilewskie, P., Brown, S., Smith, A., et al. (2011). Future Long-term Measurements of Solar Spectral Irradiance by JPSS TSIS. AGU Fall Meeting Abstracts, GC23A. ADS.

Righini, A. (2003). One century of Solar Physics in Italy 1850-1950. *Memorie della Societa Astronomica Italiana, 74*, 556. ADS.

Rosner, R., Tucker, W. H., & Vaiana, G. S. (1978). Dynamics of the quiescent solar corona. *The Astrophysical Journal, 220*, 643. DOI. ADS.

Roth, R., & Joos, F. (2013). A reconstruction of radiocarbon production and total solar irradiance from the Holocene ^{14}C and CO_2 records: implications of data and model uncertainties. *Climate of the Past, 9*, 1879. *DOI*. ADS.

Rozelot, J. P., & Damiani, C. (2012). Rights and wrongs of the temporal solar radius variability. *European Physical Journal H, 37*, 709. DOI. ADS.

Rugheimer, S., & Kaltenegger, L. (2018). Spectra of Earth-like planets through geological evolution around FGKM stars. *The Astrophysical Journal, 854*, 19. DOI. ADS.

Rust, D. M. (1994). Are flares necessary? In K. S. Balasubramaniam & G. W. Simon (Eds.), *Solar Active Region Evolution: Comparing Models with Observations*, Astronomical Society of the Pacific Conference Series, 68, 335. ADS.

Rust, D. M., & Kumar, A. (1994). Helical magnetic fields in filaments. *Solar Physics, 155(1), 69*. DOI. ADS.

Rust, D. M., & Martin, S. F. (1994). A correlation between sunspot whirls and filament type. In K.S. Balasubramaniam & G.W. Simon (Eds.), *Solar Active Region Evolution: Comparing Models with Observations*, Astronomical Society of the Pacific Conference Series, 68, 337. ADS.

Rutten, R. G. M. (1984). Magnetic structure in cool stars. VII – Absolute surface flux in CA II H and K line cores. Astronomy & Astrophysics, 130, 353. ADS.

Salabert, D., García, R. A., Jiménez, A., Bertello, L., Corsaro, E., & Pallé, P. L. (2017). Photospheric activity of the Sun with VIRGO and GOLF. Comparison with standard activity proxies. Astronomy & Astrophysics, 608, A87. DOI. ADS.

Scalo, J., Kaltenegger, L., Segura, A. G., Fridlund, M., Ribas, I., Kulikov, Y. N., et al. (2007). M stars as targets for terrestrial exoplanet searches and biosignature detection. *Astrobiology, 7*, 85. DOI. ADS.

Schatten, K. H., Wilcox, J. M., & Ness, N. F. (1969). A model of interplanetary and coronal magnetic fields. *Solar Physics, 6, 442*. DOI. ADS.

Scherrer, P. H., Bogart, R. S., Bush, R. I., Hoeksema, J. T., Kosovichev, A. G., Schou, J., et al. (1995). The Solar Oscillations Investigation – Michelson Doppler Imager. *Solar Physics, 162, 129*. DOI. ADS.

Scherrer, P. H., Schou, J., Bush, R. I., Kosovichev, A. G., Bogart, R. S., Hoeksema, J. T., et al. (2012). The Helioseismic and Magnetic Imager (HMI) investigation for the Solar Dynamics Observatory (SDO). *Solar Physics, 275, 207*. DOI. ADS.

Schmidtke, G. (2015). Extreme ultraviolet spectral irradiance measurements since 1946. *History of Geo- and Space Sciences, 6*, 3. DOI. ADS.

Seehafer, N. (1990). Electric current helicity in the solar atmosphere. *Solar Physics, 125*, 219. DOI. ADS.

Seppälä, A., Matthes, K., Randall, C. E., & Mironova, I. A. (2014). What is the solar influence on climate? Overview of activities during CAWSES-II. *Progress in Earth and Planetary Science*, *1*, 24. DOI. ADS.

Shapiro, A. I., Solanki, S. K., Krivova, N. A., Tagirov, R. V., & Schmutz, W. K. (2015). The role of the Fraunhofer lines in solar brightness variability. *Astronomy & Astrophysics, 581, A116*. DOI. ADS.

Shapiro, A. I., Solanki, S. K., Krivova, N. A., Cameron, R. H., Yeo, K. L., et al. (2017). The nature of solar brightness variations. *Nature Astronomy*, *1*, 612. DOI. ADS.

Sheeley, N. R. Jr., DeVore, C. R., & Boris, J. P. (1985). Simulations of the mean solar magnetic field during sunspot cycle 21. *Solar Physics*, *98*, 219. DOI. ADS.

Shields, A. L., Ballard, S., & Johnson, J. A. (2016). The habitability of planets orbiting M-dwarf stars. *Physical Reports*, *663*, 1. DOI. ADS.

Shiota, D., Tsuneta, S., Shimojo, M., Sako, N., Orozco Suárez, D., & Ishikawa, R. (2012). Polar field reversal observations with Hinode. *The Astrophysical Journal, 753, 157*. DOI. ADS.

Sindhuja, G., Singh, J., & Ravindra, B. (2014). Study of meridional flow using Ca-K line profiles during solar cycles 22 and 23. *The Astrophysical Journal*, *792*, 22. DOI. ADS.

Singh, J. (1988). A new technique to study the variability of the Sun. Kodaikanal Observatory Bulletins, 9, 159. ADS.

Singh, J., & Ravindra, B. (2012). Twin telescope observations of the sun at Kodaikanal Observatory. Bulletin of the Astronomical Society of India, 40, 77. ADS.

Sivaraman, K. R., Singh, J., Bagare, S. P., & Gupta, S. S. (1987). Chromospheric CA II K-line variations in the sun as a star over a solar cycle. *The Astrophysical Journal*, *313*, 456. DOI. ADS.

Skumanich, A., Smythe, C., & Frazier, E. N. (1975). On the statistical desription of inhomogeneities in the quiet solar atmosphere. I. Linear regression analysis and absolute calibration of multichannel observations of the Ca+ emission network. The Astrophysical Journal, 200, 747. DOI. ADS.

Skumanich, A., Lean, J. L., Livingston, W. C., & White, O. R. (1984). The sun as a star – three-component analysis of chromospheric variability in the calcium K line. *The Astrophysical Journal*, *282*, 776. DOI. ADS.

Smith, E. J., & Balogh, A. (2008). Decrease in heliospheric magnetic flux in this solar minimum: Recent Ulysses magnetic field observations. *Geophysical Research Letters*, *35*, 22103. DOI. ADS.

Sofia, S., & Li, L. H. (2005). Mechanisms for global solar variability. Memorie della Societa Astronomica Italiana, 76, 768. ADS.

Solanki, S. K. (1993). Smallscale solar magnetic fields – an overview. *Space Science Reviews, 63, 1*. DOI. ADS.

Solanki, S. K. (2003a). Models of solar irradiance on time scales of minutes to centuries. AGU Fall Meeting Abstracts, SH31C. ADS.Solanki, S.K. (2003b). Sunspots: An overview. The Astronomy and Astrophysics Review, 11, 153. DOI. ADS.

Solanki, S. K., & Unruh, Y. C. (1998). A model of the wavelength dependence of solar irradiance variations. Astronomy & Astrophysics, 329, 747. ADS.

Song, P. (2017). A model of the solar chromosphere: Structure and internal circulation. *The Astrophysical Journal*, *846*, 92. DOI. ADS.

Spruit, H. C. (1976). Pressure equilibrium and energy balance of small photospheric fluxtubes. *Solar Physics, 50, 269*. DOI. ADS.

Spruit, H. C. (1981). Magnetic flux tubes. NASA Special Publication, 450. ADS.

Steenbeck, M., & Krause, F. (1969). On the dynamo theory of stellar and planetary magnetic fields. I. AC dynamos of solar type. Astronomische Nachrichten, 291, 49. DOI. ADS.

Steiner, O. (2005). Radiative properties of magnetic elements. II. Center to limb variation of the appearance of photospheric faculae. Astronomy & Astrophysics, 430, 691. DOI. ADS.

Steiner, O. (2007). Photospheric processes and magnetic flux tubes. In S.S. Hasan & D. Banerjee (Eds.), *Kodai School on Solar Physics, American Institute of Physics Conference Series*, *919*, 74. DOI. ADS.

Steinhilber, F., Abreu, J. A., & Beer, J. (2008). Solar modulation during the Holocene. *Astrophysics and Space Sciences Transactions, 4, 1*. DOI. ADS.

Steinhilber, F., Beer, J., & Fröhlich, C. (2009). Total solar irradiance during the Holocene. *Geophysical Research Letter*, 36, L19704. DOI. ADS.

Stenflo, J. O. (1982). The Hanle effect and the diagnostics of turbulent magnetic fields in the solar atmosphere. *Solar Physics, 80, 209*. DOI. ADS.

Sturrock, P. A. (1966). Model of the high-energy phase of solar flares. *Nature*, *211*(5050), 695. DOI. ADS.

Svalgaard, L., Duvall, T. L. Jr., & Scherrer, P. H. (1978). The strength of the sun's polar fields. *Solar Physics, 58, 225*. DOI. ADS.

Tapping, K., & Morgan, C. (2017). Changing relationships between sunspot number, total sunspot area and F{10.7} in cycles 23 and 24. Solar Physics, 292, 73. DOI. ADS.

Tapping, K. F. (2013). The 10.7 cm solar radio flux (F10.7). Space Weather, 11, 394. DOI. ADS.

Thomas, J. H., & Weiss, N. O. (2008). *Sunspots and Starspots*. Cambridge: Cambridge University Press. ADS.

Thuillier, G., Deland, M., Shapiro, A., Schmutz, W., Bolsée, D., & Melo, S. M. L. (2012). The solar spectral irradiance as a function of the Mg ii index for atmosphere and climate modelling. *Solar Physics, 277, 245*. DOI. ADS.

Thuillier, G., Melo, S. M. L., Lean, J., Krivova, N. A., Bolduc, C., Fomichev, V. I., et al. (2014). Analysis of different solar spectral irradiance reconstructions and their impact on solar heating rates. *Solar Physics, 289, 1115*. DOI. ADS.

Tian, F., France, K., Linsky, J. L., Mauas, P. J. D., & Vieytes, M. C. (2014). High stellar FUV/NUV ratio and oxygen contents in the atmospheres of potentially habitable planets. *Earth and Planetary Science Letters*, *385*, 22. DOI. ADS.

Tlatov, A. G., Pevtsov, A. A., & Singh, J. (2009). A new method of calibration of photographic plates from three historic data sets. *Solar Physics*, *255*, 239. DOI. ADS.

Tobiska, W. K., Woods, T., Eparvier, F., Viereck, R., Floyd, L., Bouwer, D., et al. (2000). The SOLAR2000 empirical solar irradiance model and forecast tool. *Journal of Atmospheric and Solar-Terrestrial Physics*, *62*, 1233. DOI. ADS.

Tomczyk, S., Card, G. L., Darnell, T., Elmore, D. F., Lull, R., Nelson, P. G., et al. (2008). An instrument to measure coronal emission line polarization. *Solar Physics*, *247*(2), 411. DOI. ADS.

Tousey, R., Sandlin, G. D., & Purcell, J. D. (1968) On some aspects of XUV spectroheliograms. In K.O. Kiepenheuer, (Ed.), *Structure and Development of Solar Active Regions*, IAU Symposium, 35, 411. ADS.

Tritschler, A., & Uitenbroek, H. (2006). the contrast of magnetic flux concentrations at near-infrared and visible wavelengths. The Astrophysical Journal, 648, 741. DOI. ADS.

Tsuneta, S., Ichimoto, K., Katsukawa, Y., Lites, B. W., Matsuzaki, K., Nagata, S., et al. (2008). The magnetic landscape of the sun's polar region. *The Astrophysical Journal*, *688*, 1374. DOI. ADS.

Tziotziou, K., Georgoulis, M. K., & Raouafi, N.-E. (2012). The magnetic energy-helicity diagram of solar active regions. *The Astrophysical Journal*, *759*(1), L4. DOI. ADS.

Unruh, Y. C., Solanki, S. K., & Fligge, M. (1999). The spectral dependence of facular contrast and solar irradiance variations. Astronomy & Astrophysics, 345, 635. ADS.

Usoskin, I. G. (2017). A history of solar activity over millennia. *Living Reviews in Solar Physics 14, 3*. DOI. ADS.

Vaiana, G. S., Reidy, W. P., Zehnpfennig, T., van Speybroeck, L., & Giacconi, R. (1968). X-ray structures of the sun during the importance 1N Flare of 8 June 1968. *Science*, *161*(3841), 564. DOI. ADS.

van Ballegooijen, A. A., & Martens, P. C. H. (1989). Formation and eruption of solar prominences. *The Astrophysical Journal*, *343*, 971. DOI. ADS.

Vaquero, J. M., Gallego, M. C., Ruiz-Lorenzo, J. J., López-Moratalla, T., Carrasco, V. M. S., Aparicio, A. J. P., et al. (2016). Monitoring the solar radius from the royal observatory of the Spanish Navy since 1773. *Solar Physics*, *291*, 1599. DOI. ADS.

Ventura, P., Penza, V., Li, L., Sofia, S., Basu, S., & Demarque, P. (2010). 2D solar modeling. *Astrophysics and Space Science*, *328*, 295. DOI. ADS.

Vieira, L. E. A., & Solanki, S. K. (2010). Evolution of the solar magnetic flux on time scales of years to millenia. *Astronomy & Astrophysics*, *509*, A100. DOI. ADS.

Vieira, L. E. A., Solanki, S. K., Krivova, N. A., & Usoskin, I. (2011). Evolution of the solar irradiance during the Holocene. *Astronomy & AstrophysIcs*, *531*, A6. DOI. ADS.

Viereck, R., Puga, L., McMullin, D., Judge, D., Weber, M., & Tobiska, W. K. (2001). The Mg II index: A proxy for solar EUV. *Geophysical Research Letters*, *28*, 1343. DOI. ADS.

Vieytes, M. C., Mauas, P. J. D., & Díaz, R. F. (2009). Chromospheric changes in K stars with activity. *Monthly Notices of the Royal Astronomical Society, 398, 1495*. DOI. ADS.

Vieytes, M., Mauas, P., Cacciari, C., Origlia, L., & Pancino, E. (2011). Modelling chromospheric line profiles as diagnostics of velocity fields in ω Centauri red giant stars. *Astronomy & Astrophysics*, *526*, A4. DOI. ADS.

Viticchié, B., Del Moro, D., Criscuoli, S., & Berrilli, F. (2010). Imaging spectropolarimetry with IBIS. II. On the fine structure of G-band bright features. The Astrophysical Journal, 723, 787. DOI. ADS.

Vögler, A. (2005). On the effect of photospheric magnetic fields on solar surface brightness. Results of radiative MHD simulations. Memorie della Societa Astronomica Italiana, 76, 842. ADS.

Waldmeier, M. (1956). Synoptische Karten der Sonnenkorona. Mit 16 Textabbildungen. Zeitschrift fur Astrophysik, 38, 219. ADS.

Waldmeier, M., (1957). *Die sonnenkorona*. Basel: Birkhäuser, 2.

Wang, Y.-M. (2013). On the strength of the hemispheric rule and the origin of active-region helicity. *Astrophysical Journal of Letters, 775, L46*. DOI. ADS.

Wang, Y.-M., Lean, J. L., & Sheeley, N. R., Jr. (2005). Modeling the sun's magnetic field and irradiance since 1713. *The Astrophysical Journal*, *625*, 522. DOI. ADS.

Warren, H. P., Brooks, D. H., & Winebarger, A. R. (2011) Constraints on the heating of high-temperature active region loops: Observations from hinode and the solar dynamics observatory. The Astrophysical Journal, 734(2), 90. DOI. ADS.

Warren, H. P., Mariska, J. T., & Lean, J. (2001). A new model of solar EUV irradiance variability: 1. Model formulation. *Journal of Geophysical Research*, 106, 15745. *DOI. ADS.*

Webb, D. F., & Howard, T. A. (2012). Coronal mass ejections: Observations. *Living Reviews in Solar Physics*, *9*(1), 3. DOI. ADS.

White, O. R., & Livingston, W. (1978), Solar luminosity variation. II - Behavior of calcium H and K at solar minimum and the onset of cycle 21. The Astrophysical Journal, 226, 679. DOI. ADS.

White, O. R., & Livingston, W. C. (1981). Solar luminosity variation. III – Calcium K variation from solar minimum to maximum in cycle 21. The Astrophysical Journal, 249, 798. DOI. ADS.

White, O. R., Livingston, W. C., & Keil, S. L. (1992). Solar Ca II k measurements and activity cycles in solar-type stars. In R. F. Donnelly (Ed.), *Solar Electromagnetic Radiation Study for Solar Cycle 22, 160*. ADS.

Wiegelmann, T., Petrie, G. J. D., & Riley, P. (2017). Coronal magnetic field models. *Space Science Reviews, 210(1-4), 249*. DOI. ADS.

Willson, R. C., Gulkis, S., Janssen, M., Hudson, H. S., & Chapman, G. A. (1981). Observations of solar irradiance variability. *Science*, *211*(4483), 700. DOI. ADS.

Woods, T. N., Eparvier, F. G., Fontenla, J., Harder, J., Kopp, G., McClintock, W. E., et al. (2004). Solar irradiance variability during the October 2003 solar storm period. *Geophysical Research Letters*, *31*, L10802. DOI. ADS.

Wu, C.-J., Krivova, N. A., Solanki, S. K., & Usoskin, I. G. (2018). Solar total and spectral irradiance reconstruction over the last 9000 years. *Astronomy & Astrophysics*, *620*, A120. DOI. ADS.

Yeates, A. R., & Mackay, D. H. (2012). Chirality of high-latitude filaments over solar cycle 23. *Astrophysical Journal of Letters*, *753*, L34. DOI. ADS.

Yeo, K. L., Krivova, N. A., & Solanki, S. K. (2017). EMPIRE: A robust empirical reconstruction of solar irradiance variability. *Journal of Geophysical Research (Space Physics)*, *122*, 3888. DOI. ADS.

Yeo, K. L., Krivova, N. A., Solanki, S. K., & Glassmeier, K. H. (2014). Reconstruction of total and spectral solar irradiance from 1974 to 2013 based on KPVT, SoHO/MDI, and SDO/HMI observations. *Astronomy & Astrophysics*, *570*, A85. DOI. ADS.

Yeo, K. L., Solanki, S. K., Norris, C. M., Beeck, B., Unruh, Y. C., & Krivova, N. A. (2017). Solar irradiance variability is caused by the magnetic activity on the solar surface. *Physical Review Letters*, *119*(9). DOI. ADS.

Youngblood, A., France, K., Loyd, R. O. P., Brown, A., Mason, J. P., Schneider, P. C., et al. (2017). The MUSCLES treasury survey. IV. Scaling relations for ultraviolet, Ca II K, and energetic particle fluxes from M Dwarfs. *The Astrophysical Journal*, *843*, 31. DOI. ADS.

Zeeman, P. (1897). On the influence of magnetism on the nature of the light emitted by a substance. *The Astrophysical Journal, 5, 332*. DOI. ADS.

Zhang, H., Sakurai, T., Pevtsov, A., Gao, Y., Xu, H., Sokoloff, D. D., et al. 2010, A new dynamo pattern revealed by solar helical magnetic fields. *Monthly Notices of the Royal Astronomical Society*, *402*(1), L30. DOI. ADS.

Zhang, M. (2006). Helicity observations of weak and strong fields. *The Astrophysical Journal, 646(1), L85*. DOI. ADS.

Zhao, J., Bogart, R. S., Kosovichev, A. G., Duvall, J. T. L., & Hartlep, T. (2013). Detection of equatorward meridional flow and evidence of double-cell meridional circulation inside the sun. *The Astrophysical Journal*, *774*(2), L29. DOI. ADS.

4

Solar Energetic Particles

Christina M. S. Cohen[1], Gang Li[2], Glenn M. Mason[3], Albert Y. Shih[4], and Linghua Wang[5]

ABSTRACT

Solar energetic particles (SEPs) have been studied for over seven decades, initially from ground-based instrumentation and now routinely from spacecraft throughout the heliosphere. Our knowledge of the acceleration and transport of these particles from close to the Sun througzh the interplanetary medium has advanced dramatically in the last 40 years as the instrumentation and modeling capabilities have become more sophisticated. However, a full understanding of the conditions and processes that cause variability in SEP event characteristics remains elusive due to the scarcity of in situ measurements well inside 1 AU. This chapter reviews our current understanding of the properties of SEP events, how they are accelerated and transported, and the impact of those processes on event variability. It is not an exhaustive review, but provides a snapshot of the state of our understanding before the near-Sun observations from the Parker Solar Probe and Solar Orbiter missions reveal surprises that challenge this understanding.

4.1. INTRODUCTION

The particle environment of the heliosphere ranges from the solar wind at keV energies to galactic cosmic rays (GCRs) at hundreds of MeV energies and beyond. Unlike the solar wind and GCRs, solar energetic particles (SEPs) occur as episodic events and can vary in intensity by several orders of magnitude in minutes. These particles are accelerated by some of the most energetic phenomena on the Sun, flares and coronal mass ejections (CMEs), and they present a particular radiation hazard to instrumentation and humans in space. SEP events consist of ions and electrons at energies greater than a few hundred keV and tens of keV, respectively, with intensities that range from a few to $>10^5$ particles/(cm^2-sr-s-MeV/nuc) and durations from hours to days. Similarly, the variation in heavy ion composition, spectral shape, and spatial distribution is substantial.

This chapter will briefly describe the history of SEP event measurements (section 4.2) followed by descriptions of the current understanding of the general characteristics of ions observed in large, shock-accelerated SEP events (section 4.3); for ^{3}He-rich events (section 4.4); and for solar energetic electrons (section 4.5) with a focus on recent results. In addition, section 4.6 describes energetic neutral atoms and their relationship to SEPs. Finally, section 4.7 describes theories of SEP acceleration and transport.

[1] Space Radiation Laboratory, California Institute of Technology, Pasadena, California, USA
[2] Department of Space Science, University of Alabama in Huntsville, Huntsville, Alabama, USA
[3] Johns Hopkins University Applied Physics Laboratory, Laurel, Maryland, USA
[4] NASA Goddard Space Flight Center, Greenbelt, Maryland, USA
[5] School of Earth and Space Sciences, Peking University, Beijing, China

Space Physics and Aeronomy Collection Volume 1: Solar Physics and Solar Wind, Geophysical Monograph 258, First Edition.
Edited by Nour E. Raouafi and Angelos Vourlidas.

DOI: 10.1002/9781119815600.ch4

4.2. HISTORY

When SEPs with energies greater than ~500 MeV impact Earth's atmosphere, they can cause a "shower" of secondary particles resulting from the interactions and decay of particles. Among the by-products of these showers are neutrons and muons that are often measured by detectors on the ground. The first connection of a neutron monitor increase to solar flares was made in 1946 by Forbush for two events in 1942 and one in 1946 (Forbush, 1946). In his paper, he suggested that these neutron monitor increases were caused by energetic charged particles emitted by the Sun and referenced the theory of Swann for the acceleration of particles through magnetic field reconfigurations in sunspots (Swann, 1933). More details came with the investigation of the 23 February 1956 event, which was observed by six neutron monitors and analyzed in conjunction with optical and radio measurements of the associated flare (see, e.g., Meyer et al., 1956). With the launch of *Luna 1* in 1959, measurements of the energetic particle environment of space began, initially with Geiger counters and scintillators. Over the ensuing decades, instrumentation for SEPs evolved to focus on particular energy ranges and specific components such as electrons, protons, or heavy ions. Currently, electrostatic deflection is a common technique for measuring lower-energy protons and combined with time-of-flight techniques to identify heavy ions (e.g., Mason et al., 1998). At higher energies (above ~10 MeV/nuc), stacks of silicon detectors are used to measure the energy deposited by ions over a range of penetration depths, which can then be used to identify the nuclear charge, Z, and mass, M (e.g., Stone et al., 1998).

As the properties of SEP events were being studied in greater detail, so too was activity on the Sun. Flares were identified and categorized by their size and duration in a variety of wavelengths and connected to radio emission detected from the Sun. The discovery of CMEs and their correlation with flares led to an improved understanding of the variability of SEP events and the acceleration processes involved. It was generally agreed upon that there were two dominant acceleration mechanisms, one associated with flares and reconnection and the other with CME-driven shocks in the corona and solar wind (Reames, 1999). As SEP events associated with small flares that appeared to not have associated CMEs exhibit substantial compositional differences (e.g., large enhancements of ^{3}He) than those associated with large CMEs and clear shocks, the two-class paradigm was created: one class was called "impulsive" and the other "gradual" in reference to the temporal characteristic of the associated flare (Reames, 1999). In this framework, the general characteristics of the SEP event could be used to identify which acceleration mechanism was responsible.

Although the two-class system provided significant organization of the SEP events and a clearer understanding of the physics involved in creating them, some events did not seem to neatly fit into either class. This was particularly highlighted by the events of late 1997 and early 1998, which were occasionally referred to as "hybrid" events (Cohen et al., 1999; Mason, Cohen, Cummings et al., 1999). Since then the use of "gradual" and "impulsive" to identify SEP events has faded slightly, often replaced with "large"/"shock-accelerated" and "^{3}He-rich"/"flare-associated"; and "hybrid" with "large, Fe-rich." What remains fairly clear is that two dominant acceleration mechanisms are involved in the generation of SEP events (see sections 4.4.3 and 4.7), but that the characteristics of the events vary significantly depending on the conditions in which the acceleration process operates and the medium in which the particles travel before being observed.

4.3. SOLAR ENERGETIC ION VARIABILITY IN SHOCK-ACCELERATED SEP EVENTS

Large SEP events are associated with fast (≥900 km/s) and wide (≥60°) CMEs (see, e.g., Gopalswamy et al., 2008) that drive shocks in the corona and interplanetary medium. As large SEP events present a substantial space weather hazard, predicting their occurrence and properties is a national priority. Although these events have been studied in detail for nearly 40 years, with substantial emphasis on ions heavier than He in the last three decades, their large variability remains poorly understood and hampers predictive capabilities. This variability manifests itself in nearly every aspect of the event characteristics. The following sections describe the current knowledge of the temporal (section 4.3.1), spectral (section 4.3.2), compositional (section 4.3.3), and spatial (section 4.3.4) variability observed in ions of large SEP events.

4.3.1. Temporal Variations

An impulsive injection of SEPs (with a spectrum of energies) at the Sun traveling outward would be observed at 1 AU as a quick rise in intensities to a peak followed by an exponential decay. The onset time and the decay rate both depend on the energy and ionic charge state per mass (Q/M) of the particles, as rigidity plays a role in both the acceleration and transport of SEPs (Lee, 2005). Typically the accelerating shock is strongest near the Sun and weakens as it moves outward. As it weakens, the maximum energy to which the shock can accelerate particles decreases; thus, lower-energy particles will be accelerated over larger distances and longer times than their high-energy counterparts (see section 4.7 for more details). This can result in time profiles that differ with energy and do not follow the simple fast onset plus exponential decay pattern.

The strongest parameter affecting the observed time profile of an SEP event is the magnetic connection between the observer and the source region. In the case of CME-driven shocks, the nominal connection follows a Parker spiral, but the connection point on the surface of the shock moves eastward along it as the shock expands outward from the Sun. Although it is now known that the strength and orientation of the shock can vary substantially along the front and as a function of time (see, e.g., Kozarev et al., 2015; Rouillard et al., 2017), a rough organization of characteristic time profiles can still be understood through the following arguments. If the source region is close to the center of the solar disk as defined by the observer–Sun line, and the CME propagates radially outward, the magnetic connection to the observer will start at a position west of the nose of the shock, move across the nose, and continue to the eastern side. With the assumption that the nose of the shock is typically the strongest portion, and the flanks being progressively weaker, such positioning results in the observer measuring the largest portion of the SEPs accelerated by the shock. These events exhibit a reasonably fast rise time, often followed by a flattening of the time profile, peaking when the shock is near the spacecraft, followed by a decay (Cane et al., 1988).

If the source region is significantly to the west of the observer–Sun line, the magnetic connection to the observer will start near the nose of the CME-driven shock and move progressively farther east. As this results in the observer being connected to the strongest portion of the shock for a relatively short period of time while the CME is still fairly close to the Sun, these events most closely resemble the impulsive injection time profile (with a peak well before the shock reaches the distance of the observer), but are typically smaller than central meridian SEP events. At the other extreme, if the source region is east of the observer–Sun line, the magnetic connection to the western flank of the shock may not be made until the CME has propagated some distance from the Sun, and will move toward the nose of the shock. This results in a more gradual rise in the observed particle intensities, with the peak of the event being observed when the observer is connected to the nose of the shock, which may occur when the shock is past the observer. Figure 4.1 illustrates these different time profiles and their dependence on the relative positioning of the solar source and the observer (see also Cane & Lario, 2006).

For those events where the shock passes over the observer, if the shock is still strong enough to be accelerating particles, an increase in the particle intensity will be detected. This is referred to an energetic storm particle (ESP) event and is usually observed at energies less than

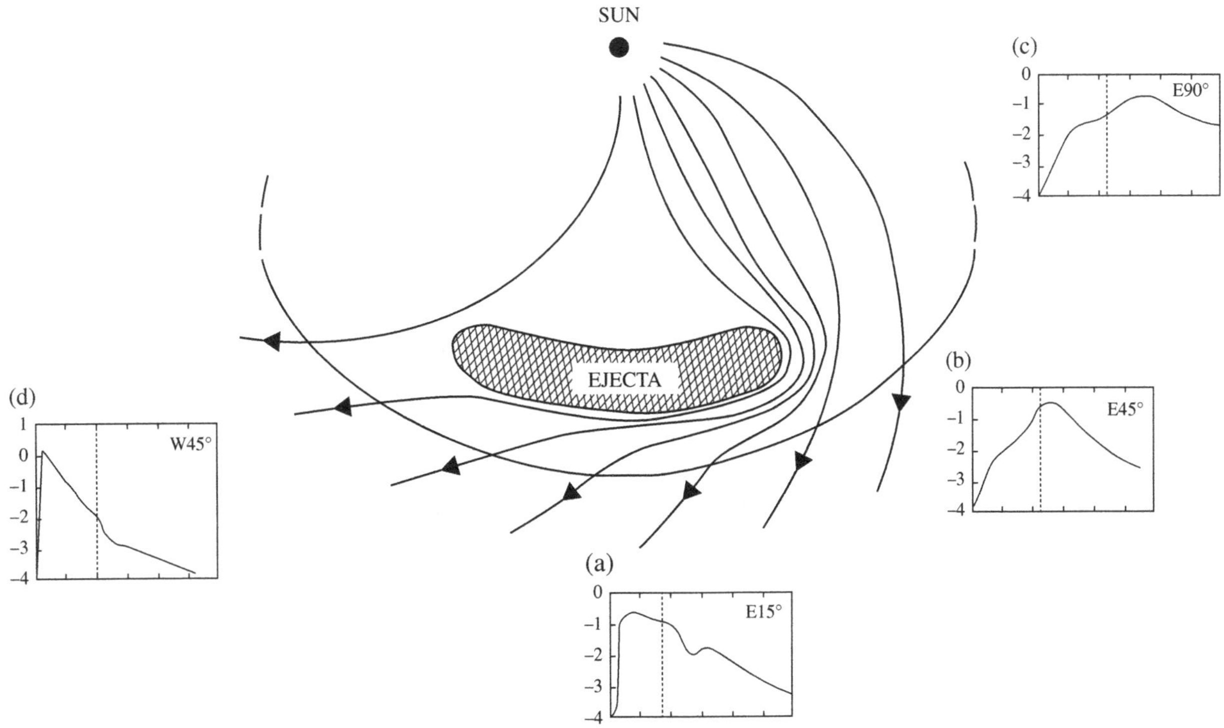

Figure 4.1 Time profiles typical for observers at different positions relative to the source region and CME propagation. (*Source*: Cane et al., 1988. © 1988 John Wiley & Sons.)

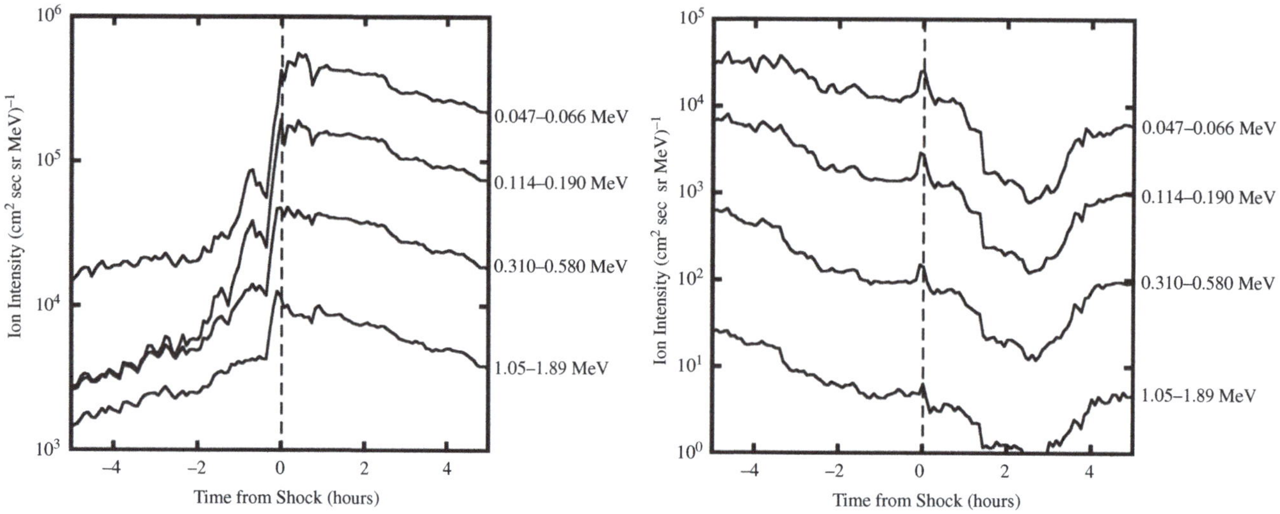

Figure 4.2 Time profiles of energetic storm particle (ESP) event types. Quasi-parallel shocks are associated with the larger events (left panel), whereas quasi-perpendicular shocks create smaller, spike-like events (right panel). (*Source*: Cohen, 2006. © 2006 John Wiley & Sons.)

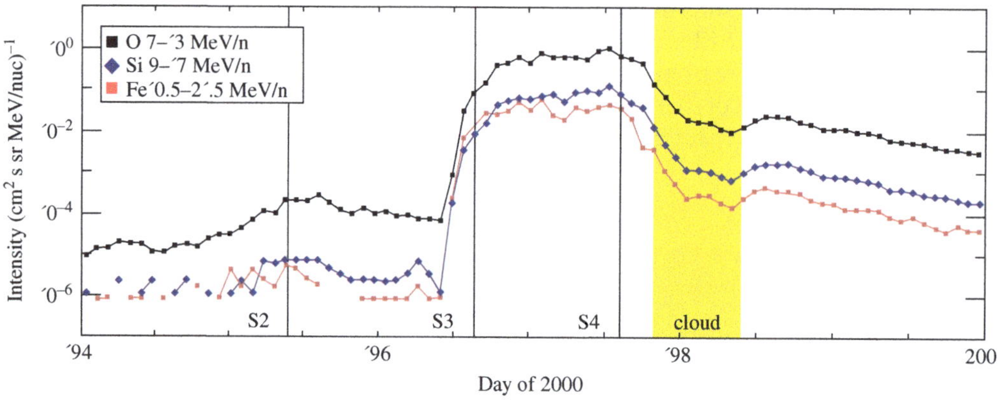

Figure 4.3 Energetic particle time profiles during the SEP event of 14 July 2000. Solid lines indicate interplanetary shock passages, and the shaded region marks the presence of an interplanetary coronal mass ejection (ICME). The particle intensities are suppressed during the ICME period. (*Source*: Smith et al., 2001. © 2001 Springer Nature.)

a few tens of MeV/nuc (Kallenrode, 1996). The duration and size of the ESP event depends not only on the strength of the shock but also on the orientation of the shock normal relative to the upstream magnetic field. A quasi-perpendicular orientation tends to result in a sharp but short-lived ESP event, whereas a quasi-parallel configuration is associated with the larger and longer ESP events (Figure 4.2; see review by Cohen, 2006, and references therein). Also common for these events is the presence of the interplanetary counterpart of the CME (referred to as an interplanetary coronal mass ejection, ICME) which drove the shock. Due to the closed magnetic structure of the ICME, SEPs cannot effectively penetrate inside the ICME, and an abrupt decrease in particle intensity is observed at the time of the ICME passage (with corresponding recovery when the observer exits the ICME); Figure 4.3.

As SEPs travel through the interplanetary medium along magnetic field lines, they can excite Alfvén waves. For large SEP events, the substantial number of energetic protons can create and amplify these waves to the point where the propagation of following SEPs is limited. This results in a relatively constant particle intensity with time apparent to the observer; the magnitude of the intensity is referred to as the "streaming limit" and has been observed to be a near constant in many SEP events (although with a radial dependence of r^{-3}; Ng et al., 2003; Reames & Ng, 2010). If an ESP event is observed in an SEP event that exhibits the streaming limit, it will likely be stronger than usual as the particles accelerated after

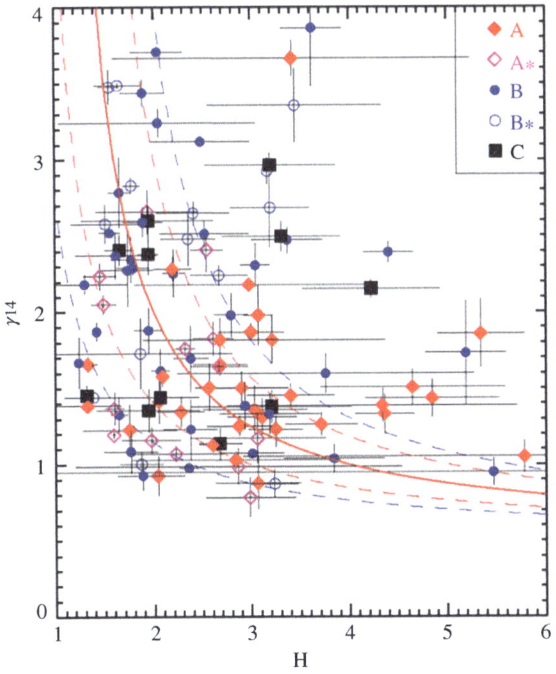

Figure 4.4 Measured spectral indices of ions observed at the time of peak intensity versus the density compression ratio of the interplanetary shock for a variety of event types. The solid curve indicates the relationship predicted by theory; dashed red and blue curves show expected uncertainties of ±25% and ±40%, respectively. The observations are not in good agreement with the theoretical predictions. (*Source*: Ho et al., 2008. © 2004 IOP Publishing.)

the streaming limit was reached are confined to the general shock region by the wave–particle interactions (Reames & Ng, 1998).

4.3.2. Spectral Variations

The standard model for particle acceleration by CME-driven shocks indicates that the resulting SEPs should exhibit a power-law spectrum (in energy) with the power-law index determined by the compression ratio of the shock (Lee, 1983). Although power-law spectra are observed over portions of the observed energy range, most often a power law at energies below ~10 MeV/nuc does not continue to higher energies. More surprisingly, the power-law index is not well predicted by the shock properties observed at 1 AU (Figure 4.4; Ho et al., 2008). This can partly be understood by the fact that the characteristics of the shock evolve as it travels from the Sun to the observer, and the particles are accelerated continuously during this time; thus, the shock properties at 1 AU may not be representative of the shock when many of the particles were accelerated. In an attempt to reduce this effect, instead of using the spectrum integrated over the

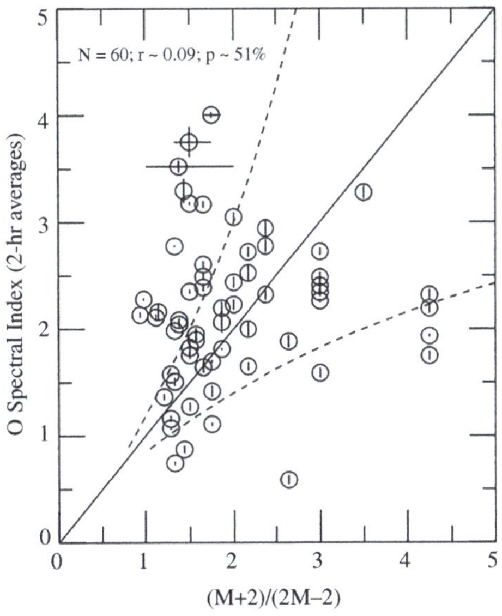

Figure 4.5 Spectral indices of oxygen during ESP events versus the theoretical value as calculated from the magnetic compression ratio, M. There is little agreement between the observed spectral indices and the theoretical predictions. (*Source*: Desai et al., 2004. © 2004 IOP Publishing.)

full SEP event, a spectrum for just the ESP portion (if there is one) can be compared to the predictions from the shock measurements. Although this improves the agreement, there is still substantial disagreement between theory and observations (Figure 4.5; Desai et al., 2004).

Two common variations of the single power-law observed in SEP events (over energy ranges of 0.1–100 MeV/n) are broken power laws and the presence of an exponential rollover at higher energies (Mewaldt et al., 2005b). Mathematical formulations used to fit these spectra are described by Band et al. (1993) and Ellison and Ramaty (1985) and correspondingly referred to as Band fits and Ellison–Ramaty fits (Figure 4.6). Although the Band fit is not rooted in any physical interpretation of the spectra, it has been shown that the maximum energy to which a shock can accelerate particles decreases as the shock weakens, and an integration of the accelerated population over the time that the shock propagates outward and degrades can result in a form similar to the Band fit (Zhao et al., 2016). Alternatively it has been suggested that transport of the SEPs through the interplanetary medium can result in a similar broken power-law form (Li & Lee, 2015). In contrast, Ellison and Ramaty justified adding the exponential rollover to the power-law form by arguing that it would be a natural result of several effects including the strength and lifetime of the shock relative to the increased particle acceleration time at higher energies as well as the increased ability of higher-energy particles to escape the shock region before enough time had passed to form a power-law distribution. Thus, the location of the "rollover" (i.e., where the exponential term dominates) could be generally related to the strength of the shock, with higher-energy rollovers resulting from stronger shocks.

Ground-level enhancement (GLE) events are a category of extreme SEP events that have sufficiently high intensities of protons with energies >500 MeV that interactions between the protons and Earth's atmosphere create enough neutrons and muons to be measured by ground-based instrumentation (Poluianov et al., 2017). Although these events are not always the largest at ~10 MeV/nuc, they typically have very hard spectra, often with a single power law extending well beyond 100 MeV/n (Mewaldt et al., 2012). Although routine measurements of SEPs beyond a few hundred MeV/nuc have not always been available, there are GLE event proton spectra extending to a few GeV for events occurring between 2006 and 2014 (Bruno et al., 2018). These typically show an exponential rollover at energies above ~100 MeV and occasionally as high as 500 MeV (Figure 4.7).

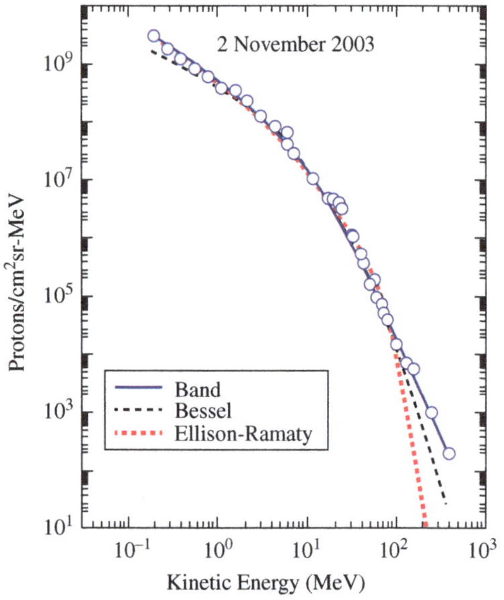

Figure 4.6 Band (solid blue curve) and Ellison–Ramaty (dashed red curve) fits to the proton spectrum from the November 2003 SEP event. Both show breaks in the spectrum; however, the Band function fits the higher-energy data better. (*Source*: Modified from Mewaldt et al., 2005b.)

4.3.3. Compositional Variations

Like the solar wind, the composition of SEP events exhibits evidence of the "first ionization potential (FIP) effect," where a plot of elemental abundances relative to O and normalized to photospheric values as a function of the elements' FIP forms a step function with the low-FIP elements being higher and a transition around 10 eV (Figure 4.8; Meyer, 1985a, 1985b). The similarity between the SEP and solar wind composition initially suggested that SEPs were likely accelerated solar wind material; however, as the abundance measurements improved, distinct differences became clear. For example, the C/O abundance ratio in the solar wind is significantly higher than in SEP events (Figure 4.9; Mewaldt et al., 2001). Current understanding is that the source of SEPs is the suprathermal ion population, which may be significantly different and more variable than the bulk solar wind (Mason et al., 2005). Not only does the tail of the solar wind contribute to this population, but so does the lowest-energy portion of prior SEP events, whether

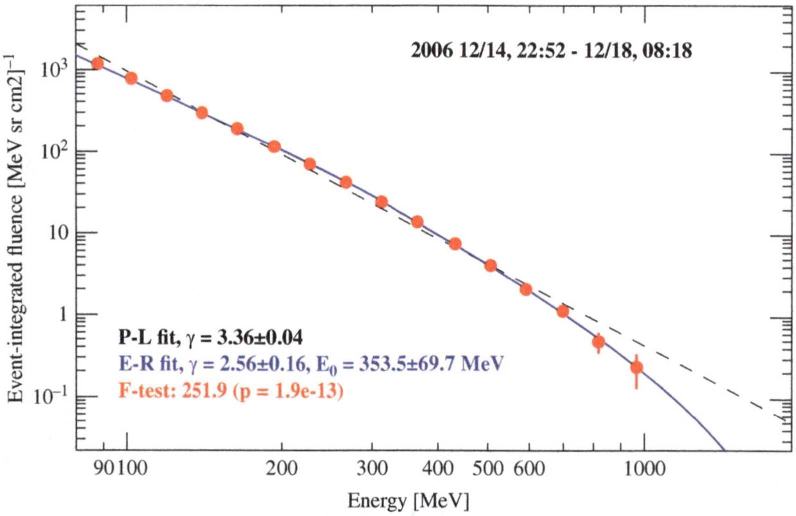

Figure 4.7 Proton spectrum from Pamela for the 14 December 2006 ground-level enhancement event shows a rollover in the spectrum at energies above ~500 MeV. (*Source*: Bruno et al., 2018. © 2018 IOP Publishing.)

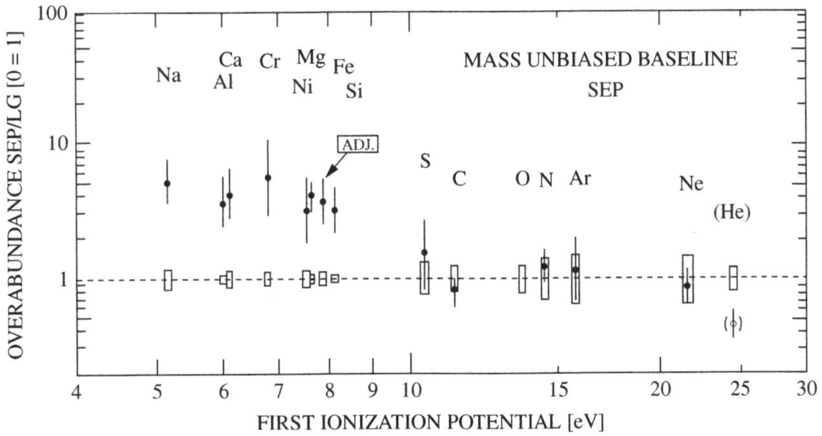

Figure 4.8 SEP abundances (relative to O), normalized to local galactic values versus the elements' first ionization potential (FIP) show a step function with the low-FIP elements being enhanced. (*Source*: Modified from Meyer, 1985a.)

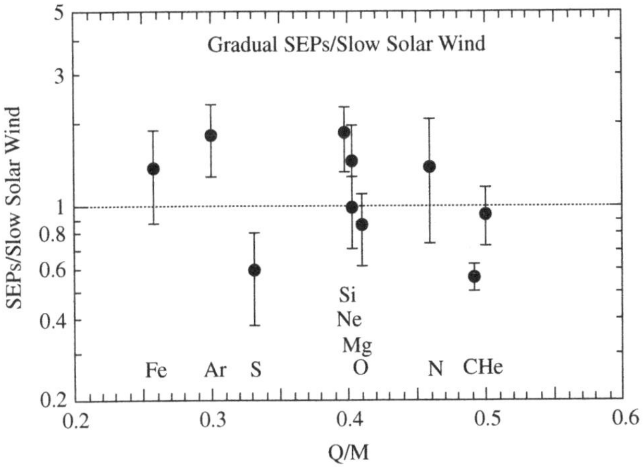

Figure 4.9 SEP abundances (relative to Si) normalized to solar wind values versus the elements' charge/mass values. That many points do not lie along the unity line indicates that the SEP and solar wind composition are distinctly different. (*Source*: From Mewaldt et al., 2001.)

shock- or flare-associated. Given the episodic nature of flares and SEP events, it is expected that the composition of the suprathermal population varies with time as well as solar distance and energy.

Characteristics of the suprathermal ions have been invoked as a possible explanation for the large Fe-rich SEP events observed in late 1997 and early 1998 (Mason, Cohen, Cummings et al., 1999; Tylka et al., 2005). One particular scenario (Tylka et al., 2005) postulates that shock acceleration generates all large SEP events; however, the resulting composition of the event depends on two aspects: (1) the characteristics of the seed population being accelerated and (2) the orientation of the shock. For the Fe-rich events, the seed population is a composite of solar wind suprathermals and flare-associated suprathermals with a harder spectrum such that the composition is energy dependent, for example, increasing Fe/O with increasing energy, and the accelerating shock is quasi-perpendicular. The theory assumes that quasi-perpendicular shocks accelerate only particles above a higher energy than quasi-parallel shocks do; thus, less of the solar wind suprathermals will be accelerated resulting in an SEP event with higher Fe/O abundance. If the shock is quasi-parallel or the flare-associated component of the seed population is missing, the resulting SEP event will have nominal composition.

Although there has been indirect evidence supporting this explanation, to date it has not be fully tested. At issue is the fact that the higher energies of the SEP event (e.g., >10 MeV/nuc) are typically accelerated close to the Sun, where no measurements are available. With the launch of the Parker Solar Probe, this will change, but not until the solar activity significantly increases.

An alternative suggestion for the creation of these hybrid events is that they are a combination of flare-associated and shock-accelerated SEPs (Cane et al., 2006, 2003). The resulting SEP composition depends on the relative strength of the two processes, with the shock acceleration dominating at lower energies, and how well the observer is connection magnetically to the flare. This explains the often-observed increase in the Fe/O abundance with increasing energy in these events and some longitudinal dependence. However, the supporting evidence for this scenario has not been clear (Cohen et al., 2017, 2013).

As discussed in the previous section, SEP spectra often exhibit breaks or rollovers. As these transitions relate to the shock's ability to confine particles, it is not surprising that the energy of the spectral breaks is element dependent. This produces energy-dependent abundance ratios (when comparing elements with different Q/M values), for example, decreasing Fe/O with increasing energy above the break energy of Fe (Figure 4.10; Cohen et al., 2007). The specific dependence of the break energies on the ion Q/M values may depend on the orientation of the shock (see section 4.7; Li et al., 2009), with stronger variability resulting from quasi-parallel shocks, as well as the turbulence ahead of the shock (which affects the ion's mean free path length). Studies have used the relative positioning of spectral breaks as a function of Q/M to infer the turbulence spectrum near the shock (Cohen et al., 2005; Mewaldt et al., 2005a).

Finally, the composition of an SEP event is also known to vary with time. The most dramatic change is typically observed at the start of the event, where ions with large rigidities (e.g., Fe) may arrive at the observer earlier than those with lower rigidities (e.g., O), which are either more efficiently contained to the shock region or experience more scattering during transport (Mason et al., 2012; Reames et al., 2000). This can be observed as a strong decrease in the Fe/O ratio as a function of time (Figure 4.11). Similar to what has been done in comparing ion spectra as a function of rigidity rather than energy/nucleon, it was found that the time profiles of higher-energy O are often very similar to that of Fe at a lower energy (picked such that the rigidity of the ions was comparable) (Mason et al., 2014). This, in turn, significantly decreases the time variability of the Fe/O abundance ratio and suggests that the transport of the particles and their temporal variation is strongly governed by the particle's rigidity.

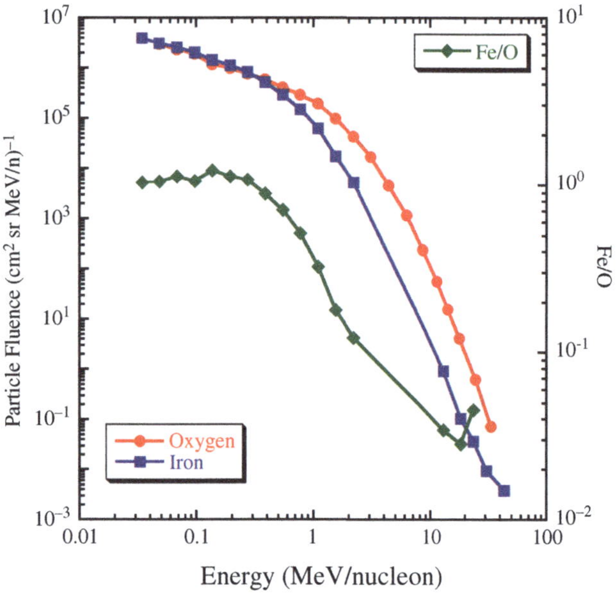

Figure 4.10 Fluences for O and Fe (left axis) versus energy exhibit different break energies which lead to an Fe/O ratio that depends strongly on energy (right axis). (*Source*: Cohen et al., 2007. © 2007 Springer Nature.)

4.3.4. Spatial Variability

The vast majority of the SEP observations are single point measurements, most often in the vicinity of Earth (including the first Lagrangian point, on the Sun–Earth line). Thus, it is challenging to characterize how SEP events vary with radius, longitude, and latitude. The first opportunity to explore this systematically was with the launch of the two *Helios* spacecraft in 1974 and 1976 (Kunow et al., 1975). By comparing SEP observations from distances as close as 0.3 AU to those made simultaneously near Earth, it was clear that, for similar magnetic connections, not only were the SEP events much larger, they were shorter in duration closer to the Sun; in some cases, multiple SEP events clearly separated at 0.3 AU had merged by the time they reached 1 AU (Figure 4.12; Wibberenz & Cane, 2006). Analysis of the relative peak intensities and fluences for a number of SEP events observed by Helios and IMP 8 suggested a radial dependence of $r^{-\partial}$, with ∂ ranging from 1.9 to 2.7 for proton peak intensities and from 1.0 to 2.1 for proton fluences (Lario et al., 2013, 2006). These values indicated a weaker radial

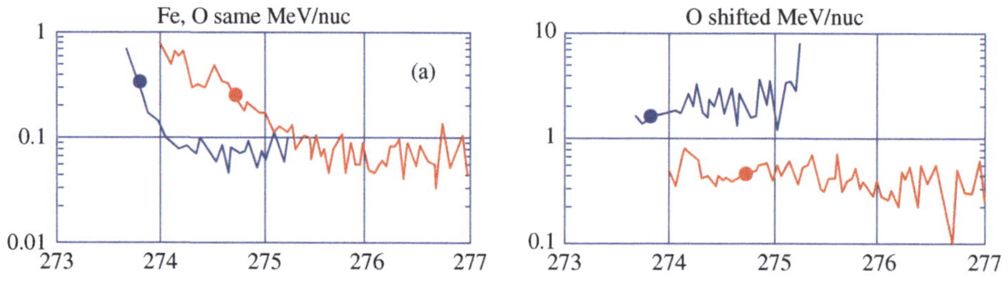

Figure 4.11 Fe/O ratios versus time as calculated at the same energy/nucleon (left panel; red curve is for 276 keV/nuc, and the blue curve is for 12 MeV/nuc) and at similar rigidities (i.e., at a "shifted" energy for oxygen; right panel). The substantial decreases with time seen in the left panel are not evident when the ratios are calculated at a common rigidity. (*Source*: Modified from Mason et al., 2012.)

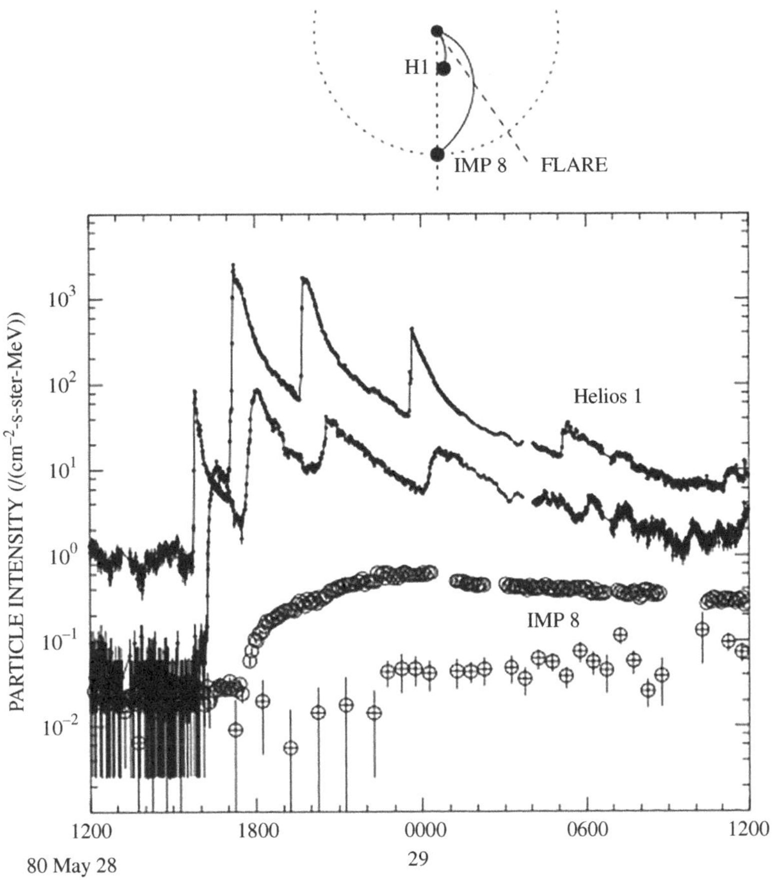

Figure 4.12 Helios and IMP 8 time profiles of energetic protons reveal that the distinct events seen at 0.3 AU have merged by the time they are measured at 1 AU. The graphic above illustrated the relative positioning of the two spacecraft. (*Source*: From Wibberenz & Cane, 2006. © 2006 IOP Publishing.)

gradient than previously expected from diffusion transport models (Hamilton et al., 1990; Ng & Reames, 1994).

Observations out of the ecliptic to date have been limited to those made by Ulysses; however, *Solar Orbiter,* launched on February 9, 2020, will reach 24° out of the ecliptic plane by 2027. *Ulysses* observed several SEP events even when at heliolatitudes of nearly 80° (Dalla et al., 2003; McKibben et al., 2003). Similar to events observed near Earth, the time profiles of the Ulysses events varied in their rise times and times to maximum. However, unlike the in-ecliptic events, where this is generally understood to be a result of the observer's magnetic connection to the shock, similar parameters did not organize the Ulysses measurements. Substantial particle transport perpendicular to the magnetic field appears to be required to explain these observations (Dalla et al., 2003; Zhang et al., 2009).

Another result from comparing the Ulysses and near-Earth observations was the discovery of periods during the decay phase of SEP events in which the intensity levels were nearly identical and decayed at the same rate despite the spacecraft being separated significantly in longitude, latitude, and radius (Lario, 2010; McKibben, 1972). Additional work has shown that this most likely results from the development of a confinement region in which the particle population is uniformly distributed and their intensities adiabatically decrease as the region expands (Reames et al., 2013, and references therein). The expectation is that the region is created by bounding magnetic structures (e.g., shocks) and that within the region the cross-field diffusion is relatively high, allowing particles access to the full volume (McKibben et al., 2001). High perpendicular diffusion was also invoked as an explanation for the reservoir observations by Zhang et al. (2009).

With the launch of the twin *STEREO* spacecraft in 2006, the possibility of measuring the same SEP event at different longitudes, but at similar radii (unlike *Helios*), became routine. Unfortunately, *STEREO* launched at the start of solar minimum, so it was ~5 years before the first multi-spacecraft SEP events were observed (Cohen et al., 2013). Over the course of solar cycle 24, >40 large SEP events were measured by at least two spacecraft, providing sufficient statistics to characterize the longitudinal dependence (Cohen et al., 2017; Richardson et al., 2014). As with many SEP characteristics, there is a wide variability in the longitudinal spread, although the average of ~45° Gaussian width in longitude is similar to that derived using the Helios data (and accounting for the radial gradient). Although some of the interpretations of single point measurements were confirmed, for example, the longitudinal dependence of an event's time profile (see section 4.3.1), there were many surprises. The 3 November 2011 event remains one of the hardest to understand; the two *STEREO* spacecraft and near-Earth satellites were roughly equally spaced around the Sun, yet the >10 MeV proton intensities were observed to rise rapidly at all three positions, with onsets within 30 min of each other (Lario et al., 2017; Richardson et al., 2014). This indicated that all 360° of the inner heliosphere was populated with SEPs within tens of minutes, requiring very efficient longitudinal transport. Additional widespread events have also been observed, both in ions and in electrons (Cohen et al., 2017; Dresing et al., 2014).

Given the success in using rigidity to organize the particle energy spectra and time profiles, it was expected that the longitudinal distribution would also be dependent on particles' Q/M values. However, in a study of H, He, O, and Fe in multi-spacecraft events, it was found that the longitude spread does not strongly depend on element but does have an energy dependence (Cohen et al., 2017). The narrower distribution of higher-energy ions may be understood in terms of a smaller longitudinal region of the CME-driven shock accelerating particles to the highest energies, or in terms of shock expansion with distance combined with a shorter acceleration period close to the Sun for the higher-energy particles.

4.4. ^{3}HE-RICH SEP EVENTS

The light isotope of helium, ^{3}He, is easily consumed in stellar cores, and so its abundance relative to ^{4}He is very low: in solar system abundances ^{3}He/^{4}He ~1.7×10^{-4} (Lodders, 2003) and in the solar wind the ratio is $4–5\times10^{-4}$ (Bochsler et al., 1990). Therefore, the unexpected discovery of large amounts of ^{3}He in some small SEP events, with ^{3}He/^{4}He ratios from tens of percent to >100%, established the existence of a distinct SEP acceleration process capable of enormously enhancing the ^{3}He isotope, unlike the situation in large SEP events (see reviews by Klecker et al., 2006a; L G Kocharov and Kocharov, 1984; Mason, 2007; Reames, 2017).

Figure 4.13 illustrates properties of ^{3}He-rich events obtained through the 1990s from remote observations and measurements of the particle properties in interplanetary space near 1 au (adapted from Benz et al., 2009; Marsden and Müller, 2011). Starting at the lower left and proceeding clockwise, the initial activity is typically an active region flare that produces X-rays in the tens of keV energy range, along with UV emission and X-ray jets (small EUV emitting plasma features). The event is usually located in the solar western hemisphere where the interplanetary magnetic field (IMF) connects to the vicinity of Earth. Kilovolt electrons escape from the region (top left panel), exciting plasma waves that emit radio waves (top right panel), whose frequency decreases as the particle bunch moves into lower-density plasma on its way out from the Sun. In the vicinity of Earth, the electron bunch arrives, showing clear velocity dispersion (lower right panel) whose timing is consistent with passage along a field line ~1.2–1.3 au in length, as expected for the Parker spiral. Hours later, low-energy (~1 MeV/nucleon) ions arrive, also showing velocity dispersion and composition greatly enriched in ^{3}He (bottom center panel) along with more modest (factors of a few times) enrichment of heavy ions. At 1 AU, the particle event is short, lasting a few hours to a day.

In the discussion below, section 4.4.1 treats source and injection properties; section 4.4.2 discusses heavy ion enrichments, and section 4.4.3 briefly reviews progress in theoretical models (see also section 4.7).

4.4.1. Source and Injection Properties

Early observations established a correlation between ^{3}He-rich SEP events and kilovolt electron events associated with type III solar radio bursts (Reames et al., 1985), and therefore with the solar activity cycle. L Wang et al. (2012) have recently examined this correlation in great detail using advanced instruments capable of detecting many more events than the earlier studies (see also section 4.5). Figure 4.14a shows a histogram of their yearly electron event occurrences (black line) along with SEP events with ^{3}He/^{4}He > 1% (red line), ^{3}He/^{4}He <1% (green line), and gradual events (blue line). The correlation between the ^{3}He-rich ion events and impulsive electron events is clearly seen on an event-by-event basis, and is also shown clearly in the yearly totals. Figure 4.14b shows the associated sunspot rates. Their survey of 1191 solar electron events from 1995 to 2005 found an occurrence rate at Earth of ~1000/year at solar maximum, which implies a whole Sun rate of ~10^4/year assuming a solar source width of ~45°. At solar minimum, the rate at Earth is ~35/year. Of these events, 98.75% were associated with type III radio bursts and 76% with ^{3}He-rich ion events. About 60% of the ion events were associated with narrow CMEs. It should be mentioned that several processes can mask the correlation between the ion and electron events; for example, the flight time for 100 keV electrons down a 1.2 AU IMF line is ~20 min, whereas 0.5 MeV/nucleon ions require ~5 hr, so if the observer at 1 AU loses magnetic connection to the solar source

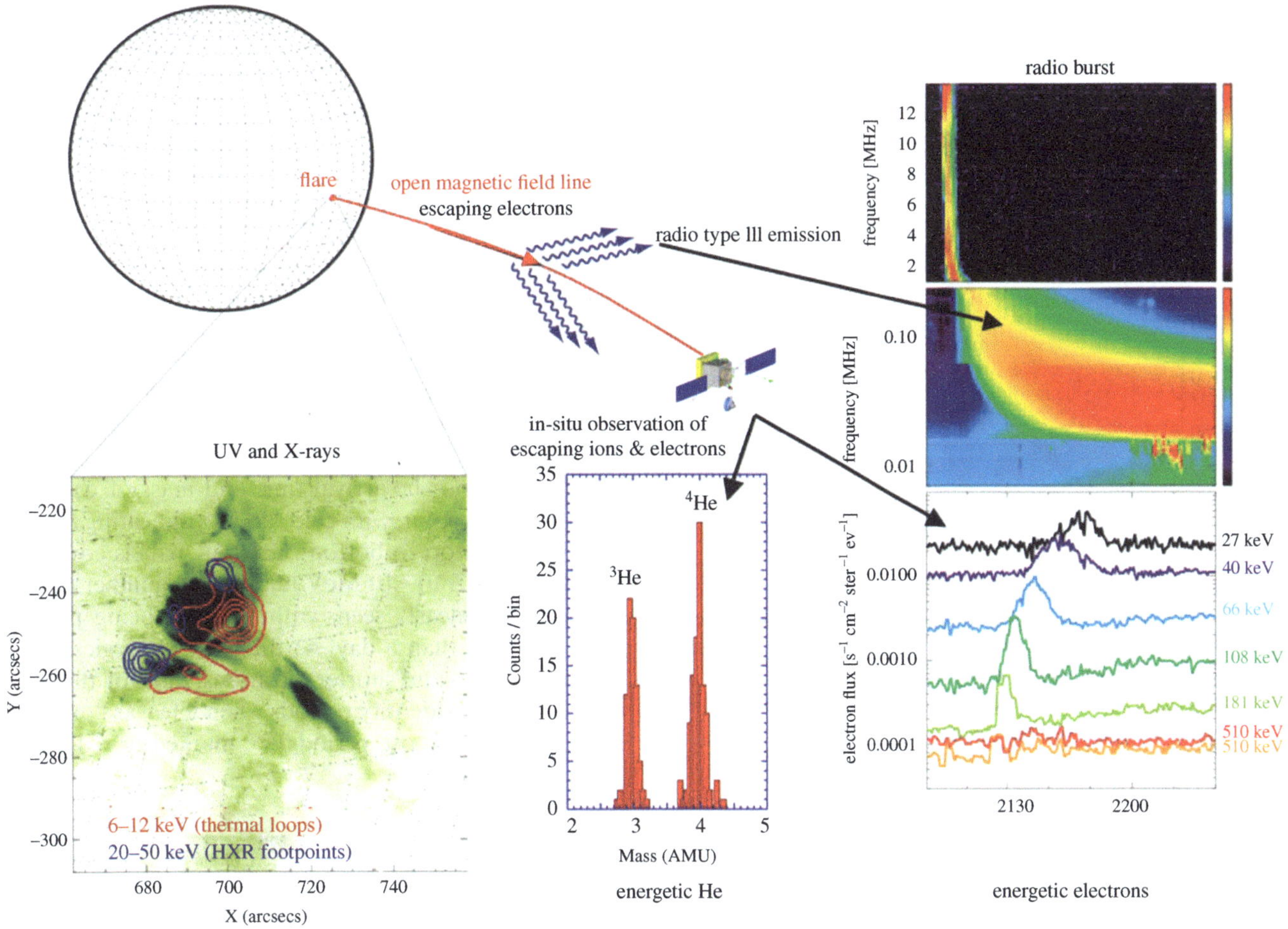

Figure 4.13 Overview of impulsive ^{3}He-rich SEP events: Clockwise from bottom left: A western-hemisphere flaring region with UV and X-ray emission and jets emits an electron beam which can be tracked from its radio burst signature, and later, by the dispersive arrival of electrons up to several hundred keV at 1 au, and still later by the dispersive arrival of ions including He with large enrichments of ^{3}He. (*Source*: Benz et al., 2009; Marsden & Müller, 2011.)

during this period, either the electrons or ions may be missed. In addition, long-lasting solar activity from CME-associated SEP events can mask the small intensity increases associated with these events. So it is quite possible that these already strong correlations are actually lower limits of the correlation.

In a detailed study of 10 electron/^{3}He-rich SEP events, L Wang et al. (2016) deduced several additional properties. Figure 4.15 shows data from a pair of events that occurred on 12 December 2002 (see also Wang et al., 2006b) from a region at N14°W35°. Figure 4.15a shows dispersive arrivals for electrons from the two injections, with sharp arrival onsets for 108 keV electrons (panel b). Figure 4.15c shows ion 1/v plots for ions from 10–70 amu; on this plot, the leading edge of the ion distributions form a reasonably straight line, consistent with travel down an IMF line of the same length at different energies. The extrapolated intersection of these leading edges down to a 1/v time of 0 marks the injection time at the Sun. Panels (d) and (e) show least squares fits to peak intensity times at different energies, with the deduced IMF field line length shown. Finally, panel (f) shows the integrated electron spectrum, which extends down to a few hundred keV. Wang et al. (2016) fitted the intensity profiles using a simple transport model that utilized the fact that in these events the electrons and ions underwent very little scattering in the interplanetary medium (IPM). They derived injection profiles for electrons and ions, with the low-energy electrons (~0.4–9 keV) injected first, the high-energy electrons (~10–300 keV) injected ~17 min later, and the ions ~75 min later. The injection profiles were extended in time, lasting ~90–160 min for the low-energy electrons, ~10–30 min for the higher-energy electrons, and ~200–550 min for the ions. Nine of the events were observed by *SOHO*/LASCO and all these had narrow (<30°) west-limb CMEs. The timing differences

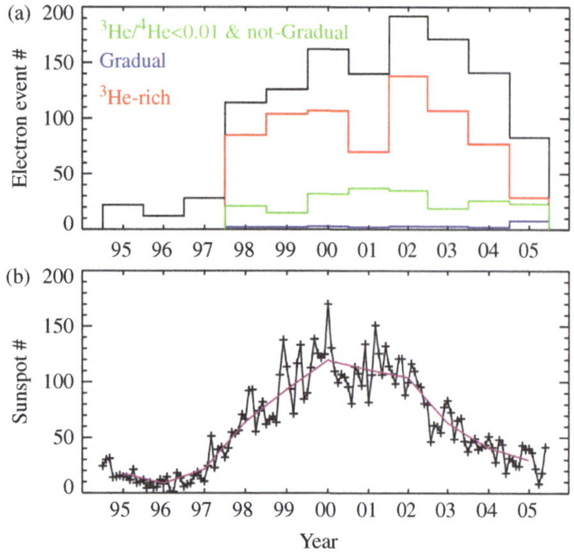

Figure 4.14 *(a):* Histogram of 1191 solar electron events (1995–2005) and their association with impulsive SEP events with ^{3}He: ^{4}He >1% (red) and <1% ^{3}He (green). (b): solar sunspot monthly (black) and yearly (pink) rates. Note the clear correlation of the particle events rates with sunspot number. (*Source*: L. Wang et al., 2012. © 2012 IOP Publishing.)

between the low- and high-energy electrons, and the ions, are puzzling because they do not appear consistent with a single source, but perhaps multiple sites exist where the low-energy electrons provide a seed population for the higher-energy electrons, and ion acceleration may occur at the flanks of the CMEs.

Early observations of ^{3}He-rich events showed only occasional cases, leaving the impression that they were rare. However, given the strong association between ^{3}He -rich ion events and impulsive electron events, it is reasonable to believe that energetic ^{3}He is often present in the IPM. Wiedenbeck and Mason (2014) have used advanced instruments on *ACE* to explore this possibility. Figure 4.16 (top panel) shows mass spectrograms for ^{3}He and ^{4}He from *ACE* instruments, plotted over a single Bartels rotation. The arrival time and mass of each ion are plotted as a point, with color coding indicating intense periods. From top to bottom, the four panels cover the energy range 0.3 to ~10 MeV/nucleon. A number of ^{3}He injections are clearly seen, with the ions being present often for more than a day. At the higher energies, there are fewer events, and shorter durations. Plots of this type were analyzed to identify periods when ^{3}He was "present" in the IPM, and those periods are marked in red in the

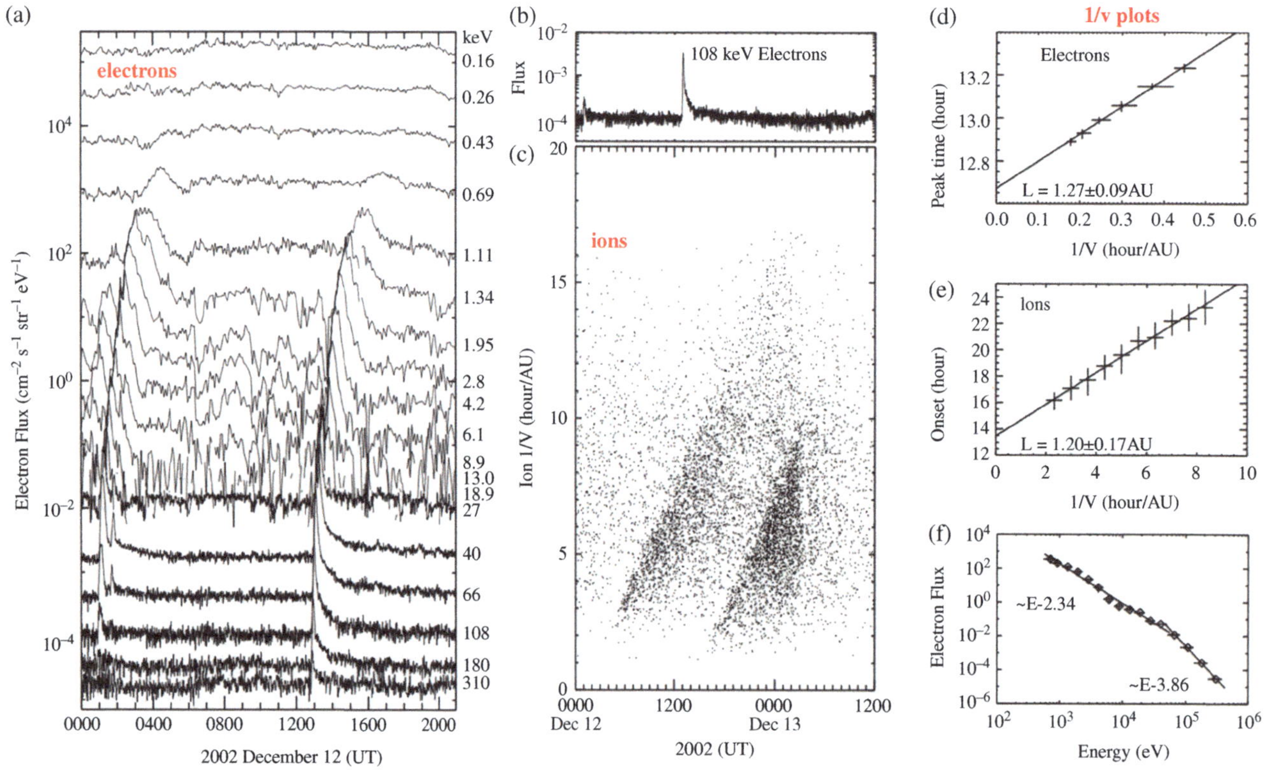

Figure 4.15 Example of 1 au electron and ion data from an impulsive SEP event: (a) dispersive arrival of electrons from two injections on 12 December 2002; (b) arrival of 108 keV electrons; (c) ion spectrogram of heavy ions (10–70 amu) from the two injections; (d and e) 1/v plots for peak intensities for electrons and heavy ions, showing a deduced IMF distance of ~1.25 au; (f) electron spectrum over a broad energy range. (*Source*: L. Wang et al., 2016. © 2016.)

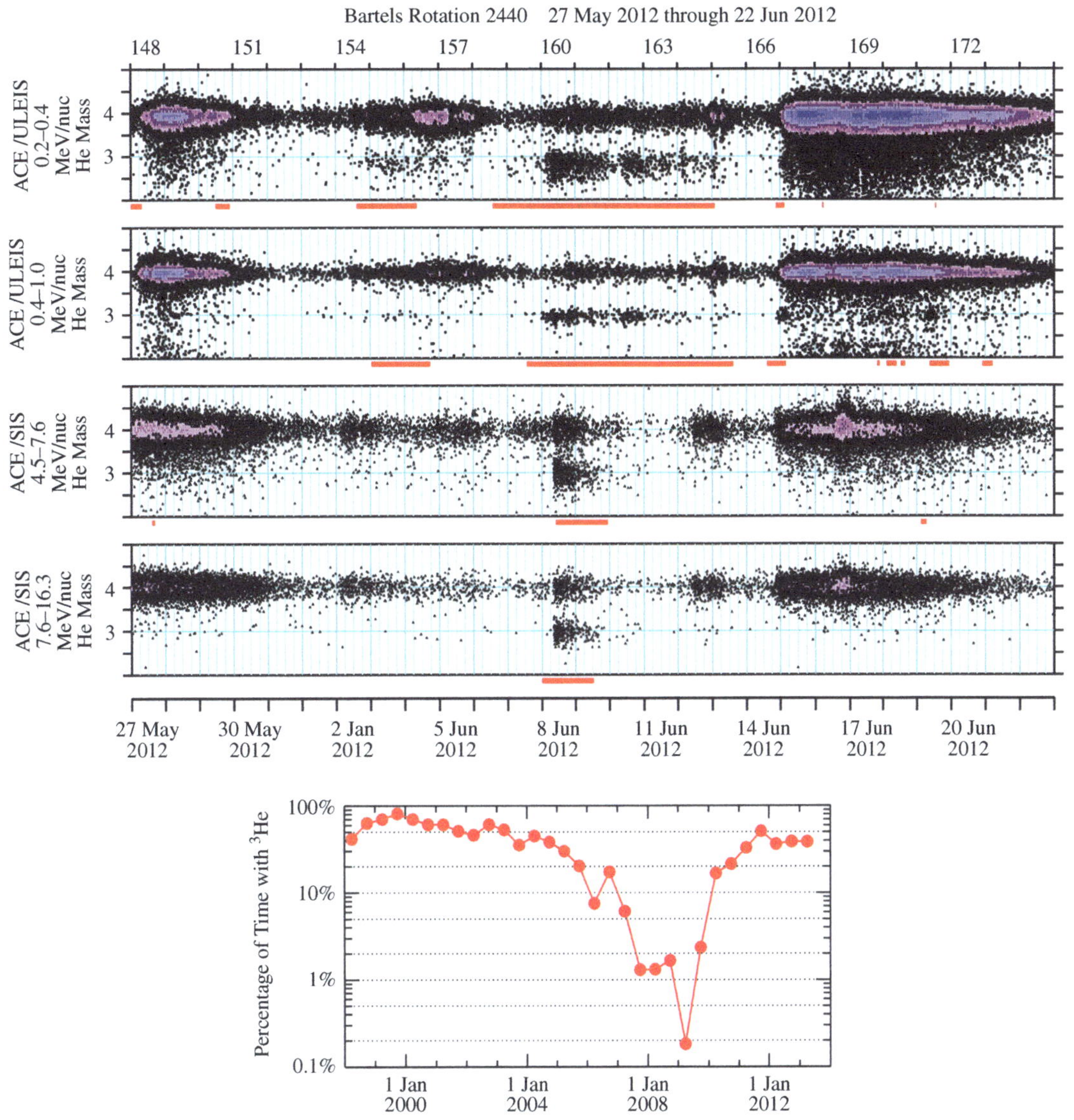

Figure 4.16 Top panel: Spectrograms of ^{3}He and ^{4}He from four energy ranges for a 27-day period in 2012 showing multiple SEP events with clearly visible increases in both ^{3}He and ^{4}He. Red lines mark intervals in which ^{3}He is considered to be "present" in the IPM. Bottom panel: Percentages of time when ^{3}He is "present" in IPM from 1998 through 2013. (*Source*: Wiedenbeck & Mason, 2014. © 2014 IOP Publishing.)

figure. For the Bartel's rotation shown, the red lines cover somewhat less than 50% of the time. The bottom panel shows the combined averages for 6-month periods from 1998 through mid-2013. The figure shows that during solar active periods, ^{3}He is present over 80% of the time, falling below 10% only for about 3 years near solar minimum. This establishes that the processes that produce ^{3}He-rich SEPs occur very frequently.

L Kocharov et al. (2008) examined time periods when there was energetic ^{3}He present for days at a time in order to constrain larger features of the solar source regions. They looked at 17 cases of multi-day-long ^{3}He-rich periods, supplemented by observations from *SOHO* to establish coronal features. All the cases were associated with slow solar wind at 1 AU, with about half of them associated with a deep well in solar wind speed followed by a fast rise. The slow solar wind is indicative of the active region, along with leading and following coronal holes (see also Zwickl et al., 1978). Bučík et al. (2014, 2015) explored this further using the *STEREO* and *ACE* spacecraft to follow individual ARs as they connected sequentially to *STEREO-B*, *ACE*, and then *STEREO-A* over periods longer than could be observed from a single location. Figure 4.17 shows an example of this analysis

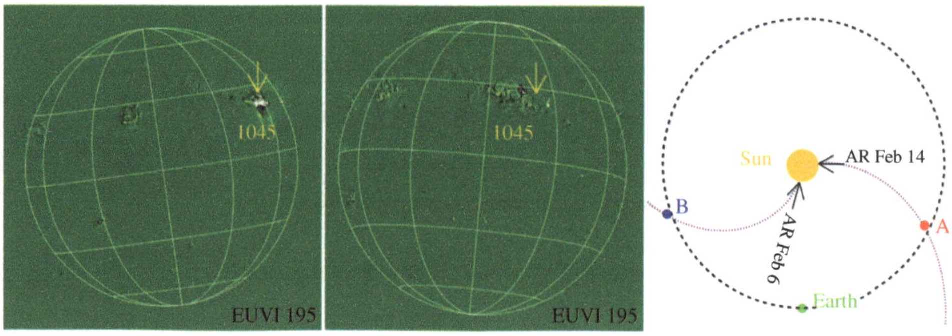

Figure 4.17 Example of a long-lived source of type III bursts and ^{3}He-rich events from AR 11045 that produced events at *STEREO-B* on 7 February 2010 and *STEREO-A* on 14 February 2010. Left and center panels show *STEREO* EUV 5-min running difference images, and the right panel shows the locations of the spacecraft and AR. (*Source*: Bučík et al., 2015. © 2015 IOP Publishing.)

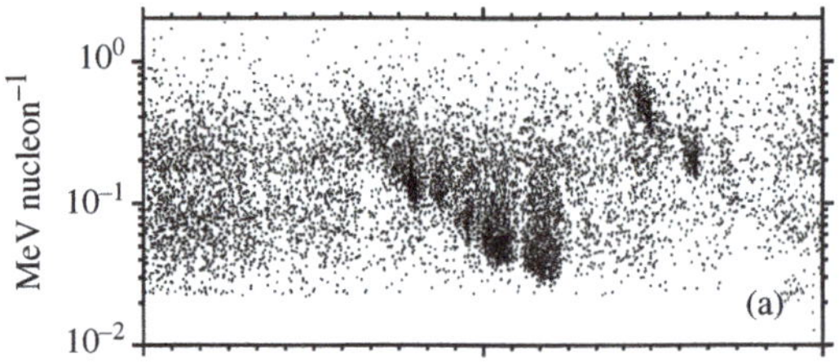

Figure 4.18 Energy versus arrival times of H-Fe from two impulsive SEP events of 9 January 1999 showing temporary periods when connectivity to the source is lost, and ions of all energies are missing. (*Source*: Modified from Giacalone et al., 2000.)

for active region (AR) 11045, which rotated between the spacecraft from 7 to 14 February 2010, producing multiple type III bursts and ^{3}He-rich events. Examining the evolution of the source regions, they found that the ^{3}He-rich events began ~1–2 days after the emergence of magnetic flux (see also Y-M Wang et al., 2006a), with subsequent ^{3}He-rich SEPs occurring after the emergence of additional magnetic flux. They concluded from potential field source surface (PFSS) models of the corona, and EUV images, that the sources were probably emerging into preexisting open field lines.

Insight into the solar source size for ^{3}He-rich SEPs has been derived from dispersionless dropouts in particle intensity during some events. Figure 4.18 shows the arrival times of ions during a two-day period in 1999 when two impulsive events occurred. The plot shows higher-energy particles arriving first, followed smoothly by the lower energies (Mazur et al., 2000). However, the arrival patterns show periods when the ions are no longer present at all energies (dropout), then reappear at all energies. The dropouts are sometimes associated with changes in the IMF directions, but usually not. The interpretation of this behavior is that there is a source region at the Sun whose field lines have the energetic ions injected into them, and then as the field lines drift outward from the Sun taking 3–4 days to reach 1 AU, they undergo mixing in the IMF due to random walk. When field lines drift past the spacecraft, they either have SEPs showing the expected velocity dispersion, or else are empty, having come from a location near but not part of the SEP source. The fact that the dropouts are the same at all energies shows that this is a geometrical effect, and not due to processes at the Sun. This random mixing is also consistent with the lack of IMF signatures at the times of the dropouts. This phenomenon is not seen in large, shock-associated events because their large source sizes fill all the field lines over a wide range of longitude, precluding the presence of such dropouts. Giacalone et al. (2000) constructed a model of this process, and showed that the relevant scale size at the Sun was the size of supergranulation cells, about 40,000 km, and that sources of this scale could produce dropouts, but much larger ones would not.

With the launch of the Solar Dynamics Observatory in 2010, high-resolution, high-cadence imaging with the Atmospheric Imaging Assembly (AIA) permitted vastly improved investigation of sources of ^{3}He-rich events. Nitta et al. (2015) studied 26 ^{3}He-rich events that were well observed on ACE, and had associated type III bursts and electron events observed on the *Wind* spacecraft. Source locations were additionally verified using EUV images from *STEREO*, which also provided critical information on regions not visible from Earth. Figure 4.19 shows a sample event from 16 May 2014. Images from the movies used for the analysis show (a) the whole solar disk at the time of the event, (b) an expanded region around the event location that includes a jet; (c) the same jet observed from *STEREO-A*, and (d) a narrow CME observed by LASCO shortly after the jet. It was found that properties of the ^{3}He-rich SEP events, such as spectral form, ^{3}He/^{4}He ratio, and highest energy particles observed, did not correlate in any significant way with the source characteristics regarding jet or CME size. Rather,

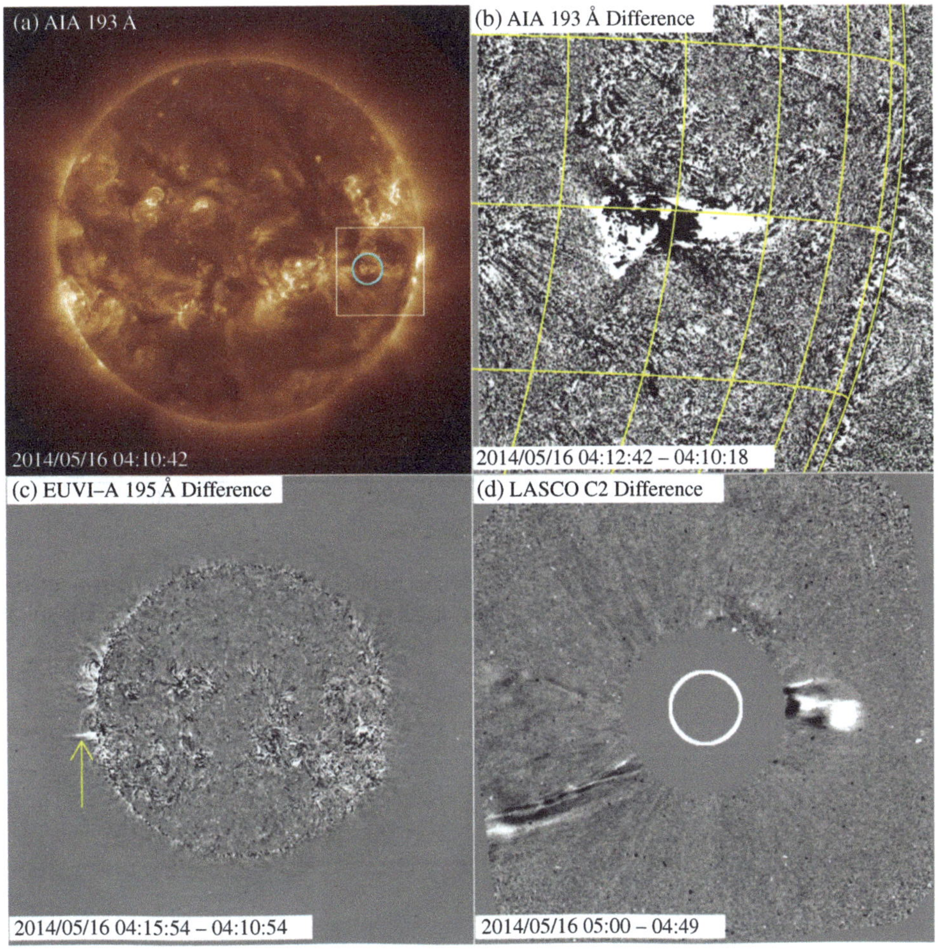

Figure 4.19 High-resolution SDO AIAA images near the time of a type III burst and associated ^{3}He-rich SEP event on 16 May 16 2014. (a) Whole disk image at 193 Å; (b) expanded area from box in preceding panel, showing a jet; (c) difference image in 195 Å from *STEREO-A* EUVI, with arrow showing the same jet; (d) LASCO C2 image shortly after the jet showing a narrow CME. (*Source*: © 2015. The American Astronomical Society. Nitta et al., 2015.)

the study confirmed previously identified correlations with type III bursts and electrons, and at the source region with a variety of motions such as eruptions and jets. Figure 4.20 explores the source locations in the present study along with two others (Nitta et al., 2006; Y-M Wang et al., 2006b). The figure shows the solar wind speed at the time of the event versus the source heliolongitude, along with a curved line showing where the expected source location would be if it followed the extrapolated Parker spiral. The width of the distribution is surprisingly large given the small source size inferred for these events and the lack of large features such as wide CMEs that might explain a broad distribution. Note that in the Nitta et al. (2006) study, a number of events are piled up at 90°, these being due to events that were consistent with being over the solar limb but without sources visible from Earth. The more recent study supplementing the Earth view with *STEREO* observations clearly shows some events with sources beyond the west limb, out to ~110°, or almost 60° away from the nominal Parker spiral connection point. This distribution is not consistent with the modeling using the PFSS and Parker spiral, and may indicate that the magnetic connection between the corona and IMF may be much more extended than previously thought. This might be related to the very puzzling cases of ^{3}He-rich events observed over a broad range of longitudes by multiple spacecraft (Wiedenbeck et al., 2013). Or in some cases the ions may be spread over a broader longitude range by large-scale structures such as those shown in Figure 4.21; however, these are unusual (Bučík et al., 2017).

Summarizing, both remote and in situ observations with vastly improved sensitivities have firmly established many properties of ^{3}He-rich SEP sources glimpsed in earlier studies, and identified fundamentally new properties such as (a) the near-ubiquitous presence of ^{3}He in the

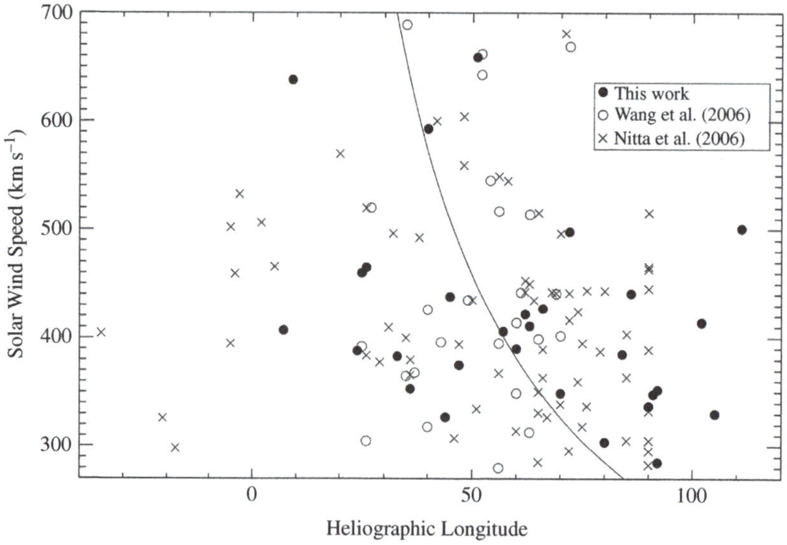

Figure 4.20 *Source* locations of ^{3}He-rich events plotted versus heliographic longitude and associated solar wind speed. The curve shows the longitude expected from the solar wind speed and nominal Parker spiral. (*Source*: Nitta et al., 2015. © 2015 IOP Publishing.)

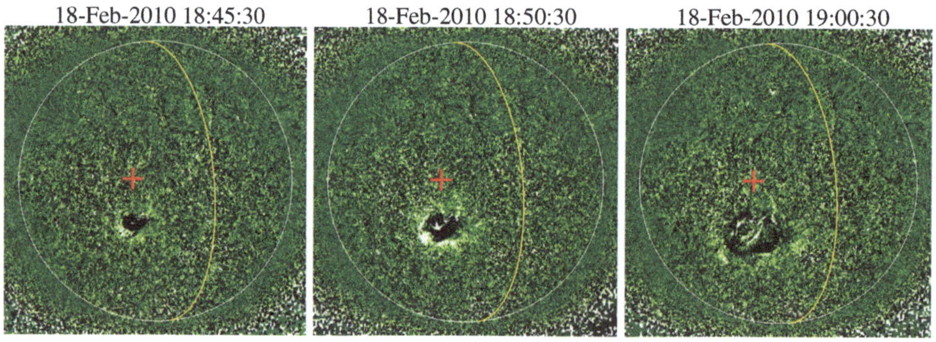

Figure 4.21 Association of ^{3}He-rich SEP event with a large-scale coronal wave seen by *STEREO-A* 195 Å EUV. L1 magnetic footpoint is shown by the red plus symbol; west limb as viewed from Earth is shown by yellow arc. (*Source*: Bučík et al., 2017.)

IPM during solar active periods, (b) the surprisingly large range of heliolongitudes over which some events are observed, (c) the association with jets and other motions in the corona, (d) the puzzling differences between electron and ion injection times and injection lengths, and (e) the deduction of a compact source size from observations of interplanetary dropouts in the events. It needs to be understood that these properties have been largely, but not entirely, developed from observations of larger ^{3}He-rich events. For example, the association with jets, although quite firm, is based on probably <100 events, whereas ^{3}He-rich SEP events more likely occur in the thousands during active periods. Some cases of ^{3}He in the IPM have been identified during solar minimum when there were no ARs in the western hemisphere, only plages (Mason et al., 2009), and for many if not most events, solar sources have not been identified. Because the ^{3}He-rich ion events last much longer than impulsive electron events at 1 AU, it is not known if seemingly continuous periods of ^{3}He presence are due to a steady source, or an unresolved series of shorter events such as impulsive electron events.

4.4.2. Properties of ^{3}He-Rich SEP Event-Averaged Electron and Ion Spectra

Event-averaged impulsive SEP electron differential energy spectra are typically steep power laws, $J \sim E^{-\gamma}$ with γ between 3.5 and 5, and some flattening at low energies (Lin, 1998; Potter et al., 1980). Figure 4.22 shows an example from the *Wind* 3DP instrument, which shows a peak in the spectrum around 1 keV, and above that power-law spectra with indices 3.0 from ~1 to 40 keV and 4.4 above that (Lin, 1998). Because the spectrum extends to very low energies before turning over, Lin et al. calculate that this portion of the spectrum must have traversed a column density of ~<9×10^{17} cm^{-2} of material (~1.5 μg) and therefore must have been accelerated at relatively high altitudes (~0.2–1.0 R_S depending on the models and activity level). However, they point out that higher-energy portions of the spectrum may have been accelerated at lower altitudes, as mentioned previously in the discussion in Wang et al. (2016).

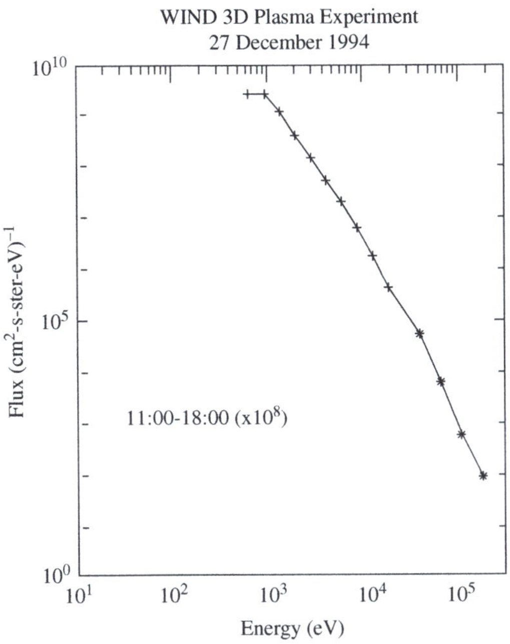

Figure 4.22 Event-averaged omnidirectional differential electron spectrum for an impulsive SEP event. Above the peak around ~1 keV the spectrum can be fit to two power laws. (*Source*: Modified from Lin, 1998.)

^{3}He-rich SEP event ion spectra show widely varying properties and unique compositional signatures. Figure 4.23 shows spectra from two impulsive events that were intense enough to allow spectral measurements up to ~10 MeV/nucleon using the high-sensitivity ACE ULEIS and SIS instruments (Mason et al., 2002a). Figure 4.23a shows one type of spectral form, differential energy power laws with spectral index γ ~1−3 at low energies, and steepening by about 1 unit above a break at a few MeV/nucleon. As mentioned in section 4.3, broken power-law spectra of this general type are often seen in large, shock-accelerated SEP events. Figure 4.23b shows the other general form of spectra observed, which are notably curved for ^{3}He and the heavy ions, but where the ^{4}He spectrum is closer to a power law. The power-law spectral forms occur more frequently in the larger events; for example, the 14-event survey in Mason et al. (2002b) selected events producing measurable spectra above several MeV/nucleon, and of this group only two events showed curved spectra. However, using the same instrument without the high-energy spectral requirement, Nitta et al. (2015) selected 26 events with no requirement on spectral shape, and of these more than half were curved spectra. This difference suggests that there may be more than one stage of acceleration, with the curved spectra being the result of an initial mechanism, and the power-law shape formed if there is an additional accelerating agent (Mason et al., 2002b).

Besides the large enrichment of ^{3}He evident in Figure 4.23, the ion spectra also show a significant enhancement of the Fe abundance relative to O, where

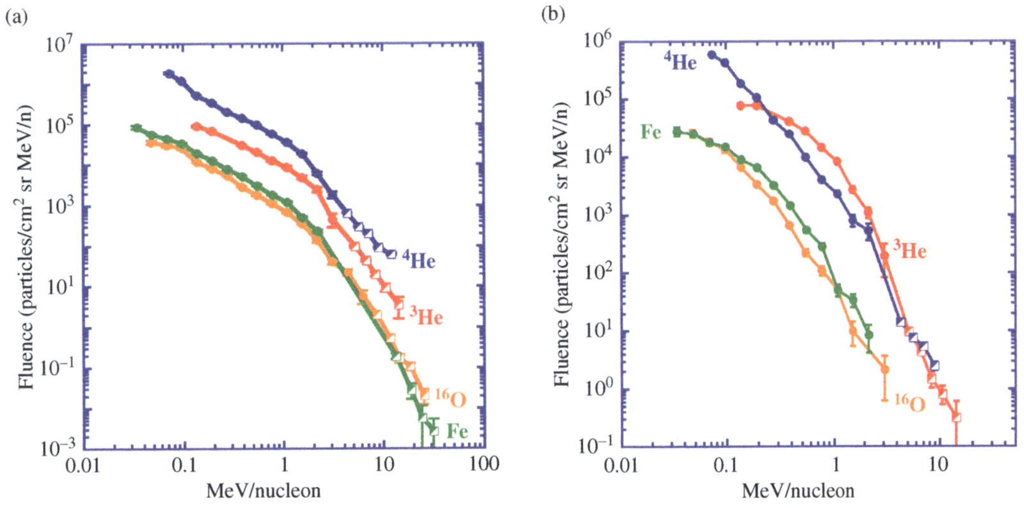

Figure 4.23 Event-averaged fluences for (a) the 9 September 1998 event, and (b) 21 March 1999 showing power-law and curved spectral forms. (*Source*: Mason et al., 2002b. © 2002 IOP Publishing.)

Table 4.1 ^{3}He-Rich SEP Event Abundances

Species	Mass	^{3}He-rich SEP abundance[a]	Gradual SEP abundance[b]	Solar System abundance[c]	^{3}He-rich SEP versus reference
^{4}He......	4	54 ± 14 [d]	57 ± 3		0.95 ± 0.24[d]
C.........	12	0.322 ± 0.003	0.465 ± 0.009		0.69 ± 0.01
N.........	14	0.129 ± 0.002	0.124 ± 0.003		1.04 ± 0.03
O.........	16	≡1.000 ± 0.006	≡1 ± 0.01		1.00 ± 0.01
Ne......	20	0.261 ± 0.003	0.152 ± 0.004		1.72 ± 0.05
Mg......	24	0.370 ± 0.003	0.196 ± 0.004		1.89 ± 0.04
Si.........	28	0.409 ± 0.004	0.152 ± 0.004		2.69 ± 0.07
S.........	32	0.118 ± 0.015	0.0318 ± 0.0007		3.70 ± 0.47
Ca.........	40	0.060 ± 0.003	0.0106 ± 0.0004		5.66 ± 0.38
Fe.........	56	0.950 ± 0.005[e]	0.134 ± 0.004	≡1	7.09 ± 0.21
	78–100	(9.6 ± 1.1) e-04		1.76 e-04	40.7 ± 4.5
	125–150	(3.1 ± 0.6) e-04		1.95 e-05	119.6 ± 23.0
	180–220	(2.2 ± 0.5) e-04		7.63 e-06	215.4 ± 49.5

[a] 385 keV nucleon^{-1} [Mason et al., 2004].
[b] 5–12 MeV nucleon^{-1} [Reames, 1995].
[c] Meteoritic, [Anders & Grevesse, 1989].
[d] Uncertainty primarily due to instrumental efficiency [see discussion in Desai et al., 2003].
[e] Multiplied by a factor of 0.92 to correct for larger mass range summed in Fe group peak.
Sources: Mason et al., 2004; Reames, 1995; Anders & Grevesse, 1989; Desai et al., 2003.

Fe/O ~1, that is, much larger than in large SEP events. Table 4.1, taken from Mason et al. (2004), shows SEP abundances from surveys of multiple events. Column 3 of Table 4.1 shows ^{3}He-rich event composition, column 4 shows higher-energy abundances from shock-accelerated SEP events, column 5 shows solar system meteoric abundances, and column 6 shows the ratio of ^{3}He-rich SEP events to the large SEP event or meteoric abundances. The enhancement factor (column 6) for ^{3}He-rich events at 385 keV/nucleon is plotted in the left panel of Figure 4.24, along with ^{3}He-rich event abundances at several MeV/nucleon (from Reames, 1995), which are very similar. The right panel shows enhancements of ions up to Z = 82, which are roughly similar to that seen at lower energies. The Fe/O enhancement in ^{3}He-rich SEPs compared to shock-accelerated events is 7.09 ± 0.21 at 385 keV/nucleon (Table 4.1) and 8.04 ± 0.42 at several MeV/nucleon (Reames, 1995).

Referring back to Figure 4.23b, the curved ^{3}He spectrum and nearly power-law ^{4}He spectrum lead to a ^{3}He/^{4}He ratio that varies with energy. However, in Figure 4.23a, the energy dependence of this ratio is much less for power-law spectra events. Figure 4.25 shows examples of the ^{3}He/^{4}He ratio for the events of (blue) 21 March 1999 (DOY 80), (red) 9 September 1998 (DOY 252), and (green) 27 September 2000 (DOY 271). The March 21 event had curved spectra similar to those in Figure 4.23b, leading to a ^{3}He/^{4}He ratio of 0.4 at 100 keV/nucleon, rising by a factor of about 10 and peaking at 1 MeV/nucleon, then falling off at higher energies. The 9 September 1998 event had broken power-law spectra as in Figure 4.23a and was reasonably constant from 100 keV/nucleon to several MeV/nucleon, after which is dropped off by a factor of ~5 around 12 MeV/nucleon. Finally, the case of 27 September 2000 was an unusual case of power-law spectra for all elements but with ^{3}He having a harder slope so that the ^{3}He/^{4}He ratio increased by a factor of ~30 from 100 keV/nucleon to ~5 MeV/nucleon and then started to drop off. It is evident from Figure 4.25 that although all three events are highly enriched in ^{3}He compared to solar wind or solar system abundances, there is no single ^{3}He/^{4}He ratio to characterize any given event because the value may vary greatly with energy. In addition, there is, however, an indication that above about 5 MeV/nucleon the ^{3}He/^{4}He ratio starts to drop off, which is probably an important clue to the differences in the acceleration mechanism for ^{3}He and ^{4}He.

Another key ion property in ^{3}He-rich SEP events is the ionization state of the energetic particles, which are much more highly stripped than those in the solar wind or large SEP events. Figure 4.26 shows histograms of ~40 ^{3}He-rich and shock-accelerated SEP events compiled by Klecker et al. (2007) from surveys of ^{3}He-rich events by DiFabio et al. (2008) and large SEP events by Klecker et al. (2006b). Compared to large events, the ionization states of other elements such as Si are also much higher in ^{3}He-rich events (Luhn et al., 1987); in fact, Si is fully stripped, or nearly so.

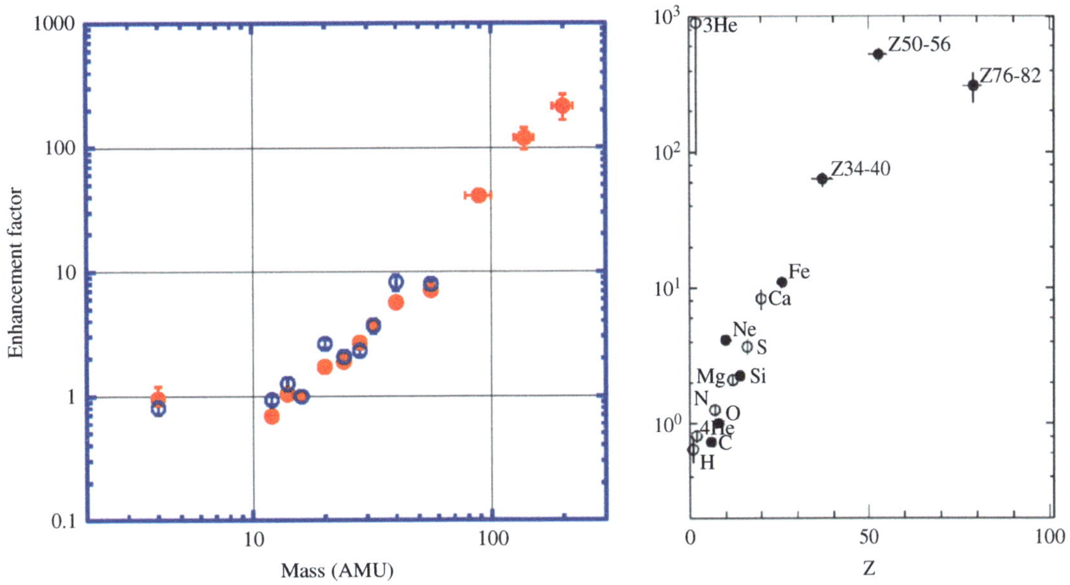

Figure 4.24 Heavy ion relative abundance enhancements compared to solar system and coronal abundances at 0.1–1.0 MeV/nucleon (left panel) and 3.3–10 MeV/nucleon (right panel). The enhancements grow to a value of a factor of several hundred at the top of the periodic table. (*Source*: Mason et al., 2004; Reames & Ng, 2004. © 2004 IOP Publishing.)

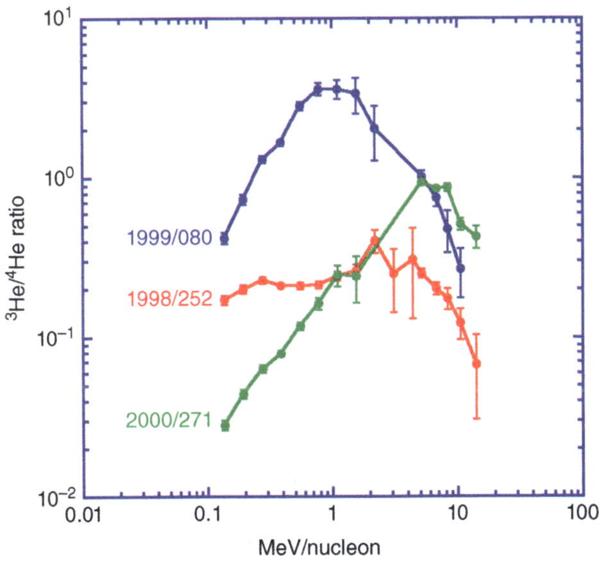

Figure 4.25 Energy dependence of $^3He/^4He$ ratio for three impulsive SEP events, showing different dependence on energy. (*Source*: Mason et al., 2002b. © 2002 IOP Publishing.)

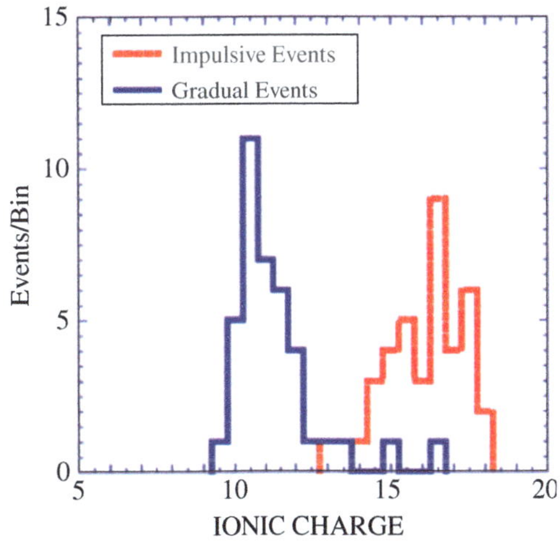

Figure 4.26 Mean ionic charge state of 0.18–0.25 MeV/nucleon Fe in ~40 impulsive events and ~40 gradual events. (*Source*: Klecker et al., 2007. © 2007 Elsevier.)

4.4.3. Theoretical Models for ^{3}He-Rich Events

Acceleration models for ^{3}He-rich SEP ions have focused on the extremely large $^3He/^4He$ ratio, the more modest heavy ion enrichments, and the high ionization states of heavy ions. Most models addressing the ^{3}He enrichment make use of its Q/M ratio, which uniquely lies between that of H and ^{4}He, whereas other heavy ions have Q/M ratios that are the same as or smaller than that for ^{4}He. Thus, ^{3}He might be able to interact with plasma wave turbulence more strongly than other ions, leading to preferential heating (Drake et al., 2009; Eichler, 2014; Fisk, 1978; L G Kocharov and Orischenko, 1983; Miller and Reames, 1996; Möbius et al., 1982; Petrosian, 2012; Temerin and Roth, 1992; Zhang, 1995). Figure 4.27 shows an example of a model of ^{3}He and ^{4}He acceleration by parallel-propagating plasma waves wherein the

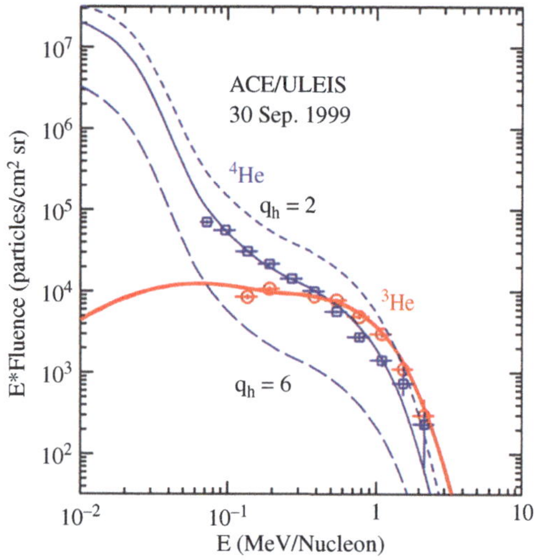

Figure 4.27 Calculated ^{3}He and ^{4}He energy spectra accelerated by stochastic turbulence, compared with the impulsive SEP event observed on 30 September 1999. (*Source*: Liu et al., 2006. © 2006 IOP Publishing.)

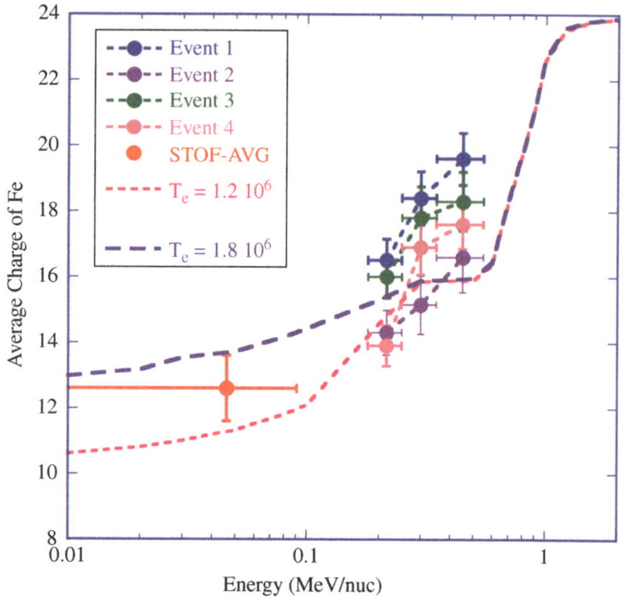

Figure 4.28 Mean ionic charge state of Fe as a function of energy for four impulsive SEP events, and calculated values from a charge-stripping model. (*Source*: Klecker et al., 2006c. © 2006 Elsevier.)

^{3}He can interact with more waves than ^{4}He and is preferentially accelerated to ~MeV energies. The figure shows calculated spectra for ^{3}He (red) and ^{4}He (blue) and also observations from a ^{3}He-rich SEP event (the dashed lines show the calculated spectra for different values of q_h, the exponent of the high wavenumber portion of the turbulence spectrum). This model fits important features of the event shown, and several others in the paper, in particular, the variation in the ^{3}He/^{4}He ratio with energy. Note that below 100 keV/nucleon, the ^{4}He dominates the ^{3}He, so the cause of the enhancement is not really that ^{3}He is preferentially heated in general, but rather that the ^{3}He reaches higher energies than the ^{4}He, and high ^{3}He/^{4}He ratios are achieved only in that range. This is quite different from a model that produces high ^{3}He/^{4}He at all energies. The rollovers of the curved ^{3}He spectra such as shown in Figure 4.23b are consistent with this picture, and immediately suggest examining particle spectra down to much lower energies than the ~100 keV/nucleon threshold of current instruments. At 1 AU, this may not be feasible because the propagation time of, for example, 20 keV/nucleon ions is about 25 hr assuming little or no scattering, and so this requires that the spacecraft be connected to the small emitting region for a long time. During this same long period, other events may take place, adding source confusion to the picture. So exploring this range from 1 AU would be difficult.

Figure 4.28 shows observations of the energy dependence of the Fe ionization state, which is seen to increase in four events from ~14–16 at 250 keV/nucleon to ~16–19 at ~500 keV/nucleon. At lower energies, the average ionization state is much lower, around 12 (Klecker et al., 2006c; L G Kocharov et al., 2000). The increase in Fe ionization states is not consistent with a single source temperature, but rather is evidence that the particles have undergone stripping during acceleration. The dashed lines show the calculated charge states for two coronal temperatures that are compatible with the low-energy Fe Q state, and then the Fe ionization state increases in the range of a few hundred keV/nucleon to higher values (Kocharov et al., 2000). The calculated higher Fe Q states, however, are seen at higher energies than the observed values, and the authors speculate that the difference in energy might be reconciled with the observation if the particles underwent adiabatic deceleration in the IPM, lowering their energy to the range below ~1 MeV/nucleon (for a more complete model, see J. J. Kartavykh et al., 2005; Y. Y. Kartavykh et al., 2007).

The enrichment of heavy ions in ^{3}He-rich SEPs has also been examined using the plasma turbulence models that preferentially accelerate the ^{3}He. Fisk (1978) pointed out that although heavy ions did not have the same Q/M ratio as ^{3}He, for partially stripped heavy ions their Q/M ratio might fall at a harmonic of the waves heating ^{3}He, and thereby have enhanced heating, although not as strong as the ^{3}He. Several problems emerged in this picture, first, the observation that ions such as Si were nearly fully stripped and could not be preferentially heated if they had the same Q/M ratio as ^{4}He. This is a major problem; however, if there was stripping during the acceleration, the difficulty might be overcome. Second, it was found

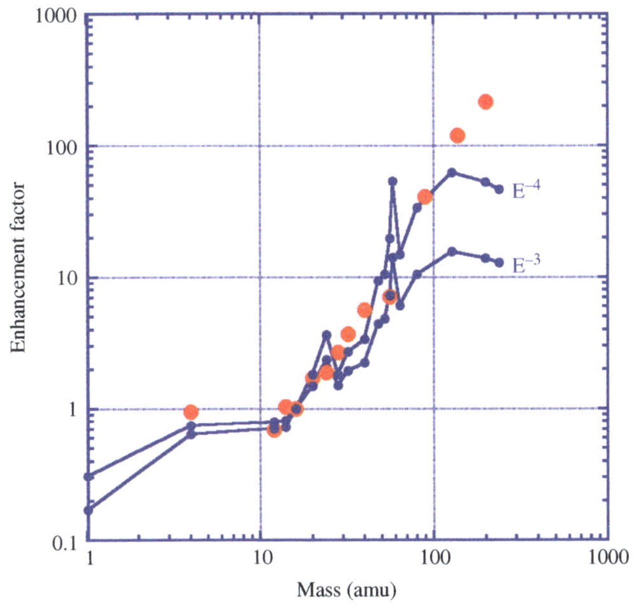

Figure 4.29 Blue points: calculated enrichment of different ions after passing through 200 μg of H gas for original power-law spectra with indices −3 and −4. Red points: Observed enhancements in 3He-rich events. (*Source*: Mason & Klecker, 2018. Licensed under CCBY 3.0.)

the size of the heavy ion enrichment was uncorrelated with the ^{3}He/^{4}He ratio (Mason et al., 1986; Reames et al., 1994). Third, the very large enhancements of ultra-heavy ions covered a very broad range of Q/M ratios, making it difficult to see how any mechanism operating in a favored Q/M range could produce the ordered enhancements observed. Although each of these difficulties might be overcome, they point to the fact that the heavy ion enrichment does not easily fit the mechanisms considered for ^{3}He enhancement.

Figure 4.29 shows an alternate approach to the heavy ion enrichment problem that is based on ionization energy loss, and the fact that as ions near the end of range below the Bragg peak, the greater electron affinity of heavier ions results in their picking up electrons and thus lowering their Q/M ratio, thereby allowing them to penetrate small amounts of material more easily (Mason & Klecker, 2018). The figure shows the enrichment that occurs if initial power laws with spectral indices 3 or 4 are passed through 200 μg of H gas. Heavier ions at lower energy/nucleon can penetrate the gas, leaving a population with the abundance enhancements shown. This process does not depend on Q/M, but rather the greater electron affinity of the heavy ions, which extends throughout the periodic table. This seed particle population enriched in heavy ions is then presumed to undergo another stage of acceleration to reach the SEP energies. If the H gas is replaced with H plasma at coronal temperatures, the required amount of material decreases by approximately a factor of 10, but the basic process remains (Mason & Klecker, 2018). The stripping implied with this approach may be compatible with the charge state energy dependence, but no detailed examination of this question has been made.

The recently launched *Parker Solar Probe* and *Solar Orbiter* missions to the inner heliosphere will provide many observations and greatly improve our understanding of ^{3}He-rich SEP events. Higher intensities will give us more detailed information on the abundance enhancements, and will make it possible to see if ultra-heavy nuclei are routinely present or are rare. Event timing will be much more precise not only due to the higher intensities, but also the much shorter travel distances from source to observer. These new data will bear directly on many of the existing puzzles described above, and may provide decisive new insights into the physical mechanisms operating in these events.

4.5. SEP ELECTRONS

Solar energetic electron events were first observed at energies above ~40 keV (Anderson & Lin, 1966; Van Allen & Krimings, 1965), with fast-rise, slow-decay temporal profiles indicative of significant scattering in the IPM. Later, events were detected down to keV energies (and even to ~0.1 keV; Gosling et al., 2003), occasionally with fast-rise, fast-decay profiles that imply essentially scatter-free propagation in the IPM (Lin, 1974). Near solar maximum, $\geq 10^2$ events year^{-1} on average occur near Earth (Lin, 1985), making these the most common solar particle acceleration phenomenon observed in the IPM.

Most solar energetic electron events are accompanied by low-energy, ~MeV nucleon^{-1} ion emissions that are highly enriched in ^{3}He (Reames et al., 1985). These electron/^{3}He-rich ion events form the class of "Impulsive" SEP events, so called because the associated soft X-ray (SXR) flares, when present, have a duration of <1 hr (Cane et al., 1986). These events often appear to fill a cone of ~30°– 60° wide in solar longitude at 1AU (sometimes as large as ≥~100°; Wiedenbeck et al., 2013), and last for from tens of minutes (for nearly scatter-free events; Figure 4.30) to many hours (for diffusive events). They are also characterized by large electron/proton (e/p) ratios, significant enhancements (factor of ~10) of heavy nuclei such as Fe, high ionization states for heavy ions, and extreme (factor of > 200) enhancements of ultra-heavy nuclei up to > 200 amu (Mason, 2007). Most solar impulsive electron events have a type III radio burst, but many events are unaccompanied by Hα flares and hard X-ray (HXR) flares, and some are even unaccompanied by SXR flares, that is, no flare-related activities at all

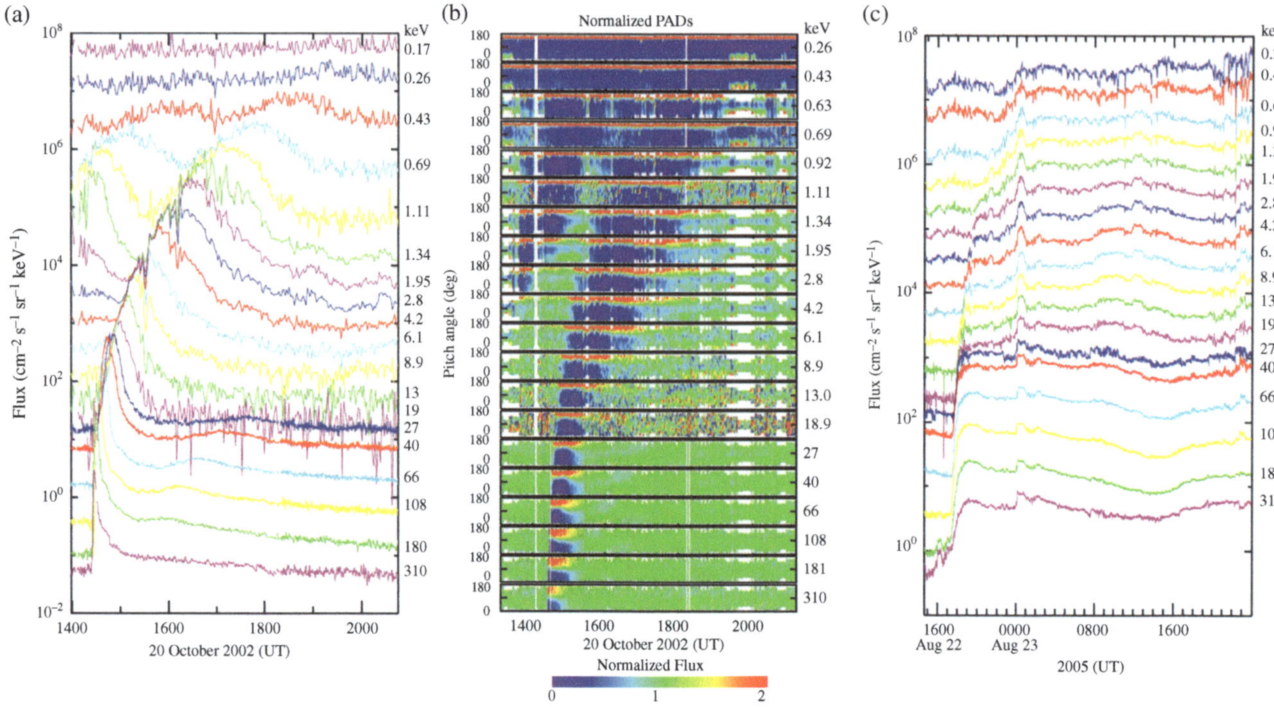

Figure 4.30 Overview of the 20 October 2002 "impulsive" electron event (a,b) and the 22 August 2005 "gradual" electron event (c). (*Source*: Wang et al., 2011, 2012. © 2011, 2012 IOP Publishing.)

(Haggerty & Roelof, 2002; Lin, 1985; Krucker et al., 1999, 2007).

Some solar energetic electron events are accompanied by large proton events that have intense fluxes of >10 MeV protons (sometimes up to tens of GeV energies), with small e/p ratios and little or no ^{3}He enhancements (Desai et al., 2006; Mason, Mazur, Dwyer et al., 1999). These electron/proton events are called "gradual" SEP events because the associated SXR flares have a duration of >1 hr. These events occur about ~10 year^{-1} at solar maximum, extend over ~100°–180° in solar longitude, last for days, and are closely associated with fast CMEs and type II radio bursts.

The following sections review characteristics of solar energetic electron events, specifically their occurrence frequency (section 4.5.1), association with solar phenomena (section 4.5.2), particle injection (section 4.5.3), energy spectra (section 4.5.4), and pitch angle distributions (section 4.5.5). The recent results presented were obtained by combining the electron measurements from solar wind plasma to 400 keV by the 3D Plasma & Energetic Particles (3DP) instrument (Lin et al., 1995) on the *Wind* spacecraft near 1 AU, with the ACE/ULEIS ion measurements (Mason et al., 1998) and with the *Wind*/WAVES radio and plasma wave measurements (Bougeret et al., 1995).

4.5.1. Occurrence Frequency of Solar Energetic Electron Events

Wang et al. (2012a) surveyed 1191 solar energetic electron events observed by *Wind*/3DP from 1995 through 2005. The event occurrence rate has a minimum of 12 year^{-1} in 1996 and a double peak at solar maximum (Figure 4.14), with a primary peak (192 year^{-1}) in 2002 and a secondary peak (162 year^{-1}) in 2000, separated by a dip (140 year^{-1}) in 2001. This solar-cycle variation is similar to that of the sunspot number.

Figures 4.31a and 4.31b show that the yearly occurrence frequency distributions of solar energetic electron events observed near the Earth fit a power law well at high fluxes but bend down at low fluxes. After correction for periods of high electron flux background, the inferred occurrence frequency of solar energetic electron events exhibits a power-law distribution: $dN/dJ \propto J^{-\gamma}$, where N represents the yearly event number, and J represents the background-subtracted peak flux. The fitted index γ ranges from ~1.08 to 1.63 (from ~1.02 to 1.38) at 40 keV (2.8 keV) for different years. It is similar/close to the frequency distributions of solar proton events with a γ of ~1.1 to 1.5 (e.g., Cliver et al., 1991), of microflares with a γ of ~1.4 to 1.6 (Christe et al., 2008a) and of CME kinetic energies with a γ of ~1 (Vourlidas et al., 2002) and ~1.7

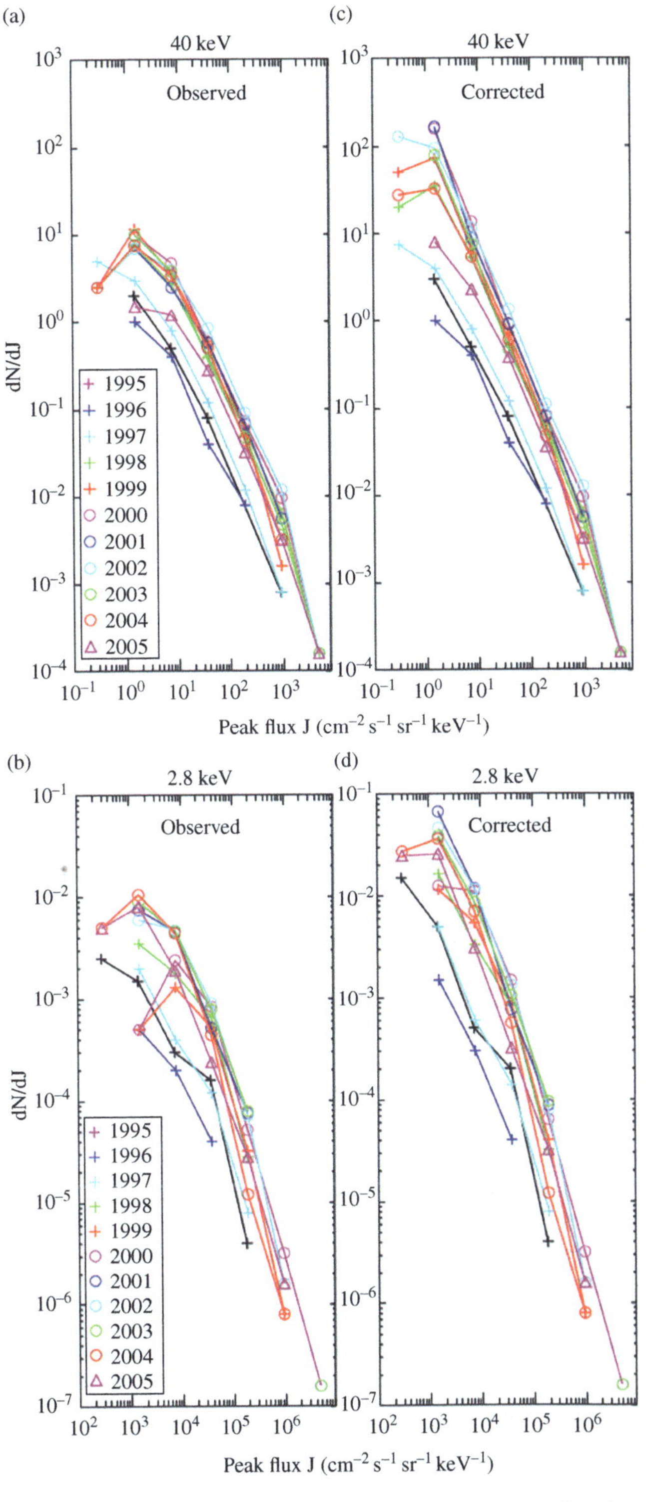

Figure 4.31 Event occurrence differential frequency distribution at 40 (top) and 2.8 (bottom) keV from 1995 to 2005. Panels (a,b) show observations and panels (c,d) show the distributions after correction for events lost due to high background fluxes. (*Source*: Wang et al., 2012. © 2012 IOP Publishing.)

(Yashiro et al., 2008), but is significantly flatter than the frequency distributions of SXR flares with a γ of ~2.2 (e.g., Belov et al., 2007; Hudson, 2007).

After correction for high-background times, the integrated occurrence rate of solar energetic electron events near the Earth is ~1000 year^{-1} at solar maximum and ~30 year^{-1} at solar minimum, up to about one order of magnitude larger than the observed occurrence rate. This implies that likely many small electron events are unable to be detected in situ near 1 AU due to high background or limited instrument sensitivity. In addition, because solar electron events typically fill a solar longitude extent of ~45° (Wang et al., 2012a), ~10^4 events/year may occur over the whole Sun at solar maximum.

4.5.2. Association with Other Solar Phenomena

Wang et al. (2012a) reported that ~99% of solar energetic electron events detected at 1 AU have a type III solar radio burst detected within 30 min by *Wind*/WAVES (~10 kHz to 14 MHz), but only ~8% have a type II radio burst (see Table 4.2). Studies of individual impulsive electron events (e.g., Lin, 1990; Wang et al., 2006) show that the Langmuir waves that scatter to produce type III radio emissions were observed in situ simultaneously with the arrival of ~2–10 keV solar electrons at 1 AU.

Near solar maximum, the inferred occurrence rate of ~10^4 solar electron events/year over the whole Sun is the same order of magnitude as the occurrence rate of type III radio bursts detected below 14 MHz by *Wind*/WAVES, indicating a one-to-one correlation between solar electron beams and interplanetary type III radio emissions.

About 76% (24%) of solar energetic electron events (when the $^3He/^4He$ measurements are available) are accompanied by low-energy ion emissions with $^3He/^4He \geq 0.01$ (<0.01) (Figure 4.14a, Table 4.2), referred to as ^{3}He-rich solar electron events (not-^{3}He-rich solar electron events). Not-^{3}He-rich solar electron events are further divided into gradual events (that are associated with gradual SXR flares) and "not-^{3}He-rich & not-gradual" events. Only ~2% of ^{3}He-rich solar electron events, as well as ~27% of "not-^{3}He-rich & not-gradual" solar electron events, are accompanied by intense proton events (peak flux > 1 cm^{-2} s^{-1} sr^{-1}) detected above 10 MeV by *GOES*; ~52% of gradual electron events are accompanied by intense >10 MeV proton events. The close correlation between solar electron events and ^{3}He-rich ion emissions suggests that the selective acceleration of ^{3}He-rich ions may be due to electron-beam-generated waves (e.g., Temerin & Roth, 1992), or due to the same mechanism that accelerates electrons.

Table 4.2 Association with Other Phenomena

			GOES SXR Flare[a]				*SOHO* west-limb CME[b]				*Wind*/WAVES Radio Bursts	
			Duration		Location		Width			No Data	Type III	Type II
	N	GOES >10 MeV proton event	≤1 hr	>1 hr	W30°–75°	Other longitudes	<50°	≥50°	Halo			
^{3}He-rich	725	15	225	26	88	53	185	171	19	101	721	13
^{3}He/^{4}He<0.01 and not-gradual	209	57	85	0	30	32	36	59	26	23	206	45
Gradual	25	13	0	25	8	7	1	10	12	2	25	12
Undetermined[c]	232	18	52	20	20	22	41	58	16	40	224	25

[a] The flare information is from the Solar Geophysical Data (SGD). (ftp://ftp.ngdc.noaa.gov/STP/SOLAR_DATA/SOLAR_FLARES/).
[b] The CME information is from the SOHO LASCO CME catalog (http://cdaw.gsfc.nasa.gov/CME_list/).
[c] With no ACE/ULEIS coverage or data.
Source: Wang et al., 2012a. © 2012 IOP Publishing.

^{3}He-rich solar electron events show a poor (~35%) association with GOES SXR flares (Wang et al., 2012a). Among these associated SXR flares, ~90% are impulsive flares; most are C-class flares and located at longitude W30°–75° that magnetically connects to the vicinity of Earth along the Parker spiral IMF. Some ^{3}He-rich electron events with no associated GOES SXR flares might have microflares that are closely associated with type III radio bursts (Christe et al., 2008b) and/or partially occulted flares (Krucker & Lin, 2008). On the other hand, ~60% of ^{3}He-rich solar electron events appear to have an associated west-limb CME measured by *SOHO*/LASCO, with a median angular width of 47° ± 36° (1σ uncertainties, excluding halo CMEs) and a median speed of 496 ± 258 km s^{-1}. Many case studies also show that some ^{3}He-rich solar electron events are associated with coronal jets and narrow CMEs that originate from western-hemisphere flaring ARs, which are next to coronal holes containing the Earth-connected open field lines (e.g., Kahler et al., 2001; Nitta et al., 2008; Pick et al., 2006; Wang et al., 2006b). These results suggest that magnetic reconnection between closed and open field lines may be involved in producing electron/^{3}He-rich SEP events.

Among "not-^{3}He-rich & not-gradual" solar electron events (Wang et al., 2012a), ~39% are associated with an impulsive GOES SXR flare; ~65% have an associated CME with a median width of 74° ± 63° (excluding halo CMEs) and a median speed of 759 ± 454 km s^{-1}. Among gradual electron events, all are associated with a gradual GOES SXR flare (according to the event definition); all have an associated CME with a median width of 111° ± 55°(excluding halo CMEs) and a median speed of 1484 ± 894 km s^{-1}.

4.5.3. Solar Electron Injections

The particle injection timing in ^{3}He-rich solar electron events carries important information for understanding the acceleration of electrons and ^{3}He-rich ions. Krucker et al. (1999) and Haggerty and Roelof (2002) found that the injection of the >25–38 keV electrons at the Sun was delayed by ~10 to 30 min after the release of type III radio bursts in the majority of solar electron events. The delayed injection of high-energy electrons may be related to propagation of large-scale coronal transient (EIT) waves (Krucker et al., 1999), propagation effects in the IPM (e.g., Cane, 2003), a second acceleration by shock waves associated with CMEs (Haggerty & Roelof, 2002; Simnett et al., 2002), or coronal magnetic restructuring in the aftermath of CMEs (Maia & Pick, 2004).

In many ^{3}He-rich solar electron events (see Figure 4.30, for example), the observed electron temporal profiles at all energies exhibit nearly symmetric rapid-rise, rapid-decay peaks, with pitch angle distributions strongly peaked antisunward along the IMF, followed by slow-decay tails at flux levels that are much lower than the peak values. Such temporal profiles imply that, in these events, most of the electrons (those in the peak) propagate through the IPM essentially scatter-free (Lin, 1974), and that the electron injection at the Sun would be nearly symmetric (Wang et al., 2006).

Wang et al. (2006a) utilized a forward-fitting model to obtain the electron injections at the Sun for highly scatter-free solar electron events detected by *Wind*/3DP at energies from ~0.4 keV to 300 keV, assuming that the injection time profile at the Sun is an isosceles triangle injection profile (with equal rise and fall time) and the

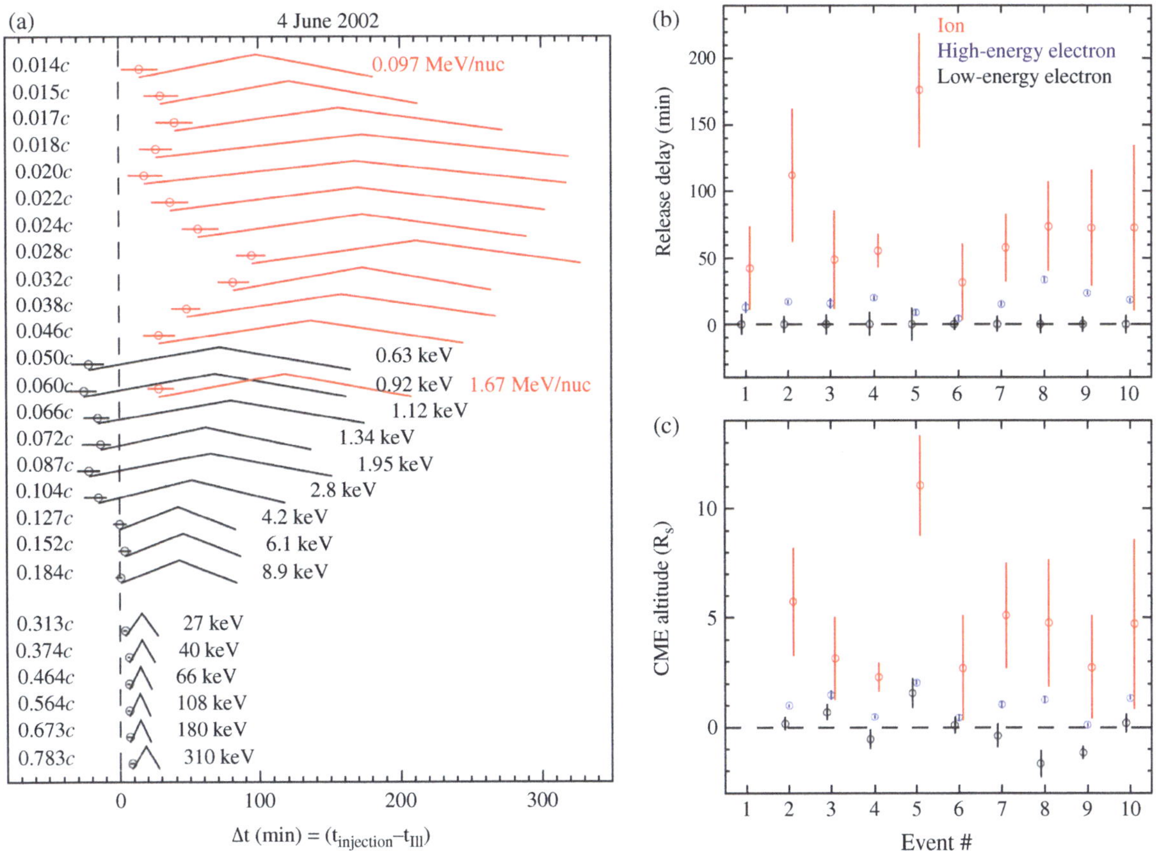

Figure 4.32 (a): Inferred solar injection timing of electrons (black) and ions (red) for the 4 June 2000 event, with the particle velocity indicated in terms of speed of light c. The x-axis shows the time in minutes, with respect to the estimated coronal release time of type III radio bursts (dashed line at $\Delta t = 0$). The triangular profiles indicate the time variation of particle injection intensities. The circles represent the start time of particle injections. (b): Particle injection delays after the start of low-energy electron injection for ten events. (c): Projected CME altitudes above the photosphere at the start of particle injections. In (b) and (c), the circles with error bars represent the start-time range of inferred solar injections for low-energy electrons (black), high-energy electrons (blue), and carbon and heavier ions (red). (*Source*: Wang et al., 2016. © 2016.)

electron path length from the Sun to 1 AU is 1.2 AU (the Parker spiral field line length for the ~400 km/s solar wind). The best-fit results show two distinct electron injections at the Sun (Figure 4.32a): a low-energy (~0.4 to 9 keV) injection that begins 9.1 ± 4.7 min earlier than the coronal release of type III radio bursts and lasts for ~50–300 min, and a high-energy (~10 to 300 keV) injection that starts 7.6 ± 1.3 min after the release of type III bursts but has durations ~5–10 times shorter. Thus, electrons of the low-energy injection are responsible for the generation of type III radio bursts, and they may provide the seed electrons for the delayed high-energy electron acceleration.

Recent injection analysis of ten good electron/^{3}He-rich SEP events (Wang et al., 2016) also suggested the presence of separate low-energy and high-energy electron injections at the Sun, followed by the ~MeV/nucleon ion (carbon and heavier) injections that start with an average delay of ~1 hr (Figure 4.32b). The ten electron/^{3}He-rich SEP events have a poor association with Hα flares or GOES SXR flares, but all have a west-limb CME observed by *SOHO*/LASCO (when its observations are available). On average, the associated CMEs appear to launch near the start of inferred low-energy electron injections, and reach altitudes of ~1.0 R_S and 4.7 R_S at the start of high-energy electron injections and ion injections (Figure 4.32c), respectively.

4.5.4. Electron Energy Spectra

For solar energetic electron events detected in situ at energies from ~ keV to 200–300 keV (Krucker et al., 2009; Lin, 1985; Wang et al., 2006), the electron peak differential

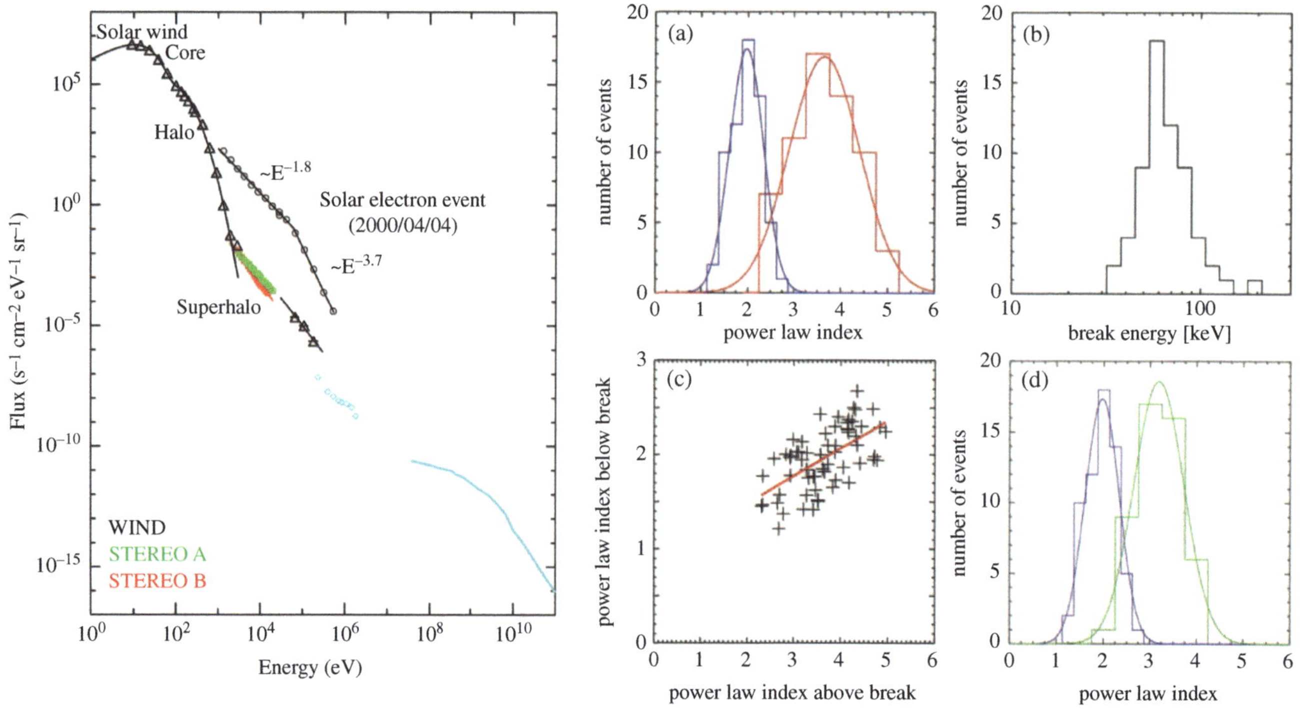

Figure 4.33 Left: Electron peak flux versus energy spectrum of solar energetic electron event (circles) on 04 April 2002 at ~1–300 keV (Krucker et al., 2009), compared with the quiet-time solar wind electron flux spectrum from ~9 eV to ~200 keV measured by *Wind* (black) and *STEREO-A* (green) and *STEREO-B* (red) on 6 December 2007. (*Source*: Modified from Wang et al., 2012.) The solid lines represent the fit by the sum of a Maxwellian and a Kappa to the solar wind core and halo distribution and the single-power-law fit to the superhalo, or the double-power-law fit to solar energetic electron data. The blue diamonds/curves represent the electron flux spectra of the interplanetary electrons at ~0.2–2 MeV (Hurford et al., 1974) and galactic cosmic rays (GCRs) at ~30 MeV–1 TeV (Lin 1974). Right: Statistical results of the fitted spectral parameters of solar energetic electron events. (*Source*: Krucker et al., 2009. © 2009 IOP Publishing.) (a) Histogram of the power-law index below δ_{low} (blue) and above δ_{high} (red) the break energy for events with full energy coverage. The distributions are fitted with a Gaussian. (b) Histogram of the break energies E_0. (c) Scatter plot of δ_{low} versus δ_{high}. (d) Histogram of δ_{low} for solar electrons events detected only below 20 keV (green) and event also detected above 25 keV (blue).

fluxes generally fit a double-power-law function with a steepening above $E_0 \sim 60$ keV (Figure 4.33):

$$J_{peak} \propto \begin{cases} E^{-\delta_{low}}, & E < E_0 \\ E^{-\delta_{high}}, & E > E_0. \end{cases}.$$

On average, the fitted spectral indices are $\delta_{low} = 1.9 \pm 0.3$ and $\delta_{high} = 3.6 \pm 0.7$, respectively, at energies below and above E_0. δ_{low} and δ_{high} appear to be linearly correlated (correlation coefficient of 0.61). Some electron events associated with large flares also show another spectral steepening observed above ~3 MeV (Lin, 1985).

For solar energetic electron events detected only at energies above 25 keV (Krucker et al., 2009), the peak flux energy spectrum mostly exhibits a spectral break below 100 keV, and a single power law above 100 keV with an average spectral index of $\delta_{high} = 3.5 \pm 0.6$, consistent with the observations of electron events with full energy coverage. For these events, measurements below 20 keV are generally dominated by solar wind suprathermal electrons (Wang et al., 2012b) and/or instrumental background.

For solar energetic electron events detected only below 20 keV (Gosling et al., 2003; Krucker et al., 2009; Potter et al., 1980), the peak flux energy spectrum generally shows a single power law with a spectral index of $\delta_{low} \sim 3-5$, much steeper than the low-energy spectrum of electron events with full energy coverage. These electron events could be the same type of events seen over the full energy range (Krucker et al., 2009), but with a steeper spectrum, or events of a different class (Potter et al., 1980).

The spectral shapes of solar energetic electron events are typically different from that of solar wind suprathermal electrons (Figure 4.33a), indicative of different origins/accelerations between these suprathermal electron populations detected in the IPM. The energy spectrum of solar energetic electron events observed in situ likely reflects the physical nature of electron acceleration and escaping processes at the Sun. At the peak of HXR flares, HXRs (produced by non-thermal electrons through bremsstrahlung) also generally exhibit a double-power-law energy spectrum with a break energy similar to that of the solar energetic electron spectrum (e.g., Dulk et al., 1992; Krucker & Lin, 2002; Lin & Schwartz, 1987). However, the solar electron spectrum observed above the break in the IPM is flatter/harder than the theoretical electron spectrum predicted from the HXR spectrum (Krucker et al., 2007, 2009), suggesting that HXR-producing electrons and escaping electrons may have different acceleration mechanisms/processes. The spectral shape of solar energetic electrons observed in situ might be also altered by energy-dependent escaping and/or propagation effects, for example, wave–particle interaction in the IPM (Kontar & Reid, 2009).

In solar energetic electron events, low-energy electrons observed in situ generally show a power-law spectrum extending down to 0.04–1 keV with no low-energy rollover (Krucker et al., 2009; Lin et al., 1996; Wang et al., 2006; Wang et al., 2012a). This spectral shape favors a source high in the corona at altitudes of ~0.1–1 R_S (Lin et al., 1996; Wang et al., 2006, 2012a), because such low-energy electrons would lose all their energy to Coulomb collisions before they can escape from low in the corona.

4.5.5. Electron Pitch Angle Distributions

As energetic electrons propagate outward from the Sun along diverging IMF lines, we expect that adiabatic motion (the invariance of $\sin^2\alpha/B$) would focus these electrons to small pitch angle (PA), α, but nonadiabatic wave–particle interactions can scatter them to large PA. Thus, in situ observations of electron PADs provide important information on local electron–wave interactions, whereas the flux versus time profiles integrate the effects of injection and propagation from the Sun to the observer. Previous measurements show that in some solar energetic electron events, the PAHM (PA width at half-maximum) of electron PADs was <13° below ~10 keV and >30°–40° above 15 keV (Lin et al., 1981; Potter et al., 1980), and the beam angular width below ~1.4 keV sometimes appears to vary with energy from ~15° to >75° (de Koning et al., 2007; Gosling et al., 2003).

Using the three-dimensional electron distributions measured by *Wind*/3DP from solar wind plasma to ~400 keV, Wang et al. (2011) examined the PADs and temporal behavior of five good ^{3}He-rich solar electron events with short time durations and low pre-event background at ~0.3–300 keV (Figure 4.30b). The electron PADs observed in situ exhibit two different behaviors at low and high energies separated by an energy transition around E_{tr} (Figure 4.34). At low energies (~0.3 keV to E_{tr}), the PAHM remains roughly constant below ~25°–30° (corresponding to an actual PAHM of 15° after taking into account the instrumental angular resolution) from onset through peak. At high energies (E_{tr} to 300 keV), the PAHM increases with both energy and time, for example, from ~30° (50°) at E_{tr} up to 85° (100°) at 300 keV at the peak (slow-tail). Thus, high-energy electrons appear to experience more PA scattering, with scattering strength increasing with energy. However, the observed electron flux temporal profiles (Figure 4.30a) retain a rapid-rise, rapid-decay peak, and the electron path length estimated from the velocity dispersion analysis is only ~4%–18% longer than the Parker spiral field length, indicating that such scattering may occur only near 1 AU.

Although the transition energy E_{tr} varies from ~10 to 40 keV from event to event, E_{tr} always corresponds to an electron gyroradius that is approximately equal to the local thermal proton gyroradius ρ_{Tp}. At the spatial scale of $\lambda \sim \rho_{Tp}$, the solar wind turbulence power spectrum ($P \propto \lambda\beta$) transitions from the inertial range ($\beta = 5/3$) above to the dissipation range ($\beta \sim 3$) below (Leamon et al., 1999). Thus, the observed electron PAD behaviors can be explained in terms of electron resonant interactions with the solar wind turbulence at $\lambda \sim \rho_e$. Low-energy electrons with $\rho_e < \rho_{Tp}$ would be weakly scattered because of weak power densities for resonant fluctuations/waves at scale $\lambda < \rho_{Tp}$ (the dissipation range), whereas high-energy electrons with $\rho_e > \rho_{Tp}$ would scatter more due to higher power densities for resonant fluctuations/waves at scale $\lambda > \rho_{Tp}$ (the inertial range).

For both fast and slow solar wind, the proton temperature Tp decreases as $r^{-0.8}$ to $r^{-1.0}$ from 0.3 to 1.0 AU (Totten et al., 1995) and as $r^{-0.5}$ to $r^{-0.7}$ beyond 1 AU (Gazis & Lazarus, 2003; Richardson et al., 1995), where r is the heliocentric distance. As solar energetic electrons propagate away from the Sun, the resonant wave scale in terms of ρ_{Tp} (i.e., $\lambda/\rho_{Tp} \sim \rho_e/\rho_{Tp}$) would gradually increase with both r and the electron energy. The heliocentric distance r_0 at which electrons start to encounter efficient resonant scattering by waves of $\lambda > \sim \rho_{Tp}$ would decrease with increasing electron energy. For example, r_0 is estimated to be ~9, 0.9, and 0.2 AU, respectively, for electrons at 2, 20, and 100 keV, assuming $T_p \propto r^{-1}$ with a value of 10 eV at 1 AU. En route to 1 AU, therefore, low-energy electrons would undergo very little scattering, whereas high-energy electrons experience efficient scattering along a distance $L = L\,(1\text{AU}) - L\,(r_0)$ that increases with the electron energy, where L is the electron path length. Because efficient scattering of high-energy electrons start to dominate only when near 1 AU, such scattering is not enough to produce the detected delay of high-energy electron injections as suggested by Cane and Erickson (2003) and Cane (2003).

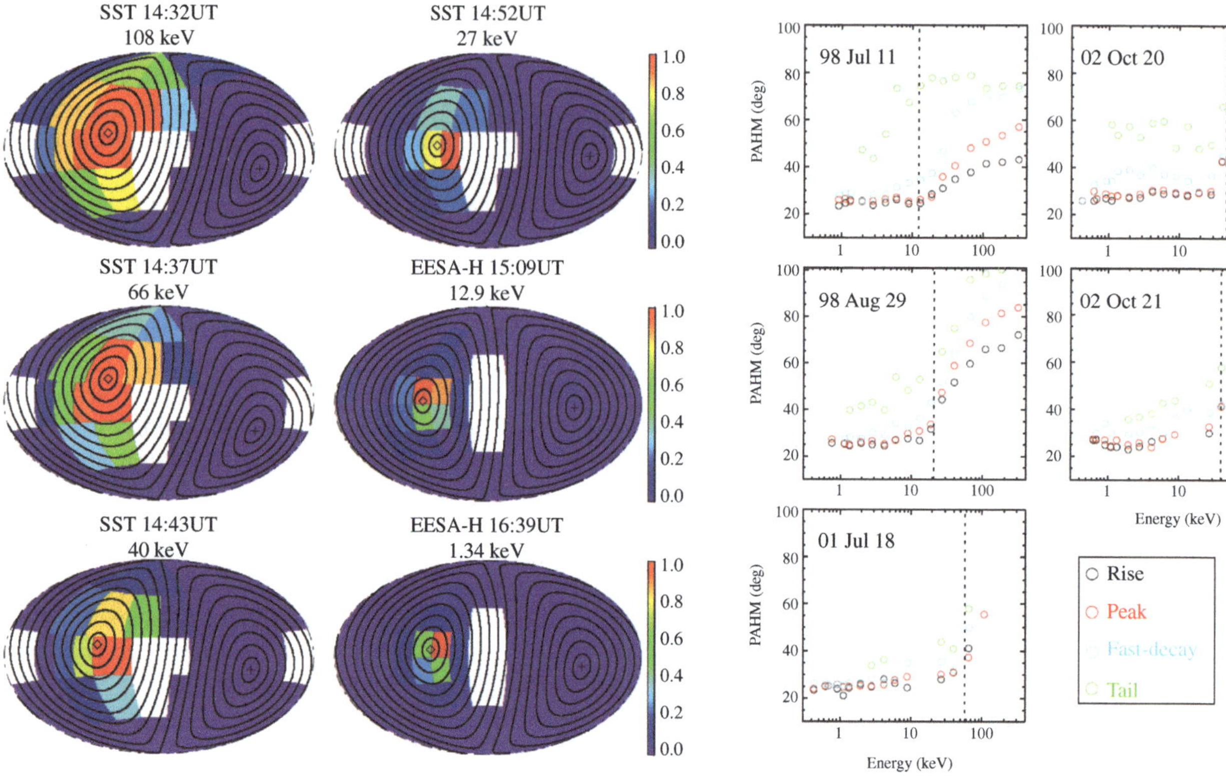

Figure 4.34 Left panels: Normalized 3D electron angular distributions measured by *Wind*/3DP at the event peak of different energies for the 20 October 2002 event. The high-background angular bins have been excluded. The cross (diamond) indicates the direction parallel (antiparallel) to the IMF. The black curves represent the iso-PA contours at every 10°. Right panels: Energy dependence of the typical PAHM at the rise (black), peak (red), fast-decay (blue), and slow-tail (green) in the five ^{3}He-rich solar electron events. The dashed lines indicate the energy of electrons with $\rho_e = \rho_{Tp}$. (*Source*: Wang et al., 2011. © 2011 IOP Publishing.)

4.6. UNDERSTANDING SEPs VIA ENERGETIC NEUTRAL ATOMS (ENAs)

Essentially all of the information about the acceleration of SEPs by CME-driven shocks comes from in situ measurements of the SEPs that escape to ~1 AU. The propagation of the SEPs along the IMF, with scattering, generally disperses the distributions in time and potentially in energy. These transport effects, as well as cross-field diffusion, mean that very little unambiguous information remains about the spatial and temporal nature of the shock acceleration. In addition, in situ measurements by a given spacecraft provide information only at a single point in space, so a global picture of the acceleration is difficult to obtain. As a result, even the most basic questions about SEP acceleration remain unanswered, including what parts of the shock are primarily responsible for SEP acceleration.

In contrast, information about the acceleration of ions by flares primarily comes from remote observations of the gamma-ray signatures produced by the accelerated ions as they interact with the solar atmosphere. These gamma-ray signatures can be studied via spectroscopy and via imaging, and most notably can be used to conclusively determine that flare-accelerated ions are associated with compact flare loops that are analogous to those containing flare-accelerated electrons (Hurford et al., 2003, 2006). In principle, SEPs traversing the corona will also produce gamma-ray signatures, but these signatures would be extremely faint (see Kahler & Ragot, 2008). Furthermore, the physics of the production of gamma-ray lines means that (a) it is not possible to directly invert the gamma-ray energies to ion energies, and (b) it is very difficult to obtain constraints on the ion energy spectrum below ~1 MeV.

As an alternative to gamma rays, ENAs provide an exciting new window of observation into SEP acceleration. An ENA is produced when a charged energetic particle undergoes a process called charge exchange, neutralizing the energetic particle by acquiring an electron from an ambient nucleus, generally a hydrogen-like or helium-like atom/ion. Figure 4.35 shows the cross sections versus energy for three charge-exchange processes. The charge-exchange process does not appreciably

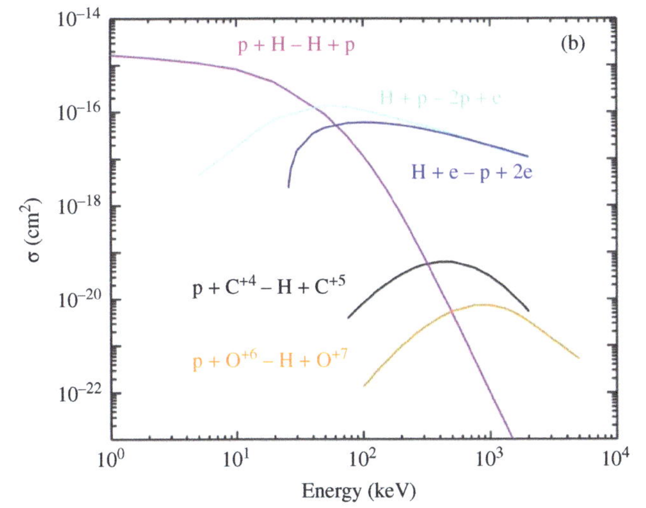

Figure 4.35 Cross sections versus energy for proton charge exchange with hydrogen (purple), C4+ ions (black), and O6+ ions (orange). Also shown are cross sections for impact ionization of a hydrogen ENA by electrons (dark blue) and protons (light blue). (*Source*: Wang et al., 2014. © 2014 IOP Publishing.)

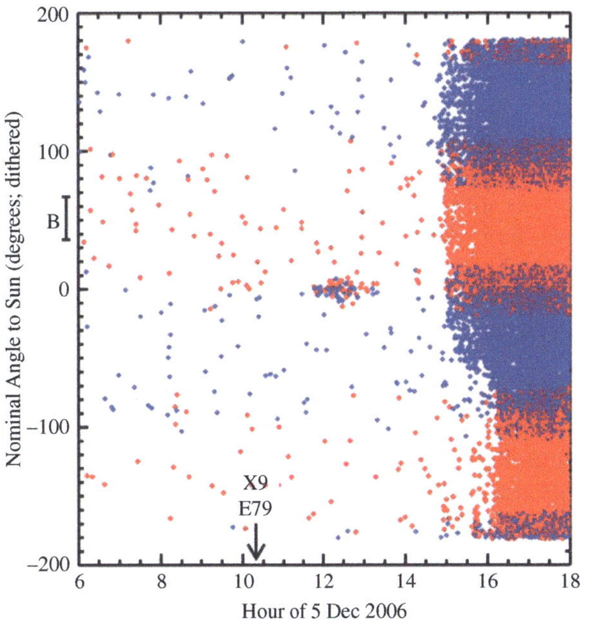

Figure 4.36 The first observation of ENAs associated with a solar eruptive event. These ~1.6–15 MeV ENAs (small clump, center) arrive at the spacecraft (red is *STEREO-A*/LET and blue is *STEREO-B*/LET) hours before the SEPs (broad swath, right), having not been scattered by the interplanetary magnetic field (IMF), and are aligned with the spacecraft–Sun line instead of with the local magnetic field (the vertical bar labeled B). (*Source*: Mewaldt et al., 2009. © 2009 IOP Publishing.)

modify either the trajectory or the energy of the initial energetic particle, but the ENA is no longer affected by magnetic fields that would guide or deflect the initial particle.

Consequently, ENA observations combine the best aspects of in situ and remote observations. Because ENAs travel in a straight line after they are produced, and cannot be scattered by the IMF, their incident direction and time of arrival can be unambiguously converted to their source on the sky and the time of production. Images over time of ENAs would reveal what parts of the CME-driven shock are responsible for SEP acceleration and when that acceleration occurs.

The following discussion details observations of solar ENAs (section 4.6.1), the associated predictions (section 4.6.2), and potential future work in this relatively new area (section 4.6.3).

4.6.1. Observation of Solar ENAs

The only known observation of solar ENAs to date was reported by Mewaldt et al. (2009), using the Low Energy Telescope (LET) instruments on the two *STEREO* spacecraft. The *STEREO*/LET instruments will detect hydrogen ENAs as energetic protons. Following the X9 GOES-class solar eruptive event on 5 December 2006, located at E79 in solar longitude, the *STEREO*/LET instruments on both *STEREO* spacecraft detected a precursor of ~1.6–15 MeV energetic protons preceding the onset of arriving SEPs by hours (Figure 4.36).

Mewaldt et al. (2009) argues that the precursor particles must be neutral particles—either ENAs or neutrons—for two reasons:

- The precursor particles arrive in a narrow angular range around the direction to the Sun, rather than showing any alignment with the direction of the IMF (Figure 4.36).
- Correcting for the time of flight of the precursor particles assuming that they are neutral and traveled the straight-line distance from the Sun to the spacecraft—rather than following the Parker spiral—results in a time profile of emission at the Sun that matches remarkably well with the time profile of X-ray emission from the flare (see figure 3 in Mewaldt et al., 2009).

Mewaldt et al. (2009) then rules out neutrons as the candidate for the neutral particles due to the lack of expected signatures from the decay of low-energy neutrons as they propagate 1 AU. Thus, ENAs are the remaining explanation.

Mewaldt et al. (2009) notes that it is unclear if these energetic protons were produced by the flare or by the CME-associated shock. The ENA time profile matches the flare time profile, but it is unknown whether the CME time profile was the same or different because there was no coronagraph imagery available for this event. Mewaldt et al. (2009) and Mewaldt et al. (2010) argue that the estimated number of flare-accelerated protons (determined via gamma-ray observations) can explain the observed number of ENAs only if the flare-accelerated protons are produced at a heliocentric radius of ~2 R_{sun} because ENAs are likely to re-ionize at lower altitudes due to collisions and photo-ionization. This altitude is much higher than where particles are anticipated to be accelerated in the standard conception for flare acceleration. Therefore, the CME shock is the most likely origin for the observed ENAs.

It should be acknowledged that Simnett (2011) has argued that *STEREO*/LET did not actually observe ENAs, but instead observed an unusual population of protons. Simnett points out that an unusual electron signature was observed by the Electron, Proton and Alpha Monitor (EPAM) instrument on the *ACE* spacecraft about 1 hr prior to the supposed ENA signature in *STEREO*/LET. In contrast to the ENA interpretation, Simnett posits the existence of a co-rotating magnetic structure containing both energetic protons (observable by *STEREO*/LET) and energetic electrons (observable by *ACE*/EPAM). The relative time delay between the two signatures is attributed to the direction of the nominal Parker spiral magnetic field and the different locations of the multiple spacecraft: near L1 for *ACE* and near Earth for the two *STEREO* spacecraft.

However, Simnett's alternative interpretation is inconsistent with the *STEREO*/LET measurements of a particle population aligned with the Earth–Sun line rather than the Parker spiral, and he acknowledges as much. It is also unclear what solar transient event would have been responsible for filling such a co-rotating magnetic structure with energetic electrons and protons. Finally, the apparent electron signature in *ACE*/EPAM is suspiciously coincident in time with the X-ray peak of the X-class flare, so the instrument may actually be responding to X-rays rather than electrons, although Simnett does argue against that possibility. Overall, the ENA interpretation of the *STEREO*/LET observations better explains the data.

4.6.2. Solar ENA Predictions

The power-law ENA energy spectrum determined by Mewaldt et al. (2009) suggests that there may exist very large numbers of ENAs at energies below what *STEREO*/LET can detect. To predict the numbers of <~1 MeV ENAs, as well as how ENAs might vary in space and time, Wang et al. (2014) constructed a model for a CME-driven shock, the SEP production by the shock, and the resulting ENAs. In addition to various assumptions about the shock parameters, they also assume that the accelerated protons are effectively trapped downstream of the shock due to strong turbulence. Importantly, they show that the predictions from the model can reproduce the *STEREO*/LET observation.

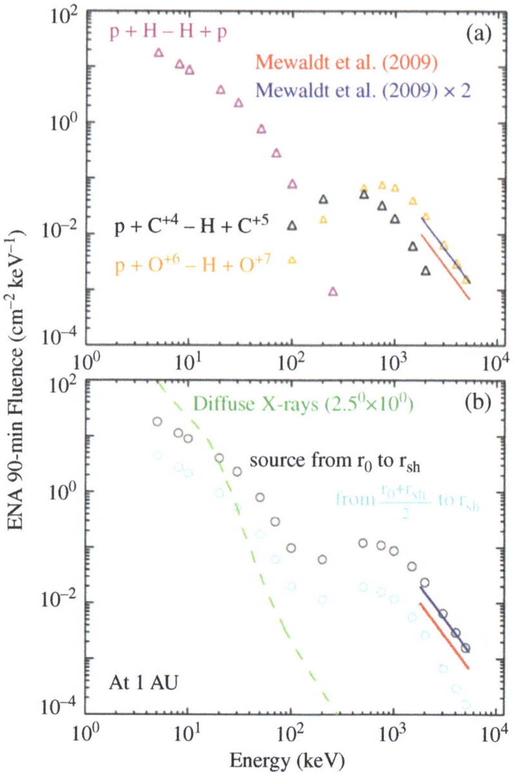

Figure 4.37 (a) Predicted ENA components from three different charge-exchange interactions. Each point is averaged over the 90 min that contains the most fluence at that energy. The red line shows the ENA spectrum reported by Mewaldt et al. (2009), and the blue line shows that the prediction is consistent within a factor of 2. (b) The total ENA spectrum (black), summed over all three interactions, is shown in comparison to a diffuse cosmic X-ray background (green), a possible source of background for some instruments. (*Source*: Wang et al., 2014. © 2014 IOP Publishing.)

As mentioned in the previous section, it is critical to consider both the production of ENAs by charge exchange and the "loss" of ENAs due to re-ionization processes. The production of ENAs decreases as the shock propagates to larger altitudes. However, the survival probability for hydrogen ENAs to reach 1 AU increases with ENA energy and source altitude, and is nearly zero within ~1.5 R_{sun}. Consequently, ENAs above ~100 keV arriving at 1 AU mostly originate from below 5 R_{sun}, whereas most of the ENAs below ~20 keV mainly originate from above 10 R_{sun}. Figure 4.37 shows the predicted ENA spectrum derived by taking the maximum ENA fluence over 90 min at each energy. That is, the lower-energy part of the spectrum reflects SEP acceleration at higher altitudes than the higher-energy part of the spectrum.

4.6.3. Future Solar ENA Observations

The power of ENA imaging to remotely study populations of energetic particles has already been demonstrated for multiple areas in the solar system, including the edge of the heliosphere by the *IBEX* mission, Earth's magnetosphere by the *IMAGE* and *TWINS* missions, and Saturn's magnetosphere by the *Cassini* mission. The observation of ENAs from the 5 December 2006 solar eruptive event is the only known observation of solar ENAs to date. Because no instrument has flown that is specifically designed to observe solar ENAs, any positive detection by past instruments would depend on multiple serendipitous factors. Importantly, given that *STEREO*/LET has no instrumental capability to discriminate between protons and neutral hydrogen, the successful measurement of the faint ENA signature depended on (a) the poorly connected nature of the event—charged particles were trapped by the IMF and thus could not reach the spacecraft until hours after the ENAs—and (b) the very low level of "background" SEPs in the heliosphere due to the lack of significant preceding solar activity. It is likely that ENAs have been produced in many events comparable to the 5 December 2006 event, but at fluxes below any available instrument's sensitivity.

A future instrument can be specifically designed to observe solar ENAs, particularly to observe the large numbers at <~1 MeV energies as predicted by Wang et al. (2014), but it is important to note that solar ENA observations have more challenging measurement requirements than non-solar ENA observations. In particular, a large effective area is required to have sufficient sensitivity to make images of where ENAs, and correspondingly SEPs, are produced as a function of energy. The production of ENAs is naturally limited to the time period of the shock travel, so sensitivity cannot be achieved by integrating over long periods of time. Fine angular resolution (e.g., a few arcmin) is required to image the spatial dependence of SEP acceleration over a CME-driven shock of ~few to >~10 R_{sun} size scales. This angular-resolution requirement may preclude the use of foils, which are commonly employed in ENA instruments to block or discriminate out non-ENA radiation. A foil in the path of ENAs may introduce too much angular blurring at the lowest energies due to scattering by the foil.

4.7. THEORY OF SEP ACCELERATION AND TRANSPORT

In this section, a few theories concerning the acceleration and transport of SEPs are presented. Section 4.7.1 describes particle acceleration in solar flares, which is most relevant to ^{3}He-rich SEP events (see also section 4.4.3), and focuses on some recent Particle-In-Cell (PIC) simulations that track particles during magnetic reconnection. With the computational infrastructure constantly improving, PIC simulations are likely to reveal more details and therefore help us better understand the underlying acceleration process. Two well-established acceleration mechanisms, the first-order and the second-order Fermi acceleration mechanisms, are described in sections 4.7.3 and 4.7.2. The first-order Fermi acceleration is also called diffusive shock acceleration (DSA), and the second-order Fermi acceleration is called stochastic acceleration (SA). SA has been proposed to be at work also in solar flares, where the acceleration process is tightly related to the plasma turbulence at the flare site. In contrast, DSA has been linked to CME-driven shocks and is believed to be at work in shock-accelerated SEP events.

Lastly, the transport of energetic particles (and cosmic rays) is discussed in section 4.7.9. The presence of a spatially varying background magnetic field is important in understanding the transport of charged energetic particles. This is because a particle's PA in a spatially varying background magnetic field is subject to magnetic focusing. Therefore, in understanding the transport of energetic particles, PA is an important consideration. However, if the turbulence field is strong enough such that particles are isotropically distributed in the phase space, then Parker's transport equation, which contains no information about particles' PA, can be applied. For example, transport of GCRs can be well described by Parker's transport equation. On the other hand, if the turbulent field is weak, such that particles' distribution has a clear PA dependence, then Parker's transport equation becomes invalid. In such a case, a focused transport equation is appropriate. The transport of SEPs is an example of a focused transport equation. Both forms of the transport equation can be cast into the form of a Fokker–Planck equation, which is mathematically equivalent to a stochastic differential equation. By casting the transport equation to a stochastic differential equation, we can obtain the solution using Monte-Carlo simulations, which again takes advantage of ever-improving computational powers.

4.7.1. Particle Acceleration Associated with Magnetic Reconnection

As explained earlier, there is compelling evidence that solar flares are driven by magnetic reconnection. During magnetic reconnection, the magnetic field reconfigures itself, and the magnetic field energy is converted to plasma heating, as well as kinetic energy of the plasma and energetic particles. Magnetic reconnection cannot occur in a perfectly conducting plasma where the conductivity $\sigma \to \infty$. The ideal Ohm's law $\mathbf{E} = -(\mathbf{v} \times \mathbf{B})/c$ implies that the magnetic flux through any surface bounded by a contour C moving with the fluid must be conserved. Therefore, magnetic field lines are frozen into the plasma (Alfvén, 1942). Magnetic reconnection emerges when we introduce a non-ideal effect to Ohm's law through a residual term **R**:

$$\mathbf{E}+(\mathrm{v}\times\mathbf{B})/c=\mathbf{R},\quad \nabla\times\mathbf{R}\neq 0. \tag{4.1}$$

The inclusion of **R** breaks the frozen-in condition and leads to the generalized Ohm's law, which contains several non-ideal terms:

$$\frac{m_e}{n_e^2}\left[\frac{\partial j}{\partial t}+\nabla\cdot\left(\mathrm{jv}+\mathrm{vj}-\frac{\mathrm{jj}}{ne}\right)\right]=\mathbf{E}+\frac{\mathrm{v}\times\mathbf{B}}{c}-\frac{\mathrm{j}\times\mathbf{B}}{nec}+\frac{1}{ne}\nabla\cdot\mathcal{P}_e-\eta\mathrm{j}. \tag{4.2}$$

Apart from the usual terms, $\mathcal{P}_e$ is the electron pressure tensor, and η is the plasma resistivity. The term on the left is due to electron inertia; the third term on the right is the Hall term; the fourth term on the right is the pressure term; and the last term is the resistive term.

The simplest reconnection model was proposed by Sweet (1958) and Parker (1957). The model considers a 2D, steady state, and incompressible plasma. "2D geometry" implies that any gradients perpendicular to the plane vanish. "Steady state" means that all physical quantities are independent of time; and "incompressible plasma" implies that the divergence of the flow velocity is zero. In the Sweet–Parker model, plasma with inflow speed v_{in} carries antiparallel magnetic fields into a current sheet with length $2L$ and thickness 2δ, resulting in bidirectional reconnection outflows v_{out}. From the continuity equation, we have

$$\frac{\upsilon_{in}}{\upsilon_{out}}=\frac{\delta}{L}. \tag{4.3}$$

For a finite resistivity η, using Ohm's law, one can rewrite the electric field as $\mathbf{E} = -(\mathbf{v}\times\mathbf{B})/c + \eta\mathbf{j}$; then, from Faraday's law:

$$\frac{\partial\mathbf{B}}{\partial t}=\nabla\times(\mathrm{v}\times\mathbf{B})+\frac{\eta c^2}{4\pi}\nabla^2\mathbf{B}. \tag{4.4}$$

For a steady state solution, one can estimate the inflow speed, $\upsilon_{in}{\sim}\eta c^2/(4\pi\delta)$ Furthermore, by requiring the inflow magnetic pressure to equal the outflow dynamic pressure, one can show that

$$\upsilon_{out}=\frac{B_0}{\sqrt{4\pi n m_i}}=V_A, \tag{4.5}$$

where V_A is the Alfvén speed of the inflow plasma. Defining the reconnection rate $\mathcal{R}$ as the ratio of the inflow speed to that of the outflow, one obtains

$$\mathcal{R}\equiv\frac{\upsilon_{in}}{\upsilon_A}=\frac{\delta}{L}=S^{-1/2}, \tag{4.6}$$

where $S = 4\pi L v_A/(\eta c^2)$ is the Lundquist number. It is the ratio of the Ohmic diffusion time $\tau_{diff} = 4\pi L^2/(\eta c^2)$ to the Alfvén time $\tau_A = L/v_A$. The corresponding reconnection time scale is

$$\tau_{rec}=(1/R)\tau_A=S^{1/2}\tau_A=\tau_{diff}^{1/2}\tau_A^{1/2}. \tag{4.7}$$

Most astrophysical plasmas have a very large S. In the solar corona, the Lundquist number is $S \sim 10^{12}$. For a flare with a spatial scale of ~ 0.001–0.01r_s, we find $\tau_A \sim 1$ s for an Alfvén speed of several thousand km/s. This yields a reconnection time scale of $\tau_{rec} \sim 10^6$ s, which is too long compared to the time scale of a flare, which is ~10^2–10^3 s. Therefore, the reconnection rate of the Sweet–Parker model is too small to explain solar flare observations. To obtain a faster reconnection rate, one obvious improvement of the Sweet–Parker model is to decrease the aspect ratio of the reconnection region. Petschek (1964) discussed such a scenario. As shown in the right panel of Figure 4.38, Petschek introduced four standing slow-mode shocks bounding a reconnection region that has an aspect ratio close to 1. In this scenario, the inflow plasma does not necessarily go through the reconnection region. It can pass through the bounding shocks and get heated. The maximum reconnection rate in Petschek's model is $\pi/(8\ln S)$. For solar corona plasma, this leads to $\tau_{rec} \sim 100$ s, which is consistent with observations.

However, the Petschek model is unstable. Numerical simulations have shown that the configuration in Petschek's model tends to evolve to that of the Sweet–Parker model (Biskamp, 1986; Malyshkin et al., 2005; Uzdensky & Kulsrud, 2000). One can relax other assumptions in the Sweet–Parker model to enhance the reconnection rate. Perhaps the most vulnerable assumption is the steady state assumption, which implies that the current sheet is stable. However, simulations have shown that an SP-like current sheet is unstable due to a tearing mode instability (e.g., Coppi et al., 1976; Furth et al., 1963) and will lead to plasmoid formation (Biskamp1986).

Formation of plasmoids can effectively decrease the aspect ratio of the reconnection region. Besides, when multiple plasmoids exist, secondary current sheets emerge when plasmoids collide. These secondary current sheets can lead to smaller plasmoids. Consequently, the configuration of plasmoids in a current sheet is fractal in nature

(Shibata & Tanuma, 2001). Furthermore, as the secondary plasmoids become smaller and smaller, eventually an MHD description becomes invalid when the length scale (here, the current sheet thickness) approaches the ion inertial length d_i (Daughton & Roytershteyn, 2012; Ji & Daughton, 2011).

The presence of multiple plasmoids is important in considering particle energization. In a sea of plasmoids, particles can be trapped within a magnetic island (staying inside an "O-point" configuration) as well as jump to neighboring plasmoids (passing through an "X-point"). Both cases can be viewed as the movement of particles from the inflow region to the outflow region of a reconnection site. Due to the change in the background plasma velocity, one expects particles to be accelerated in a process similar to that of the Fermi process. Particles also get accelerated during magnetic island coalescence.

Drake et al. (2006) formulated a transport model describing electron motion in an ensemble of plasmoids. Let F(x,y,v) = v^2 f(x,y,v) be the omnidirectional distribution function; then Drake et al. (2006) obtained

$$\nabla \cdot \mathbf{u}F - \nabla \cdot \mathbf{D} \cdot \nabla F = -\frac{\mathrm{A}}{3}\left(1-\frac{8\pi\varepsilon}{3B^2}\right)^{\frac{1}{2}} \frac{\mathrm{d}c_{\mathrm{A}x}}{\mathrm{d}y}\frac{\partial}{\partial v} vF \quad (4.8)$$

where f(x,y,v) is the particle phase space density, u is the local plasma velocity, D is the diffusion rate, and epsilon is three times the energy along the background magnetic field. With this formalism, one can show that electrons can be accelerated to non-thermal energies with a power-law spectrum and that the spectral index depends strongly on the ambient plasma beta.

Further kinetic simulations have shown that the most energetic electrons are accelerated during the process of magnetic island merging (e.g., Oka et al., 2010) instead of at X-points. More recent works using PIC simulations confirm such findings. A recent work by Li et al. (2015) found that power-law distributions can naturally emerge in a magnetically dominated environment (small plasma beta).

4.7.2. Stochastic Acceleration (SA)

SA, also known as second-order Fermi acceleration, was proposed by Fermi in 1949. In this scenario, particles are scattered randomly at a rate R_{sc} by some scattering agents (e.g., plasma turbulence). Denote the particle speed as v and the scattering agent's speed as u; then the particle gains energy at a rate proportional to $(u/v)^2 R_{sc}$. The rate is proportional to $(u/v)^2$, hence the name second-order Fermi acceleration. This is in contrast to the DSA mechanism (see the next section), which has an acceleration rate of $\sim u_{sh}/v$, where u_{sh} is the shock speed. Compared to DSA, SA requires no converging flow and therefore can operate in almost any environment, provided some scattering agents exist. SA has been long suggested as a major acceleration mechanism in solar flares, where both Alfvén waves and other forms (e.g., whistler wave) of turbulence can exist in the reconnection downstream and the reconnected closed loop. Often, a certain kind of turbulence with a broad turbulence spectrum is assumed to exist in a closed loop, and a Fokker–Planck equation is then solved in that geometry.

Assuming the gyro-radii of particles we consider are smaller than the scale of the spatial variation of the magnetic field, the gyro-phase-averaged distribution function $f(t,s,\mu,p)$, with s the spatial coordinate along the background field line, satisfies the Fokker–Planck equation,

$$\begin{aligned}\frac{\partial f}{\partial t} + \upsilon\mu\frac{\partial f}{\partial s} + \frac{\upsilon\partial B}{2B\partial s}\frac{\partial}{\partial \mu}\left[\left(1-\mu^2\right)f\right] = \frac{\partial u}{\partial s}\frac{p}{3}\left(\frac{\partial f}{\partial p}\right) \\ + \frac{1}{p^2}\frac{\partial}{\partial p}\left[p^2 \dot{p}_L f\right] + \dot{S} + \frac{1}{p^2}\frac{\partial}{\partial p}p^2\left[D_{pp}\frac{\partial f}{\partial p} + D_{p\mu}\frac{\partial f}{\partial \mu}\right] \\ + \frac{\partial}{\partial \mu}\left[D_{\mu\mu}\frac{\partial f}{\partial \mu} + D_{\mu p}\frac{\partial f}{\partial p}\right]. \quad (4.9)\end{aligned}$$

In equation. (4.9), $\dot{S}$ is a source term, and D_{pp}, $D_{\mu\mu}$, $D_{\mu p}$, and $D_{p\mu}$ are the diffusion coefficients. They describe the effects of particle interaction with turbulence and background plasma. The various $\partial/\partial s$ terms are due to the gradients of large-scale flows and magnetic fields. Equation 4.9 is very general. In studying particle acceleration at a solar flare, assumptions about the diffusion coefficients and the geometry of background flow and field need to be specified. For a recent review on particle acceleration via SA in solar flares, see Petrosian (2012). Note that in large SEP events, both flares and CME-driven shocks exist. So it is conceivable that both SA and DSA are in operation in these events. It is possible the flare-accelerated particles can serve as the seed for a second phase re-acceleration in the CME environment.

4.7.3. Diffusive Shock Acceleration (DSA)

The theory of DSA, also known as the first-order Fermi acceleration, was proposed in the late 1970s (Axford et al., 1977; Bell, 1978a,b; Blandford & Ostriker, 1978; Krymskii, 1977) to explain the observed cosmic ray spectra as a result of particle acceleration at supernova shocks. In the formalism of DSA, particles scatter back and forth between the upstream and downstream medium due to magnetic turbulence. In the solar wind, the magnetic turbulence often assumes the form of Alfvén waves. At a CME-driven shock, streaming protons can further amplify any preexisting Alfvén waves in the solar wind. Upon completing each traversal, a non-relativistic

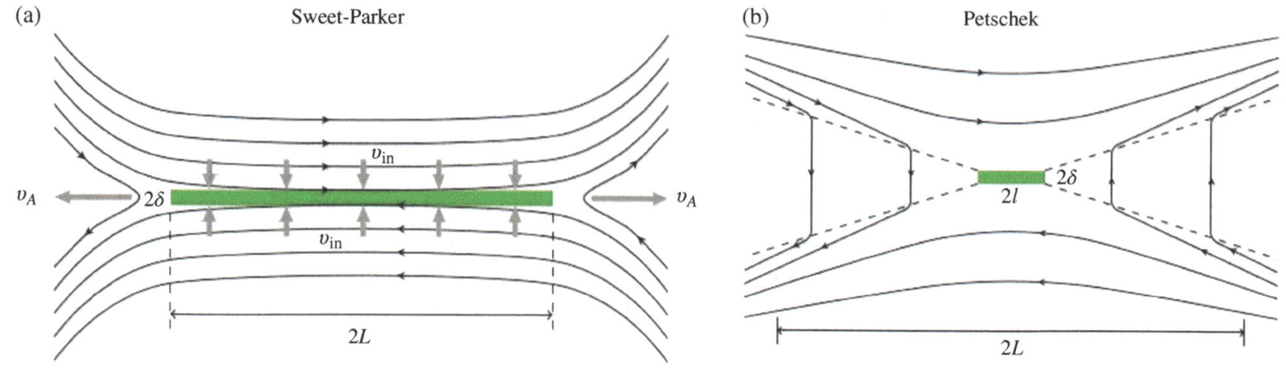

Figure 4.38 Sweet–Parker model (left) and Petschek's model (right). The blue region is the reconnection current sheet with a length $2L$ and a thickness 2δ. v_{in} is the inflow speed. v_A is the Alfvén speed. The inflow magnetic field is $B_x = \pm B_0$. The dashed lines in the right are standing slow-mode shocks, $l \ll L$ in this model.

particle will gain a momentum proportional to $m\Delta v$, where m is the particle mass, and $\Delta v = |v_2 - v_1|$ is the background velocity difference between the upstream and downstream. DSA does not differentiate different elements because the momentum gain only depends on the speed difference across the shock.

In the context of large SEP events, the CME-driven shocks have been regarded as a major acceleration site where DSA is in operation. For a propagating CME-driven shock, Alfvén waves driven by protons streaming off the shock front serve as the agent of upstream turbulence that keep particles close to the shock front. The wave intensity grows with a rate proportional to the PA anisotropy of the accelerated particle's phase space distribution function f (Lee, 1983). Consequently, the process of particle acceleration and Alfvén wave amplification are coupled together. The process is therefore intrinsically a nonlinear and time-dependent process.

To understand the physics of the acceleration process, however, we can first ignore wave amplification and assume that particle motion near the shock is fully specified by a diffusion process with the diffusion coefficient κ given. This simplification captures much of the important physics of shock-accelerated SEP events.

Consider particle acceleration at a shock. The transport equation is given by (Parker, 1965),

$$\frac{\partial f}{\partial t} + \mathbf{U}\cdot\nabla f - \nabla\cdot(\boldsymbol{\kappa}\cdot\nabla f) - \left(\frac{1}{3}\nabla\cdot\mathbf{U}\right)p\frac{\partial f}{\partial p} = 0. \quad (4.10)$$

In equation. (4.10), f is the particle distribution function, U is the background fluid velocity, t is time, p is the particle momentum, and κ is the spatial diffusion coefficient. The terms in equation. (4.10) describe convection ($U\cdot\nabla f$), diffusion ($\nabla\cdot(\kappa\cdot\nabla f)$), and particle energy change $\left((1/3)(\nabla\cdot U)p\frac{\partial f}{\partial p}\right)$, respectively. The effect of a large-scale background magnetic field B can be added as a "drift" term $V_d = (pv/3q)\nabla\times(\mathbf{B}/B^2)$ to the convection. Equation 4.10 can be cast into a "conservation form" (Gleeson & Axford, 1967),

$$\frac{\partial f}{\partial t} + \nabla\cdot\mathrm{S} + \frac{1}{p^2}\frac{\partial}{\partial p}\left(p^2 J\right) = 0, \quad (4.11)$$

$$\mathrm{S} = -\frac{p}{3}\frac{\partial f}{\partial p}\mathbf{U} - \kappa\nabla f, \quad (4.12)$$

$$J = \frac{p}{3}u\cdot\nabla f. \quad (4.13)$$

In this formalism, S represents the current in the configuration space, and J represents the current in the momentum space along the $\hat{p}$ (in analogy to $\hat{r}$ in the spherical coordinate system) direction.

When a shock is present, the transport equation. (4.10) or the set of equations (4.11) to (4.13) is solved with the matching conditions where both the distribution function f and the normal component of the current S are continuous at the shock front.

Consider a 1D parallel shock. Let x be the distance from the shock front with the upstream direction corresponding to $x > 0$ (so that $u_{up} < 0$); then the general steady state solution of f (both upstream and downstream of the shock) is

$$f = A(p) + B(p)exp\int_0^x d\zeta\,\frac{u}{\kappa}. \quad (4.14)$$

Downstream, the requirement of a finite distribution f yields

$$f_{dn} = f_0(p). \quad (4.15)$$

The subscripts (up) and (dn) denote quantities upstream and downstream of the shock, respectively. In the upstream region, assuming a seed population of $f(x,p) = h(p)$ at $(x \to \infty)$ that advects to the shock, the general form of f_{up} is

$$f_{up} = h(p) + g(p) exp \int_0^x d\zeta \frac{u}{\kappa}. \quad (4.16)$$

Applying the matching conditions and using $\beta = 3s/(s-1)$ with $s = u_{up}/u_{dn}$, one obtains (Drury, 1983)

$$f_0(p) = \beta p^{-\beta} \int_0^p q^{\beta-1} h(q) dq + C p^{-\beta} \quad (4.17)$$

where C is a constant. Setting $h(p) = 0$, we find the well-known steady state power-law solution $f_{dn}(p,x) = f_0(p) = Cp^{-\beta}$ and $f_{up}(p,x) = Cp^{-\beta} exp \int_0 x d\zeta \frac{u}{\kappa}$. Note that the index β depends only on the compression ratio of the shock.

A more interesting case is $h(p) \neq 0$. This is also more realistic because one can envision a relic population from a previous large or ^{3}He-rich SEP event. Consider first the general shape of $h(p)$. Note that $h(p)$ is not the population of the fluid particles, but the population of the energetic particles. To make this distinction, we require $h(p)$ to be defined above a certain momentum p_1, the injection momentum. As discussed above, we require $p_1/m > U$. We also require $h(p) \to 0$ when $p \to \infty$. A convenient working assumption is to treat $h(p)$ as a double power law:

$$h(p) = h_0 \begin{cases} 0 & p < p_1 \\ (p/pb)^{-\gamma_1} & p_1 \leq p \leq p_b \\ (p/pb)^{-\gamma_2} & p_b \leq p \leq p_2 \end{cases} \quad (4.18)$$

with $h_0 = h(p_b)$, the distribution function at the break momentum p_b and $\gamma_1 < \gamma_2$. With equation. (4.18), we can solve f and obtain

$$f_0(p \leq p_b) = \frac{\beta h_0}{\beta - \gamma_1} \left[\left(\frac{p}{p_b}\right)^{-\gamma_1} - \left(\frac{p}{p_b}\right)^{-\beta} \left(\frac{p_1}{p_b}\right)^{\beta-\gamma_1} \right] + h_0 \left(\frac{p_1}{p_b}\right)^{\beta-\gamma_1} \left(\frac{p}{p_b}\right)^{-\beta} \quad (4.19)$$

$$f_0(p \geq p_b) = \beta h_0 \left[\frac{1 - \left(\frac{p_1}{p_b}\right)^{\beta-\gamma_1}}{\beta - \gamma_1} + \left(\frac{p}{p_b}\right)^{-\beta} \frac{1 - \left(\frac{p}{p_b}\right)^{\beta-\gamma_2}}{\gamma_2 - \beta} \right] + h_0 \left(\frac{p_1}{p_b}\right)^{\beta-\gamma_1} \left(\frac{p}{p_b}\right)^{-\beta} \quad (4.20)$$

where the integration factor C in equation. (4.17) can be identified as

$$C = h_0 \left(\frac{p_1}{p_b}\right)^{-\gamma_1} p_1^{\beta}. \quad (4.21)$$

Clearly, if the source spectrum between $p_1 < p < p_b \sim p^{-\gamma_1}$ is harder than $p^{-\beta}$, then the accelerated particle spectrum in $p_1 < p < p_b$ is given by $p^{-\gamma_1}$; at higher momentum $p > p_b$, if γ_2 is softer than β, then the spectrum is given by $p^{-\beta}$.

Acceleration Time and Maximum Particle Energy

The steady state solution of the transport equation assumes that the shock does not vary with time and that the solution is a power law with no upper bound. However, in reality, shock parameters vary with time. Therefore the acceleration is intrinsically a time-dependent problem. Let us define the dynamic time scale through

$$t_{dyn} = \min \left\{ \frac{\mathrm{R(t)}}{\mathrm{dR(t)/dt}}, \frac{\mathrm{B(t)}}{\mathrm{dB(t)/dt}}, \frac{\mathrm{n(t)}}{\mathrm{dn(t)/dt}} \right\}. \quad (4.22)$$

Then, during t_{dyn}, the shock parameters do not vary significantly, and we can use the instantaneous steady state solution as an approximation to the solution of the time-dependent problem provided that we constrain the maximum energy by equating the dynamical time scale to the acceleration time scale:

$$f_{dn}(p) = \frac{\beta N}{4\pi p_0^3 u_{up}} (p/p_0)^{-\beta} H(p - p_0) H(p_{\max} - p), \quad (4.23)$$

where p_{max} is the maximum energy evaluated at time t, and H is the Heaviside function. Following Axford (1981 and Drury (1983), the acceleration time scale is

$$\Delta t = \beta \frac{\kappa(p)}{u_{sh}^2} \frac{\Delta p}{p}. \quad (4.24)$$

where the diffusion coefficient is given by

$$\kappa = \kappa_{up}(p) + s\kappa_{dn}(p). \quad (4.25)$$

Here κ_{up} and κ_{dn} are particle diffusion coefficients in the upstream and the downstream regions of the shock. The maximum momentum (energy) can be obtained through

$$t_{dyn} = \int_{p_1}^{p_{\max}} \beta \frac{\kappa}{u_{sh}^2} \frac{1}{p} dp. \quad (4.26)$$

In the above expression, u_{sh} is the shock speed in the upstream frame. Once the injection momentum p_1 is given, one can evaluate $_{pmax}$ from equation. (4.26).

Shock Geometry and the Injection Momentum

When considering particle acceleration at an oblique shock, the obliquity of the shock, that is, the angle between the upstream magnetic field and the shock normal, plays a crucial role. At a quasi-parallel shock (having a small shock obliquity angle), streaming particles upstream of the shock can effectively enhance the wave intensity. At a quasi-perpendicular shock, particles streaming along the field line can encounter the shock discontinuity multiple times in one gyration. In this section, we examine how the shock obliquity affects the injection momentum. We consider a planar shock, and the shock obliquity is also referred to as shock geometry in this section.

The effect of shock geometry on the injection threshold can be seen from the following argument (Giacalone and Jokipii, 1999; Zank et al., 2004, 2006): to apply the DSA formalism, the particle anisotropy should be small at the shock. The anisotropy ξ in the shock frame is defined through

$$\xi = 3\frac{|\mathbf{F}|}{4\pi p^2 f(p)}, \tag{4.27}$$

where F is the streaming flux. Assuming the shock normal is along the x-axis and the shock obliquity angle is θ_{BN}, one finds

$$\xi = \frac{3U_{up}}{\upsilon}\left\{(\beta-1)^2 + \frac{\left(\kappa_{Bohm}^2 + \kappa_{\parallel}^2\cos^2\theta_{BN}\right)\sin^2\theta_{BN}}{\left(\kappa_{\perp}\sin^2\theta_{BN} + \kappa_{\parallel}\cos^2\theta_{BN}\right)^2}\right\}^{1/2} \tag{4.28}$$

Requiring $\xi <= 1$ yields the injection speed v_1:

$$\upsilon_1 = 3U_{up}\left\{(\beta-1)^2 + \frac{\left(\kappa_{Bohm}^2 + \kappa_{\parallel}^2\cos^2\theta_{BN}\right)\sin^2\theta_{BN}}{\left(\kappa_{\perp}\sin^2\theta_{BN} + \kappa_{\parallel}\cos^2\theta_{BN}\right)^2}\right\}^{1/2} \tag{4.29}$$

At a parallel shock, this gives $v_1 = 3U_{up}/(s-1) \sim U$; at a quasi-perpendicular shock with $\lambda_{\parallel} > r_L > \lambda_{\perp}$, $v_1 = 3U_{up}/cos(\theta_{BN})$. Equation 4.29 shows that the injection speed is decided by both the parallel diffusion coefficient $\kappa_{\parallel}$ and the perpendicular diffusion coefficient $\kappa_{\perp}$.

Upstream Alfvén Waves and $\kappa_{\parallel}$

Upstream of a parallel or an oblique shock, streaming protons can excite Alfvén waves. These waves in turn interact with particles through wave–particle resonance and keep particles close to the shock front so that they can be accelerated to higher energies. In a simple 1D planar geometry, the amplification of wave action is coupled to the particle phase space distribution function f through (Bell, 1978b; Lee, 1983; Zank et al., 2000)

$$\frac{\partial A}{\partial t} + u\frac{\partial A}{\partial x} = \Gamma A,$$
$$\frac{\partial f}{\partial t} + u\frac{\partial f}{\partial x} - \frac{p}{3}\frac{\partial u}{\partial x}\frac{\partial f}{\partial p} = \frac{\partial}{\partial x}\left(\kappa_{\parallel}\frac{\partial f}{\partial x}\right), \quad (4.30), (4.31), (4.32)$$
$$\kappa_{\parallel}(p) = \frac{\kappa_0}{A(k)}\frac{B_0}{B}\frac{p^2}{E},$$

where $A(k)$ is the energy density of Alfvén waves per unit logarithmic bandwidth, and the particle wave resonance condition is given by

$$k\upsilon\mu = \Omega = \frac{QeB}{A\gamma mc}, \tag{4.33}$$

where Q is the particle charge, A the ion mass number, and c the speed of light. In the standard quasi-linear theory (QLT), the coupled system of particles and waves is related via the particle PA diffusion coefficient (Jokipii, 1966):

$$D_{\mu\mu}(p) = \frac{\pi Q^2 e^2\left(1-\mu^2\right)\upsilon}{2p^2c^2}\frac{1}{|\mu|}I\left(k = \frac{\gamma m\Omega}{p|\mu|}\right). \tag{4.34}$$

where $I(k)$, the wave intensity per unit k, is normalized through

$$\int_0^\infty I(k)dk = <\delta B^2>/2, \tag{4.35}$$

and $\delta B^2/8\pi$ is the energy density of the magnetic field fluctuations. The particle diffusion coefficient $\kappa_{\parallel}$ is related to $D_{\mu\mu}$ through

$$\kappa = \frac{\upsilon^2}{8}\int_{-1}^{+1} d\mu\frac{\left(1-\mu^2\right)^2}{D_{\mu\mu}}. \tag{4.36}$$

The wave intensity $I(k)$ satisfies the wave kinetic equation, and at the shock front it is given by

$$I\left(|k| < \gamma m|\Omega|/p_o\right) = I^e(k) + I^o(k), \tag{4.37}$$

and

$$I^e(k) = 2\pi\frac{\beta\Omega N V_A^0 p_0^{\beta-3}cos\theta_{BN}}{(\beta-4)(\beta-2)u_{up}^2}\frac{1}{k^2}\left|\frac{\gamma m\Omega}{k}\right|^{4-\beta} \tag{4.38}$$

is the wave intensity for the self-excited waves due to streaming protons; N is the injection rate, V_A^0 the Alfvén wave speed, p_0 is the injection momentum, and $I^0(k)$ is the slab component of the background solar wind MHD turbulence spectrum (Gordon et al., 1999; Lee, 1983).

When the background turbulence $I^0(k)$ is much smaller than $I^e(k)$, we can ignore it. Equation 4.38 can be rewritten as

$$I^e(k) = \frac{2\pi\beta\delta\, cos\theta_{BN}}{(\beta-4)(\beta-2)} \frac{V_A^0}{U_{up}} \frac{np_0^3}{\gamma m_p^2 \Omega_p} \left(\frac{k}{k_0}\right)^{\beta-6}, \quad (4.39)$$

where we replace the ion injection rate N by δnu_{up}. Because the waves are excited by streaming protons, we use Ω_p in the above. We see that the enhanced Alfvén waves $I^e(k)$ has a k dependence of $k^{\beta-6}$. For $\beta = 13/3$, that is, $r = 10/3$, the enhanced wave turbulence will have a Kolmogorov spectrum $\sim k^{-1.67}$. When $r > 10/3$ ($r < 10/3$), the enhanced wave turbulence has a harder (softer) spectrum than the Kolmogorov spectrum. Using $I^e(k)$ from equation. (4.44), one can obtain $\kappa_{\|}$ by numerically integrating equation. (4.42):

$$\kappa_{\|}(p) = \frac{\upsilon^2}{8}\int_{-1}^{+1} d\mu \frac{(1-\mu^2)^2}{D_{\mu\mu}} = \frac{\upsilon p^2 c^2}{\pi Q^2 e^2} \frac{I_e^{-1}(k=\Omega/\upsilon)}{(\beta-4)(\beta-2)}$$

$$= \frac{c^2}{2\pi^2 e^2}\left(\frac{A}{Q}\right)^{\beta-4} \frac{\gamma^{\beta-3} m_p \Omega_p}{\delta * n} \frac{U_{up}}{\beta V_A^0\, cos\theta_{BN}} \left(\frac{\upsilon}{\upsilon_0}\right)^{\beta-3} \quad (4.40), (4.41)$$

If we assume that the injection speed for different elements is the same, then the A/Q dependence of $\kappa_{\|}$ is given by $(A/Q)^{\beta-4}$. The factor $\gamma^{\beta-3}$ is the relativistic correction. It affects high-energy particles more than low-energy particles. From equation. (4.47), we also find that the dependence of $\kappa_{\|}$ on θ_{BN} is through $1/(\delta cos\theta_{BN})$. As θ_{BN} increases, δ quickly drops, so $\kappa_{\|}$ quickly increases. Knowing $\kappa_{\|}$, we can obtain $\lambda_{\|}$ through

$$\lambda_{\|} = 3\kappa_{\|}/\upsilon \quad (4.42)$$

Note that in the coupled equations of particle transport and wave action transport, there is no particle gyrophase; that is, the gyro-motion freedom is ignored. This implies that a particle will complete a few gyrations before it reverses its direction of motion. Therefore, we expect $r_g < \lambda_{\|}$.

Cross-Field Diffusion—the Nonlinear Guiding Center Theory

We now discuss $\kappa_\perp$, the diffusion coefficient perpendicular to the background field B_0. In studying the solar wind MHD turbulence, it was noted that the turbulent magnetic field δB is best described by a "slab+2D" model where 20% of the turbulence is in slab geometry and 80% in the 2D geometry (Bieber et al., 1994). For the 2D component, the turbulence spectrum is

$$I_{lm}^{2D}(k) = \frac{l_{2D}\delta B_{2D}^2 \Gamma(v)}{(\pi)^{3/2}\Gamma(v-1/2)} \frac{1}{\left[1+(k_\perp l_{2D})^2\right]^{s/2}} \frac{\delta(k_{\|})}{k_\perp}\left[\delta_{l,m} - \frac{k_l k_m}{k^2}\right], \quad (4.43)$$

where $l,m = x,y$ are the two directions perpendicular to B_0. Note that the $2D$ bend-over scale l_{2D} describes the 2D component of the turbulence and is independent of the slab component of the turbulence. The enhanced Alfvén waves described in the last section are of the slab in the geometry and do not affect l_{2D}. In QLT, the perpendicular diffusion is attributed to field line random walk (FLRW), and particles are thought to follow these field lines. However, numerical simulations (Mace et al., 2000; Qin et al., 2002a, 2002b) have shown that the FLRW describes only high-energy particles well. At lower energies, FLRW can underestimate $\kappa_\perp$ significantly. The first theoretical model of $\kappa_\perp$ that agrees well with numerical simulation is the Non-Linear-Guiding-Center Theory (NLGC), which was put forth by Matthaeus et al. (2003). Assuming the background magnetic field $\mathbf{B}_0$ is along the $\hat{Z}$ direction, the perpendicular diffusion coefficient $\kappa_\perp = \kappa_{xx}$ can be evaluated through Kubo's formula (Kubo, 1957):

$$\kappa_\perp = \int_0^\infty <\upsilon_x(0)\upsilon_x(\tau)> d\tau. \quad (4.44)$$

Assuming $v_x = av_z b_x/B_0$, the above equation requires evaluation of the four-point correlator ζ:

$$\zeta =<\upsilon_z(0) b_x(x(0),0)\upsilon_z(\tau) b_x(x(\tau),\tau)>. \quad (4.45)$$

If we can separate the ensemble average of the velocity from that of the magnetic field, then ζ can be replaced by

$$\zeta =<\upsilon_z(0)\upsilon_z(\tau)><b_x(x(0),0) b_x(x(\tau),\tau)>. \quad (4.46)$$

Using this assumption, Matthaeus et al. (2003) obtained

$$\kappa_\perp = \frac{a^2}{B_0^2}\int_0^\infty <\upsilon_x(0)\upsilon_x(\tau)><b_x(x(0),0) b_x(x(\tau),\tau)> d\tau. \quad (4.47)$$

Shalchi and Dosch (2008) have shown that, using the Lorentz equation, the value of a lies between 1 and 2. The correlation function $< b_x(x(0),0) b_x(x(\tau),\tau) >$, which is evaluated at the particle position, can be formally written as (see, e.g. Corrsin, 1959; Salu & Montgomery, 1977)

$$< b_x\left(x(0),0\right)b_x\left(x(\tau),\tau\right) >= \int R_{xx}\left(y,t'\right)P\left(y|t'\right)d^3y, \quad (4.48)$$

where $y = x(t') - x(0)$ is a spatial lag, $R_{xx}(y,t')$ is the magnetic correlation function $< b_x(x(0),0)b_x(x(0) + y,\tau) >$ with lag y, and $P(y|t')$ is the probability of finding a particle located at $x(t') - x(0) = y$ at $t = t'$. Using the Fourier representation of R_{xx}

$$R_{xx}\left(y,\tau\right) = \int d^3kS_{xx}\left(k\right)e^{iky}\Gamma\left(k,\tau\right), \quad (4.49)$$

where $S_{xx}(k)$ is the power density at wave number k, $\Gamma(k,\tau)$ describes the dynamic effects of the turbulence, and $\Gamma(k,\tau = 0) = 1$ at $\tau = 0$; assuming that $P(y|t')$ is given by a Gaussian distribution and approximating $< v_x(0)v_x(\tau) >$ by

$$< \upsilon_x\left(0\right)\upsilon_x\left(\tau\right) >= \left(\upsilon^2/3\right)e^{-\upsilon\tau/\lambda_\parallel}, \quad (4.50)$$

one obtains

$$\kappa_\perp = \frac{a^2\upsilon^2}{3B_0^2}\int d^3kS_{xx}\left(k\right)\int_0^\infty d\tau e^{-\upsilon\tau/\lambda_\parallel}\Gamma\left(k,\tau\right)e^{-k_\perp^2\kappa_\perp\tau - k_\parallel^2\kappa_\parallel\tau}. \quad (4.51)$$

In general, we expect the correlation to decay exponentially with τ, so $\Gamma_{xx}(k,\tau) = e^{-\gamma\tau}$ (the magnetostatic approximation corresponds to the choice of $\gamma = 0$). Carrying out the τ integral in equation. (4.57), we get the expression for $\kappa_\perp$ in NLGC theory:

$$\kappa_\perp = \frac{a^2\upsilon^2}{3B_0^2}\int \frac{S_{xx}\left(k\right)dk_xdk_ydk_z}{\frac{\upsilon^2}{\kappa_\parallel} + \left(k_x^2 + k_y^2\right)\kappa_\perp + k_\parallel^2\kappa_\parallel + \gamma\left(k\right)} \quad (4.52)$$

This is a nonlinear equation as $\kappa_\perp$ appears on both sides of the equation. It is clear from equation. (4.52) that the value of $\kappa_\perp$ depends on $\kappa_\parallel$. The dependence is twofold: first, from $v\lambda_\parallel$, which is due to the two point velocity correlation, $v^2/\kappa_\parallel$; second, from $k_\parallel{}^2\kappa_\parallel$, which arises from the assumption that the separations in both the parallel direction $\hat{Z}$ and the perpendicular direction ($\hat{x}$ and $\hat{y}$) are diffusive and $P(z|t') \sim e^{-k\|2\kappa\|t'}$.

Consider now particle acceleration at an oblique shock; the acceleration time scale is decided by the total diffusion coefficient $\kappa = \kappa_\parallel cos^2\theta_{BN} + \kappa_\perp sin^2\theta_{BN}$. In evaluating $\kappa_\parallel$, we solve self-consistently the wave action amplification equation. In evaluating $\kappa_\perp$, we use NLGC theory, which requires as an input the value of $\kappa_\parallel$. As a result, the total κ will depend on the upstream wave and the number of injected particles in a complicated manner.

4.7.4. Propagation of Energetic Particles in the Solar Wind

In all SEP events, when particles escape from the acceleration sites, their propagation in the solar wind is governed by the focused transport equation.

In a uniform and steady magnetic field, a charge particle gyrates along the background magnetic field and is tied to a single magnetic field line. Introducing inhomogeneity and turbulence to the background field causes particles to jump between neighboring field lines, a phenomenon often called cross-field diffusion. If the background field dominates the turbulent magnetic field, then a guiding center theory can be used to describe particle motion. Along the background magnetic field, particles experience PA scattering due to the turbulent field. They also respond to large-scale changes of magnetic field, conserving the adiabatic moment.

The focused transport equation (FTE) (Isenberg, 1997; Skilling, 1971), including an extra term that describes cross-field diffusion, is

$$\begin{aligned}
&\frac{\partial f}{\partial t} + \left(u_i + \upsilon\mu b_i\right)\frac{\partial f}{\partial x_i} + \left[\frac{1-3\mu^2}{2}b_ib_j\frac{\partial u_i}{\partial x_j} - \frac{1-\mu^2}{2}\frac{\partial u_j}{\partial x_j}\right. \\
&\quad \left. - \frac{\mu b_i}{\upsilon}\left(\frac{\partial u}{\partial t} + u_j\frac{\partial u_i}{\partial x_j}\right)\right]\upsilon\frac{\partial f}{\partial \upsilon} \\
&\quad + \frac{1-\mu^2}{2}\left[\upsilon\frac{\partial b_i}{\partial x_i} + \mu\frac{\partial u_i}{\partial x_i} - 3\mu b_ib_j\frac{\partial u_j}{\partial x_i}\right. \\
&\quad \left. - \frac{2b_i}{\upsilon}\left(\frac{\partial u}{\partial t} + u_j\frac{\partial u_i}{\partial x_j}\right)\right]\frac{\partial f}{\partial \mu} \\
&= \nabla\cdot\left(\kappa_\perp\cdot\nabla f\right) + \frac{\partial}{\partial \mu}D_{\mu\mu}\frac{\partial f}{\partial \mu} + S - L \quad (4.53)
\end{aligned}$$

where b_i and u_i are the components of the unit magnetic field and solar wind speed; μ is the cosine of the particle PA. S and L are source and loss terms; and the $\nabla\cdot(\kappa_\perp\cdot\nabla f)$ term describes the cross-field diffusion. The diffusion in the PA is governed by $D\mu\mu$. Equation 4.53 is solved numerically. Because a diffusion term is described by a random walk process, a natural approach to solving equation. (4.53) is through a backward stochastic differential equation.

In solving the FTE, one can choose different reference frames to perform the updates. Because the PA scattering is due to local turbulent magnetic field, we consider its effect in the local solar wind frame. The solar wind frame is not an inertial frame, so the particle's energy is not conserved in this frame. Considering the adiabatic deceleration and PA focusing effect, on the order of $(v/c)^2$, the change in particle momentum and the PA in the solar wind frame is given by (Ruffolo, 1995):

$$\begin{aligned}
\dot{p} &= -pu_{sw}\left[\frac{\sec\psi}{2L(z)}\left(1-\mu^2\right) + \cos\psi\frac{d}{dr}\sec\psi\mu^2\right] \\
\dot{\mu} &= \frac{\upsilon}{2L(z)}\left[1 + \mu\frac{u_{sw}}{\upsilon}\sec\psi - \mu\frac{u_{sw}\upsilon}{c^2}\sec\psi\right]\left(1-\mu^2\right) \\
&\quad - u_{sw}\left(\cos\psi\frac{d}{dr}\sec\psi\right)\mu\left(1-\mu^2\right) \quad (4.54), (4.55)
\end{aligned}$$

where ψ is the angle between the radial direction and the (outward-pointing) magnetic field direction, $L(z)$ is the scale length of the IMF (Roelof, 1969), and z is the arc length from the Sun along the magnetic field. For a Parker field,

$$\sec\psi = \sqrt{1+\left(\frac{\Omega_{\odot}}{u_{sw}}\right)^2 r^2}; \frac{1}{L(z)} \equiv -\frac{1}{B(z)}\frac{\partial B(z)}{\partial z}. \quad (4.56)$$

In considering the change in PA $\delta\mu$ due to the PA diffusion, one has

$$\delta\mu' = sign\left(\epsilon' - \frac{1}{2}\right) erf^{-1}(\epsilon)\sqrt{4D_{\mu\mu}\,|\,\delta t\,|} + \frac{\partial D_{\mu\mu}}{\partial\mu}\delta t, \quad (4.57)$$

where erf^{-1} is the inverse error function. Here ϵ' and ϵ are two random numbers uniformly distributed between 0 and 1. Under the framework of QLT, the PA diffusion coefficient $D\mu\mu$ is related to the turbulence power spectrum of the background plasma through

$$D_{\mu\mu} = \frac{1}{2}\frac{\langle(\delta\mu)\rangle^2}{\delta t} = \frac{\pi}{4}\left(1-\mu^2\right)\Omega_0 \frac{kP^{slab}(k)}{B^2}. \quad (4.58)$$

In the context of SEP transport in the solar wind, a Kolmogorov-type slab turbulence is often used:

$$P^{slab}(k) = A_s\lambda_c\left(\delta B_{slab}\right)^2\left(1+\left(k^2\lambda_c^3\right)\right)^{-s/2}, \text{ with } s=5/3, \quad (4.59)$$

where As is a normalization factor to ensure that the integral of $P^{slab}(k)$ over k is $\delta B_{slab}{}^2$, and λ_c is the correlation length.

Clearly, the strength of the background turbulence controls the strength of the PA scattering. Bieber et al. (1994) suggested that the solar wind turbulence is best described by a slab-2D geometry. The power $I(k)$ is

$$I(k) = \frac{<\delta B_{slab}^2>\Gamma\left(q_i/2\right)}{\sqrt{\pi}\Gamma\left(\left(q_i-1\right)/2\right)}\lambda_c\left(1+k^2\lambda_c^2\right)^{-q_i/2} \quad (4.60)$$

where q_i is the spectral indices for the inertial range, and a Kolmogorov-type spectra corresponds to $q_i = 5/3$. In equation. (4.60), we do not include the dissipation range because those waves are in resonance with particles of lower energy. The correlation length λ_c in equation. (4.60) is defined by

$$l_c = \frac{\int_0^\infty <\delta B_x(z)\delta B_x(z+l)>dl}{<\delta B_x^2>}$$
$$= \frac{\pi I^0(k=0)}{2\int_0^\infty I^0(k)dk} = \lambda_c \frac{\sqrt{\pi}\Gamma\left(\frac{q_i}{2}\right)}{\Gamma\left(\frac{q_i-1}{2}\right)} \quad (4.61)$$

The bend-over of the spectrum at low wave numbers signifies the energy-containing range. Using observations of magnetic field data from the *ACE* spacecraft, Vasquez et al. (2007) suggested that this bend-over scale is about $f_c = u_{sw}/\lambda_c = 10^{-4}$ Hz. This value of f_c is comparable to earlier Helios observations (Bavassano et al., 1982), where $f_c \sim 5 \times 10^{-4}$ Hz around 1 AU is found. Closer to the Sun, the value of f_c is somewhat larger: ~5×10^{-3} Hz (see, e.g. Tu et al., 1984), and also a recent review by Bruno and Carbone (2005), suggesting that the turbulence is actively developing from the Sun to 1 AU. The radial dependence of the solar wind turbulence power spectrum and f_c may affect the transport of SEPs. Results from models that take into account these radial dependences can be important in interpreting future observations from Parker Solar Probe and Solar Orbiter.

4.8. CONCLUDING REMARKS

Significant progress has been made over the last 70+ years in understanding the characteristics and generation of SEP events. Two dominant acceleration processes have been identified and studied in detail. However, despite the great observational advancements made since the dawn of the Space Age, a full understanding of SEP acceleration and transport is hampered by the limited in situ measurements made well inside 1 AU. Although particle acceleration at coronal/interplanetary shocks is fairly well understood, the conditions near the Sun under which this acceleration occurs are generally unknown. In contrast, the acceleration process(es) responsible for ^{3}He-rich SEP events is still under debate and not well constrained by the current observations. The lack of in situ measurements in the inner heliosphere also impedes characterization of the IPM through which the SEPs travel, making it difficult to fully understand their transport.

Thus, many questions remain unanswered: How can particles be distributed over 360° within tens of minutes of the solar event, and what are the conditions that facilitate that? Is the composition of SEP events controlled by the orientation of the shock, dependent on the suprathermal seed population, a result of a significant contribution from flare-related particles, or some combination of all three? What controls the spectral shape and strongly varying composition (e.g., ^{3}He/^{4}He, Fe/O, and ultra-heavy ion enhancements) of ^{3}He-rich SEP events? What are the conditions that result in ^{3}He-rich events, presumably from a small solar source, being spread over more than 100° in longitude?

Just as the *STEREO* mission has challenged many of the pre-launch understandings of SEP longitudinal distributions, it is anticipated the *Parker Solar Probe* (*PSP*) and *Solar Orbiter* (*SolO*) missions will be similarly transformative. The PSP, launched in August 2018, will measure the solar wind and energetic particle environment as close as

10 R_S over the course of its seven year primary mission. *SolO*, launched in February 2020, will not only explore regions into 30 R_S, but also make observations outside the ecliptic plane up to 24° in latitude. These missions will provide contextual information and characterize SEP events at a level of detail and at locations never done before. No doubt these measurements will not only begin to answer many of the outstanding questions regarding SEP acceleration and transport, but will also raise new ones.

ACKNOWLEDGMENTS

Portions of this work have been supported at JHU/APL by NASA grant NNX17AC05G/125225 and L. W. thanks NSFC for support under grants 41861134033 and 41774183.

REFERENCES

Alfvén, H. (1942). Existence of electromagnetic-hydrodynamic waves. *Nature*, *150*, 405–406.

Anders, E., & Grevesse, N. (1989). Abundances of the elements: Meteoritic and solar. *Geochimica et Cosmochimica Acta*, *53*, 197–214.

Anderson, K. A., & Lin, R. P. (1966). Observations on the propagation of solar-flare electrons in interplanetary space. *Physical Review Letters*, *16*, 1121.

Axford, W, Leer, E., & Skadron, G. (1977). Acceleration of cosmic rays by shock waves. *Proceedings of the 15th International Cosmic Ray Conference*, *11*, 132.

Band, D., Matteson, J., Ford, L., et al. (1993). BATSE observations of gamma-ray burst spectra. I. Spectral diversity. *The Astrophysical Journal*, *413*, 281.

Bavassano, B., Dobrowolny, M., Mariani, F., & Ness, N. F. (1982). Radial evolution of power spectra of interplanetary Alfvénic turbulence. *Journal of Geophysical Research*, *87*, 3617–3622.

Bell, A. R. (1978a). The acceleration of cosmic rays in shock fronts – II. *Monthly Notices of the Royal Astronomical Society*, *182*: 443–455

Bell, A. R. (1978b). The acceleration of cosmic rays in shock fronts – I. *Monthly Notices of the Royal Astronomical Society*, *182*, 147–156.

Belov, A., Kurt, V., Mavromichalaki, H., & Gerontidou, M. (2007). Peak-Size Distributions of Proton Fluxes and Associated Soft X-Ray Flares, *Solar Physics*, *246*, 457.

Benz, A. B., Krucker, S., Hurford, G. J., & Veronig, A. (2009). X-ray imaging of the flaring corona with STIX, in *Solar Orbiter Workshop 3, edited, Sorrento*, Italy.

Bieber, J. W., Matthaeus, W. H., Smith, C. W., Wanner, W., Kallenrode, M. B., & Wibberenz, G. (1994). Proton and electron mean free paths: The Palmer consensus revisited. *The Astrophysical Journal*, *420*, 294–306.

Biskamp, D. (1986). Magnetic reconnection via current sheets. *Physics of Fluids*, *29*, 1520–1531.

Blandford, R. D., & Ostriker, J. P. (1978). Particle acceleration by astrophysical shocks. *The Astrophysical Journal*, *221*, L29–29.

Bochsler, P., Geiss, J., & Maeder, A. (1990). The abundance of He-3 in the solar wind – A constraint for models of solar evolution. *Solar Physics*, *128*, 203–215. doi:10.1007/BF00154157.

Bougeret, J.-L., Kaiser, M. L., Kellogg, P. J., Manning, R., Goetz, K., Monson, S. J., et al. (1995). Waves: The Radio and Plasma Wave Investigation on the Wind Spacecraft, *Space Science Reviews*, *71*, 231.

Bruno, A., Bazilevskaya, G. A., Boezio, M., Christian, E. R., de Nolfo, G. A., Martucci, M., et al. (2018). Solar energetic particle events observed by the PAMELA mission, *The Astrophysical Journal*, *862*, 97.

Bruno, R., & Carbone, V. (2005). The solar wind as a turbulence laboratory. *Living Reviews in Solar Physics*, *2*, 4.

Bučík, R., Innes, D. E., Chen, N. H., Mason, G. M., Gómez-Herrero, R., & Wiedenbeck, M. E. (2015). Long-lived energetic particle sources on the Sun. *Journal of Physics Conference Series*, *642*, 1. doi:10.1088/1742–6596/642/1/012002.

Bučík, R., Innes, D. E., Mall, U., Korth, A., Mason, G. M., & Gómez-Herrero, R. (2014). Multi-spacecraft observations of recurrent 3He-rich solar energetic particles. *The Astrophysical Journal*, *786*, 71.

Bučík, R., Innes, D. E., Mason, G. M., & Wiedenbeck, M. E. (2017). Energy spectra of 3He-rich solar energetic particles associated with coronal waves. *Journal of Physics: Conference Series*, *767*(1). doi:10.1088/1742–6596/767/1/012002.

Cane, H. V. (2003). Near-Relativistic Solar Electrons and Type III Radio Bursts, *The Astrophysical Journal*, *598*, 1403.

Cane, H. V., & Erickson, W. C. (2003). Energetic particle propagation in the inner heliosphere as deduced from low-frequency (<100 kHz) observations of type III radio bursts, *Journal of Geophysical Research*, *108*, 1203.

Cane, H. V., & Lario, D. (2006). An Introduction to CMEs and Energetic Particles, *Space Science Reviews*, *123*, 45.

Cane, H. V., Mewaldt, R. A., Cohen, C. M. S., & Rosenvinge, von, T. T. (2006). Role of flares and shocks in determining solar energetic particle abundances, *Journal of Geophysical Research*, *111*, A06S90.

Cane, H. V., McGuire, R. E., & von Rosenvinge, T. T. (1986). Two Classes of Solar Energetic Particle Events Associated with Impulsive and Long-Duration Soft X-Ray Flares, *The Astrophysical Journal*, *301*, 448.

Cane, H. V., Reames, D. V., & Rosenvinge, von, T. T. (1988). The role of interplanetary shocks in the longitude distribution of solar energetic particles, *Journal of Geophysical Research*, *93*, 9555.

Cane, H. V., Rosenvinge, von, T. T., Cohen, C. M. S., & Mewaldt, R. A. (2003). Two components in major solar particle events, *Geophysical Research Letters*, *30*, 8017.

Christe, S., Hannah, I. G., Krucker, S., McTiernan, J., & Lin, R. P. (2008a). RHESSI Microflare Statistics. I. Flare-Finding and Frequency Distributions, *The Astrophysical Journal*, *677*, 1385.

Christe, S., Krucker, S., & Lin, R. P. (2008b). Hard X-Rays Associated with Type III Radio Bursts, *The Astrophysical Journal*, *680*, L149.

Cliver, E. W., Reames, D. V., Kahler, S. W., & Cane, H. V. (1991). Size Distributions of Solar Energetic Particle Events, In *Proceedings of 22nd International Cosmic Ray Conference*, Contrib. *A92–3680615–93*. Greenbelt, MD: NASA.

Cohen, C. M. S. (2006). Observations of Energetic Storm Particles: An Overview, *Solar Eruptions and Energetic Particles*, *165*, 275.

Cohen, C. M. S., Mason, G. M., & Mewaldt, R. A. (2017). Characteristics of Solar Energetic Ions as a Function of Longitude, *The Astrophysical Journal*, *843*, 132.

Cohen, C. M. S., Mason, G. M., Mewaldt, R. A., & Rosenvinge, von, T. T. (2013). Solar energetic particle characteristics and their dependence on longitude in solar cycle 24, *Solar Wind 13: Proceedings of the Thirteenth International Solar Wind Conference AIP*, *1539*, 151.

Cohen, C. M. S., Mewaldt, R. A., Leske, R. A., Cummings, A. C., Stone, E. C., Wiedenbeck, M. E., et al. (1999). New observations of heavy-ion-rich solar particle events from ACE. *Geophysical Research Letters*, *26*, 2697.

Cohen, C. M. S., Mewaldt, R. A., Leske, R. A., Cummings, A. C., Stone, E. C., Wiedenbeck, M. E., et al. (2007). Solar elemental composition based on studies of solar energetic particles. *Space Science Reviews*, *130*, 183.

Cohen, C. M. S., Stone, E. C., Mewaldt, R. A., Leske, R. A., Cummings, A. C., Mason, G. M., et al. (2005), Heavy ion abundances and spectra from the large solar energetic particle events of October-November 2003. *Journal of Geophysical Research*, *110*, A09S16. doi: 10.1029/2005JA011004.

Coppi, B., Galvao, R., Pellat, R, Rosenbluth, M. N., & Rutherford, P. H. (1976). Resistive internal kink modes. *Soviet Journal of Plasma Physics*, *2*, 533–535.

Dalla, S., Balogh, A., Krucker, S., Posner, A., Müller-Mellin, R., Anglin, J. D., et al. (2003). Properties of high heliolatitude solar energetic particle events and constraints on models of acceleration and propagation. *Geophysical Research Letters*, *30*, 108.

Daughton, W., & Roytershteyn, V. (2012). Emerging parameter space map of magnetic reconnection in collisional and kinetic regimes. *Space Science Reviews*, *172*, 271–282.

de Koning, C. A., Gosling, J. T., Skoug, R. M., & Steinberg, J. T. (2007). Energy dependence of electron pitch angle distribution widths in solar bursts. *Journal of Geophysical Research*, *112*, A04101.

Desai, M. I., Mason, G. M., Dwyer, J. R., Mazur, J. E., Gold, R. E., Krimigis, S. M., et al. (2003). Evidence for a suprathermal seed population of heavy ions accelerated by interplanetary shocks near 1 AU. *The Astrophysical Journal*, *588*, 1149–1162.

Desai, M. I., Mason, G. M., Gold, R. E., Krimigis, S. M., Cohen, C. M. S., Mewaldt, R. A., et al. (2006). Heavy-ion elemental abundances in large solar energetic particle events and their implications for the seed population. *The Astrophysical Journal*, *649*, 470.

Desai, M. I., Mason, G. M., Wiedenbeck, M. E., Cohen, C. M. S., Mazur, J. E., Dwyer, J. R., et al. (2004). Spectral properties of heavy ions associated with the passage of interplanetary shocks at 1 AU. *The Astrophysical Journal*, *611*, 1156.

DiFabio, R., Guo, Z., Möbius, E., Klecker, B., Kucharek, H., Mason, G. M. et al. (2008). Energy dependent charge states and their connection with ion abundances in impulsive solar energetic particle events. *The Astrophysical Journal*, *687*, 623–634.

Drake, J. F., Cassak, P. A., Shay, M. A., Swisdak, M., & Quataert, E. (2009). A magnetic reconnection mechaism for ion acceleration and abundance enhancements in impulsive flares. *The Astrophysical Journal*, *700*, L16–L20. doi:10.1088/0004-637X/700/1/L16.

Drake, J. F., Swisdak, M., Che, H., & Shay, M. A. (2006). Electron acceleration from contracting magnetic islands during reconnection. *Nature*, *443*, 553–556.

Dresing, N., Gómez-Herrero, R., Heber, B., Klassen, A., Malandraki, O., Dröge, W., et al. (2014). Statistical survey of widely spread out solar electron events observed with STEREO and ACE with special attention to anisotropies, *Astronomy & Astrophysics*, *567*, A27.

Drury, L. O. C. (1983). An introduction to the theory of diffusive shock acceleration of energetic particles in tenuous plasmas. *Reports on Progress in Physics*, *46*, 973–1027.

Dulk, G. A., Kiplinger, A. L., & Winglee, R. M. (1992). Characteristics of hard X-ray spectra of impulsive solar flares. *The Astrophysical Journal*, *389*, 756.

Eichler, D. (2014). Ultraheavy element enrichment in impulsive solar flare. *The Astrophysical Journal*, *794*, 1. doi:doi:10.1088/0004-637X/794/1/6.

Ellison, D. C., & Ramaty, R. (1985). Shock acceleration of electrons and ions in solar flares. *The Astrophysical Journal*, *298*, 400.

Fisk, L. A. (1978). 3He-rich flares: a possible explanation. *The Astrophysical Journal*, *224*, 1048–1055.

Forbush, S. E. (1946). Three unusual cosmic-ray increases possibly due to charged particles from the Sun. *Physical Review*, *70*, 771.

Furth, H. P., Killeen, J., & Rosenbluth, M. N. (1963). Finite-resistivity instabilities of a sheet pinch. *Physics of Fluids*, *6*, 459–484

Gazis, P. R., & Lazarus, A. J. (1982). Voyager observations of solar wind proton temperature: 1-10AU. *Geophysical Research Letters*, *9*, 431.

Giacalone, J., & Jokipii, J. R. (1999). The transport of cosmic rays across a turbulent magnetic field. *The Astrophysical Journal*, *520*, 204–214.

Giacalone, J., Jokipii, J. R., & Mazur, J. E. (2000). Small-scale gradients and large-scale diffusion of charged particles in the heliospheric magnetic field. *The Astrophysical Journal*, *532*, L75–L78.

Gleeson, L. J., & Axford, W. I. (1967). Cosmic rays in the interplanetary medium. *The Astrophysical Journal*, *149*, L115.

Gopalswamy, N., Yashiro, S., Akiyama, S., et al. (2008). Coronal mass ejections, type II radio bursts, and solar energetic particle events in the SOHO era. *Annales Geophysicae*, *26*, 3033.

Gordon, B. E., Lee, M. A., Möbius, E., & Trattner, K. J. (1999). Coupled hydromagnetic wave excitation and ion acceleration at interplanetary traveling shocks and Earth's bow shock revisited. *Journal of Geophysical Research*, *104*, 28263–28277.

Gosling, J. T., Skoug, R. M., & McComas, D. J. (2003). Solar electron bursts at very low energies: Evidence for acceleration in the high corona?, *Geophysical Research Letters*, *30*, 1697.

Haggerty, D. K., & Roelof, E. C. (2002). Impulsive near-relativistic solar electron events: Delayed injection with respect to solar electromagnetic emission. *The Astrophysical Journal*, *579*, 841.

Hamilton, D. C., Mason, G. M., & McDonald, F. B. (1990). The radial dependence of the peak flux and fluence in solar energetic particle events. *Proceedings of the 21st International Cosmic Ray Conference*, *5*, 237.

Ho, G. C., Lario, D., Decker, R. B., Smith, C. W., & Hu, Q. (2008). Transient shocks and associated energetic particle distributions observed by ACE during Cycle 23. *AIP Conference Proceedings*, *1039*, 184.

Hudson, H. S. (2007). The unpredictability of the most energetic solar events. *The Astrophysical Journal*, *663*, 45.

Hurford, G. J., Schwartz, R. A., Krucker, S., Lin, R. P., Smith, D. M., & Vilmer, N. (2003). First gamma-ray images of a solar flares. *The Astrophysical Journal*, *595*, L77.

Hurford, G. J., Krucker, S., Lin, R. P., Schwartz, R. A., Share, G. H., & Smith, D. M. (2006). Gamma-ray imaging of the 2003 October/November solar flares. *The Astrophysical Journal*, *644*, L93.

Isenberg, P. A. (1997). A hemispherical model of anisotropic interstellar pickup ions. *Journal of Geophysical Research*, *102*, 4719–4724.

Ji, H., & Daughton, W. (2011). Phase diagram for magnetic reconnection in heliophysical, astrophysical, and laboratory plasmas. *Physics of Plasmas*, *18*, 111207.

Jokipii, J. R. (1966). Cosmic-ray propagation. I. Charged particles in a random magnetic field. *The Astrophysical Journal*, *146*, 480.

Kahler, S. W., & Ragot, B. R. (2008). Gamma-ray emission from SEP interactions with solar wind ions. *Proceedings of the 30th International Cosmic Ray Conference*, *1*, 225.

Kahler, S. W., Reames, D. V., & Sheeley, N. R. Jr. (2001). Coronal mass ejections associated with impulsive solar energetic particle events. *The Astrophysical Journal*, *562*, 558.

Kallenrode, M.-B. (1996). A statistical survey of 5-MeV proton events at transient interplanetary shocks. *Journal of Geophysical Research*, *101*, 24393.

Kartavykh, J. J., Dröge, W., Ostryakov, V. M., & Kovaltsov, G. A. (2005). Adiabatic deceleration effects on the formation of heavy ion charge spectra in interplanetary space. *Solar Physics*, *227*, 123–135.

Kartavykh, Y. Y., Dröge, W., Klecker, B., Mason, G. M., Möbius, E., Popecki, M., & S. Krucker (2007). Evidence for a two-temperature source region in the 3He-rich solar energetic particle event of 2000 May 1st. *The Astrophysical Journal*, *671*, 947–954.

Klecker, B., Möbius, E., & Popecki, M. (2006a). Solar energetic particle charge states: An overview. *Space Science Reviews*, *124*, 289.

Klecker, B., Möbius, E., & Popecki, M. (2007). Ionic charge states of solar energetic particles: A clue to the source. *Space Science Reviews*, *130*, 273–282. doi:10.1007/s11214-007-9207-1.

Klecker, B., Möbius, E., Popecki, M., & Kistler, L. M. (2006b). *Ionic Charge States in Interplanetary Shock Related Solar Energetic Particle Events, in 2006 AGU Fall Meeting, edited, abstract id.SH41B-07*, San Francisco, CA.

Klecker, B., Möbius, E., Popecki, M., Kistler, L. M., Kucharek, H., & Hilchenbach, M. (2006c). Observation of energy-dependent ionic charge states in impulsive solar energetic particle events. *Advances in Space Research*, *38*, 493–497.

Kocharov, L., Laivola, J., Mason, G. M., Didkovsky, L., & Judge, D. L. (2008). Extended 3He-rich periods of solar energetic particles in structured solar wind. *The Astrophysical Journal Supplement*, *176*, 497–510.

Kocharov, L. G., & G. E. Kocharov (1984). 3He-rich solar flares. *Space Science Reviews*, *38*, 89–141.

Kocharov, L. G., Kovaltsov, G. A., Torsti, J., & Ostryakov, V. M. (2000). Evaluation of solar energetic Fe charge states: Effects of proton-impact ionization. *Astronomy & Astrophysics*, *357*, 716–724.

Kocharov, L. G., & Orischenko, A. V. (1983). On the mechanism of solar cosmic ray enrichment by heavy ions. *Proceedings of the 18th ICRC*, Tata Inst. Fund. Res. Mumbai, *4*, 37–40.

Kontar, E. P., & Reid, H. A. S. (2009). Onsets and spectra of impulsive solar energetic electron events observed near the Earth. *The Astrophysical Journal*, *695*, L140.

Kozarev, K. A., Raymond, J. C., Lobzin, V. V., & Hammer, M. (2015). Properties of a coronal shock wave as a driver of early SEP acceleration. *The Astrophysical Journal*, *799*, 167.

Krucker, S., Larson, D. E., Lin, R. P., & Thompson, B. J. (1999). On the origin of impulsive electron events observed at 1 AU. *The Astrophysical Journal*, *519*, 864.

Krucker, S., Kontar, E. P., Christe, S., & Lin, R. P. (2007). Solar flare electron spectra at the Sun and near the Earth. *The Astrophysical Journal*, *663*, L109.

Krucker, S., Oakley, P. H., & Lin, R. P. (2009). Spectra of solar impulsive electron events observed near Earth. *The Astrophysical Journal*, *691*, 806.

Krucker, S., & Lin, R. P. (2002). Relative timing and spectra of solar flare hard X-ray sources. *Solar Physics*, *210*, 229.

Krucker, S., & Lin, R. P. (2008). Hard X-ray emissions from partially occulted solar flares. *The Astrophysical Journal*, *673*, 1181.

Krymskii, G. (1977). Regular mechanism of charged-particle acceleration on shock-wave front. *Doklady Akademii Nauk Sssr*, *234*: 1306−1308.

Kubo, R. (1957). Statistical-mechanical theory of irreversible processes. I. General theory and simple applications to magnetic and conduction problems. *Journal of the Physical Society of Japan*, *12*: 570–586.

Kunow, H., Wibberenz, G., Green, G., Müeller-Mellin, R., Witte, M., & H. Hempe (1975). The Kiel University experiment for measuring cosmic radiation between 1.0 and 0.3 AE /E 6/, *Raumfahrtforschung*, *19*, 253.

Lario, D. (2010). Heliospheric energetic particle reservoirs: Ulysses and ACE 175-315 keV electron observations. *AIP Conference Proceedings 1216*, 625.

Lario, D., Aran, A., Gómez-Herrero, R., DresingHeber, B., Ho, G. C., Decker, R. B., et al. (2013). Longitudinal and radial dependence of solar energetic particle peak intensities: STEREO, ACE, SOHO, GOES, and MESSENGER observations. *The Astrophysical Journal*, *767*, 41.

Lario, D., Kallenrode, M.-B., Decker, R. B., Roelof, E. C., Krimigis, S. M., Aran, A., et al. (2006). Radial and longitudinal dependence of solar 4-13 MeV and 27-37 MeV proton peak intensities and fluences: Helios and IMP 8 observations. *The Astrophysical Journal*, *653*, 1531.

Lario, D., Kwon, R. Y., Riley, P., & Raouafi, N. E. (2017). On the link between the release of solar energetic particles measured at widespread heliolongitudes and the properties of the associated coronal shocks. *The Astrophysical Journal*, *847*, 103.

Leamon, R. J., Smith, C. W., Ness, N. F., & Wong, H. K. (1999). Dissipation range dynamics: Kinetic Alfvén waves and the importance of βe. *Journal of Geophysical Research*, *104*, 22331.

Lee, M. A. (1983). Coupled hydromagnetic wave excitation and ion acceleration at interplanetary traveling shocks. *Journal of Geophysical Research*, *88*, 6109–6119.

Lee, M. A. (2005). Coupled hydromagnetic wave excitation and ion acceleration at an evolving coronal/interplanetary shock. *The Astrophysical Journal*, *158*, 38.

Li, G., & Lee, M. A. (2015). Scatter-dominated interplanetary transport of solar energetic particles in large gradual events and the formation of double power-law differential fluence spectra of ground-level events during solar cycle 23. *The Astrophysical Journal*, *810*, 82.

Li, G., Zank, G. P., Verkhoglyadova, O., Mewaldt, R. A., Cohen, C. M. S., Mason, G. M., et al. (2009). Shock geometry and spectral breaks in large SEP events. *The Astrophysical Journal*, *702*, 998.

Li, X., Guo, F., Li, H., & Li, G. (2015). Nonthermally dominated electron acceleration during magnetic reconnection in a low-β PLASMA. *The Astrophysical Journal*, *811*, L24.

Lin, R. P. (1974). Non-relativistic Solar Electrons, *Space Science Reviews*, *16*, 189.

Lin, R. P. (1985). Energetic solar electrons in the interplanetary medium. *Solar Physics*, *100*, 537.

Lin, R. P. (1990). In E. R. Priest & V. Krishnan (Eds.), *Basic plasma processes on the sun* (p. 467). The Netherlands: International Astronomical Union.

Lin, R. P. (1998). WIND observations of suprathermal electrons in the interplanetary medium, *Space Science Reviews*, *86*, 61–78.

Lin, R. P. (2011). Energy release and particle acceleration in flares: Summary and future prospects. *Space Science Reviews*, *159*, 421–445. doi:10.1007/s11214–011–9801–0.

Lin, R. P., Anderson, K. A., Ashford, S., et al. (1995). A Three-Dimensional Plasma and Energetic Particle Investigation for the Wind Spacecraft, *Space Science Reviews*, *71*, 125.

Lin, R. P., & Schwartz, R. A. (1987). High spectral resolution measurements of a solar flare hard X-ray burst. *The Astrophysical Journal*, *312*, 462.

Liu, S., Petrosian, V., & Mason, G. M. (2006). Stochastic acceleration of 3He and 4He by parallel propagating plasma waves: General results. *The Astrophysical Journal*, *636*, 462–475.

Lodders, K. (2003). Solar system abundances and condensation temperatures of the elements. *The Astrophysical Journal*, *591*, 1220–1247.

Luhn, A., Klecker, B., Hovestadt, D., & Möbius, E. (1987). The mean ionic charge of silicon in 3He-rich solar flares. *The Astrophysical Journal*, *317*, 951–955.

Mace, R. L., Matthaeus, W. H., & Bieber, J. W. (2000). Numerical investigation of perpendicular diffusion of charged test particles in weak magnetostatic slab turbulence. *The Astrophysical Journal*, *538*, 192–202.

Maia, D., & Pick, M. (2004). Revisiting the origin of impulsive electron events: Coronal magnetic restructuring. *The Astrophysical Journal*, *609*, 1082.

Malyshkin, L. M., Linde, T., & Kulsrud, R. M. (2005). Magnetic reconnection with anomalous resistivity in two-and-a-half dimensions. I. Quasistationary case. *Physics of Plasmas*, *12*, 102902–102902.

Marsden, R. G., & Müller, D. (2011). Solar Orbiter, Exploring the Sun-heliosphere connection, European Space Agency.

Mason, G. M. (2007). 3He-rich solar energetic particle events. *Space Science Reviews*, *130*, 231–242. doi: 10.1007/s11214–007–9156–8.

Mason, G. M., Cohen, C. M. S., Cummings, A. C., Dwyer, J. R., Gold, R. E., Krimigis, S. M., et al. (1999a). Particle acceleration and sources in the November 1997 solar energetic particle events. *Geophysical Research Letters*, *26*, 141.

Mason, G. M., Desai, M. I., Mazur, J. E., & Dwyer, J. R. (2005). Energetic particles accelerated by shocks in the Heliosphere: What is the source material?. *AIP Conference Proceedings*, *781*, 219.

Mason, G. M., Gold, R. E., Krimigis, S. M., Mazur, J. E., Andrews, G. B., Daley, K. A., et al. (1998). The Ultra-Low-Energy Isotope Spectrometer (ULEIS) for the ACE spacecraft. *Space Science Reviews*, *86*, 409.

Mason, G. M., & Klecker, B. (2018). A possible mechanism for enriching heavy ions in 3He-rich solar energetic particle events. *The Astrophysical Journal*, *862*(7). doi:10.3847/1538–4357/aac94c.

Mason, G. M., Li, G., Cohen, C. M. S., Desai, M. I., Haggerty, D. K., Leske, R. A., et al. (2012). Interplanetary propagation of solar energetic particle heavy ions observed at 1 AU and the role of energy scaling. *The Astrophysical Journal*, *761*, 104.

Mason, G. M., Li, G., Cohen, C. M. S., Desai, M. I., Haggerty, D. K., Leske, R. A., et al. (2014). Fe enhancements in SEP onsets: Flare/CME mixture or transport effect?. *ASP Conference Series*, *484*, 137.

Mason, G. M., Mazur, J. E., & Dwyer, J. R. (1999). 3He Enhancements in large solar energetic particle events. *The Astrophysical Journal*, *525*, L133.

Mason, G. M., Mazur, J. E., & Dwyer, J. R. (2002a). A new heavy ion abundance enrichment pattern in 3He-rich solar particle events. *The Astrophysical Journal*, *565*, L51–L54.

Mason, G. M., Mazur, J. E., Dwyer, J. R., Jokipii, J. R., Gold, R. E., & Krimigis, S. M. (2004). Abundances of heavy and ultra-heavy ions in 3He-rich solar flares. *The Astrophysical Journal*, *606*, 555–564.

Mason, G. M., Nitta, N., Cohen, C. M. S., & Wiedenbeck, M. E. (2009). Solar energetic particle 3He-rich events from the nearly quiet Sun in 2007–2008. *The Astrophysical Journal*, *700*, L56. doi:10.1088/0004–1637X/1700/1081/L1056.

Mason, G. M., Reames, D. V., Klecker, B., Hovestadt, D., & von Rosenvinge, T. T. (1986). The heavy ion compositional signature in 3He-rich solar particle events. *The Astrophysical Journal*, *303*, 849–860.

Mason, G. M., Wiedenbeck, M. E., Miller, J. A., Mazur, J. E., Christian, E. R., Cohen, C. M. S., et al. (2002b). Spectral properties of He and heavy ions in 3He-rich solar flares. *The Astrophysical Journal*, *574*, 1039–1058.

Matthaeus, W. H., Qin, G., Bieber, J. W., & Zank, G. P. (2003). Nonlinear collisionless perpendicular diffusion of charged particles. *The Astrophysical Journal*, *590*, L53–L56.

Mazur, J. E., Mason, G. M., Dwyer, J. R., Giacalone, J., Jokipii, J. R., & Stone, E. C. (2000). Interplanetary magnetic field line mixing deduced from impulsive solar-flare particles. *The Astrophysical Journal*, *532*, L79-L82.

McKibben, R. B. (1972). Azimuthal propagation of low-energy solar-flare protons as observed from spacecraft very widely separated in solar azimuth. *Journal of Geophysical Research*, *77*, 3957.

McKibben, R. B., Connell, J. J., Lopate, C., Zhang, M., Anglin, J. D., Balogh, A., et al. (2003). Ulysses COSPIN observations of cosmic rays and solar energetic particles from the South Pole to the North Pole of the Sun during solar maximum. *Annales Geophysicae*, *21*, 1217.

McKibben, R. B., Lopate, C., & Zhang, M. (2001). Simultaneous observations of solar energetic particle events by IMP 8 and the Ulysses COSPIN high energy telescope at high solar latitudes. *Space Science Reviews*, *97*, 257.

Mewaldt, R. A., Cohen, C. M. S., Labrador, A. W., Leske, R. A., Desai, M. I., Looper, M. D., et al. (2005a). Proton, helium, and electron spectra during the large solar particle events of October-November 2003. *Journal of Geophysical Research*, *110*, doi: 10.1029/2005JA011038.

Mewaldt, R. A., Cohen, C. M. S., Leske, R. A., Christian, E. R., Cummings, A. C., Stone, E. C., et al. (2001). Are solar energetic particles an accelerated sample of solar wind?. *Proceedings of the 29th International Cosmic Ray Conference*, *8*, 3132.

Mewaldt, R. A., Cohen, C. M. S., Mason, G. M., Labrador, A. W., Looper, M. L., Haggerty, D. K. et al. (2005b). Solar energetic particle spectral breaks. *AIP Conference Proceedings*, *781*, 227.

Mewaldt, R. A., Looper, M. D., & Cohen, C. M. S., et al. (2012). Energy spectra, composition, and other properties of ground-level events during solar cycle 23. *Space Science Reviews*, *171*, 97.

Mewaldt, R. A., et al. (2009). STEREO observations of energetic neutral hydrogen atoms during the 2006 December 5 solar flare. *The Astrophysical Journal*, *693*, L11.

Mewaldt, R. A., Leske, R. A., Shih, A. Y., Stone, E. C., Barghouty, A. F., Cohen, C. M. S. et al. (2010). Observations and interpretations of energetic neutral hydrogen atoms from the December 5, 2006 solar event. *AIP Conference Proceedings*, *1216*, 592.

Meyer, J.-P. (1985a). The baseline composition of solar energetic particles. *The Astrophysical Journal*, *57*, 151.

Meyer, J.-P. (1985b). Solar-stellar outer atmospheres and energetic particles, and galactic cosmic rays. *The Astrophysical Journal*, *57*, 173.

Meyer, P., Parker, E. N., & Simpson, J. A. (1956). Solar cosmic rays of February, 1956 and their propagation through interplanetary space. *Physical Review*, *104*, 768.

Miller, J. A., & Reames, D. V. (1996). Heavy ion acceleration by cascading Alfvén waves in impulsive solar flares. *AIP Conference Proceedings*, *374*, 450–460.

Möbius, E., Scholer, M., Hovestadt, D., Klecker, B., & Gloeckler, G. (1982). Comparison of helium and heavy ion spectra in 3He-rich solar flares with model calculations based on stochastic Fermi acceleration in Alfvén turbulence. *The Astrophysical Journal*, *259*, 397.

Nitta, N. V., Mason, G. M., Wang, L., Cohen, C. M. S., & Wiedenbeck, M. E. (2015). Solar sources of 3He-rich solar energetic particle events in solar cycle 24. *The Astrophysical Journal*, *806*, 235.

Nitta, N. V., Mason, G. M., Wiedenbeck, M. E., et. al. (2008). Coronal jet observed by Hinode as the source of a3He-rich solar energetic particle event. *The Astrophysical Journal*, *675*, L125.

Nitta, N. V., Reames, D. V., DeRosa, M. L., Liu, Y., Yashiro, S., & Gopalswamy, N. (2006). Solar sources of impulsive solar energetic particle events and their magnetic connection to the Earth. *The Astrophysical Journal*, *650*, 438–450.

Ng, C. K., & Reames, D. V. (1994). Focused interplanetary transport of approximately 1MeV solar energetic protons through self-generated Alfven waves. *The Astrophysical Journal*, *424*, 1032.

Ng, C. K., Reames, D. V., & Tylka, A. J. (2003). Modeling shock-accelerated solar energetic particles coupled to interplanetary Alfvén waves. *The Astrophysical Journal*, *591*, 461.

Oka, M., Phan, T. D., Krucker, S., Fujimoto, M., & Shinohara, I. (2010). Electron acceleration by multi-island coalescence. *The Astrophysical Journal*, *714*, 915–926.

Parker, E. N. (1957). Sweet's mechanism for merging magnetic fields in conducting fluids. *Journal of Geophysical Research*, *62*, 509–520.

Parker, E. N. (1965). The passage of energetic charged particles through interplanetary space. *Planetary and Space Science*, *13*, 9–49.

Petrosian, V. (2012). Stochastic acceleration by turbulence. *Space Science Reviews*, *173*, 535–556. doi:10.1007/s11214-012-9900-6.

Petschek, H. (1964). *Magnetic field annihilation. NASA Spec Publ*, *50*, 425.

Pick, M., Mason, G. M., Wang, Y.-M., Tan, C., & Wang, L. (2006). Solar source regions for 3He-rich solar energetic particle events identified using imaging radio, optical, and energetic particle observations. *The Astrophysical Journal*, *648*, 1247.

Poluianov, S. V., Usoskin, I. G., Mishev, A. L., Shea, M. A., & Smart, D. F. (2017). GLE and Sub-GLE redefinition in the light of high-altitude polar neutron monitors. *Solar Physics*, *292*, 176.

Potter, D. W., Lin, R. P., & Anderson, K. A. (1980). Impulsive 2–10 keV solar electron events not associated with flares. *Astrophysical Journal Letters*, *236*, L97–L100. doi:10.1086/183206.

Qin, G., Matthaeus, W. H., & Bieber, J. W. (2002a). Perpendicular transport of charged particles in composite model turbulence: Recovery of diffusion. *The Astrophysical Journal*, *578*, L117–L120.

Qin, G., Matthaeus, W. H., & Bieber, J. W. (2002b). Subdiffusive transport of charged particles perpendicular to the large scale magnetic field. *Geophysical Research Letters*, *29*, 1048.

Reames, D. V. (1995). Coronal abundances determined from energetic particles. *Advances in Space Research*, *15*, 41.

Reames, D. V. (1999). Particle acceleration at the Sun and in the heliosphere. *Space Science Reviews*, *90*, 413.

Reames, D. V. (2017). *Solar energetic particles: A modern primer on understanding acceleration and propagation. Springer Lecture Notes in Physics*, *192*. doi:10.1007/978-3-319-50871-9.

Reames, D. V., Meyer, J. P., & von Rosenvinge, T. T. (1994). Energetic-particle abundances in impulsive solar flares. *The Astrophysical Journal Supplement*, *90*, 649–667.

Reames, D. V., & Ng, C. K. (1998). Streaming-limited intensities of solar energetic particles. *The Astrophysical Journal*, *v504*, 504, 1002.

Reames, D. V., & Ng, C. K. (2010). Streaming-limited intensities of solar energetic particles on the intensity plateau. *The Astrophysical Journal*, *723*, 1286.

Reames, D. V., & C. K. Ng (2004). Heavy-element abundances in solar energetic particle events. *The Astrophysical Journal*, *610*, 510–522.

Reames, D. V., Ng, C. K., & Tylka, A. J. (2000). Initial time dependence of abundances in solar energetic particle events. *Astrophysical Journal Letters*, *531*, L83.

Reames, D. V., Ng, C. K., & Tylka, A. J. (2013). Spatial distribution of solar energetic particles in the inner heliosphere. *Solar Physics*, *285*, 233.

Reames, D. V., von Rosenvinge, T. T., & R. P. Lin (1985). Solar 3He-rich events and nonrelativistic electron events: A new association. *The Astrophysical Journal*, *292*, 716.

Richardson, I. G., Rosenvinge, von, T. T., Cane, H. V., Christian, E. R., Cohen, C. M. S., Labrador, A. W., et al. (2014). > 25 MeV Proton events observed by the high energy telescopes on the STEREO A and B spacecraft and/or at earth during the first ~ seven years of the STEREO mission. *Solar Physics*, *289*, 3059.

Richardson, J. E., Paularena, K. I., Lazarus, A. J., & Belcher, J. W. (1995). Radial evolution of the solar wind from IMP 8 to Voyager 2, *Geophysical Research Letters*, *22*, 325.

Roelof, E. C. (1969). Propagation of solar cosmic rays in the interplanetary magnetic field. In H. Ögelman & J. Wayland (eds.), *Lectures in High-Energy Astrophysics* (p. 111) Prepared by NASA Goddard Space Flight Center. Published by the Scientific and Technical Information Division, Office of Technology Utilization, NASA, Washington, DC.

Rouillard, A. P., Lavraud, B., Génot, V., et al. (2017). *Planetary and Space Science*, *147*, 61.

Ruffolo, D. (1995). Effect of adiabatic deceleration on the focused transport of solar cosmic rays. *The Astrophysical Journal*, *442*, 861–874.

Shalchi, A., & Dosch, A. (2008). Nonlinear guiding center theory of perpendicular diffusion: Derivation from the Newton-Lorentz equation. *The Astrophysical Journal*, *685*, 971–975.

Shibata, K., & Tanuma, S. (2001). Plasmoid-induced-reconnection and fractal reconnection. *Earth Planets and Space*, *53*, 473–482.

Simnett, G. M. (2011). Energetic neutral atoms from the Sun: An alternative interpretation of a unique event. *Astronomy & Astrophysics*, *531*, A46.

Simnett, G. M., Roelof, E. M., & Haggerty, D. K. (2002). *The Astrophysical Journal*, *579*, 854.

Skilling, J. (1971). Cosmic rays in the galaxy: Convection or diffusion? *The Astrophysical Journal*, *170*, 265.

Smith, C. W., Ness, N. F., Burlaga, L. F., Skoug, R. M., McComas, D. J., Zurbuchen, T. H., et al. (2001). ACE observations of the Bastille day 2000 interplanetary disturbances. *Solar Physics*, *204*, 227.

Stone, E. C., Cohen, C. M. S., Cook, W. R., Cummings, A. C., Gauld, B., Kecman, B., et al. (1998). The solar isotope spectrometer for the advanced composition explorer. *Space Science Reviews*, *86*, 357.

Swann, W. F. G. (1933). A mechanism of acquirement of cosmic ray energies by electrons. *Journal of the Franklin Institute*, *215*, 273–279.

Sweet, P. A. (1958). The neutral point theory of solar flares. In B. Lehnert (Ed.), *Electromagnetic phenomena in Cosmic Physics* (pp. 123−134). IAU Symposium, 6.

Temerin, M., & Roth, I. (1992). The production of 3He and heavy enrichments in 3He-rich flares by electromagnetic hydromagnetic hydrogen cyclotron waves. *The Astrophysical Journal*, *391*, L105–L108.

Totten, T. L., Freeman, J. W., & Arya, S. (1995). An empirical determination of the polytropic index for the free-streaming solar wind using Helios 1 data. *Journal of Geophysical Research*, *100*, 13.

Tu, C. Y., Pu, Z. Y., & Wei, F. S. (1984). The power spectrum of interplanetary Alfvénic fluctuations: Derivation of the governing equation and its solution. *Journal of Geophysical Research*, *89*, 9695–9702.

Tylka, A. J., Cohen, C. M. S., Dietrich, W. F., et al. (2005). *The Astrophysical Journal*, *625*, 474.

Uzdensky, D. A., & Kulsrud, R. M. (2000). Two-dimensional numerical simulation of the resistive reconnection layer. *Physics of Plasmas*, *7*, 4018–4030.

Van Allen, J. A., & Krimings, S. M. (1965). Impulsive emission of 40-kev electrons from the Sun. *Journal of Geophysical Research*, *70*, 5737.

Vasquez, B. J., Smith, C. W., Hamilton, K., MacBride, B. T., & Leamon, R. J. (2007). Evaluation of the turbulent energy cascade rates from the upper inertial range in the solar wind at 1 AU. *Journal of Geophysical Research*, *112*, A07101.

Vourlidas, A., Buzasi, D., Howard, R. A., & Esfandiari, E. (2002). In J. Kuijpers (Ed.), Solar variability: From core to outer frontiers (p. 91). *ESA SP-506*; Noordwijk: ESA.

Wang, L., Krucker, S., Mason, G. M., Lin, R. P., & Li, G. (2016). The injection of ten electron/3He-rich SEP events. *Astronomy & Astrophysics*, *585*, A119, doi:10.1051/0004–6361/201527270.

Wang, L., Li, G., Shih, A. Y., Lin, R. P., & Wimmer-Schweingruber, R. F. (2014). Simulation of energetic neutral atoms from solar energetic particles. *The Astrophysical Journal*, *793*, L37.

Wang, L., Lin, R. P., & Krucker, S. (2011). Pitch-angle distributions and temporal variations of 0.3-300 keV solar impulsive electron events. *The Astrophysical Journal*, *727*, 121

Wang, L., Lin, R. P., Krucker, S., & Gosling, J. T. (2006a). Evidence for double injections in scatter-free solar impulsive electron events, *Geophysical Research Letters*, *33*, L03106.

Wang, L., Lin, R. P., Krucker, S., & Mason, G. M. (2012). A statistical study of solar electron events over one solar cycle. *The Astrophysical Journal*, *759*. doi:10.1088/0004–1637X/1759/1081/1069.

Wang, L., Lin, R. P., Salem, C., Pulupa, M., Larson, D. E., Yoon, P. H. et al. (2012b). Quiet-time interplanetary ~2-20 keV superhalo electrons at solar minimum. *The Astrophysical Journal Letters*, *753*, L23.

Wang, Y.-M., Pick, M., & Mason, G. M. (2006b). Coronal holes, jets, and the origin of 3He-rich particle events. *The Astrophysical Journal*, *639*, 495. doi:10.1086/499355.

Wibberenz, G., & Cane, H. V. (2006). Multi-spacecraft observations of solar flare particles in the inner heliosphere. *The Astrophysical Journal*, *650*, 1199.

Wiedenbeck, M. E., & G. M. Mason (2014), The interplanetary population of suprathermal ions from impulsive solar energetic particle events: solar cycle variations. *ASP Conference Series*, *484*, 225–232.

Wiedenbeck, M. E., Mason, G. M., Cohen, C. M. S., Nitta, N. V., Gómez-Herrero, R., & Haggerty, D. K. (2013). Observations of solar energetic particles from 3He-rich events over a wide range of heliographic longitudes. *The Astrophysical Journal*, *762*, 54, 1–9. doi:10.1088/0004–637X/762/1/54.

Yashiro, S., Michalek, G., & Gopalswamy, N. (2008). A comparison of coronal mass ejections identified by manual and automatic methods. *Annals of Geophysics*, *26*, 3103.

Zank, G. P., Rice, W. K. M., & Wu, C. C. (2000). Particle acceleration and coronal mass ejection driven shocks: A theoretical model. *Journal of Geophysical Research*, *105*, 25079.

Zank, G. P., Li, G., Florinski, V., Hu, Q., Lario, D., & Smith, C. W. (2006). Particle acceleration at perpendicular shock waves: Model and observations. *Journal of Geophysical Research*, *111*, A06108.

Zank, G. P., Li, G., Florinski, V., Matthaeus, W. H., Webb, G. M., & Le Roux, J. A. (2004). Perpendicular diffusion coefficient for charged particles of arbitrary energy. *Journal of Geophysical Research*, *109*, A04107.

Zhang, M., Qin, G., & Rassoul, H. (2009). Propagation of solar energetic particles in three-dimensional interplanetary magnetic fields. *The Astrophysical Journal*, *692*, 109.

Zhang, T. X. (1995). Solar 3He-rich events and ion acceleration in two stages. *The Astrophysical Journal*, *449*, 916–929.

Zhao, L., Zhang, M., & Rassoul, H. K. (2016). Double power laws in the event-integrated solar energetic particle spectrum. *The Astrophysical Journal*, *821*, 62.

Zwickl, R. D., Roelof, E. C., Gold, R. E., Krimigis, S. M., & Armstrong, T. P. (1978). Z-rich solar particle event characteristics, *The Astrophysical Journal*, *225*, 281.

5

Solar Flares and Coronal Mass Ejections

Barbara J. Thompson[1], Jiong Qiu[2], Noé Lugaz[3], and David F. Webb[4]

ABSTRACT

This chapter describes the two major types of impulsive magnetic energy release on the Sun: flares and coronal mass ejections (CMEs). The sections include a brief historical overview, a description of observational signature and detection properties, a section on theoretical considerations and outstanding problems, and a final brief section on the field of space weather and anticipated future developments in the field.

5.1. INTRODUCTION

Flares and CMEs, though often treated as separate and distinct solar phenomena, have a great deal of similarities. In fact, many researchers are coming to treat them collectively as two manifestations of a single process: large-scale magnetic reconfiguration in the solar corona.

First and foremost, both defy simple observational description. For example, when one tries to define a flare as a sudden, local, strong increase in brightness on the Sun, one has to clearly define what is meant by these descriptors, and also take into account the cases that test these general criteria.

Flares are a local brightening on the Sun, and may be observed in the photosphere, chromosphere, or corona. Where the brightening is observed depends on where the magnetic energy release occurs, and also where the energy is being deposited. The definition of "local" depends on the type of flare. Compact flares (see Figure 5.1), commonly associated with active regions, are usually scaled in megameters. However, the largest long-duration, gradual flares (see Figure 5.2), which are more common in quiet Sun filament channels, can span a solar radius or more. "Local" is defined relative to the magnetic structures undergoing reconfiguration, and as flares can occur in a variety of magnetic structures, the specification of "local" evolves quite a bit. When one starts to consider "microflares" and "nanoflares," the discussion of spatial scale becomes even more complicated. Other properties such as energy, timing, and multi-spectral brightness exhibit a similar range of scales. In Section 5.3, we discuss the observational properties of flares in greater detail.

CMEs, on the other hand, are the bulk expulsion of magnetized plasma from the corona. They are typically characterized in terms of speed, acceleration, size (such as width), and brightness. It may seem that the name "CME" is not ideal: it is the magnetic field, not the mass, that forms the basis of the ejection. The mass is part of the magnetic structure (the highly conducting plasma closely follows field lines), and most of the ways CMEs are detected on the Sun depend on how the expelled magnetic fields are filled with material. A CME's observation is dependent on the detection of an apparent massive flow, but also the manifestation of a change in magnetic field structure (see Figure 5.3). The solar wind is a continuous flow of mass from the corona; it is only when the mass contains an apparent "bulk" signature, implying a

[1]*NASA Goddard Space Flight Center, Greenbelt, Maryland, USA*

[2]*Montana State University, Bozeman, Montana, USA*

[3]*University of New Hampshire, Durham, New Hampshire, USA*

[4]*Institute for Scientific Research, Boston College, Chestnut Hill, Massachusetts, USA*

Space Physics and Aeronomy Collection Volume 1: Solar Physics and Solar Wind, Geophysical Monograph 258, First Edition.
Edited by Nour E. Raouafi and Angelos Vourlidas.

DOI: 10.1002/9781119815600.ch5

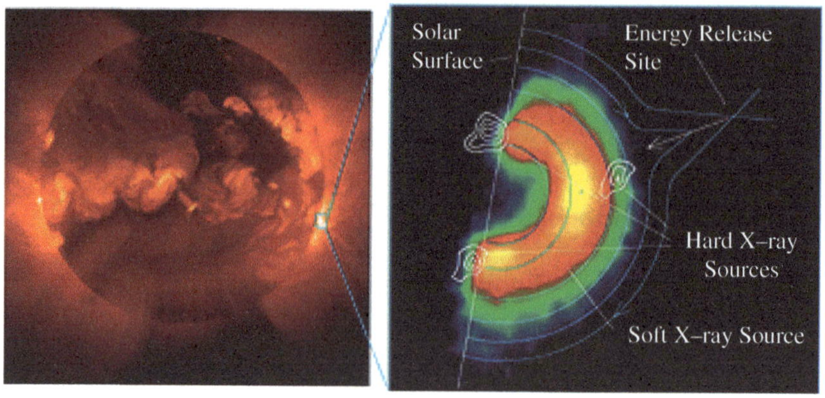

Figure 5.1 Compact flare observed in an active region off the solar limb on 13 January 1992. Left: Yohkoh Soft X-ray Telescope full-disk soft X-ray image of the Sun. The flaring region is circled. Right: Flaring region contour map of the hard X-ray emission from Yohkoh Hard X-ray Telescope, superimposed on the soft X-ray intensity map. Included are approximate magnetic field lines (with arrows indicating direction). This flare, known as the "Masuda flare" (Masuda et al., 1994), was a landmark observation because of the clarity of the coronal hard X-ray emission above flaring loops observed in profile off the solar limb. (*Source:* Maggio, 2008.)

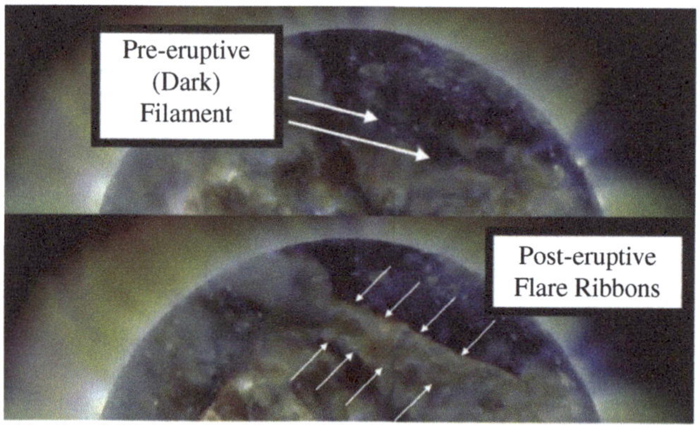

Figure 5.2 A long-duration, gradual solar flare observed in a filament channel. Above: *Solar Dynamics Observatory* (*SDO*) Atmospheric Imaging Assembly (AIA) pre-eruptive image at 02: 49 UT on 1 August 2010 (blue: 171Å, green: 193Å, red: 211Å). Below: Post-eruptive image at 13: 54 UT on 1 August 2010. The arrows in the lower image indicate the ribbons of the large-scale flare, extending across much of the visible solar surface. The dark structure in the pre-eruptive image is the filament prior to eruption (note its absence in the post-eruptive image). (*Source:* SDO/AIA/NASA.)

separate magnetic structure, that we call it a CME. However, CMEs can be small and faint, and often the observational viewpoint is not ideal; it is not uncommon for two observers to differ on whether a moving disturbance is a poorly observed CME, or if it is a different evolving structure. Again, complexity arises when one looks at small CMEs, streamer puffs, or blowouts, jets or other small-scale eruptions. CME is the term used to describe the eruption and its manifestation in inner and outer coronal observations. The term "interplanetary coronal mass ejection" (ICME) is often used to describe the interplanetary counterpart of CMEs. It has been historically used to describe the in situ measurements associated with these eruptions. With the development of heliospheric imaging with the Solar Mass Ejection Imager (*SMEI*; 2003–2011) and the *Solar TErrestrial RElations Observatory* (*STEREO*) mission's Heliospheric Imager (HI; 2007–present), and the availability of in situ measurements of interplanetary CMEs (ICMEs) in the upper corona with the *Parker Solar Probe*, the delimitation between these two terms is no longer obvious.

Forging ahead, our field has tried to pin down the key properties that observationally define both flares and CMEs. We usually start with the best examples and properties of "typical" flares and CMEs, and discuss how much these properties can vary (and why). In Section 5.3, we discuss observations in greater detail. This is important, as the two phenomena have been historically defined by their observational properties; these have evolved as our physical understanding has improved and as improved measurements became available.

Flares and CMEs both require an understanding of the buildup and release of solar magnetic energy. An understanding of the physical processes underlying flares and CMEs requires (at least) three aspects: the buildup of surplus magnetic energy, the "trigger" or initiation of the release of that energy, and the post-flare/eruptive evolution or recovery. The interplay between "buildup" and "release" is what makes flares and CMEs so energetic and impulsive; it is also what makes them so difficult to predict. The trigger, or initiation mechanism, can be extremely subtle, serving as more of a catalyst than a driver of the energy. The trigger can simply be a small tip in the balance of energy, a transition from an eruption being "not energetically favorable" to "energetically favorable." However, once the process has been initiated, the instability can grow rapidly.

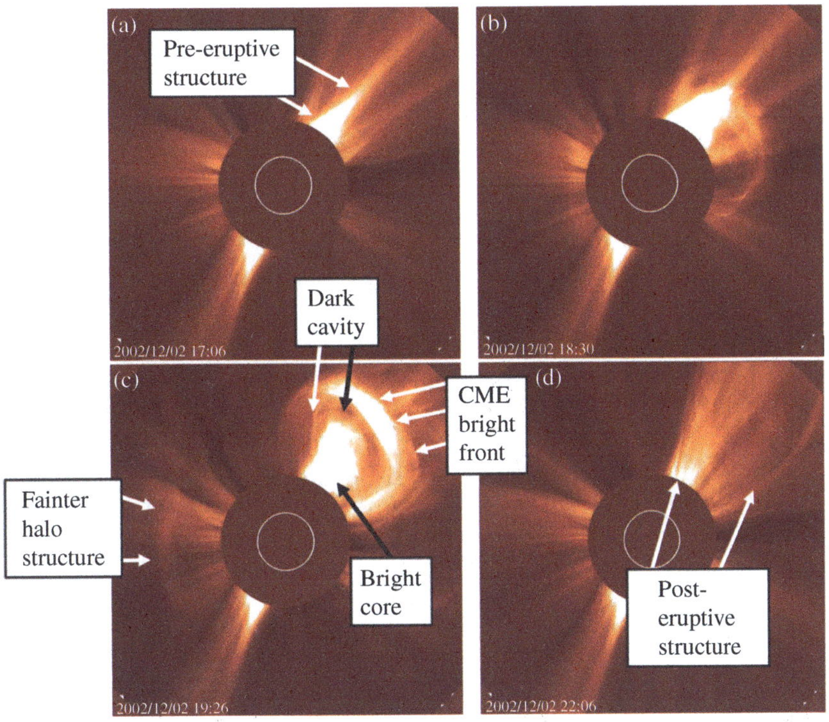

Figure 5.3 Sequence of *SOHO* LASCO C2 coronagraph images showing an eruption over several hours, with a clear loop CME (top), a fainter CME front (left), and some radial, non-eruptive streamer structures. (a) The "pre-event" image (at 17: 06 UT on 2 December 2002) shows primarily radial structures that are typical of a slowly evolving coronal structure. (b) A relatively faint structure with a "halo" component is observed, with emission observed at most angles around the disk. (c) A newer CME structure is observed, more localized to the upper-right (northwest) quadrant. This CME is much more distinguishable than the first CME. (d) A few hours later, the corona has begun the return to its original, primarily radial configuration. (*Source:* NASA.)

Both flares and CMEs can drive space weather in space and at Earth. Another aspect common to flares and CMEs is their impact on the surrounding corona, heliosphere, and planetary system. This is related, for the most part, to the "release" stage of an eruption. X-ray and extreme-ultraviolet (EUV) brightness increases from a flare can deposit energy in Earth's ionosphere and cause communication disruptions. A direct hit from a CME's magnetic field on Earth can drive strong geomagnetic storms. However, space weather is complicated by a zoo of flare- and CME-associated phenomena. These are processes driven by or triggered by flares and CMEs, such as heliospheric shock waves. Energetic particles from a shock wave are one of the most important factors when predicting space weather. The associated flare and/or CME may have no direct impact on Earth, but the shock that was sparked by an eruption can cause major disruptions, including geomagnetically induced currents and drastic changes in Earth's radiation belts. Therefore, it is not simply a question of what causes the release of energy; we must understand where the energy goes, and how it couples to other important phenomena.

Both flares and CMEs vary with the solar cycle and sunspot number. Flares and CMEs tend to be more energetic and frequent during (and just after) solar maximum, due to their role in helping the Sun liberate excess magnetic energy. During solar maximum, more magnetic flux emerges from the solar interior. The opportunities to exhibit the threefold criteria (storage, trigger, and release) are much more common. Flares and CMEs are an important means by which the Sun liberates the excess magnetic field that builds up during solar maximum, and help the global solar corona transition back to a "solar minimum" configuration. During solar maximum, CME and flare-associated space weather effects are also more common.

Flare studies and CME studies are compromised by detection thresholds. It is important to note that our historical definition of flares and CMEs using observational properties imposes several limitations. For the case of flares, they are commonly defined by their X-ray peak flux, observed against an often bright and variable X-ray solar background. During solar maximum, the corona itself may exceed smaller "A-class" and "B-class" level flares, meaning that we can only detect C-class flares and above. This does not mean that there are no A-class flares during solar maximum, it just means that classical detection methods must be augmented (a similar challenge involves flares occurring on the far side of the Sun). Similarly, the mass and magnetic structure of the solar corona are more disturbed and variable during solar maximum. A faint, slow CME that would easily be detected against a more placid solar minimum corona would be more difficult to distinguish against a more active solar maximum environment. Therefore, the statistics of less energetic CMEs may be underrepresented during solar maximum, suffering a similar fate as weaker flares.

As more sophisticated measurements became available, our ability to characterize flares and CMEs dramatically improved.

They frequently occur together, suggesting that impulsive magnetic energy releases on the Sun can exhibit varying degrees of "CME-like" properties and "flare-like" properties.

Several CME trigger mechanisms involve a flare-like reconfiguration (reconnection) of a magnetic field that transforms confined, high-pressure magnetic flux into flux that is free to expand and leave the corona. Indeed, the famous "Aly-Sturrock" conjecture (Sturrock, 1991) stated that the magnetic energy of a given force-free boundary field is maximized when the field is open, and challenged the scientific community to explain how the apparent opening of field lines in the form of a CME could happen. The flare/CME interplay changes the boundary conditions; therefore, many CME models require a flare-like reconfiguration as part of the CME process. In addition, open fields following a CME close back down and recover using flare-like magnetic reconnection processes.

From the other standpoint, many flare models involve the bulk motion and expulsion of magnetic flux in a CME-like manner. Even if the flux "blob" is not ejected fully from the Sun, it can be ejected from the flaring area and land elsewhere. Many researchers specialize in this study of "eruptive flares" and their associated "failed" eruptions, recognizing the role that flux ejection can play in flare energy release. As it turns out, most flare models contain some element of CME-like motion, and most CME models have some aspect of local flare-like energization.

They are often confused with each other. It is not at all uncommon for someone in the popular press, or even researchers in peripheral fields, to label a CME a "flare" and vice versa. And, to be honest, can we truly blame them? With the high degree of complexity (and confusion) in defining, observing, characterizing, and understanding the two phenomena, even the most experienced observer can find it a challenge to simply call it one or the other.

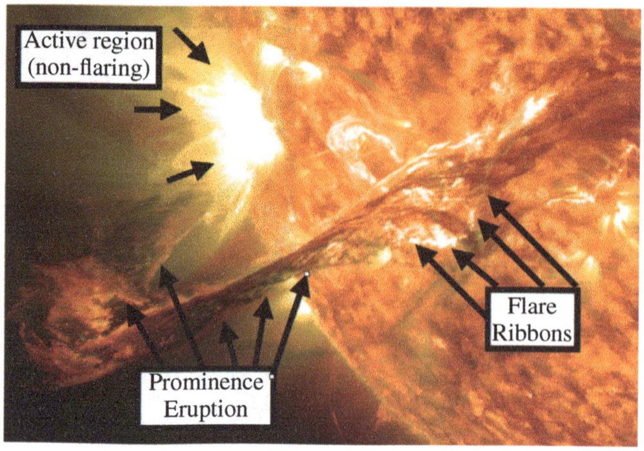

Figure 5.4 This eruption, observed by *SDO*/AIA on 31 August 2012 (orange: 304Å, yellow: 171Å), is at the time of publication one of the most common search engine results for solar flares. However, the flare-like manifestation of this event (labeled "Flare Ribbons") is one of the least remarkable aspects of the image. Undoubtedly, the attraction (and confusion) is mostly due to a non-flaring (but bright) active region, and the CME (labeled "Prominence Eruption"). (*Source:* NASA.)

Therefore, the case for treating the two collectively, as two different manifestations of a more comprehensive impulsive-magnetic-energy-release phenomenon, is compelling. Rather than specifying whether something is a "flare" or a "CME," there is a growing movement to treat the two together, and discuss how the energy release is partitioned between "flare-like" processes and "CME-like" processes. This chapter will take that route; we will discuss the "flare-like" aspect of an event, and follow with a discussion of its "CME-like" aspect. We hope the reader will enjoy the interplay between the two.

5.2. HISTORY

5.2.1. Early History

Flares and CMEs have undoubtedly been impacting Earth for over 4 billion years. However, only recently has humanity been aware of these two phenomena. The reason for this "late realization" is because their observational properties (Section 5.3) are subtle, and their effects (see Section 5.5) have become more pronounced in the technological era. The one possible exception is the aurora borealis/australis. Aurorae are a beautiful manifestation of the Sun's influence on Earth (see Figure 5.5) and do not require special instrumentation to observe.

Aurorae (*aurora borealis* indicating the North Pole, *aurora australis* being the southern complement) have been reported throughout human history. They are observable from the ground provided (a) the Sun's activity has impacted Earth's magnetosphere in a way

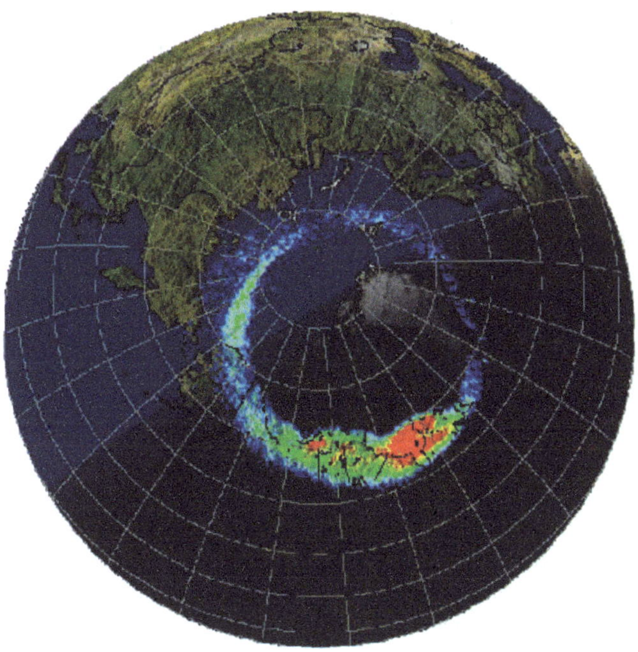

Figure 5.5 "Typical" auroral oval. Colors indicate intensity; red is strongest and blue is weakest (Holzworth & Meng, 1975), superposed over a map of Earth. Aurorae have occurred throughout Earth's history over areas that are not heavily populated. (*Source:* NASA, NSSDC.)

that drives geomagnetic activity, (b) the observer is at a latitude where aurorae are observable (primarily, high latitudes; see Figure 5.5), and (c) the aurorae are visible to the observer (nighttime conditions are ideal—cloud cover and 24-hour daylight in the summer inhibit observation).

With the realization of the very ordinary conditions under which aurorae occur, and the sporadic conditions under which they are observed, it becomes clear why these pieces of the Sun–Earth connections puzzle were so difficult to assemble. From Earth, the Sun appeared static, constant, unchanging. Aurorae appeared at night, seemed harmless, and were a reminder of the many unknown manifestations of the power of nature.

Naked-eye observations of sunspots were reported in China starting in the 10th century (Hayakawa et al., 2016). In the West, J. Fabricius was the first to publish a report of sunspots at the end of 1611, and shortly thereafter Galileo began recording sunspots using a new instrument: the telescope. This represents the dawn of Sun–Earth connections and space weather studies. Although sunspots are an excellent representation of the magnetic potential of the Sun, they are not a good indicator of how much Earth-impacting activity has transpired. The long-term sunspot record initiated by Galileo is invaluable; it helps us understand the Sun's history and variability. But sunspot number is just one part of the puzzle.

It was not until the 19th century that the evidence became clear: the Sun's activity could impact the Earth. It required two components: observations of the Sun's activity, combined with the response at Earth. The evidence began to add up. There were technological components (such as radios and telegraph machines) that provided vital clues that could be connected to the occurrence of aurorae, and there were also improved solar observations. One is left wondering how rapidly auroral (and, hence, space weather and solar activity) research would have progressed if aurorae were a commonly observed phenomenon over heavily populated areas!

To make advances in the understanding of Sun–Earth connections, humanity needed clear evidence of solar activity and of the associated response at Earth. Flares are typically observed in energetic wavelengths (X-rays and extreme ultraviolet), which are blocked by Earth's atmosphere and are only readily observable from space. CMEs are very faint compared to the Sun's emission, and are more readily observed in the outer corona, also from space. However, the outer corona is about 1 billion times fainter than the Sun itself, so a definitive CME observation is a challenge. There are reports of white light (i.e., visible spectrum) flares being observed by astronomers, and CME-like structures during solar eclipses, but there are not enough events to identify an association with any terrestrial response.

5.2.2. The "Carrington Event"

All this changed in 1859, when strong geomagnetic and space weather effects were observed soon after a major solar flare. It is also called the "Carrington event," after one of the astronomers who recorded the series of interactions (Richard C. Carrington and Richard Hodgson). On 29 August 1859, Carrington and Hodgson independently observed a flare in a large sunspot group. This was prior to X-ray and EUV observatories; this flare was so strong that it was easily observable in the visible spectrum. Less than 24 hr later (which is extremely fast, by CME standards), the effects were noticed at Earth. Aurorae were observed at nearly all latitudes over the globe, even at locations near the equator. Telegraph systems, which are prone to space weather because they use conducting wires connected over large distances, were unable to function with the geoelectric field overload; some operators reported receiving electric shocks when they attempted operation. Additional space weather effects were observed, such as strong deviations in magnetic field measurements.

The Carrington event was significant because of two key factors: the magnitude of the storm and onset of technology affected by space weather. Modern measurements that characterize storm magnitude were not available in 1859, so a number of methods have been applied to translate the magnitude of the Carrington storm in the context of more recent events (Baker et al., 2013; Ngwira et al., 2014). Although the estimates can vary, it is overwhelmingly agreed that there has not been as strong an impact on Earth since the Carrington event. It remains the benchmark against which all other major storms are judged.

There is little doubt that there were stronger storms prior to the Carrington event, but a key role was played by the emerging technology of the 19th century, including the improved international communication of the scientific community that allowed them to correlate global auroral observations. The accumulated evidence of solar observations, magnetic field measurements, auroral activity and anomalous telegraph behavior comprised an undeniable body of evidence: activity on the Sun has an impact on Earth.

In later years, measurements and technology both improved. We began to understand that it was not the solar flare, but the CME that traveled the 1.5×10^8 km (93 million miles) to Earth that was responsible for the majority of the observable impacts. Storms that were weaker than the Carrington storm had much larger impacts, because radio communications, power grids, and technology in space were much more sensitive to space weather phenomena. Still, we look to the Carrington event as a landmark event: there was no doubt that Sun–Earth connections exist, and the scientific community took up the challenge of unraveling the physics of this dynamic system.

5.2.3. Improved Observations Lead to the Modern Era

There are many measurements that led to the piecing together of the flare/CME puzzle. Not long after the Carrington event, ground-based observations of solar filaments with spectroheliographs documented their tendency to erupt (for a comprehensive review, see Parenti, 2014). Solar magnetic field measurements from the ground, developed by George Ellery Hale in 1908, have played an important role in understanding solar dynamics. However, it was the advent of observations from space that gave the scientific community access to continuous, multi-wavelength data free of atmospheric effects.

It is important to note that observations from the ground have always provided vital information as well, such as global, synoptic H-alpha imagery obtained over many decades. New ground instruments such as the Daniel K. Inouye Solar Telescope (DKIST) in Hawaii promise to yield many more advances.

As flares were observable long before CMEs were discovered, most of the early papers focused on flares as the driver of solar activity and the subsequent effects observed at Earth. A very noteworthy discovery was the identification of CMEs by Tousey (1973) using the seventh Orbiting Solar Observatory (OSO-7). Although it was known that the solar wind existed, and that it contained transient, varying structures, the definitive observation of a CME led us to investigate the causes and effects of these phenomena. It became clear that certain effects that were attributed to flares could possibly be more clearly explained by CMEs. A famous paper by J. Gosling titled the "Solar Flare Myth" (Gosling 1993) led the community to understand the tendency to believe something parallel to "ontogeny recapitulating phylogeny" – or that the discovery of one phenomenon predating another phenomenon can falsely imply a causal, hierarchical relationship between the two. Gosling argued that the community needed to reevaluate the correlations between flares and certain impacts, because some effects more clearly explained as being driven by CMEs.

Švestka (1995), however, cautioned against an overemphasis on the CME aspect of magnetic energy release. He argued that two-ribbon flares, and the study of other types of "eruptive flares," are key parts of understanding the eruption process.

Finally, Harrison (1996) led the way toward a more tempered approach, arguing that flares and CMEs are two aspects of the "same magnetic disease," and further urged against the use of simplistic terms such as "explosion" to describe the complex, interrelated process of magnetic energy liberation.

Readers are reminded that, given the long history of these studies, many excellent textbooks (e.g., Howard, 2011) and reviews are available. Fletcher et al. (2011), Hudson (2011), Benz (2017), and Webb and Howard (2012) provide more recent and general reviews of flare observations. A comprehensive account of flare physics is given by Tandberg-Hanssen and Emslie (1988), and Priest and Forbes (2000), Shibata and Magara (2011), Priest (2014) discuss in detail the magnetohydrodynamic processes and magnetic reconnection theory of solar flares. Yashiro et al. (2004a) and Vourlidas et al. (2017) cover the observed properties of CMEs from a single and multiple viewpoints, respectively. We recommend the papers by Gopalswamy and Thompson (2000) and Georgoulis et al. (2019) describing the early stages of CME development, and Howard (2006) for a historical overview of CME observations. The following sections give a brief overview of the observational signatures of flares and CMEs, with a particular focus on updating readers with observations in the recent era.

5.3. OBSERVATIONAL SIGNATURES AND DETECTION PROPERTIES

The physics of flares and CMEs is intrinsically linked; as we understand the nature of magnetic energy buildup and release, we will understand these two explosive manifestations of those processes. In this section, we briefly cover the various observations that are used to study flares and CMEs. Most of them involve remote sensing—the collection of photons in the form of images and spectra. As you will read, the scientific community relies on a diverse range of observations, and each of the observations holds vital clues about the nature of the two phenomena.

5.3.1. Flares

The most prominent observational signature of a solar flare is the impulsive enhancement of radiation across the entire electromagnetic spectrum, from radio waves with a wavelength between kilometers and millimeters, through γ rays, whose extremely short wavelength is only a small fraction of the diameter of an atom. The Carrington flare was discovered as a sudden brightening in the visible white light emission from the Sun's photosphere. Nowadays, flare emissions in many wavelength ranges are recorded. Whereas modern imaging observations of flares have revealed breathtaking details down to the scale less than one thousandth of the Sun's radius, spectroscopic observations, including polarimetry measurements, remain the astrophysicist's most powerful tool to diagnose the properties of plasma and the magnetic field in the Sun's atmosphere.

This section will focus on a few observational aspects of flares with space weather consequences, including non-thermal electrons, short-wavelength radiations, and any pertinent relationships with CMEs. Although the non-thermal electrons discussed here refer to those trapped in the flare volume, they often carry a large fraction of flare energy and play a significant role in heating the flare atmosphere, which in turn produces much enhanced EUV radiation acting on Earth's upper atmosphere. In the following, only a limited description is provided to explain the basics, and readers are referred to many existing outstanding reviews for completeness and comprehensive details on flare observations.

Time Profiles

The time history of the flare total emission at a given wavelength is called a flare light curve. Figure 5.6a shows the soft X-ray light curve in the wavelength range 1–8 Å from a large flare, observed by the Geostationary Operational Environmental Satellite (GOES) system. Flare emission at this wavelength is produced by plasma heated to more than a few million kelvin in the Sun's outer atmosphere, the corona. As a convention, this specific measurement of flare emission is used to define the magnitude of a flare: a flare is designated to be A, B, C, M, or X class when its peak soft X-ray flux is in the range of 10^{-7}, 10^{-7}–10^{-6}, 10^{-6}–10^{-5}, 10^{-5}–10^{-4}, $>10^{-4}$ watts per square meter, respectively. For example, the X8.2 flare illustrated in the figure produces 8.2×10^{-4} watts per square meter in the GOES 1-8 Å channel at its peak. Because of the logarithmic scale, X-class flares are much more energetic than C-class flares.

The soft X-ray emission of a flare is usually observed to rise rapidly on timescales on the order of minutes, and then decay more gradually, sometimes lasting for hours or even days. Traditionally, we consider the flare to be in the impulsive phase during the rise of its soft X-ray emission, and in the gradual phase after the peak of the soft X-ray emission. Significant hard X-ray and microwave emissions, which are indicators of the presence of electrons accelerated to a fraction of the speed of light, are often produced in the flare's impulsive phase (Figure 5.6b). The wavelength, or energy, of hard X-ray photons is described in units of electron volts (eV). A large flare may generate many hard X-ray photons of energy from a few tens to a few hundred keV (kilo- electron volts), as shown in the figure. Microwave emissions are produced by the interaction of fast electrons with plasma as well as the magnetic field. Here, we start to use "frequency" in units of Hz to describe the energy of microwave photons, and the figure shows the microwave emission of the large X8.2 flare at a frequency of 13 GHz (or 13 billion Hz), which carries, at its peak, a total flux of more than 10,000 solar flux units (sfu).[1]

Following the X-ray and microwave emissions, many ion lines have flare emission in the EUV wavelengths (Figure 5.6c). These are produced by the flare plasma in the corona that cools off from more than 10 million kelvin. A flare also gives off abundant visible and ultraviolet line emission as well as enhanced continuum emission, usually produced at lower temperatures (below 100,000 kelvin) from the impulsively heated chromosphere or even photosphere. Although the amount of the Sun's radiation energy in the short wavelength is rather insignificant in comparison with its visible output, a large number of EUV photons will cause severe ionospheric responses. For this reason, over the past decade, the EUV Variability Experiments (EVE; Woods et al., 2012) has been in operation to monitor the Sun's EUV output, which increases by a few orders of magnitude during flares (Chamberlin et al., 2012; Woods, 2014).

It is worth noting that many kinds of radio emissions can be produced by or in association with a flare (see review by Pick & Vilmer, 2008). Energetic particles escaping into interplanetary space cause decimetric- or metric-type radio bursts, for example, the type III radio burst by flare-produced energetic electrons. Accompanying the X8.2 flare shown in Figure 5.6, the WAVES instrument onboard the *STEREO* (Kaiser

[1] *Solar flux units describe the wave power received by a telescope detector per unit area per unit frequency range. One sfu is 10^{-22} watts per square meter per Hz.*

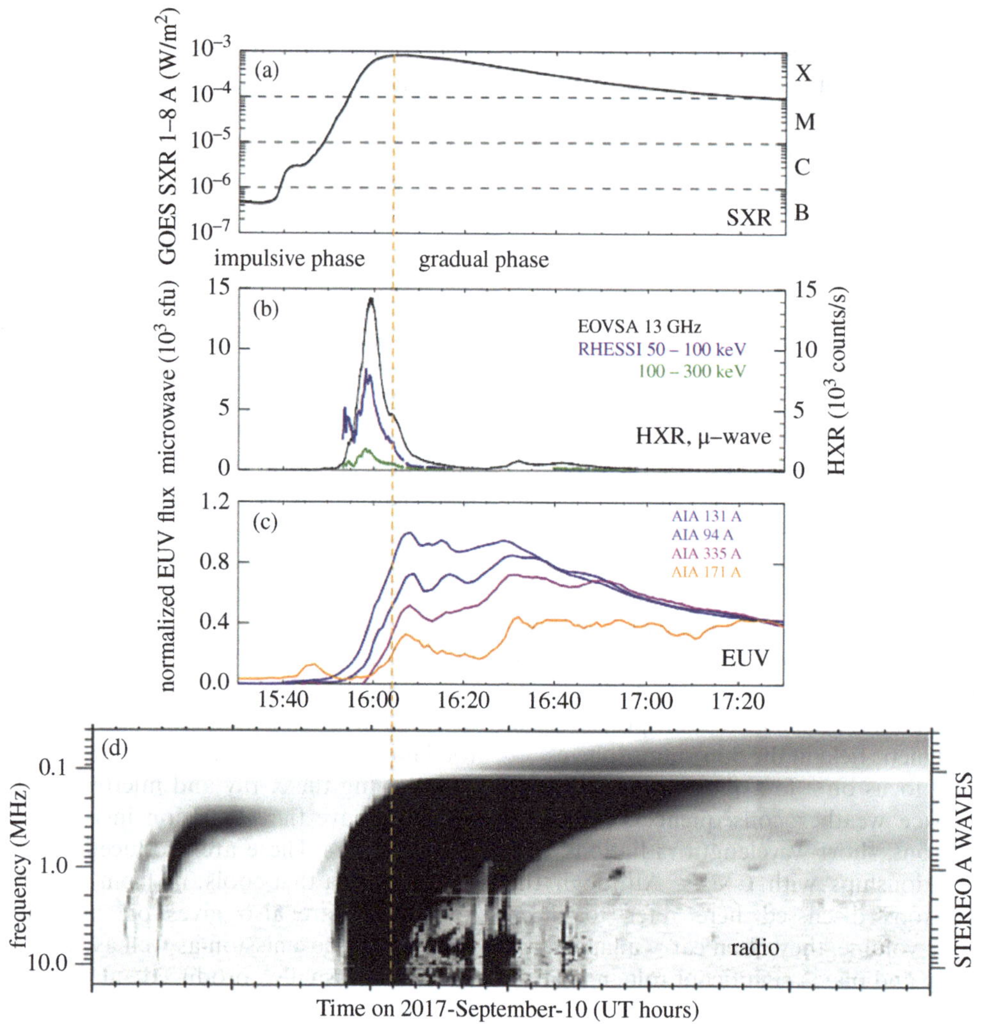

Figure 5.6 Light curves of an X8.2 flare on 10 September 2017, including (a) the soft X-ray light curve by GOES at 1–8 Å, (b) microwave light curve at 13 GHz by the Extended Owens Valley Solar Array (EOVSA; Gary et al., 2013), and hard X-ray light curves in two energy ranges by the Reuven Ramaty High Energy Solar Spectroscopic Imager (RHESSI; Lin et al., 2002), and (c) light curves of extreme-ultraviolet emissions in several passbands observed by the Atmospheric Imaging Assembly (AIA; Lemen et al., 2012) onboard the *Solar Dynamics Observatory* (*SDO;* Pesnell et al., 2012). (*Source:* Courtesy Dale Gary, Chunming Zhu; also see Gary et al. [2018].) (d): a strong type II radio perturbation is observed by the WAVES onboard the *STEREO* (Kaiser et al., 2008) satellite. (*Source:* secchirh.obspm.fr.)

et al., 2008) satellite has detected a strong type II radio burst with its frequency drifting from 10 MHz to 0.1 MHz, a signature of electrons accelerated by CME-driven shocks moving through interplanetary space. So the flare occurs together with a very fast CME, which is rapidly accelerated during the impulsive phase of the flare (Veronig et al., 2018). Such close synchronization between the impulsive rise of flare emission and CME acceleration has been illustrated in a good number of, though not all, events, suggesting a more than casual relationship between the two (e.g., Zhang et al., 2001; Gallagher et al., 2003; Qiu et al., 2004; Temmer et al., 2010; Patsourakos et al., 2010). Observations have further revealed the presence of a flux rope cavity (Long et al., 2018; see Section 5.3.2 and Figure 5.11 for CME structure), significant wave transients (Liu et al., 2018; see Section 5.3.2.2 for wave transients), solar energetic particles (SEPs; Guo et al., 2018; see Chapter 8), and ground-level enhancement events (GLEs; Gopalswamy et al., 2018; see Chapter 8). So, like most X-class flares (Yashiro et al. 2004b), the illustrated X8.2 flare is one symptom of the "same magnetic disease" (Harrison, 1996) with another symptom being a fast CME, amid many others.

Spatial and Spectral Properties

Imaging observations have shown that a flare comprises a cluster of so-called flare loops,[2] which are plasma tubes believed to be confined by the magnetic field. These are referred to as magnetic loops, largely residing in the tenuous corona with both ends anchored in the denser lower atmosphere. Therefore, in contrast to their outgoing CME sisters, flares are often described to be "closed," "confined," or "local" structures.

The aforementioned soft X-ray emission is from the coronal part of flare loops, where plasma is heated to more than ten million kelvin. When heating is attenuated and coronal plasma cools down through a sequence of decreasing temperatures, emissions of ion lines sensitive to these temperatures arise sequentially. This is shown in Figure 5.7, where a series of Fe XXI (131 Å), Fe XVIII (94 Å), Fe XVI (335 Å), Fe XIV (211 Å), Fe XII (193 Å), and Fe IX (171 Å) lines, respectively sensitive to temperatures of 10, 6, 3, 2, 2, and 1 million kelvin, become dominant emissions in an arcade of flare loops observed by AIA. More precise diagnostics of plasma temperatures can be achieved from high-resolution spectroscopic observations obtained by spectrometers such as the EVE (Warren et al., 2013; Caspi et al., 2014; also see the right panel of Figure 5.7 for some EVE spectra). These observations have helped establish our understanding of the heating and cooling sequence of coronal plasma in flare loops.

Analysis of these spectral line emissions also allows us to find the amount of plasma emitting at a given temperature (see Feldman et al., 1992; Del Zanna & Mason, 2018; for comprehensive tutorials of plasma diagnostics with EUV observations). Simply put, the more plasma is emitting, the more intense the line becomes. The temperature distribution of non-isothermal coronal plasma, which is often the case during flares, is accounted for by a property called the differential emission measure (DEM). $DEM(T) = n^2(T)dV/dT$, where n is the electron density, T the temperature, and dV the elementary volume of the plasma. The observed intensity of a specific spectral line is related to this property by $I_\lambda = \int R_\lambda(T)DEM(T)dT$, where $R\lambda(T)$ is the combined effect of the emissivity of the line, known from atomic physics (e.g., the CHIANTI atomic database by Dere et al., 1997, and later development), and the response function of the observing instrument. Advanced inversion methods have been developed to derive the DEM of flare plasma from observations made at many lines sensitive to a variety of temperatures (e.g., Hannah & Kontar, 2012; Cheung et al., 2015), and the total amount of emitting plasma is then given by $EM = \int DEM(T)dT$. Interestingly, the same technique applied to coronal dimmings (signatures of a CME; see Section 5.3.2.2) yields the emission measure that is missing from the Sun's corona (Figure 5.7), as it is carried away by an erupting CME! When we do these, it comes as a surprise that, given its "closed" and "confined" nature, the emission measure of an average flare, usually on the order of 10^{47-50} cm^{-3} (Aschwanden et al., 2015), is indeed comparable with that of a large CME (Aschwanden, 2017). Imaging observations give us an educated guess of the flare volume, and it then turns out that the average density of the flare coronal plasma is quite high, greater than that of the background flare-less corona by two orders of magnitude. Then, how does the flare corona become so dense apart from being hot?

The answer lies in the lower atmosphere—the chromospheric density is 10,000 times higher, and could easily act as a mass reservoir for the tenuous corona. The prevailing understanding is that the chromosphere is impulsively heated at the start of the flare, giving rise to a sudden increase of the pressure there, which drives upflows and therefore transports a considerable amount of dense plasma into the corona. This is evidenced by signatures of impulsive and intense optical or ultraviolet emission at the feet of flare loops, prior to substantially increased emission in the corona, as clearly illustrated in Figure 5.7.

Yet the crucial evidence is from spectroscopic observations (see review by Milligan, 2015), which have revealed plasma upflows at speeds as large as a few hundred kilometers per second, into the tenuous corona from the impulsively brightened feet of flare loops. These are accompanied by downflows into still denser atmosphere at a few tens kilometers per second, so as to maintain momentum conservation (Canfield et al., 1987; Zarro et al., 1988). The upward and downward motion, first discovered in the 1980s as the blue shift of hot lines and red shift of cool lines (e.g., Antonucci et al., 1982; Ichimoto & Kurokawa, 1984), has been called "chromospheric evaporation" and "chromospheric condensation" (Canfield, 1986; Fisher, 1989). Graham and Cauzzi (2015, Figure 5.8) have demonstrated some of the most advanced spectroscopic diagnostics of flare dynamics at a high spatial resolution (of scale around 300 km) with the recently launched *IRIS* (De Pontieu et al., 2014) that observes the Sun's lower atmosphere, the chromosphere and transition region, in the ultraviolet spectral range. Observations of this kind remind us that the advent of next-generation, diffraction-limited high-resolution telescopes with high-sensitivity spectrometers, like the Daniel K. Inouye Solar Telescope (DKIST), will certainly increase our understanding of the fundamental scale at which flare energy is released to heat the lower atmosphere and drive chromosphere evaporation (see, e.g., Jing et al., 2016, for fine-scale structures of a flare).

[2] *For consistency in the following text, we use flare "loops" to describe observationally resolvable structures having a cross section of around 1000 km, and the term "strands" or "threads" seen elsewhere may refer to still smaller-scale fine structures resolved by only a few current instruments (see Aschwanden & Peter, 2017, for a discussion of coronal loop width).*

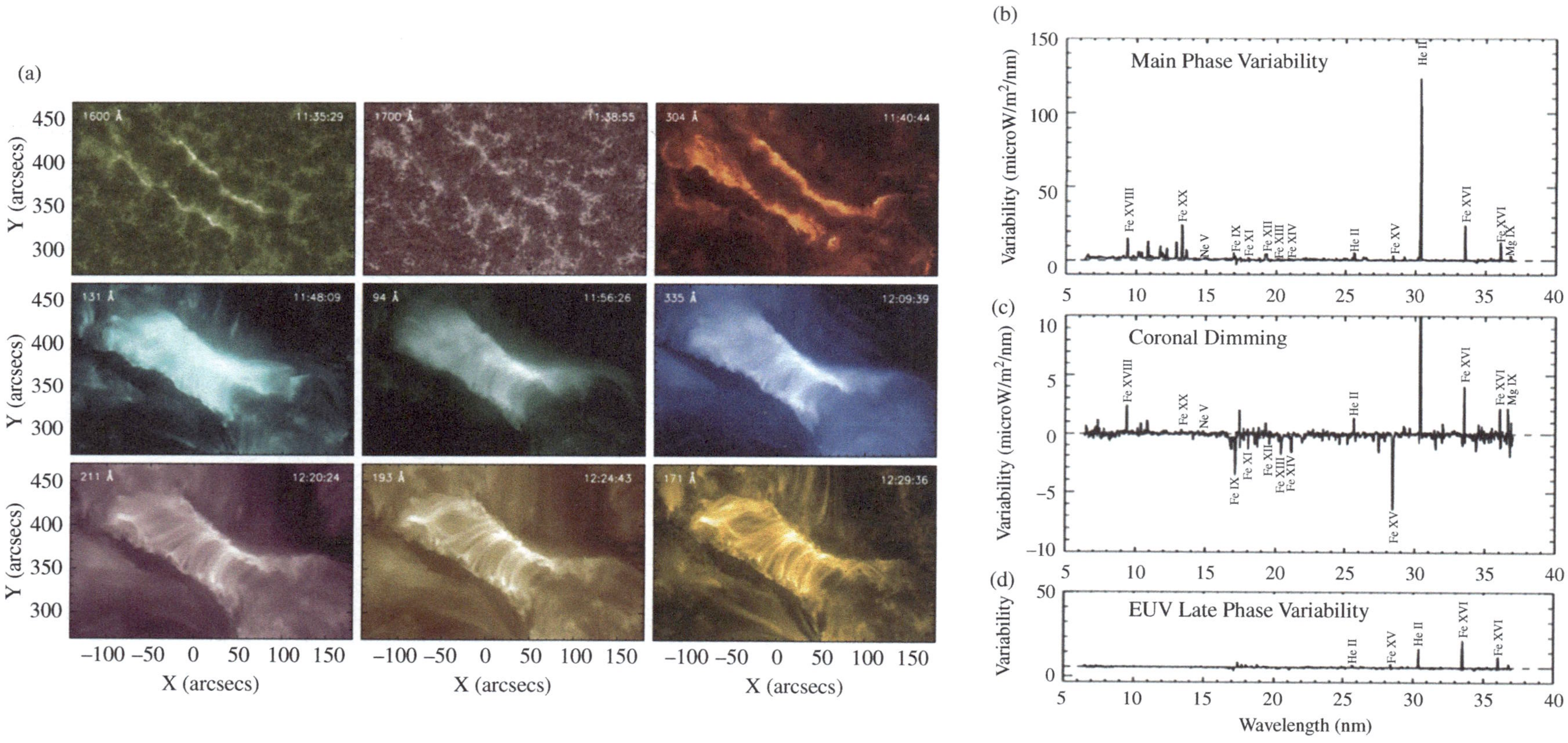

Figure 5.7 Left: An arcade of flare loops of an eruptive two-ribbon flare observed by AIA on *SDO* in nine ultraviolet and extreme-ultraviolet passbands. The top panels show enhanced ultraviolet emission in 1600 Å, 1700 Å, and 304 Å, from the heated chromosphere and transition region, outlining the feet of flare coronal loops, which are brightened subsequently in a series of iron lines sensitive to successively lower temperatures, as shown in the middle and lower panels. (*Source:* Cheng & Qiu, 2016.) Right: EUV spectra by EVE spectrometers showing flare radiation, coronal dimming, and EUV late phase in another event. (*Source:* Woods et al., 2011).

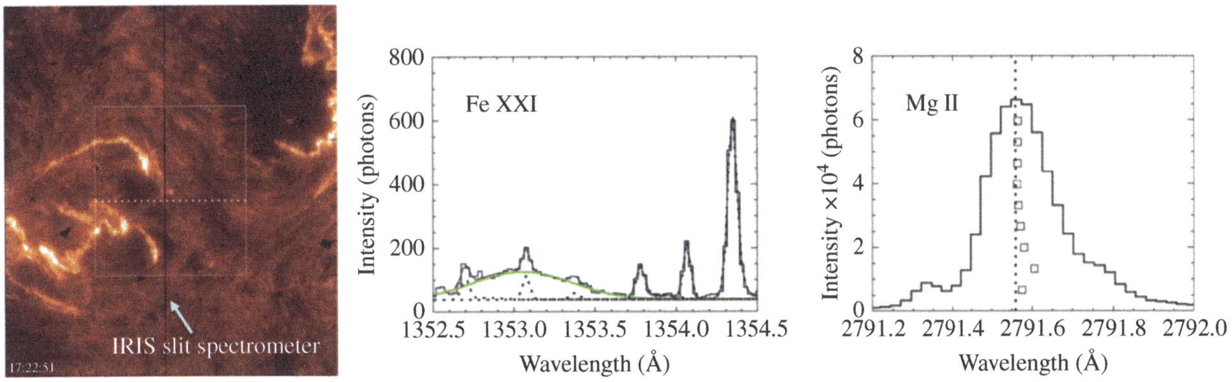

Figure 5.8 Left: Image of a flare with brightened feet of flare loops observed by the *Interface Region Imaging Spectrograph* (*IRIS*) launched in 2013. The dark vertical line shows the position of the slit spectrometer. Right: The Fe XXI and Mg II line profiles taken by the spectrometer at one location on the flare ribbon at one time. The observed hot Fe XXI line is fitted to a Gaussian profile (green), and appears to be blue-shifted from its rest wavelength 1354.1 Å, indicating hot plasma moving upward into the corona (note that dotted curves represent other lines in nearby spectral positions). The rest wavelength of the cool Mg II is at 2791.56 Å, denoted by the vertical dotted line, and black squares mark the central positions of the line from its core to the wing, which are red-shifted from the rest wavelength, indicating plasma downward motion in a complex manner in the varying depth of the chromosphere (Graham & Cauzzi, 2015).

Flare-accelerated electrons can deposit energy in the lower atmosphere. These electrons interact with the ambient plasma and magnetic field to produce hard X-ray and microwave emissions, and the manner of interaction greatly depends on the speed as well as the direction of the electrons, the latter specified by the pitch angle, the angle electrons make with magnetic field of flare loops. Kontar et al. (2011) has explained how hard X-ray observations can be used to deduce electron properties. Dulk (1985) and Bastian et al. (1998) have reviewed the physics of radio emissions in astrophysical systems in general and in solar flares in particular.

Figure 5.9 shows that energetic hard X-ray emissions, or photons of energy more than a few tens of kiloelectronvolts, are concentrated at two places: a compact source at the feet of the flare loop, and a weak yet more extended source above the top of the flare loop (Gary et al. 2018), similar to the "Masuda flare" in Figure 5.1. At both places, electrons collide with plasma, which produce thick-target (of the dense chromosphere) or thin-target (of the tenuous corona) bremsstrahlung radiation of hard X-ray photons.[3] Less energetic X-ray emission of photon energy below 20 keV, which is thermal bremsstrahlung emission, is found to outline hot flare loops, also visible in soft X-ray and EUV emissions.

Hard X-ray spectra can be used to calculate the energy distribution of electrons based on principles governing these kinds of radiations. Figure 5.9 shows that energetic hard X-ray photons exhibit a power-law-shaped spectrum, with fewer photons at higher energies. This is believed to result from the energy distribution of electrons that also follows a power law (Brown, 1971). The observed difference in the photon spectrum between the footpoint, thick-target source and the loop-top, thin-target source provides crucial diagnostics of the physical mechanisms governing the acceleration and transport of high-energy electrons (e.g., Battaglia & Benz, 2007; Krucker et al., 2008, 2011).

Fast electrons subject to the magnetic Lorentz force spiral around magnetic field lines of flare loops, and produce gyrosynchrotron radiation in microwaves (Dulk, 1985). Figure 5.9 shows the location of microwave emissions at varying frequencies from 1 to 16 GHz. In this flare, energetic electrons trapped on or above the top of the flare loop give rise to microwave emissions, and the height of the emission grows with decreasing frequency, indicating that the magnetic field strength decreases with height. Therefore, spectroscopic and imaging observations of flare microwave emissions also provide unique measurements of magnetic field in the flaring corona, which may help diagnose the dynamics of magnetic reconnection (see Section 5.4.1) and particle acceleration and transport mechanisms (Gary et al., 2013).

[3] *For this flare observed on the limb and partially occulted, it is considered that the conjugate feet of flare loops are behind the limb, so that thick-target hard X-ray emission is visible only at the feet in front of the limb.*

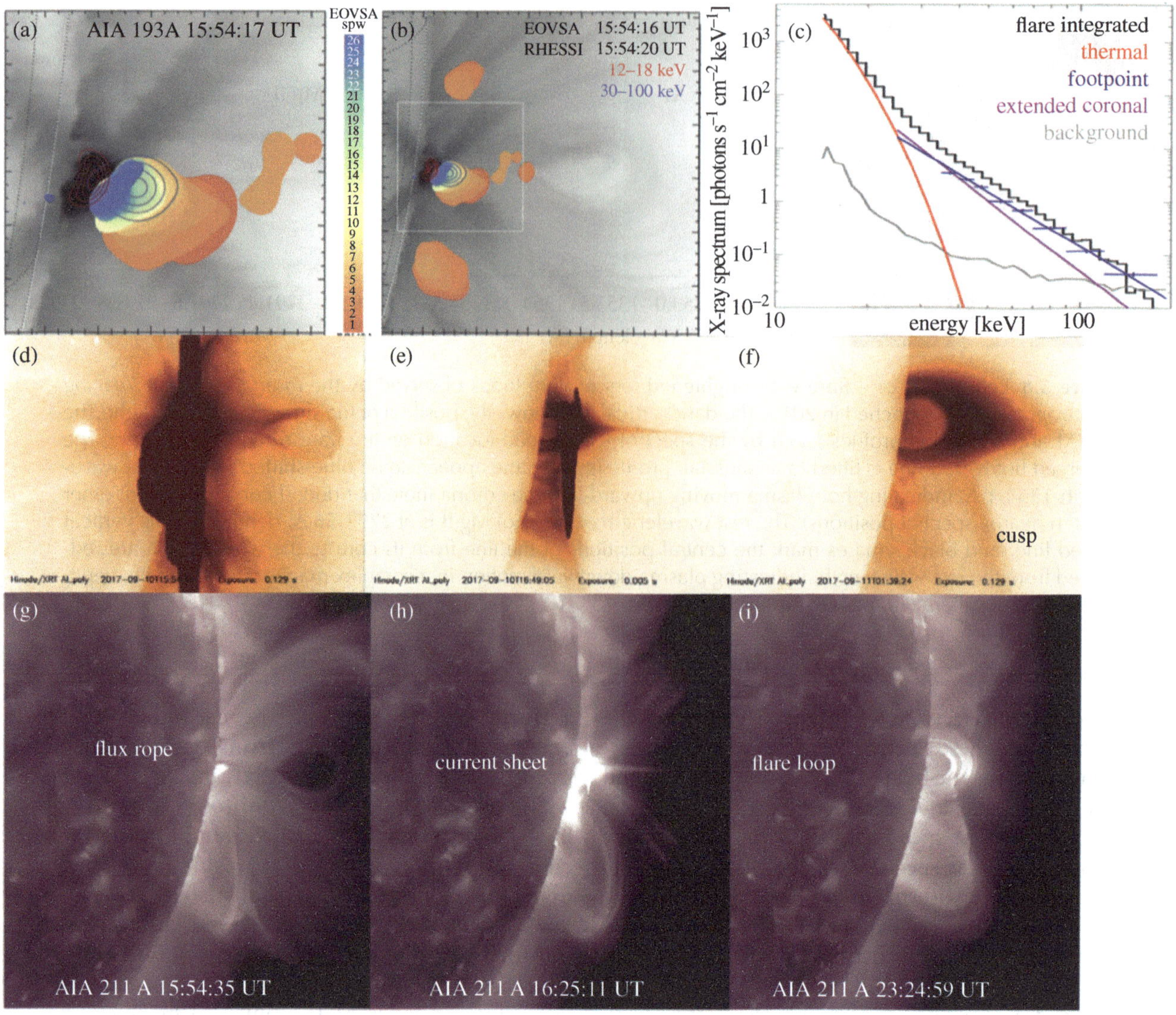

Figure 5.9 Top: X-ray emission (contours) observed by RHESSI at "soft" and "hard" energies, and microwave emission by EOVSA, in the impulsive phase during the X8.2 flare on 10 September 2017, presented in a small (a) and large (b) field of view, respectively. Background images, on reverse grayscale, are obtained from the *SDO*/AIA at Fe XII line at 193 Å, showing an erupting flux rope cavity above the flare (b). The box in (b) shows the field of view of (a) (*Source:* Gary et al., 2018). (c): The spectrum of the flare X-ray emission demonstrating electron thermal bremsstrahlung emission at lower energy, and non-thermal power-law distribution at higher energy of the footpoint source and coronal source, respectively. The solid black line shows the observed total photon spectrum, whereas the color curves show the model fit to different components of the X-ray emission (*Source:* Gary et al., 2018). Middle: soft X-ray emission by the X-ray Telescope (XRT; Golub et al., 2007) on *Hinode* (Kosugi et al., 2007) during the flare's impulsive phase (d), gradual phase (e), when a long thin structure, considered to be a current sheet, is visible trailing the erupted flux rope CME, and many hours after the explosion (f) (*Source:* Courtesy of Chunming Zhu and XRT team). Bottom (g–i): Extreme-ultraviolet images at Fe XIV line at 211 Å by *SDO*/AIA around the same times. These observations exemplify the standard eruptive flare configuration, which will be discussed in the next section.

Energetics

However particles are accelerated, and however plasma is heated, ultimately all flare energy is output in photons, by which flares are observed. Depending on the magnitude of a flare, the bolometric radiation, or the sum of electromagnetic radiation throughout the entire wavelength range, amounts to 10^{23-26} J, which is comparable to the energy of a CME (Emslie et al., 2012; Aschwanden et al., 2017; Aschwanden, 2017).[4] Note that the total energy of a CME includes the kinetic energy as well as the gravitational potential energy. In a flare or a CME, this large amount of energy is released on a timescale on the order of a few tens of minutes. In terms of the energy release rate, flares and CMEs are the most energetic events in the heliosphere.

Having understood the mechanisms of radiation of various kinds, we may go into further details of the energy distribution in a flare. In the corona, X-ray and extreme-ultraviolet spectral diagnostics tell us the temperature distribution of plasmas, so we can calculate the total thermal energy in the corona as a consequence of flare heating (Emslie et al., 2012; Aschwanden et al., 2015). The spectral diagnostic is somewhat more complex in the dense lower atmosphere. It is estimated that the lower atmosphere holds a substantial amount of heat within the budget of flare energy (e.g., Kretzschmar, 2011; Milligan et al., 2014).

Earlier, we discussed the possibility that the flare heating energy is originally carried in fast electrons. Then, how much is this energy? By integrating the electron spectrum inverted from the observed hard X-ray photon spectrum, it is found that a large flare produces a total number of 10^{34-36} electrons per second, and the total energy carried by non-thermal electrons can amount to more than one half of the total radiation energy of a flare (Emslie et al., 2012; Warmuth & Mann, 2016; Aschwanden et al., 2017). If the numbers are correct, it becomes apparent that not all of the heating is done by non-thermal electrons. Flare-accelerated ions may play a role in some large flares (Emslie et al., 2012; Aschwanden et al., 2017). In the next section, some other mechanisms will be discussed.

Magnetism

It has been well recognized that the source of the vast amount of energy, whether in flares or CMEs, is magnetic in nature. This knowledge has been corroborated by yet another critical observation: the polarimetric measurements of the magnetic field in the Sun's atmosphere. Figure 5.10 shows an eruptive M1.9 flare residing in an active region of strong and complex magnetic field, which also hosts a sigmoid (see Section 5.3.2.2). Prior to the eruption, flux emergence, rotation of the leading sunspot carrying positive magnetic field, and strong shear motion of the other polarity have been observed in the active region. These motions have built up strong vertical electric currents (Figure 5.10e), derived as the curl of the transverse magnetic field (indicated by arrows in Figure 5.10d). Through these motions, magnetic helicity, which characterizes the complexity of a magnetic field, has been injected into the active region (e.g., Chae, 2001, and many following studies). The subsequent flare and CME are found to be rooted in the areas of strong electric currents and helicity injection (James et al., 2017, 2018; Wang et al., 2019). Note that this event was observed by *SDO* as well as *STEREO-A* and *STEREO-B*, the latter having a view angle of roughly ±90° from the former. Therefore, whereas the CME magnetic flux rope (MFR) is identified in the *STEREO* coronagraph observations (Figure 5.10b, c), its footprints on the Sun's surface are identified by coronal dimming signatures (Webb et al., 2000; see Section 5.3.2.2) observed by AIA. This event illustrates that magnetic field evolution, through flux emergence, sunspot rotation, and shear motion, builds up strong electric currents and leads to the eruption (e.g., Schrijver, 2007; Liu et al., 2017, and references therein).

A large body of observational studies, with the most recent effort using machine learning methods to analyze vector magnetic field measurements (Bobra & Couvidat, 2015b, Bobra & Ilonidis, 2016b, and references therein), have been dedicated to unraveling what magnetic field characteristics cause flares and CMEs (see, e.g., recent reviews by Green et al., 2018; Georgoulis et al., 2019). In the meanwhile, a successful flare/CME model should lead to an understanding of how magnetic energy is abruptly released and converted to kinetic energy of particles, bulk motion, and heat. Such an understanding will not only improve our capability to predict space weather, but also shed light on mechanisms of energy release in other astrophysical systems like stellar flares and CMEs, corona and jets from accretion disks of black holes, binaries, and proto-stars, and even γ-ray bursts.

5.3.2. Coronal Mass Ejections

CMEs are an important aspect of coronal evolution, involving vast structures of plasma and magnetic fields that are expelled from the Sun into the heliosphere. White light coronagraphs (Figure 5.3) image CMEs as bright features moving outward from the Sun with speeds ranging from tens to >3000 km/s. They can be detected in situ

---

[4] *Readers are reminded that there are flares or flare-like radiation phenomena of much smaller scales, such as microflares, nanoflares, and bright points. The energy of the smallest of these events can be as low as 10^{17} J (Hannah et al., 2008), depending on the detection threshold.*

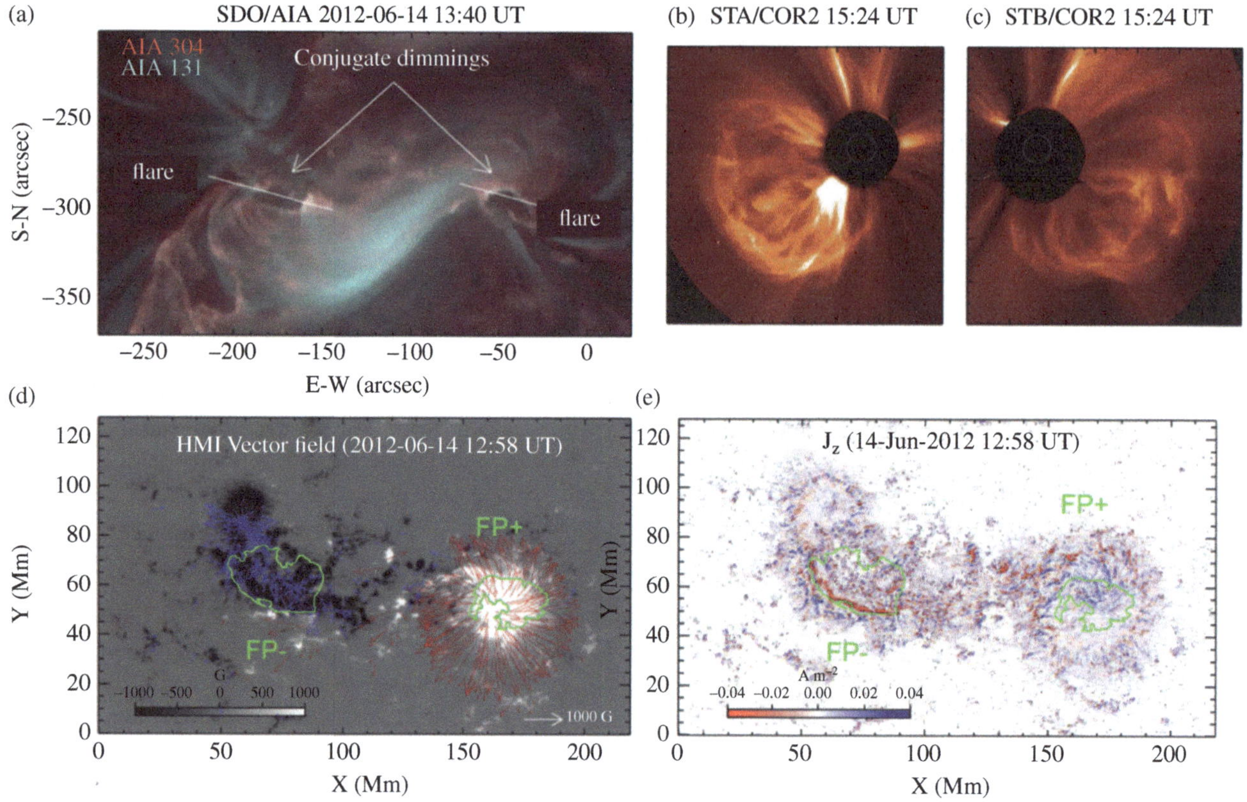

Figure 5.10 (a) An M1.9 flare observed on 14 June 2012 by *SDO*/AIA in 304 Å (red) and 131 Å (cyan) bands. The M1.9 flare is associated with a halo CME, which was captured by *STEREO*/COR2A (b) and *STEREO*/COR2B (c) from different perspectives. (d) A vector magnetogram obtained by the *SDO*/Helioseismic and Magnetic Imager [HMI Scherrer et al., 2012]; grayscale shows the longitudinal magnetic field, and arrows indicate the transverse field. (e) The vertical current density calculated from the transverse magnetic field, with the positive (negative) values colored in blue (red). Two contours (green) in (d) or (e) outline the two conjugate feet of the erupting CME magnetic flux rope, which are identified by conjugate dimmings denoted in (a). (*Source:* Wang et al., 2019; also see James et al., 2017, 2018 for this event.)

by their anomalous (compared to average solar wind) plasma and magnetic field characteristics (e.g., Zurbuchen & Richardson, 2006). Much of the plasma observed in a CME is entrained on expanding magnetic field lines, which can have the form of helical field lines with changing pitch angles, that is, an MFR. In fast CMEs, the plasma is supersonic so a detached shock forms ahead in the ambient solar wind. These shocks, in turn, can accelerate particles producing the largest SEP events (see Chapter 5) that are associated with major interplanetary disturbances and hazardous effects at Earth.

As stated previously, CMEs can be difficult to detect. Energetically, the strongest flares and CMEs are comparable: 10^{25} Joule. However, compared to the star itself, CME signatures can be very subtle. The total solar luminosity is 3.8×10^{25} Joule per second, whereas the CME frequency rarely exceeds ten per day. Even the outflow of the solar wind, integrated over the solar cycle, rivals the energy of flares and CMEs combined (Low, 2001). However, the solar wind is relatively steady, and Earth can adjust and reach an equilibrium. Conversely, impulsive transients in the solar wind driven by CMEs can deposit large amounts of energy over a relatively short period, and the concomitant response of Earth's field can drive geomagnetic currents and other space weather effects.

Even the largest CME has a small, almost inconsequential, magnetic field when compared to that of a sunspot. So how does a small ejection of flux from the solar surface have such a large impact in the corona? The reason is the large gradient in magnetic field pressure from the surface to the inner heliosphere; the magnetic pressure at the solar surface (0.5 R_{Sun}) falls off rapidly, so a CME "footnote" in the inner corona can become a dominant factor at higher altitudes (> 2 R_{Sun}). Magnetic fields on the Sun are typically tens to hundreds of gauss, and strong sunspots can exceed several thousand gauss. However, in space, the magnetic field is typically measured in nanotesla (10^{-5} gauss).

The key to detecting and observing CMEs is understanding where they have their strongest influence. In the upper corona (above 4 R_{Sun}) or the inner heliosphere (above the Alfvén surface, i.e., above about 15 R_{Sun}), local magnetic structures such as sunspots have evolved into a global structure consisting of streamers and coronal holes. The Alfvén surface is the locus where the outward speed of the accelerating solar wind passes the radial Alfvén speed; therefore, any displacement of material cannot carry information back down into the corona (e.g., DeForest et al., 2014). Thus, it is the natural outer boundary of the solar corona and the inner boundary of the heliosphere. Streamers and coronal holes are similar in that they are radially aligned, but coronal holes tend to be darker (because the plasma population is not strongly emitting in most spectral ranges, and the electron density is low), and streamers tend to be bright (plasma is more energized and emitting, and convergence of electrons in plasma flow are more effective in scattering). Both features stand in contrast to CMEs, which are more likely to take the form of easily distinguished structures (Figure 5.3). The fact that the CME is in motion, whereas the other structures are relatively static (unless disturbed by the CME), distinguishes CMEs further. However, these diagnostics are not always clear; a weak CME, or one that is not captured well by the available observations, can be difficult or ambiguous to interpret.

Usually a CME is observed via its impact on the solar atmosphere, such as evacuation of mass from the lower corona, injection of mass into the upper corona and heliosphere, dramatic magnetic field evolution, and strong interaction with the surrounding corona. Unfortunately, there are no established criteria to definitively identify a CME. Similar to flares, there are "sufficient" criteria, which when satisfied an observer may say that a CME definitely occurred. Certainly, a bright, outward-moving, loop-like structure as shown in Figure 5.3 is unambiguous evidence. However, there are no "necessary" criteria: The loop-like structure may not be observed, or the CME's direction relative to the images may make it impossible to deduce a moving structure. There are observations, and associated phenomena, that may suggest that a CME or eruption occurred.

Outer Coronal Observations

The "sufficient" observational condition for a CME is provided by a coronagraph, which is a type of imager that works as an artificial eclipse; it blocks out the Sun so that the surrounding atmosphere (which is many orders of magnitude fainter) can be observed. Coronagraph designs vary, but most extend to at least several solar radii in altitude. By this height, the solar features become more radial, and CME detection becomes less ambiguous (Figure 5.3).

Coronagraphs observe the evolution of the outer corona and the motion of density structures, typically by detecting Thomson-scattered sunlight from the free electrons in coronal and heliospheric plasma. This emission has a non-isotropic, angular dependence, which must be taken into account when interpreting the data (e.g., Billings, 1966; Vourlidas & Howard, 2006; Howard & Tappin, 2009).

The two CMEs visible in Figure 5.3 exhibit a range of features. The first CME, though fainter, is more pronounced in extent (Figure 5.3b). The primary loop of the CME is visible in the upper-right (northwest) portion of the image, but there is another front that is observed moving to the left (eastward). The latter CME also appears in the northwest (Figure 5.3c). Though more localized, is much brighter and more distinct. It has a bright outer loop, and within it a bright "core" structure.

Figure 5.11, an illustration from Hundhausen (1999), is based on a loop-like CMEs observed during 17–18 August 1980 by the coronagraph/polarimeter on the *Solar Maximum Mission*. The arrows indicate the direction of the magnetic field. The structure begins as a quasi-quiescent helmet streamer (first panel), expands as the interior magnetic cavity grows (second panel), erupts (third panel), and then evolves into a post-eruptive current sheet structure (last panel).

When compared to Figure 5.3, it is clear that some CMEs do exhibit this "classic" structure, but perhaps not all. Even authors of the earliest papers on CMEs recognized that there were two issues at play: (1) that not all CMEs had the same structure and dynamics, and (2) the angle of observation could have a strong impact on how a CME appeared in the image. In other words, was the earlier CME fainter, or was the second CME simply better aligned with the image direction to capture its full extent? Howard et al. (1985), surveying data from the Solwind coronagraph, described a large variety of structures: curved front, halo, complex, single spike, streamer blowout, and diffuse fan CMEs. The huge range of manifestations could not be explained by data ambiguities alone. However, they argued that the angle of observation was a necessary consideration. CMEs with "halo" structures (eruptions that appeared with emission surrounding the occulting disk) were most likely due to a CME traveling along the observer's line of sight (i.e., toward or away the observer), rather than a CME exploding 360° in all directions.

Later studies, based on coronagraphs viewing from multiple angles (the *STEREO* SECCHI coronagraphs [Howard et al., 2008] combined with the *SOHO* LASCO coronagraph [Brueckner et al., 1995]) confirmed that the varying CME manifestations were indeed a combination of the two effects. Vourlidas et al. (2017) surveyed thousands of eruptions, and strove to classify them more clearly with respect to their physical nature and their observed kinematics.

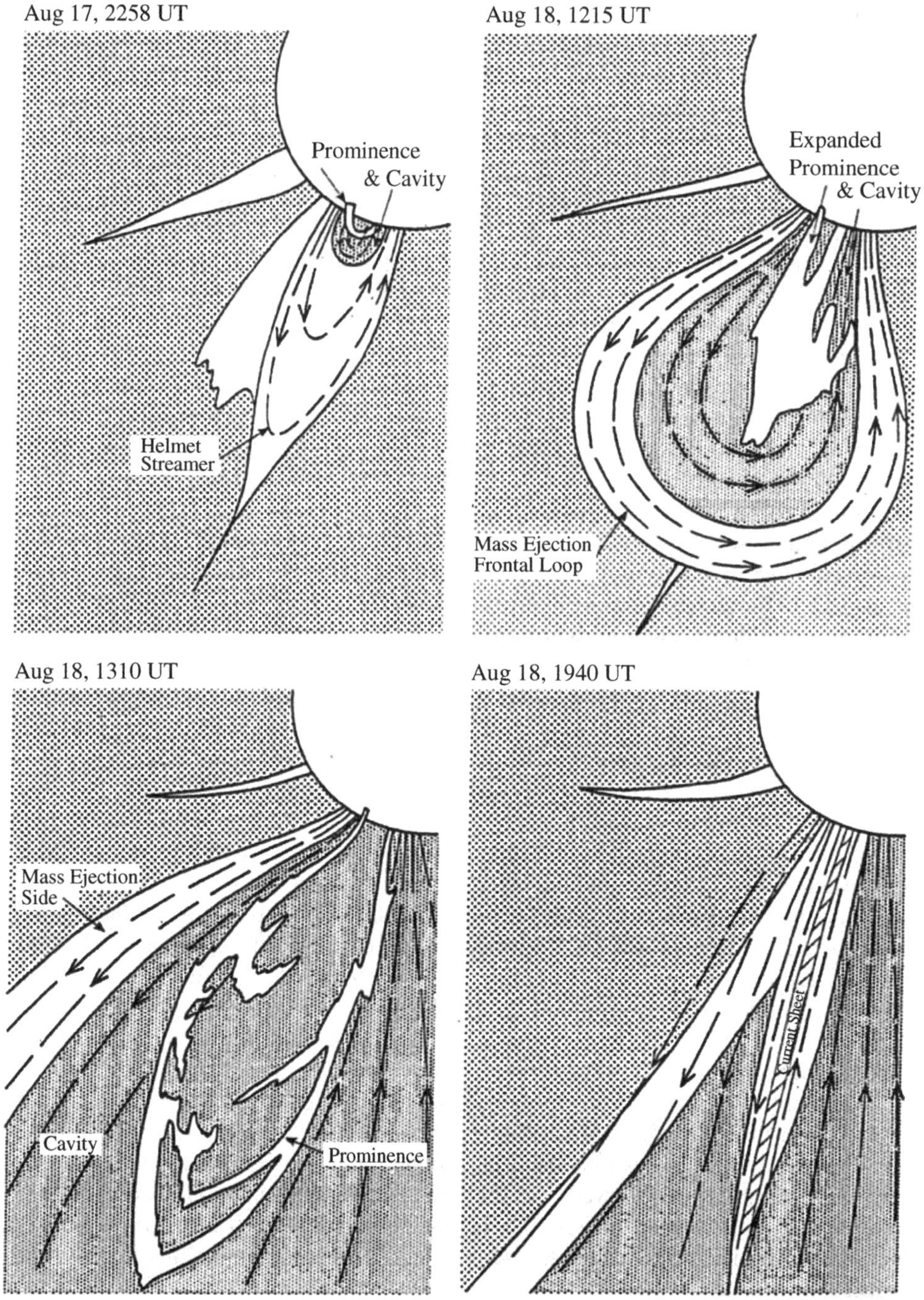

Figure 5.11 Early observations provided evidence that many CMEs have a distinctive structure, consisting of a bright loop-like front, dark cavity, and dense (bright) core. (*Source:* Figure from Hundhausen (1999), illustrating the structure of a CMEs observed during 17–18 August 1980.)

However, even multiple viewing angles are limited in their ability to deduce 3D structure. Figure 5.12 from Colaninno et al. (2013) illustrates how error or variation in identifying a point from the viewpoint of *STEREO-A* (blue lines), when combined with the ambiguity from *STEREO-B* (red lines), maps into a wide range of locations, particularly in the inner heliosphere (yellow shaded regions). The range of altitudes in the yellow shaded region is roughly proportional to the error in position relative to the two spacecraft. However, the range of possible *longitudes* is surprising to many.

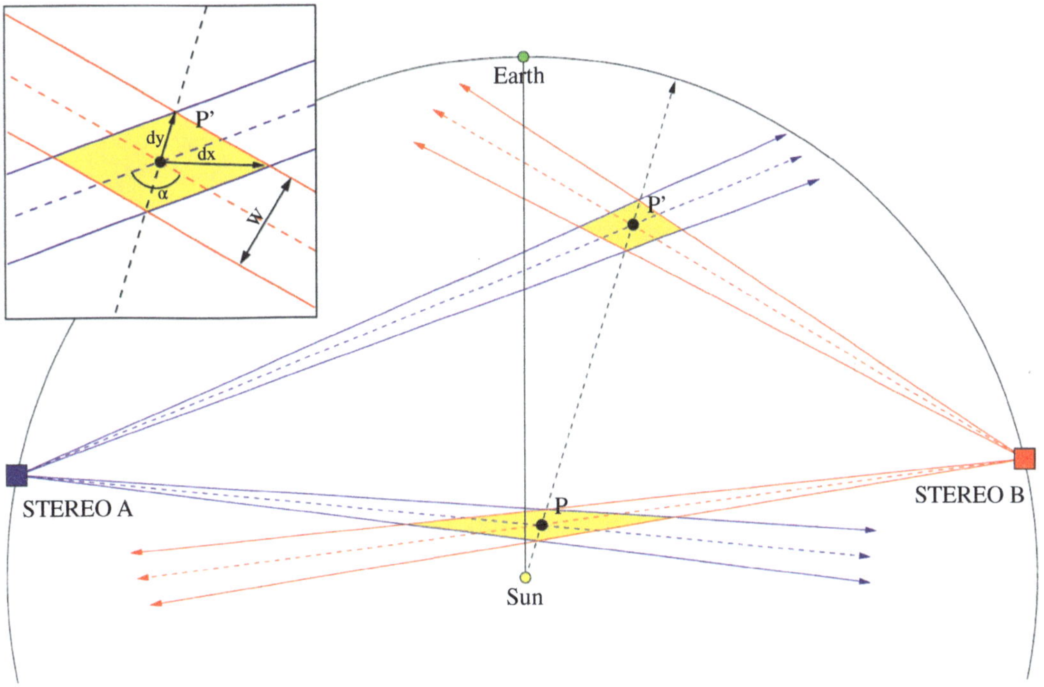

Figure 5.12 This diagram, from Colaninno et al. (2013), illustrates a difficult aspect of reconstructing a 3D CME structure using remote sensing observations. A small variation in the apparent observing angle can project to a large range of longitudes, particularly at locations close to the Sun.

Other efforts have shed light on the subtle, but important three-dimensional properties of CMEs. For example, Kwon et al. (2015) demonstrated that there were an unusually high number of "halo" CMEs if they were considered a function of the viewing angle alone. Some CMEs exhibit a halo-like structure surrounding their central emission from every direction, indicating that some halos truly were a bright structure surrounding the core CME from all directions. Thernisien et al. (2006) and Thernisien (2011) set the stage for an interpretive flux rope CME tool that allowed the users to view a CME in the context of a consistent magnetic structure.

Another type of telescope yielding outer coronal observations of CMEs is heliospheric imagers, beginning with *Helios* and the *SMEI* (Jackson et al., 2004) and culminating in the *STEREO*/SECCHI HIs (Harrison et al., 2005; Eyles et al. 2009), *Parker Solar Probe's* Wide-Field Imager for Solar Probe Plus (WISPR; Vourlidas et al., 2016) and Solar Orbiter's Heliospheric Imager (SoloHI; Howard et al., 2019). Their extended fields of view have allowed observers to track CMEs far from the Sun, study their evolution, and track their propagation through the heliosphere. Howard (2011) gives a comprehensive overview of many observations and how they relate to the physical processes behind CME physics, and Green et al. (2018) incorporates multidimensional observations for a fully three-dimensional understanding of CME origin and initiation.

Inner Coronal Observations

There are a variety of observations in the inner corona that, when combined with coronagraph observations, shed light into the origin and structure of CMEs (e.g., Thompson et al., 1999b, Gopalswamy & Thompson, 2000, Hudson & Cliver, 2001).

However, not all CMEs have inner coronal signatures (Howard & Simnett, 2008, Robbrecht et al., 2009)—the name "stealth" CME has been given to those that lack any accompanying solar surface activity. Most CMEs do have inner coronal signatures, though they have a wide variety of manifestations.

Flares, arcades, and sigmoids As flares and CMEs are related, it should not be surprising that some of the key indicators of a CME in the lower corona are provided by flare observations. Arcade formation (as shown in Figure 5.7) provides a reliable diagnostic of the evolution of magnetic structures involved in a flare and CME (Hanaoka et al., 1994; Hudson & Webb, 1997). The CME represents a topological change, and these arcades are often attributed to the re-closing of magnetic field lines and the energization of trapped plasma along the loops. In addition, Yokoyama and Shibata (1998) discussed the ability of flare reconnection to energize adjacent, previously existing arcade loops. These correspond to the soft X-ray "long-decay enhancement" (LDE) flares, which are one of the soft X-ray signatures of a CME (e.g., Hudson & Webb, 1997).

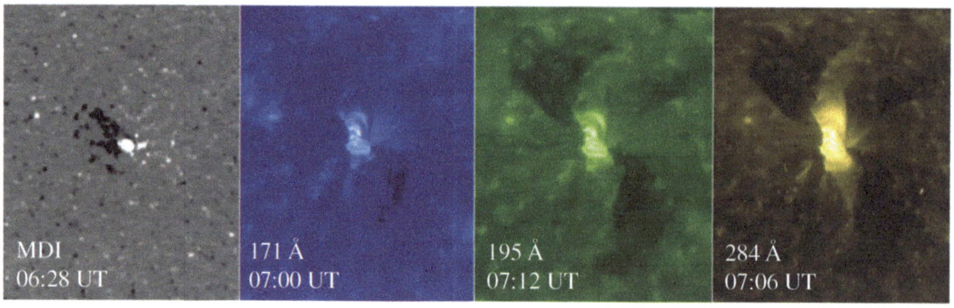

Figure 5.13 Magnetic field and EUV coronal images of a large-scale, post-eruptive dimming on 12 May 1997. First panel: *SOHO* MDI magnetic field, showing strong flux in the active region but weak prevailing unipolar flux in each of the dimming regions. The dimmings in coronal wavelengths show fairly symmetric "twin" regions. Blue: 171 Å, green: 195 Å, yellow: 284 Å. (*Source:* Figure from Thompson et al., 1998).

Canfield et al. (1999) discovered that soft X-ray "sigmoid" eruptions, named for their S-shaped structure prior to eruption, is another type of arcade signature that correlates strongly with the presence of a CME. The S-shape, which forms due to the large amount of shear built up in the magnetic field, precedes an eruption; after the eruption, the field evolves to a more relaxed, laminar arcade.

Many authors (Sheeley et al., 1975; Webb et al., 1976; Rust & Hildner, 1976; Kahler, 1992; Hudson et al., 1996) have worked toward clarifying the relationship between LDEs, accompanying H-alpha "two-ribbon" flares, sigmoids, and CMEs. The inner coronal observations confirm these results, and the flare/arcades fitting these descriptions, when observed, are a very strong indicator of a CME. The converse is not necessarily true, in that a CME can occur without the accompanying arcade formation; however, this may correspond to a more gradual reconfiguration of magnetic field which does not result in an observable enhancement.

Coronal dimmings EUV and soft X-ray dimmings of the corona, occurring on timescales ranging from minutes to hours, are another reliable signature of a CME. Such regions were called "transient coronal holes" in *Skylab* data and can appear as low coronal regions of reduced emission in EUV and soft X-ray emission.

Rust (1983) examined the brightest *Skylab* LDE events, and determined that a majority showed evidence of soft X-ray "voids" forming in association with the eruption. The voids were observed near the erupting region, either alone or in pairs, such as those shown in Figure 5.13. In general, these coronal voids lasted less than 48 hr and could decrease in emission until they were as dark as coronal holes.

Rust reasoned that because the interplanetary shocks associated with eruptions "are followed by a sustained period of high-speed solar wind, and recurrent high-speed solar wind streams originate in long-lived coronal holes (Hundhausen, 1972), then the transient solar wind speed increases might stem from transient coronal holes." Hudson et al. (1998) determined that the dimming of the soft X-ray corona associated with a CME proceeded too rapidly to be explained by radiative or conductive cooling, leading to the conclusion that the observed decrease in emission was due mostly to the outflow of material, and not just a shift in temperature. They chose the more empirical label "dimming" for this phenomenon.

Dimmings studies benefit from a direct comparison of the "transient" coronal hole dimmings with a more static coronal hole. Photometrically, the emission recorded in EUV and soft X-rays are comparable to that of a coronal hole (e.g., Zarro et al., 1999), a strong indication that the decrease in emission is primarily due to a decrease in density. In addition, Harra et al. (2007) demonstrated that coronal dimmings can exhibit Doppler shifts indicative of outflows, consistent with transient coronal hole behavior.

Sterling and Hudson (1997) demonstrated that the degree of soft X-ray dimming observed by Yohkoh SXT during a CME can account for only a fraction of the total mass of the white light CME. Thompson et al. (1998) argued that the degree of EUV dimming does not scale proportionally with the mass of the observed white light CME because a significant fraction of the CME mass source can originate from regions that are not strongly emitting in EUV or soft X-rays, such as at higher altitudes. Harrison and Lyons (2000) included spectroscopic measurements, but Aschwanden (2016) demonstrated that a full emission measure treatment of dimmings in the EUV could account for the mass deficit.

An extensive survey of coronal dimming events and their relation to CMEs was performed by Reinard and Biesecker (2008). They found mean lifetimes of 8 hr, with most disappearing within a day. Reinard and Biesecker (2009) studied CMEs with and without associated dimmings, finding that those with dimmings tended to be faster and more energetic. Bewsher et al. (2008) used SOHO/CDS and LASCO data and found a 55% association rate of dimming events with CMEs, and that most CMEs were associated with dimming events. Coronal dimmings have not been observed as frequently as other associated eruptive phenomena, but the recent, very sensitive results (e.g., Schrijver & Title, 2011; Woods et al., 2011; Mason et al., 2016) from *SDO*/AIA and *SDO*/EVE imply that dimmings are more common than previously thought. More recently, Krista and Reinard (2017) analyzed over 150 separate dimming events, and discussed their significance as the magnetic footprint of CMEs, concluding that their magnetic evolution provides essential clues of the evolution of CME eruptive topologies.

Tian et al. (2012) covered additional important factors regarding spectroscopic observations of CMEs and dimmings, shedding light on the outflow of material and mass of the evacuated material. Mason et al. (2016), Aschwanden et al. (2017), and Krista and Reinard (2017) strove to connect the statistical properties of dimmings to the properties of the associated CMEs, namely their mass and velocity, finding some good correlations. Dissauer et al. (2018) investigated the relationship of dimmings to flares, then to CMEs (Dissauer et al., 2019), combining multi-viewpoint observations from the *STEREO* mission with those along the Earth–Sun line. These 3D studies highlight the role of the magnetic field of a dimming region, allowing a closer examination of flare and CME manifestations of eruptions.

Filament eruptions The presence of an erupting filament or prominence (a filament seen off the solar limb—hereafter we will refer to either as a filament) is a fairly reliable indicator of a CME (Figure 5.4, Prominence Eruption).

As stated earlier, spectroheliograph observations of filaments began at the end of the 19th century, and they were quickly recognized to be evidence of the Sun's highly volatile nature. Filaments can be extremely compact (especially filaments in an active region) or can extend for over a solar radius in length. They can remain relatively static, or quiescent, for long periods, and as they form along magnetic inversion lines, they serve as excellent indicators of large-scale solar magnetic structure (Figure 5.14). For this reason, a great deal of effort has been dedicated to studying their formation, structure, classification, evolution, and magnetic connections (see Tandberg-Hanssen, 2013; Parenti, 2014; and Engvold, 2015). For the purpose of this chapter, we briefly discuss the role filaments play in an eruption.

Filaments appear as dark masses in Hα images, and are observed in coronal emission lines as a dark profile; the dark features are due to cool dense material shadowing the bright emission behind it. In the cooler EUV 304 Å He II wavelength (typically estimated as representing 5 - 8 $\times$ 10e-4 K), the material can be seen either in emission or absorption (Figure 5.4, Prominence Eruption). Spectrometer observations of absorption features (e.g., Kucera et al., 1998) allow an analysis of the absorption profiles and comparisons.

Filament observations show several key aspects of a CME (Schmieder et al., 1997, 2002). They trace out the base of the pre-eruptive structure, and continue to serve as an indicator of magnetic field evolution throughout the eruption process. Figure 5.15 highlights several important aspects of a filament's relationship to the CME magnetic field. A very clear example of a filament eruption and dimming is described by Webb et al. (2000), who discussed CME properties like the magnetic flux and inferred the associated large-scale magnetic structure; through studies of many events, the picture of how filaments relate to the overall structure of CMEs becomes more clear.

Wave transients Since their discovery by the EIT investigation on the *SOHO* mission (Moses et al., 1997; Thompson et al., 1998, 1999a; Thompson, 1999), *SOHO*/

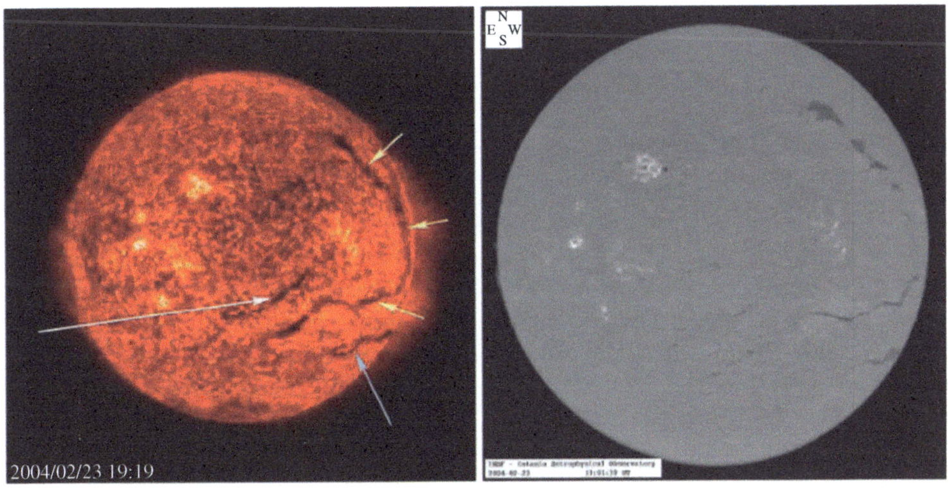

Figure 5.14 Left: *SOHO* Extreme Ultraviolet Telescope 304 Å (He II) image from 23 February 2004 showing several large filaments and their channels. Right: H-alpha image showing the same filament regions. (*Source:* Figure from Parenti, 2014.)

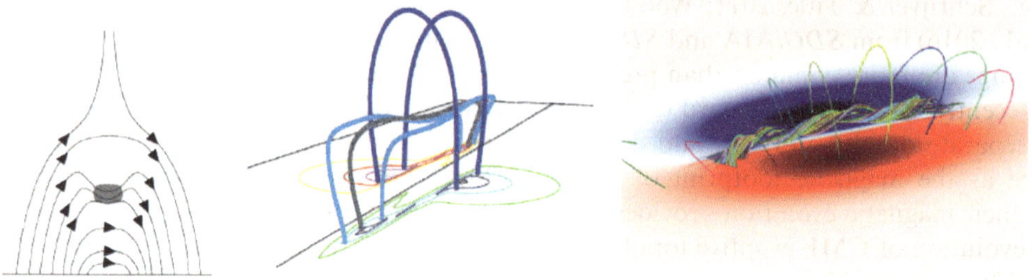

Figure 5.15 Figure 3 from Parenti (2014). Left: A simple 2D Kippenhahn–Schlüter magnetic field topology, showing the location of dense filament material relative to a larger loop structure (from Gilbert et al., 2001). Middle: A more realistic 3D sheared arcade configuration, with dipped regions supporting filament formation (from DeVore & Antiochos, 2000). Right: Flux rope configuration (from Amari et al., 2003).

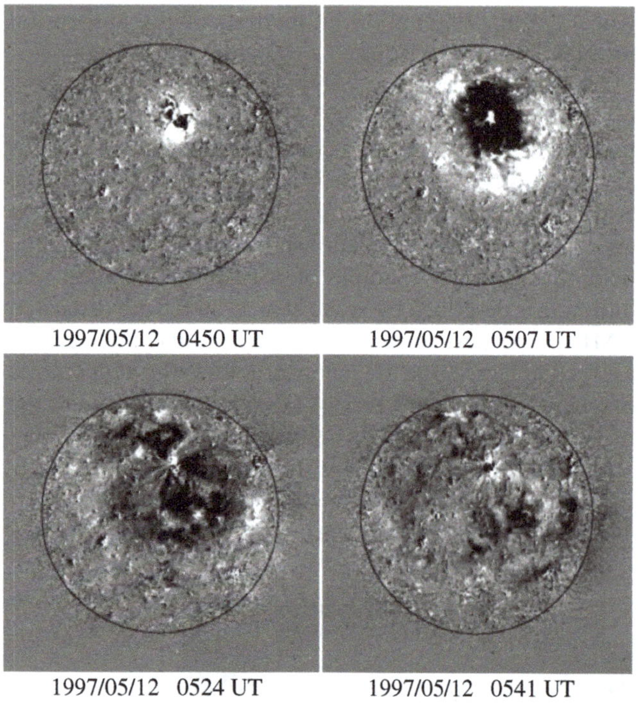

Figure 5.16 Figure from Thompson et al. (1998). Running-difference (i.e., consecutive images subtracted) of a large-scale wave observed on 12 May 1997 by *SOHO* EIT 195 Å.

EIT Waves hundreds of papers have been written on "EIT waves," which have also been called "EUV waves," large-scale coronal disturbances (LSCDs), or simply coronal wave transients (see Figure 5.16). In general, studies of this phenomenon can be assigned to at least one of these broad, non-exclusive categories:

- The physical nature and appearance of EUV waves
- Investigation of correlated phenomena, such as CMEs, flares, dimmings, and filament activity
- Probing the origin or driver of EUV waves
- Understanding the interaction with and impact on ambient coronal features.

In each of these categories, there have been major breakthroughs, primarily due to the availability of high-cadence, multi-wavelength, multi-viewpoint observations from *SDO*, *STEREO*, *Hinode*, and other data sources (for a recent review, see Long et al., 2016). Careful analysis has yielded a much-improved understanding of the EUV wave phenomenon, but it is clear that many outstanding questions remain.

When "EIT waves" were first observed, the assumption was that these phenomena were the coronal manifestation of Moreton waves (Moreton, 1960), which were seen as moving wave fronts in H-alpha images. Among others, Uchida (1968) suggested that the Moreton wave phenomenon was not purely chromospheric in origin, but was instead the "ground track" of a 3D wave front expanding in the corona. The wave shown in Figure 5.16 exhibits a progression in a series of running-difference images (each successive image subtracted from the next image): in less than an hour, the wave has covered almost the entire visible solar disk.

Over 100 of these wave transients were recorded by EIT, and they were interpreted as a signature of the impulse delivered to the ambient corona by the erupting field lines. There is evidence of lateral expansion during the very early stages of an eruption; this "kick" given to the corona is seen as a low-amplitude wave front propagating away from the erupting region (Figure 5.17). The discussion of Zhukov and Auchère (2004) examines the relationship between dimmings, waves and CMEs, and the analysis of Kwon et al. (2015) indicates that in general dimmings can map into the interior of a CME, whereas the waves generally map to the outer envelope of a CME.

Data from *STEREO* EUVI, with additional viewing angles, provided a significant improvement over prior

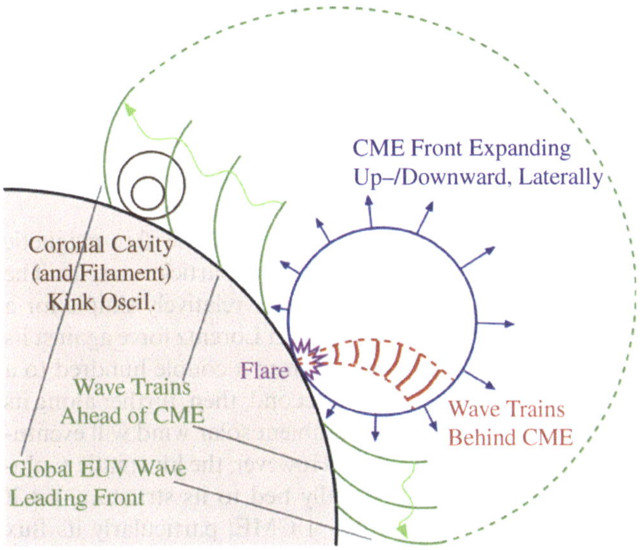

Figure 5.17 Diagram from Liu et al. (2012) summarizing the impulse from a coronal mass ejection (CME; blue arrows), the expanding coronal wave front (green), and motions associated with ongoing flare activity and an affected external filament.

observations, illuminating the complex structure and interactions of these waves (e.g., Liu et al., 2012). With these improved observations, studies of individual wave events (e.g., Long et al., 2008; Long et al., 2011) as well as comprehensive studies of multiple events (Nitta et al., 2013) have augmented early kinematic studies (Wills-Davey & Thompson, 1999; Wang, 2000), thereby elucidating the initiation and subsequent deceleration of wave transients. The extensive study by Nitta et al. (2013), however, indicated that the dynamics of waves can be extremely variable, and apart from being an indicator of the presence of a CME, there is little correlation between wave kinematic properties and CME properties.

In Situ and Other Observations

There are many other observations of CMEs that have provided vital clues as to their nature. In particular, CMEs carry into the heliosphere large amounts of coronal magnetic fields and plasma, which can be detected by remote sensing and in situ spacecraft observations. In situ signatures include transient interplanetary shocks, depressed proton temperatures, cosmic ray depressions, flows with enhanced helium abundances, unusual compositions of ions and elements, and magnetic field structures consistent with loop-like topologies (e.g., Vourlidas et al., 2012b, Webb & Howard, 2012; Webb et al., 1999).

In situ measurements provide the full plasma and magnetic field measurements at a single point in space. Statistical studies of CME properties measured by *Helios*, *MESSENGER*, *Venus Express*, *Ulysses*, and spacecraft at 1 AU (*ISEE*, *Wind*, *ACE*, *STEREO*) provide average behaviors of the radial evolution of CMEs in the inner heliosphere (Bothmer & Schwenn, 1998; Liu et al., 2005; Winslow et al., 2015; Good & Forsyth, 2016; Janvier et al., 2019), giving us one of the only ways to learn from data about the changes in size, magnetic field, and plasma properties inside a CME with distance. Conjunction studies of specific CMEs are able to reveal the evolution and case-by-case variability of individual CME events (e.g., see Nieves-Chinchilla et al., 2012; Good et al., 2015; Winslow et al., 2016). Recent reviews (e.g., see Manchester et al., 2017) discuss the evolution of CMEs in the inner heliosphere, including rotation, erosion, deflection, expansion, whereas Lugaz et al. (2017) discuss specifically interaction of multiple CMEs.

van Driel-Gesztelyi and Culhane (2009) trace the extensive body of research behind solar magnetic field measurements and their connections to interplanetary CMEs. Structured field configurations are often observed corresponding to the arrival of a CME. The field assumes the structure of a spiral or helix, and is accompanied by other signatures including strong magnetic field with low field variance, low plasma beta, and low temperature. This substructure of a CME is referred to as a magnetic cloud (Burlaga et al., 1981), and can be modeled as an MFR, a set of helical field lines with pitch angles increasing outward.

Heliospheric imagers have added wide-field viewing and tracking capabilities of CMEs in the heliosphere, the *SMEI* (2003–2011) in Earth orbit, and the *STEREO* heliospheric imagers (HIs; 2007–present) in 1 AU orbits. Images of the outer heliosphere (above coronagraph altitudes) have shown great potential in connecting coronal imagery to the distances where in situ measurements are obtained (Harrison et al., 2018). Recently, space weather models have been produced using observational CME inputs for the forecasts (see Section 5.5.1). Some insert a magnetic field structure, particularly a modeled or empirical flux rope, into the erupting CME, allowing prediction of the magnetic cloud structure at 1 AU (e.g., Kay et al., 2016; Savani et al., 2017).

There are also important studies that use radio and energetic particle observations to understand the shock properties of CMEs. Shocks driven by CMEs can accelerate particles to extremely high energies, which have important impacts on their own. Chapter 4 covers solar energetic particle events, a dramatic effect of solar eruptions.

5.4. THEORETICAL INTERPRETATIONS AND KEY PROBLEMS

Fundamentally, the Sun's magnetic field fuels flares and CMEs. Surrounding this concept, three questions have to be answered.

First, how is energy stored in the magnetic field before its explosive release into flares and CMEs? It is well accepted that the driver is down below the photosphere, where gas density is high, and convection turbulence generates a magnetic field. As the magnetic field emerges into the corona by buoyancy, it is further twisted and stressed by turbulent motions from below the photosphere. In a coronal volume, the total magnetic energy is given by $W_B = \int \frac{\vec{B}\cdot\vec{B}}{2\mu_0} dV$, where $\vec{B}$ is the magnetic field vector. It has been proved (Woltjer, 1958) that when the normal magnetic field component at the photosphere is held at fixed values (the so-called "line-tying" condition), the magnetic energy W_B is minimum if the magnetic field is current free, namely the current density $\vec{j} = \vec{\nabla} \times \vec{B} / \mu_0 = 0$. Such a magnetic field is called a potential field. Deviations from this potential field, such as when motions in the photosphere add shear or twist, will generate a current in the volume that is associated with a higher magnetic energy. The excess magnetic energy relative to the potential field energy is called the free magnetic energy, which is the energy accessible to flares and CMEs. As the current grows and free energy accumulates, at a certain critical point, an explosion occurs that restructures the field, restoring it back to a lower-energy state.

The second question is, therefore, where is this critical point that makes the energy release possible? And at this point, what is the mechanism that suddenly changes the magnetic field? Theoretically, this question has been addressed with the premise that the Sun's corona may be largely treated as an ideal conductor. This treatment means that on observed scales that are relevant to energy storage and release, plasma is tied with the magnetic field and behaves as a fluid, namely, ions and electrons moving together with the magnetic field. Therefore, a magnetohydrodynamic (MHD) approach is often used to study the evolution of the magnetic field and plasma. For the magnetic field and plasma to evolve steadily without an explosion, the forces must (mostly) balance. In the MHD framework, the net force on plasma includes the electromagnetic Lorentz force, the plasma pressure gradient, and the gravitational force. In more general terms, the critical point is reached when the system evolves into a state where no force balance can be found anymore, or when a balance, or equilibrium, is unstable, so that any small perturbation would pull the system out of the equilibrium, resulting in explosions like a CME. Very often, it turns out that local enhancement of non-ideal MHD factors such as the plasma resistivity, which are largely unimportant in the global context, can actually help lead toward the loss of equilibrium or stability. The presence of non-ideal effects breaks the ideal condition where magnetic field is frozen in plasma, and allows the system to reach a lower-energy state than in an ideal situation. This process is called magnetic reconnection, and is considered to be the governing mechanism of flare loop formation and release energy therein. Reconnection also plays a vital role in releasing CMEs. It is generally accepted that regardless of the initiation mechanism for a CME, the occurrence of magnetic reconnection removes restrictions bounding the CME (Aly, 1991), often also changing its structure, and helps expel the CME into interplanetary space.

Finally, the third question concerns how free magnetic energy is converted into plasma and particle energies. The answer may, at first glance, appear relatively simple for a CME that is accelerated by the net Lorentz force against its own gravity to attain a speed from a couple hundred to a few thousand kilometers per second; then, further along its path outward, drag by the ambient solar wind will eventually slow down a fast CME. However, the kinematic evolution of a CME is intrinsically tied to its structure, yet it remains largely elusive how a CME, particularly its flux rope structure, is formed in the first place. Is the flux rope born from below the photosphere or in the corona well before its eruption, or is it largely formed in situ during its eruption? Related to this notorious "nature or nurture" question of the CME flux rope, it is very puzzling whether and why an active region possessing sufficient free magnetic energy will be eruptive, converting a substantial amount of energy into CMEs. Here, the word "substantial" is meaningful when compared with the energy that goes into the associated flare. It becomes even more tricky if we ask how energy is converted to particle kinetic energy and plasma thermal energy in a flare. For example, charged particles, electrons or ions, can be accelerated by an electric field, by waves, or by shocks; plasma in the corona and chromosphere can be heated by fast particles, in which case plasma heating is only a secondary effect in the chain of energy conversion. But we have suggested earlier that other mechanisms may need to be invoked to heat flare plasma.

5.4.1. Magnetic Reconnection

Flares and CMEs are fueled by free magnetic energy stored in the global system. Magnetic reconnection is believed to be the primary mechanism that allows very rapid reconfiguration of the magnetic field to a lower-energy state. It forms flare loops and may also play a critical role in releasing CMEs into interplanetary space. Magnetic reconnection is a process of locally breaking the stressed magnetic field, in violation of the globally ideal condition that the magnetic field is "frozen" into the plasma and must move with it. For that to happen, an electric current sheet is formed at the place of strong magnetic field gradient, where a dissipation mechanism is invoked. The physics of magnetic reconnection with its application to the solar and heliospheric environment is very well explained by Priest and Forbes (2000), Biskamp (2000), and Birn and Priest (2007). For more recent studies, Lin et al. (2015) reviewed observational signatures and numerical simulations of magnetic reconnection in the current sheet trailing CMEs, and Janvier (2017) discussed 3D reconnection in flares and CMEs.

Back in 1947, almost a century after the discovery of the Carrington event, Giovanelli (1947) realized that flares tend to occur near sunspots at what he called the neutral point, where positive and negative magnetic fields meet, and suggested that an electric field developed there generates energetic particles responsible for observed flare radiation. In the following decades, Giovanelli's idea was developed into the theory of magnetic reconnection. Sweet (1958), Parker (1957), and Petschek (1964) established steady-state 2D reconnection models (by "steady-state," local physical quantities do not change with time). As magnetic fields of opposite sign enter the current sheet with plasma inflows v_{in} from both sides across the length of the current sheet L, due to non-ideal effects, the inflow fields of opposite signs bend, merge, and change connectivity, resulting in plasma outflows v_{out} from the other ends of the current sheet. The inflow speed v_{in} that brings in the magnetic field is therefore a measure of how fast the magnetic field is reconnected, and the dimensionless magnetic reconnection rate is henceforth defined as the ratio of this inflow speed to the outflow speed. By observing mass and energy conservation, the outflow speed is found to be roughly the Alfvén speed of the inflow magnetic field $v_{out} = v_a = B_{in} / \sqrt{\mu_0 \rho_0}$, ρ_0 being the electron density in the current sheet. In other words, the 2D reconnection rate is the inflow Alfvén Mach number.

In the Sweet–Parker model, the magnetic field (frozen-in to the plasma) flows in along a long current sheet to be dissipated by plasma resistivity η inside the current sheet (i.e., $\vec{E} = \eta \vec{j}$), and the derived reconnection rate is $M_a \equiv v_{in} / v_a = R_m^{-\frac{1}{2}}$, where R_m is the magnetic Reynolds number $R_m \equiv L v_a / \eta$. Given the very low resistivity (and therefore a high magnetic Reynolds number) of the Sun's corona, the Sweet–Parker reconnection is found to be too slow to account for the observed rapid energy release in flares. Petschek (1964) changed the inflow-outflow configuration so that magnetic field flows in only along a very small length to be reconnected. Petschek's configuration introduces a vertical magnetic field component with respect to the current sheet, which can support two pairs of slow shocks spreading outward away from the current sheet. This provides an additional mechanism to carry magnetic energy away. Petschek derived the upper limit of the reconnection rate as $M_a \sim (\ln R_m)^{-1}$. Detailed derivation and interpretation of the original Petschek model and its later development can be found in Vasyliunas (1975), Priest and Forbes (1986), and Forbes et al. (2013). The upper limit of the Petschek reconnection rate is significantly larger than the Sweet–Parker reconnection rate, and is comparable to the observed flare energy release rate. For this reason, the Sweet–Parker model is now considered as a slow reconnection model, whereas the Petschek model represents a fast reconnection model.

Around the same time Petschek published his steady-state reconnection model, Furth et al. (1963) put forward an non-steady model, showing that the Sweet–Parker long current sheet with a finite resistivity is not stable against a long wavelength perturbation. This so-called tearing mode instability will tear the current sheet into staggered X and O configurations, or a chain of magnetic islands, and the instability growth rate is proportional to $R_m^{-\frac{3}{5}}$. In a current-carrying, circular-cylinder shaped flux tube, a finite resistivity will also give rise to a resistive kink instability, whose growth rate is proportional to $R_m^{-\frac{1}{3}}$. The continuous development of these linear instabilities may lead to nonlinear growth and thinning of the current sheet, and eventually allow fast reconnection to set in (Priest & Forbes, 2000; Biskamp, 2000; Shibata & Tanuma, 2001; Pucci & Velli, 2014).

It should be noted that the above models each provide a scaling law of magnetic reconnection rate in terms of the magnetic Reynolds number defined by the resistivity due to collisions between electrons and ions. The details of the current sheet and diffusion region (where this collision is thought to be important) in these models are very complex and usually beyond analytical approaches. From the 1980s, with development of high-speed computers, numerical simulations have been conducted to examine reconnection models. These studies have found that, for fast reconnection to occur, microscopic physics below the MHD scale, treating ions and electrons as particles rather than a single fluid, has to be considered in the current sheet. These include anomalous resistivity, or other terms, apart from advective and resistive terms, in the generalized Ohm's law. There is also intriguing evidence regarding the role of "return currents" in flares, particularly relative to the role of classical or anomalous resistivity (Alaoui & Holman, 2017). Therefore, the mechanisms causing fast reconnection are still an area of active research (Biskamp, 1986, 2000; Birn et al., 2001; Birn & Priest, 2007). If the reconnection rate is no longer constrained by resistivity, it allows fast reconnection to take place in the nearly ideal solar corona. On the other hand, depending on the problem, the detailed dissipation mechanism responsible for reconnection is perhaps not important as long as fast reconnection does occur to allow global magnetic reconfiguration and consequential energy release.

A particular challenge facing studies of reconnection theory is that most reconnection models have been 2D, but reconnection is an inherently 3D process. Longcope (2005), Démoulin (2007), and Pontin (2012) have reviewed the topology of magnetic field and reconnection in three dimensions. In general, magnetic reconnection occurs at topological boundaries, called separatrices (plural of separatrix), dividing the space into domains of different magnetic field connectivities (Gorbachev & Somov, 1988). Separatrices intersect at the null point, where magnetic field vanishes. 3D separatrix structures have been extensively studied (e.g., Parnell et al., 1996). Given the continuous distribution of photospheric magnetic field, Priest and Démoulin (1995) also created the concept of quasi-separatrix-layers (QSLs) in the absence of null points. Methods to determine topological features like null points, separatrices, QSLs, and bald patches (dipped magnetic field lines) have been developed and applied to analyzing magnetic field measurements and models (Demoulin, Henoux et al., 1996; Titov et al., 2002). 3D magnetic reconnection with respect to these topological features has been studied (Lau & Finn, 1990; Priest & Titov, 1996; Aulanier et al., 2006), and flare observations have confirmed that magnetic reconnection (whose location may be mapped by flare ribbons) tends to occur at where these topological features are located (e.g., Démoulin et al., 1993; Masson et al., 2009; Savcheva et al., 2015).

Many CME eruptive models also depend heavily on the location and role of reconnection, though there is debate on what aspects are related to a flare process, and what aspects proceed independently. There are several compelling models describing the onset and early evolution of CMEs (e.g., Moore & Roumeliotis, 1992; Antiochos et al., 1999; Fan & Gibson, 2003; Lynch et al., 2005), each relating reconnecting fields to the overall eruptive topology. It is also known that a flare may accompany a CME resulting from the eruptive magnetic process rather than being a direct cause of the CME launch (Kahler, 1992; Gosling, 1993).

It has been recognized that evolution of Sun's magnetic field builds up magnetic helicity in the solar atmosphere, which characterizes the complexity of magnetic field beyond the potential field. This is discussed by Pevtsov et al. (2014). Magnetic reconnection may redistribute magnetic helicity in coronal structures—for example, van Ballegooijen and Martens (1989), Mikic and Linker (1994), Démoulin et al. (1996), Aulanier et al. (2010), and Priest and Longcope (2017) have discussed how 3D magnetic reconnection may form or change the CME structure, by which a large part of the magnetic helicity is injected into and then carried away by CMEs (Low, 1994).

In short, the 3D magnetic structure and its evolution determine where a current sheet forms and how rapidly reconnection can occur, which plays a role in determining the shape and energy of the associated flares and CMEs. The convergence of observations and models gives us a greater physical understanding of the two phenomena, and also provides a pathway for more accurate predictions (Green et al., 2018).

5.4.2. Global Configuration and Topology

Pending knowledge of specific reconnection mechanisms acting on a tiny spot inside the diffusion region, several global configurations have been envisaged for flares and/or CMEs. In an emerging flux model (Heyvaerts et al., 1977), a current sheet forms at the boundary between a magnetic field by a magnetic dipole emerging into the preexisting overlying field of opposite polarity (Figure 5.18a), where reconnection produces another set of new field lines, or flare loops.

A current sheet may also form under a rising MFR (Figure 5.18b; Forbes & Priest, 1995, Lin & Forbes, 2000), and reconnection therein generates an arcade of flare loops closing down as well as field lines curling up into the flux rope. This configuration has been known for decades as, schematically, the standard (eruptive) flare model, the Carmichael–Sturrock–Hirayama–Kopp–Pneuman (CSHKP) model (Carmichael, 1964; Sturrock, 1966;

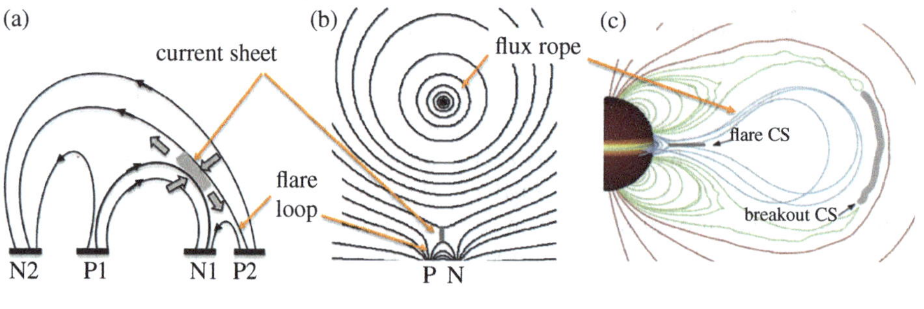

Figure 5.18 (a): Magnetic flux P1-N1 emerging into preexisting overlying flux P2-N2 of opposite directions, forming a current sheet in between, where reconnection may occur and produce new sets of field lines, or flare loops (P1-N2 and P2-N1), after Heyvaerts et al. (1977). (b): A current sheet below a flux rope, where reconnection may take place to form flare loops below the rope, and inject more flux into the flux rope, after Forbes and Priest (1995), Lin and Forbes (2000). (c): A break-out configuration showing reconnection in the break-out current sheet as well as the flare current sheet (*Source:* Karpen et al., 2012).

Hirayama, 1974; Kopp & Pneuman, 1976). The standard model describes continuous generation of flare loops at progressively higher altitude, and its many features (flux rope, trailing current sheet, loop-top hard X-rays, soft X-ray cusp, hot loops on top of cool loops, and two ribbons) have been well established by decades of observations. These are demonstrated in Figure 5.9, and summarized by Lin et al. (2015). The 3D rendition of the standard model has been developed by Demoulin, Priest et al. (1996) and Aulanier et al. (2012); also see the review by Janvier (2017).

The standard flare configuration often also accounts for filament or CME eruption; therefore, this type of flares has been called an "eruptive" flare. Moore et al. (2001) further considered that reconnection below the flux rope plays the role of "tether-cutting" to initiate the CME. In contrast, the other type of flare illustrated in Figure 5.18a has been called a "confined" flare due to its lack of apparent association with "eruptions." However, removal of the overlying flux may allow the underlying flux to further expand and even explode, corresponding to the "break-out" scenario of reconnection initiating CME eruption (Antiochos et al., 1999; Karpen et al., 2012; and Figure 5.18c).

Whereas "tether-cutting" or "break-out" reconnection may initiate CMEs, the other camp of CME initiation models rely on ideal instabilities, such as the torus instability (Bateman, 1978; Kliem & Török, 2006; Fan & Gibson, 2007; Démoulin & Aulanier, 2010) and kink instability (Newcomb, 1960; Hood & Priest, 1981; Einaudi & van Hoven, 1983; Török & Kliem, 2005; Fan & Gibson, 2007); see recent reviews from Chen (2011) and Lin et al. (2015). Most theories, however, require the ejection of an MFR irrespective of whether the MFR existed before or formed during the eruption. In other words, all CMEs would contain an MFR, which is ejected by overcoming the constraint of the so-called "strapping" field. The stronger the "strapping" field, the more it is able to prevent an MFR from erupting (e.g., Sun et al., 2015; Amari et al., 2018; and the references therein). On the other hand, Jacobs et al. (2009) argued, on the basis of MHD simulations of quadrupolar magnetic field configurations, that eruptions could occur without the formation of an MFR.

As mentioned in Section 5.3, evidence for CME-entrained flux ropes is derived from in situ measurements of magnetic clouds (Liu et al., 2008) and from the detection of striations and three-part morphologies in coronagraph images (Vourlidas, 2014; and the references therein). The statistics show that magnetic clouds comprise about 30% of CMEs detected at Earth (Gosling, 1990), and about 40% of CMEs show flux rope characteristics in coronagraph observations from a single viewpoint (Vourlidas et al., 2012a).

Another compelling aspect of global topology are studies of mid- and post-eruptive configurations. One interesting example is the study of "failed" or "partial" eruptions, where material that partially erupts and returns to the Sun can provide vital clues about the evolving eruptive magnetic field. One of the best examples, van Driel-Gesztelyi et al. (2014), documents falling material from a major filament eruption on 7 June 2011. The authors describe how tracing the filament material from their origin point to the locations observed after the eruption can be used as a diagnostic of global field reconfiguration. The filament plasma originated a single erupting active region, but was observed to fall at other locations on the Sun distant from the active region. The authors argue that these locations serve as an indicator of extended fields and regions that are also participating in the CME field evolution.

5.4.3. Energy Deposition in Flares

Electron acceleration How energetic electrons are generated remains one of the biggest puzzles in flare physics. The task of a successful theory is to address the following two findings from hard X-ray and microwave observations: first, the power-law distribution of electrons, and second, the very large number of electrons produced every second during the flare, which implies that electrons in the entire active region corona can be depleted![5]

A successful acceleration model also needs to yield the electron pitch angle distribution to help elucidate particle transport effects and understand the classic issue of electron precipitation and trapping that generate hard X-ray and microwave emissions at the flare footpoints and loop top (e.g., Melrose & Brown, 1976; Aschwanden et al., 1997; Lee & Gary, 2000; Battaglia & Benz, 2007).

Many theories have been developed to tackle these problems. Most of these theories consider the primary site of electron acceleration to be in the corona, motivated by observed hard X-ray and microwave sources at or above the flare loop top (e.g., Petrosian et al., 2002; Sui et al., 2004; Krucker et al., 2008; Gary et al., 2018; and see Figure 5.9) and also the electron time-of-flight analysis (Aschwanden et al., 1995). Some earlier models consider electron acceleration by a direct electric field in a reconnection current sheet parallel to the magnetic guide field (Holman, 1985; Benka & Holman, 1994; Litvinenko, 1996). Given the non-steady nature of magnetic reconnection that generates numerous contracting magnetic islands, electrons may likely gain energy more efficiently by interacting with these islands (Kliem et al., 2000; Drake et al., 2006; Cargill et al., 2012; Guidoni et al.,

[5] *Let us do some math here: given an active region corona of scale size $L \sim 50$ Mm and typical electron density $n \sim 10^9$ cm^{-3}, the total number of electrons is $N_{tot} \sim 10^{38}$. Recall that a large flare may produce up to 10^{36} energetic electrons per second, so the corona is depleted in a matter of two minutes.*

2016). This scenario has been invoked to interpret observations of pulsating radio bursts (e.g., Kliem et al., 2000; Aurass et al., 2009; Benz et al., 2011). More recently, quasi-periodic-pulsations (QPPs) have been found in light curves of many wavelengths than hard X-rays and microwaves (Inglis & Nakariakov, 2009; Brosius et al., 2016; Van Doorsselaere et al., 2016; for a review). The very dynamic reconnection energy release process may also generate turbulence to accelerate particles stochastically (Miller et al., 1996; Petrosian & Liu, 2004; Petrosian, 2012). Furthermore, magnetic field lines from the reconnection outflow retract or shrink: Figure 5.9 nicely demonstrates the post-reconnection evolution from the soft X-ray cusp structure (panel f) to relaxed flare loops (panel i), which is a key feature of the standard flare model (e.g., Forbes & Acton, 1996). This collapsing trap may accelerate electrons in the corona (e.g., Somov & Kosugi, 1997). The reconnection outflow colliding with earlier formed flare loops may also produce termination shocks to accelerate particles (e.g., Forbes, 1986; Tsuneta & Naito, 1998). Although rare, a few observations have been considered to provide evidence of the presence of the termination shock (Aurass & Mann, 2004; Mann et al., 2006; Chen et al., 2015). Finally, to alleviate the electron number problem, Fletcher and Hudson (2008) proposed that magnetic energy is transported by Alfvén waves to the chromosphere, where electrons are accelerated.

There are not yet deterministic winners, if at all, and production of non-thermal particles remains one of the most challenging problems in flare physics. Readers are referred to Hudson and Ryan (1995), Miller et al. (1997), Aschwanden (2002), Holman et al. (2011), and Klein and Dalla (2017) for thorough reviews of existing theories and challenges.

Flare heating The very dynamic reconnection outflow region is a place with super-hot plasma or energetic particles. From this region, energy flux will be transported predominantly along the newly reconnected flare loops and heat plasma inside. Much of the energy will deposit in the dense chromosphere, which drives chromosphere evaporation, and in this way sends mass and energy back into the corona. Within this general picture, we still do not fully understand some important details. The most critical question concerns what mechanisms heat the flare plasma. Non-thermal electrons have been the most popular candidate (Somov et al., 1981; Nagai & Emslie, 1984; Mariska et al., 1989; Emslie et al., 1992). More recently, Longcope and Guidoni (2011) and Longcope et al. (2016) have proposed to heat the flare corona by slow magnetosonic shocks from reconnection, generating super-hot emission at the flare loop top (Lin et al., 1981; Caspi & Lin, 2010). The chromosphere may be heated by thermal conduction (Gan et al., 1991), non-thermal electrons (Fisher et al., 1985a, 1985b, 1985c), X-ray, or EUV irradiation (Hawley & Fisher, 1992), or Alfvénic waves (Kerr et al., 2016; Reep & Russell, 2016; Reep et al., 2018). In each of these scenarios, what determines the properties of chromospheric evaporation, for example, explosive versus gentle evaporation (Fisher, 1989), and the transition temperature between evaporation and condensation (e.g., Milligan & Dennis, 2009)? Furthermore, it has been known for decades that flare soft X-ray and EUV emissions often decay much more slowly than conductive and radiative cooling (Withbroe, 1978; Dere & Cook, 1979; Jiang et al., 2006; Ryan et al., 2013); and more recently the EUV late phase emission has been discovered in EVE observations (Woods et al., 2011; Sun et al., 2013; and see Figure 5.7). Does the extended duration of flare emission result from successive formation of and continuous heating in a multitude of flare loops (or unresolved threads) much beyond the impulsive phase (Hori et al., 1997; Aschwanden & Alexander, 2001; Warren, 2006; Qiu & Longcope, 2016; Rubio da Costa et al., 2016; Zhu et al., 2018), or is it due to suppressed thermal conduction (Jiang et al., 2006; Wang et al., 2015; Zhu et al., 2018) or other non-equilibrium effects (see reviews by Bradshaw & Raymond, 2013; Dudík et al., 2017)?

Although reconnection is an inherently 3D process, much of the theoretical investigation of flare heating is conducted using 1D models. A major challenge in modeling flare heating is the very dynamic nature of the flare energy release and energy transport across the complex stratification of the atmosphere, particularly the sharp transition from the chromosphere to the corona. At least in the impulsive phase, equilibria or steady states cannot be assumed, posing additional difficulty in the modeling. Reale (2014) has given a comprehensive review of progress in modeling coronal heating in general, and some most recent effort particularly in heating flare loops with hydrodynamic models can be found in Reep (2014). Longcope and Klimchuk (2015, and the references therein). In the dense chromosphere, complex physical processes of absorption, emission, and scattering of photons have to be considered, by also solving radiative transfer equations. In the last decade, heating of the chromosphere has been modeled with new-generation, state-of-the-art numerical codes (Allred et al., 2005, 2015; Rubio da Costa et al., 2015; Kowalski et al., 2017). In the meantime, advanced spectroscopic observations by EIS, EVE, and IRIS with unprecedented resolution and spectral coverage have been providing a very rich diagnostic information, through a variety of spectral lines as well as continua, to help probe physical mechanisms of flare heating or physical processes involved in energy transfer in the flare atmosphere. Finally, (along-the-loop) 1D modeling having been the mainstream, a comprehensive full-dimensional modeling taking into account both the magnetic and plasma aspects of flare evolution will be the ultimate goal, and some new progress is underway (Cheung et al., 2019; and the references therein).

5.4.4. CME Mass and Energetics

For a structure containing the word "mass" in its name, the mass of CMEs is surprisingly not straightforward to measure. The main issue is again that of observability with coronagraph images providing a 2D line-of-sight representation of a 3D structure. Under a set of assumptions, including full ionization, a set ratio of heavier particles to protons, and more importantly about the geometry of the CME itself, it is possible to derive CME mass from coronagraphic images (see, e.g., Jackson & Hildner, 1978; Poland et al., 1981) following typical Thomson scattering calculation (Billings, 1966). The typical technique requires spatial integration. The excess brightness of background-subtracted calibrated images in the area of the CME is proportional to the number of excess electrons in the CME volume. The CME mass in the middle corona is found to be 10^{15} – 10^{16} g. Although CME mass has been measured starting with *Skylab* images, this was limited to the best-observed events. With the launch of *SOHO* and the improved signal-over-noise ratio of the LASCO coronagraphs, the determination of CME mass from coronagraphic images has become routine.

One of the main limitations of the mass determination is that it assumes that all the measured mass is in the plane of the sky. This results in an underestimation of the CME mass (see discussion in Vourlidas et al., 2000). On the other hand, the presence of streamers, which are often deflected during fast CMEs, may easily add spurious mass to the derived CME mass unless the CME area is carefully delimited.

A similar method can be used for heliospheric imagers (DeForest et al., 2012) following early work by Jackson and Leinert (1985) using *Helios* data. DeForest et al. (2013) found an increase by a factor of about 2 of the CME mass from 10 $R_\odot$ to 0.6 AU, confirming findings from simulations (e.g., Lugaz et al., 2005). The limitation of the mass determination can be significantly elevated by using two viewpoints, as made possible following the launch of *STEREO* in 2006. Assuming that the CME mass determined from both viewpoints must be the same, it is possible to determine both the CME mass and the direction of propagation (Colaninno & Vourlidas, 2009).

Determining the energetics of CMEs is once again not a straightforward problem. The driver of CME is certainly their magnetic field; however, coronal magnetic field measurement is still in its infancy. Once the CME mass and position have been derived from coronagraphic observations, it is straightforward to derive the CME kinetic and (gravitational) potential energies. To derive the magnetic energy, one needs to integrate the CME magnetic field over the volume of the magnetic ejecta. Even under the strong assumption of uniform magnetic field inside the ejecta, this requires knowledge of the magnetic field strength or flux inside the CME as well as the CME length perpendicular to the plane of sky. In Vourlidas et al. (2000) the authors use values of the magnetic flux measured in situ at 1 AU to derive the magnetic energy. Other approaches include using the peak acceleration and derived mass to determine the magnetic flux from Newton's second law (DeForest et al., 2012) or using solar measurements, as explained below. Vourlidas et al. (2010) determine the typical mass, kinetic and total energies of CMEs over all of solar cycle 23 from *LASCO* images, finding a median of about 10^{12} kg, 2×10^{23} J of kinetic energy, and 4×10^{23} J of total mechanical energy.

Whereas coronal magnetic fields are not typically measured, solar photospheric magnetic fields are. Therefore, the amount of reconnected magnetic flux can be derived using magnetograms combined with Hα and/or EUV measurements. One technique consists in following the area of the flare ribbons and measuring the associated magnetic flux (Qiu et al., 2007; Möstl et al., 2009). The poloidal magnetic flux in the CME is expected to be greater than the reconnected flux (as measured from ribbons), as a preexisting flux rope may exist (see, e.g., the discussion in Qiu et al., 2007). From the flux measurements, estimating the length of the flux rope (e.g., by assuming self-similar expansion), the CME magnetic energy can be derived.

The energy partition between flares and CMEs and other associated events during an "eruption" is an active area of research (Emslie et al., 2004, 2012; Feng et al., 2013). Additional work has related flare properties, such as the flux, with the CME mass or kinematics.

5.5. SPACE WEATHER AND BEYOND

5.5.1. Space Weather

Space weather is the study of how changes in space affect life and technology on Earth and in space. It requires an understanding of how solar and heliospheric changes interact with Earth's system, and it requires a partnership with "end users," that is, those whose systems are impacted by space weather processes. Like terrestrial weather, an important part of space weather is accurate forecasting, and an improved scientific understanding of the constituent processes can dramatically improve in forecast capabilities.

There are three primary areas that sum up the diversity of space weather effects: diversity of drivers (flares, CMEs, SEPs, high-speed streams), diversity of effects (such as signal interruption, spacecraft charging, orbital drag, ionospheric disturbances), and the diversity of locations (from interplanetary space to Earth's surface). For these reasons, the scientific community works together with forecasters to advance our ability to cope with space weather impacts. Some of the advances allow a better assessment of the likelihood of an eruption or disturbance. Others deal with predicting the impact after a transient is observed, and providing an accurate "nowcast" of the space environment. As our understanding of the Sun–Earth connected system improves, so will our ability to cope with the effects of our dynamic star and the impacts it can cause.

Flares and CMEs are the two main drivers of space weather activity. Every advance in our observational and theoretical understanding of flares and CMEs can be harnessed to improve our ability to predict the impact on Earth and technology in space. The other consistent cause of space weather is high-speed streams (see Chapter 7), which originate from coronal holes and travel faster than the relatively slower ambient solar wind. As the high-speed stream propagates into slower plasma, a compression region forms. Compression and the southward interplanetary magnetic field (IMF) are the main space weather drivers of heliospheric transient disturbances. Southward IMF from high-speed streams allows merging of the IMF and Earth's magnetic field and transfer of solar wind energy and mass into the magnetosphere. Such compression regions, with higher density and magnetic field than the typical solar wind, can drive geomagnetic activity at Earth and generate radiation belt activity. However, the largest and most dramatic storms are associated with CME impacts.

CMEs and their associated coronal activity cause the largest disturbances in the corona and subsequently in the heliosphere. CMEs can contain long-duration flows of southward IMF, and fast CMEs compress any southward field ahead. CMEs can drive interplanetary shocks, a key source of SEPs; see Chapter 4), which are one of the most difficult space weather effects to predict. CMEs drive space weather not only at Earth but also other planets and locations in the heliosphere and, therefore, can have a significant impact on society.

Flares are important for space weather because of their ability to produce sudden dramatic enhancements in a broad range of wavelengths. The X-ray, EUV, and radio emission from flares can drive myriad processes in Earth's atmosphere, disrupting communication, causing sudden ionospheric disturbances (SIDs), and heating of the upper layers of the atmosphere to increase drag on spacecraft.

Flare impacts are difficult to forecast because of their impulsive nature; light from the Sun takes 8 min to reach Earth, and the same light that drives space weather impacts is also used to detect flares in the first place! By the time we are aware that a flare is occurring, Earth is already experiencing space weather effects. Therefore, much of flare space weather strategy focuses on improving our ability to predict whether a region is likely to produce a flare.

Fast CMEs can reach Earth in 1–2 days, but most take at least three days to reach Earth. This may seem like a long time, but the shock-driven particles can travel much faster. The observed kinematic parameters of CMEs are typically inserted into propagation forecast models as a spherical shape, or cone. Cone models approximate initial CME parameters assuming isotropic expansion, radial propagation, and constant width of the CME. Two popular models are Wang–Sheeley–Arge (WSA)-ENLIL+Cone (Odstrcil et al., 2004), and EUHFORIA (EUropean Heliospheric FORecasting Information Asset) (Pomoell & Poedts, 2018)). The Drag-based Model (Vršnak et al., 2013) has also been extensively used with *STEREO* CME observations.

In addition, the difficulty in resolving the exact direction of propagation (see Figure 5.12), combined with the delayed availability of many critical types of data, means that it can take over a day to produce a complete CME forecast. A common strategy is ensemble modeling of CME propagation to Earth, which harnesses the ambiguity in model inputs to produce a range of possible outputs. Recent papers summarizing CME impact forecasting include: (Lee et al., 2013; Möstl et al., 2014; Mays et al., 2015; Wold et al., 2018; Verbeke et al., 2019). Verifications and validations of forecasts using the above models typically yield errors in CME arrival times on the order of 8–14 hr.

The National Oceanic and Atmospheric Administration (NOAA) Space Weather Prediction Center (SWPC) coordinates the space weather forecasts for assets in the United States. SWPC and many others have chosen (rather pragmatically) to categorize space weather impacts through a consideration of both the drivers and their effects. The three categories of "NOAA Scales" are radio blackouts, geomagnetic storms, and energetic particles/radiation. Radio blackouts are a function of flare flux; photons of higher energies deposit in Earth's upper atmosphere and ionosphere and inhibit communications, radio signals, GPS, and so on. Geomagnetic storms are the CME counterpart to flare effects. Although flare effects are due to photons, when a CME's plasma and magnetic fields arrive at Earth, they can drive a large magnetospheric reaction. The response time is longer; light takes 8 min to get to Earth, independent of wavelength, whereas the fastest CMEs take nearly a day to reach us. This can be viewed as inconvenient (many effects are distributed over time, so it is difficult to call an "all clear" forecast), or one may see the separation of effects as a scientific opportunity.

The "NOAA Scales" focus areas were chosen based on how drivers and impacts are clustered. The key to the first (radio blackouts) requires us to understand how magnetic energy stored in the Sun is converted to different emissions in the electromagnetic spectrum. As photons of different wavelengths are deposited into Earth's upper atmosphere and ionosphere, we are better able to understand how the local structure influences radio signal propagation. The second area (geomagnetic storms) requires the deliberate study of solar transients or CMEs leaving the Sun, propagating through the heliosphere, and impacting geospace. Observations in space over the last several solar activity cycles have allowed us to accumulate important evidence. It is important to understand how Earth responds to a CME, but Earth's state prior to the CME can also play a role. Nonetheless, definitive, deterministic models have shown great potential over the past decade. Finally, the third area (energetic particles and radiation) includes energetic particles in both the solar wind and Earth's radiation belts. Although SEP events can develop quickly, their associated radiation storms at Earth can last for days and can have major impacts on spacecraft at geosynchronous orbit as well as in lower orbits.

5.5.2. Exoplanet Impacts

Recent attention has focused on the impact of stellar activity on exoplanets. Like Earth, planets can feel the impact of stellar activity, and this can influence the way life evolves on exoplanets.

There have been studies of flares on other stars, and that has helped us understand stellar activity in the context of our star's activity. It has also helped us understand the range of potential behavior of Sun-like stars. X-ray and EUV radiation can heat or ionize the upper atmosphere, leading to increased atmospheric loss (Lammer et al., 2007). As the presence of a steady atmosphere is an important factor in habitability, Scalo et al. (2007) discuss how a strong planetary magnetic field may help mitigate loss effects.

We have come to understand that it is not only flares, but also CMEs that can determine this habitability. An important clue was the study by Jakosky (2015) that indicated CMEs may have affected the evolution of the Martian atmosphere. Just as space weather research at Earth has elucidated the role that CMEs and flares play, researchers are attempting to recognize the potential effects that solar transients may have in other stellar systems.

Khodachenko et al. (2007) show that in the case of a significant planetary magnetic field, the atmosphere can still be eroded when CME impacts compress the planetary magnetosphere. Kay et al. (2016) adapted their terrestrial space weather model to investigate the potential impacts of CMEs on habitability on other systems. They found that the planets in what is thought to be the "habitable zone" of magnetically active M dwarf stars may experience more extreme space weather than at Earth. Interestingly, this leads to increased flare impact as well; frequent CME impacts can lead to atmospheric erosion, leaving the surface exposed to extreme flare activity.

The first observation of a CME on another star has been recently reported (Argiroffi et al., 2019). However, Harra et al. (2016) reviewed the different observational properties of CMEs discussed in Section 5.3 and found that coronal dimmings show great promise as an indicator of CMEs on other stars. The consideration of space weather for exoplanet habitability is a fairly recent development, and many exciting new results are sure to come in the future.

5.5.3. Enabling Predictions Using Machine Learning

We close this chapter with a brief mention of an exciting new frontier in flare/CME studies. Many researchers are examining models and forecasts using new methods derived from data science, advanced statistics, and machine learning. In particular, space weather events are amenable to machine learning problems because the data are already labeled and easily used with methods such as logistic regression (e.g., flare/no flare).

Many authors (e.g., Bobra & Couvidat, 2015a) have found that machine learning methods such as neural nets, support vector machines, and random forests have been successful in improving predictive accuracy. Others have focused on impacts. For example, McGranaghan et al. (2018) address the high-latitude ionospheric impact with a data-driven discovery machine learning approach. Not only did they find improved predictive performance, but they were able to analyze the various data input parameters, or "features," to determine which played the greatest role in driving space weather effects.

Similar efforts have focused on CME prediction (e.g., Bobra & Ilonidis, 2016a) and classification (e.g., Heidrich-Meisner & Wimmer-Schweingruber, 2018). The book *Machine Learning Techniques for Space Weather* by Camporeale et al. (2018) summarizes many space weather concepts that are advanced by machine learning. Beyond simple forecasts and predictions, machine learning studies are revolutionizing our ability to identify important features, classify phenomena, and derive correlations between physical processes that have thus far eluded classical methods. The *HelioML* online book by Bobra et al. (2019) demonstrates the prospects of machine learning combined with the potential represented by the open source/open code ethic. These expanded capabilities, combined with classical methods and models, present an exciting new frontier for space weather research.

ACKNOWLEDGMENTS

The authors would like to thank Dr. Christina D. Kay for her guidance on exoplanet studies.

REFERENCES

Alaoui, M. & Holman, G. D. (2017, December). Understanding breaks in flare X-ray spectra: Evaluation of a cospatial collisional return-current model. *The Astrophysical Journal, 851*(2): 78. ISSN 1538-4357. doi: 10.3847/1538-4357/aa98de.

Allred, J. C., Hawley, S. L., Abbett, W. P., & Carlsson, M. (2005, September). Radiative hydrodynamic models of the optical and ultraviolet emission from solar flares. *Astrophysical Journal, 630*: 573–586. doi: 10.1086/431751.

Allred, J. C., Kowalski, A. F., & Carlsson, M. (2015, August). A unified computational model for solar and stellar flares. *Astrophysical Journal, 809*: 104. doi: 10.1088/0004-637X/809/1/104.

Aly, J. J. (1991). How Much Energy Can Be Stored in a Three-dimensional Force-free Magnetic Field? *Astrophyiscal Journal Letters, 375*: L61, 7 1991. doi: 10.1086/186088.

Amari, T., Luciani, J. F., Aly, J. J., Mikic, Z., & Linker, J. (2003, October). Coronal mass ejection: Initiation, magnetic helicity, and flux ropes. II. Turbulent diffusion–driven evolution. *The Astrophysical Journal, 595*(2), 1231–1250. ISSN 0004-637X. doi: 10.1086/377444.

Amari, T., Canou, A., Aly, J.-J., Delyon, F., & Alauzet, F. (2018, February). Magnetic cage and rope as the key for solar eruptions. *Nature, 554*(7691), 211–215. doi: 10.1038/nature24671.

Antiochos, S. K., DeVore, C. R., & Klimchuk, J. A. (1999, January). A model for solar coronal mass ejections. *Astrophysical Journal, 510*, 485–493. doi: 10.1086/306563.

Antonucci, E., Gabriel, A. H., Acton, L. W., Culhane, J. L., Doyle, J. G., Leibacher, J. W. et al. (1982, May). Impulsive phase of flares in soft X-ray emission. *Solar Physics, 78*, 107–123. doi: 10.1007/BF00151147.

Argiroffi, C., Reale, F., Drake, J. J., Ciaravella, A., Testa, P., Bonito, R. et al. (2019, May). A stellar flare-coronal mass ejection event revealed by X-ray plasma motions. *Nature Astronomy*, page 328. doi: 10.1038/s41550-019-0781-4.

Aschwanden, M. J. (2002, September). *Particle Acceleration and Kinematics in Solar Flares*. Kluwer Academic Publishers, Dordrecht.

Aschwanden, M. J. (2016, October). Global energetics of solar flares. IV. Coronal mass ejection energetics. *The Astrophysical Journal, 831*(1), 105. ISSN 1538-4357. doi: 10.3847/0004-637X/831/1/105.

Aschwanden, M. J. (2017, September). Global energetics of solar flares. VI. Refined energetics of coronal mass ejections. *Astrophysical Journal, 847*(1), 27. doi: 10.3847/1538-4357/aa8952.

Aschwanden, M. J. & Alexander, D. (2001, December). Flare plasma cooling from 30 MK down to 1 MK modeled from Yohkoh, GOES, and TRACE observations during the Bastille Day Event (14 July 2000). *Solar Physics, 204*, 91–120. doi: 10.1023/A:1014257826116.

Aschwanden, M. J. & Peter, H. (2017, May). The width distribution of loops and strands in the solar corona—Are we hitting rock bottom? *Astrophysical Journal, 840*(1), 4. doi: 10.3847/1538-4357/aa6b01.

Aschwanden, M. J., Schwartz, R. A., & Alt, D. M. (1995, July). Electron time-of-flight differences in solar flares. *Astrophysical Journal, 447*, 923. doi: 10.1086/175930.

Aschwanden, M. J., Bynum, R. M., Kosugi, T., Hudson, H. S., & Schwartz, R. A. (1997, October). Electron trapping times and trap densities in solar flare loops measured with COMPTON and YOHKOH. *Astrophysical Journal, 487*(2), 936–955. doi: 10.1086/304633.

Aschwanden, M. J., Boerner, P., Ryan, D., Caspi, A., McTiernan, J. M., & Warren, H. P. (2015, March). Global energetics of solar flares: II. Thermal energies. *Astrophysical Journal, 802*(1), 53. doi: 10.1088/0004-637X/802/1/53.

Aschwanden, M. J., Caspi, A., Cohen, C. M. S., Holman, G., Jing, J., Kretzschmar, M. et al. (2017, February). Global Energetics of solar flares. V. Energy closure in flares and coronal mass ejections. *Astrophysical Journal, 836*(1), 17. doi: 10.3847/1538-4357/836/1/17.

Aulanier, G., Pariat, E., Démoulin, P., & DeVore, C. R. (2006, November). Slip-running reconnection in quasi-separatrix layers. *Solar Physics, 238*(2), 347–376. doi: 10.1007/s11207-006-0230-2.

Aulanier, G., Török, T., Démoulin, P., & DeLuca, E. E. (2010, January). Formation of torus-unstable flux ropes and electric currents in erupting sigmoids. *Astrophysical Journal, 708*, 314–333. doi: 10.1088/0004-637X/708/1/314.

Aulanier, G., Janvier, M., & Schmieder, B. (2012, July). The standard flare model in three dimensions. I. Strong-to-weak shear transition in post-flare loops. *Astronomy and Astrophysics, 543*, A110. doi: 10.1051/0004-6361/201219311.

Aurass, H. and Mann, G. (2004, November). Radio Observation of Electron Acceleration at Solar Flare Reconnection Outflow Termination Shocks. *Astrophysical Journal, 615*:526–530,. doi: 10.1086/424374.

Aurass, H., Landini, F., & Poletto, G. (2009, November). Coronal current sheet signatures during the 17 May 2002 CME-flare. *Astronomy and Astrophysics, 506*(2), 901–911. doi: 10.1051/0004-6361/200912229.

Baker, D. N., Li, X., Pulkkinen, A., Ngwira, C. M., Mays, M. L., Galvin, A. B. et al. (2013, October). A major solar eruptive event in July 2012: Defining extreme space weather scenarios. *Space Weather, 11*(10), 585–591. ISSN 15427390. doi: 10.1002/swe.20097.

Bastian, T. S., Benz, A. O., & Gary, D. E. (1998, January). Radio emission from solar flares. *Annual Reviews, 36*, 131–188. doi: 10.1146/annurev.astro.36.1.131.

Bateman, G. (1978). *MHD Instabilities*. Cambridge, MA: MIT Press.

Battaglia, M. & Benz, A. O. (2007, May). Exploring the connection between coronal and footpoint sources in a thin-thick target solar flare model. *Astronomy and Astrophysics, 466*(2), 713–716. doi: 10.1051/0004-6361:20077144.

Benka, S. G. & Holman, G. D. (1994, November). A thermal/nonthermal model for solar hard X-ray bursts. *Astrophysical Journal, 435*, 469. doi: 10.1086/174829.

Benz, A. O. (2017, December). Flare observations. *Living Reviews in Solar Physics, 14*, 2. doi: 10.1007/s41116-016-0004-3.

Benz, A. O., Battaglia, M., & Vilmer, N. (2011, November). Location of decimetric pulsations in solar flares. *Solar Physics*, *273*(2), 363–375. doi: 10.1007/s11207-011-9760-3.

Bewsher, D., Harrison, R. A., & Brown, D. S. (2008, February). The relationship between EUV dimming and coronal mass ejections. *Astronomy & Astrophysics*, *478*(3), 897–906. ISSN 0004-6361. doi: 10.1051/0004-6361:20078615.

Billings, D. E. (1966). *A Guide to the Solar Corona*. New York; London: Academic Press.

Birn, J. and Priest, E. R. (2007, January). *Reconnection of Magnetic Fields*. Cambridge, UK: Cambridge University Press.

Birn, J., Drake, J. F., Shay, M. A., Rogers, B. N., Denton, R. E., Hesse, M. et al. (2001, March). Geospace Environmental Modeling (GEM) magnetic reconnection challenge. *Journal of Geophysical Research*, *106*, 3715–3720. doi: 10.1029/1999JA900449.

Biskamp, D. (1986, May). Magnetic reconnection via current sheets. *Physics of Fluids*, *29*, 1520–1531. doi: 10.1063/1.865670.

Biskamp, D. (2000, September). *Magnetic Reconnection in Plasmas*. Cambridge, UK: Cambridge University Press.

Bobra, M. G. & Couvidat, S. (2015a, January). Solar flare prediction using SDO/HMI vector magnetic field data with a machine-learning algorithm. *Astrophysical Journal*, *798*(2), 135. doi: 10.1088/0004-637X/798/2/135.

Bobra, M. G. and Couvidat, S. (2015b, January). Solar flare prediction using SDO/HMI vector magnetic field data with a machine-learning algorithm. *Astrophysical Journal*, *798*(2), 135. doi: 10.1088/0004-637X/798/2/135.

Bobra, M. G. & Ilonidis, S. (2016a, April). Predicting coronal mass ejections using machine learning methods. *Astrophysical Journal*, *821*(2), 127. doi: 10.3847/0004-637X/821/2/127.

Bobra, M. G. and Ilonidis, S. (2016b, April). Predicting coronal mass ejections using machine learning methods. *Astrophysical Journal*, *821*(2), 127. doi: 10.3847/0004-637X/821/2/127.

Bobra, M., Holdgraf, C., Mason, J., Wright, P., Díaz Baso, C. J., & Rokem, A. (2019, February). Helioml/helioml: Helioml 0.2.0 (2019-02-22). doi: https://doi.org/10.5281/zenodo.2575738.

Bothmer, V. & Schwenn, R. (1998, January). The structure and origin of magnetic clouds in the solar wind. *Annales Geophysicae*, *16*(1), 1–24. ISSN 1432-0576. doi: 10.1007/s00585-997-0001-x.

Bradshaw, S. J. & Raymond, J. (2013, October). Collisional and radiative processes in optically thin plasmas. *Space Science Reviews*, *178*(2–4), 271–306. doi: 10.1007/s11214-013-9970-0.

Brosius, J. W., Daw, A. N., & Inglis, A. R. (2016, October). Quasi-periodic fluctuations and chromospheric evaporation in a solar flare ribbon observed by Hinode/EIS, IRIS, and RHESSI. *Astrophysical Journal*, *830*(2), 101. doi: 10.3847/0004-637X/830/2/101.

Brown, J. C. (1971, July). The deduction of energy spectra of non-thermal electrons in flares from the observed dynamic spectra of hard X-ray bursts. *Solar Physics*, *18*, 489–502. doi: 10.1007/BF00149070.

Brueckner, G. E., Howard, R. A., Koomen, M. J., Korendyke, C. M., Michels, D. J., Moses, J. D. et al. (1995, December). The Large Angle Spectroscopic Coronagraph (LASCO). *Solar Physics*, *162*(1–2), 357–402. doi: 10.1007/BF00733434.

Burlaga, L., Sittler, E., Mariani, F., & Schwenn, R. (1981). Magnetic loop behind an interplanetary shock: Voyager, Helios, and IMP 8 observations. *Journal of Geophysical Research*, *860* (A8), 6673. ISSN 0148-0227. doi: 10.1029/JA086iA08p06673.

Camporeale, E., Wing, S., & Johnson, J. R. (2018). *Machine Learning Techniques for Space Weather*. Elsevier Press. ISBN 978-0-12-811788-0. doi: 110.1016/C2016-0-01976-9.

Canfield, R. C., Metcalf, T. R., Strong, K. T., & Zarro, D. M. (1987, March). A novel observational test of momentum balance in a solar flare. *Nature*, *326*, 165. doi: 10.1038/326165a0.

Canfield, R. C. (1986, January). Impulsive phase explosive dynamics. *Advances in Space Research*, *6*(6), 167–176,. ISSN 02731177. doi: 10.1016/0273-1177(86)90140-7.

Canfield, R. C., Hudson, H. S., & McKenzie, D. E. (1999, January). Sigmoidal morphology and eruptive solar activity. *Geophysical Research Letters*, *26*(6), 627–630. doi: 10.1029/1999GL900105.

Cargill, P. J., Vlahos, L., Baumann, G., Drake, J. F., & Nordlund Å. (2012, November). Current fragmentation and particle acceleration in solar flares. *Space Science Reviews*, *173*(1–4), 223–245. doi: 10.1007/s11214-012-9888-y.

Carmichael, H. (1964). A process for flares. *NASA Special Publication*, *50*, 451.

Caspi, A. & Lin, R. P. (2010, December). RHESSI line and continuum observations of super-hot flare plasma. *Astrophysical Journal*, *725*(2), L161–L166. doi: 10.1088/2041-8205/725/2/L161.

Caspi, A., McTiernan, J. M., & Warren, H. P. (2014, June). Constraining solar flare differential emission measures with EVE and RHESSI. *Astrophysical Journal Letters*, *788*(2), L31. doi: 10.1088/2041-8205/788/2/L31.

Chae, J. (2001, October). Observational determination of the rate of magnetic helicity transport through the solar surface via the horizontal motion of field line footpoints. *Astrophysical Journal Letters*, *560*(1), L95–L98. doi: 10.1086/324173.

Chamberlin, P. C., Milligan, R. O., & Woods, T. N. (2012, July). Thermal evolution and radiative output of solar flares observed by the EUV variability experiment (EVE). *Solar Physics*, *279*, 23–42. doi: 10.1007/s11207-012-9975-y.

Chen, B., Bastian, T. S., Shen, C., Gary, D. E., Krucker, S., & Glesener, L. (2015, December). Particle acceleration by a solar flare termination shock. *Science*, *350*, 1238–1242. doi: 10.1126/science.aac8467.

Chen, P. F. (2011). Coronal mass ejections: Models and their observational basis. *Living Reviews in Solar Physics*, *8*(1):1. ISSN 1614-4961. doi: 10.12942/lrsp-2011-1.

Cheng, J. X. and Qiu, J. (2016, July). The nature of CME-flare-associated coronal dimming. *Astrophysical Journal*, *825*, 37. doi: 10.3847/0004-637X/825/1/37.

Cheung, M. C. M., Boerner, P., Schrijver, C. J., Testa, P., Chen, F., Peter, H. et al. (2015, July). Thermal diagnostics with the atmospheric imaging assembly on board the solar dynamics observatory: A validated method for differential emission measure inversions. *Astrophysical Journal*, *807*, 143. doi: 10.1088/0004-637X/807/2/143.

Cheung, M. C. M., Rempel, M., Chintzoglou, G., Chen, F., Testa, P., Martnez-Sykora, J. et al. (2019, January). A comprehensive three-dimensional radiative magnetohydrodynamic simulation of a solar flare. *Nature Astronomy*, *3*, 160–166. doi: 10.1038/s41550-018-0629-3.

Colaninno, R. C., Vourlidas, A., & Wu, C. C. (2013). Quantitative comparison of methods for predicting the arrival of coronal mass ejections at earth based on multiview imaging. *Journal of Geophysical Research: Space Physics*, *118*(11), 6866–6879. doi: 10.1002/2013JA019205.

Colaninno, R. C. & Vourlidas, A. (2009, June). First determination of the true mass of coronal mass ejections: A Novel approach to using the two STEREO viewpoints. *Astrophysical Journal*, *698*(1), 852–858. doi: 10.1088/0004-637X/698/1/852.

De Pontieu, B., Title, A. M., Lemen, J. R., Kushner, G. D., Akin, D. J., Allard, B. et al. (2014, July). The Interface Region Imaging Spectrograph (IRIS). *Solar Physics*, *289*, 2733–2779. doi: 10.1007/s11207-014-0485-y.

DeForest, C. E., Howard, T. A., & McComas, D. J. (2012, January). Disconnecting open solar magnetic flux. *Astrophysical Journal*, *745*(1), 36. doi: 10.1088/0004-637X/745/1/36.

DeForest, C. E., Howard, T. A., & McComas, D. J. (2013, May). Tracking coronal features from the low corona to Earth: A quantitative analysis of the 2008 December 12 coronal mass ejection. *Astrophysical Journal*, *769*(1), 43. doi: 10.1088/0004-637X/769/1/43.

DeForest, C. E., Howard, T. A., & McComas, D. J. (2014, May). Inbound waves in the solar corona: A direct indicator of Alfvén surface location. *The Astrophysical Journal*, *787*(2), 124. ISSN 0004-637X. doi: 10.1088/0004-637X/787/2/124.

Del Zanna, G. & Mason, H. E. (2018, August). Solar UV and X-ray spectral diagnostics. *Living Reviews in Solar Physics*, *15*(1), 5. doi: 10.1007/s41116-018-0015-3.

Démoulin, P. & Aulanier, G. (2010, August). Criteria for flux rope eruption: Non-equilibrium versus torus instability. *Astrophysical Journal*, *718*, 1388–1399. doi: 10.1088/0004-637X/718/2/1388.

Démoulin, P. (2007, January). Where will efficient energy release occur in 3-D magnetic configurations? *Advances in Space Research*, *39*(9), 1367–1377. ISSN 02731177. doi: 10.1016/j.asr.2007.02.046.

Demoulin, P., van Driel-Gesztelyi, L., Schmieder, B., Hemoux, J. C., Csepura, G., & Hagyard, M. J. (1993, April). Evidence for magnetic reconnection in solar flares. *Astronomy and Astrophysics*, *271*, 292.

Demoulin, P., Henoux, J. C., Priest, E. R., & Mandrini, C. H. (1996, April). Quasi-Separatrix layers in solar flares. I. Method. *Astronomy and Astrophysics*, *308*, 643–655.

Démoulin, P., Priest, E. R., & Lonie, D. P. (1996, April). Three-dimensional magnetic reconnection without null points 2. *Application to twisted flux tubes. Journal of Geophysical Research*, *101*, 7631–7646. doi: 10.1029/95JA03558.

Dere, K. P. and Cook, J. W. (1979, April). The decay of the 1973 August 9 flare. *Astrophysical Journal*, *229*, 772–787. doi: 10.1086/157013.

Dere, K. P., Landi, E., Mason, H. E., Monsignori Fossi, B. C., & Young, P. R. (1997, October). CHIANTI – an atomic database for emission lines. *Astronomy and Astrophysics, Supplement*, *125*, 149–173. doi: 10.1051/aas:1997368.

DeVore, C. R. & Antiochos, S. K. (2000, August). Dynamical formation and stability of helical prominence magnetic fields. *The Astrophysical Journal*, *539*(2), 954–963. ISSN 0004-637X. doi: 10.1086/309275.

Dissauer, K., Veronig, A. M., Temmer, M., & Podladchikova, T. (2019, March). Statistics of coronal dimmings associated with coronal mass ejections. II. Relationship between coronal dimmings and their associated CMEs. *The Astrophysical Journal*, *874*(2), 123. ISSN 0004-637X. doi: 10.3847/1538-4357/AB0962.

Dissauer, K., Veronig, A. M., Temmer, M., Podladchikova, T., & Vanninathan, K. (2018, February). On the detection of coronal dimmings and the extraction of their characteristic properties. *The Astrophysical Journal*, *855*(2), article id. 137, 13 pp., 855. ISSN 0004-637X. doi: 10.3847/1538-4357/aaadb5.

Drake, J. F., Swisdak, M., Che, H., & Shay, M. A. (2006, October). Electron acceleration from contracting magnetic islands during reconnection. *Nature*, *443*, 553–556. doi: 10.1038/nature05116.

Dudik, J., Dzifčáková, E., Meyer-Vernet, N., Del Zanna, G., Young, P. R., Giunta, A. et al. (2017, August). Nonequilibrium processes in the solar corona, transition region, flares, and solar wind (invited review). *Solar Physics*, *292*(8), 100. doi: 10.1007/s11207-017-1125-0.

Dulk, G. A. (1985). Radio emission from the sun and stars. *Annual Reviews*, *23*, 169–224. doi: 10.1146/annurev.aa.23.090185.001125.

Einaudi, G. & van Hoven, G. (1983, October). The stability of coronal loops – Finite-length and pressure-profile limits. *Solar Physics*, *88*, 163–177. doi: 10.1007/BF00196185.

Emslie, A. G., Li, P., & Mariska, J. T. (1992, November). Diagnostics of electron-heated solar flare models. III – Effects of tapered loop geometry and preheating. *Astrophysical Journal*, *399*, 714–723. doi: 10.1086/171964.

Emslie, A. G., Kucharek, H., Dennis, B. R., Gopalswamy, N., Holman, G. D., Share, G. H. et al. (2004, October). Energy partition in two solar flare/CME events. *Journal of Geophysical Research*, *1090* (A10), A10104. ISSN 0148-0227. doi: 10.1029/2004JA010571.

Emslie, A. G., Dennis, B. R., Shih, A. Y., Chamberlin, P. C., Mewaldt, R. A., Moore, C. S. et al. (2012, November). Global energetics of thirty-eight large solar eruptive events. *Astrophysical Journal*, *759*, 71. doi: 10.1088/0004-637X/759/1/71.

Engvold, O. (2015). Description and classification of prominences. In *Solar Prominences*, pages 31–60. Springer International Publishing. doi: 10.1007/978-3-319-10416-4_2.

Eyles, C. J., Harrison, R. A., Davis, C. J., Waltham, N. R., Shaughnessy, B. M., Mapson-Menard, H. C. A. et al. (2009, February). The Heliospheric imagers onboard the STEREO mission. *Solar Physics*, *254*(2), 387–445. doi: 10.1007/s11207-008-9299-0.

Fan, Y. and Gibson, S. E. (2003, June). The emergence of a twisted magnetic flux tube into a preexisting coronal arcade. *The Astrophysical Journal*, *589*(2), L105–L108. ISSN 0004-637X. doi: 10.1086/375834.

Fan, Y. & Gibson, S. E. (2007, October). Onset of coronal mass ejections due to loss of confinement of coronal flux ropes. *Astrophysical Journal*, *668*, 1232–1245. doi: 10.1086/521335.

Feldman, U., Mandelbaum, P., Seely, J. F., Doschek, G. A., & Gursky, H. (1992, July). The potential for plasma diagnostics from stellar extreme-ultraviolet observations. *Astrophysical Journal Supplement*, *81*, 387. doi: 10.1086/191698.

Feng, L., Wiegelmann, T., Su, Y., Inhester, B., Li, Y. P., Sun, X. D. et al. (2013, February). Magnetic energy partition between the coronal mass ejection and flare from AR 11283. *The Astrophysical Journal*, *765*(1), 37. ISSN 0004-637X. doi: 10.1088/0004-637X/765/1/37.

Fisher, G. H. (1989, November). Dynamics of flare-driven chromospheric condensations. *Astrophysical Journal*, *346*, 1019–1029. doi: 10.1086/168084.

Fisher, G. H., Canfield, R. C., & McClymont, A. N. (1985a, February). Flare loop radiative hydrodynamics. V – Response to thick-target heating. VI – Chromospheric evaporation due to heating by nonthermal electrons. VII – Dynamics of the thick-target heated chromosphere. *Astrophysical Journal*, *289*, 414–441. doi: 10.1086/162901.

Fisher, G. H., Canfield, R. C., & McClymont, A. N. (1985b, February). Flare loop radiative hydrodynamics – Part Six – Chromospheric evaporation due to heating by nonthermal electrons. *Astrophysical Journal*, *289*, 425. doi: 10.1086/162902.

Fisher, G. H., Canfield, R. C., & McClymont, A. N. (1985c, February). Flare loop radiative hydrodynamics – Part Seven – Dynamics of the thick target heated chromosphere. *Astrophysical Journal*, *289*, 434. doi: 10.1086/162903.

Fletcher, L. & Hudson, H. S. (2008, March). Impulsive phase flare energy transport by large-scale alfvén waves and the electron acceleration problem. *Astrophysical Journal*, *675*, 1645–1655. doi: 10.1086/527044.

Fletcher, L., Dennis, B. R., Hudson, H. S., Krucker, S., Phillips, K., Veronig, A. et al. (2011, September). An observational overview of solar flares. *Space Science Reviews*, *159*, 19–106. doi: 10.1007/s11214-010-9701-8.

Forbes, T. G. (1986, June). Fast-shock formation in line-tied magnetic reconnection models of solar flares. *Astrophysical Journal*, *305*, 553–563. doi: 10.1086/164268.

Forbes, T. G. & Acton, L. W. (1996, March). Reconnection and field line shrinkage in solar flares. *Astrophysical Journal*, *459*, 330. doi: 10.1086/176896.

Forbes, T. G. & Priest, E. R. (1995, June). Photospheric magnetic field evolution and eruptive flares. *Astrophysical Journal*, *446*, 377. doi: 10.1086/175797.

Forbes, T. G., Priest, E. R., Seaton, D. B., & Litvinenko, Y. E. (2013, May). Indeterminacy and instability in Petschek reconnection. *Physics of Plasmas*, *20*(5), 052902. doi: 10.1063/1.4804337.

Furth, H. P., Killeen, J., & Rosenbluth, M. N. (1963, April). Finite-resistivity instabilities of a sheet pinch. *Physics of Fluids*, *6*, 459–484. doi: 10.1063/1.1706761.

Gallagher, P. T., Lawrence, G. R., & Dennis, B. R. (2003, May). Rapid acceleration of a coronal mass ejection in the low corona and implications for propagation. *Astrophysical Journal Letters*, *588*, L53–L56. doi: 10.1086/375504.

Gan, W. Q., Zhang, H. Q., & Fang, C. (1991, January). A hydrodynamic model of the impulsive phase of a solar flare loop. *Astronomy and Astrophysics*, *241*, 618–624.

Gary, D. E., Fleishman, G. D., & Nita, G. M. (2013, December). Magnetography of solar flaring loops with microwave imaging spectropolarimetry. *Solar Physics*, *288*(2), 549–565. doi: 10.1007/s11207-013-0299-3.

Gary, D. E., Chen, B., Dennis, B. R., Fleishman, G. D., Hurford, G. J., Krucker, S. et al. (2018, August). Microwave and hard X-ray observations of the 2017 September 10 solar limb flare. *Astrophysical Journal*, *863*, 83. doi: 10.3847/1538-4357/aad0ef.

Georgoulis, M. K., Nindos, A., & Zhang, H. (2019, July). The source and engine of coronal mass ejections. *Philosophical Transactions of the Royal Society of London Series A*, *377*(2148), 20180094. doi: 10.1098/rsta.2018.0094.

Gilbert, H. R., Thompson, B. J., Holzer, T. E., & Burkepile, J. T. (2001, December). A comparison of CME-associated atmospheric waves observed in coronal (19.5 nm) and chromospheric (He I 1083 nm and H-alpha 656 nm) lines. In *AGU Fall Meeting Abstracts*, volume 2001, pages SH12B–0746.

Giovanelli, R. G. (1947). Magnetic and electric phenomena in the Sun's atmosphere associated with sunspots. *Monthly Notices of the Royal Astronomical Society*, *107*, 338. doi: 10.1093/mnras/107.4.338.

Golub, L., Deluca, E., Austin, G., Bookbinder, J., Caldwell, D., Cheimets, P. et al. (2007, June). The X-ray telescope (XRT) for the Hinode mission. *Solar Physics*, *243*(1), 63–86. doi: 10.1007/s11207-007-0182-1.

Good, S. W. & Forsyth, R. J. (2016, January). Interplanetary coronal mass ejections observed by MESSENGER and Venus express. *Solar Physics*, *291*(1), 239–263. ISSN 0038-0 Good, S. W., Forsyth, R. J., Raines, J. M., 938. doi: 10.1007/s11207-015-0828-3.

Good, S. W., Forsyth, R. J., Raines, J. M., Gershman, D. J., Slavin, J. A., & Zurbuchen, T. H. (2015, July). Radial evolution of a magnetic cloud: Messenger, Stereo, and Venus express observations. *The Astrophysical Journal*, *807*(2), 177. ISSN 1538-4357.

Gopalswamy, N. & Thompson, B. J. (2000, November). Early life of coronal mass ejections. *Journal of Atmospheric and Solar-Terrestrial Physics*, *62*(16), 1457–1469. doi: 10.1016/S1364-6826(00)00079-1.

Gopalswamy, N., Yashiro, S., Mäkelä, P., Xie, H., Akiyama, S., & Monstein, C. (2018, August). Extreme kinematics of the 2017 September 10 solar eruption and the spectral characteristics of the associated energetic particles. *Astrophysical Journal Letters*, *863*(2), L39. doi: 10.3847/2041-8213/aad86c.

Gorbachev, V. S. & Somov, B. V. (1988, March). Photospheric vortex flows as a cause for two-ribbon flares – a Topological Model. *Solar Physics*, *117*(1), 77–88. doi: 10.1007/BF00148574.

Gosling, J. T. (1990). Jan Coronal mass ejections and magnetic flux ropes in interplanetary space. *Washington DC American Geophysical Union Geophysical Monograph Series*, *58*, 343–364. doi: 10.1029/GM058p0343.

Gosling, J. T. (1993, November). The solar flare myth. *Journal of Geophysical Research*, *980* (A11), 18937–18950. doi: 10.1029/93JA01896.

Graham, D. R. & Cauzzi, G. (2015, July). Temporal evolution of multiple evaporating ribbon sources in a solar flare. *Astrophysical Journal Letters*, *807*, L22. doi: 10.1088/2041-8205/807/2/L22.

Guidoni, S. E., DeVore, C. R., Karpen, J. T., & Lynch, B. J. (2016, March). Magnetic-island contraction and particle acceleration in simulated eruptive solar flares. *Astrophysical Journal*, *820*(1), 60. doi: 10.3847/0004-637X/820/1/60.

Guo, J., Dumbović, M., Wimmer-Schweingruber, R. F., Temmer, M., Lohf, H., Wang, Y. et al. (2018, August). Modeling the evolution and propagation of 10 September 2017 CMEs and SEPs arriving at Mars constrained by remote sensing and in situ measurement. *Space Weather*, *16*(8), 1156–1169. doi: 10.1029/2018SW001973.

Hanaoka, Y., Kurokawa, H., Enome, S., Nakajima, H., Shibasaki, K., Nishio, M. et al. (1994, April). Simultaneous observations of a prominence eruption followed by a coronal arcade formation in Radio, Soft X-Rays, and Hα. *Publications of the ASJ*, *46*, 205–216.

Hannah, I. G. & Kontar, E. P. (2012, March). Differential emission measures from the regularized inversion of Hinode and SDO data. *Astronomy and Astrophysics*, *539*, A146. doi: 10.1051/0004-6361/201117576.

Hannah, I. G., Christe, S., Krucker, S., Hurford, G. J., Hudson, H. S., & Lin, R. P. (2008, April). RHESSI Microflare Statistics. II. X-ray imaging, spectroscopy, and energy distributions. *Astrophysical Journal*, *677*, 704–718. doi: 10.1086/529012.

Harra, L. K., Hara, H., Imada, S., Young, P. R., Williams, D. R., Sterling, A. C. et al. (2007, November). Coronal dimming observed with hinode: Outflows related to a coronal mass ejection. *Publications of the Astronomical Society of Japan*, *590* (sp3), S801–S806. ISSN 0004-6264. doi: 10.1093/pasj/59.sp3.S801.

Harra, L. K., Schrijver, C. J., Janvier, M., Toriumi, S., Hudson, H., Matthews, S. et al. (2016, August), The characteristics of solar X-class flares and CMEs: A paradigm for stellar superflares and eruptions? *Solar Physics*, *291*(6), 1761–1782. ISSN 0038-0938. doi: 10.1007/s11207-016-0923-0.

Harrison, R. A. & Lyons, M. (2000, June). A spectroscopic study of coronal dimming associated with a coronal mass ejection. *Astronomy and Astrophysics*, *358*, 1097.

Harrison, R. A., Davis, C. J., & Eyles, C. J. (2005, January). The STEREO heliospheric imager: How to detect CMEs in the heliosphere. *Advances in Space Research*, *36*(8), 1512–1523. doi: 10.1016/j.asr.2005.01.024.

Harrison, R. A., Davies, J. A., Barnes, D., Byrne, J. P., Perry, C. H., Bothmer, V., et al. (2018, May). CMEs in the Heliosphere: I. A statistical analysis of the observational properties of CMEs Detected in the Heliosphere from 2007 to 2017 by STEREO/HI-1. *Solar Physics*, *293*(5),77. doi: 10.1007/S11207-018-1297-2.

Harrison, R. A. (1996, July). Coronal magnetic storms: A new perspective on flares and the 'solar flare myth' debate. *Solar Physics*, *166*(2), 441–444. doi: 10.1007/BF00149411.

Hawley, S. L. and Fisher, G. H. (1992, February). X-ray-heated models of stellar flare atmospheres – Theory and comparison with observations. *Astrophysical Journal Supplement*, *78*, 565–598. doi: 10.1086/191640.

Hayakawa, H., Iwahashi, K., Tamazawa, H., Isobe, H., Kataoka, R., Ebihara, Y. et al. (2016, December). East Asian observations of low-latitude aurora during the Carrington magnetic storm. *Publications of the ASJ*, *68*(6), 99. doi: 10.1093/pasj/psw097.

Heidrich-Meisner, V. & Wimmer-Schweingruber, R. F. (2018, January). Solar wind classification via k-means clustering algorithm. *Machine Learning Techniques for Space Weather*, pages 397–424. doi: 10.1016/B978-0-12-811788-0.00016-0.

Heyvaerts, J., Priest, E. R., & Rust, D. M. (1977, August). An emerging flux model for the solar flare phenomenon. *Astrophysical Journal*, *216*, 123–137. doi: 10.1086/155453.

Hirayama, T. (1974, February). Theoretical model of flares and prominences. I: Evaporating flare model. *Solar Physics*, *34*, 323–338. doi: 10.1007/BF00153671.

Holman, G. D. (1985, June). Acceleration of runaway electrons and Joule heating in solar flares. *Astrophysical Journal*, *293*, 584–594. doi: 10.1086/163263.

Holman, G. D., Aschwanden, M. J., Aurass, H., Battaglia, M., Grigis, P. C., Kontar, E. P. et al. (2011, September). Implications of X-ray observations for electron acceleration and propagation in solar flares. *Space Science Reviews*, *159*(1–4), 107–166. doi: 10.1007/s11214-010-9680-9.

Holzworth, R. H. & Meng, C. I. (1975, September). Mathematical representation of the auroral oval. *Geophysical Research Letters*, *2*(9), 377–380. doi: 10.1029/GL002i009p00377.

Hood, A. W. & Priest, E. R. (1981). Critical conditions for magnetic instabilities in force-free coronal loops. *Geophysical and Astrophysical Fluid Dynamics*, *17*, 297–318. doi: 10.1080/03091928108243687.

Hori, K., Yokoyama, T., Kosugi, T., & Shibata, K. (1997, November). Pseudo-two-dimensional hydrodynamic modeling of solar flare loops. *Astrophysical Journal*, *489*, 426–441. doi: 10.1086/304754.

Howard, R. A. (2006, October). A historical perspective on coronal mass ejections. *Washington DC American Geophysical Union Geophysical Monograph Series*, *165*, 7. doi: 10.1029/165GM03.

Howard, R. A., Sheeley, N. R. Jr., Koomen, M. J., & Michels, D. J. (1985). Coronal mass ejections: 1979–1981. *Journal of Geophysical Research: Space Physics*, *900* (A9), 8173–8191. doi: 10.1029/JA090iA09p08173.

Howard, R. A., Moses, J. D., Vourlidas, A., Newmark, J. S., Socker, D. G., Plunkett, S. P. et al. (2008, April). Sun Earth Connection Coronal and Heliospheric Investigation (SECCHI). *Space Science Reviews*, *136*(1–4), 67–115. doi: 10.1007/s11214-008-9341-4.

Howard, R. A., Vourlidas, A., Colaninno, R. C., Korendyke, C. M., Plunkett, S. P., Carter, M. T. et al. (2019, July). The solar orbiter heliospheric imager (solohi). *Astronomy and Astrophysics*, *642*, 23 pp. doi: 10.1051/0004-6361/201935202.

Howard, T. A. & Simnett, G. M. (2008). Interplanetary coronal mass ejections that are undetected by solar coronagraphs. *Journal of Geophysical Research: Space Physics*, *1130* (A8). doi: 10.1029/2007JA012920.

Howard, T. (2011). *Introduction*, pages 1–18. Springer New York, New York, NY. ISBN 978-1-4419-8789-1. doi: 10.1007/978-1-4419-8789-1_1.

Howard, T. A. & James Tappin, S. (2009, October). Interplanetary coronal mass ejections observed inÂ theÂ heliosphere: 1. review of theory. *Space Science Reviews*, *147*(1), 31–54. ISSN 1572-9672. doi: 10.1007/s11214-009-9542-5.

Hudson, H. and Ryan, J. (1995, January). High-energy particles in solar flares. *Annual Reviews*, *33*, 239–282. doi: 10.1146/annurev.aa.33.090195.001323.

Hudson, H. S. & Cliver, E. W. (2001, November). Observing coronal mass ejections without coronagraphs. *Journal of*

Geophysical Research: Space Physics, *1060* (A11), 25199–25213. ISSN 01480227. doi: 10.1029/2000JA904026.

Hudson, H. S., Acton, L. W., & Freeland, S. L. (1996, October). A long-duration solar flare with mass ejection and global consequences. *The Astrophysical Journal*, *470*, 629. ISSN 0004-637X. doi: 10.1086/177894.

Hudson, H. S., Lemen, J. R., St. Cyr, O. C., Sterling, A. C., & Webb, D. F. (1998, July). X-ray coronal changes during Halo CMEs. *Geophysical Research Letters*, *25*(14):2481–2484. ISSN 00948276. doi: 10.1029/98GL01303.

Hudson, H. S. (2011, January). Global Properties of solar flares. *Space Science Reviews*, *158*(1), 5–41. doi: 10.1007/s11214-010-9721-4.

Hudson, H. S. & Webb, D. F. (1997). Soft X-Ray Signatures of Coronal Ejections, pages 27–38. American Geophysical Union (AGU). ISBN 9781118664377. doi: 10.1029/GM099p0027.

Hundhausen, A. (1999, January). Coronal mass ejections. In K. T. Strong, J. L. R. Saba, B. M. Haisch, and J. T. Schmelz (Eds), *The many faces of the sun: a summary of the results from NASA's Solar Maximum Mission*, page 143.

Hundhausen, A. J. (1972). High-speed plasma streams and magnetic sectors. In *Coronal Expansion and Solar Wind*, pages 121–168. Springer, Berlin, Heidelberg. doi: 10.1007/978-3-642-65414-5_5.

Ichimoto, K. and Kurokawa, H. (1984, June). H-alpha red asymmetry of solar flares. *Solar Physics*, *93*, 105–121. doi: 10.1007/BF00156656.

Inglis, A. R. & Nakariakov, V. M. (2009, January). A multi-periodic oscillatory event in a solar flare. *Astronomy and Astrophysics*, *493*(1), 259–266. doi: 10.1051/0004-6361:200810473.

Jackson, B. V. & Hildner, E. (1978, November). Forerunners: outer rims of solar coronal transients. *Solar Physics*, *60*(1):155–170. doi: 10.1007/BF00152340.

Jackson, B. V. and Leinert, C. (1985, November). Helios images of solar mass ejections. *Journal of Geophysical Research*, *900* (A11), 10759–10764. doi: 10.1029/JA090iA11p10759.

Jackson, B. V., Buffington, A., Hick, P. P., Altrock, R. C., Figueroa, S., Holladay, P. E. et al. (2004, November). The Solar Mass-Ejection Imager (SMEI) mission. *Solar Physics*, *225*(1), 177–207. doi: 10.1007/s11207-004-2766-3.

Jacobs, C., Roussev, I. I., Lugaz, N., & Poedts, S. (2009, April). The internal structure of coronal mass ejections: Are all regular magnetic clouds flux ropes? *The Astrophysical Journal*, *695*(2), L171–L175. ISSN 0004-637X. doi: 10.1088/0004-637X/695/2/L171.

Jakosky, B. M. (2015, November). MAVEN GOES TO MARS. MAVEN explores the Martian Upper Atmosphere. Introduction. *Science (New York, N.Y.)*, *350*(6261), 643. ISSN 1095-9203. doi: 10.1126/science.aad3443.

James, A. W., Green, L. M., Palmerio, E., Valori, G., Reid, H. A. S., Baker, D. et al. (2017, May). On-disc observations of flux rope formation prior to its eruption. *Solar Physics*, *292*, 71. doi: 10.1007/s11207-017-1093-4.

James, A. W. Valori, G., Green, L. M., Liu, Y., Cheung, M. C. M., Guo, Y. et al. (2018, March). An observationally constrained model of a flux rope that formed in the solar corona. *Astrophysical Journal Letters*, *855*(2), L16. doi: 10.3847/2041-8213/aab15d.

Janvier, M. (2017, February). Three-dimensional magnetic reconnection and its application to solar flares. *Journal of Plasma Physics*, *83*(1), 535830101. doi: 10.1017/S0022377817000034.

Janvier, M., Winslow, R. M., Good, S., Bonhomme, E., Démoulin, P., Dasso, S. et al. (2019, February). Generic magnetic field intensity profiles of interplanetary coronal mass ejections at Mercury, Venus, and Earth from superposed epoch analyses. *Journal of Geophysical Research: Space Physics*, *124*(2), 812–836. ISSN 2169-9380. doi: 10.1029/2018JA025949.

Jiang, Y. W., Liu, S., Liu, W., & Petrosian, V. (2006, February). Evolution of the loop-top source of solar flares: Heating and cooling processes. *Astrophysical Journal*, *638*(2), 1140–1153. doi: 10.1086/498863.

Jing, J., Xu, Y., Cao, W., Liu, C., Gary, D., & Wang, H. (2016, April). Unprecedented fine structure of a solar flare revealed by the 1.6 m New Solar Telescope. *Scientific Reports*, *6*:24319. doi: 10.1038/srep24319.

Kahler, S. W. (1992, September). Solar flares and coronal mass ejections. *Annual Review of Astronomy and Astrophysics*, *30*(1), 113–141. ISSN 0066-4146. doi: 10.1146/annurev.aa.30.090192.000553.

Kaiser, M. L., Kucera, T. A., Davila, J. M., St. Cyr, O. C., Guhathakurta, M., & Christian, E. (2008, April). The STEREO mission: An introduction. *Space Science Reviews*, *136*, 5–16. doi: 10.1007/s11214-007-9277-0.

Karpen, J. T., Antiochos, S. K., & DeVore, C. R. (2012, November). The mechanisms for the onset and explosive eruption of coronal mass ejections and eruptive flares. *Astrophysical Journal*, *760*, 81. doi: 10.1088/0004-637X/760/1/81.

Kay, C., Opher, M., & Kornbleuth, M. (2016, July). Probability of CME impact on exoplanets orbiting M dwarfs and solar-like stars. *The Astrophysical Journal*, *826*(2), 195. ISSN 1538-4357. doi: 10.3847/0004-637X/826/2/195.

Kerr, G. S., Fletcher, L., Russell, A. J. B., & Allred, J. C. (2016, August). Simulations of the Mg II k and Ca II 8542 lines from an AlfvÉn wave-heated flare chromosphere. *Astrophysical Journal*, *827*(2), 101. doi: 10.3847/0004-637X/827/2/101.

Khodachenko, M. L., Lammer, H., Lichtenegger, H. I. M., Langmayr, D., Erkaev, N. V., Grießmeier, J.-M. et al. (2007, April). Mass loss of "Hot Jupiters"—Implications for CoRoT discoveries. Part I: The importance of magnetospheric protection of a planet against ion loss caused by coronal mass ejections. *Planetary and Space Science*, *55*(5), 631–642. ISSN 00320633. doi: 10.1016/j.pss.2006.07.010.

Klein, K.-L. & Dalla, S. (2017, November). Acceleration and propagation of solar energetic particles. *Space Science Reviews*, *212*(3–4), 1107–1136. doi: 10.1007/s11214-017-0382-4.

Kliem, B. & Török, T. (2006, June). Torus instability. *Physical Review Letters*, *96*(25), 255002. doi: 10.1103/PhysRevLett.96.255002.

Kliem, B., Karlický, M., & Benz, A. O. (2000, August). Solar flare radio pulsations as a signature of dynamic magnetic reconnection. *Astronomy and Astrophysics*, *360*, 715–728.

Kontar, E. P., Brown, J. C., Emslie, A. G., Hajdas, W., Holman, G. D., Hurford, G. J. et al. (2011, September). Deducing electron properties from hard X-ray observations. *Space Science Reviews*, *159*(1–4), 301–355. doi: 10.1007/s11214-011-9804-x.

Kopp, R. A. & Pneuman, G. W. (1976, October). Magnetic reconnection in the corona and the loop prominence phenomenon. *Solar Physics*, *50*, 85–98. doi: 10.1007/BF00206193.

Kosugi, T., Matsuzaki, K., Sakao, T., Shimizu, T., Sone, Y., Tachikawa, S. et al. (2007, June). The Hinode (Solar-B) mission: An overview. *Solar Physics*, *243*(1), 3–17. doi: 10.1007/s11207-007-9014-6.

Kowalski, A. F., Allred, J. C., Daw, A., Cauzzi, G., & Carlsson, M. (2017, February). The atmospheric response to high nonthermal electron beam fluxes in solar flares. I. Modeling the brightest NUV footpoints in the X1 solar flare of 2014 March 29. *Astrophysical Journal, 836(1)*, 12. doi: 10.3847/1538-4357/836/1/12.

Kretzschmar, M. (2011, June). The Sun as a Star: Observations of white-light flares. *Astronomy and Astrophysics*, *530*, A84. doi: 10.1051/0004-6361/201015930.

Krista, L. D. & Reinard, A. A. (2017, May). Statistical study of solar dimmings using CoDiT. *The Astrophysical Journal*, *839*(1), 50. ISSN 1538-4357. doi: 10.3847/1538-4357/aa6626.

Krucker, S., Battaglia, M., Cargill, P. J., Fletcher, L., Hudson, H. S., MacKinnon, A. L. et al. (2008, October). Hard X-ray emission from the solar corona. *Astronomy and Astrophysics Reviews*, *16*, 155–208. doi: 10.1007/s00159-008-0014-9.

Krucker, S., Hudson, H. S., Jeffrey, N. L. S., Battaglia, M., Kontar, E. P., Benz, A. O. et al. (2011, October). High-resolution imaging of solar flare ribbons and its implication on the thick-target beam model. *Astrophysical Journal*, *739*(2), 96. doi: 10.1088/0004-637X/739/2/96.

Kucera, T. A., Poland, A. I., Wiik, J. E., Schmieder, B., & Simnett, G. (1998, April). Helical structure in an eruptive prominence related to a CME (SUMER, CDS, LASCO). *International Astronomical Union Colloquium*, *167*, 318–321. ISSN 0252-9211. doi: 10.1017/S0252921100047825.

Kwon, R.-Y., Zhang, J., & Vourlidas, A. (2015, January). Are halo-like solar coronal mass ejections merely a matter of geometric projection effects? *Astrophysical Journal Letters*, *799*(2), L29. doi: 10.1088/2041-8205/799/2/L29.

Lammer, H., Lichtenegger, H. I.M., Kulikov, Y. N., Grießmeier, J.-M., Terada, N., Erkaev, N. V. et al. (2007, February). Coronal Mass Ejection (CME) activity of low mass M stars as an important factor for the habitability of terrestrial exoplanets. *II. CME-induced ion pick up of Earth-like exoplanets in close-in habitable zones. Astrobiology*, *7*(1), 185–207. ISSN 1531-1074. doi: 10.1089/ast.2006.0128.

Lau, Y.-T. & Finn, J. M. (1990, February). Three-dimensional kinematic reconnection in the presence of field nulls and closed field lines. *Astrophysical Journal*, *350*, 672–691. doi: 10.1086/168419.

Lee, C. O., Arge, C. N., Odstrčil, D., Millward, G., Pizzo, V., Quinn, J. M. et al. (2013, July). Ensemble modeling of CME propagation. *Solar Physics*, *285*(1–2), 349–368. ISSN 0038-0938. doi: 10.1007/s11207-012-9980-1.

Lee, J. & Gary, D. E. (2000, November). Solar Microwave bursts and injection pitch-angle distribution of flare electrons. *Astrophysical Journal*, *543*(1), 457–471. doi: 10.1086/317080.

Lemen, J. R., Title, A. M., Akin, D. J., Boerner, P. F., Chou, C., Drake, J. F. et al. (2012, January). The Atmospheric Imaging Assembly (AIA) on the Solar Dynamics Observatory (SDO). *Solar Physics*, *275*(1–2), 17–40. doi: 10.1007/s11207-011-9776-8.

Lin, J. & Forbes, T. G. (2000, February). Effects of reconnection on the coronal mass ejection process. *Journal of Geophysical Research*, *1050* (A2), 2375–2392. doi: 10.1029/1999JA900477.

Lin, J., Murphy, N. A., Shen, C., Raymond, J. C., Reeves, K. K., Zhong, J. et al. (2015, November). Review on current sheets in CME development: Theories and observations. *Space Science Reviews*, *194*(1-4), 237–302. doi: 10.1007/s11214-015-0209-0.

Lin, R. P., Schwartz, R. A., Pelling, R. M., & Hurley, K. C. (1981, December). A new component of hard X-rays in solar flares. *Astrophysical Journal Letters*, *251*, L109–L114. doi: 10.1086/183704.

Lin, R. P., Dennis, B. R., Hurford, G. J., Smith, D. M., Zehnder, A., Harvey, P. R. et al. (2002, November). The Reuven Ramaty high-energy solar spectroscopic imager (RHESSI). *Solar Physics*, *210*(1), 3–32. doi: 10.1023/A:1022428818870.

Litvinenko, Y. E. (1996, May). Particle acceleration in reconnecting current sheets with a nonzero magnetic field. *Astrophysical Journal*, *462*, 997. doi: 10.1086/177213.

Liu, W., Ofman, L., Nitta, N. V., Aschwanden, M. J., Schrijver, C. J., Title, A. M. et al. (2012). Quasi-periodic fast-mode wave trains within a Global EUV wave and sequential transverse oscillations detected by SDO/AIA. *The Astrophysical Journal*, Volume *753*, Issue 1, article id. 52, 17 pp., 753, 4 2012. ISSN 0004-637X. doi: 10.1088/0004-637X/753/1/52.

Liu, W., Jin, M., Downs, C., Ofman, L., Cheung, M. C. M., & Nitta, N. V. (2018, September). A truly global extreme ultraviolet wave from the SOL2017-09-10 X8.2+ solar flare-coronal mass ejection. *Astrophysical Journal Letters*, *864*(2), L24. doi: 10.3847/2041-8213/aad77b.

Liu, Y., Richardson, J. D., & Belcher, J. W. (2005, January). A statistical study of the properties of interplanetary coronal mass ejections from 0.3 to 5.4AU. *Planetary and Space Science*, *53*(1–3), 3–17. ISSN 00320633. doi: 10.1016/j.pss.2004.09.023.

Liu, Y., Sun, X., Török, T., Titov, V. S., & Leake, J. E. (2017, September). Electric-current neutralization, magnetic shear, and eruptive activity in solar active regions. *Astrophysical Journal Letters*, *846*, L6. doi: 10.3847/2041-8213/aa861e.

Liu, Y. C., Galvin, A. B., Simunac, K. D., Kistler, L. M., Popecki, M. A., Farrugia, C. F. et al. (2008, December). Oxygen observations by STEREO/PLASTIC in the slow solar wind. In *AGU Fall Meeting Abstracts*, volume *2008*, pages SH51B–1604.

Long, D. M., Gallagher, P. T., James McAteer, R. T., & Shaun Bloomfield, D. (2008, May). The kinematics of a globally propagating disturbance in the Solar Corona. *The Astrophysical Journal Letters*, *680*(1), pp. L81, 680: L81. ISSN 0004-637X. doi: 10.1086/589742.

Long, D. M., DeLuca, E. E., & Gallagher, P. T. (2011, September). The wave properties of coronal bright fronts observed using SDO/AIA. *The Astrophysical Journal Letters*, *741*(1), article id. L21, 6 pp., 741. ISSN 0004-637X. doi: 10.1088/2041-8205/741/1/L21.

Long, D. M., Bloomfield, D. S., Chen, P.-F., Downs, C., Gallagher, P. T., Kwon, R. Y. et al. (2016, November). Understanding the physical nature of coronal "EIT Waves". *Solar Physics*, *292*(1), 7. ISSN 0038-0938. doi: 10.1007/s11207-016-1030-y.

Long, D. M., Harra, L. K., Matthews, S. A., Warren, H. P., Lee, K.-S., Doschek, G. A. et al. (2018, March). Plasma evolution within an erupting coronal cavity. *Astrophysical Journal*, *855*(2), 74. doi: 10.3847/1538-4357/aaad68.

Longcope, D., Qiu, J., & Brewer, J. (2016, December). A reconnection-driven model of the hard X-ray loop-top source from flare 2004-Feb-26. *Astrophysical Journal*, *833*, 211. doi: 10.3847/1538-4357/833/2/211.

Longcope, D. W. & Guidoni, S. E. (2011, October). A model for the origin of high density in Looptop X-ray sources. *Astrophysical Journal*, *740*, 73. doi: 10.1088/0004-637X/740/2/73.

Longcope, D. W. & Klimchuk, J. A. (2015, November). How gas-dynamic flare models powered by petschek reconnection differ from those with ad hoc energy sources. *Astrophysical Journal*, *813*(2), 131. doi: 10.1088/0004-637X/813/2/131.

Longcope, D. W. (2005, November). Topological methods for the analysis of solar magnetic fields. *Living Reviews in Solar Physics*, *2*(1), 7. doi: 10.12942/lrsp-2005-7.

Low, B. C. (1994, May). Magnetohydrodynamic processes in the solar corona: Flares, coronal mass ejections, and magnetic helicity. *Physics of Plasmas*, *1*(5), 1684–1690. doi: 10.1063/1.870671.

Low, B. C. (2001, November). Coronal mass ejections, magnetic flux ropes, and solar magnetism. *Journal of Geophysical Research*, *106*(A11), 25141–25164. doi: 10.1029/2000JA004015.

Lugaz, N., Manchester, IV, W. B., & Gombosi, T. I. (2005, July). The evolution of coronal mass ejection density structures. *Astrophysical Journal*, *627*(2), 1019–1030. doi: 10.1086/430465.

Lugaz, N., Temmer, M., Wang, Y., & Farrugia, C. J. (2017, April). The interaction of successive coronal mass ejections: A review. *Solar Physics*, *292*(4), 64. ISSN 0038-0938. doi: 10.1007/s11207-017-1091-6.

Lynch, B. J., Gruesbeck, J. R., Zurbuchen, T. H., & Antiochos, S. K. (2005, August). Solar cycle-dependent helicity transport by magnetic clouds. *Journal of Geophysical Research: Space Physics*, *1100*(A8). ISSN 01480227. doi: 10.1029/2005JA011137.

Maggio, A. (2008, January). Non-thermal hard X-ray emission from stellar coronae. *Memorie della Società Astronomica Italiana*, *79*, 186.

Manchester, W., Kilpua, E. K. J., Liu, Y. D., Lugaz, N., Riley, P., Török, T. et al. (2017, November). The physical processes of CME/ICME evolution. *Space Science Reviews*, *212*(3-4), 1159–1219. ISSN 0038-6308. doi: 10.1007/s11214-017-0394-0.

Mann, G., Aurass, H., & Warmuth, A. (2006, August). Electron acceleration by the reconnection outflow shock during solar flares. *Astronomy and Astrophysics*, *454*(3), 969–974. doi: 10.1051/0004-6361:20064990.

Mariska, J. T., Emslie, A. G., & Li, P. (1989, June). Numerical simulations of impulsively heated solar flares. *Astrophysical Journal*, *341*, 1067–1074. doi: 10.1086/167564.

Mason, J. P., Woods, T. N., Webb, D. F., Thompson, B. J., Colaninno, R. C., & Vourlidas, A. (2016, October). Relationship of EUV irradiance coronal dimming slope and depth to coronal mass ejection speed and mass. *Astrophysical Journal*, *830*(1), 20. doi: 10.3847/0004-637X/830/1/20.

Masson, S., Pariat, E., Aulanier, G., & Schrijver, C. J. (2009, July). The nature of flare ribbons in coronal null-point topology. *Astrophysical Journal*, *700*, 559–578. doi: 10.1088/0004-637X/700/1/559.

Masuda, S., Kosugi, T., Hara, H., Tsuneta, S., & Ogawara, Y. (1994, October). A loop-top hard X-ray source in a compact solar flare as evidence for magnetic reconnection. *Nature*, *371*(6497), 495–497. doi: 10.1038/371495a0.

Mays, M. L., Taktakishvili, A., Pulkkinen, A., MacNeice, P. J., Rastätter, L., Odstrcil, D. et al. (2015, June). Ensemble modeling of CMEs Using the WSA–ENLIL+Cone model. *Solar Physics*, *290*(6), 1775–1814. ISSN 0038-0938. doi: 10.1007/s11207-015-0692-1.

McGranaghan, R. M., Mannucci, A. J., Wilson, B., Mattmann, C. A., & Chadwick, R. (2018, November). New capabilities for prediction of high-latitude ionospheric scintillation: A novel approach with machine learning. *Space Weather*, *16*(11), 1817–1846. ISSN 15427390. doi: 10.1029/2018SW002018.

Melrose, D. B. and Brown, J. C. (1976). Jul Precipitation in trap models for solar hard X-ray bursts. *Monthly Notices of the Royal Astronomical Society*, *176*, 15–30. doi: 10.1093/mnras/176.1.15.

Mikic, Z. & Linker, J. A. (1994, August). Disruption of coronal magnetic field arcades. *Astrophysical Journal*, *430*, 898. doi: 10.1086/174460.

Miller, J. A., Larosa, T. N., & Moore, R. L. (1996, April). Stochastic electron acceleration by cascading fast mode waves in impulsive solar flares. *Astrophysical Journal*, *461*, 445. doi: 10.1086/177072.

Miller, J. A., Cargill, P. J., Emslie, A. G., Holman, G. D., Dennis, B. R., LaRosa, T. N. et al. (1997, July). Critical issues for understanding particle acceleration in impulsive solar flares. *Journal of Geophysical Research*, *102*, 14631–14660. doi: 10.1029/97JA00976.

Milligan, R. O. (2015, December). Extreme ultra-violet spectroscopy of the lower solar atmosphere during solar flares (Invited review). *Solar Physics*, *290*(12), 3399–3423. doi: 10.1007/s11207-015-0748-2.

Milligan, R. O. & Dennis, B. R. (2009, July). Velocity characteristics of evaporated plasma using Hinode/EUV Imaging Spectrometer. *Astrophysical Journal*, *699*(2), 968–975. doi: 10.1088/0004-637X/699/2/968.

Milligan, R. O., Kerr, G. S., Dennis, B. R., Hudson, H. S., Fletcher, L., Allred, J. C. et al. (2014, October). The radiated energy budget of chromospheric plasma in a major solar flare deduced from multi-wavelength observations. *Astrophysical Journal*, *793*(2), 70. doi: 10.1088/0004-637X/793/2/70.

Moore, R. L. & Roumeliotis, G. (1992). Triggering of eruptive flares: Destabilization of the preflare magnetic field configuration. In *Eruptive Solar Flares*, pages 69–78. Springer Berlin Heidelberg, Berlin, Heidelberg. doi: 10.1007/3-540-55246-4_79.

Moore, R. L., Sterling, A. C., Hudson, H. S., & Lemen, J. R. (2001, May). Onset of the magnetic explosion in solar flares and coronal mass ejections. *Astrophysical Journal*, *552*(2), 833–848. doi: 10.1086/320559.

Moreton, G. E. (1960). Hα Observations of flare-initiated disturbances with velocities 1000 km/sec. *The Astronomical Journal*, *65*, 494. ISSN 00046256. doi: 10.1086/108346. URL http://adsabs.harvard.edu/cgi-bin/bib_query?1960AJ......65U.494M.

Moses, D., Clette, F., Delaboudinière, J. P., Artzner, G. E., Bougnet, M., Brunaud, J. et al. (1997, October). EIT observations of the extreme ultraviolet Sun. *Solar Physics*, *175*(2), 571–599. doi: 10.1023/A:1004902913117.

Möstl, C., Farrugia, C. J., Temmer, M., Miklenic, C., Veronig, A. M., Galvin, A. B. et al. (2009, November). Linking remote imagery of a coronal mass ejection to its in situ signatures at 1 AU. *The Astrophysical Journal*, *705*(2), L180–L185. ISSN 0004-637X. doi: 10.1088/0004-637X/705/2/L180.

Möstl, C., Amla, K., Hall, J. R., Liewer, P. C., De Jong, E. M., Colaninno, R. C. et al. (2014, May). Connecting speeds, directions and arrival times of 22 coronal mass ejections from the Sun to 1 AU. *The Astrophysical Journal*, *787*(2), 119. ISSN 0004-637X. doi: 10.1088/0004-637X/787/2/119.

Nagai, F. & Emslie, A. G. (1984, April). Gas dynamics in the impulsive phase of solar flares. I Thick-target heating by nonthermal electrons. *Astrophysical Journal*, *279*, 896–908. doi: 10.1086/161960.

Newcomb, W. A. (1960, June). Hydromagnetic stability of a diffuse linear pinch. *Annals of Physics*, *10*, 232–267. doi: 10.1016/0003-4916(60)90023-3.

Ngwira, C. M., Pulkkinen, A., Kuznetsova, M. M., & Glocer, A. (2014, June). Modeling extreme "Carrington-type" space weather events using three-dimensional global MHD simulations. *Journal of Geophysical Research: Space Physics*, *119*(6):4456–4474. ISSN 21699380. doi: 10.1002/2013JA019661.

Nieves-Chinchilla, T., Colaninno, R., Vourlidas, A., Szabo, A., Lepping, R. P., Boardsen, S. A. et al. (2012, June). Remote and in situ observations of an unusual Earth-directed coronal mass ejection from multiple viewpoints. *Journal of Geophysical Research: Space Physics*, *1170* (A6), n/a–n/a. ISSN 01480227. doi: 10.1029/2011JA017243.

Nitta, N. V., Schrijver, C. J., Title, A. M., & Liu, W. (2013, March). SDO AIA observations of large-scale coronal disturbances in the form of propagating fronts. In *SDO-3: Exploring the Network of SDO Science*, page 111.

Odstrcil, D., Pizzo, V. J., Linker, J. A., Riley, P., Lionello, R., & Mikic, Z. (2004, October). Initial coupling of coronal and heliospheric numerical magnetohydrodynamic codes. *Journal of Atmospheric and Solar-Terrestrial Physics*, *66*(15–16), 1311–1320. ISSN 13646826. doi: 10.1016/j.jastp.2004.04.007.

Parenti, S. (2014, March). Solar prominences: Observations. *Living Reviews in Solar Physics*, *11*(1), 1. ISSN 1614-4961. doi: 10.12942/lrsp-2014-1.

Parker, E. N. (1957, December). Sweet's mechanism for merging magnetic fields in conducting fluids. *Journal of Geophysical Research*, *62*, 509–520. doi: 10.1029/JZ062i004p00509.

Parnell, C. E., Smith, J. M., Neukirch, T., & Priest, E. R. (1996, March). The structure of three-dimensional magnetic neutral points. *Physics of Plasmas*, *3*, 759–770. doi: 10.1063/1.871810.

Patsourakos, S., Vourlidas, A., & Stenborg, G. (2010, December). The genesis of an impulsive coronal mass ejection observed at ultra-high cadence by AIA on SDO. *Astrophysical Journal Letters*, *724*, L188–L193. doi: 10.1088/2041-8205/724/2/L188.

Pesnell, W. D., Thompson, B. J., & Chamberlin, P. C. (2012, January). The Solar Dynamics Observatory (SDO). *Solar Physics*, *275*(1–2), 3–15. doi: 10.1007/s11207-011-9841-3.

Petrosian, V. (2012, November). Stochastic acceleration by turbulence. *Space Science Reviews*, *173*(1–4), 535–556. doi: 10.1007/s11214-012-9900-6.

Petrosian, V. & Liu, S. (2004, July). Stochastic acceleration of electrons and protons. I. Acceleration by parallel-propagating waves. *Astrophysical Journal*, *610*(1), 550–571. doi: 10.1086/421486.

Petrosian, V., Donaghy, T. Q., & McTiernan. J. M. (2002, April). Loop top hard X-ray emission in solar flares: Images and statistics. *Astrophysical Journal*, *569*(1), 459–473. doi: 10.1086/339240.

Petschek, H. E. (1964). Magnetic field annihilation. *NASA Special Publication*, *50*, 425.

Pevtsov, A. A., Berger, M. A., Nindos, A., Norton, A. A., & van Driel-Gesztelyi, L. (2014, December). Magnetic helicity, Tilt, and twist. *Space Science Reviews*, *186*(1–4), 285–324. doi: 10.1007/s11214-014-0082-2.

Pick, M. & Vilmer, N. (2008, October). Sixty-five years of solar radioastronomy: Flares, coronal mass ejections and Sun Earth connection. *Astronomy and Astrophysics Reviews*, *16*, 1–153. doi: 10.1007/s00159-008-0013-x.

Poland, A. I., Howard, R. A., Koomen, M. J., Michels, D. J., & Sheeley, Jr. N. R. (1981, January). Coronal transients near Sunspot maximum. *Solar Physics*, *69*(1):169–175. doi: 10.1007/BF00151264.

Pomoell, J. & Poedts, S. (2018, June). EUHFORIA: European heliospheric forecasting information asset. *Journal of Space Weather and Space Climate*, *8*, A35. ISSN 2115-7251. doi: 10.1051/swsc/2018020.

Pontin, D. I. (2012, July). Theory of magnetic reconnection in solar and astrophysical plasmas. *Philosophical Transactions of the Royal Society of London Series A*, *370*(1970), 3169–3192. doi: 10.1098/rsta.2011.0501.

Priest, E. (2014, May). *Magnetohydrodynamics of the Sun*. Cambridge, UK: Cambridge University Press.

Priest, E. & Forbes, T. (2000, June). *Magnetic Reconnection*. Cambridge, UK: Cambridge University Press.

Priest, E. R. & Démoulin, P. (1995, December). Three-dimensional magnetic reconnection without null points. 1. Basic theory of magnetic flipping. *Journal of Geophysical Research*, *1000* (A12), 23443–23464. doi: 10.1029/95JA02740.

Priest, E. R. & Forbes, T. G. (1986, May). New models for fast steady state magnetic reconnection. *Journal of Geophysical Research*, *91*, 5579–5588. doi: 10.1029/JA091iA05p05579.

Priest, E. R. & Longcope, D. W. (2017, January). Flux-rope twist in eruptive flares and CMEs: due to zipper and main-phase reconnection. *Solar Physics*, *292*(1), 25. doi: 10.1007/s11207-016-1049-0.

Priest, E. R. and Titov, V. S. (1996, December). Magnetic reconnection at three-dimensional null points. *Philosophical Transactions of the Royal Society of London Series A*, *354*, 2951–2992. doi: 10.1098/rsta.1996.0136.

Pucci, F. & Velli, M. (2014, January). Reconnection of quasi-singular current sheets: The "Ideal" tearing mode. *Astrophysical Journal Letters*, *780*, L19. doi: 10.1088/2041-8205/780/2/L19.

Qiu, J. & Longcope, D. W. (2016, March). Long duration flare emission: Impulsive heating or gradual heating? *Astrophysical Journal*, *820*, 14. doi: 10.3847/0004-637X/820/1/14.

Qiu, J., Wang, H., Cheng, C. Z., & Gary, D. E. (2004, April). Magnetic reconnection and mass acceleration in flare-coronal mass ejection events. *Astrophysical Journal*, *604*, 900–905. doi: 10.1086/382122.

Qiu, J., Hu, Q., Howard, T. A., & Yurchyshyn, V. B. (2007, April). On the magnetic flux budget in low-corona magnetic reconnection and interplanetary coronal mass ejections. *Astrophysical Journal*, *659*(1), 758–772. doi: 10.1086/512060.

Reale, F. (2014, July). Coronal loops: Observations and modeling of confined plasma. *Living Reviews in Solar Physics*, *11*, 4. doi: 10.12942/lrsp-2014-4.

Reep, J. W. & Russell, A. J. B. (2016, February). Alfvénic wave heating of the upper chromosphere in flares. *Astrophysical Journal Letters*, *818*(1), L20. doi: 10.3847/2041-8205/818/1/L20.

Reep, J. W., Russell, A. J. B., Tarr, L. A., & Leake, J. E. (2018, February). A Hydrodynamic model of Alfvénic wave heating in a coronal loop and its chromospheric footpoints. *Astrophysical Journal*, *853*, 101. doi: 10.3847/1538-4357/aaa2fe.

Reep, J. (2014, January). Hydrodynamic Modeling of Heating Processes in Solar Flares. PhD thesis, Rice University.

Reinard, A. A. & Biesecker, D. A. (2008, February). Coronal mass ejection–associated coronal dimmings. *The Astrophysical Journal*, *674*(1), 576–585. ISSN 0004-637X. doi: 10.1086/525269.

Reinard, A. A. & Biesecker, D. A. (2009, November). The relationship between coronal dimming and coronal mass ejection properties. *The Astrophysical Journal*, *705*(1), 914–919. ISSN 0004-637X. doi: 10.1088/0004-637X/705/1/914.

Robbrecht, E., Patsourakos, S., & Vourlidas, A. (2009, August). No trace left behind: STEREO observation of a coronal mass ejection without low coronal signatures. *Astro physical Journal*, *701*(1), 283–291. doi: 10.1088/0004-637X/701/1/283.

Rubio da Costa, F., Liu, W., Petrosian, V., & Carlsson, M. (2015, November). Combined modeling of acceleration, transport, and hydrodynamic response in solar flares. II. Inclusion of radiative transfer with RADYN. *Astrophysical Journal*, *813*, 133. doi: 10.1088/0004-637X/813/2/133.

Rubio da Costa, F., Kleint, L., Petrosian, V., Liu, W., & Allred, J. C. (2016, August). Data-driven radiative hydrodynamic modeling of the 2014 March 29 X1.0 solar flare. *Astrophysical Journal*, *827*(1), 38. doi: 10.3847/0004-637X/827/1/38.

Rust, D. M. & Hildner, E. (1976, June). Expansion of an X-ray coronal arch into the outer corona. *Solar Physics*, *48*(2), 381–387. ISSN 0038-0938. doi: 10.1007/BF00152003.

Rust, D. M. (1983, January). Coronal disturbances and their terrestrial effects. *Space Science Reviews*, *34*(1), 21. ISSN 0038-6308. doi: 10.1007/BF00221193.

Ryan, D. F., Chamberlin, P. C., Milligan, R. O., & Gallagher, P. T. (2013, November). Decay-phase cooling and inferred heating of M- and X-class solar flares. *Astrophysical Journal*, *778*, 68. doi: 10.1088/0004-637X/778/1/68.

Savani, N. P., Vourlidas, A., Richardson, I. G., Szabo, A., Thompson, B. J., Pulkkinen, A. et al. (2017, February). Predicting the magnetic vectors within coronal mass ejections arriving at Earth: 2. Geomagnetic response. *Space Weather*, *15*(2), 441–461. doi: 10.1002/2016SW001458.

Savcheva, A., Pariat, E., McKillop, S., McCauley, P., Hanson, E., Su, Y. et al. (2015, September). The relation between solar eruption topologies and observed flare features. I. Flare ribbons. *Astrophysical Journal*, *810*(2), 96. doi: 10.1088/0004-637X/810/2/96.

Scalo, J., Kaltenegger, L., Segura, A., Fridlund, M., Ribas, I., Kulikov, Y. N. et al. (2007, February). M Stars as targets for terrestrial exoplanet searches and biosignature detection. *Astrobiology*, *7*(1), 85–166. ISSN 1531-1074. doi: 10.1089/ast.2006.0125.

Scherrer, P. H., Schou, J., Bush, R. I., Kosovichev, A. G., Bogart, R. S., Hoeksema, J. T. et al. (2012, January). The Helioseismic and Magnetic Imager (HMI) investigation for the Solar Dynamics Observatory (SDO). *Solar Physics*, *275*, 207–227. doi: 10.1007/s11207-011-9834-2.

Schmieder, B, van Driel-Gesztelyi, L., Wiik, J. E., Kucera, T., Thompson, B., de Forest, C. et al. (1997, January). Prominence activity related to CME observed by SOHO, YOHKOH and ground-based observatories. In A Wilson (Ed.), *Fifth SOHO Workshop: The Corona and Solar Wind Near Minimum Activity*, volume 404 of *ESA Special Publication*, page 663.

Schmieder, B., van Driel-Gesztelyi, L., Aulanier, G., Démoulin, P., Thompson, B., De Forest, C. et al. (2002, January). Relationships between CME's and prominences. *Advances in Space Research*, *29*(10), 1451–1460. doi: 10.1016/S0273-1177(02)00211-9.

Schrijver, C. J. (2007, February). A characteristic magnetic field pattern associated with all major solar flares and its use in flare forecasting. *Astrophysical Journal Letters*, *655*, L117–L120. doi: 10.1086/511857.

Schrijver, C. J. & Title, A. M. (2011, April). Long-range magnetic couplings between solar flares and coronal mass ejections observed by SDO and STEREO. *Journal of Geophysical Research: Space Physics*, *1160* (A4), n/a–n/a. ISSN 01480227. doi: 10.1029/2010JA016224.

Sheeley, N. R., Bohlin, J. D., Brueckner, G. E., Purcell, J. D., Scherrer, V. E., Tousey, R. et al. (1975, December). Coronal changes associated with a disappearing filament. *Solar Physics*, *45*(2), 377–392. ISSN 0038-0938. doi: 10.1007/BF00158457.

Shibata, K. & Magara, T. (2011, December). Solar flares: Magnetohydrodynamic processes. *Living Reviews in Solar Physics*, *8*, 6. doi: 10.12942/lrsp-2011-6.

Shibata, K. & Tanuma, S. (2001, June). Plasmoid-induced-reconnection and fractal reconnection. *Earth, Planets, and Space*, *53*, 473–482. doi: 10.1186/BF03353258.

Somov, B. V. & Kosugi, T. (1997, August). Collisionless reconnection and high-energy particle acceleration in solar flares. *Astrophysical Journal*, *485*, 859–868. doi: 10.1086/304449.

Somov, B. V., Syrovatskii, S. I., & Spektor, A. R. (1981, September). Hydrodynamic response of the solar chromosphere to an elementary flare burst. I – Heating by accelerated electrons. *Solar Physics*, *73*, 145–155. doi: 10.1007/BF00153151.

Sterling, A. C. & Hudson, H. S. (1997, December). *Yohkoh* SXT observations of X-ray "dimming" associated with a halo coronal mass ejection. *The Astrophysical Journal*, *491*(1), L55–L58. ISSN 0004637X. doi: 10.1086/311043.

Sturrock, P. A. (1966, August). Model of the high-energy phase of solar flares. *Nature*, *211*, 695–697. doi: 10.1038/211695a0.

Sturrock, P. A. (1991). Oct Maximum energy of semi-infinite magnetic field configurations. *Astrophysical Journal*, *380*, 655. doi: 10.1086/170620.

Sui, L., Holman, G. D., & Dennis, B. R. (2004, September). Evidence for magnetic reconnection in three homologous solar flares observed by RHESSI. *Astrophysical Journal*, *612*(1), 546–556. doi: 10.1086/422515.

Sun, X., Hoeksema, J. T., Liu, Y., Aulanier, G., Su, Y., Hannah, I. G. et al. (2013, December). Hot spine loops and the nature of a late-phase solar flare. *Astrophysical Journal*, *778*, 139. doi: 10.1088/0004-637X/778/2/139.

Sun, X., Bobra, M. G., Hoeksema, J. T., Liu, Y., Li, Y., Shen, C. et al. (2015, May). Why is the great solar active region 12192 Flare-rich but CME-poor? *Astrophysical Journal Letters*, *804*, L28. doi: 10.1088/2041-8205/804/2/L28.

Švestka, Z. (1995, August). On "The Solar Flare Myth" postulated by Gosling. *Solar Physics*, *160*(1), 53–56. doi: 10.1007/BF00679093.

Sweet, P. A. (1958). The neutral point theory of solar flares. In B. Lehnert (Ed.), *Electromagnetic Phenomena in Cosmical Physics*, volume 6 of *IAU Symposium*, page 123.

Tandberg-Hanssen, E. (2013). *The Nature of Solar Prominences. Astrophysics and Space Science Library*. Springer Netherlands. ISBN 9789401733960. URL https://books.google.com/books?id=b03zCAAAQBAJ.

Tandberg-Hanssen, E. & Emslie, A. G. (1988). *The physics of solar flares*. Cambridge and New York: Cambridge University Press, 286 p.

M. Temmer, A. M. Veronig, E. P. Kontar, S. Krucker, and B. Vršnak. Combined STEREO/RHESSI Study of Coronal Mass Ejection Acceleration and Particle Acceleration in Solar Flares. *Astrophysical Journal*, *712*:1410–1420, April 2010. doi: 10.1088/0004-637X/712/2/1410.

Thernisien, A. (2011, June). Implementation of the graduated cylindrical shell model for the three-dimensional reconstruction of coronal mass ejections. *Astrophysical Journal Supplement*, *194*(2), 33. doi: 10.1088/0067-0049/194/2/33.

Thernisien, A. F. R., Howard, R. A., & Vourlidas, A. (2006, November). Modeling of flux rope coronal mass ejections. *Astrophysical Journal*, *652*(1), 763–773. doi: 10.1086/508254.

Thompson, B. J., Plunkett, S. P., Gurman, J. B., Newmark, J. S., St. Cyr, O. C., & Michels, D. J. (1998, January). SOHO/EIT observations of an Earth-directed coronal mass ejection on May 12, 1997. *Geophysical Research Letters*, *25*(14), 2465–2468. doi: 10.1029/98GL50429.

Thompson, B. J., Gurman, J. B., Neupert, W. M., Newmark, J. S., Delaboudinière, J P., St. Cyr, O. C., et al. (1999a, June). SOHO/EIT observations of the 1997 April 7 coronal transient: Possible evidence of coronal Moreton waves. *Astrophysical Journal*, *517*(2), L151–L154. doi: 10.1086/312030.

Thompson, B. J., St. Cyr, O. C., Plunkett, S. P., Gurman, J. B., Gopalswamy, N., Hudson, H. S. et al. (1999b). The correspondence of EUV and white light observations of coronal mass ejections with SOHO EIT and LASCO. In *Sun-Earth Plasma Connections*, pages 31–46. American Geophysical Union (AGU). doi: 10.1029/GM109p0031.

Tian, H., McIntosh, S. W., Xia, L., He, J., & Wang, X. (2012, March). What can we learn about solar coronal mass ejections, coronal dimmings, and extreme-ultraviolet jets through spectroscopic observations? *The Astrophysical Journal*, *748*(2), 106. doi: 10.1088/0004-637x/748/2/106.

Titov, V. S., Hornig, G., & Démoulin, P. (2002, August). Theory of magnetic connectivity in the solar corona. *Journal of Geophysical Research (Space Physics)*, *1070* (A8), 1164. doi: 10.1029/2001JA000278.

Török, T. & Kliem, B. (2005, September). Confined and ejective eruptions of Kink-unstable Flux ropes. *Astrophysical Journal Letters*, *630*, L97–L100. doi: 10.1086/462412.

Tousey, R. (1973, January). The solar corona. In *Space Research Conference*, volume *2*, pages 713–730.

Tsuneta, S. & Naito, T. (1998, March). Fermi acceleration at the fast shock in a solar flare and the impulsive loop-top hard X-ray source. *Astrophysical Journal Letters*, *495*, L67–L70. doi: 10.1086/311207.

Uchida, Y. (1968, May). Propagation of hydromagnetic disturbances in the solar corona and Moreton's wave phenomenon. *Solar Physics*, *4*(1), 30–44. ISSN 0038-0938. doi: 10.1007/BF00146996.

van Ballegooijen, A. A. & Martens, P. C. H. (1989, August). Formation and eruption of solar prominences. *Astrophysical Journal*, *343*, 971. doi: 10.1086/167766.

Van Doorsselaere, T., Kupriyanova, E. G., & Yuan, D. (2016, November). Quasi-periodic pulsations in solar and stellar flares: An overview of recent results (Invited review). *Solar Physics*, *291*(11), 3143–3164. doi: 10.1007/s11207-016-0977-z.

van Driel-Gesztelyi, L. & Culhane, J. L. (2009, April). Magnetic flux emergence, activity, eruptions and magnetic clouds: Following magnetic field from the Sun to the Heliosphere. *Space Science Reviews*, *144*(1–4), 351–381. doi: 10.1007/s11214-008-9461-x.

van Driel-Gesztelyi, L., Baker, D., Török, T., Pariat, E., Green, L. M., Williams, D. R. et al. (2014, June). Coronal magnetic reconnection driven by CME expansion—the 2011 June 7 event. *Astrophysical Journal*, *788*(1), 85. doi: 10.1088/0004-637X/788/1/85.

Vasyliunas, V. M. (1975, February). Theoretical models of magnetic field line merging. *I. Reviews of Geophysics and Space Physics*, *13*, 303–336. doi: 10.1029/RG013i001p00303.

Verbeke, C., Mays, M. L., Temmer, M., Bingham, S., Steenburgh, R., Dumbović, M. et al. (2019, January). Benchmarking CME arrival time and impact: Progress on metadata, metrics, and events. *Space Weather*, *17*(1), 6–26. ISSN 1542-7390. doi: 10.1029/2018SW002046.

Veronig, A. M., Temmer, M., & Vrsnak, B. (2008, June). High cadence observations of a global coronal wave by EUVI/STEREO. *The Astrophysical Journal*, *681*(2), L113–L116. ISSN 0004-637X. doi: 10.1086/590493.

Veronig, A. M., Podladchikova, T., Dissauer, K., Temmer, M., Seaton, D. B., Long, D. et al. (2018, December). Genesis and impulsive evolution of the 2017 September 10 coronal mass ejection. *Astrophysical Journal*, *868*(2), 107. doi: 10.3847/1538-4357/aaeac5.

Vourlidas, A., Subramanian, P., Dere, K. P., & Howard, R. A. (2000, May). Large-angle spectrometric coronagraph measurements of the energetics of coronal mass ejections. *Astrophysical Journal*, *534*(1), 456–467. doi: 10.1086/308747.

Vourlidas, A., Lynch, B. J., Howard, R. A., & Li, Y. (2012a). How many CMEs have flux ropes? Deciphering the signatures of shocks, flux ropes, and prominences in coronagraph observations of CMEs. *Solar Physics*, *8*. ISSN 0038-0938. doi: 10.1007/s11207-012-0084-8.

Vourlidas, A., Lynch, B. J., Howard, R. A., & Li, Y. (2012b, August). How many CMEs have flux ropes? Deciphering the signatures of shocks, flux ropes, and prominences in coronagraph observations of CMEs. *Solar Physics*, *284*(1), 179. ISSN 0038-0938. doi: 10.1007/s11207-012-0084-8.

Vourlidas, A. (2014, June). The flux rope nature of coronal mass ejections. *Plasma Physics and Controlled Fusion*, *56*(6), 064001. ISSN 0741-3335. doi: 10.1088/0741-3335/56/6/064001.

Vourlidas, A. & Howard, R. A. (2006, May). The proper treatment of coronal mass ejection brightness: A new methodology and implications for observations. *The Astrophysical Journal*, *642*(2), 1216–1221. doi: 10.1086/501122.

Vourlidas, A., Howard, R. A., Esfandiari, E., Patsourakos, S., Yashiro, S., and Michalek, G. (2010, August). Comprehensive analysis of coronal mass ejection mass and energy properties over a full solar cycle. *The Astrophysical Journal*, *722*(2), 1522–1538. 722:1522–1538, 8 2010. ISSN 0004-637X. doi: 10.1088/0004-637X/722/2/1522.

Vourlidas, A., Howard, R. A., Plunkett, S. P., Korendyke, C. M., Thernisien, A. F. R., Wang, D. et al. (2016, December). The Wide-Field Imager for Solar Probe Plus (WISPR). *Space Science Reviews*, *204*(1–4), 83–130. doi: 10.1007/s11214-014-0114-y.

Vourlidas, A., Balmaceda, L. A., Stenborg, G., & Lago, A. D. (2017, April). Multi-viewpoint coronal mass ejection catalog based on STEREO COR2 observations. *Astrophysical Journal*, *838*(2), 141. doi: 10.3847/1538-4357/aa67f0.

Vršnak, B., Žic, T., Vrbanec, D., Temmer, M., Rollett, T., Möstl, C. et al. (2013, July). Propagation of interplanetary coronal mass ejections: The Drag-Based Model. *Solar Physics*, *285*(1–2), 295–315. ISSN 0038-0938. doi: 10.1007/s11207-012-0035-4.

Wang, T., Ofman, L., Sun, X., Provornikova, E., & Davila, J. M. (2015, September). Evidence of thermal conduction suppression in a solar flaring loop by coronal seismology of slow-mode waves. *Astrophysical Journal Letters*, *811*, L13. doi: 10.1088/2041-8205/811/1/L13.

Wang, W., Zhu, C., Qiu, J., Liu, R., Yang, K. E., & Hu, Q. (2019, January). Evolution of a magnetic flux rope toward eruption. *Astrophysical Journal*, *871*, 25. doi: 10.3847/1538-4357/aaf3ba.

Wang, Y.-M. (2000, November). EIT waves and fast-mode propagation in the solar corona. *The Astrophysical Journal*, *543*(1), L89–L93. ISSN 0004637X. doi: 10.1086/318178.

Warmuth, A. and Mann, G. (2016, April). Constraints on energy release in solar flares from RHESSI and GOES X-ray observations. II. Energetics and energy partition. *Astronomy and Astrophysics*, *588*, A116. doi: 10.1051/0004-6361/201527475.

Warren, H. P. (2006, January). Multithread hydrodynamic modeling of a solar flare. *Astrophysical Journal*, *637*, 522–530. doi: 10.1086/497904.

Warren, H. P., Mariska, J. T., & Doschek, G. A. (2013, June). Observations of thermal flare plasma with the EUV variability experiment. *Astrophysical Journal*, *770*(2), 116. doi: 10.1088/0004-637X/770/2/116.

Webb, D. F., Krieger, A. S., & D. M. Rust (1976, May). Coronal X-ray enhancements associated with H? filament disappearances. *Solar Physics*, *48*(1), 159–186. ISSN 0038-0938. doi: 10.1007/BF00153342.

Webb, D. F., St. Cyr, O. C., Plunkett, S. P., Howard, R. A., & Thompson, B. J. (1999, May). The characteristics and geoeffectiveness of SOHO-LASCO Halo CMEs. In *American Astronomical Society Meeting Abstracts #194*, volume 194 of *American Astronomical Society Meeting Abstracts*, page 17.03.

Webb, D. F., Cliver, E. W., Crooker, N. U., St. Cry, O. C., & Thompson, B. J. (2000, April). Relationship of halo coronal mass ejections, magnetic clouds, and magnetic storms. *Journal of Geophysical Research*, *1050*(A4), 7491–7508. doi: 10.1029/1999JA000275.

Webb, D. F. and Howard, T. A. (2012). Coronal mass ejections: Observations. *Living Reviews in Solar Physics*, *9*(1), 3. ISSN 1614-4961. doi: 10.12942/lrsp-2012-3.

Wills-Davey, M. J. and Thompson, B. J. (1999, December). Observations of a propagating disturbance in TRACE. *Solar Physics*, *190*, 467–483. doi: 10.1023/A:1005201500675.

Winslow, R. M., Lugaz, N., Philpott, L. C., Schwadron, N. A., Farrugia, C. J., Anderson, B. J. et al. (2015, August). Interplanetary coronal mass ejections from MESSENGER orbital observations at Mercury. *Journal of Geophysical Research: Space Physics*, *120*(8), 6101–6118. ISSN 21699380. doi: 10.1002/2015JA021200.

Winslow, R. M., Lugaz, N., Schwadron, N. A., Farrugia, C. J., Yu, W., Raines, J. M. et al. (2016, July). Longitudinal conjunction between MESSENGER and STEREO A: Development of ICME complexity through stream interactions. *Journal of Geophysical Research: Space Physics*, *121*(7), 6092–6106. ISSN 21699380. doi: 10.1002/2015JA022307.

Withbroe, G. L. (1978, October). The thermal phase of a large solar flare. *Astrophysical Journal*, *225*, 641–649. doi: 10.1086/156524.

Wold, A. M., Leila Mays, M., Taktakishvili, A., Jian, L. K., Odstrcil, D., & MacNeice, P. (2018, March). Verification of real-time WSAâ˘ENLIL+Cone simulations of CME arrival-time at the CCMC from 2010 to 2016. *Journal of Space Weather and Space Climate*, *8*, A17. ISSN 2115-7251. doi: 10.1051/swsc/2018005.

Woltjer, L. (1958, June). A theorem on force-free magnetic fields. *Proceedings of the National Academy of Science*, *44*, 489–491. doi: 10.1073/pnas.44.6.489.

Woods, T. N. (2014, September). Extreme ultraviolet late-phase flares: Before and during the solar dynamics observatory mission. *Solar Physics*, *289*, 3391–3401. doi: 10.1007/s11207-014-0483-0.

Woods, T. N., Hock, R., Eparvier, F., Jones, A. R., Chamberlin, P. C., Klimchuk, J. A. et al. (2011, October). Tobiska. New solar extreme-ultraviolet irradiance observations during flares. *Astrophysical Journal*, *739*, 59. doi: 10.1088/0004-637X/739/2/59.

Woods, T. N., Eparvier, F. G., Hock, R., Jones, A. R., Woodraska, D., Judge, D. et al. (2012, January). Extreme Ultraviolet Variability Experiment (EVE) on the Solar Dynamics Observatory (SDO): Overview of science objectives, instrument design, data products, and model developments. *Solar Physics*, *275*, 115–143. doi: 10.1007/s11207-009-9487-6.

Yashiro, S., Gopalswamy, N., Michalek, G., St. Cyr, O. C., Plunkett, S. P., Rich, N. B. et al. (2004a, July). A catalog of white light coronal mass ejections observed by the SOHO spacecraft. *Journal of Geophysical Research (Space Physics)*, *1090* (A7):A07105. doi: 10.1029/2003JA010282.

Yashiro, S., Gopalswamy, N., Michalek, G., St. Cyr, O. C., Plunkett, S. P., Rich, N. B. et al. (2004b, July). A catalog of white light coronal mass ejections observed by the SOHO spacecraft. *Journal of Geophysical Research (Space Physics)*, *1090* (A7), A07105. doi: 10.1029/2003JA010282.

Yokoyama, T. & Shibata, K. (1998, February). A two-dimensional magnetohydrodynamic simulation of chromospheric evaporation in a solar flare based on a magnetic reconnection model. *Astrophysical Journal*, *494*(1), L113–L116. doi: 10.1086/311174.

Zarro, D. M., Canfield, R. C., Strong, K. T., & Metcalf, T. R. (1988 January). Explosive plasma flows in a solar flare. *Astrophysical Journal*, *324*, 582–589. doi: 10.1086/165919.

Zarro, D. M., Sterling, A. C., Thompson, B. J., Hudson, H. S., & Nitta, N. (1999, August). SOHO EIT observations of extreme-ultraviolet "dimming" associated with a halo coronal mass ejection. *Astrophysical Journal Letters*, *520*(2), L139–L142. doi: 10.1086/312150.

Zhang, J., Dere, K. P., Howard, R. A., Kundu, M. R., & White, S. M. (2001, September). On the temporal relationship between coronal mass ejections and flares. *Astrophysical Journal*, *559*, 452–462. doi: 10.1086/322405.

Zhu, C., Qiu, J., & Longcope, D. W. (2018, March). Two-phase heating in flaring loops. *Astrophysical Journal*, *856*, 27. doi: 10.3847/1538-4357/aaad10.

Zhukov, A. N. & Auchère, F. (2004, November). On the nature of EIT waves, EUV dimmings and their link to CMEs. *Astronomy & Astrophysics*, *427*(2), 705–716. ISSN 0004-6361. doi: 10.1051/0004-6361:20040351.

Zurbuchen, T. H. & Richardson, I. G. (2006, November). In-situ solar wind and magnetic field signatures of interplanetary coronal mass ejections. *Space Science Reviews*, *123*(1–3), 31–43. ISSN 0038-6308. doi: 10.1007/s11214-006-9010-4.

6

Fine-Scale Features of the Sun's Atmosphere: Spicules and Jets

Alphonse C. Sterling

ABSTRACT

We present an overview of the subset of fine-scale features in the Sun's atmosphere consisting of spicules, coronal jets, and similar dynamic features. We consider older and newer observations, and theories for chromospheric spicules and coronal jets. We then discuss the possibility that there is a continuum of jet-like features ranging from spicules (and/or near-spicule-size-scale events) to large-scale CME-producing eruptions, all driven by similar magnetic processes operating on differing corresponding size scales. We also consider the connection between these features and some other solar atmospheric phenomena. Future observational and theoretical studies will help clarify further the nature of these solar events and elucidate possible connections between them.

6.1. INTRODUCTION

A striking aspect of a typical image of sufficient resolution showing the Sun's atmosphere above the photosphere is the amount of structure present. Frequently this structure is in the form of transient and dynamic features that resemble geysers, in that they grow to be longer than (sometimes many times longer than) their widths. These are jets of solar plasma that shoot out from near the photosphere into the higher atmospheric layers. They are seen on the size scale of less than an arcsecond to tens of arcseconds. These fascinating jet-like features, which make up key components of different regions of the solar atmosphere, are the topic of this section.

We will anchor our discourse on two primary jet features: spicules and coronal jets. Spicules are part of the chromosphere. They are comparatively small features, with widths near the limits of most ground-based instruments and extending only into the bottom of the corona (widths ~0″.5, heights ~5″–10″). They have been primarily observed at visible wavelengths such as Hα and Ca II. Chromospheric spicules have lifetimes of a few minutes, and occur in vast numbers over essentially all of the Sun's surface and seemingly in all solar regions (perhaps avoiding sunspots). Thus, they are a key constituent of the solar chromosphere. Similar features appear at ultraviolet (UV) and extreme ultraviolet (EUV) wavelengths. For a long time it was not known whether these made up a hot component of (chromospheric) spicules, or if they are entirely different entities; thus, these features are usually described with an adjective, that is, UV spicules or EUV spicules, to differentiate them from the chromospheric spicules. Recent investigations, however, now show that at least some of the UV spicules are hotter components of chromospheric spicules (section 6.2.3). Figure 6.1 shows examples of spicules at the solar limb observed in Ca II by the *Hinode* spacecraft.

Coronal jets, as the name indicates, are coronal features, primarily seen at EUV and X-ray wavelengths; thus, they are largely observable only from space (except for their chromospheric components). They have a size scale larger than those of spicules, with widths of ~5″–10″, and heights of a few tens or hundreds of arcseconds. They typically last for 10 min or so, and occur in coronal

NASA Marshall Space Flight Center, Huntsville, Alabama, USA

Space Physics and Aeronomy Collection Volume 1: Solar Physics and Solar Wind, Geophysical Monograph 258, First Edition.
Edited by Nour E. Raouafi and Angelos Vourlidas.

DOI: 10.1002/9781119815600.ch6

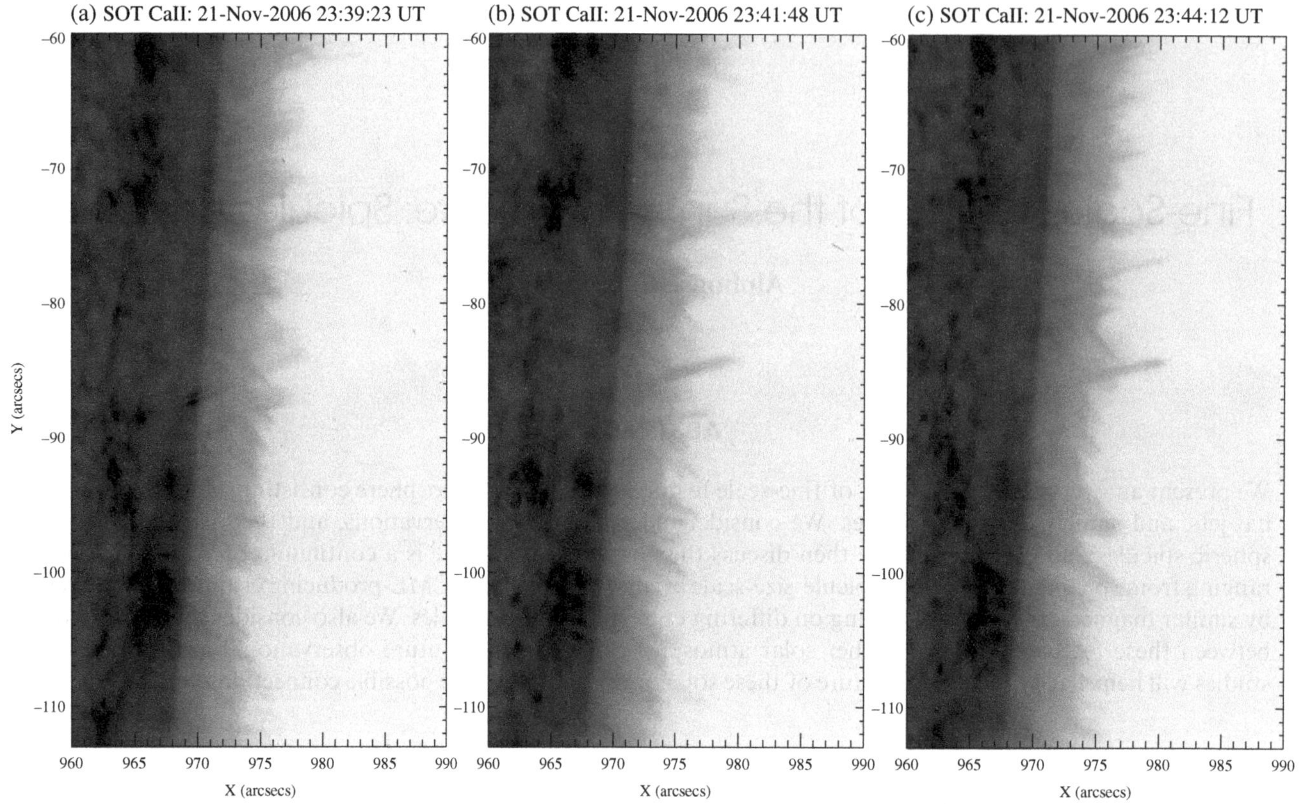

Figure 6.1 Examples of chromospheric spicules observed at the solar west limb with the *Hinode* spacecraft in Ca II. (a)–(c) Dynamic changes can be seen in many of the spicules that reach across the limb over the course of panels (a) through (c). Intensity colors are reversed, so that bright features appear dark and vice versa. (*Source*: Data from this period were analyzed by De Pontieu, McIntosh, et al., 2007 and by Zhang et al., 2012.)

holes, quiet Sun, and in the periphery of active regions (ARs). Figure 6.2 shows examples observed in X-rays in a polar coronal hole (Figure 6.2a), and near an AR (Figure 6.2b).

Following a discussion of these two basic features (spicules and jets), we will briefly discuss features of size scales between spicules and jets, the question of whether there is a connection between features in these two broad categories, and the relationship of jet-like features with other solar activity. These fields are very extensive, very active, and fast changing. Here we describe aspects of our understanding of spicules and coronal jets as of early 2019.

6.2. SOLAR SPICULES

Because they are observable in the visible portion of the spectrum, spicules have been studied for a long time. They were described by Secchi (1877), and have been studied since then by many workers. There are always many spicules on the Sun, unlike other more transient visible-from-the-ground solar features, such as sunspots. Their near-resolution-limit visibility left even their basic properties a challenge to define, likely adding to their allure in some circles. The study of spicules, however, seems to have hit a trough during the decades of the 1970s through the early 2000s, during which time space-based observations were popularizing other areas of solar research, areas that had previously been totally inaccessible from the ground, such as observations at UV and shorter wavelengths. Although some spicule-like features were studied in the UV during this period (e.g., Dere et al., 1983), the spatial resolution of many of the space-based observations were too coarse to resolve spicules. Improvements in ground-based capabilities, direct observations of the chromosphere from space with the *Hinode* satellite, and the advent of continuous, high-resolution observations of the lower corona and the transition region resulted in a "spicule-observation renaissance" from about 2000. Here we first consider spicules observations during the earlier period, and then we will consider the later renaissance period.

The literature on spicules is extensive, and we are not attempting a full review here. Beckers (1968) and Beckers (1972) are two highly cited review papers covering the earlier studies. Sterling (2000) updates some of the

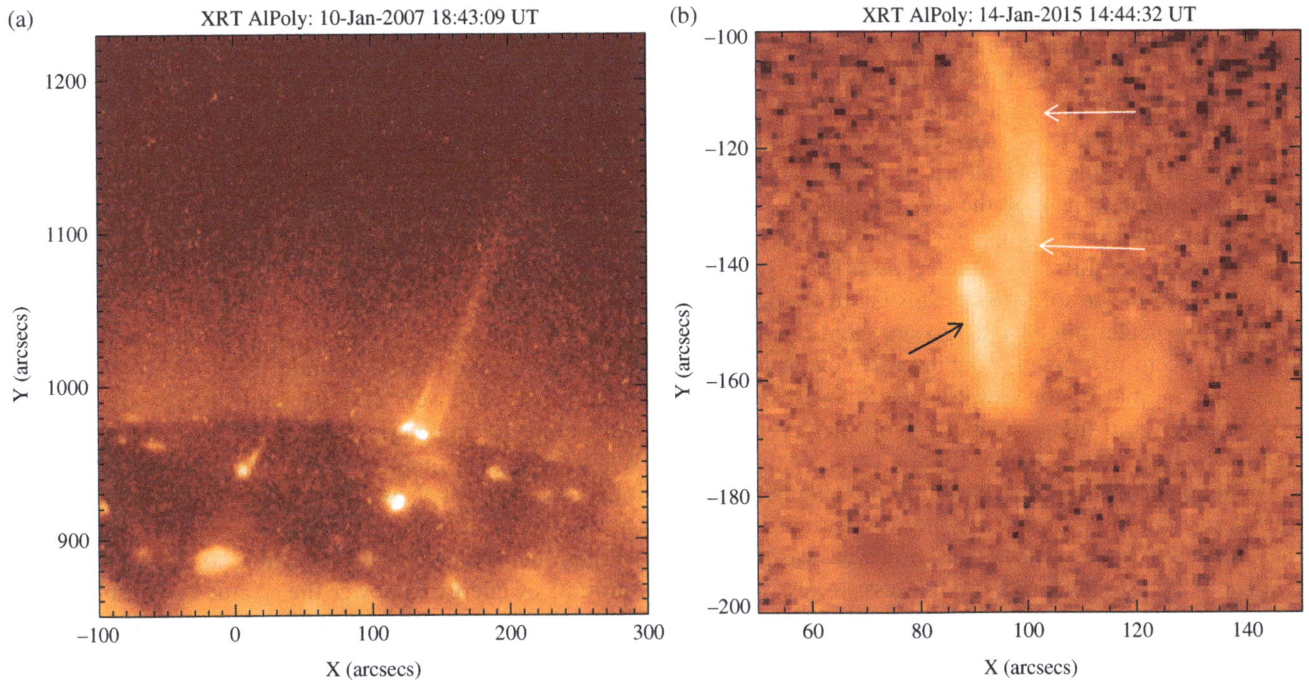

Figure 6.2 Examples of X-ray images of solar coronal jets observed with the *Hinode*/X-ray telescope (XRT), with (a) showing the north polar coronal hole limb region (taken with the XRT "Al Poly" filter), and (b) showing a close-up of a jet in an active region on the disk (taken with the "Be thin" filter); white arrows point to the spire. The typical morphology for the jets is a spire shooting away from the Sun, with a bright base at the bottom of the spire. Often one side of the base is particularly bright (e.g., the left side of the base in (b); black arrow), a feature we refer to as a "jet bright point" (JBP; following Sterling et al., 2015). (*Source*: The jets in (a) were studied by Cirtain et al., 2007 and that in (b) was studied in Sterling et al., 2017.)

observational discussion and reviews theoretical ideas for spicules since the Beckers papers, with an emphasis on the results of numerical modeling. Tsiropoula et al. (2012) provide an updated review emphasizing observations, and Zaqarashvili and Erdléyi (2009) review waves in and on spicules.

6.2.1. Earlier Spicule Observations

Spicules were detected with the advent of Hα observations of the chromosphere in the late 1800s and early 1900s, and were also studied in white-light eclipse photographs. Given adequate atmospheric seeing and sufficient spatial resolution, they appear prominently in emission against the dark background of space at the solar limb. Their presence gives a fine-scale nonuniformity to the Hα limb. When those early observations were of sufficient cadence, individual spicules were seen to move out from the photosphere and reach into the corona, and then to either fall back to the photosphere or to fade before falling.

Due to their small size, which rendered them near the edge of detectability with the instruments of those times, it was a challenge even to characterize consistently the properties of spicules. Various works (e.g., Lippincott, 1957; Rush & Roberts, 1954) resulted in some average values for spicules, such as maximum heights of 6500–9500 km, widths of ~300–1500 km (e.g., Lynch et al., 1973; Nishikawa, 1988), and lifetimes of 1–10 min (Bray & Loughhead, 1974; Lippincott, 1957). Other studies yielded spicule temperatures of 5000–15,000 K and densities of ~3×10^{-13} g cm^{-3}; there was much discussion of spicule rise velocities, with a typical quoted value being ~25 km s^{-1} (see summaries in, e.g., Beckers, 1968, 1972).

There has also been consideration of whether spicules always fall back to the solar surface. The upshot is that although both rising and falling motions were reported, a percentage of them fade from view (in the chromospheric line of the observation) after reaching their maximum heights. That reported percentage varied by a large amount, perhaps depending on the region in which the spicules occurred and/or the specific observation circumstances. For example, Lippincott (1957) reports that (only) 58% of her sample of 77 spicules appear to descend or fade from the top, whereas Pasachoff et al. (2009) report that ~70% of 40 QS spicules they observed fade. Suematsu et al. (1995) report that 80% could be followed through complete up-and-down cycles for spicules they observed on-disk in a region of enhanced network. Basic considerations regarding mass flux, however, show that something like 99% of the material that goes upward in spicules must return to the surface. That is, at least most of the fading spicule must return to the surface. This is because a calculation using the above properties for spicules, and assuming that, say, 30% of the upward material escapes the Sun results in an upward mass flux ~100 times that observed in the solar wind. This led to the suspicion that many spicules might undergo sufficient heating during their ascent to stop emitting in visible chromospheric lines. And comparing the upward mass flux with observed UV downflows in the magnetic network helped fuel this suspicion (Pneuman & Kopp, 1978).

To understand the driving source of spicules, it was important to know what was going on at their bases as they started to rise upward. That, however, was not (and still is not even today) trivial to address. A key difficulty is that spicules are best seen at the limb. But because they live only for a few minutes, it is not possible to follow the same spicule seen at the limb as it rotates onto the solar disk. (So, for example, if we observe a spicule at the east limb, it will be gone in 5 min, and so when the base of that particular spicule is on the disk—several hours or a day later—the spicule, and the fine-scale magnetic flux arrangement at its base, will be long gone.) Moreover, in contrast to spicules always emitting when viewed at the limb, disk observations showed spicule-like features (i.e., of similar physical dimensions and roughly similar lifetimes) that were either absorbing or emitting: these features are respectively called dark and bright mottles, with use of the term "mottle" emphasizing that a one-to-one correspondence with spicules had not been established unambiguously (some workers use the term spicule instead of mottle for the on-disk features also, e.g., Zirin, 1988). Near-limb mottles were seen as limb spicules in careful studies (Koutchmy & Macris, 1971). Nevertheless, a key finding from Hα on-disk studies of the spicule-like mottles is that they are rooted in the magnetic network. This supports the notion that the spicules have a strong magnetic connection, in that they are at the least channeled along the field. Features resembling mottles that are more horizontal than vertical are referred to as *fibrils* (Foukal, 1971b, 1971a; Zirin, 1988).

Meanwhile, observations in UV (Dere et al., 1983) showed features that resemble spicules, in that they grow in length to become long and thin like spicules, but are much hotter (~10^5 K) than the chromospheric spicules. And an unanswered question was that of the relationship between these *UV spicules* and normal (i.e., chromospheric) spicules.

Also, some EUV features (Bohlin et al., 1975; Dere et al., 1989; Harrison et al., 1997; Loboda & Bogachev, 2019; Withbroe et al., 1976; Xia et al., 2005) were described as larger-than-normal spicules, and hence called *macrospicules*. Moore et al. (1977) and LaBonte (1979) identified some macrospicules with surge-like eruptions. Several observations support that macrospicules spin (Kamio et al., 2010; Pike & Mason, 1998; Sterling et al., 2010).

6.2.2. Earlier Spicule Models

Because the roots of on-disk spicule-like features were seen to congregate at the boundaries of the magnetic network, it was clear very early on that spicules are dependent upon the magnetic field. But what drives them? Whatever it is, it must be a very common aspect of the lower solar atmosphere in order to produce spicules in the observed copious quantities.

Also, the mechanism must be able to supply sufficient amounts of energy to power the spicules. A simple calculation shows that the required amount of gravitational energy, based on the average numbers presented above, is ~10^{24}–10^{25} erg for an individual spicule (e.g., Sterling, 2000). This energy can easily be supplied by photospheric motions, given that individual photospheric granules have kinetic energy flux that is about two orders of magnitude larger than this. On the other hand, the rate at which this energy must be supplied into the area of the base of the spicule (assuming a lifetime of 5 min and a radius of 300 km, say) is 10^6–10^7 erg cm^{-2} s^{-1}; this is a substantial value, in the sense that it is comparable to the energy fluxes required to heat the corona. If, however, we assume that spicules cover only about 1% of the Sun's surface (which seems reasonable, based on the density of mottles on the disk), then these energy fluxes would *not* be sufficient for coronal heating.

This reasoning has been used in the past to argue that spicules are not capable of heating the corona. More recent studies, however (e.g., Falconer et al., 2003; Moore et al., 2011; Zaqarashvili & Erdléyi, 2009), hold that the energy from spicules might indeed be large enough to power the corona if one takes into consideration the energy in spicule oscillations, for example, Alfvén waves, which some recent observations suggest is substantial. And in fact the question of whether spicules heat the corona is a topic still under active debate (Bradshaw & Klimchuk, 2015; De Pontieu et al., 2011; Klimchuk, 2012; Klimchuk & Bradshaw, 2014; Madjarska et al., 2011).

There are several excellent summaries of discussions of spicule models up to the early 1970s (e.g., Athay, 1976; Beckers, 1968, 1972; Bray & Loughhead, 1974). From the later 1970s, numerical methods were becoming more practical due to improvements in computing power, resulting in detailed numerical models of spicules (reviewed by Sterling, 2000).

A series of models assumed that hydrodynamic waves and/or shocks channeled by the magnetic field generated spicules (e.g., Blake & Sturrock, 1985; Cheng, 1992; Hollweg, 1982; Shibata et al., 1982; Sterling & Hollweg, 1988; Suematsu et al., 1982).

A subset of these models (e.g., Cheng, 1992; Hollweg, 1982; Hollweg et al., 1982; Sterling & Hollweg, 1988; Sterling & Mariska, 1990) relied on *rebound shocks* to create the spicule. In the 1D simulations, these shocks result from a pulse of ~1 km s^{-1} (say, induced by granule motions) at the base of a vertical magnetic flux tube. The pulse propagates upward and nonlinearly evolves into a series of shocks (the rebound shocks, formed when some of the material raised by a shock falls and rebounds off the denser material below). These shocks then lift the transition region, and also shock-heat the chromospheric material below the transition region. This raised and heated chromospheric material was proposed to be the spicule. The simulated non-steady rise of the spicule top could match early reports that some spicules appear to rise in jerks (Bray & Loughhead, 1974). There were, however, several difficulties with this model, such as the question of whether there was a proper accounting for the radiative losses. Also, it was not certain whether spicules showed the non-steady upward motion predicted as the rebound shocks successively strike the bottom side of the transition region at (roughly) the acoustic cutoff frequency of the chromosphere. (More recent observations from *Hinode*, e.g., De Pontieu, McIntosh, et al., 2007, do *not* show spicules to have such drastic non-ballistic motions, and so the earlier reports of jerky trajectories may have been due to terrestrial atmospheric effects.)

Another series of these models (Shibata et al., 1982; Shibata & Suematsu, 1982; Suematsu et al., 1982) has a similar initial atmospheric setup, but relies on an initial pulse that is stronger than in the rebound shock model. Among the predictions for a spicule resulting from this process was that a base brightening would be expected at the spicule start time, due to the large initial energy input. Observations in Hα of Suematsu et al. (1995), however, did not find such base brightenings at the expected times.

Other models (Shibata et al., 1982; Sterling et al., 1991, 1993) considered whether an initial energy deposition in the chromosphere (instead of the photosphere as in many other simulations) could drive spicules. This resulted in model spicules with rise trajectories closer to ballistic (and hence similar to later observations), but still predicted base brightenings that do not appear to match the observations (e.g., Suematsu et al., 1995). A later study (Guerreiro et al., 2013) also shows that this idea has difficulties matching observations.

Yet another focus was models based on Alfvén waves entering the base of a vertical flux tube that nonlinearly couple to other MHD modes to generate the spicules (Kudoh & Shibata, 1999; Hollweg et al., 1982; Mariska & Hollweg, 1985). More recent versions of this idea are continuing today (e.g., Cranmer & Woolsey, 2015, and Iijima & Yokoyama, 2017, which is discussed in section 2.4).

A conclusion of the Sterling (2000) review was that a resolution of the spicule problem was still outstanding, with progress required on the observational front to guide theoretical studies. Some such progress has come over the subsequent two decades.

6.2.3. More Recent Spicule Observations

With its launch in 2006, the *Hinode* spacecraft (Kosugi et al., 2007) provided the first long-term high-resolution observations of the solar chromosphere from space, with its 50 cm Solar Optical Telescope (SOT; Tsuneta et al., 2008).

"Type I, "Type II," and "Classical" Spicules

SOT limb views revealed spicules with characteristics quite different from those discussed previously. From observations at the limb in Ca II with *Hinode*/SOT, along with earlier work reported in De Pontieu et al. (2004), De Pontieu, McIntosh et al. (2007) concluded that there were two different spicule populations, which they called type I and type II spicules.

Describing the latter of these first: Type II spicules are prevalent in coronal holes and quiet Sun; they have relatively high velocities (30–150 km s^{-1}), lifetimes of 10–150 s, and usually fade from view without falling back to the surface (a later study by Skogsrud et al., 2015, shows that often a faint trace of the falling type II spicules can be seen with sufficient image-intensity enhancement). Type I spicules seem to be most prevalent in and around ARs. They have upward velocities of 15–40 km s^{-1}, lifetimes of 3–10 min, and are seen to rise and fall back to the surface (some of these values are from an updated summary in Pereira et al., 2014).

Even prior to this usage of the term "Type I spicules," small jet-like features were identified in plage regions with high-resolution ground-based images, sometimes observed in conjunction with UV images and given various designations, including "dynamic fibrils" (e.g., Hansteen et al., 2006). Extensive observations (Berger et al., 1999; De Ponteiu et al., 1999; De Pontieu et al., 2003), and both 2D (e.g., Hansteen et al., 2006) and (e.g., Heggland et al., 2011; Martínez-Sykora et al., 2009) 3D MHD simulations, support the idea that these features (along with at least some quiet Sun mottles; Rouppe van der Voort et al., 2007) are driven by acoustic waves. Most of these show oscillations close to the acoustic cutoff frequency (Hollweg, 1982), with additional input from *p*-mode waves on sufficiently inclined flux tubes

(De Pontieu et al., 2004; Suematsu, 1990). These studies provided insight into investigations of the *Hinode*-era "type I" spicules.

This type I/type II designation for spicules, however, has not been universally accepted, with Zhang et al. (2012) maintaining that there is no fundamental difference between spicules in the different regions. Pereira et al. (2012), however, contest the Zhang et al. (2012) findings, and present counterarguments.

(Another possible point of confusion with the terms "type I" and "type II" spicules is that the same terminology was used earlier by Beckers (1968), where his meaning is completely *different* from type I and type II as used by De Pontieu et al! Beckers' type I and type II refer to spicules with Ca line widths that are, respectively, relatively more broad and less broad; see section 3.3.1.4 in Beckers (1968) for details. In this work, and in most of the discussions in the literature on spicules in recent years, the terms refer to the spicules as defined by De Pontieu, McIntosh, et al., 2007.)

Most of the earlier ground-based observations were in Hα, and therefore there are inherent differences from the spicules observed in Ca II by SOT. And moreover, the specific Ca II filter of SOT is different from that used in ground-based Ca II observations. Therefore, it is not clear—at least not to this author—that SOT sees the same component of spicules as that described in the historical literature on spicules from ground-based observations. (And indeed, there could be differences in what was observed and described by different ground-based observations.) That is, the SOT Ca II passband may be seeing limited aspects of the spicules described earlier. In order to be clear when discussing the newer observations in terms of the older ground-based observations (and newer observations taken in the manner of the older observations from the ground), we call the spicules seen in ground-based observations from prior to approximately the year 2000 (and hence, including the Beckers' era spicules) "*classical spicules*." This terminology was used in Sterling, Harra, and Moore (2010), and also by Pereira et al. (2013).

Assuming for the moment that there are two types of spicules (that is, assuming for the moment that De Pontieu, McIntosh, et al., 2007— rather than Zhang et al., 2012—are correct in saying that there are two types of spicules), we might ask which of the two types corresponds to the spicules observed over the first 100 years of spicule studies? I maintain that the classical spicules are type II spicules, via the following rationale.

Earlier (i.e., classical) investigations are either unclear on whether spicules exist over plage regions, or they say that they are absent over plages (Michard, 1974; Zirin, 1974, 1988). Moreover, the tallest, most pronounced spicules studied during that earlier period were those in coronal holes (Rabin & Moore 1980). (Gaizauskas, 1984, does report spicules at the *borders* of an active region, but does not talk about spicules over the main part of the AR itself.) It is true that most of those observations were with various resolutions, time cadences, and likely varying quality of atmospheric seeing, and therefore the physical properties derived from those earlier periods would indeed be subject to question. But the location on the Sun in which they occurred would be quite reliable. The classical spicules were generally observed in quiet Sun and coronal holes. Because that is precisely where the type II spicules are positively observed, *most classical spicules must correspond to the type II spicules*. Pereira et al. (2013) also present arguments that type II spicules look like classical spicules in many aspects, if the SOT images are degraded in time and cadence to mimic the ground-based circumstances. For example, in the original data, the mean maximum velocity of their type II spicules was 60.0 km s^{-1}. In their degraded data, this was reduced to 25.8 km s^{-1}, which is close to the reported values for classical spicules. This procedure, however, did not show as high a percentage of spicules falling back to the surface as quoted in the classical-spicule literature; however, the classical results are mainly derived from Hα rather than from Ca II as used by SOT, and so the comparison is tricky in many respects.

Type I spicules, on the other hand, are reported to be seen primarily in plage regions. Thus, type I spicules likely were rarely, if ever, observed during the classical period. Indeed, using *Hinode*/SOT images, Anan et al. (2010) found lengths of spicules in a plage region to average only ~1000 km, which is substantially shorter than those found in quieter regions. Therefore, type I spicules were probably not reported in classical observations because their shorter extent in plage regions made them more difficult to detect. (Shibata et al., 1982, discuss the question of the near-absent appearance of spicules in plage regions in terms of a wave-pulse model.)

So in summary: Classical spicules are those discussed by Beckers (1968) and Beckers (1972), meaning seen from the ground mostly in Hα (with the inherent limitations of ground-based observations), mainly in CHs and QS regions, with upward velocities of 25 km s^{-1} and other properties as described above. Type I and type II spicules are as defined by De Pontieu, McIntosh, et al. (2007): type IIs (observed in SOT/ Ca II) dominate in CHs and QS, have upward velocities of ~30–150 km s^{-1}, generally fade without falling, and so on, as described above. Type I spicules occur in or near ARs, have upward velocities of ~15–40 km s^{-1}, show both up-and-down motions, and so on, as described above.

And if there are two types of spicules, there is supporting evidence that type II spicules correspond most closely to the classical spicules. The average quoted speeds and lifetimes of type I spicules is closer to the classical average values than are those of type II spicules, but as just described, the match between the classical and type II average values are closer when the quality of the images of the type II spicules is degraded, as presented by Pereira et al. (2013).

(Several works equate type I spicules with classical spicules, but as I have just argued, this association is not correct; or at least, it is unclear and confusing. Some papers that make this association are Scullion et al., 2011, Tsiropoula et al., 2012, Klimchuk, 2012, and Klimchuk & Bradshaw, 2014; in none of these cases does this association have consequences for the main themes of those works.)

Other Aspects of Newer Spicule Observations

We briefly touch on a few more aspects of spicule observations, with an emphasis on insights gained from newer studies.

Because classical observations had long indicated that some spicules observed in chromospheric spectral lines fade, it was long suspected that spicules are heated as they rise (e.g., Pneuman & Kopp, 1978; Sterling, 1998; Sterling & Hollweg, 1984). Such evolution from cooler to warmer spicules has received strong support in recent years by combining observations from SOT with those from the Atmospheric Imaging Array (AIA) (De Pontieu et al., 2011) and IRIS (De Pontieu et al., 2014; Pereira et al., 2014; Skogsrud et al., 2015), showing a progression of emission from cooler (chromospheric) spectral lines to hotter (UV and EUV) lines. These observations also showed definitively that at least some chromospheric and transition-region spicules have a one-to-one correspondence with each other. Several other studies have also considered the question of whether spicules have counterparts at transition region and coronal temperatures (e.g., Jiao et al., 2015; Samanta et al., 2015; Tian et al., 2014).

There also has been suspicion from classical observations that spicules display twisting motions, due to some of the observed spectral lines appearing tilted when the spectral slit is placed normal to the length of the extended spicules, consistent with the two sides of the spicule moving in opposite directions (Beckers, 1968; Livshits, 1967; Pasachoff et al., 1968; Rompolt, 1975). Newer observations now provide supporting evidence that as least some spicules twist (De Pontieu et al., 2012; Suematsu et al., 2008). In addition, some UV spicules similarly likely show twist (De Pontieu, Rouppe van der Voort, et al., 2014). Zaqarashvili and Erdléyi (2009) review oscillations and waves in spicules.

Spicules are essentially invisible in SOT Ca II on-disk observations (Beck et al., 2013), unless extremely close to the limb (Sterling, Moore, & DeForest, 2010). Improvements in ground-based techniques, however, reveal new features that are suspected of being counterparts to some of the limb-observed high-speed spicules; these include so-called "straws" (Rutten, 2007), "rapid redshifted excursions" (RREs), and "rapid blueshifted excursions" (RBEs; e.g., Langangen et al., 2008; Rouppe van der Voort et al., 2009; Sekse et al., 2012; Sekse, Rouppe van der Voort, & De Pontieu, 2013; Sekse, Rouppe van der Voort, De Pontieu et al. 2013). This is an exciting, but involved, topic that we do not discuss further here.

6.2.4. More Recent Spicule Models

Since the review of earlier spicule models by Sterling (2000), one of the ideas for spicule generation that has garnered much attention is that they result from the leakage of photospheric *p*-mode oscillations into the upper atmosphere. This concept differed from some of the earlier pulse models (e.g., Hollweg, 1982; Suematsu et al., 1982) in that those earlier models considered only pulses of frequency below the chromospheric acoustic cutoff frequency, which corresponds to a period of about 220 s in the chromosphere. This means that *p*-mode waves, of periods ~300 s, would be expected to be evanescent in the chromosphere. De Pontieu et al. (2004) showed that using an input spectrum of *p*-modes in a numerical simulation could reproduce model spicules with characteristics that closely matched spicules observed in an active region. It turns out that—in the simulation—the initially evanescent chromospheric waves driven by the *p*-mode oscillations can propagate far enough into the chromosphere to reach a layer where they again become propagating, and *p*-mode oscillations on a steeply enough inclined-from-the-vertical flux tube propagate instead of being evanescent. These "leaked" *p*-modes waves result in the spicules in the model. Suematsu (1990) conducted a preliminary investigation of *p*-mode waves on inclined flux tubes. The idea received increased attention later, in particular after De Pontieu, McIntosh, et al. (2007) argued that many of the spicules they refer to as "type I" are produced by such a mechanism.

Observations of dynamic fibrils (De Pontieu, Hansteen, et al., 2007; Hansteen et al., 2006; also see Marsh, 1976), and at least some quiet Sun mottles (Rouppe van der Voort et al., 2007) show oscillations both near the acoustic cutoff frequency and the *p*-mode frequency, consistent with wave models. De Pontieu, McIntosh, et al. (2007) provide evidence that the spicules they refer to as type II behave differently, and are thus likely to be driven by a different mechanism.

Modeling features with characteristics of type II spicules has been more challenging. For example, simulations of Martínez-Sykora et al. (2011) and Martínez-Sykora et al. (2013) using a 3D MHD code produce model spicules with the reported type II properties, but not with a frequency of occurrence needed to match the true Sun; spicules are rare in that model but common on the Sun.

Much more promising are simulations of Martínez-Sykora et al. (2017), who find that, with the addition of ambipolar diffusion processes, they can produce copious type II spicules in a model solar atmosphere. Their simulations suggest that spicules result when "magnetic tension is amplified and transported upward through interactions between ions and neutrals or ambipolar diffusion."

Other recent spicule models include that of Iijima and Yokoyama (2017), which drives spicules via Alfvén twists at the spicule base, following on to the work of Hollweg et al. (1982) and Kudoh and Shibata (1999). Their resulting simulated spicules look realistic, and so their work supports the hypothesis that this process may explain some spicules. In order to reproduce spicules of large enough heights, however, their simulations appear to require motions near the photosphere of 5–10 km s^{-1} (see figure 7 of Iijima & Yokoyama, 2017). Historically, typical reported photospheric motions due to granules are only ~1 km s^{-1}, for example, Foukal (2013), and so an interesting question is whether improved high-resolution observations of granules, intergranular flows, and/or intergranular magnetic elements might show the required velocities.

Using a different approach, Judge et al. (2011) suggest that type II spicules might be warped current sheets that only appear to be long and thin fast-moving features, due to a kind of illusion when the sheets are viewed in superposition. Alfvén fluctuations in the sheets driven from below would produce the apparent motions. They argue that this could explain the sudden appearance of some spicules along their full length. An alternative explanation however for this sudden appearance, one that does not rely on sheets, is that it results from torsional motions of spicules suddenly becoming Doppler shifted into the observed wave band (Sekse, Rouppe van der Voort, & De Pontieu, 2013; Sekse, Rouppe van der Voort, De Pontieu et al. 2013).

Sterling and Moore (2016) present arguments that spicules work like miniature versions of the larger-scale coronal jets, building on an idea presented earlier by Moore (1990). (Also see Falconer et al., 2003, and Moore et al., 2011.) We will revisit this suggestion in section 6.4.2, after discussing coronal jets in the next section.

The field of spicule research has again become extremely rich, and many other studies not discussed here have been carried out in recent years. Moreover, active observational and theoretical studies based on past and current data are still underway. Continued improvements in observational capabilities—for example, the Daniel K. Inouye Solar Telescope (DKIST), as well as new telescopes at Big Bear Solar Observatory and in China—should allow for new large strides in our understanding of spicules.

6.3. CORONAL JETS

One of the more dramatic early features observed with the soft X-ray telescope (SXT; Tsuneta et al., 1991) on the *Yohkoh* spacecraft (launched in 1991) were dynamic jets shooting out from the lower atmosphere into the corona. These SXT observations motivated jet studies over the following two decades. Similar to the situation with spicules however, there have been two eras of jet studies, albeit on a much shorter time frame than the over 100 years of spicule studies (although surges, at least some of which are cool components of coronal jets, have been long observed from the ground). After the initial *Yohkoh*-era work, new observations with the X-ray telescope (XRT) on *Hinode* (launched in 2006), and with instruments on the *Solar Dynamics Observatory* (*SDO*, launched in 2010), spearheaded new understanding of jets. On *SDO* the key instruments for jet studies were its AIA (Lemen et al., 2012), and its Helioseismic and Magnetic Imager (HMI; Scherrer et al., 2012). The presentation below will cover both of these periods. We will, however, emphasize developments in the later era, as the earlier period has been covered in much greater detail in the extensive review of Raouafi et al. (2016).

6.3.1. Earlier Jet Studies: Observations

There were indications of the nature of coronal jets prior to SXT (Raouafi et al., 2016), but they were only fully appreciated with the new long-term and nearly continuous high cadence and resolution capabilities of SXT (Shibata et al., 1992; Strong et al., 1992). Shimojo et al. (1996, 1998) examined the properties of the jets in detail, finding values that generally covered wide ranges, for example, lengths ranging from ~10^4–10^5 km, widths of 5000–10^5 km, velocities from tens of km s^{-1} to over 1000 km s^{-1} (averaging 200 km s^{-1}), and lifetimes covered from minutes to hours; Shimojo and Shibata (2000) found jet temperatures to range over 3–8 MK, with an average of 5.6 MK. In some cases, the wide ranges are a consequence of the variable cadence and spatial resolution used in the SXT observations. For example, although SXT was capable of high cadences of a few seconds, it also had large gaps in coverage, due to the satellite day-night cycle and other factors. Jets in ARs can sometimes occur in rapid succession (Sterling et al., 2016), but these might be inferred as a single jet without the benefit of continuous coverage at sufficient cadence ($\lesssim$1min). If, for example, the footpoints of those multiple jets were slightly offset from each other, it might be concluded from low-cadence observations that there is one exceptionally broad jet occurring from that area, whereas higher cadences could reveal two or more jets.

A common aspect of essentially all jets is that they have a bright point, or, in more detail, a small bright loop or arcade of loops, at the base of the spire that makes up the extended part of the jet. In many jets, this bright point is located asymmetrically off to one side of the spire, rather than directly beneath the spire. For convenience we will call this brightening the jet base's jet bright point (JBP; cf. Sterling et al., 2015). The JBP is frequently more prominent in X-ray images than in, say, EUV, indicating that it is substantially hotter than other parts of the jet and the nearby corona.

Jets observed with SXT are, of course, X-ray jets. Moreover, Shimojo et al. (1996) found that most SXT jets (68%) occurred near ARs. As we will see shortly (section 6.3.3), coronal jets (not all of them necessarily detectable in X-rays) are also plentiful in coronal holes and quiet Sun. The SXT bias toward the AR jets is likely due to the temperature response of SXT (Tsuneta et al., 1991), which tends to see hotter plasmas than does, for example, XRT on *Hinode* (section 6.3.3 below).

Following the initial X-ray observations of coronal jets by *Yohkoh*, they have been observed in EUV by several different instruments. Several instruments on the *Solar and Heliospheric Observatory* (*SOHO*, launched in 1995) satellite contributed to jet studies. The *SOHO*/Extreme Ultraviolet Imaging Telescope (EIT) took images at four different EUV wavelengths: 171, 193, 284, and 304 Å. Its cadence was 12 min, which somewhat limited its ability to study the evolution of jets because this is comparable to the lifetime of many jets (Savcheva et al., 2007).

6.3.2. Earlier Jet Studies: Theories

To create brightenings in the corona of the intensity of jets requires substantial energy, and from early on it was suspected that they are powered by magnetic reconnection. Shibata et al. (1992) recognized the JBP as a clue to the cause of the jet. They proposed that reconnection between a newly emerging magnetic bipole and a preexisting ambient far-reaching magnetic field was the source of the jet.

Figure 6.3 shows a schematic of this proposed emerging-flux mechanism. Here, the background field is of a single polarity (negative in this schematic), as in, say, a coronal hole region. When a bipole emerges into this region, clearly the polarity of one of the bipole's two polarities will match that of the background, and the other pole will be of the opposing polarity. As emergence continues, a current sheet develops at the interface between the bipole and the open field, and eventually magnetic reconnection can occur along this sheet.

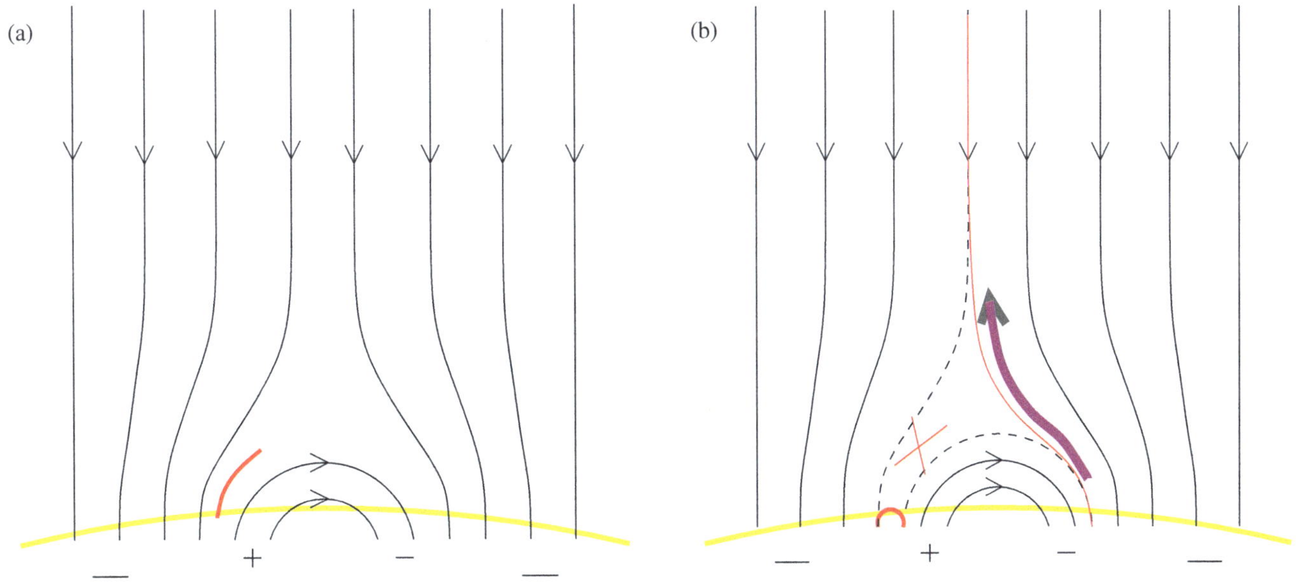

Figure 6.3 This is a schematic representation of the emerging flux model for coronal jets proposed by Shibata et al. (1992). They suggested that a jet results when (a) a bipolar field emerges from beneath the solar surface, forming a current sheet where the field lines are oppositely directed. (b) Continued emergence results in reconnection at the current sheet (red X), with one reconnection product being a small loop (red semicircle), which was proposed to represent the jet bright point (JBP); and the other being a new open field line (red line), along which the jet spire flows outward (purple arrow). Dotted lines represent the field that reconnected to give the JBP and the new open field. (*Source*: Sterling et al. 2015, based on ideas presented in Shibata et al. 1992.)

A result would be two post-reconnection products, one of which would be a new open field line along which reconnection-heated jet spire material could flow. A second product would be a new loop that is smaller than the emerging-bipole loops (at least in the geometry of Figure 6.3). Shibata et al. (1992) suggested that the new small loop was the bright point (later called the JBP). Moreover, in addition to the flows induced by reconnection heating, a slingshot effect would assist in propelling the spire material outward, as the open field reconnection product snaps away from the reconnection point due to magnetic tension.

At the time it was suspected that surges in ARs resulted from the same type of emerging-flux mechanism (Heyvaerts et al., 1977, and references therein). And, in fact, subsequent observations showed that surge-like cool-material ejections can also accompany X-ray jets (Canfield et al., 1996).

Yokoyama and Shibata (1995, 1996) performed numerical simulations of the emerging-flux idea, resulting in features that resembled observed jets. They also found that by assuming that the emerging-flux loop runs into an overlying coronal field that is horizontal rather than vertical or oblique, they could reproduce features that look like "two-sided loop jets," which had also been observed (Shibata et al., 1994; Yokoyama & Shibata, 1995, 1996).

Over the next 20 years, there were several simulations that included refinements to this basic idea (e.g., Archontis & Hood, 2013; Fang et al., 2014; Miyagoshi & Yokoyama, 2004; Moreno-Insertis & Galsgaard, 2013; Nishizuka et al., 2008; Yokoyama & Shibata, 2001; also see the Raouafi et al., 2016, review for a complete discussion of these earlier models).

A different set of numerical simulations were a variation on this emerging-flux idea, motivated, at least in part, by the fact that several jets and jet-like features are observed to spin (section 6.3.5). These simulations (Pariat et al., 2009, 2010, 2015) initiate the jet by adding an Alfvénic shear to the base of the jet, leading to an outward eruption and jet flows.

6.3.3. Later Jet Studies: Observations

A new era of jet studies began with the *Hinode* and *SDO* satellites. *Hinode* carried XRT (Golub et al., 2007), the follow-up to *Yohkoh*'s SXT. Because of its Sun-synchronous orbit, *Hinode* could carry out observations for extended periods of time without day-night cycle interruptions. XRT has increased spatial resolution over SXT (1″ pixels compared to SXT's 2″.5 pixels). Its spectral response (Narukage et al., 2011) extended to "softer" (less energetic) X-rays than SXT, in particular with its Al-poly, Al-mesh, and C-poly filters, so that it could effectively detect plasmas in the ~1–2 MK range. Such temperatures were too cool for easy detection by SXT (under comparatively quiet conditions, SXT could detect plasmas of ~1.5 MK; Sterling, 1999).

A set of early dedicated observations by XRT looking at the north polar region revealed that coronal jets are prominent and frequent in solar polar coronal holes (Cirtain et al., 2007), occurring at an estimated rate of 60/day in the two polar coronal holes (PCHs) (Savcheva et al., 2007). These PCH jets could reach lengths of ~50,000 km, and had lifetimes of about 10 min. There seem to have been few studies of PCH regions with SXT, but one such study (with its AlMgMn "thick filter," sensitive to temperatures near 3–5 MK) did possibly detect jet brightenings (Koutchmy et al., 1997).

Studies of the temperatures of coronal hole jets provide a possible explanation for lack of many SXT PCH jet observations. Pucci et al. (2013) used the ratio of intensities in different XRT filters for one PCH jet and *STEREO* EUV filter intensity ratios for a second PCH jet to determine jet temperatures using the filter-ratio method. Similarly, Paraschiv et al. (2015) used the intensity ratios among different XRT filters to determine the temperatures of 18 PCH jets. Both investigations gave temperatures in the range 1.5–2.0 MK. These (comparatively) low temperatures could, at least in part, explain the paucity of reported SXR jets in PCHs, and also the comparatively low SXT detection of jets outside of ARs (with 68% occurring near ARs as noted above) from Shimojo et al. (1996). Moreover, studies using the *Hinode*/EIS spectrometer yield higher temperatures, 2–3 MK, for jets occurring in ARs (Chifor et al., 2008; Lee et al., 2013; Matsui et al., 2012; Mulay et al., 2016, 2017a, 2017b).

Standard and Blowout Jets

By investigating a substantial number of jets in X-rays, Moore et al. (2010, 2013) observed that they could be categorized by their morphology into roughly two different types. To begin with, all jets have a long spire extending into the corona, and a broad base region that brightens, with the spire emanating from that base region. In one type of jet, the basic pattern is that the spire of the jet remained narrow throughout its life, and the JBP was the brightest part of the base throughout the life of the jet, and was located off to one side of the spire. Here, "narrow" can be defined in terms of the size of the base; so in this type of jet, when viewed in X-rays, the spire always remains narrow compared to that jet's base.

A second pattern seen in the data was that the jet spire would start out narrow, but then the spire would broaden so that it became about as wide as the base of the jet. Also, the brightening at the base was often not confined to one location on the side of the spire, but instead the entire base brightens to an intensity rivaling or exceeding that of the initial JBP.

Moore et al. (2010, 2013) called the narrow and broad jets respectively "standard jets" and "blowout jets" for the reasons we now present. (We alert the reader, however, that these explanations are no longer believed by the primary authors who defined those terms to be fully valid! Their current view of the explanations for standard and blowout jets is presented in section 6.3.4.2.)

Moore et al. (2010) argued that the jets that remained narrow in X-rays were produced by the emerging-flux mechanism, whereby the emerging bipole undergoes reconnection with ambient field and produces a spire and JBP as in Figure 6.3, and throughout the lifetime of the jet the emerging-flux arcade at the base of the jet remains intact. Moore et al. (2010) called these narrow jets "standard jets," because they were thought to follow the standard picture for jets presented in Shibata et al. (1992).

By contrast, Moore et al. (2010) suspected that the broad-spire jets started out in the same fashion as the standard jet, with the flux emerging into the corona with an ambient field, but that the emerging (or already emerged) bipole becomes destabilized and erupts explosively. In this case, early in the eruption, reconnections would occur between the bipole and the ambient field, leading to the initial narrow spire and JBP from that reconnection just as in the standard jet case. After that (however still well within the ~10 min lifetime of the jet), the bipole's eruption continues, and the entire body of the erupting bipole would result in brightening over the entire width of the base region, through reconnections with the ambient field. Reconnections in the legs of the erupting bipole would result in brightenings similar to larger-scale flare brightenings, illuminating the entire base region. Because in this scenario the key element is the emerging (or emerged) bipole that erupts out and away from the solar surface, Moore et al. (2010) called these broader-spire jets "blowout jets." They found that, for 109 X-ray jets over those two studies, 53 were classified as standard and 50 as blowout, with 6 being classified as ambiguous.

Although many jets clearly fit into one or the other category, for several others jets the category determination—based on whether a spire is broad or narrow and other morphological factors—are likely somewhat subjective. Nonetheless, the nearly 50-50 split from the Moore et al. studies shows that both types of jets occur frequently, at least in PCH regions.

Complementary observations in EUV are consistent with this basic dichotomy into standard and blowout jets; in fact, the EUV observations were used as further evidence in developing this story. EUV 304 Å observations show erupting filament/prominence material, especially when observing at the limb. Moore et al. (2010) used EUV 304 Å observations from *STEREO* of four XRT X-ray jets, and Moore et al. (2013) used EUV 304 Å observations from AIA to observe all of the XRT X-ray jets of that study. In total, they found that 29 out of 32[1] of the blowout jets (as identified in X-rays) had a cool 304 Å component; this supports the notion that an emerging or emerged bipole could be erupting, as suggested in the blowout-jet scenario first presented in Moore et al. (2010).[1]

Several other papers investigated blowout jets, including Hong et al. (2017), Chandra et al. (2017), Zhu et al. (2017), Joshi et al. (2017), Shen et al. (2017), and Joshi et al. (2018).

6.3.4. Later Jet Studies: Theories; the Cause of Jets Revisited

Adams et al. (2014) studied a jet in an on-disk coronal hole using *SDO* AIA EUV and HMI magnetic-field observations, with the goal of finding the flux emergence that was hypothesized to be the agent driving the jet. They did not, however, find any such emergence; that is, their data did not support the hypothesis (pictured in Figure 6.3) that an emerging or recently emerged bipole was the source of the jet. Instead, the source of the jet appeared to be a small-scale filament that erupted to form the jet spire, and the jet bright point (JBP) that occurred at the side of the jet appeared to be a scaled-down version of a flare that accompanies large-scale filament eruptions. Moreover, watching the long-term evolution of the magnetograms showed that the jet occurred at a location where opposing polarity magnetic elements converged and canceled, leading to the jet.

Thus, Adams et al. (2014) concluded that this jet was produced when magnetic flux cancelation resulted in a small-scale filament eruption that evolved to form the jet spire, the JBP being the small-scale flare accompanying the small-scale filament eruption. This was quite different from what was expected based on the emerging-flux idea first presented about 20 years earlier, and Adams et al. (2014) say explicitly that their "jet is a newly recognized variety of blowout jet," different from that proposed in Moore et al. (2010).

Other studies, it turns out, had seen various elements of this conclusion. Several workers suggested that their observed jet(s) could be due to small-scale eruptions (e.g., Hong et al., 2014; Nisticò et al., 2009; Raouafi et al., 2010). Shen et al. (2012) very nicely show an erupting small-scale filament leading to a jet. And going back much further, although there was no identification with coronal jets, Moore (1990) suggested that miniature filament eruptions produced macrospicules, and Wang et al. (2000) found that miniature filament eruptions are common in the quiet Sun. Several studies also found that flux cancelation accompanied jets in some events or circumstances (Hong et al., 2014; Innes & Teriaca, 2013; Shen et al., 2012; Young & Muglach, 2014a, 2014b).

Other instruments contributing to jet studies from this period include *SOHO*/CDS, SUMER, and UVCS, and also the EUV *TRACE* telescope. We refer the reader to Raouafi et al. (2016) for detailed discussions of jet results from these instruments.

Minifilament Eruption Model for Coronal Jets

Sterling et al. (2015) used the data set of Moore et al. (2013) of near-limb PCH X-ray jets, and examined 20 of these jets in all of the AIA channels. (Moore et al., 2013, had inspected AIA 304 Å data, but not that of other channels.) They found that *all* of the jets were consistent with having resulted from eruptions of minifilaments, with the JBP being a miniature flare accompanying the minifilament's eruption. Figure 6.4 shows one of the 20 jets Sterling et al. (2015) found to result from minifilament eruptions. Earlier studies had shown minifilaments erupting to form jets in studies of limited numbers of events; Figure 6.5 shows an example of a minifilament eruption producing a jet studied by Shen et al. (2012). Sterling et al. (2015) showed that all jets in a sample of 20 were formed by this process, and suggested that all jets were made this way.

Figure 6.6 shows the schematic scenario they presented for this minifilament eruption model; this figure is slightly modified from the original version in Sterling et al. (2015), to reflect the findings of Moore et al. (2018). Figure 6.6a shows a double-bipole field embedded in an ambient coronal field having a single magnetic polarity and extending approximately vertically from the surface. This view is a good approximation to the expected situation in a coronal hole region, but it also applies to other regions of the Sun. For example, the ambient field might be part of a coronal loop that is large compared to the double bipole, the requirement being that the loop is large compared to the double bipole shown in the figure. (More precisely, the loop has to be substantially larger than its base, and large enough that reflections from the far side of the loop are of no significant consequence to the jet's development; if these conditions hold, then the picture of Figure 6.6 can be considered apropos without modification.)

In 2D cross section, Figure 6.6 shows a double bipole embedded in the base of the ambient field. This setup would naturally occur on the Sun if an opposite-polarity field were embedded in the base of the unipolar ambient field. In 3D, that opposite-polarity field would bloom out into an anemone-type configuration (Shibata et al., 2007), so that opposite-polarity flux connects to the surrounding ambient flux; Figure 6.6 shows a 2D cross-sectional cut of that 3D anemone structure. In this cross-sectional view, one side of the bipole is more compact (the right-hand side in Figure 6.6a) and more non-potential than the other, likely due to asymmetric photospheric flows and flux cancelation (section 6.3.5). Magnetic shear and twist would build up in that stressed side. It is well known that large-scale filaments form along polarity inversion lines in sheared field; the location between M2 and M3 in Figure 6.6 is such a sheared-field location, and this is just where we observe miniature filaments existing prior to the jets (section 6.3.5).

Figure 6.6b shows the next stage. At some point (again due to the processes discussed in section 6.3.5), the magnetic field enveloping the minifilament becomes destabilized and starts to erupt. Prior to erupting, the field enveloping the minifilament was either a sheared magnetic arcade structure, or a magnetic flux rope. In either case, at the latest it is a flux rope soon after the eruption starts. As it erupts outward from the surface, magnetic reconnections occur at two different locations. One location is the field below the erupting minifilament; that is, the "legs" of the erupting minifilament field come together below the minifilament as it erupts away from the photospheric magnetic neutral line. This reconnection, which we call *internal reconnection* (because it is internal to the erupting minifilament field) is identical to the reconnection that occurs beneath erupting large-scale filaments resulting in solar flares, in the standard model for solar eruptions, sometimes called the CSHKP model after Carmichael (1964), Sturrock (1966), Hirayama (1974), and Kopp and Pneuman (1976) (also see, e.g., Shibata et al., 1995, Moore et al., 2001). In this case, the result is a miniature flare, which we identify as the JBP, below an erupting flux-rope field wrapped around a cool-material minifilament.

The other reconnection that occurs as the minifilament flux rope erupts takes place in the region where the flux rope field contacts the ambient field, a location that is an approximate magnetic null point. We call this *external reconnection*, as it occurs on the external side of the minifilament flux rope. This is identical to the reconnection referred to as "breakout reconnection" (Antiochos, 1998; Antiochos et al., 1999), or "interchange reconnection" (Crooker et al., 2002; Crooker & Owens, 2012). This reconnection results in two products: one is an additional new open, vertical field, and the second is a new closed field on the larger bipole. (Sterling & Moore, 2001, show similar external reconnection in a large-scale eruption; compare that paper's figure 5 with Figure 6.6 here.)

Figure 6.6c shows the jet forming as the minifilament flux rope field reconnects and mixes in with the ambient field. This results in formation of the spire, by both reconnection heating of the ambient coronal plasma at the site of the external reconnection, and by flows due to chromospheric "evaporation" that occur in standard flare models. There is also likely a "whipping effect" that propels chromospheric material onto the newly opening external-reconnection field line as it snaps to vertical

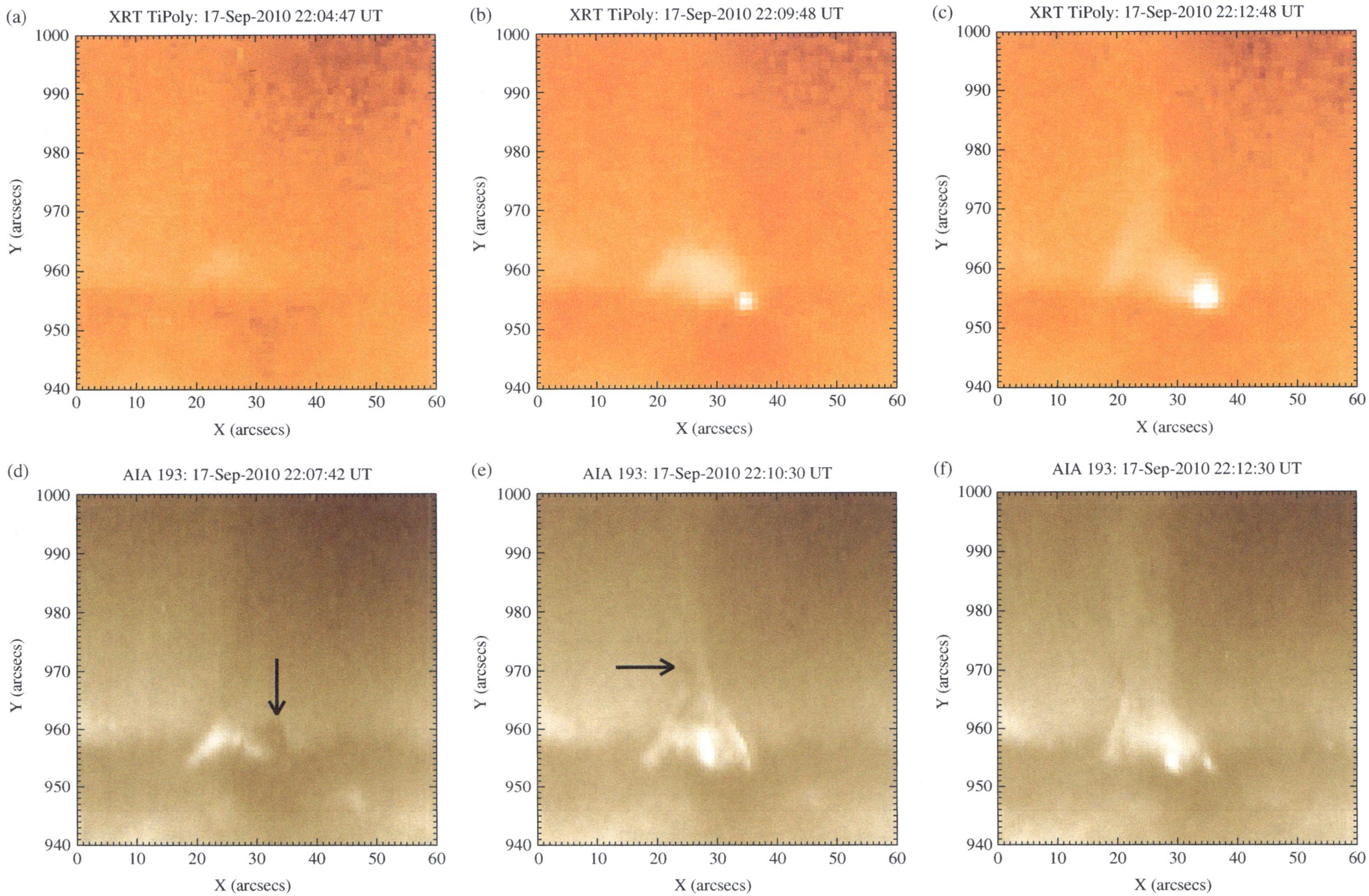

Figure 6.4 Observed X-ray jet near at the solar limb in (a) X-rays from Yohkoh/XRT and (b) in 193 Å EUV from AIA on *SDO*. EUV frequently shows dark (presumably cool) absorbing filament material ("minifilament") erupting and traveling out along the spire, and the JBP appearing at the location from which the minifilament erupted. Here, the JBP is visible as the bright point-like features in (b) that grows larger and brighter in (c). The arrow in (d) shows the minifilament emanating from the location where the JBP forms; in (e), the erupting minifilament is moving out along the jet spire, and is almost gone in (f). (*Source*: Sterling et al., 2015.)

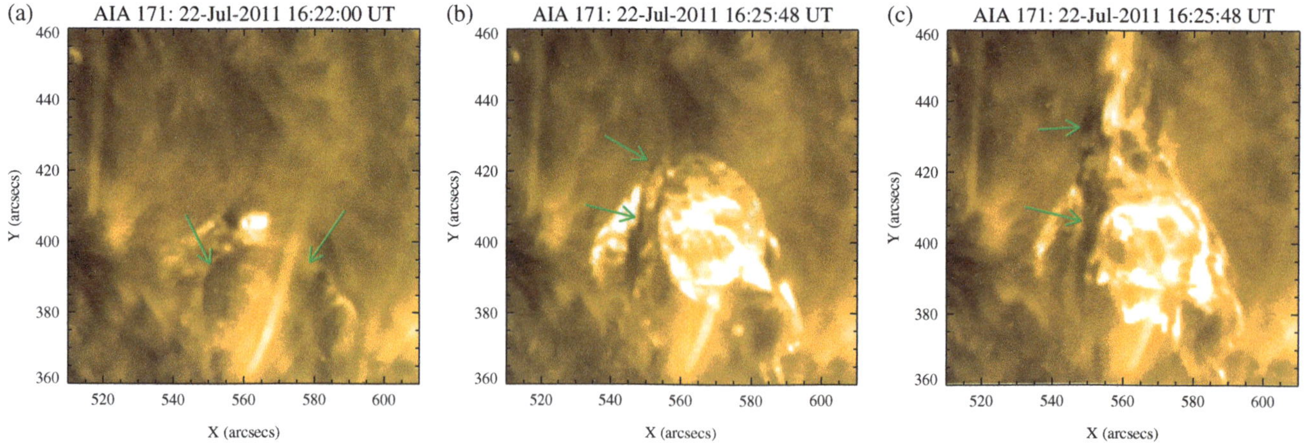

Figure 6.5 Minifilament observed erupting in *SDO*/AIA 171 Å, where the arrows in the panels point to the minifilament, which shows up in absorption. (a) The minifilament is starting to expand upward, and the accompanying brightenings are still minimal. (b) Strong brightenings are developing beneath the erupting filament, forming the JBP and auxiliary brightenings. (c) As the minifilament eruption continues, the spire starts to form above the bright base. (*Source*: This event was studied extensively in Shen et al., 2012.)

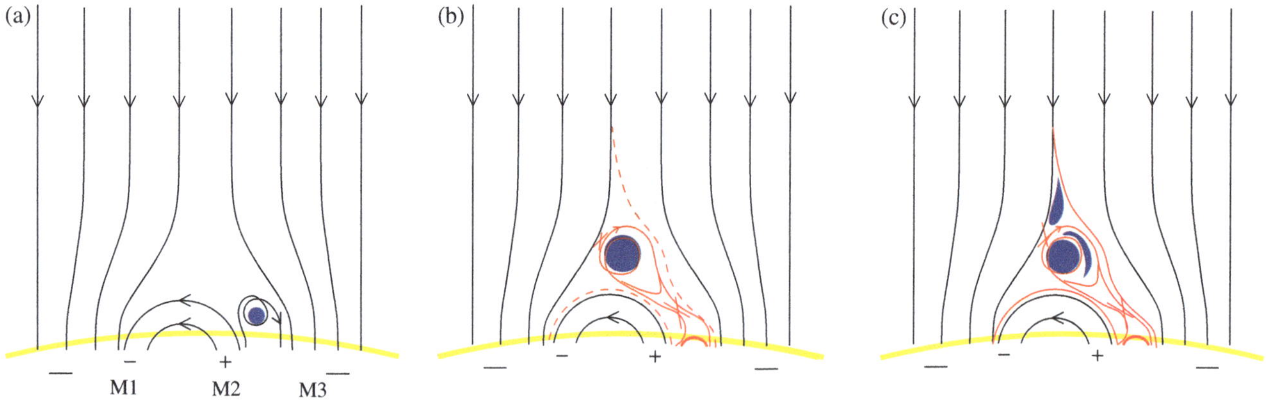

Figure 6.6 Schematic representation of the minifilament eruption model for coronal jets. Sterling et al. (2015) proposed that jets result from locations where in panel (a) a miniature filament (*minifilament*; blue circle) is initially in a bipolar anemone region embedded inside of a surrounding open (or far-reaching) field. This shows a cross section of a 3D anemone structure, where one side of the structure has increased stress (the smaller—more compact-loop field, rooted between the locations M2 and M3 labeled in Panel (a)) that contains the cool-material minifilament (blue circle). (b) Upon eruption of the minifilament, the field surrounding the cool minifilament reconnects with adjacent open field via *external reconnection* (upper red X), producing a new closed loop over the larger loop (between M1 and M2), and a new open field, along which the jet spire flows outward. Meanwhile, *internal reconnection* occurs among the legs of the erupting minifilament field (lower red X), producing a miniature flare, similar to the formation of a typical solar flare in the wake of larger filament eruptions as described by the CSHKP standard model for solar eruptions; here, the miniature flare in Panels (b) and (c) (small red semicircle between locations M2 and M3 of Panel (a)) is identified with the JBP. (c) If the minifilament eruption progresses far enough into the open-field region, enough of the outer envelope of the erupting minifilament field can be eroded away by the external reconnection for the cool minifilament material to flow outward along the spire as part of the jet. Dotted lines in Panel (b) represent a field soon to be made by the external reconnection. (*Source*: Sterling et al., 2015. © 2015, Springer Nature.) (Panel (b) is modified from that in Sterling et al. 2015, with the addition of red dotted lines showing more explicitly that it is uncertain whether the external or internal reconnection occurs first. Moore et al. 2018 discuss the order of occurrence of these two reconnections.)

following the reconnection (Shibata & Uchida, 1986). In this way, a jet spire naturally forms with the JBP off to the side of the base.

Modified View of Standard and Blowout Jets

Among the 20 events studied in Sterling et al. (2015), 14 were classified as blowout jets, 5 were standard jets, and 1 was ambiguous. They concluded that, rather than via the emerging-flux picture of Figure 6.3, the blowout jets instead work as described in section 6.3.4.1 and in Figure 6.6. In this case, the erupting minifilament comes from the base bipolar region, as originally proposed in Moore et al. (2010), except that now there is no longer an assumption that the erupting bipole is an emerging (or recently emerged) bipole. Sterling et al. (2015) did not directly address the cause of minifilament formation and eruption, pointing out that both flux emergence and cancelation would be candidates for triggering the eruption (although they indicated that the observational evidence up until that date had been favoring cancelation).

Blowout jets first appear as a narrow-spire standard jet, and then evolve into a broader jet, where the spire can be as large as, or even larger than, the jet's base. How does this match up with the minifilament eruption model? During the early stages of the jet formation (Figure 6.6b), the erupting minifilament flux rope will have only external reconnection with a few of the ambient vertical field lines; this will produce the observed narrow spire. Also, during the minifilament's eruption, internal reconnections will form the JBP as the brightest part of the base, as observed during the early stages of blowout jets. Later in the jet development, it is observed that the spire becomes broader, and the base immediately below the spire becomes brighter. This fits the minifilament eruption model, as the spire becomes broader as the erupting minifilament flux rope plows deeper into the ambient "open" coronal field, making more and more new "open" field lines through external reconnection. Moreover, the movement of the flux rope is away from the JBP with time (at least in the simplified 2D picture of Figure 6.6), indicating that the spire should grow in a direction away from the JBP. That is, the spire drift should be away from the JBP rather than toward the JBP; this is in agreement with the trend of spire drift in observed jets (Savcheva et al., 2009). (In contrast, the emerging-flux model predicts a spire drift toward the JBP with time; Moreno-Insertis & Galsgaard, 2013; Sterling et al., 2015.) These same external reconnections also progressively add more and more bright reconnection-product loops to the large lobe immediately beneath the spire, increasing its intensity in agreement with the observations.

What about the narrow-spire jets; the so-called "standard jets"? For the five cases among the jets in Sterling et al. (2015) that fit this description, there was still evidence for presence of an erupting minifilament. But for those events the minifilaments were often faint, and in none of the cases was there a very obvious minifilament eruption completely away from the Sun (it is unclear whether there was a weak ejection from the Sun, but even if so, they were weaker than those seen in typical blowout jets). Instead, the minifilament's eruption was largely arrested prior to reaching the upper corona, or likely only part of the minifilament leaked out onto the vertical field. Thus, the minifilament eruptions in the standard jet cases seem to correspond to failed or partially failed filament eruptions that have been observed in larger-scale filaments (e.g., Ji et al., 2003, Li & Ding, 2017; Sterling et al., 2011).

Thus, for standard jets, the eruption would stop at about the stage pictured in Figure 6.6b, before the full filament flux rope has a chance to escape into the upper corona (and also, observations indicate it usually stops before much of the cool filament material reaches the spire, that is, prior to the time depicted in Figure 6.6c). In comparing this with the observed properties of standard jets, the external reconnection would form a hot spire, but the erupting minifilament would not progress far enough into the ambient open field for much, if any, of the cool minifilament material to flow out along the open field. This would explain why in the observations the spire remains narrow, and why most of the standard jets would not show a cool component of 304 Å material. Meanwhile, the internal reconnection would form the JBP, but the lack of extensive external reconnection would limit the intensity of the jet base immediately beneath the spire to modest levels, so that the JBP intensity is more likely to dominate the base-region intensity throughout the life of the standard jet, as observed.

Therefore, in this view, both standard jets and blowout jets are the result of minifilament eruptions, but in the former case the eruption is likely more confined, whereas in the latter case the minifilament erupts fully ("blows out"). This picture renders the name "standard jet" a misnomer, because these jets are no longer thought to obey the previously held standard emerging-flux picture (Figure 6.3). So the term "standard jet" should be thought of as a morphological description of jets, where the spire remains narrow throughout the jet's life, and the brightness of the broader part of the base region does not rival that of the JBP. For blowout jets, the morphological description of a narrow spire growing into a broader spire also holds, but the cause is now suggested to be the minifilament eruption picture of Figure 6.6 (Sterling et al., 2015), rather than an eruptive version of the emerging-flux picture of Figure 6.3 (Moore et al., 2010; Shibata et al., 1992).

A note of caution: The translation of these properties of standard and blowout jets to EUV-observed jets should be done with care, as the terms *standard* and *blowout* were originally made in the context of the X-ray appearance. So, for example, it is possible that in standard jets the JBP does not stand out prominently compared to the rest of the jet base in some EUV channels, as it would in X-rays. On the other hand, a clear and explosive eruption of a minifilament into the far corona observed in EUV would be expected to have properties of a blowout jet when observed in X-rays.

Other examples of minifilament eruptions causing jets include the descriptions in Hong et al. (2016), Zhang et al. (2016), and Hong et al. (2017).

6.3.5. The Magnetic Cause of Jets

Understanding why jets occur requires observing them on the solar disk. As mentioned in section 6.3.4, there had been several observations of on-disk jets. These had found either cancelation, or emergence and cancelation occurring below jets (Hong et al., 2014; Innes & Teriaca, 2013; Shen et al., 2012; Young & Muglach, 2014a, 2014b). Even earlier studies also saw jets occurring where one pole of an emerging bipole underwent cancelation, and the jet occurring from the cancelation site (Chae et al., 1999; Liu & Kurokawa, 2004).

In retrospect however, none of these earlier studies found conclusive evidence for jets occurring with emergence in the absence of cancelation, with some of those studies concluding that cancelation was the primary trigger of the jet (e.g., Young & Muglach, 2014a, 2014b). Following up on those studies, along with the work of Adams et al. (2014), using AIA and HMI data, Panesar, Sterling, Moore, et al. (2016) looked at ten quiet Sun coronal jets, and Panesar, Sterling, and Moore (2018) looked at 13 on-disk coronal hole coronal jets. They found that all of these jets resulted from erupting minifilaments, and that magnetic flux cancelation occurred at the minifilament location in the hours preceding minifilament eruption (Figure 6.7).

Panesar, Sterling, Moore, et al. (2016) and Panesar, Sterling, and Moore (2018) measured the rate of flux cancelation for their jets. Their procedure was to look at the evolution of the magnetic field in the region around where the cancelation occurred over the hours leading up to the cancelation. Among the canceling elements, one of the polarities—the majority polarity—was usually dominant in the region, whereas the other polarity—the minority polarity—was less prevalent. Panesar et al. tracked the flux inside only one of those polarities, usually the minority polarity, because it was easier to isolate inside a prescribed box as the cancelation occurred. This isolation is important when determining the total integrated flux variation with time. If instead the flux is not sufficiently isolated, then any flows of flux into or out of the box over time would confuse the results: in that case, it would be very difficult to determine whether an increase or decrease in the measured flux value was due to a true emergence or cancelation on the one hand, or due to a flow into or out of the bounding box's boundary over time on the other hand. More sophisticated methods also exist for tracking flux changes in regions (Green et al., 2003). The above-mentioned Panesar et al. (2016, 2018) studies found that flux cancels at a rate of 0.5–2.0×10^{18} Mx s^{-1} and 0.9–4.0×10^{18} Mx s^{-1} in the coronal-hole and quiet-Sun jet cases, respectively. (Sterling et al., 2018, provide a table summarizing these and other cancelation rates.)

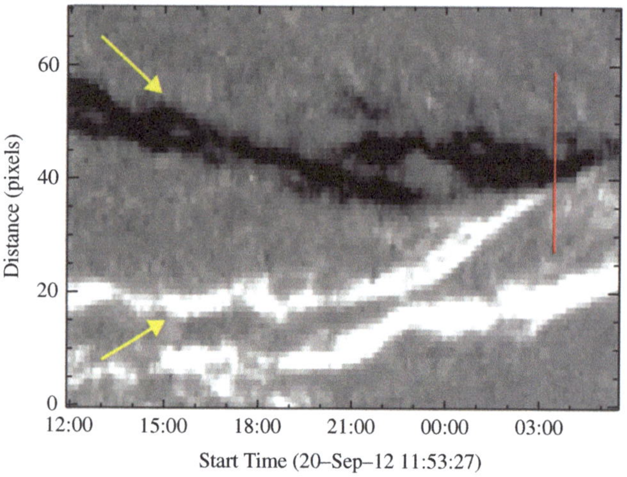

Figure 6.7 Plot showing magnetic evolution as a function of time of positive- (white) and negative- (black) polarity fields at the location where a coronal jet occurred on 21 September 2012. These magnetic data are from the *SDO*/HMI magnetograph, which has pixels of 0″.5. At the start of the plot, the polarity elements (arrows) are about a distance of 15″ apart, and over the next 15 hr flow together, eventually converging. At the time of convergence (red line), a minifilament eruption occurred, producing the observed coronal jet. Panesar, Sterling, Moore, et al. (2016) discuss this event, and nine others like it, in detail. (*Source*: Panesar, Sterling, Moore, et al., 2016.)

For the 10 jets of Panesar, Sterling, Moore, et al. (2016), Panesar et al. (2017) examined how and when the minifilaments that erupted to produce the jets formed. They found that, in all cases, the minifilaments themselves formed via flux cancelation. Following its creation, the minifilament was present for a period ranging from 1.5 hr to two days prior to erupting to drive the jet. Based on these findings then, the complete picture for the 10 jets of Panesar, Sterling, Moore, et al. (2016) is that an episode of flux cancelation forms a minifilament; then, either continued flux cancelation, or a second episode of flux cancelation, leads to eruption of the minifilament and formation of the jet via the process shown in Figure 6.6. They also showed that this process can result in formation of multiple minifilaments, that subsequently erupt to form a series of homologous jets.

In the author's view, then, several of these recent studies (Adams et al., 2014; Moore et al., 2018; Panesar, Sterling, Moore, et al., 2016; Panesar et al., 2017; Panesar, Sterling, & Moore, 2018; Sterling et al., 2015) suggest the following picture for jets in quiet regions and coronal holes. Photospheric flows drive together minority- and majority-polarity flux patches, or flux emergence drives one pole of the emerging flux into a pre-exisiting opposite-polarity flux patch. This results in flux cancelation and a buildup of a magnetic flux rope at the canceling neutral line. The process that is responsible for causing large-scale filaments to form (Parenti, 2014; van Ballegooijen & Martens, 1989) works in this situation too, resulting in formation of the minifilament along the sheared field/flux rope. (We do not attempt to differentiate between highly sheared non-potential field and a bona fide flux rope in the pre-eruption phase.) The presence of the minority flux patch results in the anemone geometry of Figure 6.6a, where the flux rope and minifilament form at the location indicated in that figure as a result of the photospheric flows and flux cancelation. Continued cancelation, or a new episode of cancelation, then results in destabilization of the minifilament, its eruption, and jet formation, as shown in Figures 6.6b and 6.6c.

Even though the above scenario holds that flux cancelation is the key agent leading to minifilament formation and eruption, this does not rule out a role for flux emergence in the jet process. In fact, several of the events in the Panesar et al. studies, as well as other jet studies (e.g., Chae et al., 1999; Huang et al., 2012; Liu & Kurokawa, 2004; Liu et al., 2011) find that the jets occur only after one pole of the emerging-flux bipole undergoes cancelation with neighboring field, as allowed in the above scenario. Thus, at least in these cases, the flux emergence plays an auxiliary role to the jet process, but the main process is flux cancelation.

There are, however, also reports of jets without flux cancelation. Kumar et al. (2018) present an example of a jet that originates from an equatorial coronal hole location that shows no obvious flux emergence or cancelation at the time of the jet. There are, however, oppositely directed motions of opposite-polarity flux before and near the time of the jet, and they argue that the magnetic shear from these shearing motions is responsible for the jet, consistent with the numerical modeling of Pariat et al. (2009). Thus, this may be a jet without direct cancelation. (Clear cancelation *does* occur within ~20″ of the jet location, but this cancelation is away from, and—seemingly—not connected to, the main neutral line of the jet.) Thus, other than shearing motion, another possibility for onset of this event are that there is an (unseen) magnetic connection between the jet location and the location of that nearby flux cancelation. And, of course, there is always the possibility that flux cancelation is occurring at the jet site at a level too weak to be detected by HMI (although similar statements could be made arguing that undetected flux emergence or shearing motions cause jets too).

Mulay et al. (2016) looked at 20 jets and their magnetic setting. They concluded that 10 of the jets resulted from flux cancelation, 4 of them resulted from flux emergence followed by cancelation, and that 6 of them were from flux emergence regions. A difference however between these Mulay et al. (2016) jets, from the jets in the Panesar, Sterling, Moore, et al. (2016), Panesar, Sterling, and Moore (2018), and Kumar et al. (2018) studies, is that the Mulay et al. events occurred near the edge of ARs, rather than in quiet Sun or CHs. Because activity in ARs is often more rapid and complex than in quieter solar regions, flux emergence and cancelation often occur in close proximity to each other, complicating determination of which is the key mechanism in the episode being studied. Similarly, Sterling et al. (2016) observed that emergence and cancelation were concurrently occurring in the AR jets that they examined (they also found cancelation in all but one of their events). Shen et al. (2012) also found a jet near an AR to occur in a region of both emergence and cancelation at its base. In any case, the Mulay et al. (2016) study confirms that flux cancelation is the primary agent for jet production in many jets. That study also confirms the importance of continued investigations into the magnetic source of jets, and also (perhaps) the need for cross-comparisons and/or calibrations between studies carried out by different workers or research groups. Future studies will help ascertain and/or confirm whether jets can occur in the absence of flux cancelation in some cases. (Section 6.3.6 below discusses AR jets further.)

Jets frequently show clear twisting motions during their rise, with recent examples including Patsourakos et al. (2008), Moore et al. (2015), Innes (2016), Joshi et al. (2018), Bogdanova et al. (2018), Miao et al. (2018), Liu et al. (2018), Panesar, Sterling, and Moore (2016), and Schmieder et al. (2013). Such twisting could be indicative of a preexisting twisted field on which the jet spire travels, a twisting of the field with time, or an untwisting of a previously twisted field. The detailed study of Moore et al. (2015) of 14 jets extending from PCH regions out to beyond 2.2 solar radii strongly supports the last explanation. In the minifilament eruption model view, this is consistent with the pre-eruption flux tube holding the minifilament material being twisted prior to eruption. This twist would be released then the erupting minifilament undergoes the external reconnection with the external field, via a process described by Shibata and Uchida (1986). Thus, in this view, the twisting would be due to an *unwinding* of a previously twisted erupting minifilament flux rope.

6.3.6. AR Jets

Jets in ARs would be expected to be more energetic versions of the jets in QS and CH regions. This assumption, however, leads to some factors that are not so easily understood. Chief among these is that some AR jets can be very prominent in X-rays and EUV, but lack an obvious accompanying erupting minifilament. So the question arises: is the same physics occurring without an erupting minifilament, or is a substantially different process responsible for the coronal jets occurring in ARs?

To begin with, the jets in ARs that we are concerned with here are strictly coronal jets, such as those appearing in X-rays (e.g., Shibata et al., 1992; Shimojo et al., 1996). These jets appear at the edge of ARs. Shimojo et al. (1996) found that about 90% of AR jets with an X-ray bright point (by which they meant a point-like brightening unresolved by XRT, and so likely the same feature Sterling et al., 2015, call the JBP) occurred in the western part of the AR, and they add that such jets "tend to appear at the western edge of the preceding spot." (Smaller "penumbral jets," which occur near sunspots, will not be discussed here; see, e.g., Katsukawa et al., 2007; Tiwari et al., 2016; Tiwari et al., 2018.)

Since the ground-breaking work of *Yohkoh*, there have been many studies of jets in ARs, but often without a mention of (mini)filaments (e.g., Guo et al., 2013; Kim et al., 2007), whereas other studies did not mention them explicitly but did discuss a connection with a surge (e.g., Zhang & Ji, 2014).

In their study of 20 jets in EUV with AIA, Mulay et al. (2016) found that they occurred at the periphery of ARs, and did not mention examples of minifilament eruption as causing the jets. They did point out that all of their jets were accompanied by Hα surges. But although Wang and Liu (2012) found that some surges that made circular-ribbon flares came from erupting filaments, more generally the connection between surge material and filaments has not been clear (e.g., Roy, 1973). Therefore, for many AR jets, the connection between the AR jets and minifilament eruption is not as obvious as it is in many quiet Sun and coronal hole jets.

In an attempt to understand whether AR jets behave in the manner described in Figure 6.6, that is, in the way that quiet Sun and CH jets appear to operate, Sterling et al. (2016) looked at a series of jets from a single AR, using EUV data obtained from AIA and *STEREO*, magnetic data from HMI, and X-ray data from the Soft X-ray Imager (SXI). They found that some of the jets (three of which they examined in detail) showed clear minifilament eruption as causing the jets, and that these eruptions displayed slow-rise and fast-rise phases that are frequently seen in typical (larger-scale) filament eruptions. Thus, they concluded that these jets followed the pattern of Figure 6.6, and thus of coronal jets in non-AR regions. All the jets occurred at magnetic neutral lines.

In addition, two of the AR jets examined in detail by Sterling et al. (2016) and that had clear erupting minifilaments also showed strong ejections in AIA 304 Å images, consistent with surges and sprays. Moreover, the footpoint brightenings corresponding to location M1 in Figure 6.6a appeared as semicircular patterns in the images, which is similar to emissions seen in a large percentage of surges (cf. Öhman, 1972). This suggests that many or all surges might be formed as in Figure 6.6, consistent with the findings for a set of surges by Wang and Liu (2012).

Several of the AR jets of Sterling et al. (2016), however, did *not* show clear minifilament eruptions. A series of these jets developed faster than the jets of that same paper that did show clear minifilament eruptions. In a follow-up study, Sterling et al. (2017) called such faster-developing jets "violent jets," and explored several of them in detail for a different AR, using data from AIA, XRT, *IRIS*, and HMI. They again found that several of the violent jets did not have prominent minifilament eruptions, but they instead had very thin ($\lesssim 2''$) filament "strands" that erupted away from the jet region, similar to features seen by Schmieder et al. (2013). Moreover, these eruptions took place at neutral lines where clear flux cancelation was occurring, similar to what is found in many QS and CH jets. In the earlier study of Sterling et al. (2016), all but one of the jets occurred where there was clear canceling flux. In the one exception, the jet occurred at the neutral line of an emerging-flux element. (In retrospect, we now suspect that that jet occurred where that emerging, or emerged, bipole was canceling at its own main neutral line during the later stages of the emerged bipole's development, similar to the larger-scale filament eruptions in Sterling et al., 2018). Table 6.1 of Sterling et al. (2018) gives the average rate of flux cancelation measured in these studies. More generally, the overall magnetic setup for all of the AR jets of both Sterling et al. (2016) and Sterling et al. (2017) was as depicted in Figure 6.6. These observations support the idea that the AR jets are largely identical to quiet Sun and coronal hole jets.

However, it is not clear why some AR jets are accompanied by negligible visible minifilaments. It is possible that a magnetic flux rope is erupting from the location of the expected minifilament, but that field merely does not contain enough cool material to appear as a cool minifilament, similar to how large-scale eruptions can occur without an accompanying filament eruption. One more puzzle with many of the violent jets is that the larger magnetic lobe (between locations M1 and M2 in Figure 6.6a) adjoining the location of the minifilament eruption (or of the expected minifilament field eruption, when the field is not marked with cool material), can far exceed the intensity from the expected JBP location (between locations M2 and M3 of Figure 6.6a); that is, in QS and CH jets, the JBP prominently stands out as a bright part of the base, at least early in the jet development, but sometimes the JBP location intensity remains subdued in AR jets. Sterling et al. (2017) speculated that this could be due to secondary eruptions triggered by the eruption of the first jet, similar to what is observed in some large-scale eruptions (e.g., Schrijver et al., 2013; Sterling et al., 2014; Török et al., 2011). These aspects of AR jets, in comparison to QS and CH jets, are yet to be fully understood.

In regard to the magnetic source of AR jets, although Sterling et al. (2016) and Sterling et al. (2017) found that all but one of their jets occurred at sites of unmistakable flux cancelation, in several cases these cancelations occurred in flux emergence regions, where one pole of the emerging flux ran into preexisting opposite-polarity flux, with the jet occurring along the so-formed neutral line. In the 20 AR jets they examined, Mulay et al. (2016) report that 10 occurred due to flux cancelation, and 4 occurred at sites of emergence and then cancelation; thus, 14 out of 20 were due to flux cancelation. They further report, however, that 6 of the 20 occurred with flux emergence without cancelation. We point out that in ARs the magnetic field can be very dynamic, and it can be difficult to ascertain with certainty whether flux emergence is truly occurring without cancelation (where that cancelation might be between one of the poles and nearby opposite-polarity field, or perhaps at the neutral line at the center of the emerging bipole itself). Further studies of jets, both inside and outside of ARs, should be carried out to confirm the magnetic source of jets.

Although we do not detail spectroscopic studies of jets here (e.g., Chifor et al., 2008; Tian et al., 2012; see Raouafi et al., 2016, for more citations), there is hope that the resulting information on twisting motions, mass flows, temperatures, and densities, can help clarify the observed properties of jets in all solar regions and give further insight into their driving mechanism.

6.3.7. Jet Numerical Models: Recent Progress

Raouafi et al. (2016) reviewed in detail earlier jet numerical simulation studies. Subsequently, Liu et al. (2016) presented numerical simulations based on the emerging-flux model, extending the simulations of Fang et al. (2014) to describe a nearly concurrent pair of jets (twin jets). Ni et al. (2017) also used the flux emergence model to explore blobs in jets resulting from the plasmoid and Kelvin–Helmholtz instabilities. As discussed above, the question arises as to how often flux emergence might be the primary trigger of jets. But even if the jet trigger is something different, some of the physics described by these models should still be relevant to aspects of jet phenomena.

Several studies have not explicitly relied on flux emergence for producing jets. One such set of studies by Pariat and colleagues (Pariat et al., 2009, 2010, 2015, mentioned earlier) assumes an anemone field embedded inside an ambient background field, where the system is energized via rotational motion of the anemone field inside the surrounding ambient field. Eventual reconnection between the flux systems results in energy release and jets along the ambient field extending vertically above the bipole. Karpen et al. (2018) and Roberts et al. (2018) extended these studies to consider, among other effects, the manifestation of the resulting jets in the far corona. Szente et al. (2017) use a variation of the same model, whereby the twisting anemone is located beneath the photosphere, and report close matches with the physical structure, dynamics, and emission of observed EUV and X-ray jets.

So far only one set of studies has explicitly attempted to simulate the minifilament eruption scenario for jets. Wyper et al. (2017) use an initial simulated magnetic setup similar to that of Figure 6.6a, and they are able to reproduce the process depicted in the subsequent panels of Figure 6.6, therefore providing numerical support for the concept, including the jet spire and the representation of the JBP as a miniature flare beneath the erupting minifilament flux rope. Wyper, DeVore, and Antiochos (2018) and Wyper, DeVore, Karpen, et al. (2018) extend the simulations to differing environmental circumstances. They refer to their simulations as a "breakout model for jets" because the reconnection at the magnetic null point, between the bipole field and the ambient overlying field (and where the external reconnection occurs at the upper red X in Figure 6.6b), is identical to the breakout model setup developed for larger-scale eruptions (Antiochos, 1998; Antiochos et al., 1999).

To date, no model has self-consistently produced jets based on minifilament eruptions triggered by magnetic flux cancelation. From the preceding discussion, several recent observations suggest that this is how at least a large fraction of jets operate. Some recent studies offer hope that such simulations might be possible in the near future (Galsgaard et al., 2019; Lionello et al., 2016; Syntelis et al., 2019).

AR jets (and probably, to some extent, non-AR jets as well) can be a source of accelerated particles. See Raouafi et al. (2016) for discussions. Examples of more recent work include Nitta et al. (2015) and Glesener and Fleishman (2018).

6.4. JET-LIKE PHENOMENA ON DIFFERING SIZE SCALES

6.4.1. Coronal Jets and Large-Scale Eruptions

If coronal jets are truly small-scale versions of larger-scale eruptions, then we expect the processes that trigger jets perhaps also to trigger large-scale eruptions. (In this case, we are referring to large-scale flux rope eruptions that produce a flare in tandem with a CME, or a confined flare eruption without a CME.) Because of the substantial evidence that converging and canceling fields trigger small-scale eruptions that cause jets, we can question whether they also trigger the large-scale eruptions. A difference, however, between jets and the larger eruptions is that the magnetic-field structures that produce jets (specifically, minifilaments) form only hours to days before onset of typical jets, based on the work of Panesar et al. (2017). In contrast, the buildup to large-scale eruptions can persist for several weeks before the eruptions occur, and therefore longer than a two-week disk-passage period. Therefore, it is much easier to follow the complete life span of the magnetic structures of small-scale jets compared to larger-scale erupting regions.

In order to determine with more confidence whether large-scale eruptions mimic coronal jets (i.e., the formation and dynamics of minifilaments that erupt to form jets), Sterling et al. (2018) looked at two CME-producing ARs, where each of those ARs was small enough to be followed from the time of its birth until the time of its CME-producing eruption within a single disk-side passage. They found that, in both cases, the regions emerged as bipoles, spread apart, and then the two polarities approached each other again, and then the eruptions occurred. In both cases the entire evolution lasted roughly five days. As the two poles converged, they clearly underwent cancelation at the neutral line at the center of the respective bipoles. This apparently built flux ropes (or sheared-field filament channels), along which filaments formed. These filaments erupted, producing CMEs. In one of the regions, an additional filament formed on a neutral line between one of the bipoles spreading during emergence and an exterior ambient opposite-polarity field; that filament also erupted. Other than this, the regions were magnetically isolated, so that they had little other interaction with the surrounding field. This behavior is similar to that of jet magnetic elements, in that in both the (mini)filament is formed by canceling fields, and the eruption is later triggered by canceling fields.

Some flux rope CMEs occur from eruptions that do not have an obvious erupting cool-material filament; in that case, the field still erupts and launches a flux rope outward, but for whatever reason a filament is not formed on the field prior to its eruption. Thus, we might expect that some jets also occur without a discernible cool minifilament (e.g., Sterling et al., 2017).

Coronal Jets and CMEs

Thus, the erupting minifilament in jets resembles the eruptions that cause CMEs in larger events. In jets, the erupting minifilament can also produce ejections that extend into the heliosphere. Here we discuss two different mechanisms for CME formation from jets.

One mechanism produces the narrow features that are called "white-light jets" as observed in space-borne coronagraphs. These are also called "narrow CMEs," as they typically have angular widths (measured from the Sun center) of less than 10° or 15°, with different studies using different criteria; this compares with an average width of ~44° for all CMEs (Gopalswamy et al., 2009). There are several other examples of such narrow CMEs from jets (e.g., Bemporad et al., 2005; Dobrzycka et al., 2003; Gilbert et al., 2001; Hong et al., 2011; Moore et al., 2015; Nisticò et al., 2009; Shen et al., 2012; Sterling et al., 2016; Wang et al., 1998; Wang & Sheeley, 2002; Yashiro et al., 2003). Wang et al. (1998) showed that the source of these jets on the Sun was coronal jets (which they observed with EIT). Moore et al. (2015) showed that the coronal jets that resulted in narrow CMEs in *SOHO*/LASCO coronagraph images, tended to be the coronal jets that displayed the strongest untwisting motions. Moreover, they showed evidence that the jet twisting motions propagated onto the white-light jets, manifesting as swaying-like dynamics of those coronagraph structures.

In terms of the minifilament eruption model, this white-light-jet twisting can be understood as discussed in section 6.3.5, that is, by imagining that the erupting minifilament contains twist, and that twist is then transferred onto the ambient open field via the external reconnection of Figures 6.6b and 6.6c. Furthermore, this provides a ready explanation for the difference between a white-light jet and a conventional CME: In the white-light jet, the erupting flux rope (that is, the magnetic-flux-rope shell surrounding the erupting cool-material minifilament) contains a relatively small amount of total flux, and all of this flux is transferred to the ambient open field; that is, the minifilament flux rope is totally consumed, or "eaten up," by the ambient field. In contrast, for a typical CME, the flux contained in the erupting flux rope is large enough (compared to the surrounding field) that a substantial portion of the flux rope can escape the corona intact, appearing as the bubble portion of the CME core.

The second mechanism we describe here for producing CMEs from jets results in what have been called "streamer-puff" CMEs (Alzate & Morgan, 2016; Bemporad et al., 2005; Panesar, Sterling, & Moore, 2016). In this case, a jet occurs at one side of the base of a coronal streamer, and a loop in the "helmet" of the streamer is blown out. Bemporad et al. (2005) provided an explanation for these features, but this explanation was updated by Panesar, Sterling, and Moore (2016) to mesh with the minifilament eruption picture of jets. According to this explanation, instead of undergoing external reconnection with an ambient open field as in the case of the narrow CMEs, an erupting twisted minifilament field instead undergoes external reconnection with the streamer-helmet loop field, rendering the streamer loop unstable so that it erupts outward as the streamer-puff CME.

6.4.2. Smaller-Scale Jets: Jetlets and Spicules(?)

Sterling and Moore (2016) suggested that there might be a power-law relationship for eruptions of filament-like structures, with CME-producing large-scale filament eruptions on the large end, extending down through coronal jets, and possibly continuing down to spicules on the small end. We emphasize that erupting-filament-like features have not been observed in spicules, and therefore extending this concept to spicules rests on the speculative assumption that spicules result from eruptions of hitherto unseen *microfilaments*. With that caveat, Sterling and Moore (2016) plotted estimates for the number of spicules, coronal jets, and CME-producing eruptions that are on the Sun at a given time, against the size (observed or estimated) of the erupting filament-like feature. This produces the plot of Figure 6.8, which we show with a linear fit line overlaid. For the estimate of the number of spicules, they used published numbers of spicule occurrences from Athay (1959) and Lynch et al. (1973). A more recent study by Judge and Carlsson (2010), however, suggests that these values may be severely underestimated, and estimate that there are ~2×10^7 spicules on the Sun at any given time. Therefore the uncertainty in the spicule counts is very high, and the plotted point in Figure 6.8 should be regarded as a minimum value. Therefore, it is possible that only some percentage of spicules is jet-like in nature. Those spicules would be analogous to coronal jets, and the formation of those spicules would be just as presented in Figure 6.6, but scaled down to occur only in the photosphere and chromosphere. Moreover, these spicules would be expected to form at very weak neutral lines, and likely where magnetic cancelation is occurring. But how big or small is the percentage of spicules formed this way is currently not known, and possibly can only be addressed with better observations, such as with the upcoming DKIST telescope.

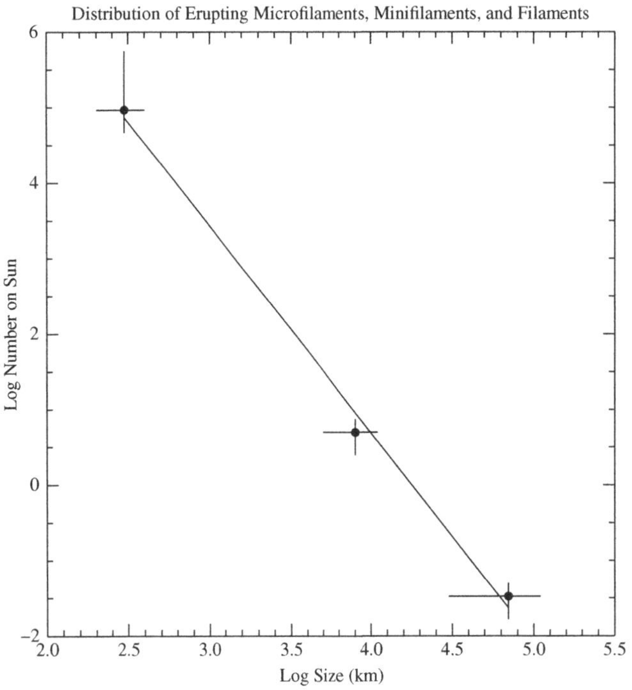

Figure 6.8 Plot of estimated observed number of erupting filament-like features on the Sun at any given time (vertical axis) as a function of the size of the feature. The rightmost point represents CME-producing filament eruptions; the middle point represents observed coronal jets, most or all of which likely result from minifilament eruptions; and the leftmost point represents spicules, under the still speculative assumption that they result from eruptions of postulated *microfilaments*, of size similar to spicule observed widths, and with a number-on-Sun value matching estimates of Athay (1959) (and with the top of that point's vertical bar matching the higher estimate due to Lynch et al., 1973). As mentioned in the text, the number of spicules may be underestimated, perhaps severely so. The line is a least-squares fit to the three points. *Source*: This figure is from Sterling and Moore (2016); see that paper for further details.

There is, however, evidence for jet-like features that, although not as small as spicules, are smaller than typical jets but still big enough to be (often just marginally) resolved with current instruments. These features were called *jetlets* by Raouafi and Stenborg (2014). They identified them as transient jet-like features occurring at the sites of quasi-random magnetic cancelations at the bases of plumes in coronal holes. They proposed that the jetlets, along with plume transient bright points (PTBPs), help sustain the plumes over their lifetime of several days.

Panesar, Sterling, Moore, Tiwari, et al. (2018) confirmed the presence of jetlets in plumes, but also found that they are prevalent in non-plume chromospheric network regions too. Hence, they refer to the features with the more general term network jetlets; these are similar to or the same as the network jets of Tian et al. (2014; also see, e.g., Kayshap et al., 2018; Narang et al., 2018). Using AIA/EUV and IRIS/UV images, for 10 measured jetlets, Panesar, Sterling, Moore, Tiwari, et al. (2018) found them to have lengths of ~30,000 km, spire widths of 3000 km, and durations of 3 min. They also found that 9 of them occur at sites of magnetic cancelations; although the origin of the tenth was uncertain, they speculate that it too was due to (unresolved) flux cancelation. Panesear, Sterling, Moore, Tiwari, et al. (2018) suggest that network jetlets might be small-scale analogs of larger-scale coronal jets.

6.5. CORONAL JETS AND OTHER PHENOMENA

We close by briefly addressing other important topics on jets that are not addressed in detail here.

6.5.1. Jets and Plumes

As noted in section 6.4.2, Raouafi and Stenborg (2014) suggested that jetlets and PTBPs might sustain plumes over their several-day lifetime. Even earlier however, Raouafi et al. (2008) and Raouafi et al. (2010) used *STEREO*/SECCHI and *Hinode*/XRT data to present evidence that the formation of plumes is preceded by coronal jets occurring at the same locations. Pucci et al. (2014) also found that a long-lived plume they studied was preceded by a coronal jet. (Also, however, see Avallone et al., 2018.) Although the idea that jets are critical for plume formation requires further verification (Raouafi & Stenborg, 2014), it is an important consideration in understanding the source of plume material, and also the magnetic environment at the base of plumes. See Poletto (2015) for further discussions.

6.5.2. Jets and Macrospicules

Pike and Harrison (1997) claimed that X-ray jets and EUV macrospicules were manifestations of the same phenomenon (cf. Kiss et al., 2017). Subsequent observations support this view, at least for some X-ray jets (Curdt et al., 2012; Moore et al., 2010, 2013; Sterling, Harra, & Moore, 2010). Combined with the minifilament eruption view of X-ray jets, this supports the hypothesis that macrospicules are the result of miniature filament eruptions, as proposed by Moore (1990).

6.5.3. Jets and Coronal Heating

Clearly, coronal jets supply energy to the corona, but so far it is unclear whether jets (on various size scales) can supply enough energy to explain general coronal heating. Individual jets are estimated to contribute something like 10^{26}–10^{27} erg for coronal hole jets, and 10^{28}–10^{29} erg for AR jets (Pucci et al., 2013; Shimojo & Shibata, 2000; Sterling et al., 2017, 2018). Considering coronal hole jets in aggregate, Paraschiv et al. (2015) estimated the amount of heating due to jets to be $\sim 2\times10^3$ erg cm^{-2} s^{-1} (comparable to values obtained by Yu et al., 2014), which is far below, for example, the rate of 6×10^5 erg cm^{-2} s^{-1} required for coronal holes (Withbroe & Noyes, 1977). Poletto et al. (2014) and Szente et al. (2017) also found energy rates from jets less than that required to sustain coronal heating, based on the estimated contribution of coronal jets alone. Moore et al. (2011) argue, however, that the heating rate could be sufficient for coronal heating, provided that the jet process continues down to events the size scale of spicules, and if spicules result from erupting magnetic bipoles as described in that Moore et al. (2011) paper.

6.6. THE FUTURE

The study of solar jet-like structures is clearly rich and fascinating, with these features forming fundamental components of the solar atmosphere. As demonstrated above, more work is required to confirm what mechanism/mechanisms is/are driving spicules, coronal jets, and other features such as jetlets. New instrumentation in the near future, such as DKIST and upcoming high-resolution space missions, will play a crucial role in unraveling these mysteries.

ACKNOWLEDGMENTS

This work was supported by funding from the Heliophysics Division of NASA's Science Mission Directorate through the Heliophysics Guest Investigators (HGI) Program, and the *Hinode* Project. We thank an anonymous referee for useful comments. We thank Ronald L. Moore for many invaluable discussions. We also thank R. L. Moore and Navdeep K. Panesar for carefully reading and commenting on the manuscript.

NOTE

1. Moore et al. (2013) contains a typo in section 5, second paragraph, where it says "all 29 blow-out X-ray jets displayed a cool component." This should instead say: "29 out of 32 blow-out X-ray jets displayed a cool component." This typo has no consequences for the subsequent discussions of that paper.

REFERENCES

Adams, M., Sterling, A. C., Moore, R. L., & Gary, G. A. (2014). A small-scale eruption leading to a blowout macrospicule jet in an on-disk coronal hole. *Astrophysical Journal*, *783*, 11. doi: 10.1088/0004-637X/783/1/11.

Alzate, N., & Morgan, H. (2016). Jets, coronal "puffs," and a slow coronal mass ejection caused by an opposite-polarity region within an active region footpoint. *Astrophysical Journal*, *823*, 129. doi: 10.3847/0004-637X/823/2/129.

Anan, T., Kitai, R., Kawate, T., Matsumoto, T., Ichimoto, K., Shibata, K., et al. (2010). Spicule dynamics over a plage region. *Publications of the Astronomical Society of Japan*, *62*, 871. doi: 10.1093/pasj/62.4.871.

Antiochos, S. K. (1998). The magnetic topology of solar eruptions. *Astrophysical Journal*, *502*, L181. doi: 10.1086/311507.

Antiochos, S. K., DeVore, C. R., & Klimchuk, J. A. (1999). A model for solar coronal mass ejections. *Astrophysical Journal*, *510*, 485. doi: 10.1086/306563.

Archontis, V., & Hood, A. W. (2013). A numerical model of standard to blowout jets. *Astrophysical Journal*, *769*, 21. doi: 10.1088/2041-8205/769/2/L21.

Athay, R. G. (1959). The number of spicules in the middle chromosphere. *Astrophysical Journal*, *129*, 164. doi: 10.1086/146603.

Athay, R. G. (1976). *The solar chromosphere and corona: Quiet* sun. Dordrecht, Holland: D. Reidel Publ. Co.

Avallone, E. A., Tiwari, S. K., Panesar, N. K., Moore, R. L., & Winebarger, A. (2018). Critical magnetic field strengths for solar coronal plumes in quiet regions and coronal holes? *Astrophysical Journal*, *861*, 111. doi: 10.3847/1538-4357/aac82c.

Beck, C., Rezaei, R., & Puschmann, K. G. (2013). Can spicules be detected at disc centre in broad-band Ca II H filter imaging data? *Astronomy and Astrophysics*, *556*, 127. doi: 10.1007/978-94-010-1715-2.

Beckers, J. M. (1968). Solar spicules (invited review paper). *Solar Physics*, *3*, 367. doi: 10.1007/BF00171614.

Beckers, J. M. (1972). Solar spicules. *Annual Review of Astronomy and Astrophysics*, *10*, 73. doi: 10.1146/annurev.aa.10.090172.000445.

Bemporad, A., Sterling, A. C., Moore, R. L., & Poletto, G. (2005). A new variety of coronal mass ejection: Streamer puffs from compact ejective flares. *Astrophysics! Journal*, *635*L, 189. doi: 10.1086/499625.

Berger, T. E., De Ponteiu, B., Schrijver, C. J., & Title, A. M. (1999). High-resolution imaging of the solar chromosphere/corona transition region. *Astrophysical Journal*, *519*, L97. doi: 10.1086/312088.

Blake, M. L., & Sturrock, P. A. (1985). Spicules and surges. *Astrophysical Journal*, *290*, 359. doi: 10.1086/162993.

Bogdanova, M., Zhelyazkov, I., Joshi, R., & Chandra, R. (2018). Solar jet on 2014 april 16 modeled by kelvin-helmholtz instability. *New Astronomy*, *63*, 75. doi: 10.1016/j.newast.2018.03.001.

Bohlin, J. D., Vogel, S. N., Purcell, S. N., Sheeley, N. R., Tousey, R., & Van Hoosier, M. E. (1975). A newly observed solar feature - macrospicules in He II 304 A. *Astrophysical Journal*, *197*, L133. doi: 10.1086/181794.

Bradshaw, S. J., & Klimchuk, J. A. (2015). Chromospheric nanoflares as a source of coronal plasma. ii. repeating nanoflares. *Astrophysical Journal*, *811*, 129. doi: 10.1088/0004-637X/811/2/129.

Bray, R. J., & Loughhead, R. E. (1974). *The solar chromosphere*. London: Chapman and Hall.

Canfield, R. C., Reardon, K. P., Leka, K. D., Shibata, K., Yokoyama, T., & Shimojo, M. (1996). H-alpha surges and X-ray jets in AR 7260. *Astrophysical Journal*, *464*, 1016. doi: 10.1086/177389.

Carmichael, H. (1964). The physics of solar flares. In W. N. Hess (Ed.), *Proceedings of the aas-nasa symposium held 28-30 October, 1963 at the goddard space flight center, greenbelt, md* (p. 451). Washington, DC: National Aeronautics and Space Administration, Science and Technical Information Division.

Chae, J., Qiu, J., Wang, H., & Goode, P. R. (1999). Active region loops observed with SUMER on board the SOHO. *Astrophysical Journal*, *513*, L75. doi: 10.1086/308640.

Chandra, R., Mandrini, C. H., Schmieder, B., Joshi, B., Cristiani, G. D., Cremades, H., et al. (2017). Blowout jets and impulsive eruptive flares in a bald-patch topology. *Astronomy and Astrophysics*, *598*, 41. doi: 10.1051/0004-6361/201628984.

Cheng, Q. Q. (1992). Fluid motions in the solar atmosphere – Part two – A spicule model with emission lines. *Astronomy and Astrophysics*, *266*, 549.

Chifor, C., Young, P. R., Isobe, H., Mason, H. E., Tripathi, D., Hara, H., et al. (2008). An active region jet observed with Hinode. *Astronomy and Astrophysics*, *481*, 57. doi: 10.1051/0004-6361:20079081.

Cirtain, J. W., Golub, L., Lundquist, L., van Ballegooijen, A., Savcheva, A., Shimojo, M., et al. (2007). Evidence for Alfvén waves in solar X-ray jets. *Science*, *318*, 1580. doi: 10.1126/science.1147050.

Cranmer, S. R., & Woolsey, L. N. (2015). Driving solar spicules and jets with magnetohydrodynamic turbulence: Testing a persistent idea. *Astrophysical Journal*, *812*, 71. doi: 10.1088/0004-637X/812/1/71.

Crooker, N. U., Gosling, J. T., & Kahler, S. W. (2002). Reducing heliospheric magnetic flux from coronal mass ejections without disconnection. *Journal of Geophysical Research*, *107*, 1028. doi: 10.1029/2001JA000236.

Crooker, N. U., & Owens, M. J. (2012). Interchange reconnection: Remote sensing of solar signature and role in heliospheric magnetic flux budget. *Space Science Reviews*, *172*, 201. doi: 10.1007/s11214-011-9748-1.

Curdt, W., Tian, H., & Kamio, S. (2012). Explosive events: Swirling transition region jets. *Solar Physics*, *280*, 417. doi: 10.1007/s11207-012-9940-9.

De Pontieu, B., Carlsson, M., Rouppe van der Voort, L. H. M., Rutten, R. J., Hansteen, V. H., & Watanabe, H. (2012). Ubiquitous torsional motions in type II spicules. *Astrophysical Journal*, *752L*, 12. doi: 10.1088/2041-8205/752/1/L12.

De Pontieu, B., Erdélyi, R., & de Wijn, A. G. (2003). Intensity oscillations in the upper transition region above active region plage. *Astrophysical Journal*, *595*, L63. doi: 10.1086/378843.

De Pontieu, B., Erdélyi, R., & James, S. P. (2004). Solar chromospheric spicules from the leakage of photospheric oscillations and flows. *Nature*, *430*, 536. doi: 10.1038/nature02749.

De Pontieu, B., Hansteen, V. H., Rouppe van der Voort, L., van Noort, M., & Carlsson, M. (2007). High-resolution observations and modeling of dynamic fibrils. *Astrophysical Journal*, *655*, 624. doi: 10.1086/509070.

De Pontieu, B., McIntosh, S., Hansteen, V. H., Carlsson, M., Schrijver, C. J., Tarbell, T. D., et al. (2007). A tale of two spicules: The impact of spicules on the magnetic chromosphere. *Publications of the Astronomical Society of Japan*, *59*, 655. doi: 10.1093/pasj/59.sp3.S655.

De Pontieu, B., McIntosh, S. W., Carlsson, M., Hansteen, V. H., Tarbell, T. D., Boerner, P., et al. (2011). The origins of hot plasma in the solar corona. *Science*, *331*, 55. doi: 10.1126/science.1197738.

De Pontieu, B., Rouppe van der Voort, L., McIntosh, S. W., Pereira, T. M. D., Carlsson, M., Hansteen, V., et al. (2014). On the prevalence of small-scale twist in the solar chromosphere and transition region. *Science*, *346*, 1255732. doi: 10.1126/science.1255732.

De Pontieu, B., Title, A. M., Lemen, J. R., Kushner, G. D., Akin, D. J., Allard, B., et al. (2014). The interface region imaging spectrograph (IRIS). *Solar Physics*, *289*, 2733. doi: 10.1007/s11207-014-0485-y.

De Ponteiu, B., Berger, T. E., Schrijver, C. J., & Title, A. M. (1999). Dynamics of transition region "moss" at high time resolution. *Solar Physics*, *190*, 419. doi: 10.1023/A:1005220606223.

Dere, K. P., Bartoe, J.-d. F., & Brueckner, G. E. (1983). Chromospheric jets – possible extreme-ultraviolet observations of spicules. *Astrophysical Journal*, *267L*, 65. doi: 10.1086/184004.

Dere, K. P., Bartoe, J.-D. F., Brueckner, G. E., Cook, J. W., Socker, D. G., & Ewing, J. W. (1989). UV observations of macrospicules at the solar limb. *Solar Physics*, *119*, 55. doi: 10.1007/BF00146212.

Dobrzycka, D., Raymond, J. C., Biesecker, D. A., Li, J., & Ciaravella, A. (2003). Ultraviolet spectroscopy of narrow coronal mass ejections. *Astrophysical Journal*, *588*, 586. doi: 10.1086/374047.

Falconer, D. A., Moore, R. L., Porter, J. G., & Hathaway, D. H. (2003). Solar coronal heating and the magnetic flux content of the network. *Astrophysical Journal*, *593*, 549. doi: 10.1086/376359.

Fang, F., Fan, Y., & McIntosh, S. W. (2014). Rotating solar jets in simulations of flux emergence with thermal conduction. *Astrophysical Journal*, *789*, 19. doi: 10.1088/2041-8205/789/1/L19.

Foukal, P. (1971a). Hα fine structure and the chromospheric field. *Solar Physics*, *20*, 298. doi: 10.1007/BF00159759.

Foukal, P. (1971b). Morphological relationships in the chromospheric Hα fine structure. *Solar Physics*, *19*, 59. doi: 10.1007/BF00148824.

Foukal, P. V. (2013). *Solar astrophysics* (3rd ed.). Weinhelm, Germany: Wiley-VCH.

Gaizauskas, V. (1984). Suspended spicules associated with the enhanced bright network in an active region. *Solar Physics*, *93*, 257. doi: 10.1007/BF02270838.

Galsgaard, K., Madjarska, J. S., Mackay, D. H., & Mou, C. (2019). Eruptions from quiet sun coronal bright points. ii. non-potential modelling. *Astronomy and Astrophysics*, *623*, 78. doi: 10.1051/0004-6361/201834329.

Gilbert, H. R., Serex, E. C., Holzer, T. E., MacQueen, R. M., & McIntosh, P. S. (2001). Narrow coronal mass ejections. *Astrophysical Journal*, *550*, 1093. doi: 10.1086/319816.

Glesener, L., & Fleishman, G. D. (2018). Electron acceleration and jet-facilitated escape in an M-class solar flare on 2002 August 19. *Astrophysical Journal*, *867*, 84. doi: 10.3847/1538-4357/aacefe.

Golub, L., Deluca, E., Austin, G., Bookbinder, J., Caldwell, D., Cheimets, P., et al. (2007). The X-ray telescope (XRT) for the Hinode mission. *Solar Physics*, *243*, 63. doi: 10.1007/s11207-007-0182-1.

Gopalswamy, N., Yashiro, S., Michalek, G., Stenborg, G., Vourlidas, A., Freeland, S., et al. (2009). The SOHO/LASCO CME catalog. *Earth, Moon, & Planets*, *104*, 295. doi: 10.1007/s11038-008-9282-7.

Green, L. M., Démoulin, P., Mandrini, C. H., & Van Driel-Gesztelyi, L. (2003). How are emerging flux, flares and CMEs related to magnetic polarity imbalance in MDI data? *Solar Physics*, *215*, 307. doi: 10.1023/A:1025678917086.

Guerreiro, N., Carlsson, M., & Hansteen, V. (2013). Numerical simulations of spicule acceleration. *Astrophysical Journal*, *766*, 128. doi: 10.1088/0004-637X/766/2/128.

Guo, Y., Démoulin, P., Schmieder, B., Ding, M. D., Vargas Domnguez, S., & Liu, Y. (2013). Recurrent coronal jets induced by repetitively accumulated electric currents. *Astronomy and Astrophysics*, *555*, A19. doi: 10.1051/0004-6361/201321229.

Hansteen, V. H., De Pontieu, B., Rouppe van der Voort, L., van Noort, M., & Carlsson, M. (2006). Dynamic fibrils are driven by magnetoacoustic shocks. *Astrophysical Journal*, *647L*, 73. doi: 10.1086/507452.

Harrison, R. A., Fludra, A., Pike, C. D., Payne, J., Thompson, W. T., Poland, A. I., et al. (1997). High-resolution observations of the extreme ultraviolet Sun. *Solar Physics*, *170*, 123. doi: 10.1023/A:1004913326580.

Heggland, L., Hansteen, V. H., De Pontieu, B., & Carlsson, M. (2011). Wave propagation and jet formation in the chromosphere. *Astrophysical Journal*, *743*, 142. doi: 10.1088/0004-637X/743/2/142.

Heyvaerts, J., Priest, E. R., & Rust, D. M. (1977). An emerging flux model for the solar flare phenomenon. *Astrophysical Journal*, *216*, 123. doi: 10.1086/155453.

Hirayama, T. (1974). Theoretical model of flares and prominences I: Evaporating flare model. *Solar Physics*, *34*, 323. doi: 10.1007/BF00153671.

Hollweg, J. V. (1982). On the origin of solar spicules. *Astrophysical Journal*, *257*, 345. doi: 10.1086/159993.

Hollweg, J. V., Jackson, S., & Galloway, D. (1982). Alfvén waves in the solar atmosphere. III – nonlinear waves on open flux tubes. *Solar Physics*, *75*, 35. doi: 10.1007/BF00153458.

Hong, J., Jiang, Y., Yang, J., Bi, Y., Li, H., Yang, B., et al. (2014). Coronal bright points associated with minifilament eruptions. Astrophysical Journal, *796*, 73. doi: 10.1088/0004-637X/796/2/73.

Hong, J., Jiang, Y., Yang, J., Yang, B., Xu, Z., & Xiang, Y. (2016). Mini-filament eruption as the initiation of a jet along coronal loops. *Astrophysical Journal*, *830*, 60. doi: 10.3847/0004-637X/830/2/60.

Hong, J., Jiang, Y., Yang, Y., Li, H., & Xu, Z. (2017). Minifilament eruption as the source of a blowout jet, C-class flare, and type-iii radio burst. *Astrophysical Journal*, *835*, 35. doi: 10.3847/1538-4357/835/1/35.

Hong, J., Jiang, Y., Zheng, R., Yang, J., Bi, Y., & Yang, B. (2011). A micro coronal mass ejection associated blowout extreme-ultraviolet jet. *Astrophysical Journal*, *738L*, 20. doi: 10.1088/2041-8205/738/2/L20.

Huang, Z., Madjarska, M. S., Doyle, J. G., & Lamb, D. A. (2012). Coronal hole boundaries at small scales. iv. SOT view. Magnetic field properties of small-scale transient brightenings in coronal holes. *Astronomy and Astrophysics*, *548*, 62. doi: 10.1051/0004-6361/201220079.

Iijima, H., & Yokoyama, T. (2017). A three-dimensional magnetohydrodynamic simulation of the formation of solar chromospheric jets with twisted magnetic field lines. *Astrophysical Journal*, *848*, 38. doi: 10.3847/1538-4357/aa8ad1.

Innes, D., & Teriaca, L. (2013). Quiet sun explosive events: Jets, splashes, and eruptions. *Solar Physics*, *282*, 453. doi: 10.1007/s11207-012-0199-y.

Innes, D. E., Bučik, R., Guo, L.-J., & Nitta, N. (2016). Observations of solar X-ray and EUV jets and their related phenomena. *Astronomische Nachrichten*, *337*, 1024. doi: 10.1002/asna.201612428.

Jiao, F., Xia, L., Bo, L., Huang, Z., Li, X., Chandrashekhar, K., et al. (2015). Sources of quasi-periodic propagating disturbances above a solar polar coronal hole. *Astrophysical Journal*, *809*, L17. doi: 10.1088/2041-8205/809/1/L17.

Ji, H., Wang, H., Schmahl , E.J., Moon, Y.-J., Jiang, Y. (2003). Observations of the failed eruption of a filament. *Astrophysical Journal*, *595*, 2, L135-L138. doi: 10.1086/378178.

Joshi, B., Thalmann, J. K., Mitra, P. K., Chandra, R., & Veronig, A. M. (2017). Observational and model analysis of a two-ribbon flare possibly induced by a neighboring blowout jet. *Astrophysical Journal*, *851*, 29. doi: 10.3847/1538-4357/aa9564.

Joshi, N. C., Nishizuka, N., Filippov, B., Magara, T., & Tlatov, A. G. (2018). Flux rope breaking and formation of a rotating blowout jet. *MNRAS*, *476*, 1286. doi: 10.1093/mnras/sty322.

Judge, P. G., & Carlsson, M. (2010). On the solar chromosphere observed at the limb with Hinode. *Astrophysical Journal*, *719*, 469. doi: 10.1088/0004-637X/719/1/469.

Judge, P. G., Tritschler, A., & Chye Low, B. (2011). Thermal fine structure and magnetic fields in the solar atmosphere: Spicules and fibrils. *Astrophysical Journal*, *730L*, 4. doi: 10.1088/2041-8205/730/1/L4.

Kamio, S., Curdt, W., Teriaca, L., Inhester, B., & Solanki, S. K. (2010). Observations of a rotating macrospicule associated with an X-ray jet. *Astronomy and Astrophysics*, *510*, 1. doi: 10.1051/0004-6361/200913269.

Karpen, J. T., DeVore, C. R., Antiochos, S. K., & Pariat, E. (2018). Reconnection-driven coronal-hole jets with gravity and solar wind. *Astrophysical Journal*, *834*, 62. doi: 10.3847/1538-4357/834/1/62.

Katsukawa, Y., Berger, T. E., Ichimoto, K., Lites, B. W., Nagata, S., Shimizu, T., et al. (2007). Small-scale jetlike features in penumbral chromospheres. *Science*, *318*, 1594. doi: 10.1126/science.1146046.

Kayshap, P., Murawski, K., Srivastava, A. K., & Dwivedi, B. N. (2018). Rotating network jets in the quiet Sun as observed by IRIS. *Astronomy and Astrophysics*, *616*, A99. doi: 10.1051/0004-6361/201730990.

Kim, Y.-H., Moon, Y.-J., Park, Y.-D., Sakurai, T., Chae, J., Cho, K. S., et al. (2007). Small-scale X-ray/EUV jets seen in Hinode XRT and TRACE. *Publications of the Astronomical Society of Japan*, *59*, 763. doi: 10.1093/pasj/59.sp3.S763.

Kiss, T. S., Gyenge, N., & Erdélyi. (2017). Systematic variations of macrospicule properties observed by SDO/AIA over half a decade. *Astrophysical Journal*, *835*, 47. doi: 10.3847/1538-4357/aa5272.

Klimchuk, J. A. (2012). The role of Type II spicules in the upper solar atmosphere. *Journal of Geophysical Research*, *117*, A12102. doi: 10.1029/2012JA018170.

Klimchuk, J. A., & Bradshaw, S. J. (2014). Are chromospheric nanoflares a primary source of coronal plasma? *Astrophysical Journal*, *791*, 60. doi: 10.1088/0004 -637X/791/1/60.

Kopp, R. A., & Pneuman, G. W. (1976). Magnetic reconnection in the corona and the loop prominence phenomenon. *Solar Physics*, *50*, 85. doi: 10.1007/BF00206193.

Kosugi, T., Matsuzaki, K., Sakao, T., Shimizu, T., Sone, Y., Tachikawa, S., et al. (2007). The Hinode (Solar-B) mission: An overview. *Solar Physics*, *243*, 3. doi: 10.1007/s11207-007-9014-6.

Koutchmy, S., Hara, H., Suematsu, Y., & Reardon, K. (1997). SXR coronal flashes. *Astronomy and Astrophysics*, *320L*, 33.

Koutchmy, S., & Macris, C. (1971). Observation in the wing of the Hα line and identification of the spicular structure near the solar limb. *Solar Physics*, *20*, 295. doi: 10.1007/BF00159758.

Kudoh, T., & Shibata, K. (1999). Alfvén wave model of spicules and coronal heating. *Astrophysical Journal*, *514*, 493. doi: 10.1086/306930.

Kumar, P., Karpen, J. T., Antiochos, S. K., Wyper, P. F., DeVore, C. R., & DeForest, C. E. (2018). Evidence for the magnetic breakout model in an equatorial coronal-hole jet. *Astrophysical Journal*, *854*,155. doi: 10.3847/1538-4357/aaab4f.

LaBonte, B. J. (1979). Activity in the quiet Sun. I - Observations of macrospicules in H-alpha and D3. *Solar Physics*, *61*, 283. doi: 10.1007/BF00150413.

Langangen, O., De Pontieu, B., Carlsson, M., Hansteen, V. H., Cauzzi, G., & Reardon, K. (2008). Search for high velocities in the disk counterpart of Type II spicules. *Astrophysical Journal*, *679*, 167L. doi: 10.1086/589442.

Lee, K. S., Innes, D. E., Moon, Y.-J., Shibata, K., Lee, J.-Y., & Park, Y.-D. (2013). Fast extreme-ultraviolet dimming associated with a coronal jet seen in multiwavelength and stereoscopic observations. *Astrophysical Journal*, *766*, 1L. doi: 10.1088/0004-637X/766/1/1.

Lemen, J. R., Title, A. M., Akin, D. J., Boerner, P. F., Chou, C., Drake, J. F., et al. (2012). The atmospheric imaging assembly (AIA) on the solar dynamics observatory (SDO). *Solar Physics*, *275*, 17. doi: 10.1007/s11207-011-9776-8.

Li, Y., & Ding, M. D. (2017). Plasma brightenings in a failed solar filament eruption. *Astrophysical Journal*, *838*, 15L. doi: 10.3847/1538-4357/aa6348.

Lionello, R., Török, T., Titov, V. S., Leake, J. E., Mikić, Z, Linker, J. A., et al. (2016). The contribution of coronal jets to the solar wind. *Astrophysical Journal Letters*, *831L*, 2. doi: 10.3847/2041-8205/831/1/L2.

Lippincott, S. L. (1957). Chromospheric spicules. *Smithsonian Contribution to Astrophysics*, *2*, 15.

Liu, C., Deng, N., Liu, R., Ugarte-Urra, I., Wang, S., & Wang, H. (2011). A standard-to-blowout jet. *Astrophysical Journal*, *735*, L18. doi: 10.1088/2041-8205/735/1/L18.

Liu, J., Erdélyi, R., Wang, Y., & Liu, R. (2018). Untwisting jets related to magnetic flux cancellation. *Astrophysical Journal*, *852*, 10. doi: 10.3847/1538-4357/aa992d.

Liu, J., Fang, F., Wang, Y., McIntosh, S. W., Fan, Y., & Zhang, Q. (2016). On the observation and simulation of solar coronal twin jets. *Astrophysical Journal*, *817*, 126. doi: 10.3847/0004-637X/817/2/126.

Liu, Y., & Kurokawa, H. (2004). On a surge: Properties of an emerging flux region. *Astrophysical Journal*, *610*, 1136. doi: 10.1086/421715.

Livshits, M. A. (1967). Line widths in spicules. *Soviet Astronomy-AJ*, *10*, 570.

Loboda, I. P., & Bogachev, S. A. (2019). What is a macrospicule? *Astrophysical Journal*, *871*, 230. doi: 10.3847/1538-4357/aafa7a.

Lynch, D. K., Beckers, J. M., & Dunn, R. B. (1973). A morphological study of solar spicules. *Solar Physics*, *30*, 63L. doi: 10.1007/BF00156173.

Madjarska, M. S., Vanninathan, K., & Doyle, J. G. (2011). Can coronal hole spicules reach coronal temperatures? *Astronomy and Astrophysics*, *532*, L1. doi: 10.1051/0004-6361/201116735.

Mariska, J. T., & Hollweg, J. V. (1985). Alfvénic pulses in the solar atmosphere. *Astrophysical Journal*, *296*, 746. doi: 10.1086/163491.

Marsh, K. A. (1976). The lifetime and evolution of fibrils. *Solar Physics*, *50*, 37. doi: 10.1007/BF00206188.

Martínez-Sykora, J., De Pontieu, B., Hansteen, V. H., Rouppe van der Voort, L., Carlsson, M., & Pereira, T. M. D. (2017). On the generation of solar spicules and Alfvénic waves. *Science*, *356*, 1269. doi: 10.1126/science.aah5412.

Martínez-Sykora, J., De Pontieu, B., Leenaarts, J., Pereira, T. M. D., Carlsson, M., Hansteen, V., et al. (2013). A detailed comparison between the observed and synthesized properties of a simulated Type II spicule. *Astrophysical Journal*, *771*, 66. doi: 10.1088/0004-637X/771/1/66.

Martínez-Sykora, J., Hansteen, V., De Pontieu, B., & Carlsson, M. (2009). Spicule-like structures observed in three-dimensional realistic magnetohydrodynamic simulations. *Astrophysical Journal*, *701*, 1569. doi: 10.1088/0004-637X/701/2/1569.

Martínez-Sykora, J., Hansteen, V., & Moreno-Insertis, F. (2011). On the origin of the Type II spicules: Dynamic three-dimensional MHD simulations. *Astrophysical Journal*, *736*, 9. doi: 10.1088/0004-637X/736/1/9.

Matsui, Y., Yokoyama, T., Kitagawa, N., & Imada, S. (2012). Multi-wavelength spectroscopic observation of extreme-ultraviolet jet in AR 10960. *Astrophysical Journal*, *759*, 15. doi: 10.1088/0004-637X/759/1/15.

Miao, Y., Liu, Y., Li, H. B., Shen, Y., Yang, S., Elmhamdi, A., et al. (2018). A blowout jet associated with one obvious extreme-ultraviolet wave and one complicated coronal mass ejection event. *Astrophysical Journal*, *869*, 39. doi: 10.3847/1538-4357/aaeacl.

Michard, R. (1974). Spicules and their surroundings. In R. G. Athay (Ed.), *Proceedings from IAU symposium no. 56 held at surfer's paradise, qld., Australia, 3-7 September 1973. edited by R. Grant Athay. International Astronomical Union. Symposium No. 56* (p. 3). Dordrecht; Boston: Reidel.

Miyagoshi, T., & Yokoyama, T. (2004). Magnetohydrodynamic simulation of solar coronal chromospheric evaporation jets caused by magnetic reconnection associated with magnetic flux emergence. *Astrophysical Journal*, *614*, 1042. doi: 10.1086/423731.

Moore, R. L. (1990). Hallmarks of the magnetic field in the solar atmosphere - structure, evolution, heating, and flaring. *Società Astronomica Italiana, Memorie (MmSAI)*, *61*, 317.

Moore, R. L., Cirtain, J. W., Sterling, A. C., & Falconer, D. A. (2010). Dichotomy of solar coronal jets: Standard jets and blowout jets. *Astrophysical Journal*, *720*, 757. doi: 10.1088/0004-637X/720/1/757.

Moore, R. L., Sterling, A. C., Cirtain, J. W., & Falconer, D. A. (2011). Solar X-ray jets, type-II spicules, granule-size emerging bipoles, and the genesis of the heliosphere. *Astrophysical Journal*, *731L*, 18. doi: 10.1088/2041-8205/731/1/L18.

Moore, R. L., Sterling, A. C., Falconer, D. A., & Robe, D. (2013). The cool component and the dichotomy, lateral expansion, and axial rotation of solar X-ray jets. *Astrophysical Journal*, *769*, 134. doi: 10.1088/0004-637X/769/2/134.

Moore, R. L., Sterling, A. C., Hudson, H. S., & Lemen, J. R. (2001). Onset of the magnetic explosion in solar flares and coronal mass ejections. *Astrophysical Journal*, *552*, 833. doi: 10.1086/320559.

Moore, R. L., Sterling, R. L., & Falconer, D. A. (2015). Magnetic untwisting in solar jets that go into the outer corona in polar coronal holes. *Astrophysical Journal*, *806*, 11. doi: 10.1088/0004-637X/806/1/11.

Moore, R. L., Sterling, R. L., & Panesar, N. K. (2018). Onset of the magnetic explosion in solar polar coronal X-ray jets. *Astrophysical Journal*, *859*,3. doi: 10.3847/1538-4357/aabe79.

Moore, R. L., Tang, F., Bohlin, J. D., & Golub, L. (1977). H-alpha macrospicules – identification with EUV macrospicules and with flares in X-ray bright points. *Astrophysical Journal*, *218*, 286. doi: 10.1086/155681.

Moreno-Insertis, F., & Galsgaard, K. (2013). Plasma jets and eruptions in solar coronal holes: A three-dimensional flux emergence experiment. *Astrophysical Journal*, *771*, 20. doi: 10.1088/0004-637X/771/1/20.

Mulay, S. M., Del Zanna, G., & Mason, H. (2017a). Cool and hot emission in a recurring active region jet. *Astronomy and Astrophysics*, *606*, 4. doi: 10.1051/0004-6361/201730429.

Mulay, S. M., Del Zanna, G., & Mason, H. (2017b). Temperature and density structure of a recurring active region jet. *Astronomy and Astrophysics*, *598*, 11. doi: 10.1051/0004-6361/201628796.

Mulay, S. M., Tripathi, D. T., Del Zanna, G., & Mason, H. (2016). Multiwavelength study of 20 jets that emanate from the periphery of active regions. *Astronomy and Astrophysics*, *589*, A79. doi: 10.1051/0004-6361/201527473.

Narang, N., Arbacher, R. T., Tian, H., Banerjee, D., Cranmer, S. R., DeLuca, E. E., et al. (2018). Statistical study of network jets observed in the solar transition region: A comparison between coronal holes and quiet-sun regions. *Solar Physics*, *291*, 1129. doi: 10.1007/s11207-016-0886-1.

Narukage, N., Sakao, T., Kano, R., Hara, H., Shimojo, M., Bando, T., et al. (2011). Coronal-temperature-diagnostic capability of the Hinode/X-ray telescope based on self-consistent calibration. *Solar Physics*, *269*, 169. doi: 10.1007/s11207-010-9685-2.

Ni, L., Zhang, Q.-M., Murphy, N. A., & Lin, J. (2017). Blob formation and ejection in coronal jets due to the plasmoid and Kelvin-Helmholtz instabilities. *Astrophysical Journal*, *841*, 27. doi: 10.3847/1538-4357/aa6ffe.

Nishikawa, T. (1988). Spicule observations with high spatial resolution. *Publications of the Astronomical Society of Japan*, *40*, 613.

Nishizuka, N., Shimizu, M., Nakamura, T., Otsuji, K., Okamoto, T. J., Katsukawa, Y., et al. (2008). Giant chromospheric anemone jet observed with Hinode and comparison with magnetohydrodynamic simulations: Evidence of propagating Alfvén waves and magnetic reconnection. *Astrophysical Journal*, *683*, 83. doi: 10.1086/591445.

Nisticò, G., Bothmer, V., Patsourakos, S., & Zimbardo, G. (2009). Characteristics of EUV coronal jets observed with STEREO/SECCHI. *Solar Physics*, *259*, 87. doi: 10.1007/s11207-009-9424-8.

Nitta, N. V., Mason, G. M., Wang, L., Cohen, C. M. S., & Wiedenbeck, M. E. (2015). Solar sources of ^{3}He-rich solar energetic particle events in solar cycle 24. *Astrophysical Journal*, *806*, 235. doi: 10.1088/0004-637X/806/2/235.

Öhman, Y. (1972). Remark on rotational motions in flares and prominences. *Solar Physics*, *23*, 134. doi: 10.1007/BF00153896.

Panesar, N. K., Sterling, A. C., & Moore, R. L. (2016). Homologous jet-driven coronal mass ejections from solar active region 12192. *Astrophysical Journal*, *822L*, 7. doi: 10.3847/2041-8205/822/2/L23.

Panesar, N. K., Sterling, A. C., & Moore, R. L. (2017). Magnetic flux cancellation as the origin of solar quiet-region pre-jet minifilaments. *Astrophysical Journal*, *844*, 131. doi: 10.3847/1538-4357/aa7b77.

Panesar, N. K., Sterling, A. C., & Moore, R. L. (2018). Magnetic flux cancelation as the trigger of solar coronal jets in coronal holes. *Astrophysical Journal*, *853*, 189. doi: 10.3847/1538-4357/aaa3e9.

Panesar, N. K., Sterling, A. C., Moore, R. L., & Chakrapani, P. (2016). Magnetic flux cancelation as the trigger of solar quiet-region coronal jets. *Astrophysical Journal*, *832L*, 7. doi: 10.3847/2041-8205/832/1/L7.

Panesar, N. K., Sterling, A. C., Moore, R. L., Tiwari, S. K., De Pontieu, B., & Norton, A. A. (2018). IRIS and SDO observations of solar jetlets resulting from network-edge flux cancelation. *Astrophysical Journal*, *868L*, 27. doi: 10.3847/2041-8213/aaef37.

Paraschiv, A. R., Bemporad, A., & Sterling, A. C. (2015). Physical properties of solar polar jets. A statistical study with Hinode XRT data. *Astronomy and Astrophysics*, *579*, 96. doi: 10.1051/0004-6361/201525671.

Parenti, S. (2014). Solar prominences: Observations. *Living Reviews in Solar Physics*, *11*, 1. doi: 10.12942/lrsp-2014-1.

Pariat, E., Dalmasse, K., DeVore, C. R., Antiochos, S. K., & Karpen, J. T. (2015). Model for straight and helical solar jets. I. parametric studies of the magnetic field geometry. *Astronomy and Astrophysics*, *573*, 130. doi: 10.1051/0004-6361/201424209.

Pariat, E. A., K., S., & DeVore, C. R. (2009). A model for solar polar jets. *Astrophysical Journal*, *691*, 61. doi: 10.1088/0004-637X/691/1/61.

Pariat, E., Antiochos, S. K., & DeVore, C. R. (2010). Three-dimensional modeling of quasi-homologous solar jets. *Astrophysical Journal*, *714*, 1762. doi: 10.1088/0004 -637X/714/2/1762.

Pasachoff, J. M., Jacobson, W. A., & Sterling, A. C. (2009). Limb spicules from the ground and from space. *Solar Physics*, *260*, 59. doi: 10.1007/s11207-009-9430-x.

Pasachoff, J. M., Noyes, R. W., & Beckers, J. M. (1968). Spectral observations of spicules at two heights in the solar chromosphere. *Solar Physics*, *5*, 131. doi: 10.1007/BF00147962.

Patsourakos, S., Pariat, E., Vourlidas, A., Antiochos, S. K., & Wuelser, J. P. (2008). Stereo SECCHI stereoscopic observations constraining the initiation of polar coronal jets. *Astrophysical Journal*, *680*, L73. doi: 10.1086/589769.

Pereira, T. M. D., De Pontieu, B., & Carlsson, M. (2012). Quantifying spicules. *Astrophysical Journal*, *759*, 18. doi: 10.1088/0004-637X/759/1/18.

Pereira, T. M. D., De Pontieu, B., & Carlsson, M. (2013). The effects of spatiotemporal resolution on deduced spicule properties. *Astrophysical Journal*, *764*, 69. doi: 10.1088/0004-637X/764/1/69.

Pereira, T. M. D., De Pontieu, B., Carlsson, M., Hansteen, V., Tarbell, T. D., Lemen, J., et al. (2014). An Interface Region Imaging Spectrograph first view on solar spicules. *Astrophysical Journal*, *792*, L15. doi: 10.1088/2041-8205/792/1/L15.

Pike, C. D., & Harrison, R. A. (1997). EUV observations of a macrospicule: Evidence for solar wind acceleration? *Solar Physics*, *175*, 457. doi: 10.1023/A: 1004987505422.

Pike, C. D., & Mason, H. E. (1998). Rotating transition region features observed with the SOHO Coronal Diagnostic Spectrometer. *Solar Physics*, *182*, 333. doi: 10.1023/A:1005065704108.

Pneuman, G. W., & Kopp, R. A. (1978). Downflow in the supergranulation network and its implications for transition region models. *Solar Physics*, *57*, 49. doi: 10.1007/BF00152043.

Poletto, G. (2015). Solar coronal plumes. *Living Reviews in Solar Physics*, *12*, 7. doi: 10.1007/lrsp-2015-7.

Poletto, G., Sterling, A. C., Pucci, S., & Romoli, M. (2014). The contribution of X-ray polar blowout jets to the solar wind mass and energy. In B. Schmieder, J.-M. Malherbe, & S. T. Wu. (Eds.), *Nature of prominences and their role in space weather* (p. 239). Cambridge: Cambridge University Press. doi: 10.1017/S1743921313011046.

Pucci, S., Poletto, G., Sterling, A. C., & Romoli, M. (2013). Physical parameters of standard and blowout jets. *Astrophysical Journal*, *776*, 16. doi: 10.1088/0004 -637X/776/1/16.

Pucci, S., Poletto, G., Sterling, A. C., & Romoli, M. (2014). Birth, life, and death of a solar coronal plume. *Astrophysical Journal*, *793*, 86. doi: 10.1088/0004 -637X/793/2/86.

Rabin, D., & Moore, R. L. (1980). Coronal holes, the height of the chromosphere, and the origin of spicules. *Astrophysical Journal*, *241*, 394. doi: 10.1086/158352.

Raouafi, N.-E., Georgoulis, M. K., Rust, D. M., & Bernasconi, P. N. (2010). Micro-sigmoids as progenitors of coronal jets: Is eruptive activity self-similarly multi-scaled? *Astrophysical Journal*, *718*, 981. doi: 10.1088/0004-637X/718/2/981.

Raouafi, N. E., Patsourakos, S., Pariat, E., Young, P. R., Sterling, A. C., Savcheva, A., et al. (2016). Solar coronal jets: Observations, theory, and modeling. *Space Science Reviews*, *201*, 1. doi: 10.1007/s11214-016-0260-5.

Raouafi, N. E., Petrie, G. J. D., Norton, A. A., Henney, C. J., & Solanki, S. K. (2008). Evidence for polar jets as precursors of polar plume formation. *Astrophysical Journal*, *682*, L137. doi: 10.1086/591125.

Raouafi, N. E., & Stenborg, G. (2014). Role of transients in the sustainability of solar coronal plumes. *Astrophysical Journal*, *787*, 118. doi: 10.1088/0004-637X/787/2/118.

Roberts, M. A., Uritsky, V. M., DeVore, C. R., & Karpen, J. T. (2018). Simulated encounters of the parker solar probe with a coronal-hole jet. *Astrophysical Journal*, *866*, 14. doi: 10.3847/1538-4357/aadb41.

Rompolt, B. (1975). Spectral features to be expected from rotational and expansional motions in fine solar structures. *Solar Physics*, *41*, 329. doi: 10.1007/BF00154070.

Rouppe van der Voort, L. H. M., De Pontieu, B., Hansteen, V. H., Carlsson, M., & van Noort, M. (2007). Magnetoacoustic shocks as a driver of quiet-sun mottles. *Astrophysical Journal*, *660L*, 169. doi: 10.1086/518246.

Rouppe van der Voort, L., Leenaarts, J., de Pontieu, B., Carlsson, M., & Vissers, G. (2009). On-disk counterparts of Type II spicules in the Ca II 854.2 nm and Hα lines. *Astrophysical Journal*, *705*, 272. doi: 10.1088/0004-637X/705/1/272.

Roy, J.-R. (1973). The dynamics of solar surges. *Solar Physics*, *32*, 139. doi: 10.1007/BF00152734.

Rush, J. H., & Roberts, W. O. (1954). Recent studies of chromospheric spicules. *Australian Journal of Physics*, *7*, 230. doi: 10.1071/PH540230.

Rutten, R. J. (2007). Observing the solar chromosphere. In P. Heinzel, I. Dorotovic, & R. J. Rutten (Eds.), *The Physics of chromospheric plasmas asp conference series, Proceedings of the conference held 9-13 October, 2006 at the University of Coimbra in Coimbra, Portugal. San Francisco* (p. 27). San Francisco: Astronomical Society of the Pacific.

Samanta, T., Pant, V., & Banerjee, D. (2015). Propagating disturbances in the solar corona and spicular connection. *Astrophysical Journal*, *815*, L16. doi: 10.1088/2041-8205/815/1/L16.

Savcheva, A., Cirtain, J., Deluca, E. E., Lundquist, L. L., Golub, L., Weber, M., et al. (2007). A study of polar jet parameters based on Hinode XRT observations. *Publications of the Astronomical Society of Japan*, *59*, 771. doi: 10.1093/pasj/59.sp3.S771.

Savcheva, A., Cirtain, J. W., DeLuca, E. E., & Golub, L. (2009). Does a polar coronal hole's flux emergence follow a hale-like law? *Astrophysical Journal*, *702*, L32. doi: 10.1088/0004-637X/702/1/L32.

Scherrer, P. H., Schou, J., Bush, R. I., Kosovichev, A. G., Bogart, R. S., Hoeksema, J. T., et al. (2012). The helioseismic and magnetic imager (HMI) investigation for the solar dynamics observatory (SDO). *Solar Physics*, *275*, 207. doi: 10.1007/s11207-011-9834-2.

Schmieder, B., Guo, Y., Moreno-Insertis, F., Aulanier, G., Yelles Chaouche, L., Nishizuka, N., et al. (2013). Twisting solar coronal jet launched at the boundary of an active region. *Astronomy and Astrophysics*, *559*, A1. doi: 10.1051/0004-6361/201322181.

Schrijver, C. J., Title, A. M., Yeates, A. R., & DeRosa, M. L. (2013). Pathways of large-scale magnetic couplings between solar coronal events. *Astrophysical Journal*, *773*, 93. doi: 10.1088/0004-637X/773/2/93.

Scullion, E., Erdélyi, R., Fedun, V., & Doyle, J. G. (2011). The response of a three-dimensional solar atmosphere to wave-driven jets. *Astrophysical Journal*, *743*, 14. doi: 10.1088/0004-637X/743/1/14.

Secchi, A. (1877). *Le soleil* (Vol. 2). Paris: Gauthier-Villars.

Sekse, D. H., Rouppe van der Voort, L., & De Pontieu, B. (2012). Statistical properties of the disk counterparts of type II spicules from simultaneous observations of rapid blueshifted excursions in Ca II 8542 and Hα. *Astrophysical Journal*, *752*, 108. doi: 10.1088/0004-637X/752/2/108.

Sekse, D. H., Rouppe van der Voort, L., & De Pontieu, B. (2013). On the temporal evolution of the disk counterpart of Type II spicules in the quiet Sun. *Astrophysical Journal*, *764*, 164. doi: 10.1088/0004-637X/764/2/164.

Sekse, D. H., Rouppe van der Voort, L., De Pontieu, B., & Scullion, E. (2013). Interplay of three kinds of motion in the disk counterpart of Type II spicules: Upflow, transversal, and torsional motions. *Astrophysical Journal*, *769*, 44. doi: 10.1088/0004-637X/769/1/44.

Shen, Y., Liu, Y., Su, J., & Deng, Y. (2012). On a coronal blowout jet: The first observation of a simultaneously produced bubble-like CME and a jet-like CME in a solar event. *Astrophysical Journal*, *745*, 164. doi: 10.1088/0004-637X/745/2/164.

Shen, Y., Liu, Y. D., Su, J., Qu, Z., & Tian, Z. (2017). On a solar blowout jet: Driving mechanism and the formation of cool and hot components. *Astrophysical Journal*, *851*, 67. doi: 10.3847/1538-4357/aa9a48.

Shibata, K., Ishido, Y., Acton, L. W., Strong, K. T., Hirayama, T., Uchida, Y., et al. (1992a). Observations of X-ray jets with the Yohkoh soft X-ray telescope. *Publications of the Astronomical Society of Japan*, *44*, L173.

Shibata, K., Masuda, S., Shimojo, M., Hara, H., Yokoyama, T., Tsuneta, S., et al. (1995). Hot-plasma ejections associated with compact-loop solar flares. *Astrophysical Journal Letters*, *451*, L83.

Shibata, K., Nakamura, T., Matsumoto, T., Otsuji, K., Okamoto, T. J., Nishizuka, N., et al. (2007). Chromospheric anemone jets as evidence of ubiquitous reconnection. *Science*, *318*, 1591. doi: 10.1126/science.1146708.

Shibata, K., Nishikawa, T., Kitai, R., & Suematsu, Y. (1982). Numerical hydrodynamics of the jet phenomena in the solar atmosphere. II - surges. *Solar Physics*, *77*, 121. doi: 10.1007/BF00156100.

Shibata, K., Nitta, N., Matsumoto, R., Tajima, T., Yokoyama, T., Hirayama, T., et al. (1994). Two types of interaction between emerging flux and coronal magnetic field. In Y. Uchida, T. Watanabe, K. Shibata, & H. S. Hudson (Eds.), *X-ray solar physics from Yohkoh. frontiers science series, Proceedings of the International Symposium on the Yohkoh scientific results, held February 23-25, 1993, Sagamihara, Kanagawa, Japan* (p. 27). Tokyo, Japan: University Academy Press.

Shibata, K., Nozawa, S., & Matsumoto, R. (1992). Magnetic reconnection associated with emerging magnetic flux. *Astronomical Society of Japan Publications* (ISSN 0004-6264), *44*(3), 265–272.

Shibata, K., & Suematsu, Y. (1982). Why are spicules absent over plages and long under coronal holes. *Solar Physics*, *78*, 333. doi: 10.1007/BF00151612.

Shibata, K., & Uchida, Y. (1986). A magnetodynamic mechanism for the formation of astrophysical jets. II - Dynamical processes in the accretion of magnetized mass in rotation. *Solar Physics*, *178*, 379.

Shimojo, M., Hashimoto, S., Shibata, K., Hirayama, T., Hudson, H. S., & Acton, L. W. (1996). Statistical study of solar X-ray jets observed with the YOHKOH soft X-ray telescope. *Publications of the Astronomical Society of Japan*, *48*, 123. doi: 10.1093/pasj/48.1.123.

Shimojo, M., & Shibata, K. (2000). Physical parameters of solar X-ray jets. *Astrophysical Journal*, *542*, 1100. doi: 10.1086/317024.

Shimojo, M., Shibata, K., & Harvey, K. L. (1998). Magnetic field properties of solar X-ray jets. *Solar Physics*, *178*, 379. doi: 10.1023/A:1005091905214.

Skogsrud, H., Rouppe van der Voort, L., De Pointieu, D., & Pereira, T. M. D. (2015). On the temporal evolution of spicules observed with IRIS, SDO, and Hinode. *Astrophysical Journal*, *806*, 170. doi: 10.1088/0004-637X/806/2/170.

Sterling, A. C. (1998). Alfvénic resonances on ultraviolet spicules. *Astrophysical Journal*, *508*, 916. doi: 10.1086/306433.

Sterling, A. C. (1999). Electron temperatures of a late-phase solar active region from YOHKOH BCS and SXT observations. *Astronomy and Astrophysics*, *346*, 995.

Sterling, A. C. (2000). Solar spicules: A review of recent models and targets for future observations – (invited review). *Solar Physics*, *196*, 79. doi: 10.1023/A: 1005213923962.

Sterling, A. C., Harra, L. K., & Moore, R. L. (2010). Fibrillar chromospheric spicule-like counterparts to an extreme-ultraviolet and soft X-ray blowout coronal jet. *Astrophysical Journal*, *722*, 1644. doi: 10.1088/0004-637X/722/2/1644.

Sterling, A. C., & Hollweg, J. V. (1984). Alfvénic resonances on solar spicules. *Astrophysical Journal*, *285*, 843. doi: 10.1086/162563.

Sterling, A. C., & Hollweg, J. V. (1988). The rebound shock model for solar spicules – dynamics at long times. *Astrophysical Journal*, *327*, 950. doi: 10.1086/166252.

Sterling, A. C., & Mariska, J. T. (1990). Numerical simulations of the rebound shock model for solar spicules. *Astrophysical Journal*, *349*, 647. doi: 10.1086/168352.

Sterling, A. C., Mariska, J. T., Shibata, K., & Suematsu, Y. (1991). Numerical simulations of microflare evolution in the solar transition region and corona. *Astrophysical Journal*, *381*, 313. doi: 10.1086/170653.

Sterling, A. C., & Moore, R. L. (2001). EIT crinkles as evidence for the breakout model of solar eruptions. *Astrophysical Journal*, *560*, 1045. doi: 10.1086/322241.

Sterling, A. C., & Moore, R. L. (2016). A microfilament-eruption mechanism for solar spicules. *Astrophysical Journal*, *828*, L9. doi: 10.3847/2041-8205/828/1 L9.

Sterling, A. C., Moore, R. L., & DeForest, C. E. (2010b). Hinode solar optical telescope observations of the source regions and evolution of "type II" spicules at the solar polar limb, astrophysical journal. *Astrophysical Journal*, *714L*, 1. doi: 10.1088/2041-8205/714/1/L1.

Sterling, A. C., Moore, R. L., Falconer, D. A., & Adams, M. (2015). Small-scale filament eruptions as the driver of X-ray jets in solar coronal holes. *Nature*, *523*, 437. doi: 10.1038/nature14556.

Sterling, A. C., Moore, R. L., Falconer, D. A., & Knox, J. M. (2014). New aspects of a lid-removal mechanism in the onset of an eruption sequence that produced a large solar energetic particle (SEP) event. *Astrophysical Journal*, *788*, L20. doi: 10.1088/2041-8205/788/2/L20.

Sterling, A. C., Moore, R. L., Falconer, D. A., Panesar, N. K., Akiyama, S., Yashiro, S., et al. (2016). Minifilament eruptions that drive coronal jets in a solar active region. *Astrophysical Journal*, *821*, 100. doi: 10.3847/0004-637X/821/2/100.

Sterling, A. C., Moore, R. L., Falconer, D. A., Panesar, N. K., & Martinez, F. (2017). Solar active region coronal jets. II. Triggering and evolution of violent jets. *Astrophysical Journal*, *844*, 28. doi: 10.3847/1538-4357/aa7945.

Sterling, A. C., Moore, R. L., & Freeland, S. L. (2011). Insights into filament eruption onset from solar dynamics observatory observations. *Astrophysical Journal*, *731L*, 3. doi: 10.1088/2041-8205/731/1/L3.

Sterling, A. C., Moore, R. L., & Panesar, N. K. (2018). Magnetic flux cancelation as the buildup and trigger mechanism for CME-producing eruptions in two small active regions. *Astrophysical Journal*, *864*, 68. doi: 10.3847/1538-4357/aad550.

Sterling, A. C., Shibata, K., & Mariska, J. T. (1993). Solar chromospheric and transition region response to energy deposition in the middle and upper chromosphere. *Astrophysical Journal*, *407*, 778. doi: 10.1086/172559.

Strong, K. T., Harvey, K., Hirayama, T., Nitta, N., Shimizu, T., & Tsuneta, S. (1992). Observations of the variability of coronal bright points by the soft X-ray telescope on Yohkoh. *Publications of the Astronomical Society of Japan*, *44*, L161.

Sturrock, P. A. (1966). Model of the high-energy phase of solar flares. *Nature*, *211*, 695. doi: 10.1038/211695a0.

Suematsu, Y. (1990). Influence of photospheric 5-minute oscillations on the formation of chromospheric fine structures. In Y. Osaki & H. Shibahashi (Eds.), *Progress of seismology of the sun and stars, Proceedings of the OJI international seminar held at Hakone, Japan, 11-14 December 1989. Lecture Notes in Physics, vol. 367* (p. 211). Berlin Heidelberg New York: Springer-Verlag.

Suematsu, Y., Ichimoto, K., Katsukawa, Y., Shimizu, T., Okamoto, T., Tsuneta, S., et al. (2008). High resolution observations of spicules with Hinode/SOT. In S. A. Matthews, J. M. Davis, & L. K. Harra (Eds.), *First results from Hinode ASP conference series, vol. 397, Proceedings of the conference held 20–24 August, 2007, at Trinity College Dublin, Dublin, Ireland* (p. 27). San Francisco: Astronomical Society of the Pacific.

Suematsu, Y., Shibata, K., Nishikawa, T., & Kitai, R. (1982). Numerical hydrodynamics of the jet phenomena in the solar atmosphere. I - Spicules. *Solar Physics*, *75*, 99. doi: 10.1007/BF00153464.

Suematsu, Y., Wang, H., & Zirin, H. (1995). High-resolution observation of disk spicules. I. Evolution and kinematics of spicules in the enhanced network. *Astrophysical Journal*, *450*, 411. doi: 10.1086/176151.

Syntelis, P., Priest, E. R., & Chitta, L. P. (2019). A cancellation nanoflare model for solar chromospheric and coronal heating. II. 2D theory and simulations. *Astrophysical Journal*, *872*, 32. doi: 10.3847/1538-4357/aafaf8.

Szente, J., Toth, G., Manchester, W. B., IV, van der Holst, B., Landi, E., Gombosi, T. I., et al. (2017). Coronal jets simulated with the global Alfvén wave solar model. *Astrophysical Journal*, *834*, 123. doi: 10.3847/1538-4357/834/2/123.

Tian, H., DeLuca, E. E., Cranmer, S. R., De Pontieu, B., Peter, H., Martnez-Sykora, J., et al. (2014). Prevalence of small-scale jets from the networks of the solar transition region and chromosphere. *Science*, *346*(1255711). doi: 10.1126/science.1255711.

Tian, H., McIntosh, S. W., Xia, L., He, J., & Wang, X. (2012). What can we learn about solar coronal mass ejections, coronal dimmings, and extreme-ultraviolet jets through spectroscopic observations? *Astrophysical Journal*, *748*, 106. doi: 10.1088/0004-637X/748/2/106.

Tiwari, S. K., Moore, R. L., De Pontieu, B., Tarbell, T. D., Panesar, N. K., Winebarger, A. R., et al. (2018). Evidence of twisting and mixed-polarity solar photospheric magnetic field in large penumbral jets: IRIS and Hinode observations. *Astrophysical Journal*, *869*, 147. doi: 10.3847/1538-4357/aaf1b8.

Tiwari, S. K., Moore, R. L., Winebarger, A. R., & Alpert, S. E. (2016). Transition-region/coronal signatures and magnetic setting of sunspot penumbral jets: Hinode (SOT/FG), HI-C, and SDO/AIA observations. *Astrophysical Journal*, *816*, 92. doi: 10.3847/0004-637X/816/2/92.

Török, T., Panasenco, O., Titov, V. S., Mikić, Z., Reeves, K. K., Velli, M., et al. (2011). A model for magnetically coupled sympathetic eruptions. *Astrophysical Journal*, *739*, L63. doi: 10.1088/2041-8205/739/2/L63.

Tsiropoula, G., Tziotziou, K., Kontogiannis, I., Madjarska, M. S., Doyle, J. G., & Suematsu, Y. (2012). Solar fine-scale structures. I. Spicules and other small-scale, jet-like events at the chromospheric level: Observations and physical parameters. *Space Science Reviews*, *169*, 181. doi: 10.1007/s11214-012-9920-2.

Tsuneta, S., Acton, L., Bruner, M., Lemen, J., Brown, W., Caravalho, R., et al. (1991). The soft X-ray telescope for the Solar-A mission. *Solar Physics*, *136*, 37. doi: 10.1007/BF00151694.

Tsuneta, S., Ichimoto, K., Katsukawa, Y., Nagata, S., Otsubo, M., Shimizu, T., et al. (2008). The solar optical telescope for the Hinode mission: An overview. *Solar Physics*, *249*, 167. doi: 10.1007/s11207-008-9174-z.

van Ballegooijen, A. A., & Martens, P. C. H. (1989). Formation and eruption of solar prominences. *Astrophysical Journal*, *343*, 971. doi: 10.1086/167766.

Wang, H., & Liu, C. (2012). Circular ribbon flares and homologous jets. *Astrophysical Journal*, *760*, 101. doi: 10.1088/0004-637X/760/2/101.

Wang, J., Li, W., Denker, C., Lee, C., Wang, H., Goode, P. R., et al. (2000). Minifilament eruption on the quiet sun. I. observations at Hα central line. *Astrophysical Journal*, *530*, 1071. doi: 10.1086/308377.

Wang, Y.-M., & Sheeley, N. R. Jr. (2002). Coronal white-light jets near sunspot maximum. *Astrophysical Journal*, *575*, 542. doi: 10.1086/341145.

Wang, Y.-M., Sheeley, N. R., Socker, J., G., D., Howard, R. A., Brueckner, G. E., Michels, D. J. et al. (1998). Observations of correlated white-light and extreme-ultraviolet jets from polar coronal holes. *Astrophysical Journal*, *508*, 899. doi: 10.1086/306450.

Withbroe, G. L., Jaffe, D. T., Foukal, P. V., Huber, M. C. E., Noyes, R. W., Reeves, E. M., et al. (1976). Extreme-ultraviolet transients observed at the solar pole. *Astrophysical Journal*, *203*, 528. doi: 10.1086/154108.

Withbroe, G. L., & Noyes, R. W. (1977). Mass and energy flow in the solar chromosphere and corona. *Annual Review of Astronomy and Astrophysics*, *15*, 363. doi: 10.1146/annurev.aa.15.090177.002051.

Wyper, P. F., Antiochos, S. K., & DeVore, C. R. (2017). A universal model for solar eruptions. *Nature*, *544*, 452. doi: 10.1038/nature22050.

Wyper, P. F., DeVore, C. R., & Antiochos, S. K. (2018a). Breakout model for solar coronal jets with filaments. *Astrophysical Journal*, *852*, 98. doi: 10.3847/1538-4357/aa9ffc.

Wyper, P. F., DeVore, C. R., Karpen, J. T., Antiochos, S. K., & Yeates, A. R. (2018b). A model for coronal hole bright points and jets due to moving magnetic elements. *Astrophysical Journal*, *864*, 165. doi: 10.3847/1538-4357/aad9f7.

Xia, L. D., Popescu, M. D., Doyle, J. G., & Giannikakis, J. (2005). Time series study of EUV spicules observed by SUMER/SoHO. *Astronomy and Astrophysics*, *438*, 1115. doi: 10.1051/0004-6361:20042579.

Yashiro, S., Gopalswamy, N., Michalek, G., & Howard, R. A. (2003). Properties of narrow coronal mass ejections observed with LASCO. *Advances in Space Research*, *32*, 2631. doi: 10.1016/j.asr.2003.03.018.

Yokoyama, T., & Shibata, K. (1995). Magnetic reconnection as the origin of X-ray jets and Hα surges on the Sun. *Nature*, *375*, 42. doi: 10.1038/375042a0.

Yokoyama, T., & Shibata, K. (1996). Numerical simulation of solar coronal X-ray jets based on the magnetic reconnection model. *Publications of the Astronomical Society of Japan*, *48*, 353. doi: 10.1093/pasj/48.2.353.

Yokoyama, T., & Shibata, K. (2001). Magnetohydrodynamic simulation of a solar flare with chromospheric evaporation effect based on the magnetic reconnection model. *Astrophysical Journal*, *549*, 1160. doi: 10.1086/319440.

Young, P. R., & Muglach, K. (2014a). A coronal hole jet observed with Hinode and the solar dynamics observatory. *Publications of the Astronomical Society of Japan*, *66*, 12. doi: 10.1093/pasj/psu088.

Young, P. R., & Muglach, K. (2014b). Solar dynamics observatory and Hinode observations of a blowout jet in a coronal hole. *Solar Physics*, *289*, 3313. doi: 10.1007/s11207-014-0484-z.

Yu, H.-S., Jackson, B. V., Buffington, A., Hick, P. P., Shimojo, M., & Sako, N. (2014). The three-dimensional analysis of Hinode polar jets using images from LASCO C2, the STEREO COR2 coronagraphs, and SMEI. *Astrophysical Journal*, *784*, 166. doi: 10.1088/0004-637X/784/2/166.

Zaqarashvili, T. V., & Erdléyi, R. (2009). Oscillations and waves in solar spicules. *Solar Physics*, *149*, 355. doi: 10.1007/s11214-009-9549-y.

Zhang, Q. M., & Ji, H. S. (2014). A swirling flare-related EUV jet. *Astronomy & Astrophysics*, *561*, A134. doi: 10.1051/0004-6361/201322616.

Zhang, Q. M., Li, D., Ning, Z. J., Su, Y. N., Ji, H. S., & Guo, J. (2016). Explosive chromospheric evaporation in a circular-ribbon flare. *Astrophysical Journal*, *827*, 27. doi: 10.3847/0004-637X/827/1/27.

Zhang, Y. Z., Shibata, K., Wang, J. X., Mao, X. J., Matsumoto, T., Liu, Y., et al. (2012). Revision of solar spicule classification. *Astrophysical Journal*, *750*, 16. doi: 10.1088/0004-637X/750/1/16.

Zhu, X., Wang, H., Cheng, X., & Huang, C. (2017). A solar blowout jet caused by the eruption of a magnetic flux rope. *Astrophysical Journal*, *844*, 20. doi: 10.3847/2041-8213/aa8033.

Zirin, H. (1974). The magnetic structure of plages. *Solar Physics*, *56*, 161. doi: 10.3847/0004-637X/827/1/27.

Zirin, H. (1988). *Astrophysics of the sun*. Cambridge and New York: Cambridge University Press.

7

Solar Interior

Mark Linton[1], Mausumi Dikpati[2], and Rachel Howe[3]

ABSTRACT

This chapter introduces the reader to the solar interior, in particular the convection zone. The first section explores the solar cycle and the dynamo models that have been studied to explain this cycle. The second section explores helioseismology observations and analysis of the solar interior, and reviews the fundamental knowledge that has been gained from these studies. The third and final section reviews observations and theory of magnetic fields emerging from the convection zone into the solar corona.

7.1. SOLAR DYNAMO

7.1.1. Solar Dynamo Models: Overall Context

The Sun's magnetic field is very complex, with fields observed in the photosphere showing spatial variations over a wide range of scales, from the smallest resolvable scales (~100 km) to the size of the Sun itself (~x 10^5–10^6 km). The smallest scales vary on time scales of minutes, whereas the global scales can take months to years to change much. Table 7.1 presents a few examples of the hierarchy of magnetic field structures and evolutionary patterns.

The generation, maintenance, and evolution of solar magnetic fields, in random as well as in systematic fashions, are most likely due to a "dynamo" operating inside the Sun. A dynamo is a process by which the magnetic field in an electrically conducting fluid is maintained against Ohmic dissipation. In astrophysical objects, there can always be a dynamo whenever the plasma consists of seed magnetic fields and flow fields. The Sun, consisting of plasma, can have its own dynamo. But to this day, there is no unified, single solar dynamo model that can explain all solar magnetic features.

Although in a pure hydrodynamic (HD) situation there is evidence that large-scale motions develop from small-scale ones through inverse cascading (Miesch et al., 2000), the situation is more complicated in the magnetohydrodynamic (MHD) case. There exists observational evidence of both cascading and inverse cascading; for example, the breakup of large active regions is a cascading phenomenon, and their coalescence to form the large-scale, diffuse magnetic fields (Leighton, 1964) is an inverse cascade.

However, the organization of the small-scale magnetic fields into large-scale patterns, which cyclically evolve with an 11-year periodicity, has not yet been established theoretically. Most likely two different types of dynamos, the small-scale type and the large-scale type, are operating in the Sun coexisting together—each type is responsible for producing certain aspects of the solar magnetic features. Small-scale turbulent dynamos are primarily responsible for the generation and evolution of small-scale magnetic fields in the short term and in a more random fashion, whereas the generation and systematic

[1] *Space Science Division, Naval Research Laboratory, Washington, DC, USA*
[2] *High Altitude Observatory, Boulder, Colorado, USA*
[3] *School of Physics and Astronomy, University of Birmingham, Birmingham, United Kingdom*

Space Physics and Aeronomy Collection Volume 1: Solar Physics and Solar Wind, Geophysical Monograph 258, First Edition.
Edited by Nour E. Raouafi and Angelos Vourlidas.

DOI: 10.1002/9781119815600.ch7

Table 7.1 Hierarchy of Magnetic Field Structures and Evolutionary Patterns

Hierarchy in length scales	Hierarchy in flux and/or field strength	Hierarchy in temporal variations
Network fields: ~ 100 km size	Diffuse fields outside sunspots: A few tens of Gauss	Small-scale mixed polarity Turbulent fields Random features
Sunspots, ephemeral regions, plage: ~ 10,000 – 30,000 km diameter	Plage: A few hundreds of Gauss	Butterfly diagram Polar reversal, Cyclic features
Large-scale, unipolar regions: ~100,000 km extent	Spots, active regions, network fields: A few thousand Gauss	Active longitudes Persistent features

evolution (in a cyclic fashion with 11-year periodicity) of large-scale magnetic fields is due to global dynamos.

7.1.2. Small-Scale Turbulent Dynamos and Their Observational Signatures

Numerical simulations of local dynamo action in a 3D box, by Emonet et al. (2001) (see also Cattaneo, 1999; Emonet et al., 2001; Meneguzzi & Pouquet, 1989), have indicated that a turbulent flow interacting with a seed magnetic field, which can be either a random field or an imposed uniform vertical field, can produce dynamo action and generate small-scale magnetic fields. Depending on the flow fields, the generated magnetic fields can correspond to intra-network fields (in the case of granular scale flows), and to ephemeral regions and network fields (in the case of supergranular scale flows).

A few among many observational signatures of the small-scale surface dynamo action in the Sun are: magnetic flux evolution in plage—observed in high-resolution G-band (Berger & Title, 1996), the presence of weak magnetic field of ~1 Gauss over 1 arcs^2 in the quiet Sun—observed from Zeeman splitting in high-resolution infra-redlines (Lin & Rimmele, 1999), generation of mixed-polarity network fields in the quiet Sun—observed in the Solar and Heliospheric Observatory/Michelson Doppler Imager (SOHO/MDI: Scherrer et al., 1995) magnetogram images (Schrijver et al., 1997), the existence of unsigned, weak (5 − 15 Gauss) fields in the photosphere, chromospheric spicules—observed using the Hanle depolarization effect (Stenflo, 1999; Trujillo Bueno et al., 2005), and the existence of the small-scale, turbulent magnetic structures of sizes below the mean free path of a photon—derived from the observed asymmetry in the Stokes V profile (Sánchez Almeida & Lites, 2000).

The continuous churning of small-scale magnetic fields in the granular lanes outside the sunspot can be compared with the simulation output of Emonet et al. (2001; see their figure 1).

7.1.3. Global Dynamo Models and Solar Cycle

The existence of a "solar cycle" has been known since the middle of the 19th century. Panels (a) and (b) of Figure 7.1 show the classical representations of this cycle. Figure 7.1a is the so-called "butterfly diagram," which is a time–latitude plot of the fractional area covered by sunspots, with the northern and southern hemispheres averaged together. The pattern is rather regular, with each new cycle starting in the neighborhood of 30° and ending close to the equator. We can see that the cycle period averages about 11 years, but there is a variance of more than one year, shorter and longer. As seen in Figure 7.1b, solar cycle amplitudes as measured by sunspot area vary by a factor of 2–3 from cycle to cycle, not randomly, but with an envelope of cycle amplitudes that varies over longer time scales than a single cycle. Among other manifestations of the solar cycle are the polar field reversals every 11 years, evolution of "active longitudes," and variation in the global coronal structure.

There is a general consensus that the solar cycle is produced by a global MHD dynamo operating in the solar convection zone and the solar tachocline just below. General dynamo theory began in the 1930s, motivated mainly by the existence of Earth's magnetic field, thought to be generated by a dynamo operating in the liquid interior. Dynamo models with specific application to the Sun first appeared in Parker (1955b). Solar dynamo models have advanced greatly since then, but there are still many challenges to overcome in the quest to build a dynamo model that simulates the most prominent features of a solar cycle, and allows prediction of amplitude and other properties of future cycles.

We describe in this chapter the current state of global solar dynamo modeling, including various classes of models, the important mathematical tools used to build them, what has been accomplished to date, and what outstanding challenges remain to be answered to make further progress. There are several excellent, recent reviews,

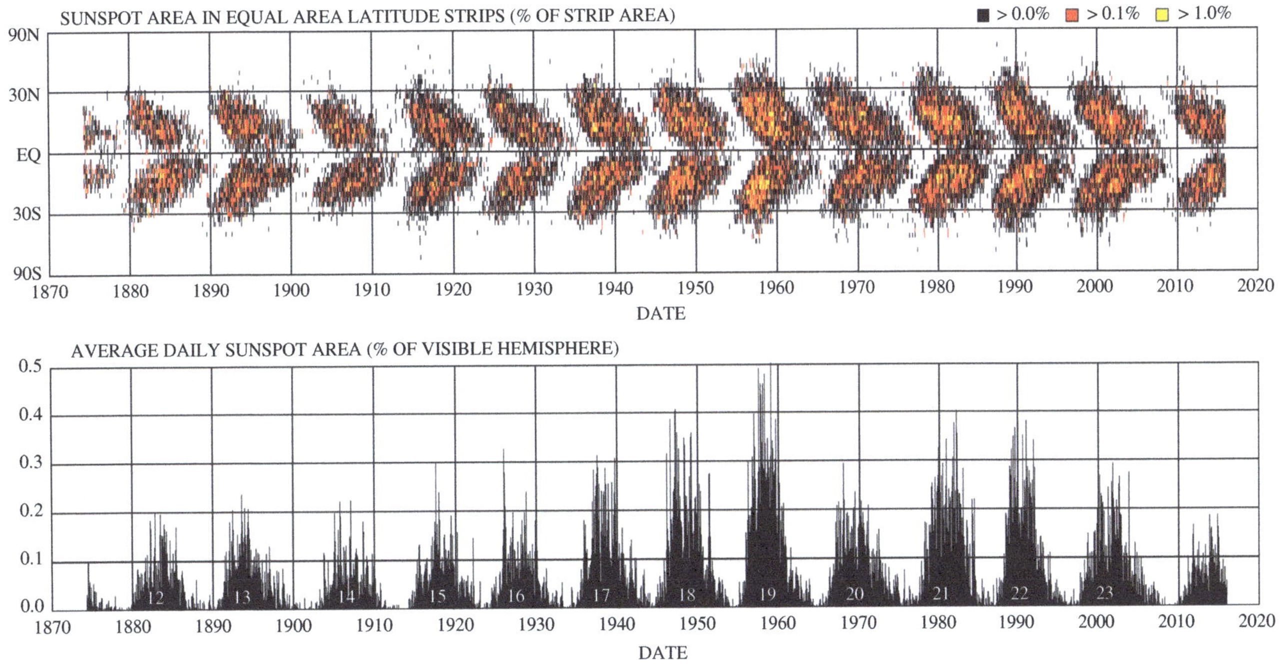

Figure 7.1 Upper: Time–latitude diagram of spot area; Lower: Total spot area as a function of time (Hathaway, 2015). (*Source*: Daily sunspot area averaged over individual solar rotations, Hathaway NASA/ARC 2016/10. Public domain.)

such as Charbonneau (2010, 2014) for solar cycle models, and Hathaway (2015) and Petrie (2015) for observations, which the readers are encouraged to consult. This review is primarily from a perspective of what needs to be done to go forward, in order to motivate the younger generation in this field of research.

Many stars have dynamo-maintained magnetic fields, some similar to the Sun, others quite different, some with magnetic cycles, some not. Solving the solar dynamo problem has obvious value for understanding stellar dynamos generally; conversely, successfully modeling the properties of stellar dynamos will have great value for understanding the solar cycle problem.

Ever since the first solar dynamo model developed by Parker (1955b), solar dynamo research has proceeded following two different approaches in parallel (see Charbonneau, 2010, for a detailed review). The first one uses the mean-field formalism (Steenbeck et al., 1966), which solves the magnetic induction equation to obtain the solution for the large-scale, mean magnetic field, by approximating small-scale motions and magnetic fields. The equatorially antisymmetric helical part of the correlation between small-scale motions and magnetic fields gives rise to the so-called "α-effect," and the mirror-symmetric non-helical part is the turbulent diffusivity (η_T). These models are computationally fast and useful for simulating solar activity features, but one of the limitations is that they rely on the parametric form of α-effect processes and turbulent magnetic diffusivity, and hence rely on capturing the physics of these processes. Many solar dynamo models have been built over the past 70 years using the mean-field formalism, primarily with the aim of reproducing the sunspot butterfly diagram, the most well-known characteristic of the solar cycle.

In the second approach, full 3D magnetic induction equations are solved along with the flow and energy equations including thermodynamics under the anelastic approximation (Brun et al., 2004; Gilman, 1983; Glatzmaier, 1985). These models have recently advanced to the level of sophistication that they produce cycles of solar-like cycle length (Fan & Fang, 2014; Racine et al., 2011), as well as intricate details, such as wreathe-like formations of toroidal fields (Brown et al., 2010; Nelson et al., 2013). These models also have their limitations. Even though they solve the full set of MHD equations in 3D with the aim of resolving all processes relevant for dynamo action, the contributions from subgrid scale processes are parameterized. Furthermore, it has been a challenge to reach the right parameter regime in terms of Reynolds and Rossby numbers. Currently this class of models does not produce solar-like differential rotation if solar-type rotation and luminosity are used (see, e.g., Fan & Fang, 2014; Gastine et al., 2014), indicating a too high Rossby number, that is, too fast convective flows.

Overview of Past Accomplishments

Successful modeling of the global fluid flow and magnetic fields in a planetary atmosphere or ocean, or the convection zone of a rotating star, generally was begun with relatively simple models that make no attempt to model all the small-scale processes of the whole system, but rather focus on those physical processes thought to be important for that system. For example, models for instability of global flows in Earth's atmosphere began to be developed in the 1940s (Charney, 1947; Eady, 1949). It took another 20 years to develop models that had enough physical processes and enough realism and spatial resolution to plausibly model the atmospheric circulation (Trenberth, 1992). Since then, such models have become vastly more detailed and realistic, simulating and the predicting global circulation in ever greater detail.

Dynamo models for solar cycles are developing in an analogous way, but are at a much earlier stage. Parker's model contained very relevant physical processes (in particular, shearing of magnetic field lines by differential rotation and twisting and lifting of field lines by helical turbulence), but certainly was not realistic enough to simulate the primary features of a particular solar cycle. But now such models do exist, and they can and have been calibrated to major features of solar cycles. Despite the complexity and realism of current global models for the general circulation of Earth's atmosphere, such models still must parameterize important physical processes that are too small in spatial scale to be calculated explicitly everywhere on the globe. The situation is the same for modeling solar cycles with a dynamo model. Currently calibrated solar cycle simulation models are virtually always in two dimensions, latitude and radius, and can simulate only axisymmetric solar cycle features—which, fortunately, describe a large part of what constitutes a cycle. In these models, all non-axisymmetric processes are parameterized with axisymmetric representations. As we discuss in later sections, the most successful of these models in simulating solar cycles have been the so-called Babcock–Leighton flux-transport (BLFT) dynamos.

So-called "full 3D" solar dynamo models exist (Fan & Fang, 2014; Gilman, 1983; Lawson et al., 2015) that do include departures from axisymmetry; these models have recently produced cyclic dynamo solutions, but they have yet to be calibrated for the Sun. Furthermore, they too do not spatially resolve energetically important scales of motion and magnetic fields in the Sun, and so still rely heavily on averages of and parameterizations of unresolved processes. This limitation on spatial resolution will not be overcome for many generations of computing power growth, and hence, despite their higher spatial resolution, including variations in longitude, the "full 3D" dynamos will continue to be used to average over processes occurring on smaller spatial scales for the foreseeable future.

New Challenges

The previous section described in broad terms where solar cycle modeling using dynamo models currently stand. Here we outline what the major challenges are to making meaningful progress toward realistic simulation of actual solar cycles.

The most obvious, and perhaps most important, challenge is to move from 2D to 3D solar cycle simulation models, in order to correctly simulate global but longitude-dependent solar cycle features such as magnetic flux emergence, active longitudes, and solar sector structure, which has so much influence in organizing and modulating the solar wind and interplanetary magnetic field that transmits the effects of solar activity to Earth, particularly to its upper atmosphere. Because the non-axisymmetric solar cycle features to be simulated are themselves global in scale, it is not clear whether the full 3D convective dynamo simulation approach, which would spatially resolve a whole spectrum of convection, is either warranted or practical. Perhaps it may be more productive to generalize the BLFT dynamo models to include the most important global 3D effects (see, e.g., Miesch & Dikpati, 2014).

There is incomplete information available about key components of solar dynamo models, particularly meridional circulation (see Section 7.2.6). Observations of meridional circulation are good near the surface, but much more uncertain for deeper layers. Helioseismic methods are generating some possible profiles, but they show large variations for different analysis techniques and time periods. It will be necessary to develop, in parallel with the 3D dynamo model itself, better theories for meridional circulation, which take account of what we do know about flow in the convection zone, such as the differential rotation there. This knowledge can be used to constrain meridional circulation profiles that are possible.

Modern data assimilation (DA) techniques, in the early stages of being included in solar dynamo models, will provide a powerful tool for inferring the form and amplitude of meridional circulation with depth and latitude (Dikpati et al., 2014, 2016a, 2016b; Jouve et al., 2015; Hung et al., 2015), using so-called observing system simulation experiments (OSSEs). Even more important, DA will provide the computational framework for optimizing the use of observations of solar velocities and magnetic fields to initialize and generate the best possible simulations of actual solar cycles, including global 3D effects. More advanced use of such methods has been enormously useful in improving simulations and predictions of global atmospheric flows. Even though it is true that solar data come much more from the photosphere and above than from the solar interior, it is far more important to use the available solar data for assimilation by implementing modern DA methods to estimate the spatiotemporal profiles of unknown ingredients for the regions where we lack observational data. In this context, we recall the statement of Kalnay—the use of the model forecast is essential to achieve four-dimensional DA. The model transports information from data-rich to data-poor regions, and it provides a complete estimation of the four-dimensional state of the atmosphere. The inclusion of the data-rich northern hemisphere in the assimilation scheme has greatly improved the forecasts in the southern hemisphere (Kalnay, 2003).

The kinematic system can be the place to start, because we do not need a model to simulate differential rotation, because it is nearly constant, but crucial 3D dynamical processes, such as the role of global MHD tachocline and convection zone dynamics, can be included as part of input of solar flow fields.

7.1.4. Solar Cycle Observations to Be Modeled by a Cyclic Dynamo

Figure 7.1 characterizes only sunspot patterns and statistics. But there is a very important relationship between sunspot fields and other surface magnetic fields, the so-called "poloidal" fields, especially those near the poles. Figure 7.2 shows this relationship. Blue and yellow shading represent opposite polarity fields. We see the low-latitude butterfly diagram, but we also see rather rapid migration of poloidal fields toward the poles (the steeply slanted streaks). One sign of field predominates in this migration, opposite in the northern and southern hemispheres, and opposite in each succeeding cycle. We can clearly see that this migration reverses the sign of the polar field near the maximum of the sunspot cycle. The polar field retains its sign in each hemisphere until the next maximum.

Figure 7.3 shows a trace of the polar field amplitude for cycles 21 to the present (see also Petrie, 2015). We see that the strength of this polar field itself varies from cycle to cycle, and for the past few cycles has shown a downward trend. The transition from one polarity to the other takes place over a fairly small fraction of a cycle, followed by a longer period over which the field is often (but not always) nearly constant. There are differences in timing of polar field reversal between north and south, and occasionally there can be multiple field reversals—but always an odd number, so that the basic pattern of field reversal is sustained. It is clear from Figure 7.2 that a mechanism for producing different polar field amplitudes in succeeding cycles is variations in the amount of surface magnetic flux that migrates to the poles in each cycle. If the new cycle is weaker than the previous one, then there is less new flux available to reverse the polar field, so when it does reverse, the new polar field peak is likely to be weaker than the previous one (and vice versa).

Some properties of sunspot cycles, such as amplitude and phase, can vary substantially between the northern and southern hemispheres. If one correlates the cycle peak amplitude between north and south, the correlation coefficient is about 0.5 for three-month running averages, rising to over 0.8 for running averages exceeding 1.5 years (20 solar rotations). Thus, on short time scales, the amount of activity can be very different between

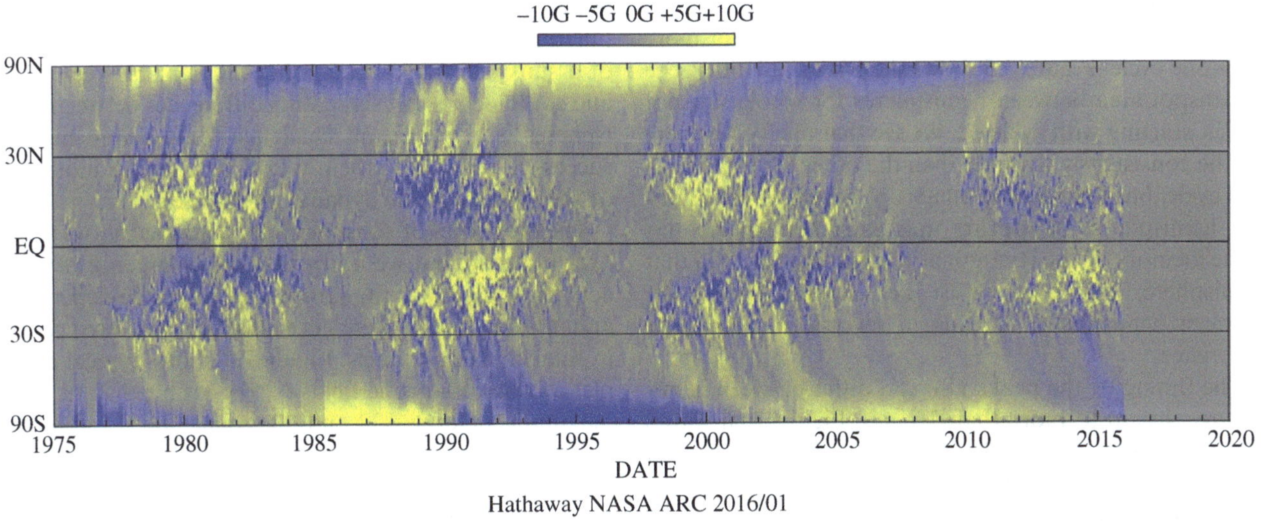

Figure 7.2 Magnetic butterfly diagram, constructed from longitudinally averaged radial magnetic fields obtained from Kitt Peak and SOHO measurements. Blue and yellow shading denote opposite polarities. Note the clear presence of polar field reversals and migration of flux to the poles (the nearly vertical streaks). (*Source*: Daily sunspot area averaged over individual solar rotations, Hathaway NASA/ARC 2016/10. Public domain.)

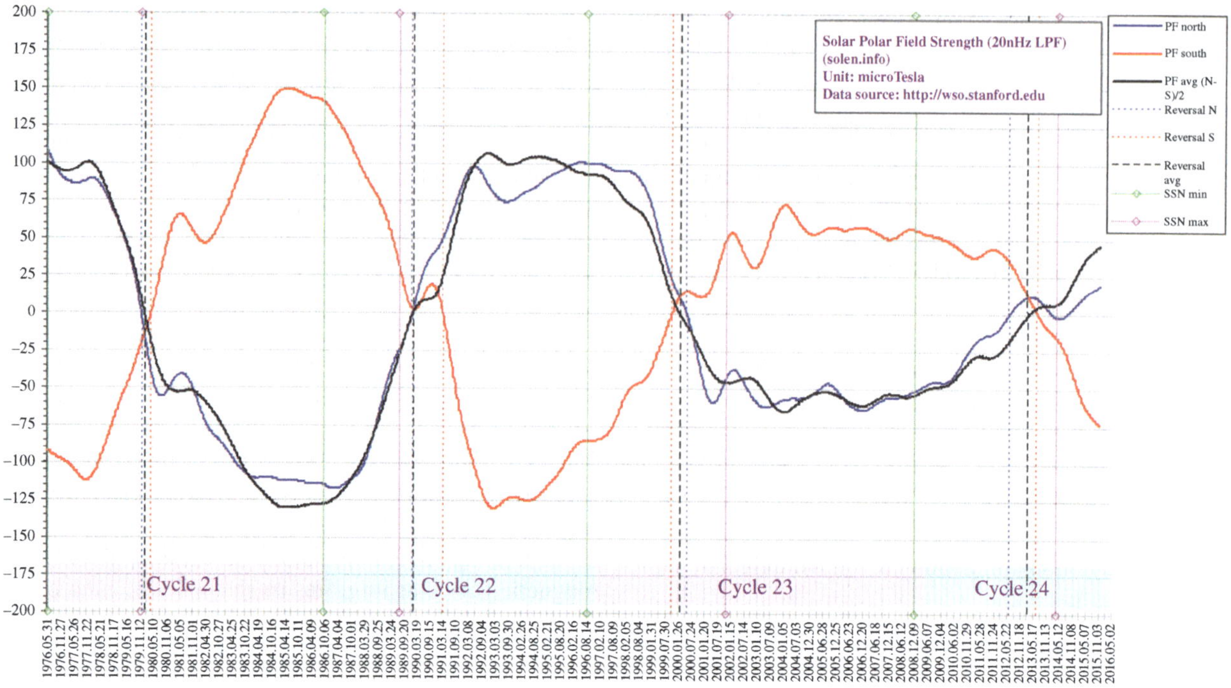

Figure 7.3 Polar field amplitudes for cycles 21–24 from the Wilcox Solar Observatory (WSO). Shown are north and south poles separately and together. (*Source*: Data from Hoeksema and WSO website.)

hemispheres. But even over a whole cycle there are substantial differences. Figure 7.4 shows the hemispheric sunspot number, with the excess of one hemisphere over the other shaded in green or red, depending on which hemisphere has more activity. We see that for several successive years one hemisphere has consistently more flux than the other.

The same sign of the difference often persists for more than one cycle. Figure 7.5 shows the fractional difference in sunspot area between hemispheres for whole sunspot cycles, starting with cycle 12. We see that one hemisphere can be consistently stronger than the other for more than one cycle, but followed by an abrupt switch to the opposite hemisphere. It appears that each hemisphere has some memory of the strength of the previous cycle in that hemisphere, but there must be enough connection between hemispheres to cause a reversal in amplitude difference.

The timing of the peaks also can differ between hemispheres, by up to more than 2 years. Figure 7.6a shows this clearly. In fact, it can be argued that the northern and southern hemispheres avoid having synchronized activity peaks. By contrast, the timing of the minima in the two hemispheres varies by no more than a year, and is usually less. So some interaction between hemispheres tends to re-synchronize the cycles by the end of a cycle. This has to occur during the declining phase, because, from Figure 7.6b, the ascending phase is always much shorter in one hemisphere than the other, commonly by substantially more than a year.

The hemisphere that ascends faster typically persists for more than one cycle. These statistics also indicate that there is a relatively weak but not insignificant interaction between the hemispheres that keeps the timing of cycle minima fairly well in phase, but with each hemisphere pursuing its own pace within the cycle. Thus the "shape" of a sunspot cycle can be rather different between north and south. Correctly simulating differences between hemispheres is a significant challenge for dynamo models; success in simulating them would be a strong indication of the validity of the dynamo model used.

The properties of sunspot cycles described above are all axisymmetric, but we know that the emergence of solar activity in a cycle is fundamentally longitude dependent, and this longitude dependence must be taken into account in solar dynamo models. Figure 7.7 shows typical examples of solar magnetic fields on the visible disk near the maximum of cycles 22 and 23. Again, blue and yellow shading denote opposite field polarities. In Figure 7.7 we clearly see Hale's polarity law, which states that the leading and following parts of each active region have opposite polarities in the northern and southern hemispheres, which polarities reverse from one cycle to the next. We also see that all active regions by eye are tilted with respect to latitude circles, such that the follower polarities are closer to the poles, on average, than are the leader

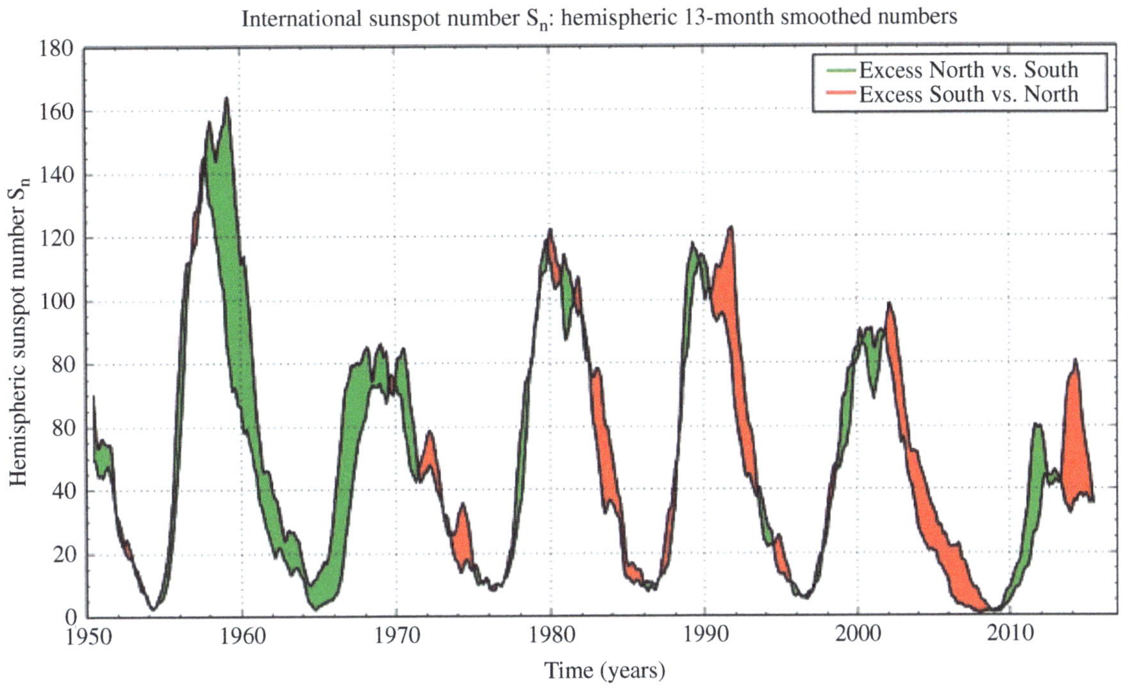

Figure 7.4 Excess hemispheric sunspot number, north (green) over south (red), showing that differences in phase between hemispheres can persist for multiple cycles. (*Source*: SILSO data/image, Royal Observatory of Belgium, Brussels.)

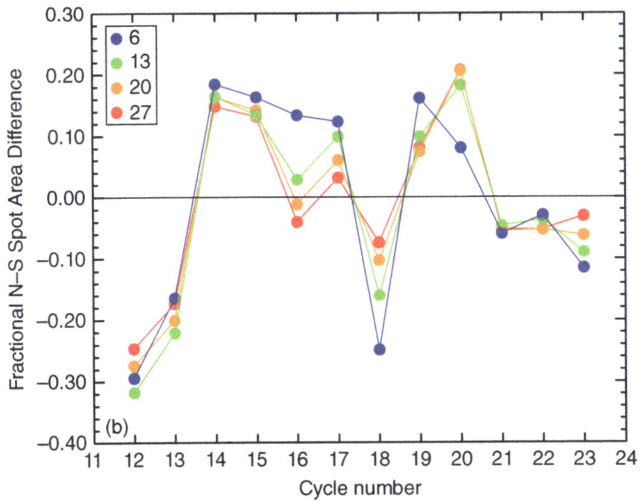

Figure 7.5 Fractional difference between spot area in the north and south. Shows that the difference can be up to 30% of the average between hemispheres, and persist for multiple cycles. Color key with numbers denotes the length of the running average used, in solar rotations, showing that the basic result is evident for all averaging intervals. (*Source*: Dikpati et al., 2007. © 2007, Springer Nature.)

polarities. This tilt, called Joy's law, is crucial for determining the sign of the net magnetic flux that migrates to the poles to reverse the polar fields seen in Figure 7.3. All dynamo models must take this tilt into account, either by including it in parameterizations of surface poloidal flux emergence (2D models) or by actually calculating it, in a way that is calibrated to observed tilts.

The average amount of this tilt is now well known from observations of active regions (see, e.g., figure 1 of Wang et al., 2015). The angle is greatest at the start of the cycle, decreasing nearly linearly to the end. There are some differences between cycles, but on average these are relatively small. Different data sets give similar results. Particularly when coupled with statistics for the latitude centroid of sunspot occurrence, the combination of tilt angle, centroid, and cycle amplitude can give us good estimates of emerged surface poloidal flux available for migration to the poles to reverse the polar fields.

There are other measures of longitude dependence in surface magnetic fields that are important for dynamo models to simulate. These involve especially the tendency for new magnetic flux to emerge in the same longitude band for many months, the so-called active longitudes. These persistent features are known to lead to global longitude-dependent magnetic structures in the corona and interplanetary medium, the so-called magnetic sectors.

Finally, it is known that solar cycles can seem to virtually disappear for several decades, an extreme variation in the envelope of cycle amplitudes. The so-called Maunder minimum is the most recent solar manifestation of this phenomenon, for which there were very few spots seen on the Sun from about 1645 to 1715. There is some evidence

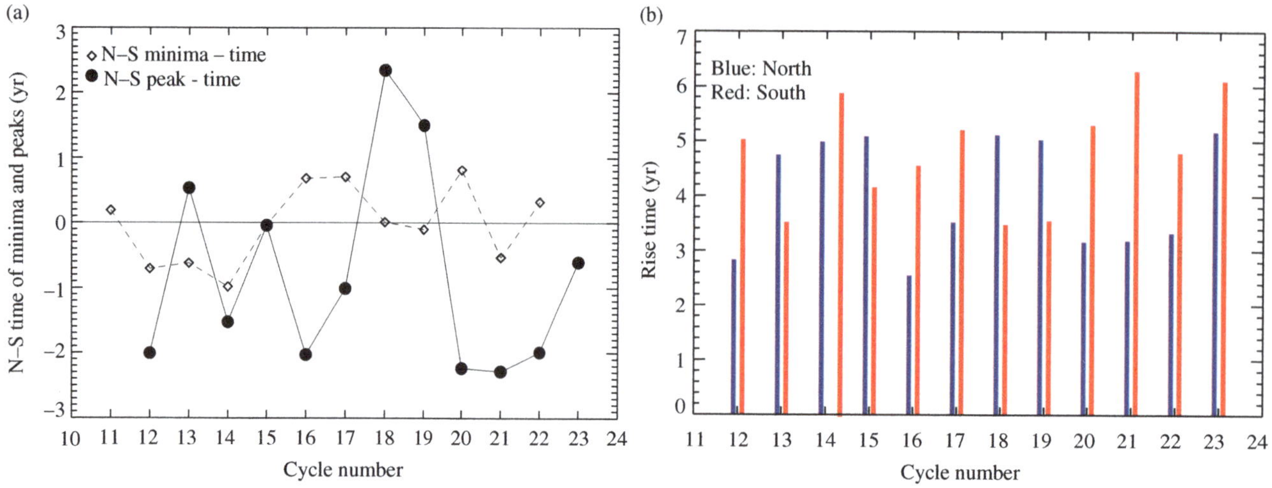

Figure 7.6 Difference between north and south in timing of peaks and minima. Frame (a) shows that the difference in timing of cycle peaks is usually more than one year, often more than two years. By contrast, the timings of minima differ by less than one year. The difference in these timing differences comes from the differences in rise time of a cycle between north and south, which is virtually always greater than one year, and sometimes more than two years. One hemisphere always jumps ahead of the other at the beginning of a new cycle. (*Source*: Dikpati et al., 2007. © 2007, Springer Nature.)

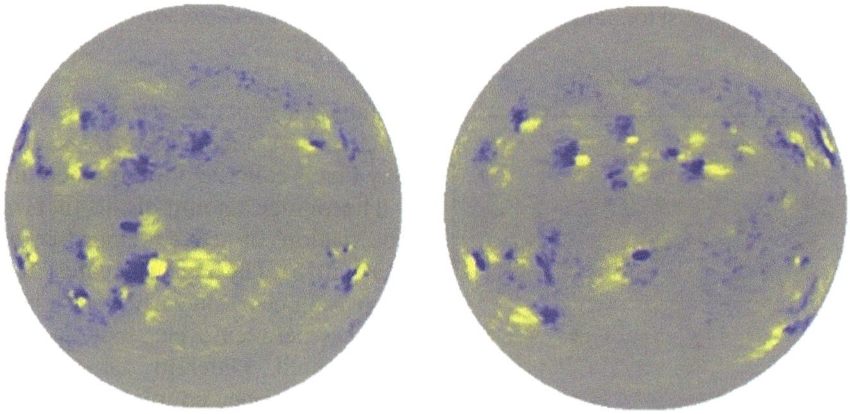

Figure 7.7 Typical synoptic magnetograms from cycles 22 and 23. Hale's polarity laws and Joy's laws are very evident. (*Source*: Hathaway NASA/ARC.)

that the cycle continued, but at a very low amplitude. Simulating these multi-cycle minima is a particularly difficult challenge for dynamo models, particularly from actual solar data.

7.1.5. Recent Models and Results

Brief History of Solar Dynamo Model Developments

A brief history of developments in solar dynamo theory is diagrammed in Figure 7.8. The modern era of solar dynamos really began with Parker's "lifting and twisting" dynamo, which showed how a combination of helical flow and differential rotation could produce a "dynamo wave" that propagated in latitude with time and which, with the right choice of helical flow and differential rotation, could roughly simulate the "butterfly diagram" seen in Figure 7.1. This model was followed by more heuristic models of Babcock (1961) and Leighton (1964, 1969) that used observed patterns and migration of fields to create magnetic cycles. The rise of "mean-field" dynamo theory from the German school (Rädler & Rheinhardt, 2007) gave solar dynamo theory much more rigor and connection to MHD turbulence concepts.

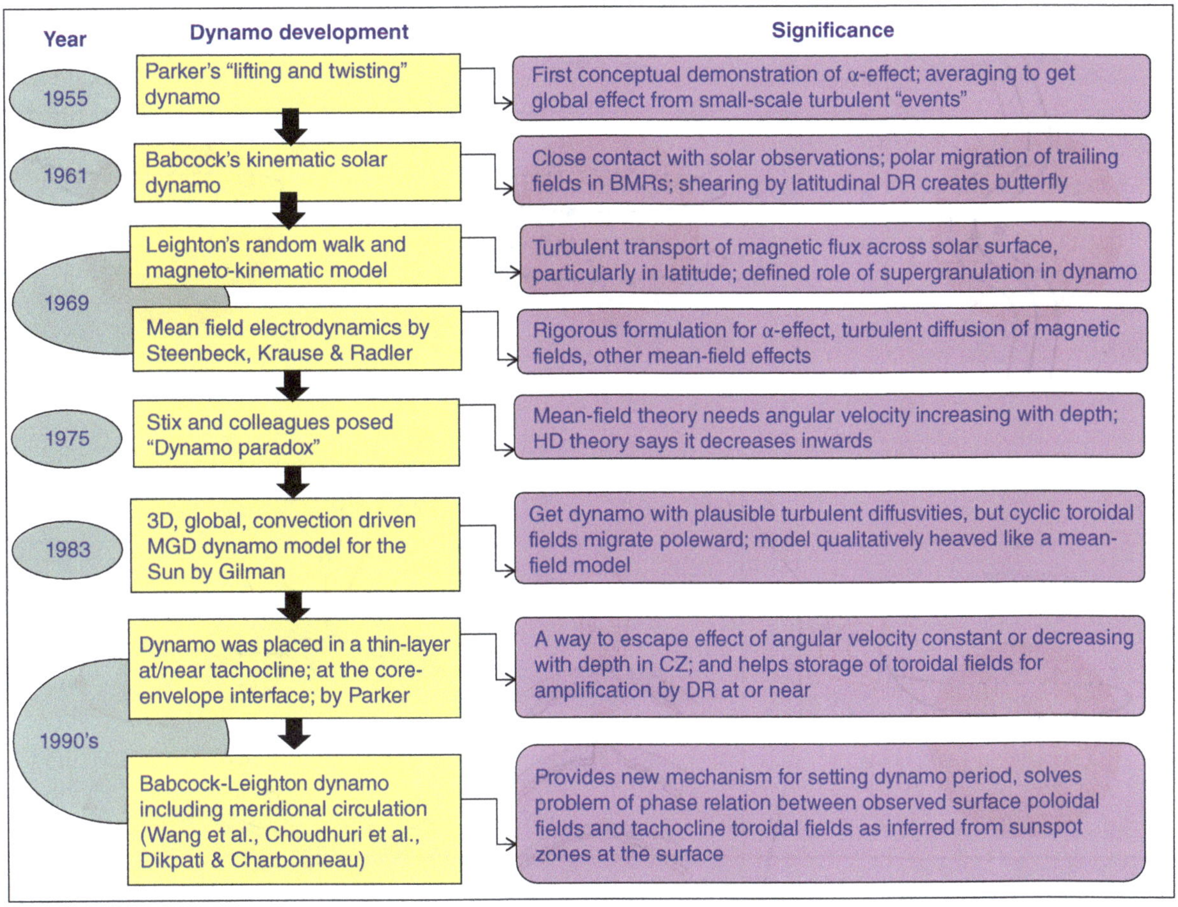

Figure 7.8 Flowchart of the history of development of solar cycle dynamo models through the 1990s. (*Source:* Data from Parker, 1955b. Provided by the SAO/NASA Astrophysics Data System.)

Mean-field dynamos gave plausible cycles and butterfly diagrams provided the angular velocity increased inward in the bulk of the convection zone, but this picture was overturned by the discoveries of helioseismology (see Section 7.2). And the first full 3D convectively driven solar dynamo of Gilman (1983) gave plausible differential rotation but an "anti-solar" butterfly diagram in which toroidal fields migrated toward the poles. These developments led to shifting the focus of the dynamo from the bulk of the convection zone to its base. The problem of reversed butterfly diagrams was overcome by the demonstration of a new mechanism for toroidal field migration, provided by meridional circulation (Choudhuri et al., 1995; Dikpati & Charbonneau, 1999; Durney, 1995; Wang et al., 1991). Thus was born the so-called Babcock–Leighton flux-transport dynamos, which have proved so far to be the most successful dynamos for simulating solar cycles (Belucz et al., 2015; Dikpati & Gilman, 2001; Dikpati et al., 2004; Guerrero & Muñoz, 2004; Hotta & Yokoyama, 2010; Muñoz-Jaramillo et al., 2009; Nandy & Choudhuri, 2001). Extensive references on BLFT dynamos can be found in Charbonneau (2010). These are 2D (latitude-radius) models, but efforts have begun to generalize them to 3D.

Figure 7.9 shows a schematic diagram that depicts how this dynamo works, from Dikpati and Gilman (2006). It begins with the shearing of an antisymmetric poloidal field by the latitudinal differential rotation (frames a,b) to produce a toroidal field near the base of the convection zone where the turbulent magnetic diffusivity is relatively low. This is followed (frames c,d) by toroidal flux loops rising to the surface while twisting, creating new poloidal loops there (frames e,f), which are tilted with respect to a latitude circle. These then contribute to a new axisymmetric poloidal field (frame g) that is carried toward the poles by meridional circulation and then down into the convection zone to its base (frame h), which has the opposite sign relative to the original poloidal field (frame a), from which a new toroidal field, also of opposite sign, is generated, leading to a repeat of the whole process, with all magnetic field signs reversed.

In actual dynamo simulations with the flux-transport dynamo equations, if they are done with a full spherical shell with no symmetry conditions imposed at the

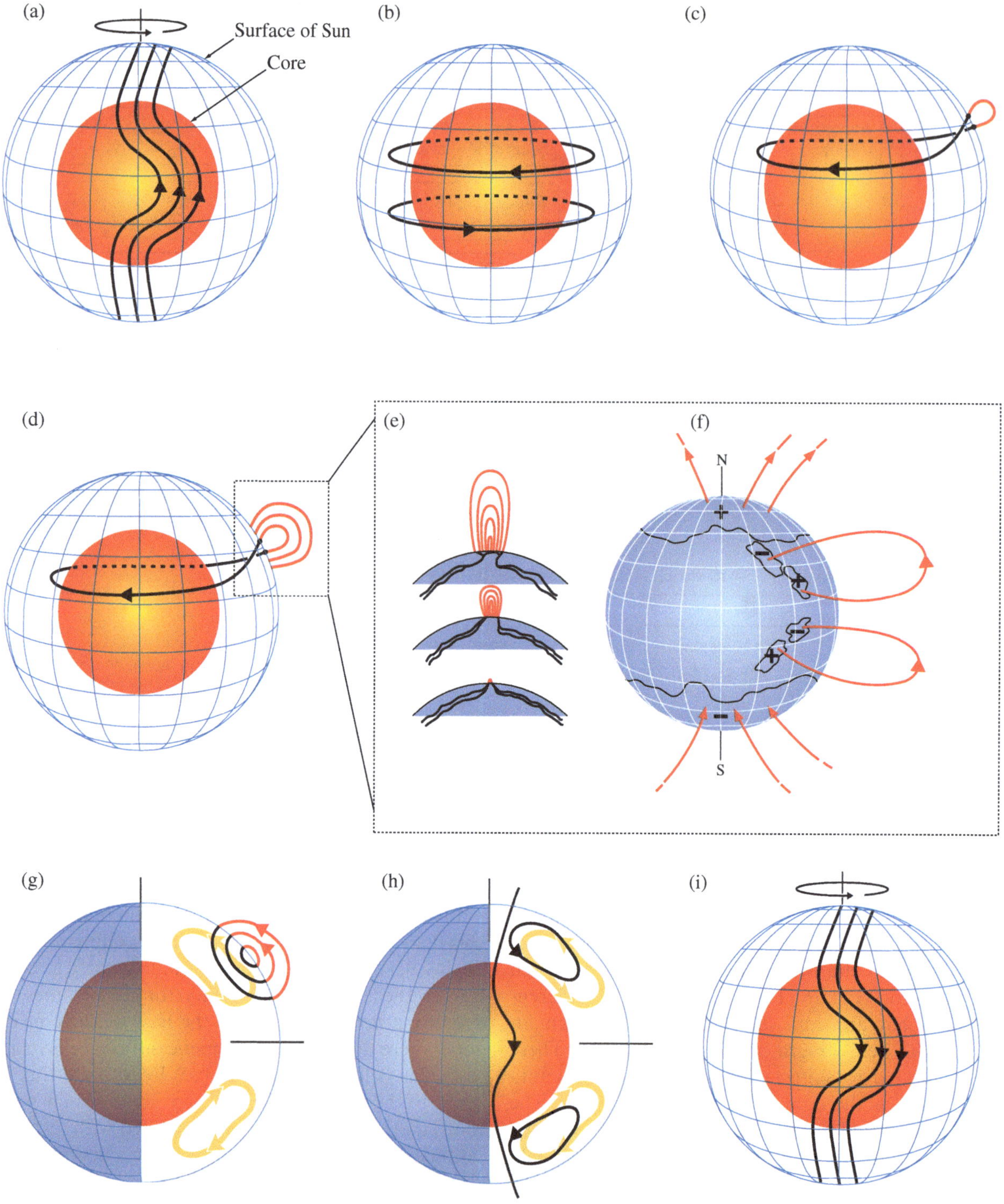

Figure 7.9 Schematic sequence of induction and transport processes contained in Babcock–Leighton flux-transport dynamo models. (*Source*: Dikpati & Gilman, 2006. © 2006, IOP Publishing.)

equator, there is no guarantee that the model will select the correct field symmetry for the Sun. In fact, various calculations showed a tendency to pick the opposite, or "quadrupole" symmetry. Dikpati and Gilman (2001) looked into this question in detail, and found that even if the simulation was started with the correct solar symmetry, if only the surface Babcock–Leighton mechanism for forming surface poloidal flux from emerging toroidal fields (Figure 7.9d–f) was present, within a few hundred years of simulation the symmetry switched over to the quadrupole type. In this situation, a butterfly diagram remains plausible, but Hale's polarity law is violated. Dikpati and Gilman (2001) showed that if there was a second mechanism for generating a poloidal from a toroidal field, such as the lifting and twisting (α-effect) due to global hydrodynamic instability of differential rotation in the tachocline, the original dipole type symmetry was retained. The effect of the bottom α-effect preserves the original symmetry to provide a booster to the poloidal field brought down from surface polar regions, which is then swept toward the equator before it can dissipate, getting in synch with the other hemisphere to create poloidal field lines that cross the equator rather than closing there within each hemisphere. The equatorward flowing meridional circulation is a key factor, because it brings a stronger poloidal field from high latitudes to merge near the equator with its opposite-hemisphere counterpart. Without the bottom α-effect, there is little poloidal field available to merge with its opposite-hemisphere counterpart through the equator to form a dipole structure, which can produce an antisymmetric toroidal field after being sheared by the differential rotation.

Symmetry selection processes have been studied by many others later on (Belucz et al., 2015; Bonanno et al., 2002; Chatterjee et al., 2004; Hotta & Yokoyama, 2010), who proposed and tested additional possible mechanisms to ensure the correct symmetry is selected.

Calibrated and Benchmarked Models

It makes sense to calibrate a dynamo model to solar cycle properties only if the model gives at least the correct dominant symmetry of the magnetic fields about the equator. Figure 7.10 shows a calibrated butterfly diagram, in which the spot-producing toroidal fields, taken from the tachocline, are shown as contours, the surface poloidal field as light and dark shading. We see that there is a strong resemblance between Figures 7.10 and 7.2. The shape of both field patterns is about right, the polar fields reverse near the maximum in toroidal fields, and reach maximum between sunspot cycles. Eight versions of this class of model, using different numerical algorithms, have been benchmarked against each other for certain parameter choices in common, and have been found to agree very closely (see Jouve et al., 2008).

Solar Cycle Simulation and Prediction

The quality of the calibration suggested that the flux-transport dynamo model could be used to attempt dynamo-based simulations of particular solar cycles, and even the first dynamo-based prediction of a solar cycle. Dikpati and colleagues (Dikpati & Gilman, 2006; Dikpati et al., 2006) used the same calibrated flux-transport dynamo model that generated Figure 7.10 to simulate the peaks of cycles 12–23, and predict the peak amplitude of the then future cycle, cycle 24. For cycles 12–23, they achieved a correlation between the observed and simulated peaks in excess of 0.94 for all turbulent magnetic diffusivities chosen, up to $3 \times 10^{11} cm^2 s^{-1}$. To perform these calculations, data "nudging" was used to drive the model with the observed time history of surface poloidal flux. Thus, surface magnetic fields from previous cycles led to the amplitude of the next cycle. When the northern and southern hemispheres were simulated separately (Dikpati et al., 2007), the model produced larger observed differences between hemispheres. The correlation between the simulated and observed peaks was somewhat lower, but still in excess of 0.8. Simulations of this length were possible because for the turbulent magnetic diffusivities chosen, the "memory" of the model was about two sunspot cycles. In the diffusivity range chosen, the dynamo was definitely in the advection-dominated regime. However, the duration of the model's memory may not be settled yet, as Yeates et al. (2008) produced only one sunspot cycle.

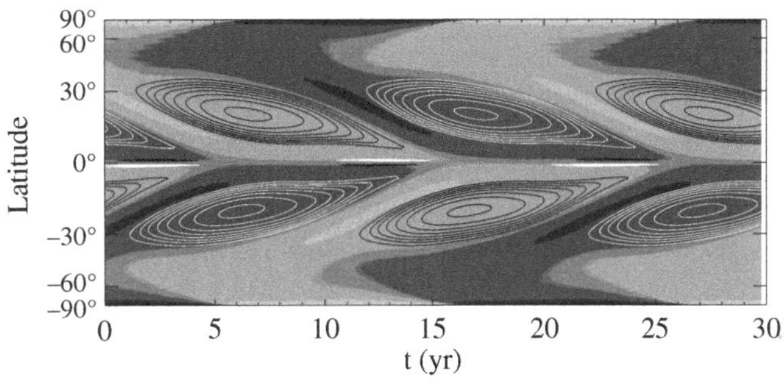

Figure 7.10 Calibrated butterfly diagram from a Babcock–Leighton flux-transport dynamo. Shading is for the surface poloidal field (light and dark shading opposite polarities); contours are for the toroidal field amplitude at the bottom of the dynamo domain. This diagram closely resembles the observed diagram shown in Figure 7.2. The narrow black/white streaks at the equator near the end of each cycle are the weak remains of the toroidal field from the previous cycle, which are slow to cancel across the equator because of the low turbulent magnetic diffusivity at the bottom. (*Source*: Based on Jouve et al., 2008.)

Unfortunately, the observed peak in cycle 24 has turned out to be substantially lower than predicted by Dikpati and colleagues. Also using a flux-transport dynamo model, but higher turbulent magnetic diffusivities and input of surface poloidal fields only near the poles and only near solar minimum, Choudhuri et al. (2007) predicted a much lower peak for cycle 24, actually lower than observed. Their model is in the diffusion-dominated regime, with a much shorter memory.

There is currently no consensus as to what the turbulent magnetic diffusivity is in the convection zone, so it is hard to choose among models that use different values.

But there is a problem with high-diffusivity models as true self-contained dynamo models, in that it can be shown by scale analysis that for diffusivities in excess of 10^{12}cm^{2}s^{-1}, such as suggested by mixing-length theory applied to the Sun, and as used by Choudhuri et al. (2007), it would take a very unrealistically large surface poloidal source or α-effect to sustain a dynamo at all.

One way to see this is by calculating the critical dynamo number P for the model, which must be exceeded for a dynamo to be sustained, following, for example, Stix (1976). The original argument is for α-ω dynamo models, but it holds equally as well for flux-transport dynamos. $P = \alpha_0 \Delta\omega \lambda^3 r_\odot^3 / \eta_T^2$, in which α_0 is the α-effect, $\Delta\omega$ a measure of the difference in rotation across the shell, λ the shell thickness in fractions of its radius, $r_\odot$ the solar radius, and η_T the turbulent magnetic diffusivity. It is well established that P should exceed 3×10^3 for sustained dynamo action. For a mixing-length amplitude diffusivity of ~3×10^{12} cm^{2}s^{-1}, this requires an α_0 of 1.5×10^3 ms^{-1}, which is completely unrealistic for any scale of solar convection anywhere in the convection zone, or for any surface poloidal source. By contrast, an η_T of 10^{11} cm^{2}s^{-1} would require an α_0 of 1.5 ms^{-1}, a much more plausible value. In fact, this is what was argued by Choudhuri et al. (1995) in their very first Babcock–Leighton flux-transport dynamo model, and therefore a value of 3 ms^{-1} was carefully considered. However, Choudhuri and colleagues developed the so-called high-diffusivity dynamo models by choosing a high diffusivity to poloidal fields, but a low diffusivity for toroidal field, along with the use of an unrealistically high α-effect, to obtain a sustained dynamo solution. It remains confusing whether a dynamo model with high and low diffusivities applied to respectively the poloidal and toroidal fields can be called a diffusion-dominated model or a dynamo with diffusion-dominated poloidal field and advection-dominated toroidal field.

If one uses the unrealistically high surface poloidal source, with high diffusivity, the calibration is negated because the resulting dynamo period is much too short. The Choudhuri et al. (2007) prediction model gets around these problems because it is being forced at the top using magnetic data from previous cycles. But the Sun does not have that freedom; it must be a self-contained calibrated dynamo to begin with.

In any case, it is more important to analyze reasons why the advection-dominated flux-transport dynamo-based prediction for the cycle 24 amplitude has not succeeded. These have to do with the simplifying assumptions that went into the early models. In particular, the early prediction models used a steady single-celled meridional circulation, but we now know that the circulation amplitude and profile both vary with time within a solar cycle. Even allowing for just a change in the high-latitude boundary of the primary meridional circulation flow toward the poles is enough to explain the two-year-longer duration for cycle 23 compared to cycles 21 and 22 (Dikpati et al., 2010). In the earlier cycles, this cell reached about 65° latitude, whereas in cycle 23 most of the time it reached all the way to the poles. The longer conveyor belt in cycle 23 led to a longer cycle, because it took longer for the surface poloidal fields in active latitudes to be transported to the poles and down to the bottom of the convection zone. Figure 7.11 shows a plot of the relationship between the cycle period, peak circulation speed, and the latitude of the poleward boundary of the circulation.

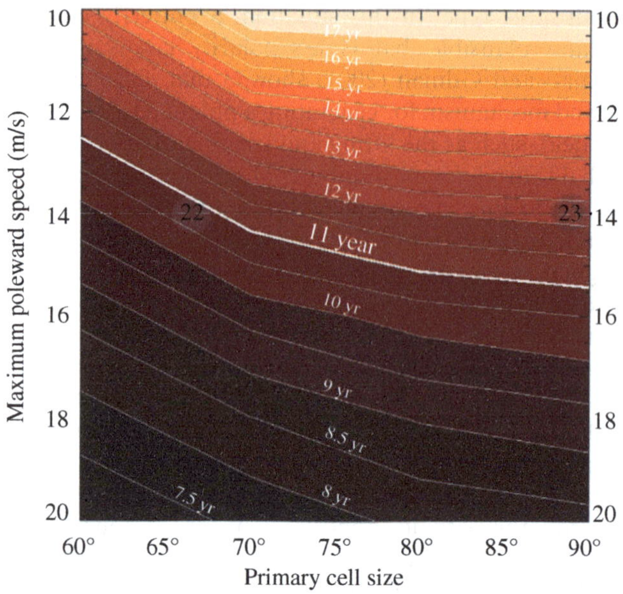

Figure 7.11 Relationship between latitude of poleward boundary of primary meridional circulation and period of magnetic cycles produced by Babcock–Leighton flux-transport dynamo. (*Source*: Dikpati et al., 2010. © 2010, John Wiley and Sons.)

Recently evidence has been found of a possible second, reversed meridional circulation cell below the primary cell that flows poleward, carrying the surface poloidal flux to the poles to reverse the polar field. Although there exist some debates about the existence of the weak reverse cell (Kholikov et al., 2014b), if such a two-celled pattern persists in the Sun, then the calibration of flux-transport dynamo models to the Sun is completely destroyed. An example is given in Figure 7.12, from Belucz et al. (2015). Because in this case the bottom circulation is toward the poles rather than the equator, as it would be with a single circulation cell with depth, the butterfly diagram constructed from the bottom toroidal fields is completely reversed from the real Sun. If this is what is happening in the Sun, then the Sun must not be a Babcock–Leighton flux-transport dynamo at all, and there must be another paradigm shift in solar dynamo theory. But observations of the second cell with depth are quite uncertain, and so it is premature to conclude that such a shift is needed.

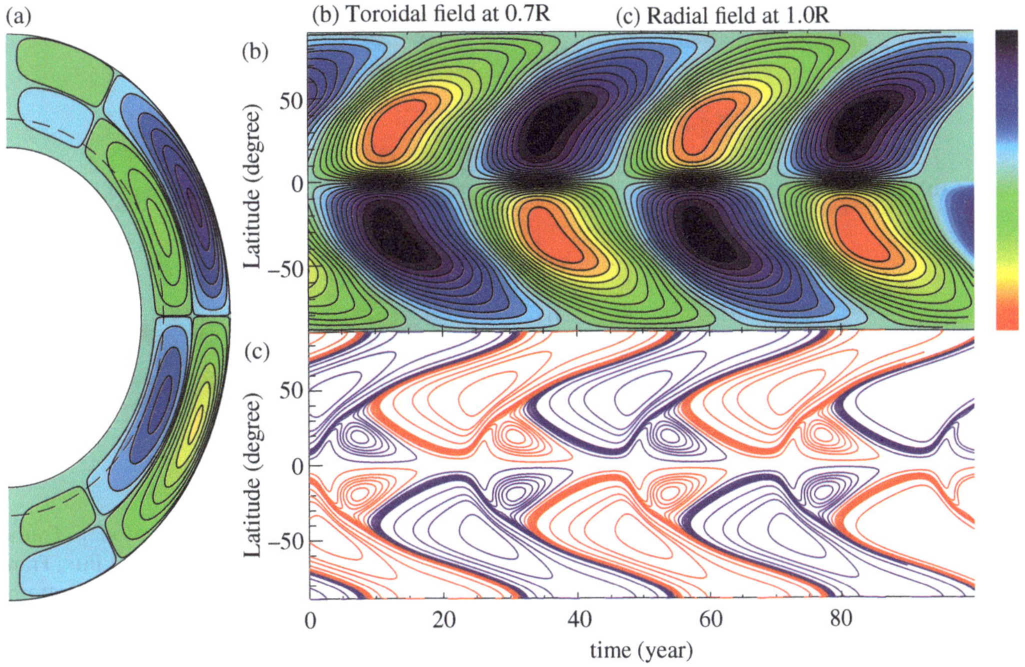

Figure 7.12 Butterfly diagrams (frames b,c) from flux-transport dynamo containing two primary meridional circulation cells stacked in depth (frame a). It is evident that two cells in depth leads to a non-solar (reversed) butterfly diagram, particularly for the toroidal (sunspot) fields. This is because the toroidal field at the bottom where it is strongest is being carried toward the poles rather than the equator. (*Source*: Belucz et al., 2015. © 2015, IOP Publishing.)

Beyond changes in meridional circulation, several other factors not previously accounted for in solar dynamo models may also contribute to the lack of success in solar cycle prediction models. Asymmetries between the northern and southern hemispheres are substantial and virtually always present; the early models treated a single hemisphere. There is still uncertainty about the profile of turbulent magnetic diffusivity with depth. The data nudging scheme used in early prediction models did not make use of much of the available solar observations, which could change the predictions substantially. As discussed in Section 7.1.3.2, modern DA methods need to be incorporated into the models to use all available data. Finally, a substantial part of the observed magnetic fields and patterns of flux emergence is longitude dependent, and all the prediction models proposed up to now have been axisymmetric. For example, creating active region tilts is strongly longitude dependent as well as highly variable in time. Active longitudes, although varying on longer time scales, are obviously also functions of longitude.

Beginnings of Application of DA

Application of DA methods to the solar dynamo problem is in its very early stages. As a precursor, Dikpati and Anderson (2012) have found from numerical experiments that the "response time" of a Babcock–Leighton flux-transport dynamo model to a change in meridional circulation amplitude is about six months. This is a key time scale for designing an optimum DA scheme, because the time interval for updating data needs to be long enough that the model has time to start responding to the change. In a more recent study, Dikpati et al. (2014) have carried out OSSEs that show it is possible to reconstruct time changes in meridional circulation from the magnetic fields generated by the dynamo, using a relatively limited number of "observations," synthetically generated from the model. This kind of reconstruction will be necessary (see also Hung et al., 2015) because there do not now exist consistent observations of meridional circulation below the solar surface, including near the bottom of the convection zone and in the tachocline, where the circulation plays a key role in moving the dynamo-generated toroidal field.

Flux Emergence and Active Longitudes

Work has begun on incorporating longitude-dependent flux emergence into flux-transport dynamos. For example, using a "spotmaker" recipe that generates statistically appropriate active region tilts, Miesch and Dikpati (2014) have created a 3D Babcock–Leighton type dynamo that generates a time sequence of longitude-dependent domains of surface flux similar to those in the original

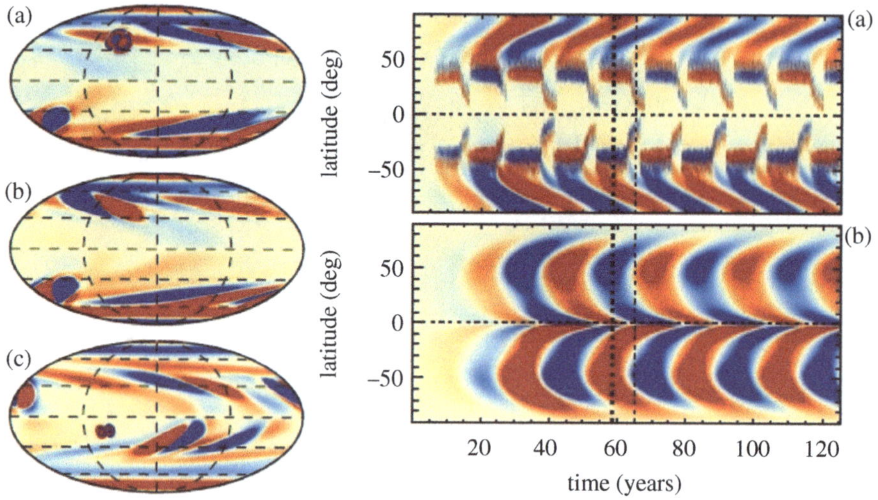

Figure 7.13 Early results from a 3D Babcock–Leighton flux-transport dynamo model (BASH model). Frames (a)–(c) on the left show typical evolution of the surface radial field originating from newly emerged flux. Frames (a) and (b) on the right show, respectively, time–latitude butterfly diagrams of the surface poloidal and bottom toroidal fields for a 120 year model run. (*Source*: Miesch & Dikpati, 2014. © 2014, IOP Publishing.)

Leighton model, as seen in Figure 7.13a. Averaging in longitude and time over this sequence creates a surface poloidal flux source for the 2D Babcock–Leighton flux-transport dynamo. Solutions from this dynamo contain plausible butterfly diagrams for surface radial field and toroidal field near the base, seen respectively in Figures 7.13a and 7.13b.

The longitude locations where new magnetic flux emerges appear not to be random; certain locations are repeated sources for new flux, which can persist for many months, even years. Dikpati and Gilman (2005) proposed a theory for these "active longitudes" in terms of global hydrodynamic (HD) and MHD instabilities occurring in the solar tachocline. Because the upper boundary of the tachocline can deform, instability of the combined latitudinal differential rotation and toroidal field there creates "bulges" of the tachocline into the convection zone above that could be favored locations from which toroidal flux can rise through the convection zone to the photosphere according to the physics of flux emergence (see Section 7.3). This kind of effect must be incorporated into global flux-transport dynamo models for generalizing them to 3D.

Recently McIntosh et al. (2015) have shown that many manifestations of solar activity exhibit global periodicities with periods in the range 6–18 months. This range of periods may also be due to global dynamics and MHD of the solar tachocline. Dikpati et al. (2018b) showed, using a nonlinear MHD shallow water model of unstable latitudinal differential rotation in the tachocline, that there is a well-defined nonlinear oscillation between energy in the differential rotation and in the low longitudinal wave number "Rossby waves" that gain energy from it. When these waves extract enough energy from the differential rotation, it ceases to be unstable, and subsequently extracts energy from the disturbances, growing until it becomes unstable again. In this way, the differential rotation and perturbation energies oscillate out of phase with each other, with the total energy of the system remaining virtually constant. Similar oscillations should still occur when toroidal fields are added; we would expect that the maximum amount of new flux would emerge at the surface when the perturbations and therefore the bulges have maximum amplitude.

7.1.6. Outstanding Challenges to Be Explored

The overriding priority for the 21st century solar cycle dynamo models is to build a 3D calibrated dynamo. 3D is essential for simulating the substantial departures from axisymmetry seen in typical synoptic magnetograms produced starting in the 1950s, and now available many times per solar rotation from both the Heliospheric and Magnetic Imager (HMI; Scherrer et al., 2012) onboard the Solar Dynamics Observatory (SDO) and from ground-based observing systems such as the National Solar Observatory's Synoptic Optical Long-term Investigations of the Sun (SOLIS) facility and the Mount Wilson Observatory (MWO). Calibration, including for longitude-dependent features, is essential for determining that the model simulates typical solar field patterns through a solar cycle and that the model can be used to simulate past cycles as well as applied to forecasting future cycles.

In principle, there are at least two approaches that could be taken. One is to use available full 3D convectively driven MHD dynamo models; the other is to generalize an already calibrated 2D Babcock–Leighton flux-transport model to include longitude-dependent fields and flows. The latter approach is advantageous over the former, because no current full 3D dynamo model has been successfully calibrated for the Sun.

What form should this generalization take? It should be strongly influenced by solar observations. For example, the typical solar synoptic magnetogram shows photospheric magnetic fields have truly global patterns, with high amplitude in low longitudinal wave numbers m. Only a few low m's can capture much of the important departures from the axisymmetric or $m = 0$ part of the field. In addition, as we go above the photosphere into the corona and beyond, the lowest m's dominate even more in the magnetic structure. Solar "sectors" generally are well represented by just $m = 1, 2$ modes. For these reasons, the model that is built should allow for picking just the lowest wave numbers, and allow experimentation to determine how few are needed. Therefore, fast Fourier transform representation of interactions between modes should be avoided in the formulation.

It is straightforward to expand the kinematic induction equation to include longitude dependence of this type. In the linear case, one simply gets separate equations for the amplitude of each longitudinal wave number, including the original 2D $m = 0$ mode. Linear solar dynamo models of this type were first studied in the early 1970s by Stix (1971), but have not been pursued much since. In these models, the flow fields included were still axisymmetric, as was the included parameterized helical turbulence. For future models, we have strong motivation, and theoretical support, for going far beyond this limited 3D model, to include velocity fields that are themselves dependent on longitude. The theoretical support is provided by the substantial theory of global HD and MHD in the solar tachocline that has been developed over the past 20 years (Cally, 2001; Cally, 2003; Cally et al., 2003; Dikpati, 2012; Dikpati & Gilman, 1999, 2001; Dikpati et al., 2017, 2018a, 2018b; Garaud, 2001; Gilman & Fox, 1997). This theory also favors the existence of global helical velocity and magnetic field disturbances in the tachocline, for which the same low longitudinal wave numbers $m = 1, 2$ are the most unstable, and therefore the most likely to dominate in the tachocline and layers above. The only exception to that is if the toroidal field present there is confined to a narrow (<10°) band of latitudes. In this case, somewhat higher m's are also excited.

In concentric spherical shells, starting from the tachocline, the model would include 3D helical flow from global tachocline instabilities, from which bulges into the convection zone above would be generated. In the bulk of the convection zone, there would be differential rotation, meridional circulation, global helical flows, and buoyantly rising magnetic flux (see, e.g., Section 7.3). Near the outer boundary there would be a Babcock–Leighton surface poloidal field source, generated from newly emerged flux from active longitude locations.

Unstable modes in the tachocline can have at least three important effects on the dynamo. First, they can "imprint" longitude-dependent patterns on the bottom of the convection zone, particularly in magnetic field, which can be transmitted to the photosphere where they can be seen. Second, they can provide limited longitudinal bands where upward bulges in the tachocline can appear, which, when they coincide with the presence of strong toroidal fields, can provide favored sites for toroidal flux tubes that become buoyant and rise to the photosphere. Third, they also are a powerful source of kinetic helicity for the dynamo itself, supplementing the Babcock–Leighton type surface poloidal source in creating, amplifying, and evolving the dynamo's poloidal field. All three of these effects could be important in the solar dynamo, so it is essential that their effects be included in the model.

It is feasible to incorporate all three of these effects into a global 3D flux-transport dynamo model. The imprinting of magnetic and velocity patterns at the bottom of the dynamo domain follows directly from including nonaxisymmetric flow and fields from solutions to the linear instability equations, with assigned amplitudes. Amplitudes for such modes can be calculated directly from nonlinear MHD models for global tachocline instabilities (Dikpati et al., 2018b).

This 3D class of model will make much heavier use of solar observations, particularly at the surface; incorporating all this data into the model in a sensible way will require sophisticated DA algorithms, most likely of the sequential type, to update the model integrations. The choice of update interval will be constrained by the time resolution of the observed data as well as by the "response" time of the dynamo model to changes in an input, such as meridional circulation. Experience so far with simpler 2D models with DA point to an update interval of 15–30 days, which is consistent with the time cadence of the solar observations to be used.

The initial 3D models should be kinematic; they should solve just the induction equation. In later versions, it should be possible, and desirable, to couple the induction equation to equations that calculate the global tachocline instability modes. The toroidal field produced by the dynamo as a function of time can be incorporated into linear and nonlinear models of tachocline instability to calculate how the unstable modes evolve. Then these can be included in computational updates of the induction equation, which include all changes in the dynamo inputs.

Although mean-field dynamo models have been quite successful in simulating important solar cycle magnetic features, they still rely on assumptions and parametric representations of certain physical processes they cannot resolve spatially. Full 3D convective dynamo simulations for the Sun, starting with Gilman (1983), and continuing to this day by Toomre and colleagues with the various anelastic spherical harmonic (ASH) codes (Augustson et al., 2015; Browning et al., 2006), as well as by Racine et al. (2011) and Fan & Fang (2014), have produced cycles, but are yet to be calibrated with solar cycle features. In addition, there is a third major track of model developments that avoid the complications of spherical geometry and work in boxes (Cattaneo & Tobias, 2014; Nigro et al., 2017; Pongkitiwanichakul et al., 2016; Rempel, 2014; Tobias et al., 2008, 2011; Tobias & Cattaneo, 2013), focusing in greater detail on important processes contributing to dynamo action, including rotation, shear, convection, and various types of turbulence, including magnetorotational turbulence. These studies allow development of greater physical insight into the interplay of various important physics, which may lead to better estimates of various parameters used in mean-field models, as well as better understanding of full 3D spherical shell MHD dynamo simulations. All three of these tracks are likely to contribute to the "ultimate," or physically most realistic, solar cycle dynamo model.

Once a 3D global solar dynamo as described above is produced, which calibrates well to the Sun, it will be possible to extend such a model into the corona to calculate dynamo-produced coronal structures with longitude dependence that can be compared with observed coronal structures. Then it should be possible to relate these structures to features such as coronal holes, polar structures, sectors, and high-speed solar wind streams (see Chapters 1–6). Finally, if the model calibrates well to the Sun, it can be tested for its ability to predict both longitude-averaged and longitude-dependent features of future solar cycles, using real solar data.

In closing, we judge that enormous progress has been made in modeling the solar dynamo, including simulating important features of solar cycles. But even now there is much to do before there is a consensus that the solar cycle problem has been "solved." It remains a very exciting challenge, one for which powerful tools available now should help us reach that goal.

7.2. HELIOSEISMOLOGY

7.2.1. Basic Principles

A wealth of information comes from the layers of the Sun's atmosphere, but how can we study what is happening below the opaque photosphere? Helioseismology is one of the few ways in which observers can probe the Sun's interior, using sound waves that travel through it.

The surface of the Sun oscillates with a period—or rather, many periods—of around five minutes, corresponding to a frequency of about 3 mHz. Deubner (1975) showed that the power in "five-minute oscillations" of the solar surface, when plotted in the wave-number–frequency plane, revealed a structure that identified them as nonradial acoustic eigenmodes of the solar interior (see Figure 7.14 for a more modern example). The cavity in which these "*p*-modes" (so-called because their restoring force is pressure) propagate is bounded by the solar

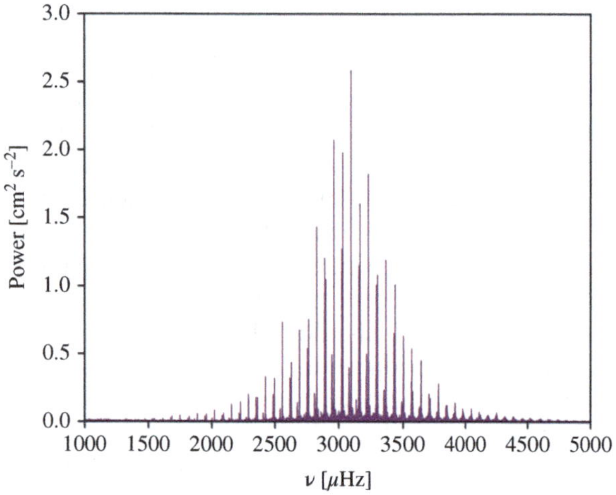

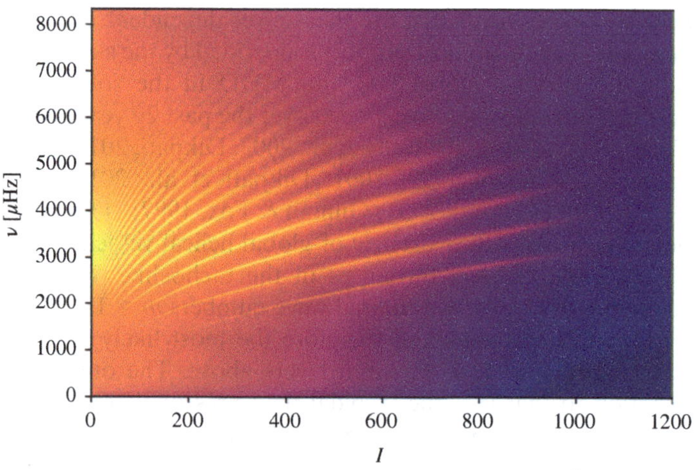

Figure 7.14 Left: Sun-as-a-Star power spectrum from the Birmingham Solar Oscillations Network (BiSON) network. Right: Rotation-corrected, *m*-averaged power spectrum in the $l - \nu$ plane, from one day of Global Oscillations Network Group (GONG) observations. Each ridge corresponds to a single radial order *n*. (*Source*: Original image courtesy GONG/NSO/AURA.)

surface and by the depth at which inward-propagating waves are refracted back toward the surface. The depth of this inner turning point varies monotonically with ν/L, where ν is the frequency, and L is related to the degree l by $L \equiv \sqrt{l(l+1)}$; the radial modes ($l = 0$) penetrate to the core, whereas modes of $l = 100$ penetrate only about $0.05R_\odot$ below the surface. Both "local" and "global" modes of oscillation exist; the global modes are those whose lifetime is longer than the time to make a complete circuit of the Sun, so that they can form a coherent global standing-wave pattern, whereas local modes—in general, the modes of shorter wavelength—are damped out more quickly and are coherent only over a small region of the solar surface. Both are used in helioseismology.

The acoustic modes can be described by spherical harmonics of degree l, azimuthal order m, and radial order n. Modes with $n = 0$ are a special case known as f-modes; these are surface gravity waves confined to the outer layers of the Sun. In the case of perfect spherical symmetry, modes of the same l and n but different m would be degenerate, but this degeneracy is lifted both by rotation and by the more subtle violation of spherical symmetry caused, for example, by the latitudinal distribution of magnetic activity. This makes it possible to use inversion techniques to infer a two-dimensional (2D) profile of the interior rotation, as we will discuss below.

7.2.2. Observations

The acoustic modes are observed by monitoring velocity—with Doppler techniques—or brightness variations at the photosphere, either in integrated light ("Sun as a star") or using imaging. Sun-as-a-star observations are limited to the lowest-degree modes, mostly those of $l \le 3$. Early results came from relatively short series of observations from a variety of single instruments—see, for example, the summary given by Howe (2009), of which perhaps the best known is that operated at the Big Bear Solar Observatory from 1986 to 1990 by Libbrecht and coworkers (e.g., Libbrecht & Woodard, 1990). For ground-based observations, it is preferable to use a global network of observing sites, to reduce the systematic errors introduced by interruptions. The Birmingham Solar Oscillations Network (BiSON: Hale et al., 2016) has been operating as a six-site network for low-degree helioseismic observations since 1992, with sparser observations stretching back into the late 1970s; Figure 7.14 (left) shows a sample BiSON spectrum. For medium- and (later) high-degree observations (Figure 7.14 (right)), the GONG; Harvey et al., 1996) has been in operation since 1995. Another strategy for obtaining near-continuous data is to observe from space. MDI (Scherrer et al., 1995) observed medium-degree oscillations from *SOHO* almost continuously from 1996 to 2011, with a few months of observations at high degree each year. MDI was then superseded by the HMI (Scherrer et al., 2012) instrument onboard SDO, which has been in use since 2010.

7.2.3. Inversions

The principle on which helioseismic inversions are based is that each observation d_i (e.g., a frequency, rotational splitting, or rotational splitting coefficient) represents a spatially weighted average of the quantity of interest over the solar interior, with an added error term ε_i.

Rotation Inversions

In the case of a rotation inversion, this can be written in the form

$$d_i = \int_0^{R_\odot}\int_0^{\pi} K_i(r,\theta)\Omega(r,\theta)\,dr\,d\theta + \varepsilon_i, \tag{7.1}$$

where $R_\odot$ is the solar radius, r and θ are respectively the distance from the center of the Sun in units of the solar radius and the colatitude, and K is the kernel, a spatial weighting function that depends on the solar model being used (Cuypers, 1980; Hansen et al., 1977). Schou et al. (1994) write the 2d-rotation kernel of an (n, l, m) mode as

$$\begin{aligned} K_{nlm}(r,\theta) = \frac{m}{I_{nl}}\Bigg\{ & \xi_{nl}(r)\left[\xi_{nl}(r) - \frac{2}{L}\eta_{nl}(r)\right]P_l^m(x)^2 \\ & + \frac{\eta_{nl}(r)^2}{L^2}\Bigg[\left(\frac{dP_l^m}{dx}\right)^2(1-x^2) \\ & - 2P_l^m\frac{dP_l^m}{dx}x + \frac{m^2}{1-x^2}P_l^m(x)^2\Bigg]\Bigg\}\rho(r)\,r\sin\theta, \end{aligned} \tag{7.2}$$

with

$$I_{nl} = \int_0^{R_\odot}\left[\xi_{nl}(r)^2 + \eta_{nl}(r)^2\right]\rho(r)r^2dr, \tag{7.3}$$

$x = \cos\theta$, and $L^2 = l(l + 1)$, where the radial and horizontal displacements are ξ_{nl} and $L^{-1}\eta_{nl}$, and $\rho(r)$ is the density. P_m^l is an associated Legendre function. Figure 7.15 shows some sample kernels.

The inferred rotation rate $\bar{\Omega}(r_0,\theta_0)$ at the location (r_0, θ_0) can be written as

$$\bar{\Omega}(r_0,\theta_0) = \sum_{i=1}^{M} c_i(r_0,\theta_0)\,d_i, \tag{7.4}$$

where the c_i's are weighting coefficients to be determined by the inversion. We can substitute Equation (7.1) into the right-hand side of Equation 7.4 to obtain

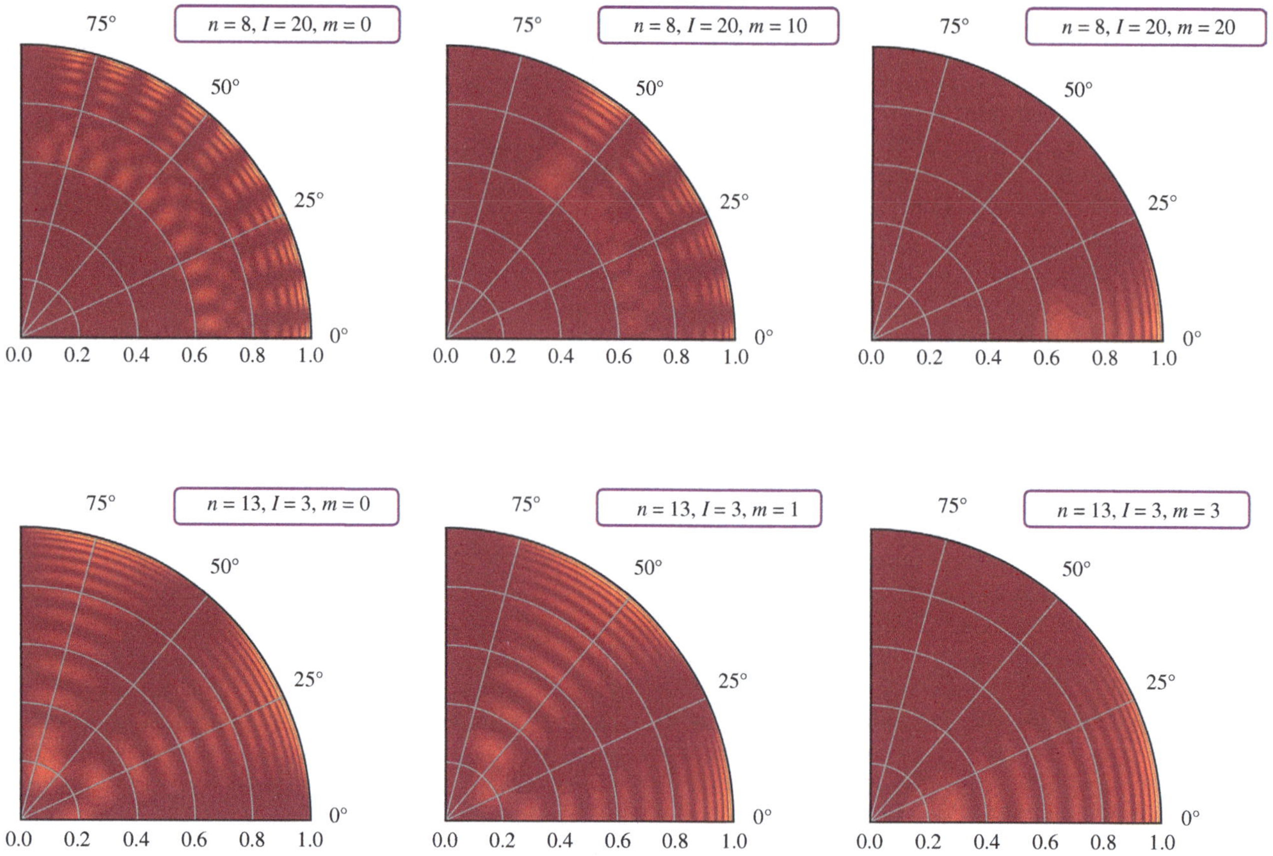

Figure 7.15 Sample rotation kernels based on Equation 7.2.

$$\bar{\Omega}(r_0,\theta_0)=\int_0^{R_\odot}\int_0^{\pi}K(r_0,\theta_0,r,\theta)\Omega(r,\theta)\,dr\,d\theta+\varepsilon_i, \quad (7.5)$$

where

$$K(r_0,\theta_0,r,\theta)\equiv\sum_{i=1}^{M}c_i(r_0,\theta_0)K_i(r,\theta) \quad (7.6)$$

is called the averaging kernel for location (r_0,θ_0). The averaging kernels do not depend on d_i, but are affected by the uncertainties in d_i, as these are used to weight the data in the inversion calculation.

The formal uncertainty on the input profile is given by

$$\sigma^2\left[\Omega(r_0,\theta_0)\right]=\sum_i\left[c_i(r_0,\theta_0)\sigma_i\right]^2. \quad (7.7)$$

where σ_i's are the uncertainties on the input data, provided that these are uncorrelated and normally distributed.

There are several possible approaches to solving the basic inversion problem. One commonly used method is "regularized least squares" (RLS), which uses, essentially, a least-squares fit with a penalty term to impose smoothness to find the rotation profile that best fits the input data (Schou et al., 1994). Another approach is "subtractive optimally localized averaging" (SOLA) (Backus & Gilbert, 1968, 1970; Pijpers & Thompson, 1992, 1994), where the quantity to be minimized is the difference between the actual averaging kernels and a target kernel. For examples of other methods, see Schou et al. (1998). The localization is never perfect, and there is always a trade-off between error and resolution; as only modes of lower degree penetrate into the deep interior, the resolution that can be achieved by the inversions is poorer at greater depth. For more details, see Howe (2009).

It is important to remember that these global inversion techniques always average the northern and southern hemispheres; to study the two separately requires local helioseismology or other more mathematically sophisticated techniques.

Structure Inversions

When helioseismologists talk about "structure," they generally mean the density and sound speed profiles of the solar interior, or quantities related to these. The problem of inverting for such quantities is quite complex; for a review, see Basu (2016). In practice, structure inversions are carried out using not the "raw" mode frequencies but

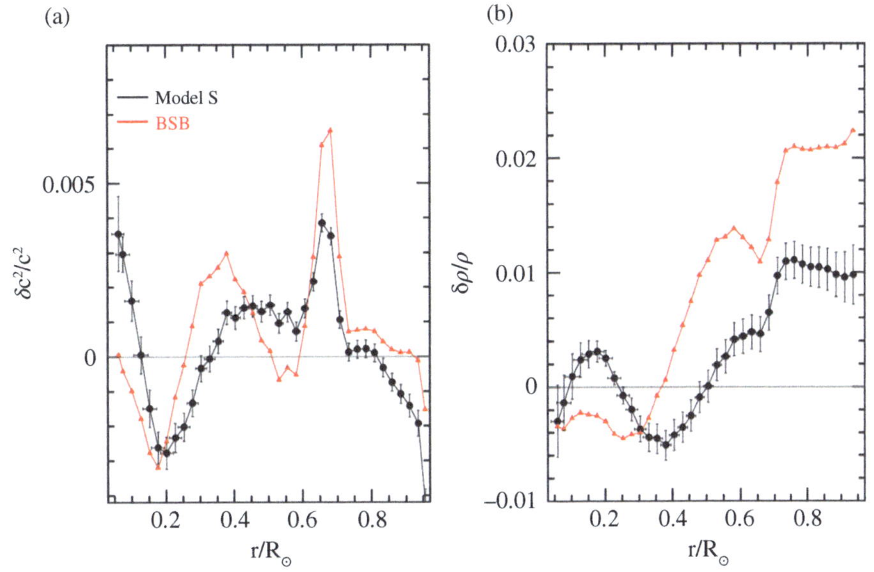

Figure 7.16 (a) The relative differences of the squared sound speed between the Sun and two other solar models, Model S and model BSB. (b) The relative density differences between the Sun and the two models. (*Source*: From Schmitt & Basu, 2015. © 2015 IOP Publishing.)

the difference between the observed frequencies and those predicted by a model. It is usually necessary to remove a frequency-dependent "surface term" that arises because the near-surface layers are not well modeled (see, e.g., Ball & Gizon, 2014, and the references therein). Once this term has been taken into account, the remaining differences between observation and model are very small—less than one part in a thousand, as seen, for example, in Figure 7.16. "Model S" of Christensen-Dalsgaard et al. (1996) is a particularly good match to the data and is therefore popular with helioseismologists; unfortunately, since the model was developed there have been changes to the calculation of element abundances in the Sun that have worsened the agreement between models and data (Bahcall et al., 2005), and the discrepancy has proved difficult to reconcile. Basu and Antia (2008) and Basu et al. (2009) have suggested that the problem may lie with the revised element abundances.

7.2.4. Solar Cycle Variations

It has been known since quite early in the history of helioseismology that the frequencies—and other properties—of the *p*-modes vary with the solar cycle. Woodard and Noyes (1985) looked at the mode frequencies in Sun-as-a-Star observations from the Active Cavity Radiation Monitor instrument onboard the Solar Maximum Mission and found that they were significantly lower at solar minimum than at solar maximum. The result was confirmed by Pallé et al. (1989) and Elsworth et al. (1990) using ground-based instruments that were to become part of the BiSON network. Around the same time, Libbrecht and Woodard (1990) found variations in medium-degree observations from the Big Bear Solar Observatory over the years 1986–1990. The medium-degree results showed that not only did the frequencies increase with increasing activity, but that higher-frequency modes experienced greater changes than those at lower frequency, following the variation of the so-called "mode inertia." This is a quantity defined by Christensen-Dalsgaard (1986) as being equal to

$$E_{nl} = \frac{\int_0^R \left[\xi_r^2(R) + l(l+1)\xi_t^2(R)\right]\rho r^2 \mathrm{d}r}{4\pi M\left[\xi_r^2(R) + l(l+1)\xi_t^2(R)\right]}, \qquad (7.8)$$

with ξ_r and ξ_t being the radial and tangential components of the displacement, M the total mass of the Sun, and R its photospheric radius. The fact that the frequency changes scale in this way shows that they are concentrated in the near-surface layers; higher-degree modes with shallower lower turning points, or higher-frequency modes with shallower upper turning points, are most strongly affected.

The variation of mode frequencies with activity is not only temporal; it also reflects the geometry of surface activity. In medium-degree modes, the even-order rotational splitting coefficients vary with time in a way that correlates with the coefficients of a Legendre polynomial expansion of the latitudinal distribution of surface activity (e.g., Antia et al., 2001; Howe et al., 2002, and the references therein). Howe et al. (2002) used a one-dimensional

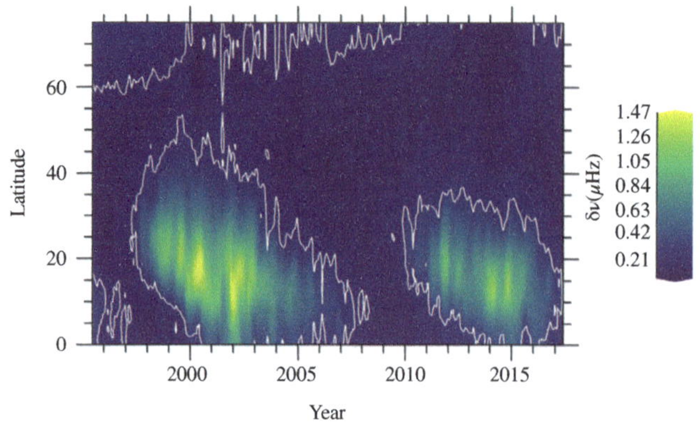

Figure 7.17 Latitudinal distribution of frequency shifts from GONG, averaged over modes with $40 \le l \le 80$ and $9 \le n \le 11$, based on the analysis of Howe et al. (2002). The white contour represents the 5 G level of the north–south averaged unsigned longitudinal magnetic field strength from National Solar Observatory synoptic maps. (*Source*: Modified from Howe et al., 2002.)

inversion technique to map the latitudinal distribution of the m-dependent p-mode frequency shifts from GONG observations and showed that they mapped closely to the (north–south symmetrized) magnetic butterfly diagram, as shown in Figure 7.17.

For local modes, Hindman et al. (2000) showed that the mode frequencies in "ring-diagram analysis"—which uses the three-dimensional (3D) power spectrum of the oscillations in a small patch of the solar surface (Hill, 1988)—increased in patches corresponding to active regions, and Rajaguru et al. (2001) also found increased ring-diagram frequencies in active regions, with larger shifts for higher frequencies.

Although the activity-related changes in mode frequencies are relatively easy to detect, their effect seems to be largely confined to the outer couple of percent of the solar radius, and in inversions it is dominated by the surface term (e.g., Antia et al., 2001). It is much more challenging to find signatures of changing magnetic activity deeper in the interior. Eff-Darwich et al. (2002b) placed an upper limit of 3×10^{-5} on any fractional change in the interior sound speed. The subtle changes found at the base of the convection zone by Baldner and Basu (2008) are just within this limit.

7.2.5. Rotation

The interior rotation of the Sun can be divided into five regimes, working from the surface inward: the near-surface shear layer, the bulk of the convection zone, the tachocline, the radiative interior, and the core (see Figure 7.18). The maximum rotation rate at the equator corresponds to a speed of about 330 ms^{-1}. We consider each of these regimes in turn.

Near-Surface Shear

The rotation rate increases inward in about the outer 5% of the solar radius. An early helioseismic detection of an inward-increasing rotation gradient was made by Deubner et al. (1979) using high-degree observations that probed about the upper 20 Mm of the convection zone. There were also hints that such a shear layer might exist in non-helioseismic observations, with a higher rotation rate seen in direct Doppler measurements at the photosphere than from tracking surface features (e.g., Brown et al., 1989), which might be explained by the idea that features such as sunspots are anchored in a faster-rotating layer below the surface. The shear layer was clearly seen in the early global rotation inversion results from GONG (Thompson et al., 1996) and MDI (Kosovichev et al., 1997); in both cases, the shear apparently reversed at high latitudes.

Because the shear layer is not fully resolved using medium-degree p-modes alone, it is best studied using f-modes, as was done by Corbard and Thompson (2002) with MDI data. They found that the rotation rate decreased away from the rotation axis at a rate of about -400 nHz/$R_\odot$ at low latitudes, but this slope decreased to a small value by 30° latitude and possibly reversed sign at latitudes above that (but only in the outermost 5 Mm). However, they pointed out that the high-latitude result depended on the handful of modes with $l \ge 250$. They also found that the rotation rate varied almost linearly with depth, which is unexpected, as considerations of conservation of momentum would suggest the rotation rate should vary with the inverse square of the distance from the rotation axis. Barekat et al. (2014), using HMI data and re-analyzed MDI data, confirmed the low-latitude results of Corbard and Thompson (2002), but found that the sign of the shear remained negative up to at least 60° latitude.

The Bulk of the Convection Zone

Like the surface rotation, the solar rotation in the bulk of the convection zone is differential—fastest near the equator and slowest near the poles. Below the near-surface shear layer, the rotation profile is almost—but not quite—constant along radial lines. The contours of constant rotation in fact make an angle of about 35° to the rotation axis over a wide range of latitudes (Gilman & Howe, 2003; Howe et al., 2005); this may be due to Coriolis forces acting on the meridional flow.

The Tachocline

Below the convection zone lies a relatively thin region of strong radial shear known as the tachocline, marking the transition from differential rotation to the more or less rigid rotation of the radiative interior. This region is not fully resolved in inversions, meaning that the

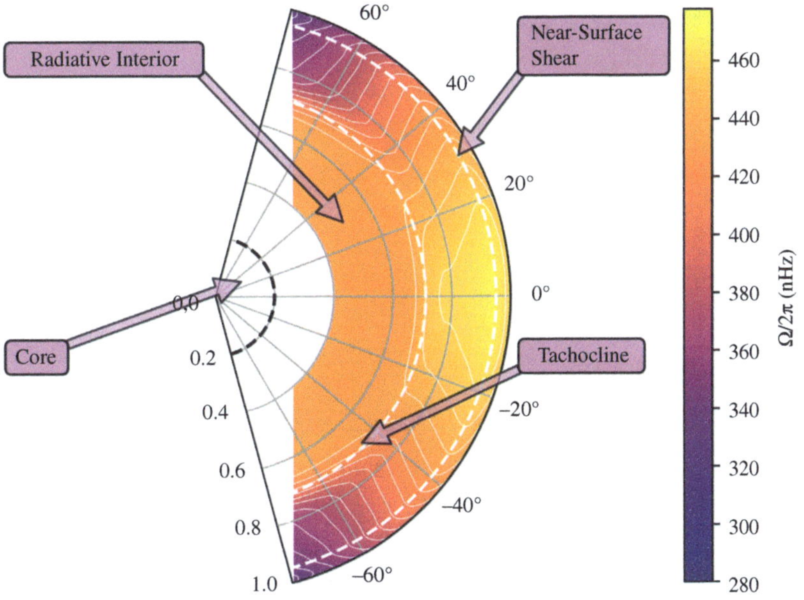

Figure 7.18 Solar interior rotation profile, based on the average of 2dRLS inversions of 45 72-day periods of HMI data from 2010 to 2018. The white contours are at intervals of 10 nHz. The dashed arcs represent the boundary of the core at 0.2 $R_\odot$, the base of the convection zone at 0.71 $R_\odot$, and the base of the near-surface shear layer at 0.95 $R_\odot$.

transition appears wider and smoother in inversion results than it is in reality, so forward calculations or other strategies are needed to estimate its thickness. Table 2 of Howe (2009) summarizes the work of various authors to parameterize the shape of the tachocline at equatorial latitudes. The consensus is that the tachocline covers a few percent of the solar radius and is centered slightly below the base of the convection zone (0.713 $R_\odot$) as determined from structure inversions (Basu & Antia, 1997). Basu and Antia (2001) also found that the tachocline is prolate—shallower at higher latitudes—a result more recently confirmed by Antia and Basu (2011).

The Radiative Interior

Below the tachocline, the latitudinal and depth resolution of the inversion results is relatively poor, but this region appears to rotate rigidly; Eff-Darwich et al. (2002a) found a rate of approximately 435 nHz with no significant differential rotation between 0.4 and 0.7$R_\odot$.

Core Rotation

Only the radial ($l = 0$) modes penetrate to the very center of the Sun, and these carry no information about the rotation. The $l = 1$ modes do have their lower turning points within the nuclear-burning core of the Sun, but even these modes spend the majority of their time closer to the surface, so estimating the core rotation is not as simple as measuring the $l = 1$ rotational splitting. It is difficult to localize an averaging kernel below about 0.25 $R_\odot$ (e.g., Chaplin et al., 1999), and to date it has not been possible to put strong constraints on the core rotation rate using *p*-modes. It would be helpful if we could measure the solar *g*-modes (modes where the restoring force is gravity, and which have their greatest amplitude in the interior and a very small amplitude at the surface), but there are no independently confirmed *g*-modes observations for the Sun, though the results of García et al. (2007) are intriguing. Using observations from the Global Oscillations at Low Frequencies (GOLF) instrument onboard *SOHO*, they found a periodic structure in the low-frequency spectrum consistent with *g*-mode peaks favoring a faster rotation rate than that in the rest of the radiative interior. However, this result remains controversial.

The Torsional Oscillation

The solar rotation at the surface is not constant over time; it shows a pattern of migrating bands of faster and slower rotation (or in other words, zonal flows that are prograde and retrograde in the frame of the overall solar rotation) that is closely associated with the migration of the active latitudes toward the equator over the course of a solar cycle (Figure 7.19). The variation in the rotation rate at the equator corresponds to a speed change of a few meters per second. This pattern was first identified, and dubbed the "torsional oscillation" by Howard and LaBonte (1980) in direct Doppler observations from the MWO. The bands were first detected by helioseismic means by Kosovichev and Schou (1997) using *f*-mode

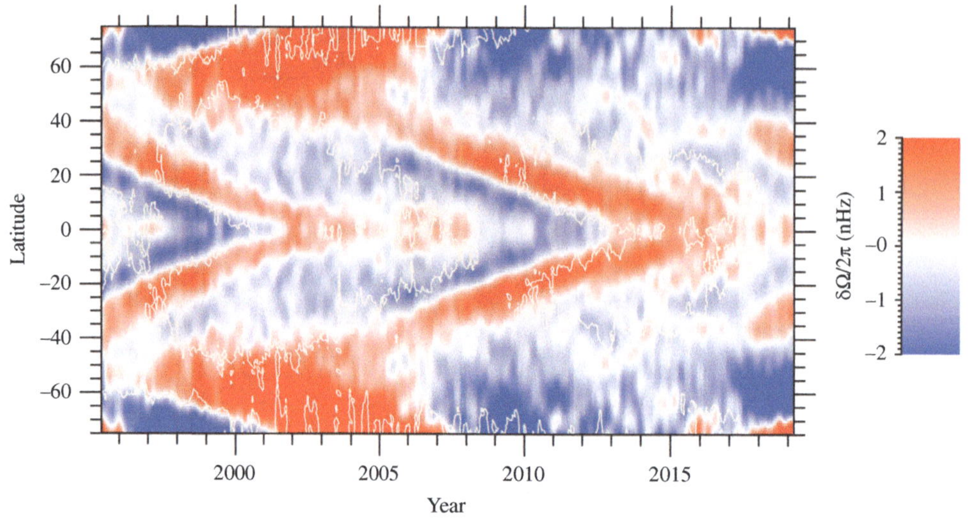

Figure 7.19 Zonal-flow residuals at 0.99 $R_\odot$ from 2dRLS inversions of GONG, MDI and HMI data after subtraction of a temporal mean at each latitude. The white contours mark the 5 G level of the longitudinal magnetic field strength from synoptic maps produced by the National Solar Observatory. (*Source*: Color map from Kovesi, 2015).

splittings from MDI. As data from GONG and MDI accumulated during the rising phase of Solar Cycle 23, it became possible to detect the migrating flows using *p*-mode data. This revealed that the flows penetrated to at least 0.92 $R_\odot$, or about 30 Mm, below the photosphere (Antia & Basu, 2000; Howe et al., 2000; Toomre et al., 2000; Vorontsov et al., 2002). The helioseismic observations also revealed the presence of a poleward-propagating branch at latitudes above about 45° that appears to penetrate through the whole depth of the convection zone (Antia & Basu, 2001). As more data were collected and the signal-to-noise of the observations improved, it became apparent that the pattern may be coherent through the whole of the convection zone, with the signal near the base of the convection zone seeming to lead that at the surface by about two years—or, to look at it another way, the phase of the pattern is constant along lines of constant rotation rather than lines of constant latitude (Howe et al., 2005).

Around solar minimum, when the main bands of faster-than average rotation are close to the equator, an acceleration in the near-surface zonal flows at mid-latitudes appears to precede the appearance of new cycle magnetic activity. During the unusually long minimum following Solar Cycle 23, the new equatorward branch associated with Cycle 24 was visible, but its migration toward the equator was delayed (Howe et al., 2009). As noted by Howe et al. (2011), the new cycle activity appeared only when the flow belt reached a latitude of about 25°, as had been seen in previous cycles.

During Solar Cycle 24, the poleward-propagating branch of the torsional oscillation pattern appeared much weaker, but this appearance was shown by Howe et al. (2013) to be partly due to a slowdown in the underlying rotation rate at higher latitudes, which may be related to the weaker polar fields during this cycle (Rempel, 2012). In the most recent observations (Howe et al., 2018), the beginnings of the equatorward branch for Cycle 25 can be clearly seen.

7.2.6. Meridional Flow

The Sun has a well-known pattern of meridional flow from the equatorial regions toward the poles, as found by Duvall (1979) using the Stanford magnetograph and by LaBonte and Howard (1982) in Mount Wilson Doppler observations at the surface. This flow plays an important part in the solar magnetic cycle (e.g., Wang et al., 1989); the flow carries magnetic flux from decaying active regions poleward, until at solar maximum the sign of the polar field reverses. Conventional global helioseismic techniques cannot measure this flow, but it can be accessed by local helioseismic techniques, at least in the outer layers of the convection zone. Giles et al. (1997) used "helioseismic tomography"—a variant of the time-distance technique (Duvall et al., 1993), which measures the travel time of waves between pairs of points on the solar surface—to measure the longitudinally averaged flow in the outer 4% of the solar radius, and found that the speed of the flow was approximately constant over that depth range. Subsequent results (e.g., González Hernández et al., 1999; Haber et al., 2000; Zhao & Kosovichev, 2004) confirmed that the flow in the outer convection zone is poleward up to at least 50 deg latitude and down to a depth of 30 Mm. Zhao & Kosovichev (2004) also reported a time-varying component to the meridional flow, with flows converging toward the magnetic activity belt above 12 Mm, and diverging flows below 18 Mm. Zhao et al. (2014) again found flows converging toward the activity belts, and also reported that the meridional flow speed above latitude 30Â° was slower when following-polarity flux was being transported and faster when leading-polarity flux was being transported. Similar solar cycle variations of the meridional flow have been seen in other work, for example, by Hathaway and Rightmire (2011), who studied the surface flows by correlation tracking of features in MDI magnetograms and found, in addition to the inflows into the activity belts, that cells of equatorward flow appeared at higher latitudes, in the southern hemisphere during the rising phase of Solar Cycle 23 and in the northern hemisphere during its declining phase. González Hernández et al. (2011) also reported on modulation of the meridional flow by the activity cycle from ring-diagram analysis of GONG data; an interesting feature of this work is that, as with the zonal flows, the changes in the meridional flow pattern start to appear in advance of the new cycle activity.

The meridional "return flow" that must exist (so that mass does not pile up at the poles), as well as any other complex interior structure of the meridional flow, is much more challenging to observe. Several attempts to probe the meridional flow deep in the convection zone have been made in recent years, particularly since the center-to-limb correction (Zhao et al., 2012) was introduced. Several of these attempts produced somewhat inconsistent results: Zhao et al. (2013), using time-distance analysis on two years of HMI data, Schad et al. (2013) using an analysis of perturbations to global modes caused by the meridional flow, and Kholikov et al. (2014a) and Jackiewicz et al. (2015)—both using GONG data—all found some indications of a multi-cell structure, but Rajaguru and Antia (2015) did not find strong evidence of a multi-cell flow in their analysis of four years of GONG data. More recently, using seven years of HMI data, Chen and Zhao (2017) found a two-cell structure in each hemisphere, with an equatorward flow in the middle of the convection zone and poleward flows above and below (Figure 7.20), in broad agreement with the earlier findings of Zhao et al. (2013). On the other hand, Mandal et al. (2018) and Liang et al. (2018) reported results favoring a single-cell structure with the return flow close to the base of the convection zone, although the latter work did find an anomalous two-cell structure in the northern hemisphere during Solar Cycle 24. It is probably still too early to say that the question of the interior meridional flow has been settled.

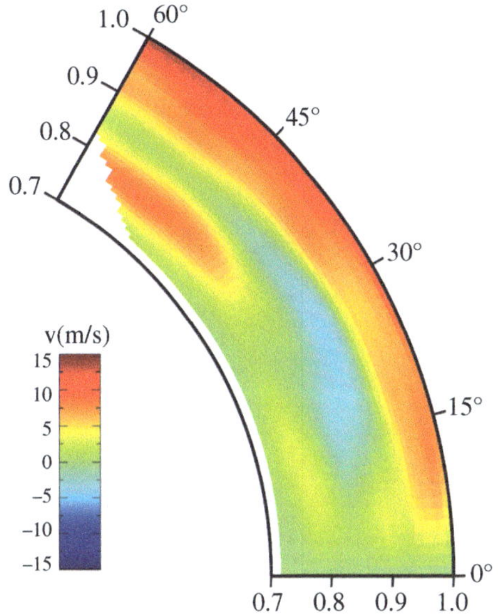

Figure 7.20 North–south average of meridional flow profile. (*Source*: figure 8 from Chen & Zhao, 2017. © 2017 IOP Publishing.)

7.3. FLUX EMERGENCE

7.3.1. Observations of Emerging Flux

Early Observations of Sunspots

A sunspot is a darkening of the Sun's surface, corresponding to a reduction in temperature at the surface from ~5700–6500 degrees Kelvin (K) outside sunspots to ~3500–4500 K in sunspots. This darkening is due to the strong concentration in sunspots of magnetic flux. The magnetic field, concentrated to several kilogauss (kG) in sunspots, inhibits the convective motion of the plasma within that field, and so prevents the efficient transport of energy by convection through the convection zone just below the sunspot, where the magnetic field is particularly strong. Rather, energy here is transported by radiation, which is less efficient in convection zone conditions, and so the surface temperature above these regions is lower. The dark, central regions of these spots are referred to as the sunspot umbrae. Fully formed sunspots are surrounded by a brighter, striated region called the penumbra. The images below, from Tiwari et al. (2015), show the white light (continuum intensity) of a well-formed sunspot (Figure 7.21), as well as the temperature, magnetic field strength, magnetic field inclination from the vertical, and line-of-sight velocity in and around the spot (Figure 7.22).

Sunspots largely appear within ±40° in latitude from the solar equator, and last for several days to a few months. They usually appear as pairs of positive and negative magnetic field polarity, suggesting that they are

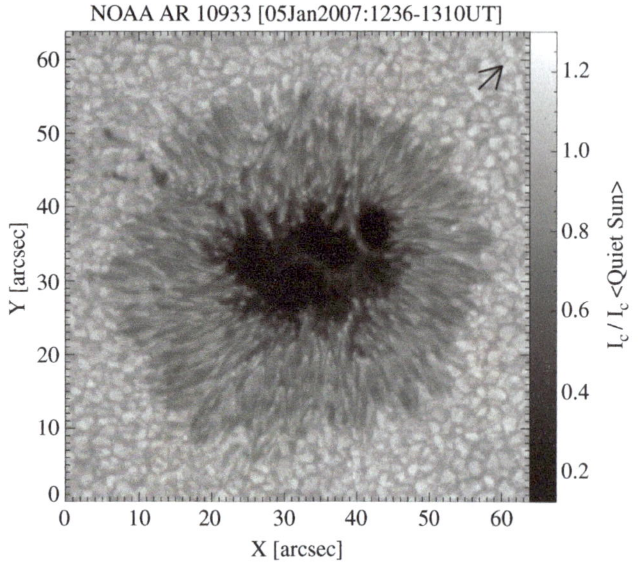

Figure 7.21 White-light observation of a sunspot (National Oceanographic and Atmospheric Administration active region, NOAA AR, 10933), taken by the Solar Optical Telescope (SOT) on board the *Hinode* satellite. (*Source*: From Tiwari et al., 2015. Public Domain.)

caused by the emergence of an Ω-shaped rope of magnetic flux through the photosphere: where one part of the emerged rope intersects the photosphere, a positive vertical field is measured, and where the other part intersects the photosphere, a negative vertical field is measured. Bright loops of magnetic field are observed between these two spots in extreme ultraviolet and X-ray observations of the solar atmosphere, completing the picture of a rope of magnetic field arching from one sunspot to the other.

Coyne (1991) presents an informative review of early sunspot observations. The first mention of sunspot observations is from the fourth century BCE by Teophrastus of Athens (Coyne, 1991). Naked eye observations of sunspots were then recorded in China, and to some extent in Japan and Korea, starting in the second century BCE (Bray & Loughhead, 1964; Coyne, 1991; Yau & Stephenson, 1988). In Europe, telescopic observations of sunspots date to around 1611, when a number of astronomers observed and reported them. The review by Coyne (1991) of these earliest observations credits Galileo Galilei, Johann Goldschmid, Christopher Schneier, Thomas Hariott, and Francesco Sizzi with these first observations. The review by Morfill et al. (1991) credits Schwabe (1843) with the first

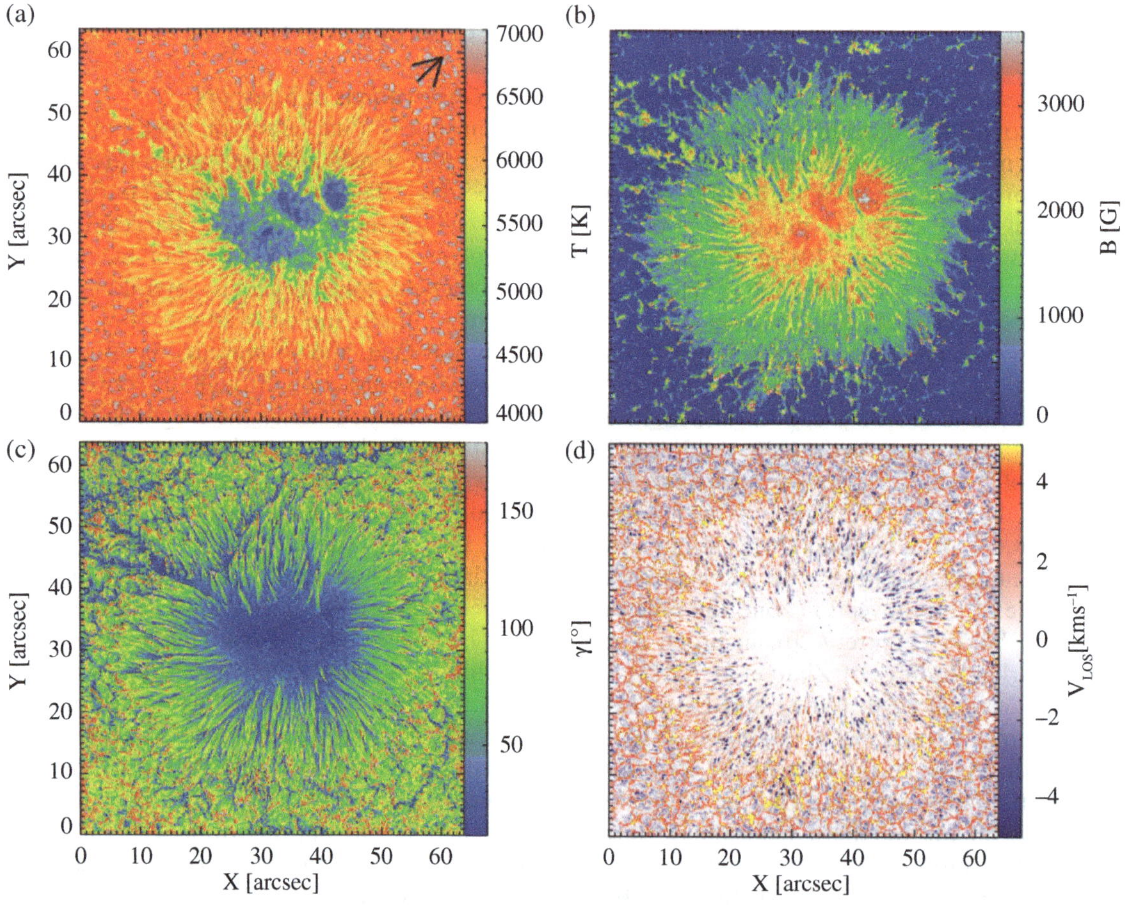

Figure 7.22 (a) Inferred temperature, (b) magnetic field strength, (c) inclination of magnetic field vector to the solar surface, and (d) line-of-sight velocity for NOAA AR 10933. Data taken by *Hinode*/SOT. (*Source*: From Tiwari et al., 2015. Public Domain.)

reported observation indicating that there is a sunspot cycle, as discussed in Section 7.1. Within this cycle are several key features of sunspot emergence and evolution. See Zirin (1988) for a review of these solar cycle features, briefly summarized below.

Sunspot Latitude

The latitude at which sunspots emerge decreases during the cycle, starting with an average latitude of about ±30° early in the cycle, and ending near the equator (Zirin, 1988). This was first reported by Carrington, and is referred to as Spörer's law (Carrington, 1858) and is discussed in more detail in Section 7.1. The top panel of Figure 7.1 shows this effect: concentrations of sunspots emerge around ±30° latitude at the beginning of each cycle with subsequent spots emerging closer to the equator as the cycle progresses.

Sunspot Magnetic Field

Hale (1912) used the Zeeman effect (Zeeman, 1897) to determine that sunspots have strong magnetic fields. The Zeeman effect is the process whereby a magnetically sensitive ionized atom in a magnetic field will emit linear and circularly polarized light, with the relative characteristics of the different polarizations depending on the direction and strength of the magnetic field. Panels (b) and (c) of Figure 7.22 show the resulting magnetic field strength and magnetic field inclination with respect to the vertical for a spot observed by Tiwari et al. (2015). Hale used his Zeeman effect measurements to show that sunspot pairs are aligned so that the leading spot (in the sense of solar rotation) is of one magnetic polarity and the following spot is of the opposite polarity. Critically, Hale et al. (1919) then found that the polarity of the leading spots is same for the majority of spot pairs in one hemisphere (northern or southern), and is the opposite for the majority of spot pairs in the other hemisphere. This polarity rule then flips with each subsequent solar cycle (see Section 7.1).

Sunspot Tilt

An additional observed rule for sunspot emergence is the tilt of the sunspots relative to the equator. As was pointed out in Section 7.1, although a sunspot pair from leading to following spot is aligned largely parallel to the equator, there is on average a tilt with respect to the equator, with the leading spot lying closer to the equator. This tilt varies from about 11° at ±30° latitude to about 3° near the equator (see Hale et al., 1919), and is referred to as Joy's law (Hale & Nicholson, 1925). Numerous measurements of this variation in tilt with latitude have been made over the years: see Wang and Sheeley (1989), Stenflo and Kosovichev (2012), and McClintock et al. (2014), as well as the review by Pevtsov et al. (2014). Figure 7.23 shows one such recent measurement of sunspot tilt versus latitude by Tlatova et al. (2018).

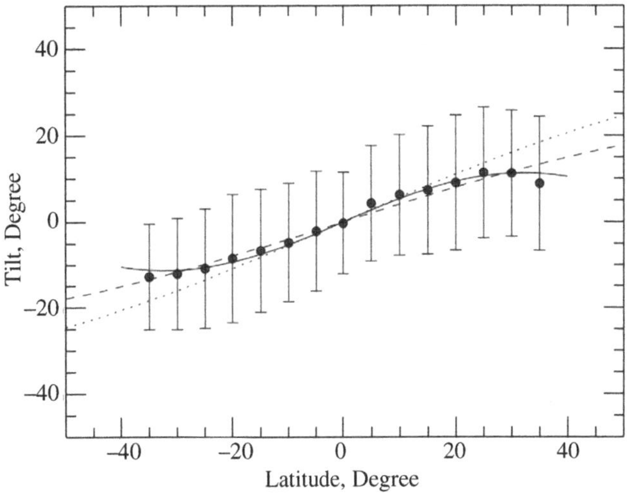

Figure 7.23 Measured tilt of solar magnetic sunspot pairs, taken from Mount Wilson Observatory measurements from 1917 to 2017. The filled circles show the measured mean tilts, and the error bars show the measured standard deviations. The dashed line shows the fit to this data, and the dotted line shows a fit from the earlier study of Stenflo and Kosovichev (2012). (*Source*: From Tlatova et al., 2018. © 2018 Springer Nature.)

Sunspot Complexity

Although sunspots in general appear as a positive and negative polarity pair, or a positive and negative polarity group of spots, and obey the leading-following (Hale) and tilt (Joy) laws outlined above, sunspot regions, also referred to as active regions, can sometimes form in more complex configurations. Hale et al. (1919) developed a classification system, often referred to as the Mount Wilson classification, to describe these types of spots. Single spots with no corresponding opposite spot are termed α spots. These spots obviously have a magnetic field that connects to the opposite magnetic field at the photosphere—but in this case that field is weaker and more dispersed, and not strong enough to form a sunspot. The standard configuration with a pair of opposite polarity spots is termed β. Spots with an "irregular distribution of polarities" (Zirin, 1988) are termed γ, and spots with "umbrae of opposite polarity within a single penumbra" are termed δ (Künzel, 1960). These different types can overlap and combine together and so extended classifications, generally consisting of combinations of these α, β, γ, and δ classifications, have been developed and used over the years. Jaeggli and Norton (2016) analyzed the statistics of active region classifications from 1992 to 2015 and found that ~20% of active regions are classified as α or $\alpha\gamma$ spots, ~75% are classified as β or $\beta\gamma$ spots, and the remaining 5% are some form of δ-spots.

δ-Spots

The flaring activity generated above sunspots will be the focus of Chapter 5, and we focus briefly here on the contribution that δ-spots make to strong solar flares.

The complexity associated with δ-spots has been shown to indicate a stronger likelihood of eruption (Künzel, 1960). These regions often have an inverted polarity relative to Hale's law, and Tanaka (1980) showed that 90% of such inverted δ-spot regions are highly active. Zirin and Liggett (1987) and Sammis et al. (2000) investigated the flare activity of these different types of spots. From this and similar studies, they found that the more complex spots, in particular those with a δ classification, either on its own or combined with another classification, were responsible for the majority of large, or so-called X-class, flares. The knowledge of this increased activity has fostered a number of theoretical and modeling studies to explain this, as discussed below.

Twist and Rotation of Sunspot Fields

Kurokawa (1991) analyzed observations of sunspot emergence with the Domeless Solar Telescope at the Hida Observatory of Kyoto University. By studying the time evolution of structures visible in the chromosphere, 1000 km or so above the photosphere, they found evidence that magnetic fields emerge as a set of twisted and sheared loop structures. This argues that emerging structures can be in the form of so-called flux ropes; that is, that they consist of a set of field lines wrapping around a single axis, forming a tight helix. Pevtsov et al. (1994) analyzed vector magnetic field measurements taken with the Stokes Polarimeter at the Mees Solar Observatory to measure the effective twist of active regions. They measured the ratio of the vertical current to the vertical magnetic field at the photosphere. This gives a measure, called α, which is related to the inverse wavelength with which field lines twist about each other. They found that locally this twist measure can reach $5 \times 10^{-7}/m$, indicating a twist wavelength on the order of 10,000 km. Pevtsov et al. (1995) and Longcope et al. (1998) then surveyed α based on the coronal geometry of field lines for each active region. They found that the average α over each active region lies in the range $\pm 2 \times 10^{-8}/m$, and increases moderately with latitude. Luoni et al. (2011) analyzed the vertical magnetic field structure of active regions, and found that the tongue-like structure often seen in emerging active regions, which distorts the bipolar structure into more of a yin-yang shape, can be well reproduced by the emergence of a twisted, arched structure through the photosphere. If the twist in such a torus goes to zero, the tongues disappear, and the emerging region returns to the classic, symmetric bipolar structure.

Leka et al. (1996) mapped out the rotation of pair of emerging sunspot groups about their center of mass. They studied an emerging, flaring sunspot region with the Imaging Vector Magnetograph at the Mees Solar Observatory. By following the location of footpoints of the two primary concentrations of magnetic polarity in the photosphere as a function of time, they were able to generate a series of time versus photospheric location maps of these footpoints. These maps indicated that the observed sunspot regions were emerging with a large-scale helical rotation, or writhe (a rotation of the axis of a tube about itself).

López Fuentes et al. (2000, 2003) found significant rotation of some sunspot regions over longer timescales as well. Following sunspot regions over the course of several months (i.e., including several passes around the back side of the Sun), they found net rotations of these regions about their center of masses by up to 180°, as shown in Figure 7.24. They found that the majority of the sunspot regions they observed had the same sign of twist (rotation of the magnetic field of a flux tube about its axis) as of rotation, indicating that the rotation was likely due to convective motions writhing the flux ropes and adding twist to them as they rose to the surface. This is consistent with the Σ effect predicted by Longcope and Klapper (1997; see discussion below in Section 7.3.2.1).

Emergence Rate and Separation Distance

Chintzoglou and Zhang (2013) extended the distance–time reconstruction of Leka et al. (1996) to generate 3D maps of entire active regions, such as the one shown in Figure 7.25 for NOAA AR 11158 (SDO/HMI). This map shows 100 hr of photospheric magnetic field measurements, with time plotted on the vertical axis and 2D measurements of the line-of-sight magnetic field on the solar surface plotted on the horizontal axes. Isosurfaces of magnetic field strength are drawn for 400 G in the top panel, and for 1100 G in the bottom panel. These isosurfaces are surfaces drawn such that the volume enclosed within the surface has field strength larger than the isosurface magnetic field strength and the volume outside that surface has a field smaller than that level. The resulting map illustrates both the separation of positive from the negative emerging regions with time, and the internal coalescence of each of these two structures with time, particularly for the 400 G isosurface level. Note that this presents a useful way to visualize the variation in horizontal location and extent of the emerging field with time, and helps to infer the magnetic topology that is transferred through the photosphere, but the result should not be taken as a reconstruction of the 3D shape of the resulting coronal field. This is in part because vertical emergence flows do not factor into this reconstruction, so the apparent vertical extent of the structure may vary from one location to the next (thus, the vertical axis here is labeled with time, rather than with height), and in part because the expansion of the emerging field into the corona once it passes through the photosphere and chromosphere will likely reshape and reconfigure the field.

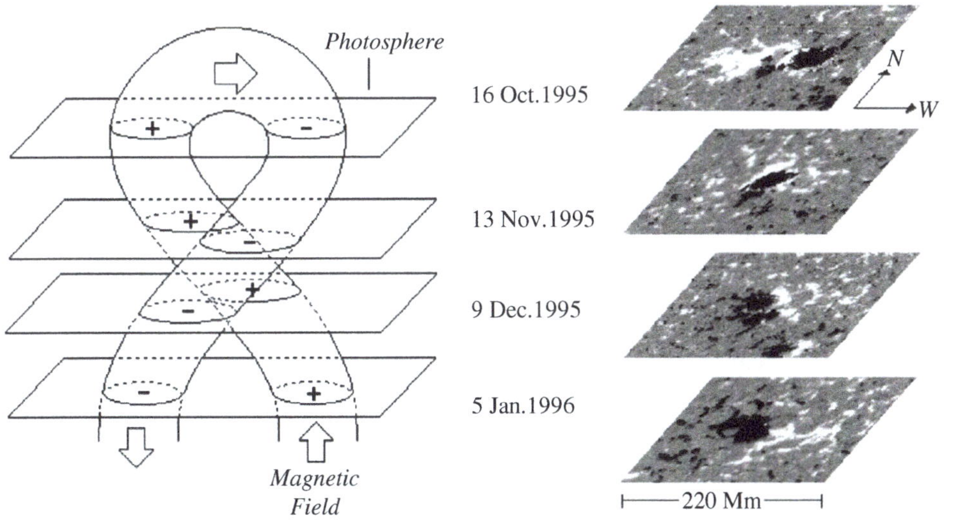

Figure 7.24 Observation of the long-term (three-month) evolution of the line-of-sight magnetic field measurements of NOAA AR 7912. This shows how the magnetic fields of the region rotate by ~180° over this time frame. The cartoon to the left shows an inferred geometry for the sunspot region that could reproduce this behavior. (*Source*: López Fuentes et al., 2000. © 2000 IOP Publishing.)

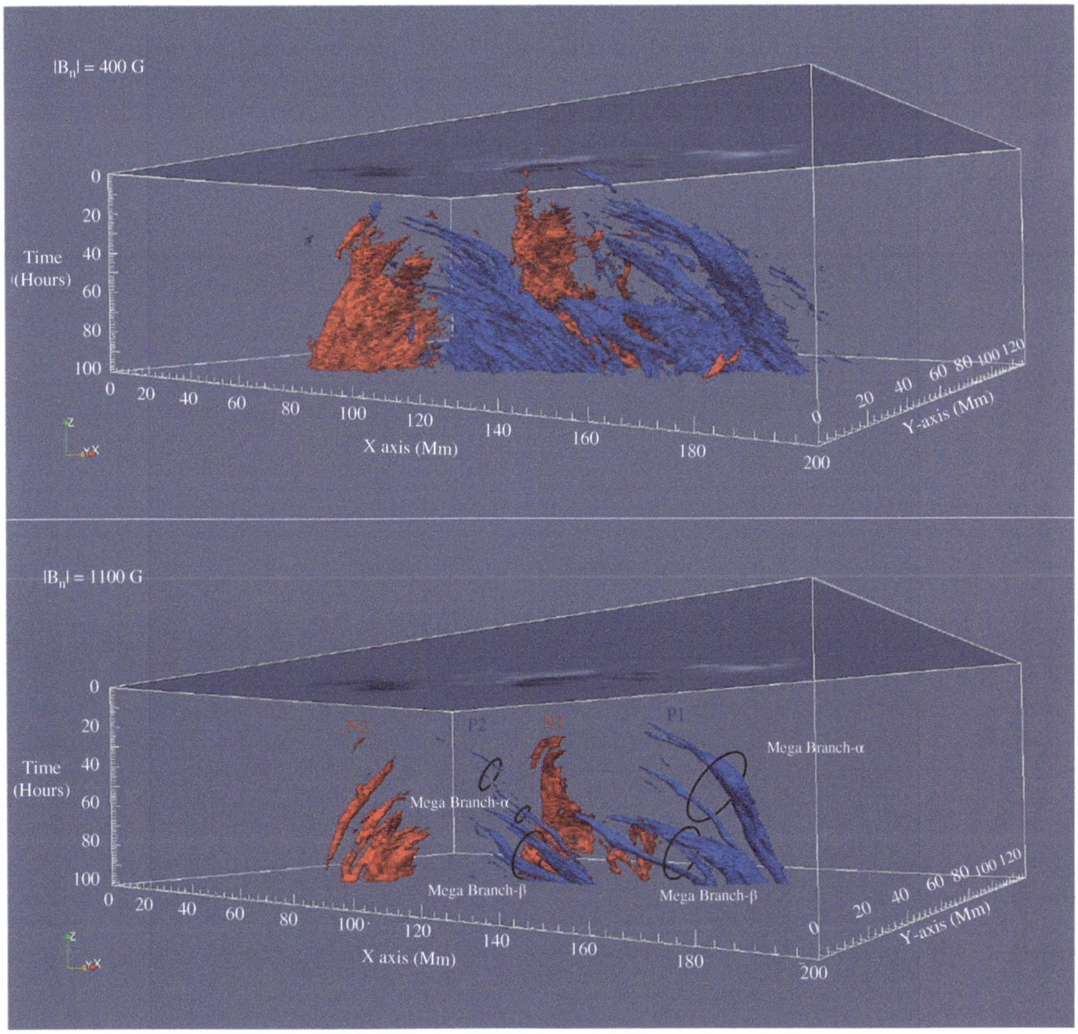

Figure 7.25 Height–time reconstruction of the line-of-sight component of the magnetic field in NOAA AR 11158 over the course of 100 hr during its emergence. The top panel shows isosurfaces at 400 G and the bottom panel shows isosurfaces at 1100 G, with the positive field in blue and the negative field in red. This gives a 3D rendering of the time history of the photospheric field, and gives one an idea of the geometry of the structure that rises through the photosphere during this flux emergence. (*Source*: From Chintzoglou and Zhang, 2013. © 2013 IOP Publishing.)

McClintock and Norton (2016) investigated sunspot separation as a function of sunspot area. They found that the separation of the opposite polarity spots is about 40,000 km for small regions when their spot area is at its maximum, the separation is about 45,000 to 50,000 km for midsize regions at that stage in their evolution, and large regions reach about 60,000 km at that stage. All regions then proceeded to move further apart in the 2–3 days following peak spot area, with large regions reaching 70,000 to 80,000 km in separation. Here, small areas are defined as regions with maximum umbral areas less than 45 MSH (millions of a solar hemisphere, MSH = 3 million square kilometers), midsize regions are between 45 MSH and 90 MSH, and large regions are larger than 90 MSH. Interestingly, all regions have a separation distance of ~40,000 km early in their evolution. McClintock and Norton (2016) hypothesize that this may be due to the influence of so-called supergranular convection cells, which have a similar diameter, as originally proposed by Frazier (1972). This proposal had been countered by Harvey and Martin (1973), who discovered that the emerging fields they studied did not collect at supergranular boundaries, as would be expected if the emergence was driven by these flows. However, these recent results do suggest that even if supergranules are not directly responsible for emerging flux, they may play a role in setting the scale. For example, the convective dynamo simulations by Nelson et al. (2013) find that large-scale convective rolls cause fields to detach from toroidal wreaths of magnetic field on those convection scales and then rise to the surface. Determining what sets this scale is a critical remaining question of dynamo and flux emergence theory, and poses a promising task for comparisons of observation and modeling.

Norton et al. (2017) then investigated the rate at which flux emerges in small to midsized bipolar active regions, studying in depth the evolution of 10 active regions whose emergence was fully captured on the visible solar disk. This study found that flux emerged at a mean rate of $\sim 7 \times 10^{19}$ Mx/hr to generate a mean peak active region flux of $\sim 6 \times 10^{21}$ Mx. They present a comprehensive review of other, complementary flux emergence rate studies that preceded theirs, and summarize these various results in Figure 7.26. This figure shows the flux emergence rate versus the peak flux for all of these studies, as well as the same measure for a number of flux emergence simulations (see the discussion of flux emergence simulations in Section 7.3.2). Together, these various studies cover three orders in magnitude of peak flux. Both this composite of various studies and the study of Otsuji et al.

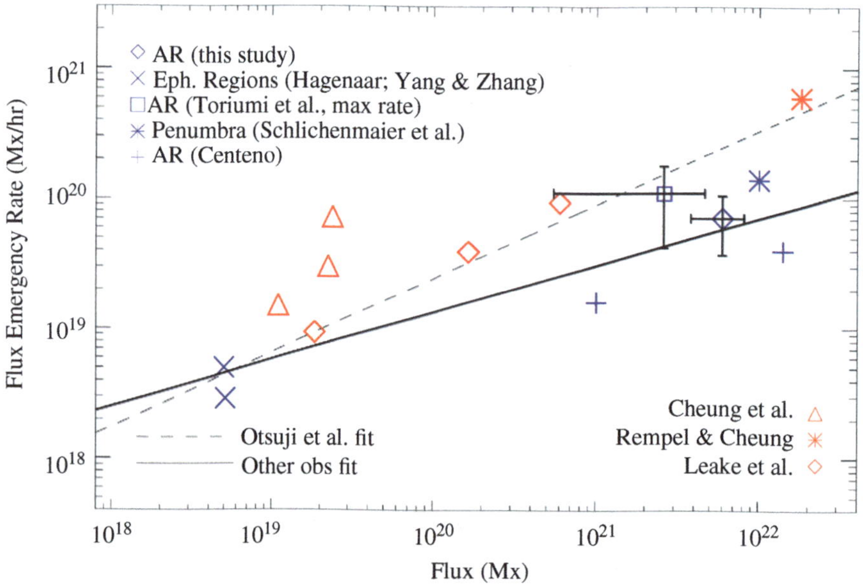

Figure 7.26 Magnetic flux emergence rate versus peak magnetic flux from a number of studies. Observations are represented by blue symbols: the Norton et al. (2017) study of 10 emerging regions is indicated by the diamond, the *x* symbols show data from Hagenaar (2001) and Yang and Zhang (2014), the square shows data measured by Toriumi et al. (2014a), and the plus symbol shows data measured by Centeno (2012). Simulations are represented by red symbols: the triangles show simulations from Cheung et al. (2007, 2008), the asterisks show simulation results from Rempel and Cheung (2014), and the diamonds show simulation data based on Leake et al. (2017). The dashed line shows a fit to observations made by Otsuji et al. (2011), whereas the solid line shows a fit to all of the observed data shown on this plot, made by Norton et al. (2017). (*Source*: From Norton et al., 2017. © 2017 IOP Publishing.)

(2011) find that flux emerges at a rate that depends on the peak flux to a power: $d\Phi / dt \propto \Phi^{B}_{max}$. Norton et al. (2017) found a power-law index of B = 0.36, whereas Otsuji et al. (2011) found a power-law index of B = 0.57. This again raises interesting questions about the flux emergence process. That is, what sets these emergence rates, and why are they well described by a power-law function of the total emerging flux?

Critical to the rate of emergence, and possibly to the form that emergence takes, is the speed at which active region magnetic flux rises to the surface prior to emergence. This is difficult to measure, as photon signatures are not detectable below the surface, and so helioseismology methods must be used to infer rise speed. Ilonidis et al. (2011) reported a helioseismological detection of active region flux at about 65 Mm below the surface one to two days prior to the eventual surface detection of the associated flux. This implied a rise speed for that last 65 Mm of ~500 m/s. This is consistent with the rise speeds reported by earlier helioseismology studies (see, e.g., the review by Kosovichev et al., 2000), but conflicts with the later analysis of Birch et al. (2016), which found a maximum rise speed of 150 m/s at 20 Mm below the photosphere. Further study is needed to more accurately determine the magnitude and range of magnetic flux rise speed prior to active region formation, both to elucidate the dynamics occurring in the convection zone, and to impose constraints on the rise speeds that should be explored in numerical models (see Section 7.3.2).

7.3.2. Theory and Modeling of Flux Emergence

Thin Flux Tube Studies

The emergence of magnetic fields to form sunspots at the solar photosphere is the most obvious manifestation of the solar magnetic field. However, the mechanisms that generate these magnetic fields in the convection zone below and the means by which they rise to the surface and emerge are critically important to understanding these surface manifestations. For a more extensive review of the theory of magnetic field evolution in the convection zone and of magnetic flux emergence, see Fan (2009) and Cheung and Isobe (2014). The predominant theories explaining the sources of these magnetic fields are described in the review of solar dynamo theories in Section 7.1. As discussed there, most dynamo models predict that the source of sunspot fields lies at the base of the convection zone in a toroidal band of field stored in the shear layer at that depth. Therefore, significant research efforts have been invested in modeling how these magnetic fields rise to the surface to form sunspots. As discussed above, observations indicate that many sunspot pairs can be explained by invoking the emergence of a coherent loop of magnetic field from the convection zone through the photosphere. The logical question to investigate then was how a coherent loop of magnetic field would rise from the base of the convection zone to the surface. Until recently, this was prohibitively expensive to model computationally for fully resolved 3D flux loops. However, significant early progress was made by invoking the so-called thin flux tube approximation. Spruit (1981) used this approximation to develop a set of one-dimensional thin flux tube equations. This model first assumes that the radius of the flux rope being studied is significantly smaller than other length scales of importance to the problem, in particular the local curvature of the flux tube axis, and the local length scale of the stratified pressure gradients of the convection zone (i.e., the pressure scale height). Second, it assumes that the internal structures of the flux rope, for example, how the magnetic field and the gas pressure profiles vary over the cross section of the tube, can be represented by an average of those structures over the cross section. This model then gives a set of one-dimensional differential equations for the motion of the flux tube axis through the stratified background convection zone, and for the evolution of the velocity parallel to the tube axis, the magnetic field parallel to the axis, the density, and the temperature (Spruit, 1981).

A critical feature of this, and of more sophisticated 2D and 3D models of magnetic flux tube dynamics in the convection zone, is that these flux tubes are naturally buoyant (Parker, 1955a). This follows from the fact that magnetic fields exert forces on plasmas, and in particular, a magnetic field has an effective pressure that is proportional to the square of the magnetic field. Thus, for a magnetic flux tube to be in pressure equilibrium with its surroundings, the plasma plus magnetic pressure inside the flux tube must be equal to the plasma pressure outside the flux tube. The plasma pressure inside the tube is therefore reduced with respect to the plasma pressure outside the tube. If one assumes that (given enough time) a flux tube settles into temperature equilibrium with its surroundings, then the ideal gas law, where pressure is proportional to the product of the temperature and density, requires that the flux tube be less dense than its surroundings. A less dense flux tube will therefore be buoyant and will start to rise up through the convection zone. Assuming adiabatic expansion of the tube as it rises into the lower-pressure layers of an approximately adiabatically stratified convection zone, the tube remains buoyant and can in theory rise all the way to the surface if no other forces disturb its rise.

These one-dimensional models had significant success in explaining many of the large-scale features of sunspot emergence described above. In particular, by including the effect of solar rotation, they showed how the Coriolis effect will alter the trajectory and shape of rising flux tubes. These simulations explored how flux tubes rising

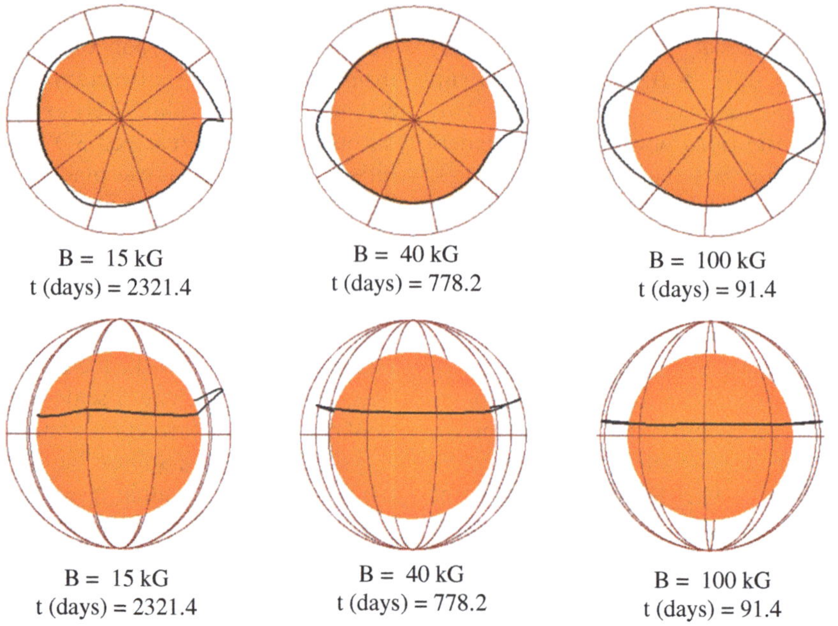

Figure 7.27 Figures showing the shape of a rising thin flux tube in a 3D convection zone, without convection, from simulations by Weber et al. (2011). The solid black line shows the thin flux tube, at a time when its apex nears the top of the convection zone, for three simulations: an initial magnetic field strength of 15 kG (left panels), 40 kG (center panels), and 100 kG (right panels). The orange sphere in each image shows a sphere just above the base of the convection zone. In the top three panels, the viewpoint is from the solar north pole, with the Sun rotating counterclockwise. In the bottom three panels, the viewpoint is from the equator. (*Source*: Figure reproduced from Weber et al., 2011).

with different initial field strengths, and therefore buoyancies, and from different latitudes can explain the observed latitudinal ranges over which sunspots emerge (Fan et al., 1993; Schüssler et al., 1994). D'Silva and Choudhuri (1993) and Fan et al. (1994) showed how rising flux tubes will be rotated out of their initial plane of symmetry by the Coriolis effect. That is, for flux tubes initially lying parallel to the equatorial plane, the leading side of the rising loop will rotate toward the equator, whereas the following side of the rising loop will rotate away from the equator. This introduces a tilt with respect to the equator of the tubes at their apex that will have a left-handed sense in the northern hemisphere and a right-handed sense in the southern hemisphere, causing the leading spots in both hemispheres to emerge closer to the equator than the following spots. D'Silva and Choudhuri (1993) and Fan et al. (1994) showed that the magnitude of the tilt will increase with emergence latitude, in agreement with the observations. A number of authors then predicted that flows will be set up along the rising flux tube axis, counter to the sense of solar rotation, with the result that the magnetic field strength in the leading spot will be more concentrated, and therefore stronger, than that in the following spot (Caligari et al., 1995; Fan et al., 1993; Moreno-Insertis et al., 1994). This may explain the observed asymmetries in leading versus following spots, where the leading spot is often more concentrated and coherent than the following spot. An example of several of these thin flux tube simulations is shown in Figure 7.27, from Weber et al. (2011), which shows snapshots of rising tubes encircling the Sun from simulations starting with 15 kG, 40 kG, and 100 kG field strengths. The top three panels in this figure are shown looking down from solar north, so that solar rotation is counterclockwise from this perspective. This perspective shows how the leading side of the loop slopes gradually up to the surface, whereas the following side of the loop has a steeper slope up to the surface, as the counter-rotational flow pushes the loop in toward this following side.

Subsequent studies then focused on the effects that convection and turbulence have on such thin flux tubes as they pass through the convection zone. Longcope and Klapper (1997) derived a set of equations to describe how the internal twist of a flux tube, that is, the rotation of flux tube's field lines about its axis, would be modified if the flux tube itself were "writhed," that is, distorted into a helical shape. Longcope et al. (1998) then investigated how thin flux tubes would be modified by this effect, called the "Σ effect," when exposed to the turbulent spectrum present in the convection zone. They found that the

Σ effect can explain the magnitude and spread in the observed twist measured in sunspots at the photosphere, as well as the dependence of the average twist on latitude (Pevtsov et al., 1995). Weber et al. (2011, 2013) then took the analysis further by simulating the evolution of thin flux tubes, modeled by the equations derived in Spruit (1981), as they move through a 3D realization of the dynamic convection zone, simulated with the ASH code (Miesch et al., 2006). With this, Weber et al. (2011, 2013) were able to study how the emergence latitude and the tilt of the emerging regions depended on the initial field strength. A key conclusion from this study was that flux tubes with initial field strengths of 40–50 kG were best able to reproduce Joy's law and the observed asymmetries between leading and following spots.

Two-Dimensional and 3D Buoyancy Studies

In addition to the behavior of the large-scale features of buoyant magnetic flux tubes, the internal structure and dynamics of these flux tubes are also important, in part because they can modify the large-scale dynamics. Schüssler (1979) explored the effect of hydrodynamic forces on 2D flux tubes. In addition to applying a drag on the buoyant rise of these flux tubes, as described in Parker (1979), the counter-rotating eddies generated by the flow of fluid around a tube as it rises through the convection zone can break it up into a pair of tubes. Longcope et al. (1996) followed up on this study with a series of 2D simulations, one of which is shown in Figure 7.28, exploring how quickly such a flux tube will break up, and showed that the previously buoyant flux tube will lose its buoyancy once it breaks up, as the pair of counter-rotating flux tubes generated by this splitting exert a downward drag on each other.

Emonet and Moreno-Insertis (1998) then explored how magnetic twist in a buoyant 2D flux tube could maintain the tube's coherence against these destructive vortex flows. Such twisted flux tubes are often referred to as flux ropes. Here the twist field, wrapping around the flux rope axis in a helical manner, acts as a set of hoops holding the flux rope together via the inward directed tension force of these twisted fields. Their simulations showed that the transition from coherent to incoherent tubes occurs when the Alfvén speed of the twist field exceeds the terminal rise speed of the flux tube, that is, the speed at which the buoyancy force equals the drag force. Figure 7.29 shows one of these simulations where a significant section of the flux tube maintains its coherence against the destructive eddies: here, the outer parts of the flux rope with weaker field strength and therefore smaller Alfvén speed are sloughed off by the flow, but the inner, stronger parts of the flux rope maintain their coherence and continue to rise.

Abbett et al. (2000, 2001) then explored several 3D effects that can also help buoyant convection zone flux

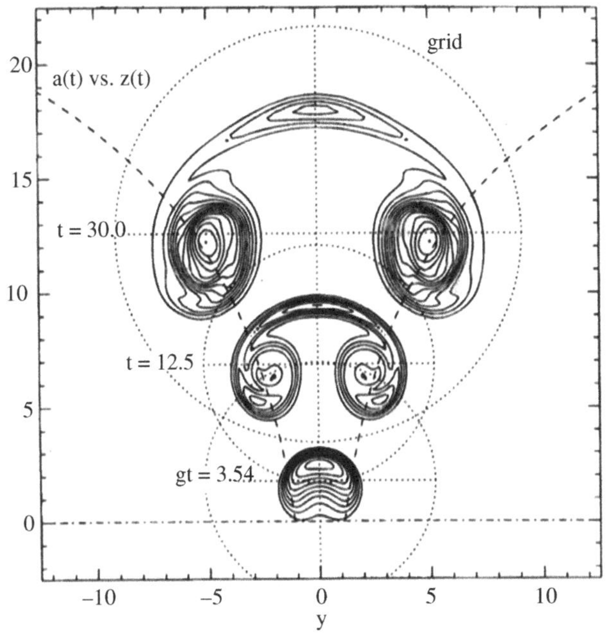

Figure 7.28 Two-dimensional simulation of a rising magnetic flux tube. The solid lines show contours of constant under-density (buoyancy), generated by the displacement of plasma by magnetic pressure in the flux tube. The three snapshots at t = 3.54, 12.5, and 30.0 show how the flux tube cross section is split into two counter-rotating flux tubes, joined by a thin section of flux tube arching over them. (*Source*: From Longcope et al., 1996. © 1996 IOP Publishing.)

tubes maintain their coherence against these hydrodynamic forces. Abbett et al. (2000) simulated the rise of an initially horizontal flux tube where only a segment near the center of the length of the flux tube was made buoyant. This flux tube, as shown in Figure 7.30 then rises with an Ω shape. They found that this tube will remain coherent if it has sufficient twist (left panel), but that it will break into two tubes if there is no twist in the tube (right panel), as it did in the 2D simulations of Schüssler (1979) and Longcope et al. (1996). Interestingly, they found that the tubes remain more coherent in all cases in 3D Ω-shaped tubes relative to equivalent 2D rising tubes, and that the coherence increases as the radius of curvature of the buoyant Ω-loop decreases. The authors conclude that this coherence for short buoyant sections was due to the twisting up of tubes during breakup, which effectively adds the stabilizing magnetic twist component explored by Emonet and Moreno-Insertis (1998). Abbett et al. (2001) then followed up this study with an exploration of the effects of the Coriolis force on rising flux tubes. Here, they found that the extra flows induced in the flux tube by the Coriolis force could also keep a rising, 3D flux tube coherent.

Jouve and Brun (2007, 2009), Fan (2008), and Jouve et al. (2013) carried out a series of global-Sun simulations

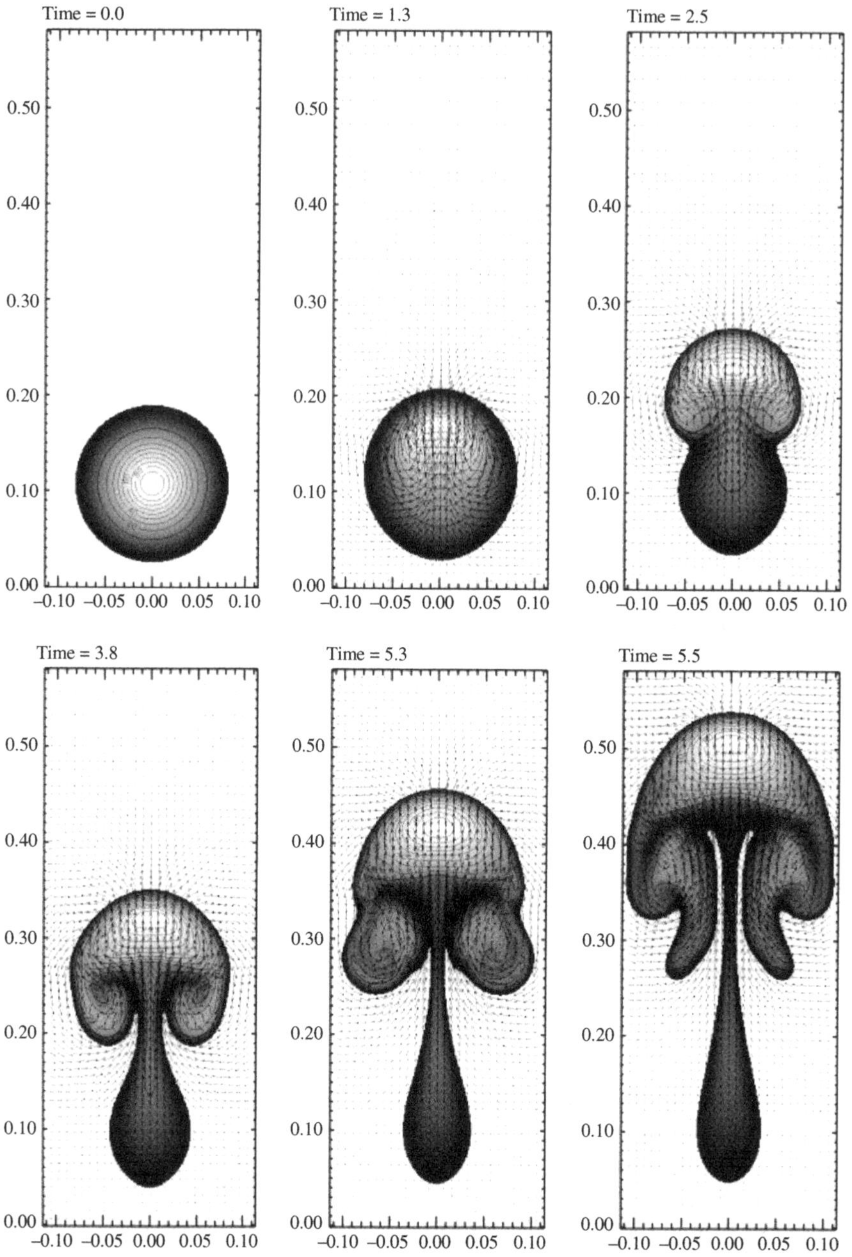

Figure 7.29 Two-dimensional simulation of the rise of a buoyant, twisted flux tube through the stratified convection zone. The axial magnetic field strength is shown by the grayscale and the contour lines. The vectors show the local velocity. The twist in the tube keeps the center of the tube, where the twist tension is strongest, coherent during its rise. The outer parts of the flux tube, with weaker twist tension, are still pulled from the flux tube by the fluid flowing past the tube. (*Source*: From Emonet & Moreno-Insertis, 1998. © 1998 IOP Publishing.)

exploring how a 3D twisted flux tubes initially circling the Sun at the base of the convection zone would rise buoyantly toward the surface. Jouve and Brun (2007) and Fan (2008) studied the rise of flux tubes in a stable (non-convecting) solar interior, and confirmed that a minimum twist is required in these tubes for them to successfully reach the surface. Fan (2008) found that for even relatively high twist (relative to what is observed at the surface), tubes can lose about half of their flux as they rise to the surface. Several studies were then carried out to explore the effects of preexisting convection on these rising flux tubes. Figure 7.31, from Jouve and Brun (2009), shows a time sequence for one of these simulations. Here, over the course of 10 days, one can see how most of the flux rope rises to the surface, but also how the rope is significantly distorted by the convective upflows and downflows

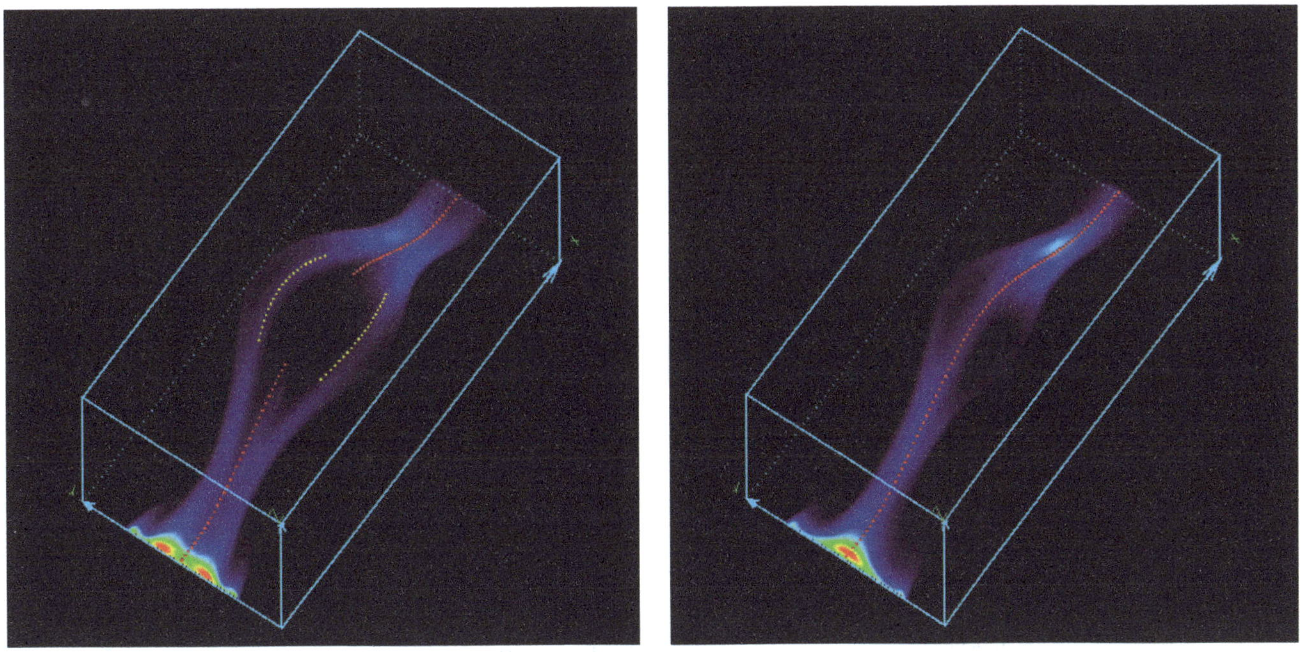

Figure 7.30 Snapshots from two simulations of 3D buoyant flux tubes rising through a stratified convection zone. The left panel shows a tube with no twist which breaks into two as it rises. The right panel shows a flux tube with a finite twist that stays coherent as it rises. (*Source*: Figure reproduced from Abbett et al., 2000.)

impinging upon it. In particular, the segment of the flux rope lying in an upflow (red arrow) rises faster and farther than the rest of the rope, whereas the segment lying below a downflow (black arrow) rises more slowly. These results are consistent with similar simulations carried out by Fan et al. (2003) on the dynamics of buoyant 3D flux tubes in a convecting, Cartesian atmosphere.

Emergence Through the Photosphere

Even after magnetic flux makes it to the photospheric surface, it still must undergo a further instability to emerge into the corona. This is because the photosphere is the transition between the unstably stratified convection zone and the stably stratified chromosphere and corona. For a review of these emergence instabilities and associated phenomena, see Archontis (2012). A flux tube that is buoyant enough to rise to the photosphere will lose its buoyancy there and stall just below the photosphere until enough flux builds up there to push it through the photosphere. Shibata et al. (1989) performed an early investigation of this mechanism, simulating the 2D evolution of a horizontal sheet of magnetic field that rises up from the high convection zone to the photosphere, and then later emerges to either populate a magnetic field-free corona, or to interact with a preexisting coronal field. Matsumoto et al. (1993, 1998) then extended this study to three dimensions, first exploring the interaction of multiple emerging flux bundles, and then exploring the formation of multiple active regions from a single, helically kinking flux tube, as shown in Figure 7.32.

Magara (2001) simulated the 2D emergence of the cross section of a flux tube, showing that the tube pinches down at the surface, with the bottom half of the flux tube piling up below the surface, and the top half of the flux tube emerging and expanding into the corona. Fan (2001) explored a similar scenario in 3D, simulating the emergence of a buoyant segment of a flux tube. This study showed that even in 3D, the bottom half of a twisted flux tube is trapped below the photosphere during emergence. Fan (2001) found that this is because dense convection zone material is trapped on the concave-up field lines that lie below the axis of a twisted flux tube. The result, shown in Figure 7.33, is that the axis of the flux rope lies below the surface, with the two legs of the flux rope moving off to the side to form a pair of sunspot-like magnetic field concentrations. Archontis et al. (2004) pointed out that the magnetic interchange instability (Acheson, 1979; Schubert, 1968) is a useful way to diagnose when a bundle of flux trapped below the photosphere will emerge. This instability is similar to the Rayleigh–Taylor instability, where a low-density plasma lies underneath and supports a higher-density plasma, but in this case, the lower-density plasma is a low-density, high-magnetic-field plasma. Toriumi and Yokoyama (2010, 2011) explored this effect in detail, focusing on how initially buoyant magnetic fields will be halted at the photosphere, and then in a second

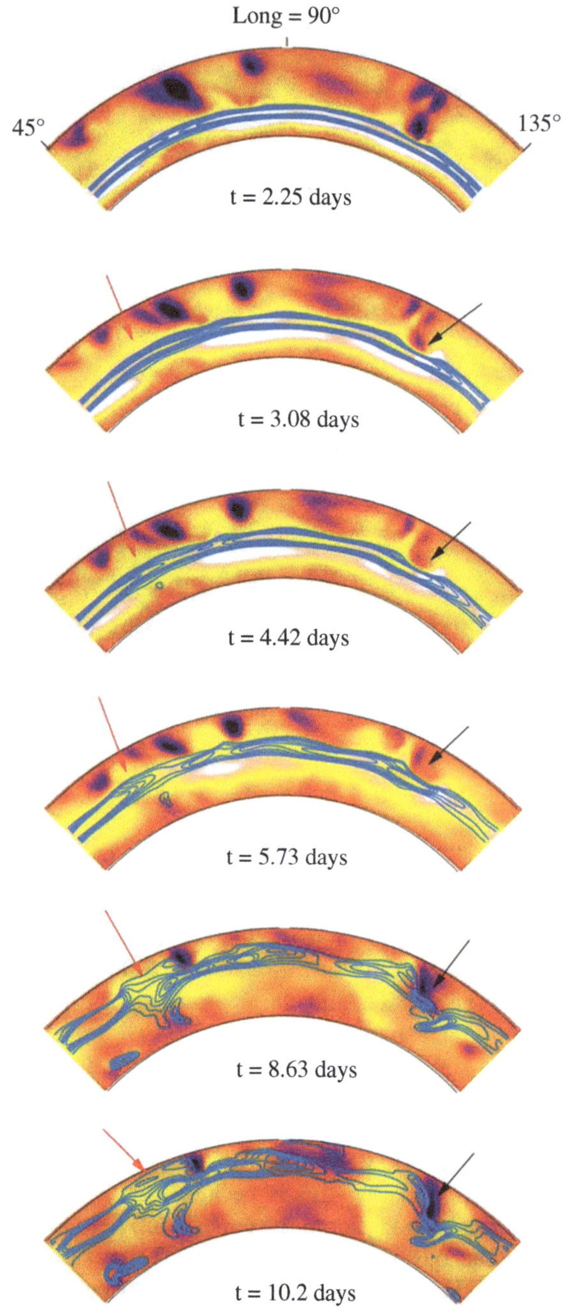

Figure 7.31 Simulation of a flux tube rising through a rotating convection zone over the course of ten days. The color scale shows radial flow: blue for downflow and yellow for upflow. The blue contour lines show contours of magnetic energy, indicating the location of the magnetic flux tube. The black arrow indicates a downflow, and the red arrow indicates an upflow. (*Source*: From Jouve & Brun, 2009. © 2009 IOP Publishing.)

stage will become buoyant again and emerge into the corona. Hood et al. (2009) and MacTaggart and Hood (2009) explored scenarios that would allow more of the flux tube cross section to emerge. They hypothesized that the emergence of a tube with a significantly arched axis could allow the convection zone plasma to drain out of an emerging flux tube, even below the axis. The idea here was that the downward arch of the tube itself would be larger than the upward arch of the field lines within the tube, and so there would be no concave-up field lines to trap dense plasma during the emergence. To test this, they simulated the emergence of an initially toroidal twisted flux tube, with a higher curvature than tested before. These simulations showed that the axis of the flux tube then easily emerges, confirming their hypothesis.

The question of why some fields emerge and some do not is important for understanding the process of sunspot formation, and for understanding how much of the magnetic field populating the convection zone eventually manifests itself in the corona. Murray et al. (2006) and Murray and Hood (2008) attacked this problem by carrying out a parameter study investigating the effect of magnetic field strength and magnetic twist on the efficiency of emergence. They showed that flux tubes with larger field strength are better able to emerge, as do flux ropes with larger twist. Lower field strength or twist in a buoyant flux tube results in it being trapped for a longer time below the photosphere, and sometimes these tubes are unable to emerge. Murray et al. (2006) postulated that this failure to emerge at low field strength was due to the field strength being too weak to excite the interchange instability discussed above (Acheson, 1979). Similarly, for weak twist, this instability is not excited, now because the weak twist is ineffective at keeping the magnetic field coherent in the face of high-pressure fluid and plasma flow forces, and so the field is dispersed and its strength ends up too weak even when it initially was quite strong. Archontis et al. (2013) then revisited this experiment and found that flux tubes with a significantly lower twist than the limit established by Murray et al. (2006) could emerge, but that they emerge on much longer timescales, and in a less coherent fashion. Low twist emergence, shown in Figure 7.34, proceeds via the emergence of two side lobes of the arched tube, whereas higher twist tubes first emerge at the apex of the Ω loop.

Partial Ionization

Whereas the convection zone and the overlying corona are hot enough that the plasma there is nearly entirely ionized, the thin layer, about 1000 km thick, of the photosphere and chromosphere between them is cool enough that neutral plasma can be the dominant species. Several efforts have investigated how these neutral species affect emerging flux. The primary mechanism expected to be in play is the extra diffusion term from the neutral species interacting with the ionized species. The neutral species can flow across magnetic field lines, whereas the ionized species are bound to magnetic field lines, so a friction is

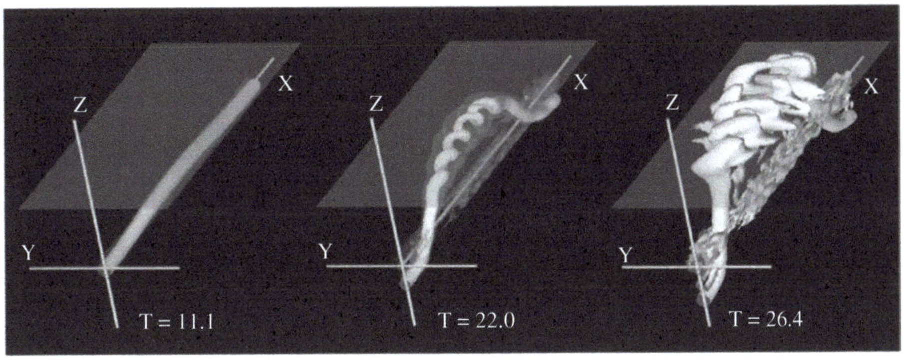

Figure 7.32 Simulation of the rise and emergence of a kinking magnetic flux rope. The gray plane in each panel is the solar surface. Two isosurfaces are shown in each panel, one for weak field (magnetic pressure over gas pressure is 0.04) and one for strong field (magnetic pressure over gas pressure is 0.36). This shows the tube both arching in an Ω shape and kinking as it rises and emerges. (*Source*: Figure reproduced from Matsumoto et al., 1998.)

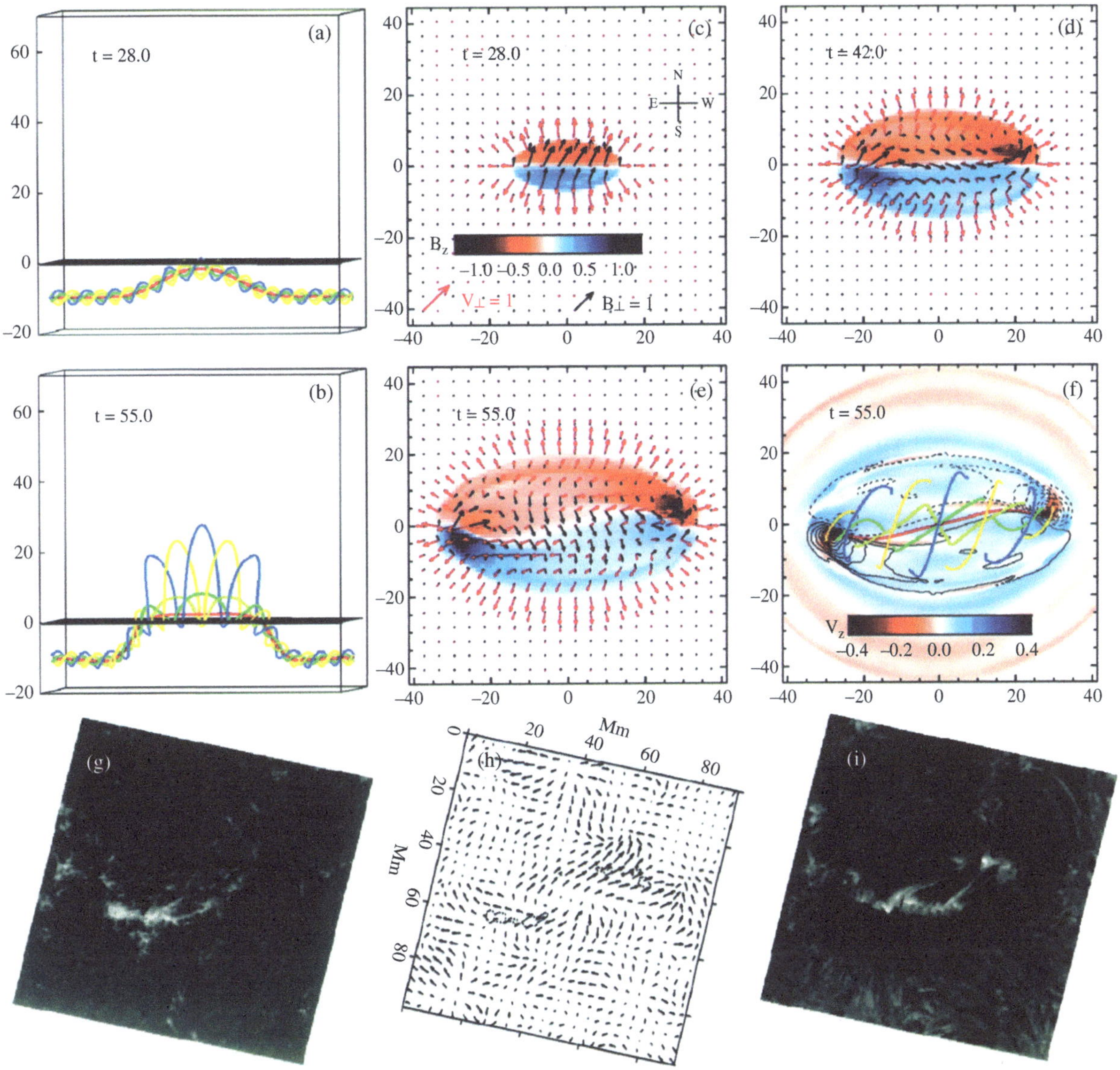

Figure 7.33 Simulation (top two rows) of the rise and emergence of an Ω loop formed by a buoyant magnetic flux rope in the solar convection zone. Field lines are shown in Panels (a) and (b), showing how the concave-up field lines below the flux rope axis are trapped at the photosphere (horizontal plane). Panels (c–e) show the vertical magnetic field at the photosphere during this emergence (color scale) as well as the horizontal magnetic field (black vectors) and the horizontal velocity (red vectors). Panel (f) shows the same field lines as Panel (b) but from above the photosphere. Panels (g–i) show an example of the observed features of an emerged active region: the line-of-sight magnetic field (g), the horizontal velocity field (h), and an image in the hydrogen α wavelength (i), from Strous et al. (1996). (*Source*: Figure reproduced from Fan, 2001.)

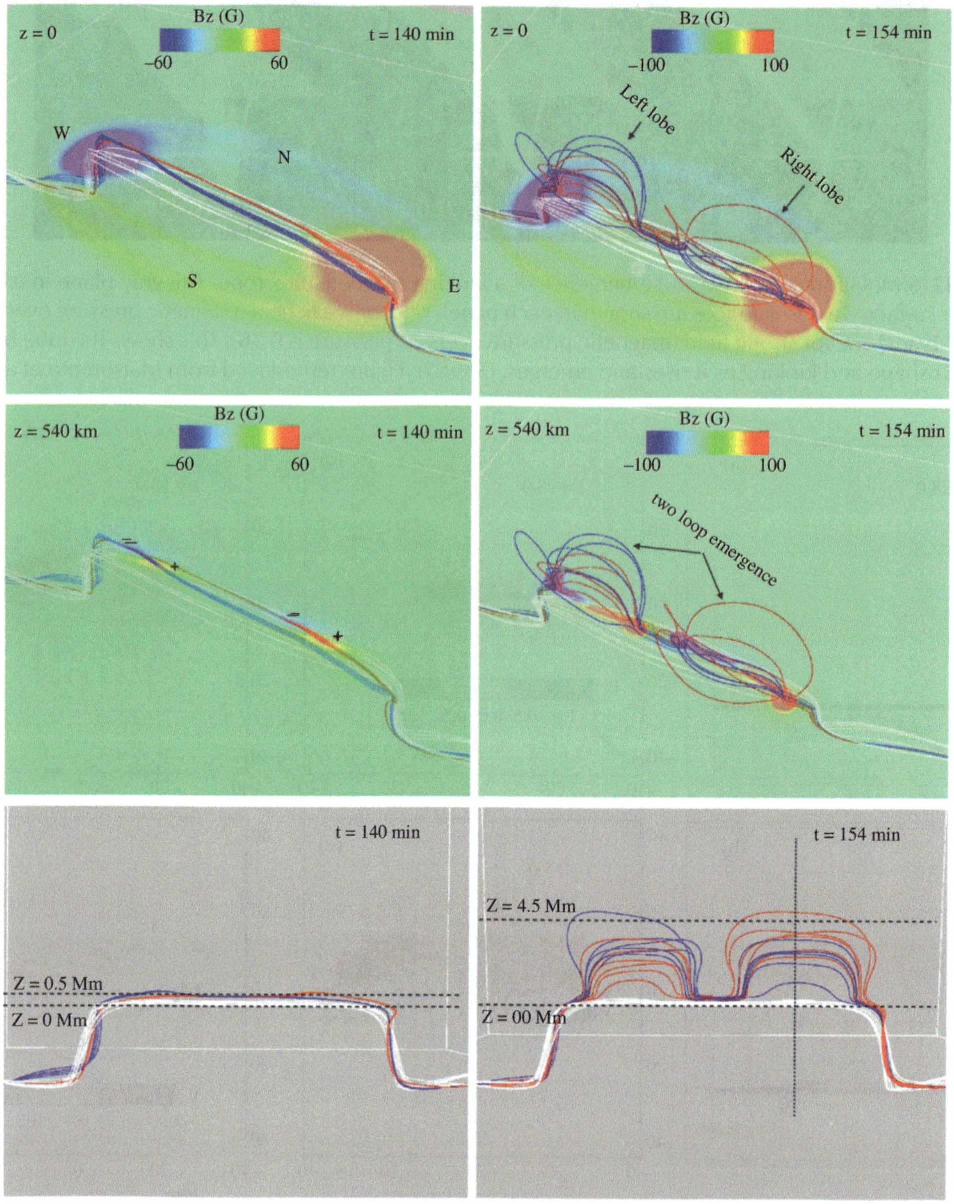

Figure 7.34 Simulation of the emergence of a very low twist flux rope into the corona. The left panels show the flux rope after it has reached and piled up at the photosphere (t =140 min). The right panels show the flux rope after two lobes have emerged (t = 154 min). The vertical magnetic field is shown by the color scale: in the top panels, this is shown at the solar surface (z = 0), and in the center panels, this is shown 540 km above the surface. The field lines are plotted as the colored lines. The bottom panel shows the side view. (*Source*: Figure reproduced from Archontis et al., 2013.)

generated between the two, resulting in the so-called Pedersen resistivity (Cowling, 1956; see, e.g., Leake et al., 2014a, for a review of this resistivity, and of the related Cowling resistivity). Leake and Arber (2006) and Arber et al. (2007) included this resistive term in a set of flux emergence simulations, and found that the extra resistivity allows fields to emerge more quickly, while also reducing the magnetic forces in the emerging structures. Leake and Linton (2013) later showed that in 2D configurations (where the plasma cannot drain along the tube's axis), this diffusion term provides an efficient way for excess convection zone plasma to drain from an emerging flux rope. Martínez-Sykora et al. (2015) then showed that the extra Pedersen conductivity plays an important role in heating the chromosphere during flux emergence, partially overcoming the excessive cooling seen in the chromosphere in fully ionized flux emergence simulations, such as that reported in Archontis et al. (2004).

δ-Spots

As discussed above, most active regions emerge to form well-separated pairs of sunspots, aligned nearly parallel to the equator, but a small number emerge in a compact, highly rotated formation called a δ-spot formation. Correspondingly, most flux emergence studies aim to explain the classic bipolar, well-separated regions. But flux emergence studies also provide a promising way to explore the formation of δ-spots. Following the hypotheses of Tanaka (1991) that these spots may be formed by the helical deformation of an emerging flux rope, Linton et al. (1996, 1998, 1999) and Fan et al. (1998, 1999) explored the implications of the helical kink instability for δ-spot characteristics. These studies explored the helical deformation that acts on twisted flux tubes with a twist in excess of the kink stability threshold (see, e.g., Shafranov, 1957; Kruskal et al., 1958). Linton et al. (1999) simulated the formation of a kinked, knotted structure in a convection zone environment, and argued that if this structure were to emerge into the corona, it would form a compact sunspot, due to its knotted structure, and that it would rotate significantly with respect to the usual Hale orientation, due to the helical writhe of the kink. Fan et al. (1999) showed that this compact distorted structure would naturally form, even for initially stable structures, as a flux rope rose through the convection zone and expanded, as predicted by Linton et al. (1996). Matsumoto et al. (1998), Takasao et al. (2015), Toriumi and Takasao (2017), and Knizhnik et al. (2018) carried this study further, exploring the photospheric and coronal signatures of kinking flux tubes when they were allowed to emerge through the photosphere. Figure 7.35, from Takasao et al. (2015), shows how the emerging structure is indeed compact and rotated, with a high level

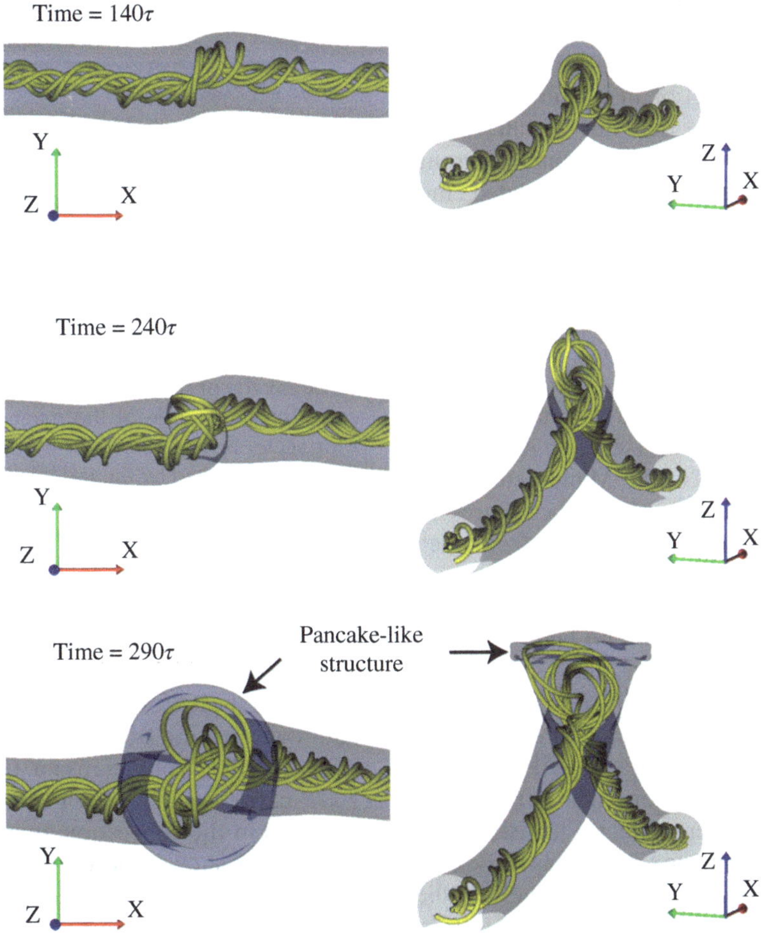

Figure 7.35 Simulation of the emergence of a kinking flux rope from the convection zone to form a δ-spot like active region. Isosurfaces are shown in each panel at a value of half of the initial peak magnetic field strength in the tube. The yellow lines show magnetic field lines within the tube. Times 140τ and 240τ show how the rise of the tube in an initial Ω shape is distorted into a helical, writhed shape by the kink instability. The bottom panels, at 290τ, show the structure as it hits and emerges through the surface in a compact, rotated configuration, consistent with δ-spot characteristics. (*Source*: Figure reproduced from Takasao et al., 2015.)

of twist, consistent with many δ-spot structures. Toriumi et al. (2014b), Fang and Fan (2015), Toriumi and Takasao (2017), and Jouve et al. (2018) explored alternate scenarios for forming these complex structures wherein multiple flux ropes interact with each other during their rise through the convection zone, or where multiple parts of a single flux rope collide with each other prior to emergence. These studies also found a good correspondence between key aspects of these structures and common types of δ-spots.

Emergence into Coronal Fields

A number of explorations have focused on how emerging fields interact with preexisting coronal fields. Heyvaerts et al. (1977) proposed and explored the theoretical consequences of a mechanism by which newly emerging magnetic flux would interact with preexisting coronal fields to generate a reconnecting current sheet and drive flaring, plasma heating, and particle acceleration. Yokoyama and Shibata (1995) explored such a mechanism numerically by simulating the emergence of a loop from a horizontal sheet of magnetic flux into an inclined coronal field, finding that this initiates reconnection and jet-like flows up into the corona. Archontis et al. (2004) and Galsgaard et al. (2005) simulated emergence into a horizontal overlying field, finding that although much of the flux rope is trapped below the photosphere, sufficient flux emerges to generate current sheets and reconnection at the interface between the emerging flux and the coronal field. This interaction of emerging fields with both a field-free corona and with overlying coronal fields has since led to numerous models of coronal flares and eruptions (see, e.g., Archontis & Török, 2008; Leake et al., 2014b; Manchester et al., 2004). Chapter 5 explores in more detail these flares and eruptions resulting from the emergence of magnetic fields into the corona.

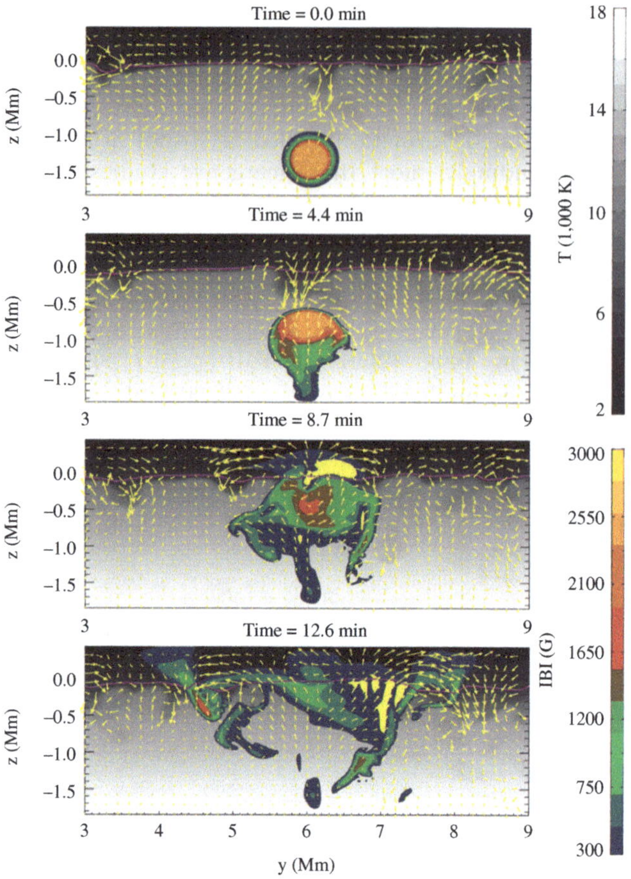

Figure 7.36 Two-dimensional cuts from a 3D simulation of the rise of a twisted magnetic flux tube through convective flows just below the photosphere. Temperature is shown by the grayscale. The magnitude of the magnetic field is shown by the color scale, flows are shown by the yellow vectors, and the surface is indicated by the red line. Time advances from top to bottom, showing how the buoyant tube rises to the surface and emerges, all the while being torn apart by the convective flows. (*Source*: Figure reproduced from Cheung et al., 2007.)

Effects of Radiation and Convection on Emergence

Convective motions, as discussed above, can play an important role in the evolution of rising flux tubes. Dorch et al. (2001) first investigated the interaction of buoyant, rising flux tubes with surrounding convective motions. They found that even for a flux rope that in theory had sufficient twist to remain coherent during its rise, convective flows strip off fields from the flux rope as it rises. Fan et al. (2003) simulated a similar scenario, and established that the magnetic energy density of buoyant flux tubes must be roughly three times the kinetic energy of convective downflows for the tubes to successfully rise toward the surface.

A number of studies have investigated how the transition from the convecting atmosphere below the photosphere to the stable chromosphere and corona affects and interacts with emerging flux. To achieve this transition, radiative effects must be included in simulations, so that the loss of heat from the photosphere can generate the required transition from a convectively unstable to a convectively stable zone across a thin layer. Cheung et al. (2007) carried out one such study, investigating both the dynamics in the convection zone and during emergence. Figure 7.36 shows one of these simulations where the tube is significantly distorted by the convective motions, but still reaches the surface to emerge. Cheung et al. (2007) and Cheung et al. (2008) then showed that these emerging fields can disrupt the convective flow pattern seen at the surface, consistent with observations of emerging flux (see also Martínez-Sykora et al., 2008, 2009; Tortosa-Andreu and Moreno-Insertis, 2009; Fang et al., 2012). Stein et al. (2011) showed that such emerging flux

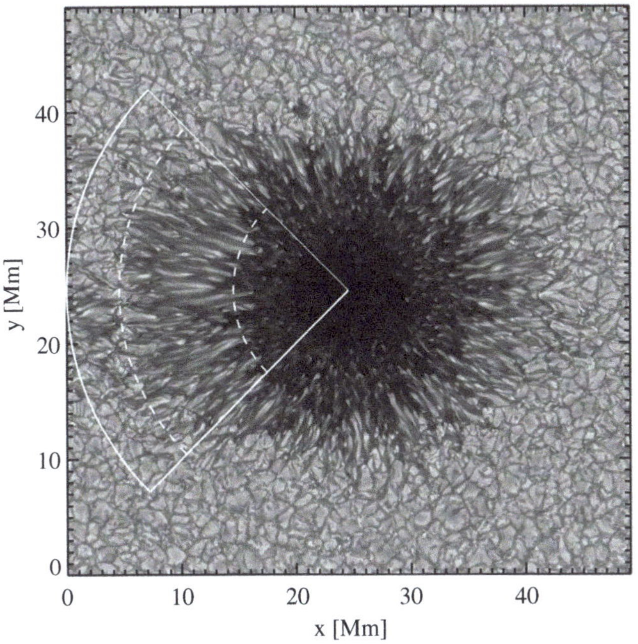

Figure 7.37 Simulated white-light emission from modeled sunspot formation in a convecting atmosphere. These simulations by Rempel et al. (2009a,b), Rempel (2011), where a magnetic field is added to a convecting atmosphere to generate a sunspot, produce realistic umbral and penumbral structure and emission. (*Source*: Figure reproduced from Rempel, 2011.)

can concentrate at the surface during its emergence to form small-scale, high-field strength regions often seen in observations, and referred to as pores. See Stein (2012b) and Stein (2012a) for reviews of the interactions of convection and magnetic fields.

These earlier convective emergence simulations focused on smaller-scale fields, which disrupted the convection but did not form sunspot-like concentrations. Larger-scale simulations then allowed for sunspot formation to be explored. Figure 7.37 shows simulated white-light intensity emission from a sunspot simulated by Rempel et al. (2009a, 2009b) and analyzed in Rempel (2011). This sunspot structure is generated via the ad hoc insertion of a predominantly vertical magnetic field into a simulation in which convection has previously been set up. The structure and emission of the resulting sunspot umbra and penumbra display a remarkable resemblance to sunspot observations, such as that shown in Figure 7.21. Figure 7.38 shows a simulated surface magnetogram and a vertical cut of the magnetic field strength below that magnetogram for a more self-consistently formed sunspot pair that is generated by the emergence into the corona of a toroidal flux tube injected at 8,000 km below the photosphere, at the bottom boundary of a convecting simulation. Rempel and Cheung (2014) then extended this type of simulation to investigate both the formation of a sunspot via convective flux emergence and the decay

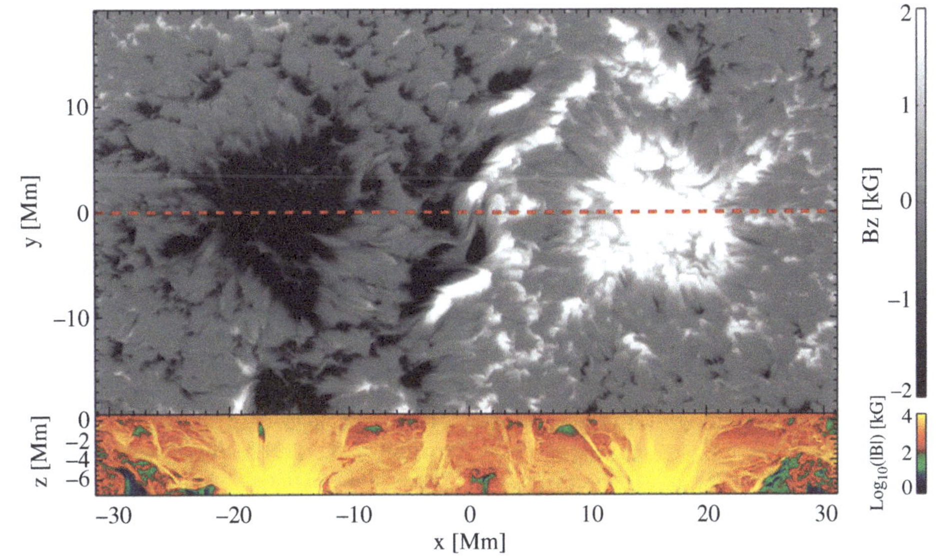

Figure 7.38 Synthetic magnetogram from a simulation of the formation of an active region sunspot pair from the emergence of magnetic flux from the convection zone. The grayscale shows the vertical magnetic field at the photosphere. The color scale shows the log of the magnetic field strength in a cut into the convection zone, perpendicular to the photosphere, made where the dashed red line crosses the photospheric plot. (*Source*: Figure reproduced from Cheung et al., 2010.)

of the sunspot via convective turbulence. Critically, in this case the torus injected at the bottom boundary had zero twist, but its field still rose to the surface to form a sunspot pair. In this case, the field was caught up in strong updrafting convection flows, and so those flows carried the flux to the surface. This contrasts with the non-convecting cases discussed in Section 7.3.2.2 where an untwisted tube is torn apart as it rises through a largely static atmosphere and with the convecting examples discussed above, where convection that flows counter to the rise of flux trying to emerge tears it apart rather than enabling emergence. Clearly, convection can both enable and inhibit flux emergence, and determining when and how often it will play either of these roles is a promising avenue for further research.

A statistical study of how magnetic fields are distributed across the photosphere in emerging active regions by Dacie et al. (2016, 2017) found that convective simulations such as these accurately reproduce the observed spectral distributions of magnetic fields during their evolution, whereas simulations of flux emergence without convection are only able to match magnetic field distributions once the emergence is mature. These convective simulations are remarkable in that they can reproduce statistical measures of sunspot characters such as this, and that they can produce magnetic field and white-light emission signatures that, to the eye, are largely indistinguishable from observed sunspot signatures.

These various explorations, from one-dimensional models focusing on the large-scale evolution of flux tubes, to 2D and 3D models focusing on the interaction of flux tubes with the plasma surrounding them and the internal reaction of flux tubes to their own twist, to highly sophisticated models investigating the effects of convection and radiation loss on emerging flux tubes have all contributed critical knowledge to our understanding of how magnetic fields emerge into the solar corona and form sunspots and active regions. Many important questions remain, and arguably all of these tools will have something to contribute to advancing our understanding to the next stage.

ACKNOWLEDGMENTS

Funding for M. Dikpati was provided by the National Center for Atmospheric Research, which is sponsored by the National Science Foundation. R. Howe acknowledges computing support from the National Solar Observatory and the University of Birmingham. Funding for M. G. Linton was provided by the NASA Living with a Star and Heliophysics Supporting Research programs and by the Office of Naval Research. SDO data was provided courtesy of SDO (NASA) and the HMI and AIA consortium. WSO data was provided via the website http://wso.stanford.edu at courtesy of J. T. Hoeksema. The Wilcox Solar Observatory is currently supported by NASA. This work utilizes data obtained by the GONG program, managed by the National Solar Observatory, which is operated by AURA, Inc., under a cooperative agreement with the National Science Foundation. The data were acquired by instruments operated by the Big Bear Solar Observatory, High Altitude Observatory, Learmonth Solar Observatory, Udaipur Solar Observatory, Instituto de Astrofísica de Canarias, and Cerro Tololo Interamerican Observatory. BiSON is funded by the UK Science and Technology Facilities Council (STFC). This review has made use of NASA's Astrophysics Data System.

M. G. Linton thanks Aimee Norton for useful feedback on Section 7.3.1.

REFERENCES

Abbett, W. P., Fisher, G. H., & Fan, Y. (September 2000). The three-dimensional evolution of rising, twisted magnetic flux tubes in a gravitationally stratified model convection zone. *Astrophysical Journal*, *540*, 548–562. doi: 10.1086/309316.

Abbett, W. P., Fisher, G. H., & Fan, Y. (2001, January). The effects of rotation on the evolution of rising omega loops in a stratified model convection zone. *Astrophysical Journal*, *546*, 1194–1203. doi: 10.1086/318320.

Acheson, D. J. (1979, May). Instability by magnetic buoyancy. *Solar Physics*, *62*, 23–50. doi: 10.1007/BF00150129.

Antia, H. M., & Basu, S. (2011, July). Revisiting the solar tachocline: Average properties and temporal variations. *The Astrophysicsal Journal Letters*, *735*, L45. doi: 10.1088/2041-8205/735/2/L45.

Antia, H. M., Basu, S., Hill, F., Howe, R., Komm, R. W., & Schou, J. (2001, November). Solar-cycle variation of the sound-speed asphericity from GONG and MDI data 1995-2000. *Monthly Notices of the Royal Astronomical Society*, *327*, 1029–1040. doi: 10.1046/j.1365-8711.2001.04819.x.

Antia, H. M., & Basu, S. (2000). Temporal variations of the rotation rate in the solar interior. *The Astrophysical Journal*, *541*, 442–448. doi: 10.1086/309421.

Antia, H. M., & Basu, S. (2001). Temporal variations of the solar rotation rate at high latitudes. *The Astrophysical Journal Letters*, *559*, L67–L70. doi: 10.1086/323701.

Arber, T. D., Haynes, M., & Leake, J. E. (2007, September). Emergence of a flux tube through a partially ionized solar atmosphere. *Astrophysical Journal*, *666*, 541–546. doi: 10.1086/520046.

Archontis, V. (2012, July). Magnetic flux emergence and associated dynamic phenomena in the Sun. *Philosophical Transactions of the Royal Society of London Series A*, *370*, 3088–3113. doi: 10.1098/rsta.2012.0001.

Archontis, V., & Török, T. (2008, December). Eruption of magnetic flux ropes during flux emergence. *Astronomy & Astrophysics*, *492*, L35–L38. doi: 10.1051/0004-6361:200811131.

Archontis, V., Moreno-Insertis, F., Galsgaard, K., Hood, A., & O'Shea, E. (2004, November). Emergence of magnetic flux from the convection zone into the corona. *Astronomy & Astrophysics*, *426*, 1047–1063. doi: 10.1051/0004-6361:20035934.

Archontis, V., Hood, A. W., & Tsinganos, K. (2013, November). The emergence of weakly twisted magnetic fields in the Sun. *Astrophysical Journal*, *778*, 42. doi: 10.1088/0004-637X/778/1/42.

Augustson, K., Brun, A. S., Miesch, M., & Toomre, J. (2015, August). Grand minima and equatorward propagation in a cycling stellar convective dynamo. *Astrophysical Journal*, *809*, 149. doi: 10.1088/0004-637X/809/2/149.

Babcock, H. W. (1961, March). The topology of the Sun's magnetic field and the 22-year cycle. *Astrophysical Journal*, *133*, 572. doi: 10.1086/147060.

Backus, G., & Gilbert, F. (1968, October). The Resolving Power of Gross Earth Data. *Geophysical Journal*, *16*, 169–205. doi: 10.1111/j.1365-246X.1968.tb00216.x.

Backus, G., & Gilbert, F. (1970, March). Uniqueness in the inversion of inaccurate gross Earth data. *Philosophical Transactions of the Royal Society of London Series A*, *266*, 123–192. doi: 10.1098/rsta.1970.0005.

Bahcall, J. N., Basu, S., Pinsonneault, M., & Serenelli, A. M. (2005, January). Helioseismological Implications of Recent Solar Abundance Determinations. *The Astrophysical Journal*, *618*, 1049–1056. doi: 10.1086/426070.

Baldner, C. S., & Basu, S. (2008, October). Solar cycle related changes at the base of the convection zone. *The Astrophysical Journal*, *686*, 1349–1361. doi: 10.1086/591514.

Ball, W. H., & Gizon, L. (2014, August). A new correction of stellar oscillation frequencies for near-surface effects. *Astronomy and Astrophysics*, *568*, A123. doi: 10.1051/0004-6361/201424325.

Barekat, A., Schou, J., & Gizon, L. (2014, October). The radial gradient of the near-surface shear layer of the Sun. *Astronomy and Astrophysics*, *570*, L12. doi: 10.1051/0004-6361/201424839.

Basu, S., & Antia, H. M. (1997, May). Seismic measurement of the depth of the solar convection zone. *Monthly Notices of the Royal Astronomical Society*, *287*, 189–198. doi: 10.1093/mnras/287.1.189.

Basu, S., & Antia, H. M. (2001, June). A study of possible temporal and latitudinal variations in the properties of the solar tachocline. *Monthly Notices of the Royal Astronomical Society*, *324*, 498–508. doi: 10.1046/j.1365-8711.2001.04364.x.

Basu, S., & Antia, H. M. (2008, March). Helioseismology and solar abundances. *Physics Reports*, *457*, 217–283,. doi: 10.1016/j.physrep.2007.12.002.

Basu, S., Chaplin, W. J., Elsworth, Y., New, R., & Serenelli, A. M. (2009, July). Fresh insights on the structure of the solar core. *The Astrophysical Journal*, *699*, 1403–1417. doi: 10.1088/0004-637X/699/2/1403.

Basu, S. (2016, August). Global seismology of the Sun. *Living Reviews in Solar Physics*, *13*(1), 2. ISSN 1614-4961. doi: 10.1007/s41116-016-0003-4. URL https://doi.org/10.1007/s41116-016-0003-4.

Belucz, B., Dikpati, M., & Forgács-Dajka, E. (2015, June). A Babcock-Leighton solar dynamo model with multi-cellular meridional circulation in advection- and diffusion-dominated regimes. *Astrophysical Journal*, *806*, 169. doi: 10.1088/0004-637X/806/2/169.

Berger, T. E., & Title, A. M. (1996, May). On the dynamics of small-scale solar magnetic elements. *Astrophysical Journal*, *463*, 365. doi: 10.1086/177250.

Birch, A. C., Schunker, H., Braun, D. C., Cameron, R., Gizon, L., Lo ptien, B., et al. (2016, July). A low upper limit on the subsurface rise speed of solar active regions. *Science Advances*, *2*(7), e1600557–e1600557. doi: 10.1126/sciadv.1600557.

Bonanno, A., Elstner, D., Rüdiger, G., & Belvedere, G. (2002, August). Parity properties of an advection-dominated solar alpha 2 Omega-dynamo. *Astronomy & Astrophysics*, *390*, 673–680. doi: 10.1051/0004-6361:20020590.

Bray, R. J., & Loughhead, R. E. (1964). *Sunspots*. Dover, New York.

Brown, B. P., Browning, M. K., Brun, A. S., Miesch, M. S., & Toomre, J. (2010, March). Persistent magnetic wreaths in a rapidly rotating Sun. *Astrophysical Journal*, *711*, 424–438. doi: 10.1088/0004-637X/711/1/424.

Brown, T. M., Christensen-Dalsgaard, J., Dziembowski, W. A., Goode, P., Gough, D. O., & Morrow, C. A. (1989, August). Inferring the Sun's internal angular velocity from observed p-mode frequency splittings. *The Astrophysical Journal*, *343*, 526–546. doi: 10.1086/167727.

Browning, M. K., Miesch, M. S., Brun, A. S., & Toomre, J. (2006, September). Dynamo action in the solar convection zone and tachocline: Pumping and organization of toroidal fields. *Astrophysical Journal Letters*, *648*, L157–L160. doi: 10.1086/507869.

Brun, A. S., Miesch, M. S., & Toomre, J. (2004, October). Global-scale turbulent convection and magnetic dynamo action in the solar envelope. *Astrophysical Journal*, *614*, 1073–1098. doi: 10.1086/423835.

Caligari, P., Moreno-Insertis, F., & Schüssler, M. (1995). Emerging flux tubes in the solar convection zone. i. Asymmetry, tilt and emergence latitude. *Astrophysical Journal*, *441*, 886–902.

Cally, P. S. (2001, April). Nonlinear evolution of 2d tachocline instabilities. *Solar Physics*, *199*, 231–249. doi: 10.1023/A:1010390814663.

Cally, P. S. (2003, March). Three-dimensional magneto-shear instabilities in the solar tachocline. *Monthly Notices of the Royal Astronomical Society*, *339*, 957–972. doi: 10.1046/j.1365-8711.2003.06236.x.

Cally, P. S., Dikpati, M., & Gilman, P. A. (2003, January). Clamshell and tipping instabilities in a two-dimensional magnetohydrodynamic tachocline. *Astrophysical Journal*, *582*, 1190–1205. doi: 10.1086/344746.

Carrington, R. (1858, November). On the distribution of the solar spots in latitudes since the beginning of the year 1854, with a map. *Monthly Notices of the Royal Astronomical Society*, *19*, 1.

Cattaneo, F. (1999, April). On the origin of magnetic fields in the quiet photosphere. *Astrophysical Journal Letters*, *515*, L39–L42. doi: 10.1086/311962.

Cattaneo, F., & Tobias, S. M. (2014, July). On large-scale dynamo action at high magnetic Reynolds number. *Astrophysical Journal*, *789*, 70. doi: 10.1088/0004-637X/789/1/70.

Centeno, R. (2012, November). The naked emergence of solar active regions observed with SDO/HMI. *Astrophysical Journal*, *759*, 72. doi: 10.1088/0004-637X/759/1/72.

Chaplin, W. J., Christensen-Dalsgaard, J., Elsworth, Y., Howe, R., Isaak, G. R., Larsen, R. M., et al. (1999, September). Rotation of the solar core from BiSON and LOWL frequency observations. *Monthly Notices of the Royal Astronomical Society*, *308*, 405–414. doi: 10.1046/j.1365-8711.1999.02691.x.

Charbonneau, P. (2010, September). Dynamo models of the solar cycle. *Living Reviews in Solar Physics*, *7*, 3. doi: 10.12942/lrsp-2010-3.

Charbonneau, P. (2014, August). Solar dynamo theory. *Annual Reviews of Astronomy and Astrophysics*, *52*, 251–290. doi: 10.1146/annurev-astro-081913-040012.

Charney, J. G. (1947, October). The dynamics of long waves in a baroclinic westerly current. *Journal of Atmospheric Sciences*, *4*, 136–162. doi: 10.1175/1520-0469(1947)004⟨0136:TDOLWI⟩2.0.CO;2.

Chatterjee, P., Nandy, D., & Choudhuri, A. R. (2004, December). Full-sphere simulations of a circulation-dominated solar dynamo: Exploring the parity issue. *Astronomy & Astrophysics*, *427*, 1019–1030. doi: 10.1051/0004-6361:20041199.

Chen, R., & Zhao, J. (2017, November). A comprehensive method to measure solar meridional circulation and the center-to-limb effect using time-distance helioseismology. *The Astrophysical Journal*, *849*, 144. doi: 10.3847/1538-4357/aa8eec.

Cheung, M. C. M., & Isobe, H. (2014, July). Flux emergence (theory). *Living Reviews in Solar Physics*, *11*, 3.

Cheung, M. C. M., Schüssler, M., & Moreno-Insertis, F. (2007, May). Magnetic flux emergence in granular convection: Radiative MHD simulations and observational signatures. *Astronomy & Astrophysics*, *467*, 703–719. doi: 10.1051/0004-6361:20077048.

Cheung, M. C. M., Schüssler, M., Tarbell, T. D., & Title, A. M. (2008, November). Solar surface emerging flux regions: A comparative study of radiative MHD modeling and Hinode SOT observations. *Astrophysical Journal*, *687*, 1373–1387. doi: 10.1086/591245.

Cheung, M. C. M., Rempel, M., Title, A. M., & Schüssler, M. (2010, September). Simulation of the formation of a solar active region. *Astrophysical Journal*, *720*, 233–244. doi: 10.1088/0004-637X/720/1/233.

Chintzoglou, G., & Zhang, J. (2013, February). Reconstructing the subsurface three-dimensional magnetic structure of a solar active region using SDO/HMI observations. *Astrophysical Journal Letters*, *764*, L3. doi: 10.1088/2041-8205/764/1/L3.

Choudhuri, A. R., Schüssler, M., & Dikpati, M. (1995, November). The solar dynamo with meridional circulation. *Astronomy & Astrophysics*, *303*, L29.

Choudhuri, A. R., Chatterjee, P., & Jiang, J. (2007, March). Predicting solar cycle 24 with a solar dynamo model. *Physical Review Letters*, *98*(13), 131103. doi: 10.1103/PhysRevLett.98.131103.

Christensen-Dalsgaard, J. (1986). Theoretical aspects of helio- and asteroseismology. In D. O. Gough, (Ed.), *NATO Advanced Science Institutes (ASI) Series C* (Vol. *169*, pp. 23–53). *NATO Advanced Science Institutes (ASI) Series C.*

Christensen-Dalsgaard, J., Dappen, W., Ajukov, S. V., Anderson, E. R., Antia, H. M., Basu, S., et al. (1996, May). The current state of solar modeling. *Science*, *272*, 1286–1292. doi: 10.1126/science.272.5266.1286.

Corbard, T., & Thompson, M. J. (2002, February). The subsurface radial gradient of solar angular velocity from MDI f-mode observations. *Solar Physics*, *205*, 211–229. doi: 10.1023/A:1014224523374.

Cowling, T. G. (1956). The dissipation of magnetic energy in an ionized gas. *Monthly Notices of the Royal Astronomical Society*, *116*, 114. doi: 10.1093/mnras/116.1.114.

Coyne, G. V., S. J. (1991). Sunspots: The historical background. In C. P. Sonett, M. S. Giampapa, & M. S. Matthews (Eds.), *The Sun in time* (pp. 3–10). The University of Arizona Press, Tucson, Arizona.

Cuypers, J. (1980, September). On the calculation of the frequency splitting of adiabatic nonradial stellar oscillations by slow differential rotation. *Astronomy and Astrophysics*, *89*, 207,.

Dacie, S., Démoulin, P., van Driel-Gesztelyi, L., Long, D. M., Baker, D., Janvier, M., et al. (2016, December). Evolution of the magnetic field distribution of active regions. *Astronomy & Astrophysics*, *596*, A69. doi: 10.1051/0004-6361/201628948.

Dacie, S., van Driel-Gesztelyi, L., Démoulin, P., Linton, M. G., Leake, J. E., MacTaggart, D., et al. (2017, October). Field distribution of magnetograms from simulations of active region formation. *Astronomy & Astrophysics*, *606*, A34,. doi: 10.1051/0004-6361/201730767.

Deubner, F.-L. (1975). Observations of low wavenumber nonradial eigenmodes of the Sun. *Astronomy and Astrophysics*, *44*, 371–375.

Deubner, F.-L., Ulrich, R. K., & Rhodes, E. J., Jr. (1979, February). Solar p-mode oscillations as a tracer of radial differential rotation. *Astronomy and Astrophysics*, *72*, 177–185.

Dikpati, M. (2012, February). Nonlinear evolution of global hydrodynamic shallow-water instability in the solar tachocline. *Astrophysical Journal*, *745*, 128. doi: 10.1088/0004-637X/745/2/128.

Dikpati, M., & Anderson, J. L. (2012, September). Evaluating potential for data assimilation in a flux-transport dynamo model by assessing sensitivity and response to meridional flow variation. *Astrophysical Journal*, *756*, 20. doi: 10.1088/0004-637X/756/1/20.

Dikpati, M., & Charbonneau, P. (1999, June). A Babcock-Leighton flux transport dynamo with solar-like differential rotation. *Astrophysical Journal*, *518*, 508–520. doi: 10.1086/307269.

Dikpati, M., & Gilman, P. A. (1999, February). Joint instability of latitudinal differential rotation and concentrated toroidal fields below the solar convection zone. *Astrophysical Journal*, *512*, 417–441. doi: 10.1086/306748.

Dikpati, M., & Gilman, P. A. (2001). September flux-transport dynamos with α-effect from global instability of tachocline differential rotation: A solution for magnetic parity selection in the Sun. *Astrophysical Journal*, *559*, 428–442. doi: 10.1086/322410.

Dikpati, M., & Gilman, P. A. (2005, December). A shallow-water theory for the Sun's active longitudes. *Astrophysical Journal Letters*, *635*, L193–L196,. doi: 10.1086/499626.

Dikpati, M., & Gilman, P. A. (2006, September). Simulating and predicting solar cycles using a flux-transport dynamo. *Astrophysical Journal*, *649*, 498–514. doi: 10.1086/506314.

Dikpati, M., de Toma, G., Gilman, P. A., Arge, C. N., & White, O. R. (2004, February). Diagnostics of polar field reversal in solar cycle 23 using a flux transport dynamo model. *Astrophysical Journal*, *601*, 1136–1151,. doi: 10.1086/380508.

Dikpati, M., de Toma, G., & Gilman, P. A. (2006, March). Predicting the strength of solar cycle 24 using a flux-transport dynamo-based tool. *Geophysical Research Letters*, *33*, L05102. doi: 10.1029/2005GL025221.

Dikpati, M., Gilman, P. A., de Toma, G., & Ghosh, S. S. (2007, September). Simulating solar cycles in northern and southern hemispheres by assimilating magnetic data into a calibrated flux-transport dynamo. *Solar Physics*, *245*, 1–17. doi: 10.1007/s11207-007-9016-4.

Dikpati, M., Gilman, P. A., de Toma, G., & Ulrich, R. K. (2010, July). Impact of changes in the Sun's conveyor-belt on recent solar cycles. *Geophysical Research Letters*, *37*, L14107. doi: 10.1029/2010GL044143.

Dikpati, M., Anderson, J. L., & Mitra, D. (2014, August). Ensemble Kalman filter data assimilation in a Babcock-Leighton solar dynamo model: An observation system simulation experiment for reconstructing meridional flow speed. *Geophysical Research Letters*, *41*, 5361–5369. doi: 10.1002/2014GL061077.

Dikpati, M., Anderson, J. L., & Mitra, D. (2016a, September). Data assimilation in a solar dynamo model using ensemble Kalman filters: Sensitivity and robustness in reconstruction of meridional flow speed. *Astrophysical Journal*, *828*, 91. doi: 10.3847/0004-637X/828/2/91.

Dikpati, M., Mitra, D., & Anderson, J. L. (2016b, October). Role of response time of a Babcock-Leighton solar dynamo model in meridional flow-speed reconstruction by EnKF data assimilation. *Advances in Space Research*, *58*, 1589–1595,. doi: 10.1016/j.asr.2016.08.004.

Dikpati, M., Cally, P. S., McIntosh, S. W., & Heifetz, E. (2017, November). The origin of the "seasons" in space weather. *Scientific Reports*, *7*, 14750. doi: 10.1038/s41598-017-14957-x.

Dikpati, M., Belucz, B., Gilman, P. A., & McIntosh, S. W. (2018a, August). Phase speed of magnetized rossby waves that cause solar seasons. *Astrophysical Journal*, *862*, 159. doi: 10.3847/1538-4357/aacefa.

Dikpati, M., McIntosh, S. W., Bothun, G., Cally, P. S., Ghosh, S. S., Gilman, P. A., et al. (2018b, February). Role of interaction between magnetic Rossby waves and Tachocline differential rotation in producing solar seasons. *Astrophysical Journal*, *853*, 144,. doi: 10.3847/1538-4357/aaa70d.

Dorch, S. B. F., Gudiksen, B. V., Abbett, W. P., & Nordlund, Å. (2001, December). Flux-loss of buoyant ropes interacting with convective flows. *Astronomy & Astrophysics*, *380*, 734–738. doi: 10.1051/0004-6361:20011397.

D'Silva, S., & Choudhuri, A. R. (1993). A theoretical model for tilts of bipolar magnetic regions. *Astronomy & Astrophysics*, *272*, 621–633.

Durney, B. R. (1995, September). On a Babcock-Leighton dynamo model with a deep-seated generating layer for the toroidal magnetic field. *Solar Physics*, *160*, 213–235. doi: 10.1007/BF00732805.

Duvall, T. L., Jr. (1979, August). Large-scale solar velocity fields. *Solar Physics*, *63*, 3–15. doi: 10.1007/BF00155690.

Duvall, T. L., Jr., Jefferies, S. M., Harvey, J. W., & Pomerantz, M. A. (1993, April). Time-distance helioseismology. *Nature*, *362*, 430–432. doi: 10.1038/362430a0.

Eady, E. T. (1949, August). Long waves and cyclone waves. *Tellus Series A*, *1*, 33–52. doi: 10.1111/j.2153-3490.1949.tb01265.x.

Eff-Darwich, A., Korzennik, S. G., & Jiménez-Reyes, S. J. (2002a, July). Inversion of the internal solar rotation rate. *The Astrophysical Journal*, *573*, 857–863. doi: 10.1086/340747.

Eff-Darwich, A., Korzennik, S. G., Jiménez-Reyes, S. J., & Pérez Hernández, F. (2002b, November). An upper limit on the temporal variations of the solar interior stratification. *The Astrophysical Journal*, *580*, 574–578. doi: 10.1086/343063.

Elsworth, Y., Howe, R., Isaak, G. R., McLeod, C. P., & New, R. (1990, May). Variation of low-order acoustic solar oscillations over the solar cycle. *Nature*, *345*, 322–324. doi: 10.1038/345322a0.

Emonet, T., & Moreno-Insertis, F. (1998, January). The physics of twisted magnetic tubes rising in a stratified medium: Two-dimensional results. *Astrophysical Journal*, *492*, 804–821. doi: 10.1086/305074.

Emonet, T., Cattaneo, F., & Weiss, N. O. (2001). Magnetoconvection. In D. Chossat, D. Armbruster, & I. Oprea (Eds.), *Dynamo and dynamics: A mathematical challenge* (pp. 173–180). Springer, Netherlands,.

Fan, Y. (2001, June). The emergence of a twisted Ω-tube into the solar atmosphere. *Astrophysical Journal Letters*, *554*, L111–L114. doi: 10.1086/320935.

Fan, Y. (2008, March). The Three-dimensional evolution of buoyant magnetic flux tubes in a model solar convective envelope. *Astrophysical Journal*, *676*, 680–697. doi: 10.1086/527317.

Fan, Y. (2009, December). Magnetic fields in the solar convection zone. *Living Reviews in Solar Physics*, *6*, 4. doi: 10.12942/lrsp-2009-4.

Fan, Y., & Fang, F. (2014, July). A simulation of convective dynamo in the solar convective envelope: Maintenance of the solar-like differential rotation and emerging flux. *Astrophysical Journal*, *789*, 35. doi: 10.1088/0004-637X/789/1/35.

Fan, Y., Fisher, G. H., & DeLuca, E. E. (1993). The origin of morphological asymmetries in bipolar active regions. *Astrophysical Journal*, *405*, 390–401.

Fan, Y., Fisher, G. H., & McClymont, A. N. (1994). Dynamics of emerging active region flux loops. *Astrophysical Journal*, *436*, 907–928.

Fan, Y., Zweibel, E. G., Linton, M. G., & Fisher, G. H. (1998, September). The rise of kink-unstable magnetic flux tubes in the solar convection zone. *Astrophysical Journal Letters*, *505*, L59–L63. doi: 10.1086/311597.

Fan, Y., Zweibel, E. G., Linton, M. G., & Fisher, G. H. (1999, August). The rise of kink-unstable magnetic flux tubes and the origin of δ-configuration sunspots. *Astrophysical Journal*, *521*, 460–477. doi: 10.1086/307533.

Fan, Y., Abbett, W. P., & Fisher, G. H. (2003, January). The dynamic evolution of twisted magnetic flux tubes in a three-dimensional convecting flow. I. Uniformly buoyant horizontal tubes. *Astrophysical Journal*, *582*, 1206–1219. doi: 10.1086/344798.

Fang, F., & Fan, Y. (2015, June). δ-Sunspot formation in simulation of active-region-scale flux emergence. *Astrophysical Journal*, *806*, 79. doi: 10.1088/0004-637X/806/1/79.

Fang, F., Manchester, W., IV, Abbett, W. P., & van der Holst, B. (2012, January). Dynamic coupling of convective flows and magnetic field during flux emergence. *Astrophysical Journal*, *745*, 37. doi: 10.1088/0004-637X/745/1/37.

Frazier, E. N. (1972, September). The magnetic structure of arch filament systems. *Solar Physics*, *26*(1), 130–141. doi: 10.1007/BF00155113.

Galsgaard, K., Moreno-Insertis, F., Archontis, V., & Hood, A. (2005, January). A Three-dimensional Study of Reconnection, Current Sheets, and Jets Resulting from Magnetic Flux Emergence in the Sun. *Astrophysical Journal Letters*, *618*, L153–L156. doi: 10.1086/427872.

Garaud, P. (2001, June). Latitudinal shear instability in the solar tachocline. *Monthly Notices of the Royal Astronomical Society*, *324*, 68–76. doi: 10.1046/j.1365-8711.2001.04245.x.

García, R. A., Turck-Chièze, S., Jiménez-Reyes, S. J., Ballot, J., Pallé, P. L., Eff-Darwich, A., et al. (2007, June). Tracking solar gravity modes: The dynamics of the solar core. *Science*, *316*, 1591. doi: 10.1126/science.1140598.

Gastine, T., Yadav, R. K., Morin, J., Reiners, A., & Wicht, J. (2014, February). From solar-like to antisolar differential rotation in cool stars. *Monthly Notices of the Royal Astronomical Society*, *438*, L76–L80. doi: 10.1093/mnrasl/slt162.

Giles, P. M., Duvall, T. L., Scherrer, P. H., & Bogart, R. S. (1997, November). A subsurface flow of material from the Sun's equator to its poles. *Nature*, *390*, 52–54. doi: 10.1038/36294.

Gilman, P. A. (1983, October). Dynamically consistent nonlinear dynamos driven by convection in a rotating spherical shell. II – Dynamos with cycles and strong feedbacks. *Astrophysical Journal Supplement*, *53*, 243–268. doi: 10.1086/190891.

Gilman, P. A., & Fox, P. A. (1997, July). Joint instability of latitudinal differential rotation and toroidal magnetic fields below the solar convection zone. *Astrophysical Journal*, *484*, 439–454. doi: 10.1086/304330.

Gilman, P. A., & Howe, R. (2003, February). Meridional motion and the slope of isorotation contours. In H. Sawaya-Lacoste (Ed.), *GONG+ 2002. Local and global helioseismology: The present and future* (Vol. 517, pp. 283–285). ESA Special Publication.

Glatzmaier, G. A. (1985, April). Numerical simulations of stellar convective dynamos. II – Field propagation in the convection zone. *Astrophysical Journal*, *291*, 300–307. doi: 10.1086/163069.

González Hernández, I., Patrón, J., Bogart, R. S., & T. SOI Ring Diagram Team (1999, January). Meridional flows from ring diagram analysis. *The Astrophysical Journal Letters*, *510*, L153–L156. doi: 10.1086/311811.

González Hernández, I., Komm, R., Kholikov, S., Howe, R., & Hill, F. (2011, January). Meridional flow observations: Implications for the current flux transport models. *GONG-SoHO 24: A New Era of Seismology of the Sun and Solar-Like Stars*, *271*(1), 012073. doi: 10.1088/1742-6596/271/1/012073.

Guerrero, G. A., & Muñoz, J. D. (2004, May). Kinematic solar dynamo models with a deep meridional flow. *Monthly Notices of the Royal Astronomical Society*, *350*, 317–322. doi: 10.1111/j.1365-2966.2004.07655.x.

Haber, D. A., Hindman, B. W., Toomre, J., Bogart, R. S., Thompson, M. J., & Hill, F. (2000, March). Solar shear flows deduced from helioseismic dense-pack samplings of ring diagrams. *Solar Physics*, *192*, 335–350. doi: 10.1023/A:1005235610132.

Hagenaar, H. J. (2001, July). Ephemeral regions on a sequence of full-disk Michelson Doppler imager magnetograms. *Astrophysical Journal*, *555*, 448–461. doi: 10.1086/321448.

Hale, G. E. (1912). The 150-Foot tower telescope of the Mount Wilson solar observatory. *The* Publications *of the Astronomical Society of the Pacific*, *24*, 223.

Hale, G. E., & Nicholson, S. B. (1925, November). The law of Sun-spot polarity. *Astrophysical Journal*, *62*, 270. doi: 10.1086/142933.

Hale, G. E., Ellerman, F., Nicholson, S. B., & Joy, A. H. (1919). The magnetic polarity of Sun-spots. *Astrophysical Journal*, *49*, 153.

Hale, S. J., Howe, R., Chaplin, W. J., Davies, G. R., & Elsworth, Y. P. (2016). Performance of the birmingham solar-oscillations network (bison). *Solar Physics*, *291*(1), 1–28,. ISSN 0038-0938. doi: 10.1007/s11207-015-0810-0. URL http://dx.doi.org/10.1007/s11207-015-0810-0.

Hansen, C. J., Cox, J. P., & van Horn, H. M. (1977, October). The effects of differential rotation on the splitting of nonradial modes of stellar oscillation. *The Astrophysical Journal*, *217*, 151–159. doi: 10.1086/155564.

Harvey, J. W., Hill, F., Hubbard, R., Kennedy, J. R., Leibacher, J. W., Pintar, J. A., et al. (1996). The Global Oscillation Network Group (GONG) project. *Science*, *272*, 1284–1286.

Harvey, K. L., & Martin, S. F. (1973, October). Ephemeral active regions. *Solar Physics*, *32*(2), 389–402. doi: 10.1007/BF00154951.

Hathaway, D. H. (2015, September). The solar cycle. Living Reviews in Solar Physics, *12*, 4. doi: 10.1007/lrsp-2015-4.

Hathaway, D. H., & Rightmire, L. (2011, March). Variations in the axisymmetric transport of magnetic elements on the Sun: 1996-2010. *Astrophysical Journal*, *729*, 80. doi: 10.1088/0004-637X/729/2/80.

Heyvaerts, J., Priest, E. R., & Rust, D. M. (1977, August). An emerging flux model for the solar phenomenon. *Astrophysical Journal*, *216*, 123–137.

Hill, F. (1988, October). Rings and trumpets – Three-dimensional power spectra of solar oscillations. *Astrophysical Journal*, *333*, 996–1013. doi: 10.1086/166807.

Hindman, B., Haber, D., Toomre, J., & Bogart, R. (2000, March). Local fractional frequency shifts used as tracers of magnetic activity. *Solar Physics*, *192*, 363–372. doi: 10.1023/A:1005283302728.

Hood, A. W., Archontis, V., Galsgaard, K., & Moreno-Insertis, F. (2009, September). The emergence of toroidal flux tubes from beneath the solar photosphere. *Astronomy & Astrophysics*, *503*, 999–1011. doi: 10.1051/0004-6361/200912189.

Hotta, H., & Yokoyama, T. (2010, February). Importance of surface turbulent diffusivity in the solar flux-transport dynamo. *Astrophysical Journal*, *709*, 1009–1017. doi: 10.1088/0004-637X/709/2/1009.

Howard, R., & Labonte, B. J. (1980). The Sun is observed to be a torsional oscillator with a period of 11 years. *The Astrophysical Journal Letters*, *239*, L33–L36. doi: 10.1086/183286.

Howe, R., Christensen-Dalsgaard, J., Hill, F., Komm, R. W., Larsen, R. M., Schou, J., et al. (2000). Deeply penetrating banded zonal flows in the solar convection zone. *The Astrophysical Journal Letters*, *533*, L163–L166. doi: 10.1086/312623.

Howe, R., Komm, R. W., & Hill, F. (2002, December). Localizing the solar cycle frequency shifts in global p-modes. *The Astrophysical Journal*, *580*, 1172–1187. doi: 10.1086/343892.

Howe, R., Christensen-Dalsgaard, J., Hill, F., Komm, R., Schou, J., & Thompson, M. J. (2005, December). Solar convection-zone dynamics, 1995-2004. *The Astrophysical Journal*, *634*, 1405–1415. doi: 10.1086/497107.

Howe, R., Christensen-Dalsgaard, J., Hill, F., Komm, R., Schou, J., & Thompson, M. J. (2009, August). A note on the torsional oscillation at solar minimum. *The Astrophysical Journal Letters*, *701*, L87–L90. doi: 10.1088/0004-637X/701/2/L87.

Howe, R., Hill, F., Komm, R., Christensen-Dalsgaard, J., Larson, T. P., Schou, J., et al. (2011, January). The torsional oscillation and the new solar cycle. *Journal of Physics Conference Series*, *271*(1), 012074,. doi: 10.1088/1742-6596/271/1/012074.

Howe, R., Christensen-Dalsgaard, J., Hill, F., Komm, R., Larson, T. P., Rempel, M., et al. (2013, April). The high-latitude branch of the solar torsional oscillation in the rising phase of cycle 24. *The Astrophysical Journal Letters*, *767*, L20. doi: 10.1088/2041-8205/767/1/L20.

Howe, R., Hill, F., Komm, R., Chaplin, W. J., Elsworth, Y., Davies, G. R., et al. (2018, July). Signatures of solar cycle 25 in subsurface zonal flows. *The Astrophysical Journal Letters*, *862*, L5. doi: 10.3847/2041-8213/aad1ed.

Howe, R. (2009, February). Solar interior rotation and its variation. Living Reviews in Solar Physics, 6(1), 1. ISSN 1614-4961. doi: 10.12942/lrsp-2009-1. URL https://doi.org/10.12942/lrsp-2009-1.

Hung, C. P., Jouve, L., Brun, A. S., Fournier, A., & Talagrand, O. (2015, December). Estimating the deep solar meridional circulation using magnetic observations and a dynamo model: A variational approach. *Astrophysical Journal*, *814*, 151. doi: 10.1088/0004-637X/814/2/151.

Ilonidis, S., Zhao, J., & Kosovichev, A. (2011, August). Detection of emerging sunspot regions in the solar interior. *Science*, *333*(6045), 993. doi: 10.1126/science.1206253.

Jackiewicz, J., Serebryanskiy, A., & Kholikov, S. (2015, June). Meridional flow in the solar convection zone. II. Helioseismic inversions of GONG data. *The Astrophysical Journal*, *805*, 133. doi: 10.1088/0004-637X/805/2/133.

Jaeggli, S. A., & Norton, A. A. (2016, March). The magnetic classification of solar active regions 1992-2015. *Astrophysical Journal Letters*, *820*, L11. doi: 10.3847/2041-8205/820/1/L11.

Jouve, L., & Brun, A. S. (2007, December). 3-D non-linear evolution of a magnetic flux tube in a spherical shell: The isentropic case. *Astronomische Nachrichten*, *328*, 1104,. doi: 10.1002/asna.200710887.

Jouve, L., & Brun, A. S. (2009, August). Three-dimensional nonlinear evolution of a magnetic flux tube in a spherical shell: Influence of turbulent convection and associated mean flows. *Astrophysical Journal*, *701*, 1300–1322. doi: 10.1088/0004-637X/701/2/1300.

Jouve, L., Brun, A. S., Arlt, R., Brandenburg, A., Dikpati, M., Bonanno, A., et al. (2008, June). A solar mean field dynamo benchmark. *Astronomy & Astrophysics*, *483*, 949–960,. doi: 10.1051/0004-6361:20078351.

Jouve, L., Brun, A. S., & Aulanier, G. (2013, January). Global dynamics of subsurface solar active regions. *Astrophysical Journal*, *762*, 4. doi: 10.1088/0004-637X/762/1/4.

Jouve, L., Gastine, T., & Lignières, F. (2015, March). Three-dimensional evolution of magnetic fields in a differentially rotating stellar radiative zone. *Astronomy & Astrophysics*, *575*, A106. doi: 10.1051/0004-6361/201425240.

Jouve, L., Brun, A. S., & Aulanier, G. (2018, April). Interactions of twisted Ω-loops in a model solar convection zone. *Astrophysical Journal*, *857*, 83. doi: 10.3847/1538-4357/aab5b6.

Kalnay, E. (2003). *Atmospheric modeling, data assimilation and predictability*. Cambridge Univ. Press, Cambridge,.

Kholikov, S., Serebryanskiy, A., & Jackiewicz, J. (2014a, April). Meridional flow in the solar convection zone. I. Measurements from GONG data. *The Astrophysical Journal*, *784*, 145. doi: 10.1088/0004-637X/784/2/145.

Kholikov, S., Serebryanskiy, A., & Jackiewicz, J. (2014b, April). Meridional flow in the solar convection zone. I. Measurements from GONG data. *Astrophysical Journal*, 784(2), 145. doi: 10.1088/0004-637X/784/2/145.

Knizhnik, K. J., Linton, M. G., & DeVore, C. R. (2018, September). The role of twist in kinked flux rope emergence and delta-spot formation. *Astrophysical Journal*, *864*, 89. doi: 10.3847/1538-4357/aad68c.

Kosovichev, A. G., Schou, J., Scherrer, P. H., Bogart, R. S., Bush, R. I., Hoeksema, J. T., et al. (1997). Structure and rotation of the solar interior: Initial results from the MDI medium-L program. *Solar Physics*, *170*, 43–61. doi: 10.1023/A:1004949311268.

Kosovichev, A. G., Jr. Duvall, T. L. _Jr., & Scherrer, P. H. (March 2000). Time-distance inversion methods and results – (Invited review). *Solar Physics*, *192*, 159–176. doi: 10.1023/A:1005251208431.

Kosovichev, A. G., & Schou, J. (1997). Detection of zonal shear flows beneath the Sun's surface from f-mode frequency splitting. *The Astrophysical Journal Letters*, *482*, L207–L210. doi: 10.1086/310708.

Kovesi, P. (2015, September). Good colour maps: How to design them. *arXiv e-prints*, art. arXiv:1509.03700,.

Kruskal, M. D., Johnson, J. L., Gottlieb, M. D., & Goldman, L. M. (1958). Hydromagnetic Instability in a Stellarator. *Phys. Fluids*, *1*, 421.

Künzel, H. (1960). Die Flare-Häufigkeit in Fleckengruppen unterschiedlicher Klasse und magnetischer Struktur. *Astronomische Nachrichten*, *285*, 271.

Kurokawa, H. (1991). Optical observations of flare-productive flux emergence. *Lecture Notes in Physics*, *387*, 39.

Labonte, B. J., & Howard, R. (1982, October). Solar rotation measurements at Mount Wilson. III – Meridional flow and limbshift. *Solar Physics*, *80*, 361–372. doi: 10.1007/BF00147982.

Lawson, N., Strugarek, A., & Charbonneau, P. (2015, November). Evidence of active MHD instability in EULAG-MHD simulations of solar convection. *Astrophysical Journal, 813*, 95. doi: 10.1088/0004-637X/813/2/95.

Leake, J. E., & Arber, T. D. (2006, May). The emergence of magnetic flux through a partially ionised solar atmosphere. *Astronomy & Astrophysics, 450*, 805–818. doi: 10.1051/0004-6361:20054099.

Leake, J. E., & Linton, M. G. (2013). February Effect of Ion-Neutral Collisions in Simulations of Emerging Active Regions. *Astrophysical Journal, 764*, 54,. doi: 10.1088/0004-637X/764/1/54.

Leake, J. E., DeVore, C. R., Thayer, J. P., Burns, A. G., Crowley, G., Gilbert, H. R., et al. (2014a, November). Ionized plasma and neutral gas coupling in the Sun's chromosphere and Earth's ionosphere/thermosphere. *Space Science Reviews, 184*, 107–172. doi: 10.1007/s11214-014-0103-1.

Leake, J. E., Linton, M. G., & Antiochos, S. K. (2014b, May). Simulations of emerging magnetic flux. II. The formation of unstable coronal flux ropes and the initiation of coronal mass ejections. *Astrophysical Journal, 787*, 46. doi: 10.1088/0004-637X/787/1/46.

Leake, J. E., Linton, M. G., & Schuck, P. W. (2017, April). Testing the accuracy of data-driven MHD simulations of active region evolution. *Astrophysical Journal, 838*, 113. doi: 10.3847/1538-4357/aa6578.

Leighton, R. B. (1964, November). Transport of magnetic fields on the Sun. *Astrophysical Journal, 140*, 1547. doi: 10.1086/148058.

Leighton, R. B. (1969, April). A magneto-kinematic model of the solar cycle. *Astrophysical Journal, 156*, 1. doi: 10.1086/149943.

Leka, K. D., Canfield, R. C., McClymont, A. N., & van Driel Gesztelyi, L. (1996). Evidence for current-carrying emerging flux. *Astrophysical Journal, 462*, 547.

Liang, Z.-C., Gizon, L., Birch, A. C., Duvall, T. L., & Rajaguru, S. P. (2018, November). Solar meridional circulation from twenty-one years of SOHO/MDI and SDO/HMI observations. Helioseismic travel times and forward modeling in the ray approximation. *Astronomy and Astrophysics, 619*, A99. doi: 10.1051/0004-6361/201833673.

Libbrecht, K. G., & Woodard, M. F. (1990, June). Solar-cycle effects on solar oscillation frequencies. *Nature, 345*, 779–782. doi: 10.1038/345779a0.

Lin, H., & Rimmele, T. (1999, March). The granular magnetic fields of the quiet Sun. *Astrophysical Journal, 514*, 448–455. doi: 10.1086/306925.

Linton, M. G., Longcope, D. W., & Fisher, G. H. (1996, October). The helical kink instability of isolated, twisted magnetic flux tubes. *Astrophysical Journal, 469*, 954. doi: 10.1086/177842.

Linton, M. G., Dahlburg, R. B., Fisher, G. H., & Longcope, D. W. (1998, November). Nonlinear evolution of kink-unstable magnetic flux tubes and solar δ-spot active regions. *Astrophysical Journal, 507*, 404–416. doi: 10.1086/306299.

Linton, M. G., Fisher, G. H., Dahlburg, R. B., & Fan, Y. (1999, September). Relationship of the multimode kink instability to δ-spot formation. *Astrophysical Journal, 522*, 1190–1205. doi: 10.1086/307678.

Longcope, D. W., & Klapper, I. (1997). The effect of convection zone turbulence on a rising flux tube. *Astrophysical Journal, 488*, 443.

Longcope, D. W., Fisher, G. H., & Arendt, S. (1996). The evolution and fragmentation of rising magnetic flux tubes. *Astrophysical Journal, 464*, 999.

Longcope, D. W., Fisher, G. H., & Pevtsov, A. A. (1998, November). Flux-tube twist resulting from helical turbulence: The Σ-effect. *Astrophysical Journal, 507*, 417–432. doi: 10.1086/306312.

López Fuentes, M. C., Démoulin, P., Mandrini, C. H., & van Driel-Gesztelyi, L. (2000, November). The counterkink rotation of a non-hale active region. *Astrophysical Journal, 544*, 540–549. doi: 10.1086/317180.

López Fuentes, M. C., Démoulin, P., Mandrini, C. H., Pevtsov, A. A., & van Driel-Gesztelyi, L. (2003, January). Magnetic twist and writhe of active regions. On the origin of deformed flux tubes. *Astronomy & Astrophysics, 397*, 305–318. doi: 10.1051/0004-6361:20021487.

Luoni, M. L., Démoulin, P., Mandrini, C. H., & van Driel-Gesztelyi, L. (2011, May). Twisted flux tube emergence evidenced in longitudinal magnetograms: Magnetic tongues. *Solar Physics, 270*, 45–74. doi: 10.1007/s11207-011-9731-8.

MacTaggart, D., & Hood, A. W. (2009, November). On the emergence of toroidal flux tubes: general dynamics and comparisons with the cylinder model. *Astronomy & Astrophysics, 507*, 995–1004. doi: 10.1051/0004-6361/200912930.

Magara, T. (2001, March). Dynamics of emerging flux tubes in the Sun. *Astrophysical Journal, 549*, 608–628. doi: 10.1086/319073.

Manchester, W., IV, Gombosi, T., DeZeeuw, D., & Fan, Y. (2004, July). Eruption of a buoyantly emerging magnetic flux rope. *Astrophysical Journal, 610*, 588–596. doi: 10.1086/421516.

Mandal, K., Hanasoge, S. M., Rajaguru, S. P., & Antia, H. M. (2018, August). Helioseismic inversion to infer the depth profile of solar meridional flow using spherical born kernels. *The Astrophysical Journal, 863*, 39. doi: 10.3847/1538-4357/aacea2.

Martínez-Sykora, J., Hansteen, V., & Carlsson, M. (2008, May). Twisted flux tube emergence from the convection zone to the corona. *Astrophysical Journal, 679*, 871–888. doi: 10.1086/587028.

Martínez-Sykora, J., Hansteen, V., & Carlsson, M. (2009, September). Twisted flux tube emergence from the convection zone to the corona. II. Later states. *Astrophysical Journal, 702*, 129–140. doi: 10.1088/0004-637X/702/1/129.

Martínez-Sykora, J., De Pontieu, B., Hansteen, V., & Carlsson, M. (2015, April). The role of partial ionization effects in the chromosphere. *Philosophical Transactions of the Royal Society of London Series A, 373*, 20140268–20140268. doi: 10.1098/rsta.2014.0268.

Matsumoto, R., Tajima, T., Shibata, K., & Kaisig, M. (1993, September). Three-dimensional magnetohydrodynamics of the emerging magnetic flux in the solar atmosphere. *Astrophysical Journal, 414*, 357–371. doi: 10.1086/173082.

Matsumoto, R., Tajima, T., Chou, W., Okubo, A., & Shibata, K. (1998, January). Formation of a kinked alignment of solar active regions. *Astrophysical Journal Letters, 493*, L43–L46. doi: 10.1086/311116.

McClintock, B. H., & Norton, A. A. (2016, February). Tilt angle and footpoint separation of small and large bipolar sunspot regions observed with HMI. *Astrophysical Journal, 818*, 7. doi: 10.3847/0004-637X/818/1/7.

McClintock, B. H., Norton, A. A., & Li, J. (2014, December). Re-examining sunspot tilt angle to include anti-hale statistics. *Astrophysical Journal*, *797*, 130. doi: 10.1088/0004-637X/797/2/130.

McIntosh, S. W., Leamon, R. J., Krista, L. D., Title, A. M., Hudson, H. S., Riley, P., et al. (2015, April). The solar magnetic activity band interaction and instabilities that shape quasi-periodic variability. *Nature Communications*, *6*, 6491. doi: 10.1038/ncomms7491.

Meneguzzi, M., & Pouquet, A. (1989, August). Turbulent dynamos driven by convection. *Journal of Fluid Mechanics*, *205*, 297–318. doi: 10.1017/S0022112089002041.

Miesch, M. S., & Dikpati, M. (2014, April). A three-dimensional Babcock-Leighton solar dynamo model. *Astrophysical Journal Letters*, *785*, L8. doi: 10.1088/2041-8205/785/1/L8.

Miesch, M. S., Brun, A. S., & Toomre, J. (2006, April). Solar differential rotation influenced by latitudinal entropy variations in the tachocline. *Astrophysical Journal*, *641*, 618–625. doi: 10.1086/499621.

Miesch, M. S., Elliott, J. R., Toomre, J., Clune, T. L., Glatzmaier, G. A., & Gilman, P. A. (2000, March). Three-dimensional spherical simulations of solar convection. I. Differential rotation and pattern evolution achieved with laminar and turbulent states. *Astrophysical Journal*, *532*(1), 593–615. doi: 10.1086/308555.

Moreno-Insertis, F., Caligari, P., & Schuessler, M. (1994, August). Active region asymmetry as a result of the rise of magnetic flux tubes. *Solar Physics*, *153*, 449–452. doi: 10.1007/BF00712518.

Morfill, G. E., Scheingraber, H., & Voges, W. (1991). Sunspot number variations: Stochastic or chaotic. In C. P. Sonett, M. S. Giampapa, & M. S. Matthews (Eds.), *The Sun in time* (pp. 30–58). The University of Arizona Press, Tucson, Arizona.

Muñoz-Jaramillo, A., Nandy, D., & Martens, P. C. H. (2009, June). Helioseismic data inclusion in solar dynamo models. *Astrophysical Journal*, *698*, 461–478,. doi: 10.1088/0004-637X/698/1/461.

Murray, M. J., & Hood, A. W. (2008, February). Emerging flux tubes from the solar interior into the atmosphere: effects of non-constant twist. *Astronomy & Astrophysics*, *479*, 567–577. doi: 10.1051/0004-6361:20078852.

Murray, M. J., Hood, A. W., Moreno-Insertis, F., Galsgaard, K., & Archontis, V. (2006, December). 3D simulations identifying the effects of varying the twist and field strength of an emerging flux tube. *Astronomy & Astrophysics*, *460*, 909–923. doi: 10.1051/0004-6361:20065950.

Nandy, D., & Choudhuri, A. R. (2001, April). Toward a mean field formulation of the Babcock-Leighton type solar dynamo. I. α-Coefficient versus Durney's double-ring approach. *Astrophysical Journal*, *551*, 576–585. doi: 10.1086/320057.

Nelson, N. J., Brown, B. P., Brun, A. S., Miesch, M. S., & Toomre, J. (2013, January). Magnetic Wreaths and cycles in convective dynamos. *Astrophysical Journal*, *762*, 73. doi: 10.1088/0004-637X/762/2/73.

Nigro, G., Pongkitiwanichakul, P., Cattaneo, F., & Tobias, S. M. (2017, January). What is a large-scale dynamo? *Monthly Notices of the Royal Astronomical Society*, *464*, L119–L123. doi: 10.1093/mnrasl/slw190.

Norton, A. A., Jones, E. H., Linton, M. G., & Leake, J. E. (2017, June). Magnetic flux emergence and decay rates for preceder and follower sunspots observed with HMI. *Astrophysical Journal*, *842*, 3. doi: 10.3847/1538-4357/aa7052.

Otsuji, K., Kitai, R., Ichimoto, K., & Shibata, K. (2011, October). Statistical study on the nature of solar-flux emergence. *Publications of the Astronomical Society of Japan*, *63*, 1047–1057,. doi: 10.1093/pasj/63.5.1047.

Pallé, P. L., Regulo, C., & Roca Cortes, T. (1989, October). Solar cycle induced variations of the low L solar acoustic spectrum. *Astronomy and Astrophysics*, *224*, 253–258.

Parker, E. N. (1955a, March). The formation of sunspots from the solar toroidal field. *Astrophysical Journal*, *121*, 491. doi: 10.1086/146010.

Parker, E. N. (1955b, September). Hydromagnetic dynamo models. *Astrophysical Journal*, *122*, 293. doi: 10.1086/146087.

Parker, E. N. (1979, June). Sunspots and the physics of magnetic flux tubes. II. Aerodynamic drag. *Astrophysical Journal*, *230*, 914–923.

Petrie, G. J. D. (2015, November). Solar magnetism in the polar regions. *Living Reviews in Solar Physics*, *12*, 5. doi: 10.1007/lrsp-2015-5.

Pevtsov, A. A., Canfield, R. C., & Metcalf, T. R. (1994, April). Patterns of helicity in solar active regions. *Astrophysical Journal Letters*, *425*, L117–L119. doi: 10.1086/187324.

Pevtsov, A. A., Canfield, R. C., & Metcalf, T. R. (1995). Latitudinal variation of helicity of photospheric magnetic fields. *Astrophysical Journal Letters*, *440*, L109.

Pevtsov, A. A., Berger, M. A., Nindos, A., Norton, A. A., & van Driel-Gesztelyi, L. (2014, December). Magnetic helicity, tilt, and twist. *Space Science Reviews*, *186*, 285–324. doi: 10.1007/s11214-014-0082-2.

Pijpers, F. P., & Thompson, M. J. (1992, September). Faster formulations of the optimally localized averages method for helioseismic inversions. *Astronomy and Astrophysics*, *262*, L33–L36.

Pijpers, F. P., & Thompson, M. J. (1994, January). The SOLA method for helioseismic inversion. *Astronomy and Astrophysics*, *281*, 231–240.

Pongkitiwanichakul, P., Nigro, G., Cattaneo, F., & Tobias, S. M. (2016, July). Shear-driven Dynamo Waves in the Fully Nonlinear Regime. *Astrophysical Journal*, *825*, 23. doi: 10.3847/0004-637X/825/1/23.

Racine, É., Charbonneau, P., Ghizaru, M., Bouchat, A., & Smolarkiewicz, P. K. (2011, July). On the Mode of dynamo action in a global large-eddy simulation of solar convection. *Astrophysical Journal*, *735*, 46. doi: 10.1088/0004-637X/735/1/46.

Rädler, K.-H., & Rheinhardt, M. (2007, April). Mean-field electrodynamics: critical analysis of various analytical approaches to the mean electromotive force. *Geophysical and Astrophysical Fluid Dynamics*, *101*, 117–154. doi: 10.1080/03091920601111068.

Rajaguru, S. P., & Antia, H. M. (2015, November). meridional circulation in the solar convection zone: Time-distance helioseismic inferences from four years of HMI/SDO observations. *The Astrophysical Journal*, 813, 114. doi: 10.1088/0004-637X/813/2/114.

Rajaguru, S. P., Basu, S., & Antia, H. M. (2001, December). Ring diagram analysis of the characteristics of solar oscillation modes in active regions. *The Astrophysical Journal*, *563*, 410–418. doi: 10.1086/323780.

Rempel, M. (2011, March). Penumbral fine structure and driving mechanisms of large-scale flows in simulated sunspots. *Astrophysical Journal*, *729*(1), 5.

Rempel, M. (2012, May). High-latitude solar torsional oscillations during phases of changing magnetic cycle amplitude. *The Astrophysical Journal Letters*, *750*, L8. doi: 10.1088/2041-8205/750/1/L8.

Rempel, M. (2014, July). Numerical simulations of quiet Sun magnetism: On the contribution from a small-scale dynamo. *Astrophysical Journal*, *789*, 132. doi: 10.1088/0004-637X/789/2/132.

Rempel, M., & Cheung, M. C. M. (2014, April). Numerical simulations of active region scale flux emergence: From spot formation to decay. *Astrophysical Journal*, *785*, 90. doi: 10.1088/0004-637X/785/2/90.

Rempel, M., Schüssler, M., Cameron, R. H., & Knölker, M. (2009a, July). Penumbral structure and outflows in simulated sunspots. *Science*, *325*(5937), 171. doi: 10.1126/science.1173798.

Rempel, M., Schüssler, M., & Knölker, M. (2009b, January). Radiative magnetohydrodynamic simulation of sunspot structure. *Astrophysical Journal*, *691*, 640–649. doi: 10.1088/0004-637X/691/1/640.

Sammis, I., Tang, F., & Zirin, H. (2000, September). The dependence of large flare occurrence on the magnetic structure of sunspots. *Astrophysical Journal*, *540*, 583–587. doi: 10.1086/309303.

Sánchez Almeida, J., & Lites, B. W. (2000, April). Physical properties of the solar magnetic photosphere under the MISMA Hypothesis. II. Network and internetwork fields at the disk center. *Astrophysical Journal*, *532*, 1215–1229. doi: 10.1086/308603.

Schad, A., Timmer, J., & Roth, M. (2013, December). Global helioseismic evidence for a deeply penetrating solar meridional flow consisting of multiple flow cells. *The Astrophysical Journal Letters*, *778*, L38. doi: 10.1088/2041-8205/778/2/L38.

Scherrer, P. H., Schou, J., Bush, R. I., Kosovichev, A. G., Bogart, R. S., Hoeksema, J. T., et al. (2012, January). The Helioseismic and Magnetic Imager (HMI) investigation for the Solar Dynamics Observatory (SDO). *Solar Physics*, *275*, 207–227. doi: 10.1007/s11207-011-9834-2.

Scherrer, P. H., Bogart, R. S., Bush, R. I., Hoeksema, J. T., Kosovichev, A. G., Schou, J., et al. (1995). The solar oscillations investigation – Michelson Doppler Imager. *Solar Physics*, *162*, 129–188.

Schmitt, J. R., & Basu, S. (2015, August). Modeling the asteroseismic surface term across the HR diagram. *The Astrophysical Journal*, *808*, 123,. doi: 10.1088/0004-637X/808/2/123.

Schou, J., Christensen-Dalsgaard, J., & Thompson, M. J. (1994, September). On comparing helioseismic two-dimensional inversion methods. *The Astrophysical Journal*, *433*, 389–416. doi: 10.1086/174653.

Schou, J., Antia, H. M., Basu, S., Bogart, R. S., Bush, R. I., Chitre, S. M., et al. (1998, September). Helioseismic studies of differential rotation in the solar envelope by the solar oscillations investigation using the Michelson Doppler Imager. *The Astrophysical Journal*, *505*, 390–417. doi: 10.1086/306146.

Schrijver, C. J., Title, A. M., van Ballegooijen, A. A., Hagenaar, H. J., & Shine, R. A. (1997, September). Sustaining the quiet photospheric network: The balance of flux emergence, fragmentation, merging, and cancellation. *Astrophysical Journal*, *487*, 424–436. doi: 10.1086/304581.

Schubert, G. (1968, March). The stability of toroidal magnetic fields in stellar interiors. *Astrophysical Journal*, *151*, 1099. doi: 10.1086/149508.

Schüssler, M. (1979). Magnetic buoyancy revisited: Analytical and numerical results for rising flux tubes. *Astronomy & Astrophysics*, *71*, 79–91.

Schüssler, M., Calgari, P., Ferriz-Maz, A., & Moreno-Insertis, F. (1994). Instability and eruption of magnetic flux tubes in the solar convection zone. *Astronomy & Astrophysics*, *281*, L69–L72.

Schwabe, H. (1843). Die Sonne. *Astronomische Nachrichten*, *20*, 283.

Shafranov, V. D. (1957). On the stability of a cylindrical gaseous conductor in a magnetic field. *Journal of Nuclear Energy II*, *5*, 86,.

Shibata, K., Tajima, T., Steinolfson, R. S., & Matsumoto, R. (1989, October). Two-dimensional magnetohydrodynamic model of emerging magnetic flux in the solar atmosphere. *Astrophysical Journal*, *345*, 584–596. doi: 10.1086/167932.

Spruit, H. C. (1981). Motion of magnetic flux tubes in the solar convection zone and chromosphere. *Astronomy & Astrophysics*, *98*, 155–160.

Steenbeck, M., Krause, F., & Rädler, K. H. (1966, April). Berechnung der mittleren LORENTZ-Feldstärke für ein elektrisch leitendes Medium in turbulenter, durch CORIOLIS-Kräfte beeinflußter Bewegung. *Zeitschrift Naturforschung Teil A*, *21*, 369. doi: 10.1515/zna-1966-0401.

Stein, R. F. (2012a, July). Magneto-convection. *Philosophical Transactions of the Royal Society of London Series A*, *370*, 3070–3087. doi: 10.1098/rsta.2011.0533.

Stein, R. F. (2012b, July). Solar surface magneto-convection. *Living Reviews in Solar Physics*, 9, 4. doi: 10.12942/lrsp-2012-4.

Stein, R. F., Lagerfjärd, A., Nordlund, Å., & Georgobiani, D. (2011, February). Solar flux emergence simulations. *Solar Physics*, *268*, 271–282. doi: 10.1007/s11207-010-9510-y.

Stenflo, J. O. (1999). Solar magnetism and the second solar spectrum: Future directions. In K. N. Nagendra & J. O. Stenflo (Eds.), *Solar polarization* (pp. 1–16). Kluwer, Dodrecht.

Stenflo, J. O., & Kosovichev, A. G. (2012, February). Bipolar magnetic regions on the Sun: Global analysis of the SOHO/MDI data set. *Astrophysical Journal*, *745*, 129. doi: 10.1088/0004-637X/745/2/129.

Stix, M. (1971, July). A non-axisyinmetric a-effect dynamo. *Astronomy & Astrophysic*, *13*, 203.

Stix, M. (1976, March). Differential rotation and the solar dynamo. *Astronomy & Astrophysics*, *47*(2), 243–254.

Strous, L. H., Scharmer, G., Tarbell, T. D., Title, A. M., & Zwaan, C. (1996, February). Phenomena in an emerging active region. I. Horizontal dynamics. *Astronomy & Astrophysics*, *306*, 947.

Takasao, S., Fan, Y., Cheung, M. C. M., & Shibata, K. (2015, November). Numerical study on the emergence of kinked flux tube for understanding of possible origin of δ-spot regions. *Astrophysical Journal*, *813*, 112. doi: 10.1088/0004-637X/813/2/112.

Tanaka, K. (1980). Active region and flare predictions. In R.F. Donnelly (Ed.), *Solar-terrestrial predictions proceedings* (Vol. 3, p. C1). NOAA-ERL.

Tanaka, K. (1991). Studies on a very flare-active δ group: Peculiar δ spot evolution and inferred subsurface magnetic rope structure. *Solar Physics*, *136*, 133–149,.

Thompson, M. J., Toomre, J., Anderson, E. R., Antia, H. M., Berthomieu, G., Burtonclay, D., et al. (1996, May). Differential rotation and dynamics of the solar interior. *Science*, *272*, 1300–1305. doi: 10.1126/science.272.5266.1300.

Tiwari, S. K., van Noort, M., Solanki, S. K., & Lagg, A. (2015, November). Depth-dependent global properties of a sunspot observed by Hinode using the Solar Optical Telescope/Spectropolarimeter. *Astronomy & Astrophysics*, *583*, A119. doi: 10.1051/0004-6361/201526224.

Tlatova, K., Tlatov, A., Pevtsov, A., Mursula, K., Vasil'eva, V., Heikkinen, E., et al. (2018, August). Tilt of sunspot bipoles in solar cycles 15 to 24. *Solar Physics*, *293*, 118. doi: 10.1007/s11207-018-1337-y.

Tobias, S. M., & Cattaneo, F. (2013, May). Shear-driven dynamo waves at high magnetic Reynolds number. *Nature*, *497*, 463–465,. doi: 10.1038/nature12177.

Tobias, S. M., Cattaneo, F., & Brummell, N. H. (2008, September). Convective dynamos with penetration, rotation, and shear. *Astrophysical Journal*, *685*, 596–605,. doi: 10.1086/590422.

Tobias, S. M., Cattaneo, F., & Brummell, N. H. (2011, February). On the generation of organized magnetic fields. *Astrophysical Journal*, *728*, 153. doi: 10.1088/0004-637X/728/2/153.

Toomre, J., Christensen-Dalsgaard, J., Howe, R., Larsen, R. M., Schou, J., & Thompson, M. J. (2000). Time variability of rotation in solar convection zone from soi-mdi. *Solar Physics*, *192*, 437–448.

Toriumi, S., & Takasao, S. (November 2017). Numerical simulations of flare-productive active regions: δ-sunspots, Sheared polarity inversion lines, energy storage, and predictions. *Astrophysical Journal*, *850*, 39. doi: 10.3847/1538-4357/aa95c2.

Toriumi, S., & Yokoyama, T. (2010, May). Two-step emergence of the magnetic flux sheet from the solar convection zone. *Astrophysical Journal*, *714*, 505–516,. doi: 10.1088/0004-637X/714/1/505.

Toriumi, S., & Yokoyama, T. (2011, July). Numerical experiments on the two-step emergence of twisted magnetic flux tubes in the Sun. *Astrophysical Journal*, *735*, 126,. doi: 10.1088/0004-637X/735/2/126.

Toriumi, S., Hayashi, K., & Yokoyama, T. (2014a, October). Statistical analysis of the horizontal divergent flow in emerging solar active regions. *Astrophysical Journal*, *794*, 19. doi: 10.1088/0004-637X/794/1/19.

Toriumi, S., Iida, Y., Kusano, K., Bamba, Y., & Imada, S. (2014b, September). Formation of a flare-productive active region: Observation and numerical simulation of NOAA AR 11158. *Solar Physics*, *289*(9), 3351–3369.

Tortosa-Andreu, A., & Moreno-Insertis, F. (2009, November). Magnetic flux emergence into the solar photosphere and chromosphere. *Astronomy & Astrophysics*, *507*, 949–967. doi: 10.1051/0004-6361/200912394.

Trenberth, K. E. (1992). *Climate System Modeling*. Cambridge Univ. Press, Cambridge.

Trujillo Bueno, J., Merenda, L., Centeno, R., Collados, M., & Landi Degl'Innocenti, E. (2005, February). The Hanle and Zeeman Effects in Solar Spicules: A Novel Diagnostic Window on Chromospheric Magnetism. *Astrophysical Journal Letters*, *619*, L191–L194. doi: 10.1086/428124.

Vorontsov, S. V., Christensen-Dalsgaard, J., Schou, J., Strakhov, V. N., & Thompson, M. J. (2002). Helioseismic measurement of solar torsional oscillations. *Science*, *296*, 101–103. doi: 10.1126/science.1069190.

Wang, Y.-M., & Sheeley, N. R., Jr. (1989, March). Average properties of bipolar magnetic regions during sunspot cycle 21. *Solar Physics*, *124*, 81–100,. doi: 10.1007/BF00146521.

Wang, Y.-M., Nash, A. G., & Sheeley, N. R., Jr. (1989, August). Magnetic flux transport on the Sun. *Science*, *245*, 712–718,. doi: 10.1126/science.245.4919.712.

Wang, Y.-M., Sheeley, N. R., Jr., & Nash, A. G. (1991, December). A new solar cycle model including meridional circulation. *Astrophysical Journal*, *383*, 431–442,. doi: 10.1086/170800.

Wang, Y.-M., Colaninno, R. C., Baranyi, T., & Li, J. (2015, January). Active-region Tilt angles: Magnetic versus white-light determinations of Joy's law. *Astrophysical Journal*, *798*, 50. doi: 10.1088/0004-637X/798/1/50.

Weber M. A., Fan, Y., & Miesch, M. S. (2011, November). The rise of active region flux tubes in the turbulent solar convective envelope. *Astrophysical Journal*, *741*, 11. doi: 10.1088/0004-637X/741/1/11.

Weber, M. A., Fan, Y., & Miesch, M. S. (2013, October). Comparing simulations of rising flux tubes through the solar convection zone with observations of solar active regions: constraining the dynamo field strength. *Solar Physics*, *287*, 239–263. doi: 10.1007/s11207-012-0093-7.

Woodard, M. F., & Noyes, R. W. (1985, December). Change of solar oscillation eigenfrequencies with the solar cycle. *Nature*, *318*, 449–450,.

Yang, S., & Zhang, J. (2014, January). Properties of solar ephemeral regions at the emergence stage. *Astrophysical Journal*, *781*, 7. doi: 10.1088/0004-637X/781/1/7.

Yau, K. K. C., & Stephenson, F. R. (1988, June). A revised catalogue of Far Eastern observations of sunspots (165 BC to AD 1918). *Royal Astronomical Society, Quarterly Journal*, *29*, 175–197.

Yeates, A. R., Nandy, D., & Mackay, D. H. (2008, January). Exploring the physical basis of solar cycle predictions: Flux transport dynamics and persistence of memory in advection-versus diffusion-dominated solar convection zones. *Astrophysical Journal*, *673*, 544–556. doi: 10.1086/524352.

Yokoyama, T., & Shibata, K. (1995, May). Magnetic reconnection as the origin of X-ray jets and Hα surges on the Sun. *Nature*, *375*, 42–44,. doi: 10.1038/375042a0.

Zeeman, P. (1897, May). On the influence of magnetism on the nature of the light emitted by a substance. *Astrophysical Journal*, *5*, 332. doi: 10.1086/140355.

Zhao, J., & Kosovichev, A. G. (2004, March). Torsional oscillation, meridional flows, and vorticity inferred in the upper convection zone of the Sun by time-distance helioseismology. *The Astrophysical Journalre*, *603*, 776–784, doi: 10.1086/381489.

Zhao, J., Nagashima, K., Bogart, R. S., Kosovichev, A. G., & Duvall, T. L., Jr. (2012, April). Systematic center-to-limb variation in measured helioseismic travel times and its effect on inferences of solar interior meridional flows. *The Astrophysical Journal Letters*, *749*, L5. doi: 10.1088/2041-8205/749/1/L5.

Zhao, J., Bogart, R. S., Kosovichev, A. G., Duvall, T. L., Jr., & Hartlep, T. (2013, September). Detection of equatorward meridional flow and evidence of double-cell meridional circulation inside the Sun. *The Astrophysical Journal Letters*, *774*, L29. doi: 10.1088/2041-8205/774/2/L29.

Zhao, J., Kosovichev, A. G., & Bogart, R. S. (2014, July). Solar meridional flow in the shallow interior during the rising phase of cycle 24. *Astrophysical Journal Letters*, *789*, L7. doi: 10.1088/2041-8205/789/1/L7.

Zirin, H. (1988). *Astrophysics of the Sun*. Cambridge Univ. Press, Cambridge.

Zirin, H. & Liggett, M. A. (1987). Delta spots and great flares. *Solar Physics*, *113*, 267–83.

INDEX

Space Physics and Aeronomy Collection Volume 1: Solar Physics and Solar Wind, Geophysical Monograph 258, First Edition.
Edited by Nour E. Raouafi and Angelos Vourlidas.

DOI: 10.1002/9781119815600.index

Printed in the USA/Agawam, MA
September 17, 2026

910661.048